AF760456

ACOUSTIC WAVES – FROM MICRODEVICES TO HELIOSEISMOLOGY

Edited by **Marco G. Beghi**

Acoustic Waves – From Microdevices to Helioseismology
Edited by Marco G. Beghi

Published by InTech
Janeza Trdine 9, 51000 Rijeka, Croatia

Copyright © 2011 InTech
All chapters are distributed under the Creative Commons Attribution 3.0 license, which allows users to download, copy and build upon published articles even for commercial purposes, as long as the author and publisher are properly credited, which ensures maximum dissemination and a wider impact of our publications. After this work has been published by InTech, authors have the right to republish it, in whole or part, in any publication of which they are the author, and to make other personal use of the work. Any republication, referencing or personal use of the work must explicitly identify the original source.

As for readers, this license allows users to download, copy and build upon published chapters even for commercial purposes, as long as the author and publisher are properly credited, which ensures maximum dissemination and a wider impact of our publications.

Notice
Statements and opinions expressed in the chapters are these of the individual contributors and not necessarily those of the editors or publisher. No responsibility is accepted for the accuracy of information contained in the published chapters. The publisher assumes no responsibility for any damage or injury to persons or property arising out of the use of any materials, instructions, methods or ideas contained in the book.

Publishing Process Manager Sandra Bakic

Technical Editor Teodora Smiljanic

Cover Designer InTech Design Team

Additional hard copies can be obtained from orders@intechopen.com

Acoustic Waves – From Microdevices to Helioseismology, Edited by Marco G. Beghi
p. cm.
ISBN 978-953-307-572-3

We are IntechOpen, the world's leading publisher of Open Access books

Built by scientists, for scientists

4,100+
Open access books available

116,000+
International authors and editors

120M+
Downloads

151
Countries delivered to

Our authors are among the
Top 1%
most cited scientists

12.2%
Contributors from top 500 universities

CLARIVATE ANALYTICS BOOK CITATION INDEX INDEXED

WEB OF SCIENCE™

Selection of our books indexed in the Book Citation Index
in Web of Science™ Core Collection (BKCI)

Interested in publishing with us?
Contact book.department@intechopen.com

Numbers displayed above are based on latest data collected.
For more information visit www.intechopen.com

For EU product safety concerns:
IN TECH d.o.o., Prolaz Marije Krucifikse Kozulić 3, 51000 Rijeka, Croatia,
info@intechopen.com or visit our website at intechopen.com.

Meet the editor

Dr. Marco G. Beghi graduated in 1979 in Nuclear Engineering at Politecnico di Milano, Italy. He spent one year at University of California, Los Angeles. In 1984 he became research fellow at the Department of Nuclear Engineering, Politecnico di Milano, and in 2003 associate professor of Condensed Matter Physics. He was member of government bodies of Politecnico: Board of Administrators and Academic Senate. Presently he is in the Micro- and Nanostructured Materials Laboratory of the Department of Energy, Politecnico. Since 1991 he has been teaching Experimental Physics, Condensed Matter Physics, and Technology of Nuclear Materials, to undergraduate and graduate students of Nuclear and Materials Engineering. His experimental research concerns the physics of materials. He analysed the mechanical behaviour of metals in terms of dislocation dynamics and irreversible thermodynamics. He then worked on thin films and their properties, exploiting Brillouin spectrometry to measure the elastic properties. He co-authored over seventy peer reviewed publications.

Contents

Preface

The subject of acoustic waves might easily be considered a mature one, quite specialized, with narrow and circumscribed fields of interest and of application. The present book is an evidence of the opposite: it witnesses how the concept of acoustic wave, a collective displacement of matter which perturbs an equilibrium configuration, is a pervasive concept, which emerges in very different fields. This type of phenomena can be analyzed from different points of view, it can be exploited in different ways, and is the object of active investigations. The present book, far from pretending to give an exhaustive overview of the subject, offers instead a sampling of various points of view, of applications, and of research objectives which are actively pursued.

It must first be remembered that acoustic waves are supported by all the forms of matter: solids, liquids, gases and plasmas. And if similarities among the different phenomena are deep enough for them to deserve the same name, nevertheless the peculiarities connected to the various media are significant. Although the range of involved length and time scales is huge, going from sub-micrometric layers exploited in microdevices to seismic waves propagating in the Sun's interior, the more profound peculiarities of the various cases concern the very heart of the phenomena, namely the type of forces which, in different types of media, tend to restore the equilibrium configuration.

These phenomena can be approached under different points of view. A first type of approach aims at a better comprehension of phenomena. Many aspects of acoustic waves are nowadays well understood, but the investigation is obviously never ending. A line of research aims at the theoretical exploration, also by relatively sophisticated mathematical analyses, of various aspects of phenomena whose basic laws are well established. Concerning acoustic waves in elastic solids, Huang recalls the characters of such waves in homogeneous isotropic media. Then he exploits recent computational tools to analyze the modifications occurring in media which are periodically inhomogeneous, like composite materials. Alshits, Lyubimov & Radowicz investigate instead the elastic waves in solids which are homogeneous but anisotropic, like single crystals. They show that the addition of a dissipative term to the elasto-dynamic equations has consequences which go far beyond the intuitive introduction of a damping. This term can modify the same topology of the slowness surface, inducing a splitting of acoustic axes. Homentcovschi & Miles review and reformulate in an operational way the 'acoustic analogy' theory which describes how noise is generated in the interaction of gas flow with stationary or mobile bodies; the application of this approach to a range of technologies (jets, propellers, aircrafts) is easily imagined. Khater & Hassan consider various nonlinear evolution equations which are well established in plasma physics and fluid dynamics, and which admit wave solutions, either periodic waves or solitary waves. They seek exact solutions, which helps to understand phenomena more than purely numerical solutions. Karmakar considers various aspects involved in perturbations of plasmas, from ion acoustic excitation to turbulence, and focuses on the effects of the inertia of electrons, which is much smaller than that of ions but is not always completely negligible. He then combines various arguments to give a picture of solar wind plasma, which needs the description of the solar surface boundary.

A second type of approach exploits acoustic waves as probes to gain information about the properties or the behavior of a system. Beghi revises various methods based on acoustic waves which aim at the elastic characterization of materials, namely of thin films. Gao & Pan consider a specific problem of significant technical relevance for the oil and gas industry: the drillability of rocks, and in particular its anisotropy. They shows how the outcome of laboratory acoustic tests correlates with the drilling properties of rocks. Gu & Choi consider instead the acoustic emission from rotating machinery, and show how it can be exploited for the early detection of faults. Burgholzer and coworkers focus on the photoacoustic imaging technique, and in particular on the image reconstruction to achieve the tomographic capability: they analyze methods to compensate for ultrasound attenuation in the media being observed.

Since acoustic waves are relatively a non invasive probe, they can be exploited also on delicate materials and on biological systems. Nassar presents various applications to delicate systems in the agro-industry, like cheese undergoing ripening, for which dedicated low frequency sensors had to be developed. Erzegen and co-workers characterize the performance of the multi-resonant thickness shear mode sensor, exploited with a genetic algorithm for data processing: this type of sensor is devoted to the characterization of biological interfaces. Finally, two chapters present measurements performed in vivo. Michalski, Dziewiszek & Bochnia discuss the performance of phase sensitive techniques to characterize non linear systems, and show how these techniques can be applied to cochlear microphonics to study ear behavior. Mota, Carvalho & Barja present photoacoustic measurements performed on human skin, to characterize the skin itself, and the transdermal drug delivery.

A completely different system is found in the ionosphere, where acoustic-gravity waves are found. Koucka & Mosna show how the ionogram technique can be exploited to investigate the ionosphere, in particular exploiting the waves excited by the shadow of an eclipse.

A third type of approach exploits acoustic waves to perform some kind of manipulation. Foldyna shows how acoustic transducers and waveguides can be exploited to generate and control pulsating water jets, which can be used as machining tools. Zinovev, Veryovkin & Pellin discuss the Laser Induced Acoustic Desorption technique to vaporize solid material to be analyzed by mass spectrometry. This technique is less prone to induce modifications of the analyte than the more widespread MALDI technique, although these authors show by some experiments that the operational mechanism is still not well understood. At the other extreme, that of high intensity laser pulses, Bykovsky & Senatsky demonstrate how stimulated Brillouin scattering can generate shock waves, able to induce permanent modifications of the materials, like phase changes and cracks. Finally, Branski considers the problem of active vibration control of beams, and investigates the optimal distribution of actuators to perform such a control.

A fourth type of approach exploits the properties of acoustic waves to design various types of devices, mainly micro devices. The most widely exploited type of device has a simple basic structure: a substrate, at least one layer of piezoelectric material, an interdigitated transducer (IDT) operating as an emitter, and another one operating as receiver. These type of devices were originally introduced as delay lines and filters, and were then developed also for other purposes.

Before discussing this type of device, it must be remembered that other types of devices also exist. Kreft & Blick discuss applications of surface acoustic waves to quantum electronics, made possible by devices like quantum dots and by the interaction of surface acoustic waves with the electron gas. This type of device is nanomechanical, and also exploits IDT, with acoustic waveguides to match their acoustic impedance to that of nanomechanical devices. The chapter by Cuadrado-Laborde and co-workers considers instead the in-fiber photonic devices, and the acousto-optic modulator which obtained exciting traveling or standing acoustic waves by a piezoelectric actuator. This way, a dynamic and controllable modulation of the fiber properties is obtained by the acousto-optic effect. They review a wide variety of configurations, showing how different devices can be obtained, including Q-switched lasers and mode locking lasers.

Returning to the most widespread type of microdevice, its interest is witnessed by the numerous chapters devoted to it. Naumenko reviews the most common design configurations, and presents detailed analyses of their behavior. Hoang presents the most adopted method of analysis of such devices, based on equivalent circuits and the so called Mason model. The method is adopted also in other chapters, and Hoang gives a detailed introduction of the method itself, also presenting the applications to basic configurations. Rocas & Collado analyze, for these devices, the various sources which can introduce a non perfectly linear behavior, leading to 3rd order intermodulation distortion. Most of the devices of this type exploit longitudinal acoustic waves, or surface waves polarized in the plane normal to the surface. Waves transversally polarized in the plane of the device surface are less considered. Yanagitani compares the performances of the two types of operation, showing the possible advantages of transversal waves. Avramov performs a similar comparison, between surface waves of the Rayleigh type and of the transverse type, for devices which are polymer coated to act as gas sensors. Some chapters focus instead on the production and the characterization of various materials which are of interest for the production of this type of devices. Buck considers ultranano crystalline diamond: diamond is the acoustically fastest material, which allows operation at the highest frequencies. Both Kar & Bose and Caliendo consider AlN layers, a piezoelectric material whose properties are interesting under several respects. Caliendo also considers multilayers, including platinum and sapphire layers.

As mentioned above, the various approaches documented in this book represent a sampling of the wide spectrum of methods and techniques involving acoustic waves. This book is offered to the scientific community in the hope of promoting a cross fertilization of ideas and of approaches.

Marco G. Beghi
Politecnico di Milano, Energy Department and NEMAS Center,
Milano,
Italy

Part 1

Theoretical and Numerical Investigations of Acoustic Waves

1

Analysis of Acoustic Wave in Homogeneous and Inhomogeneous Media Using Finite Element Method

Zi-Gui Huang
Department of Mechanical Design Engineering, National Formosa University
Taiwan

1. Introduction

Even though the propagation of elastic/acoustic waves in inhomogeneous and layered media has been an active research topic for many decades already, new problems and challenges continue to be posed even up to now. In fact, during the last few years, renewed interests have been witnessed by researchers in the various fields of acoustics, such as acoustic mirrors, filters, resonators, waveguides, and other kinds of acoustic devices, in relation to wave propagation in periodic elastic media. In acoustics and applied mechanics, these developments have been triggered by the need for new acoustic devices in order to obtain quality control of elastic/acoustic waves.

What sort of material can allow us to have complete control over the elastic/acoustic wave's propagation? We would like to discuss and answer this question in this chapter. It is well known that the successful applications of photonic band-gap materials have hastened the related researches on phononic band-gap materials. *Analysis of Acoustic Wave in Homogeneous and Inhomogeneous Media Using Finite Element Method* explores the theoretical road leading to the possible applications of phononic band gaps. It should quickly bring the elastic/acoustic professionals and engineers up to speed in this field of study where elastic/acoustic waves and solid-state physics meet. It will also provide an excellent overview to any course in elastic/acoustic media.

Previous research on photonic crystals (Johnson & Joannopoulos, 2001, 2003; Joannopoulos et al., 1995; Leung & Liu, 1990) has sparked rapidly growing interest in the analogous acoustic effects of phononic crystals and periodic elastic structures. The various techniques for band structure calculations were introduced (Hussein, 2009). There are many well-known methods of calculating the band structures of photonic and phononic crystals in addition to the reduced Bloch mode expansion method: the plane-wave expansion (PWE) method (Huang & Wu, 2005; Kushwaha et al., 1993; Laude et al., 2005; Tanaka & Tamura, 1998; Wu et al., 2004 ; Wu & Huang, 2004), the multiple-scattering theory (MST) (Leung & Qiu, 1993; Kafesaki & Economou, 1999; Psarobas & Stefanou, 2000; Wang et al., 1993), the finite-difference (FD) method (Garica-Pabloset et al., 2000; Sun & Wu, 2005; Yang, 1996), the transfer matrix method (Pendry & MacKinnon, 1992), the meshless method (Jun et al., 2003), the multiple multipole method (Moreno et al., 2002), the wavelet method (Checoury & Lourtioz, 2006; Yan & Wang, 2006), the pseudospectral method (Chiang et al., 2007), the finite element method (FEM) (Axmann & Kuchment, 1999; Dobson, 1999; Huang & Chen,

2011; Wu et al., 2008), the mass-in-mass lattice model (Huang & Sun, 2010), and the micropolar continuous modeling (Salehian & Inman, 2010).
Many studies on phononic band structures from the past decade use the PWE, MST, and FD methods to analyze the frequency band gaps of bulk acoustic waves (BAW) in composite materials or phononic band structures. Studies adopting the PWE method investigate the dispersion relations and the frequency band-gap feathers of the BAW and surface acoustic wave (SAW) modes. Other studies use the layered MST to study the frequency band gaps of bulk acoustic waves in three-dimensional periodic acoustic composites and the band structures of phononic crystals consisting of complex and frequency-dependent Lame' coefficients. Other researchers applied the finite-difference time-domain method to predict the precise transmission properties of slabs of phononic crystals and analyze the mode coupling in joined parallel phononic crystal waveguides.
The techniques for tuning frequency band gaps of elastic/acoustic waves in phononic crystals are very important, and remain exciting research topics in the physics community. The filling fraction, rotation of noncircular rods, different cuts of anisotropic materials, and the temperature effect all produce large frequency band gaps in the BAW and SAW modes of periodic structures. A previous review paper (Burger et al., 2004) discusses the technique used to optimize the unit cell material distribution, achieving the largest possible band gap in photonic crystals for a given cell symmetry. Studies over the past decade focus on the theoretical and numerical analysis of phononic structures based on circular or square cylinders embedded in background materials. In this case, the PWE method can easily calculate the dispersion relations by constructing the structural functions with Bessel or Sinc functions. However, research on the more complicated problem of waves in the reticular and other special periodic band structures has not started until recently.
This chapter uses the 2D and 3D finite element methods to discuss the wave velocities of isotropic and anisotropic materials in homogeneous media. It also considers the tunable band gaps of acoustic waves in two-dimensional phononic crystals with reticular geometric structures (Huang & Chen, 2011). The concept of adopting a reticular geometric structure comes from the variations of similar geometry in bio-structural reticular formation and fibers. The PWE method used to calculate the structural functions of densities and elastic constants cannot numerically analyze the Gibbs phenomenon. Therefore, this chapter adopts the FEM to discuss this special periodic band structure. Changing the filling fraction, scale parameters, and rotating angles of reticular geometric structures can tune the frequency band gaps of mixed polarization modes. This technique is suitable for analyzing the phenomenon of frequency band gaps in special band structures.

2. Theory

In this chapter, based on the theorems of solid-state physics and the finite element method with Bloch calculations, equation of motion of the acoustic modes in two-dimensional inhomogeneous media, phononic band structures, are derived and discussed in detail. In the beginning, the concepts of the real space and **k** space are introduced while the Brillouin zone is also addressed in the text. Generalized techniques of Bloch calculations in finite element method are used to analyze the acoustic modes in two-dimensional homogeneous and inhomogeneous media, phononic band structures, consisting of materials with general anisotropy. The mixed and transverse polarization modes and quasi-polarization modes are investigated in the text.

2.1 Real space and k space

It is well-known that the analysis of wave motion in infinite periodic structures is difficult in real space. For dealing with the periodic structures, the Fourier series and Bloch's theorem are used to expand the periodic parameters such as the density, material constants, displacement fields, or potential. Regarding to the transformation of the real space and **k** space, the reciprocal lattice vectors (RLVs) are adopted from the solid-state physics. In general, we consider a three-dimensional phononic crystal with primitive lattice vectors $\mathbf{a}_1$, $\mathbf{a}_2$, and $\mathbf{a}_3$. The complete set of lattice vectors is written as $\{\mathbf{R} \mid \mathbf{R} = l_1\mathbf{a}_1 + l_2\mathbf{a}_2 + l_3\mathbf{a}_3\}$, where l_1, l_2, and l_3 are integers. The associated primitive reciprocal lattice vectors $\mathbf{b}_1$, $\mathbf{b}_2$, and $\mathbf{b}_3$ are determined by (Kittel, 1996)

$$\mathbf{b}_i = 2\pi \frac{\varepsilon_{ijk}\mathbf{a}_j \times \mathbf{a}_k}{\mathbf{a}_1 \cdot (\mathbf{a}_2 \times \mathbf{a}_3)}, \tag{1}$$

where ε_{ijk} is the three-dimensional Levi-Civita completely antisymmetric symbol. The complete set of reciprocal lattice vectors is written as $\{\mathbf{G} \mid \mathbf{G} = N_1\mathbf{b}_1 + N_2\mathbf{b}_2 + N_3\mathbf{b}_3\}$, where N_1, N_2, and N_3 are integers. Figure 1 shows the primitive unit cell in two-dimensional real space while the Fig. 2 shows the relationship between the real space and **k** space. A property between the primitive lattice vectors and associated primitive reciprocal lattice vectors is $\mathbf{b}_i \cdot \mathbf{a}_j = 2\pi\delta_{ij}$, where δ_{ij} is the kronecker symbol. Note that the associated primitive reciprocal lattice vectors are constructed as **k** space from the concept of crystal diffraction.

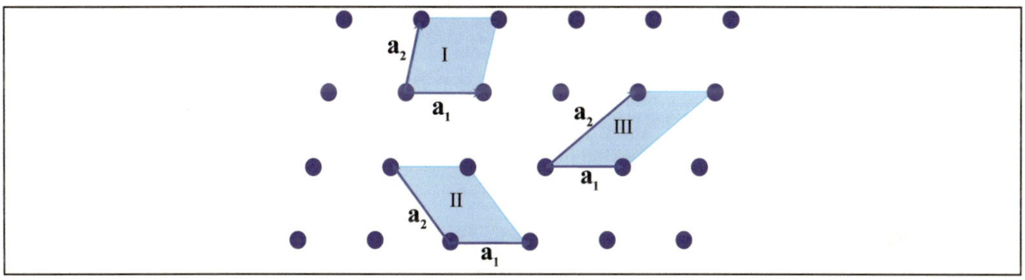

Fig. 1. Primitive unit cell in real space

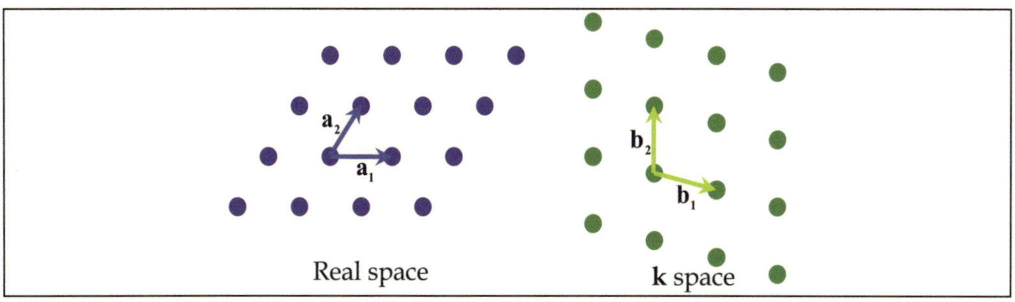

Fig. 2. Relationship between the real space and **k** space

We will find that, in following sections, the discrete translational symmetry of a phononic crystal allows us to classify the elastic/acoustic waves with a wave vector **k**. The propagating modes can be written in "Bloch form," consisting of a plane wave modulated by a function that shares the periodicity of the lattice (Joannopoulos et al., 1995):

$$\mathbf{P}_{\mathrm{k}}(\mathbf{r}) = e^{i\mathbf{k}\cdot\mathbf{r}}\mathbf{u}_{\mathrm{k}}(\mathbf{r}) = e^{i\mathbf{k}\cdot\mathbf{r}}\mathbf{u}_{\mathrm{k}}(\mathbf{r}+\mathbf{R}). \tag{2}$$

The important feature of the Bloch states is that different values of **k** do not necessarily lead to different modes. It is clear that a mode with wave vector **k** and a mode with wave vector **k+G** are the same mode, where **G** is a reciprocal lattice vector. The wave vector **k** serves to specify the phase relationship between the various cells that are described by **u**. If **k** is increased by **G**, then the phase between cells is increased by **G**·**R**, which we know is $2\pi n$ ($n = l_1N_1+l_2N_2+ l_3N_3$ is an integer) and not really a phase difference at all. So incrementing **k** by **G** results in the same physical mode. This means that we can restrict our attention to a finite zone in reciprocal space in which we *cannot* get from one part of the volume to another by adding any **G**. All values of **k** that lie outside of this zone, by definition, can be reached from within the zone by adding **G**, and are therefore redundant labels shown in Fig. 3. This zone is the so-called *Brillouin zone*.

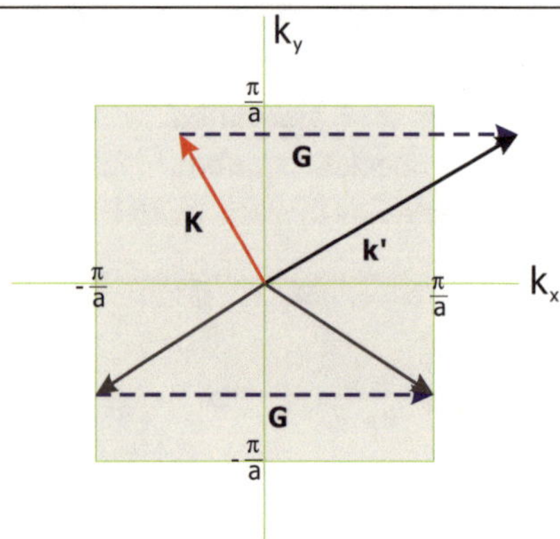

Fig. 3. All values of **k** that lie outside of this zone, by definition, can be reached from within the zone by adding **G**

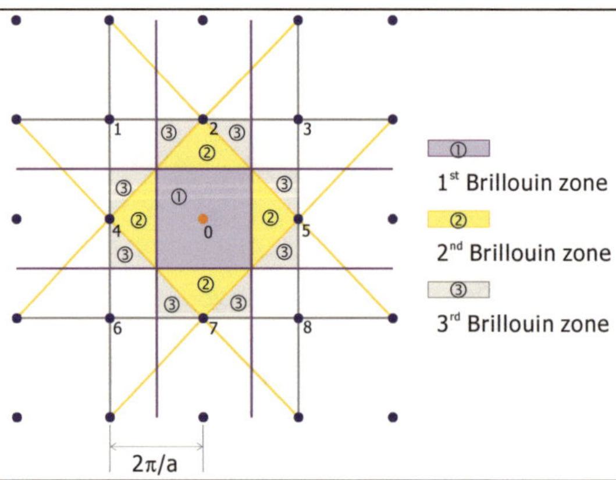

Fig. 4. Brillouin zones in a square lattice

By the periodicity of the reciprocal lattice, any reciprocal lattice point which represents a wave vector **k** outside the first Brillouin zone can be found a corresponding point in the first Brillouin zone. Therefore, the wave vectors **k** can always be confined in the first Brillouin zone. In the square lattice, only the wave vectors **k** in the region of the first Brillouin zone between $-\pi/a$ to π/a (the lattice constant is a) need to be considered. The Fig. 4 shows the *first*, *second*, and *third Brillouin zones*. For more details, it is best to consult the first few chapters of a solid-state physics text, such as Kittel, 1996, or consult the appendix of popular photonic text like Joannopoulos et al. 1995 and Johnson & Joannopoulos, 2001, 2003.

2.2 Equation of motion

This section provides a brief introduction of the theory of analyzing acoustic wave propagation in inhomogeneous media like as phononic band structures. The theory in this chapter can also be used to discuss acoustic wave propagation in homogeneous media because a homogeneous medium is symmetric with respect to any periodicity.

In an inhomogeneous linear elastic medium with no body force, the equation of motion of the displacement vector $\mathbf{u}(\mathbf{r},t)$ can be written as

$$\rho(\mathbf{r})\ddot{u}_i(\mathbf{r},t) = \partial_j[C_{ijmn}(\mathbf{r})\partial_n u_m(\mathbf{r},t)], \tag{3}$$

where $\mathbf{r}=(\mathbf{x},z)=(x,y,z)$ is the position vector, t is the time variable, and $\rho(\mathbf{r})$ and $C_{ijmn}(\mathbf{r})$ are the position-dependent mass density and elastic stiffness tensor, respectively. The following discussion considers a periodic structure consisting of a two-dimensional periodic array (x-y plane) of material A embedded in a background material B shown in Fig. 5. It is noted that when the properties of materials A and B tend to coincide, the homogeneous case is recovered.

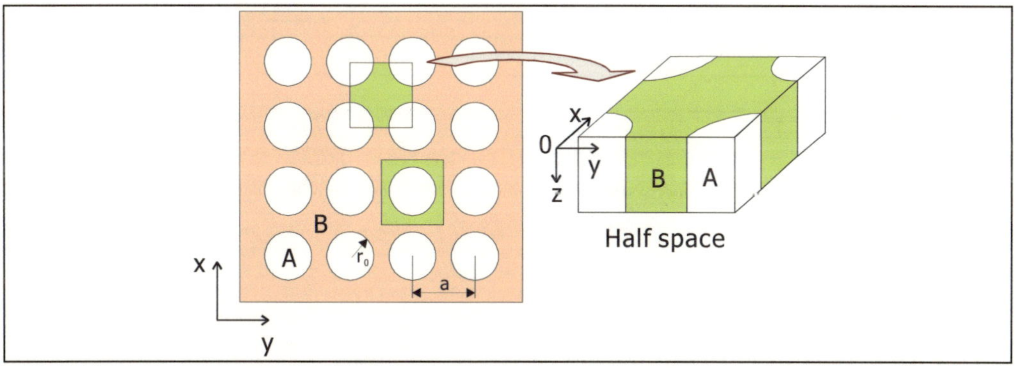

Fig. 5. Periodic structures with square lattice. When the properties of materials A and B tend to coincide, the homogeneous case is recovered

To calculate the dispersion diagrams of periodic structures, this study uses COMSOL Multiphysics software to apply the Bloch boundary condition to the unit cell domain in the FEM method. Based on the periodicity of phononic crystals, the displacement and stress components in the periodic structure are expressed as follows:

$$u_i(\mathbf{x},t) = e^{i\mathbf{k}\cdot\mathbf{x}}U_i(\mathbf{x},t), \tag{4}$$

$$\sigma_{ij}(\mathbf{x},t)=e^{i\mathbf{k}\cdot\mathbf{x}}T_{ij}(\mathbf{x},t), \tag{5}$$

where $\mathbf{k}=(k_1,k_2)$ is the Bloch wave vector, and $i=\sqrt{-1}$; $U_i(\mathbf{x},t)$ and $T_{ij}(\mathbf{x},t)$ are periodic functions that satisfy the following relation (Tanaka et al., 2000):

$$U_i(\mathbf{x}+\mathbf{R},t)=U_i(\mathbf{x},t), \tag{6}$$

$$T_{ij}(\mathbf{x}+\mathbf{R},t)=T_{ij}(\mathbf{x},t), \tag{7}$$

where $\mathbf{R}$ is a lattice translation vector with components of R_1 and R_2 in the x and y directions. The relationships between the original variables $u_i(\mathbf{x},t)$, $\sigma_{ij}(\mathbf{x},t)$, $u_i(\mathbf{x}+\mathbf{R},t)$, and $\sigma_{ij}(\mathbf{x}+\mathbf{R},t)$ about the Bloch boundary conditions are characterized as:

$$u_i(\mathbf{x}+\mathbf{R},t)=e^{i\mathbf{k}\cdot(\mathbf{x}+\mathbf{R})}U_i(\mathbf{x}+\mathbf{R},t)=e^{i\mathbf{k}\cdot\mathbf{R}}e^{i\mathbf{k}\cdot\mathbf{x}}U_i(\mathbf{x},t)=e^{i\mathbf{k}\cdot\mathbf{R}}u_i(\mathbf{x},t), \tag{8}$$

$$\sigma_{ij}(\mathbf{x}+\mathbf{R},t)=e^{i\mathbf{k}\cdot(\mathbf{x}+\mathbf{R})}T_{ij}(\mathbf{x}+\mathbf{R},t)=e^{i\mathbf{k}\cdot\mathbf{R}}e^{i\mathbf{k}\cdot\mathbf{x}}T_{ij}(\mathbf{x},t)=e^{i\mathbf{k}\cdot\mathbf{R}}\sigma_{ij}(\mathbf{x},t). \tag{9}$$

The Bloch calculations in this study record the variation of the displacements, stress fields, and eigen-frequencies as the wave vector increases. By using the FEM, the unit cell is meshed and divided into finite elements which connect by nodes, and is used to obtain the eigen-solutions and mechanical displacements. The types of finite elements used in this chapter are the default element types, Lagrange-quadratic, in COMSOL Multiphysics. In order to simulate the dispersion diagrams, the wave vectors are condensed inside the first Brillouin zone in the square lattice. According to the above theories, the results of dispersion relations in a band structure along the $\Gamma-\mathrm{X}-\mathrm{M}-\Gamma$ are characterized and presented in the following sections.

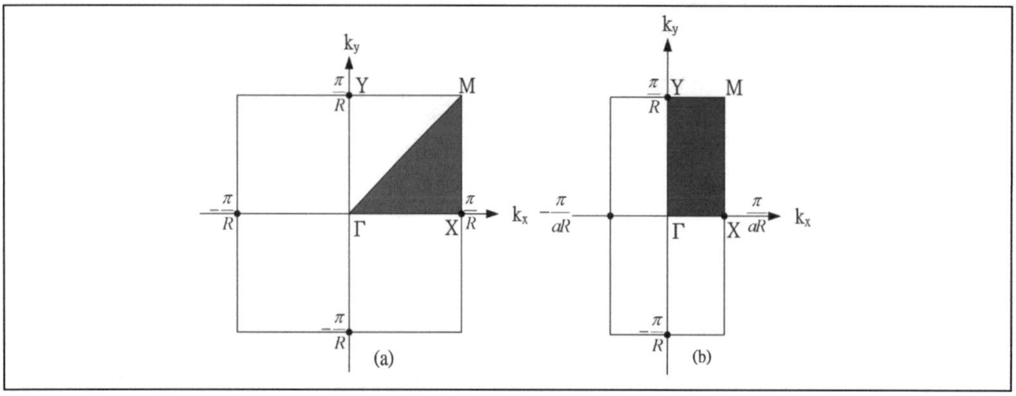

Fig. 6. Brillouin regions of the square and rectangular lattices

This chapter considers a periodic homogeneous medium with square lattice and phononic structures with square and rectangular lattices. These lattices consist of periodic structures that form two-dimensional lattices with lattice spacing R (square lattice) and lattice spacing aR (rectangular lattice). The term a is a scale from 0.1 to 2.0 in this chapter. The periodic structures are parallel to the z-axis. Figures 6(a) and 6(b) illustrate the Brillouin regions of the square lattice and rectangular lattice, respectively. In the square lattice, Fig. 6(a) shows the irreducible part of the Brillouin zone, which is a triangle with vertexes Γ, X, and M. Similarly, Fig. 6(b) shows the irreducible part of the Brillouin zone of a rectangular lattice due to the geometric anisotropy, which is a rectangle with vertexes Γ, X, M, and Y, and the same as discussing the material anisotropy (Wu et al., 2004).

The finite element method divides a unit cell with a three-dimensional model into finite elements connected by nodes. The FEM obtains the eigen-solutions and contours of a mode shape. To simulate the dispersion diagrams, the wave vectors are condensed inside the first Brillouin zone in the square and rectangular lattices. Using the theories above, the following section presents the results of dispersion relations in a band structure for the $\Gamma-X-M-\Gamma$ square lattice or isotropic materials, and $\Gamma-X-M-Y-\Gamma$ rectangular lattice or anisotropic materials. Note that the 2D FEM model calculates the dispersion relations of mixed polarization modes, while the 3D FEM model describes the dispersion relations of mixed and transverse polarization modes.

3. Acoustic wave in homogeneous media

It can be noted that a homogeneous medium is symmetric with respect to any periodicity, and it can be shown that the results for an infinite homogeneous medium can be cast in the form appropriate for a periodic medium. In this section, we introduce the mixed polarization modes and transverse polarization modes in a homogeneous medium. Displacement fields (polarizations) are also investigated and used to distinguish the different modes in the dispersion relations. The aluminum and quartz are adopted for examples and discussed in the section. The wave velocities of different propogating modes are also observed and discussed.

3.1 Isotropic medium

In Fig. 5, when the properties of materials A and B tend to coincide, the homogeneous case is recovered. Consider a periodic structure consisting of aluminum (Al) circular cylinders embedded in a background material of Al forming a two-dimensional square lattice with lattice spacing R. It means this is a homogeneous medium in a 3D FEM model. Figure 7 shows the dispersion relations along the boundaries of the irreducible part of the Brillouin zone $\Gamma-X-M-\Gamma$. The vertical axis is the frequency (Hz) and the horizontal axis is the reduced wave vector $k^*=kR/\pi$. Here, k is the wave vector along the Brillouin zone. The Young's modulus E, Poisson's ratio ν, and density ρ of the material Al utilized in this example are E=70 GPa, ν=0.33, and ρ=2700 kg/m^3.

As the elastic waves propagate along the x axis, the nonvanishing displacement fields of the shear horizontal mode (SH), shear vertical mode (SV), and longitudinal mode (L) are u_y, u_z, and u_x respectively. It is noted that wave velocity $c_{S,L}=d\omega/dk=2R*m_{S,L}$, so the slopes of dispersion curves in the $\Gamma-X$ section of Fig. 7 are exactly the straight lines and can be explained as the wave velocities of shear (S) and longitudinal (L) modes. Here, $m_{S,L}$ are the slopes of shear and longitudinal modes in Fig. 7. It is noted that the wave velocities of shear horizontal mode and shear vertical mode are the same in an isotropic material. From the results in Fig. 7, the wave velocities of shear and longitudinal modes are 3119 and 6174 m/s. As we know, the wave velocities of shear and longitudinal modes in an isotropic material can be obtain from

$$c_S = \sqrt{\frac{E}{\rho}\frac{1}{2(1+\nu)}} = 3122 \quad m/s, \tag{10}$$

$$c_L = \sqrt{\frac{E}{\rho}\frac{(1-\nu)}{(1+\nu)(1-2\nu)}} = 6031 \quad m/s. \tag{11}$$

Note that the FEM method can easily describe the mode characteristics. Figure 8 shows the vibration mode shapes of unit cell for shear and longuitudinal modes in X point. In this example, Fig. 8(a) is a shear horizontal mode with mode vibrating displacement along the y direction when the wave propagates along the x direction ($\Gamma - X$ direction). Also, Fig. 8(b) is a shear vertical mode with mode vibrating displacement along z direction, and Fig. 8(c) is a longitudinal mode with mode vibrating displacement along x direction. The arrows shown in Fig. 8 are the polarizations.

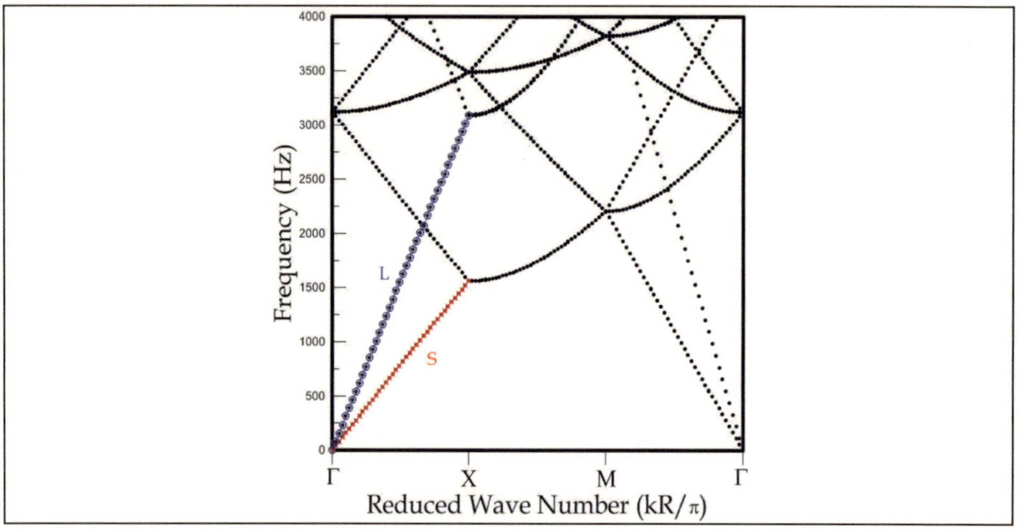

Fig. 7. The dispersion relations of homogeneous and isotropic material Al along the boundaries of the irreducible part of the Brillouin zone $\Gamma - X - M - \Gamma$

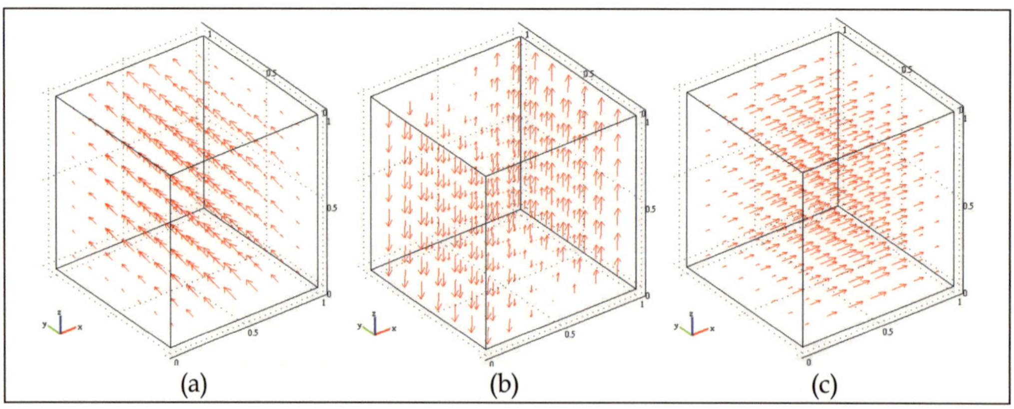

Fig. 8. (a) shear horizontal mode (b) shear vertical mode, (c) longitudinal mode in the Al

3.2 Anisotropic medium

Similarly, the method in this chapter is used to discuss the wave velocities of acoustic modes in an anisotropic material. Consider a periodic structure consisting of quartz circular cylinders embedded in a background material of quartz forming a two-dimensional square lattice with lattice spacing R. This is also a homogeneous medium. The quartz is a piezoelectric and anisotropic material. The density ρ =2651 kg/m^3. The elastic constants, piezoelectric constants, and relative permittivity of quartz utilized in this example are shown in Tables 1-3. The piezoelectric material, quartz, is a complete structural-electrical material, and thus all piezoelectric material properties were defined and entered into the FEM model. Figure 9 shows the dispersion relations along the boundaries of the irreducible part of the Brillouin zone $\Gamma-X-M-Y-\Gamma$ due to the material anisotropy. In the calculations, the x-y plane is parallel to the (001) plane and the x axis is along the [100] direction of quartz. The vertical axis is the frequency in Hz unit and the horizontal axis is the reduced wave vector.

86.7362	6.98527	11.9104	17.9081	0	0
6.98527	86.7362	11.9104	-17.9081	0	0
11.9104	11.9104	107.194	0	0	0
17.9081	-17.9081	0	57.9428	0	0
0	0	0	0	57.9492	17.9224
0	0	0	0	17.9224	39.9073

Table 1. The elastic constants of quartz in GPa unit

-0.19543	0.19543	0	-0.1212	0	0
0	0	0	0	0.12127	0.19558
0	0	0	0	0	0

Table 2. The piezoelectric constants of quartz in C/m^2 unit

4.4093	0	0
0	4.4092	0
0	0	4.68

Table 3. The relative permittivity of quartz

Shown in $\Gamma-X$ section of Fig. 9, the cross symbols represent the quasi shear horizontal (quasi-SH) mode. The square symbols represent the quasi shear vertical (quasi-SV) mode and the open circle symbols represent the quasi longitudinal (quasi-L) mode. The wave velocities of quasi-SH, quasi-SV, and quasi-L modes along x axis are 3306, 5116, and 5741 m/s. Similarly, The wave velocities of quasi-SH, quasi-SV, and quasi-L modes along y axis ($\Gamma-Y$ section) are 3922, 4311, and 6009 m/s respectively.

Figure 10 also shows the vibration mode shapes of unit cell for quasi-SH, quasi-SV, and quasi-L modes in X point. The arrows shown in Fig. 10 are the polarizations. In this example, the quasi-longitudinal and quasi-transverse waves are almost indistinguishable from the truly longitudinal and truly transverse waves of Fig. 8.

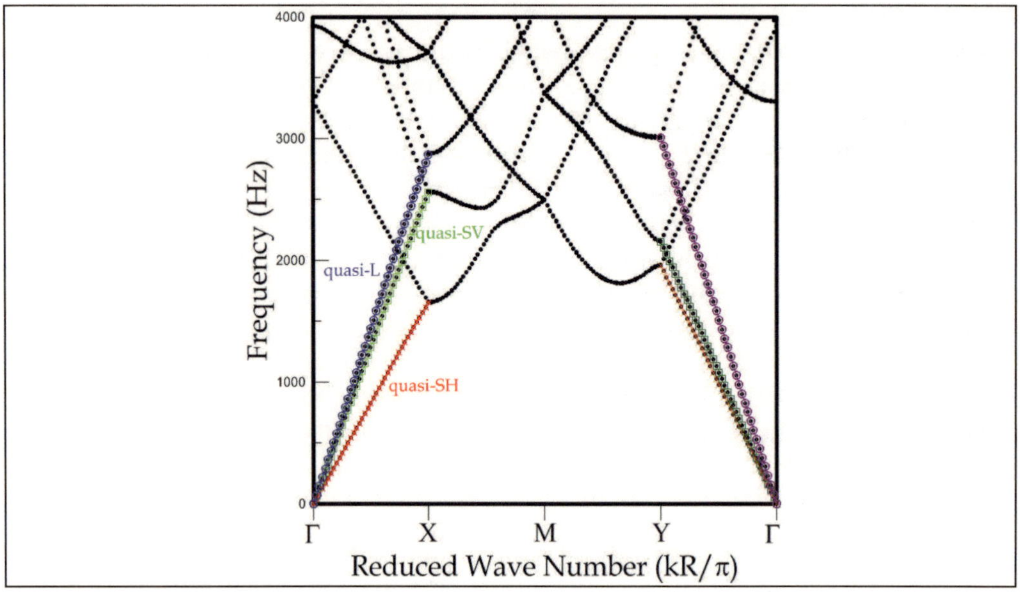

Fig. 9. The dispersion relations of homogeneous material quartz along the boundaries of the irreducible part of the Brillouin zone $\Gamma - X - M - Y - \Gamma$

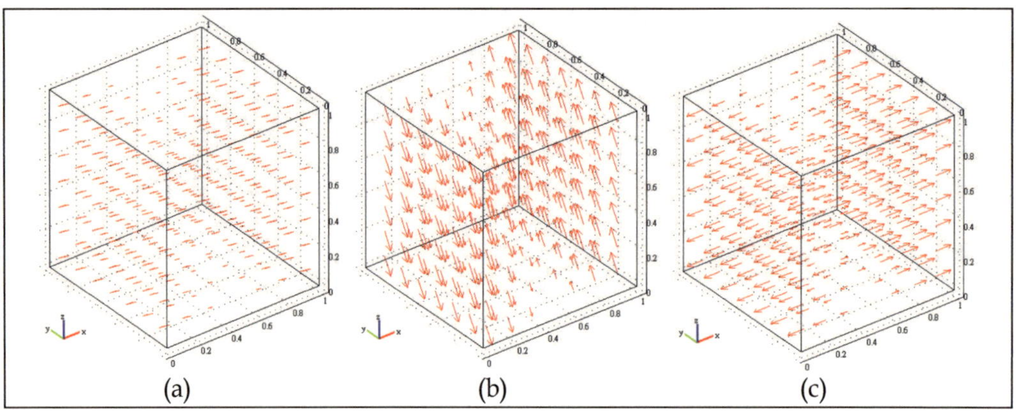

Fig. 10. (a) quasi shear horizontal mode (b) quasi shear vertical mode, (c) quasi longitudinal mode in the quartz

From the discussion, it shows that the method adopted in this chapter can be used to discuss the wave propagations in isotropic and anisotropic media.

4. Acoustic wave in inhomogeneous media

Previous studies on photonic crystals raise the exciting topic of phononic crystals. This section presents the results of acoustic waves in inhomogeneous media, Al/Ni periodic structures and phononic crystals with reticular geometric structures. It also discusses the tunable band gaps in the acoustic waves of two-dimensional phononic crystals with reticular geometric structures using the 2D and 3D finite element methods. This section calculates and discusses the band gap variations of the bulk modes due to different sizes of reticular geometric structures. Results show that adjusting the orientation of the reticular geometric structures can increase or decrease the total elastic band gaps for mixed polarization modes.

4.1 Periodic structure with two media

It is necessary and worthy to provide evidence supporting the FEM method's (COMSOL Multiphysics) ability to perform Bloch calculations with two media. This chapter compares the dispersion relations of Al/Ni band structure using the PWE method with the results of using the FEM method. Consider a phononic structure consisting of Al circular cylinders embedded in a background material of Ni to form a two-dimensional square lattice with lattice spacing R. Figure 11 shows the dispersion relations along the boundaries of the irreducible part of the Brillouin zone in Fig. 6(a) with filling ratio 0.6. The vertical axis represents the normalized frequency $\omega^* = \omega R / C_t$ and the horizontal axis represents the reduced wave number $k^* = kR / \pi$. Here, C_t and k are the shear velocity of Ni and the wave vector along the Brillouin zone, respectively. The Young's modulus E, Poisson's ratio ν , and density ρ of the material Ni utilized in this example are E=214 GPa, ν =0.336, and ρ =8905 kg/m³.

The diamond symbols represent the dispersion relations of the transverse polarization modes (shear vertical modes), and the cross symbols represent the mixed polarization modes (shear horizontal mode coupled with longitudinal mode) in the PWE method. The open circles represent the dispersion relations of all modes in the FEM method with a 3D model. The results of the FEM method match well with those of the PWE method. In the similar cases, when the differences of mass densities and elastic constants between the two periodic materials are larger, the convergence of the PWE method is slower and costs more CPU time.

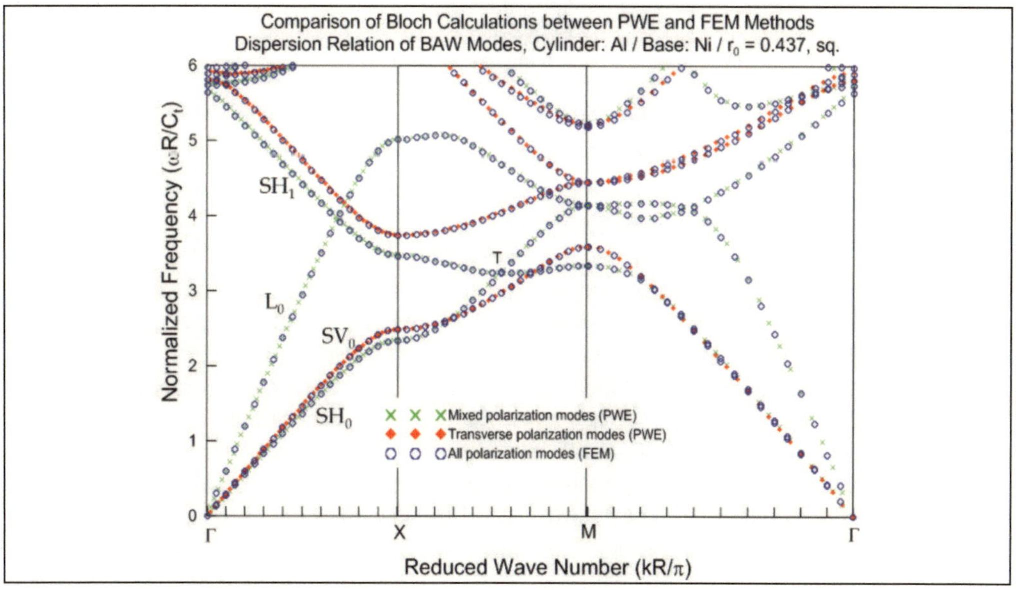

Fig. 11. Comparison of Bloch calculations between the PWE and FEM methods

As the elastic waves propagate along the x axis, the nonvanishing displacement fields of the shear horizontal mode, shear vertical mode, and longitudinal mode are u_y, u_z, and u_x respectively. For the sequence modes appear, the modes are always the same. When representing the whole wave vector space by the first Brillouin zone alone, they appear as further branches from higher Brillouin zones. In this example, the phase velocities of the SV_0 mode (diamond symbols) are larger than those of the SH_0 mode. The boundary of the Brillouin zone X-M of Fig. 11 represents the dispersion of the bulk waves with propagating direction varied 0 deg~ 45 deg counterclockwise away from the x direction.

4.2 Periodic structure with single medium

Figure 12(a) depicts a two-dimensional phononic crystal with the reticular geometric structures of square lattice. These reticular structures are parallel to the z-axis. In a perfect two-dimensional phononic crystal, the periodic structure is constant in the z direction and the size of the structure is infinite in the x and y directions. To analyze the dispersion relations of all bulk acoustic modes in this band structure, the FEM should consider the 3D model in Fig. 12(c). The dimensions of the unit cell in Fig. 12(a) are c=d=0.8R and R=h=1 in the calculations.

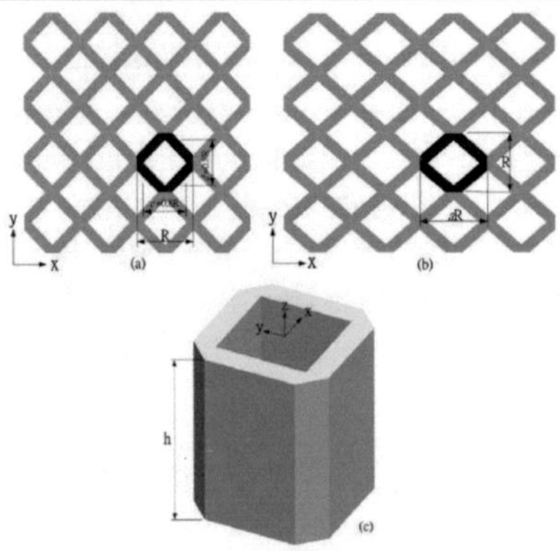

Fig. 12. (a) square lattice with lattice spacing R and (b) rectangular lattice with lattice spacing aR along x-axis and R along y-axis, (c) a unit cell with reticular structures in a 3D FEM model

The material of the reticular structures in the unit cell in this chapter is aluminum. Figure 12(c) shows a diagram of the unit square lattice in a 3D FEM model. The periodicity of phononic crystals along the z direction is used to calculate the dispersion relations of the mixed and transverse polarization modes. The types of finite elements used for the 2D and 3D cases are the default element types, Lagrange-Quadratic, in COMSOL Multiphysics. Figure 13 shows the dispersion relations of the mixed and transverse polarization modes along the boundaries of the irreducible part of the Brillouin zone in Fig. 6(b) with the scales R=h=1, c=0.8, and a=1.2. The horizontal axis represents the reduced wave number along $\Gamma-X-M-Y-\Gamma$ and the vertical axis represents the frequency (Hz). Note that this band structure shows no full band gap of the mixed and transverse polarization modes. Adopting the 2D FEM model to discuss the mixed polarization modes in this kind of band structure shows that there is only one full frequency band gap in Fig. 13, located at 3311 ~ 3400 Hz. Figure 13 compares the 3D and 2D FEM models. Open circles represent the dispersion relations of mixed polarization modes in the 2D FEM model, while solid circles represent the results of all bulk modes in the 3D FEM model. Figure 14 shows the eigenmode shapes with 4×4 supercell of total displacements for M_1 and M_2 modes indicated in Fig. 13. These figures clearly show the phenomena of wave localizations in this reticular geometric structure. Note that the FEM method can easily describe the mode characteristics. In this chapter, M_1 is a shear horizontal mode with mode vibrating displacement along the y direction when the wave propagates along the x direction ($\Gamma - X$ direction). Also, M_2 is a shear vertical mode with mode vibrating displacement along z direction, and it does not couple with the mixed polarization modes.

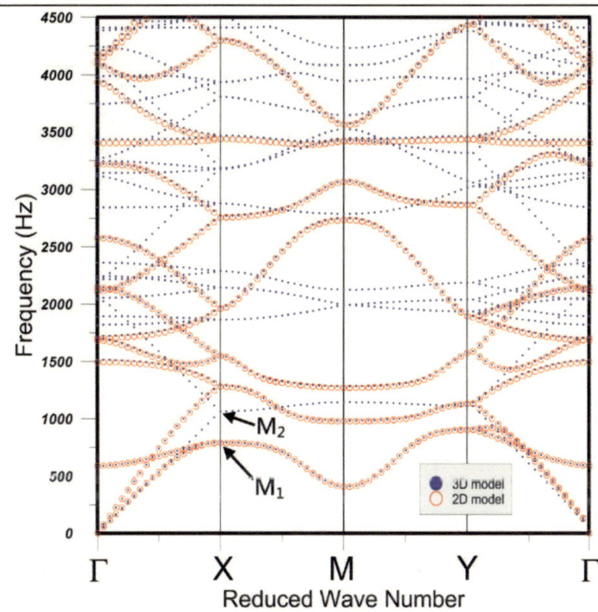

Fig. 13. The dispersion relations of the mixed and transverse polarization modes along the boundaries of the irreducible part of the Brillouin zone with the scales R=h=1, c=0.8, and a=1.2

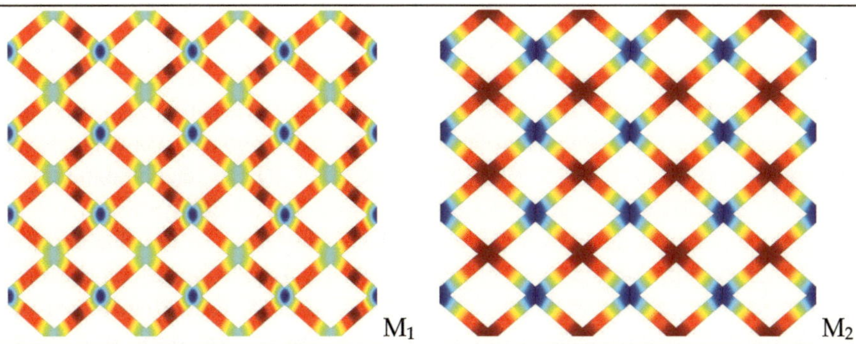

Fig. 14. The eigenmode shapes with 4×4 supercell of total displacements for M_1 and M_2 modes indicated in Fig. 13

The following discussion addresses several parameters of the reticular geometric in this chapter. First, the effect of filling fraction is discussed when the parameters c=d varied from 0.1 to 0.9 in Fig. 12(a). Figure 15 shows the distribution of the total band gaps of mixed polarization modes, in which only one total band gap appears at approximately 3560 ~ 3736 Hz in c=d=0.8. The horizontal axis represents the parameter c, and the vertical axis represents frequency (Hz). Figure 15 also shows the 2D diagrams of the reticular geometric structures with c=d=0.1, 0.5, and 0.8.

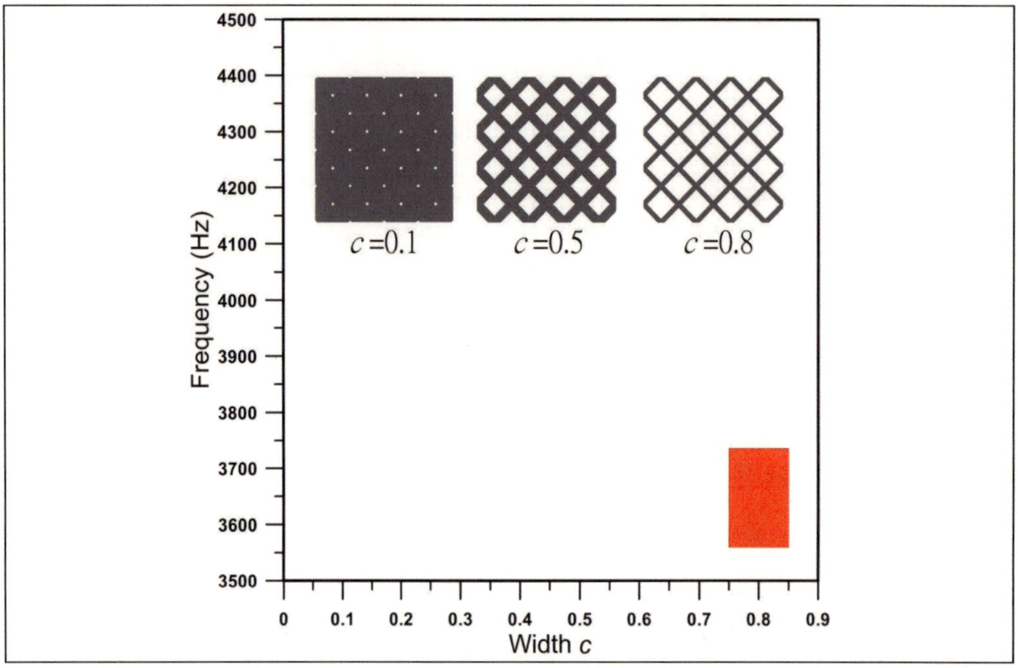

Fig. 15. The band gap width with parameters c=d varying from 0.1 to 0.9 when the vertical range is selected from 3500 to 4500 Hz

On the other hand, the scale a in Fig. 12(b) varies from 0.1 to 2.0 along the x direction and the width of the unit cell along y direction remains 1.0 in the Bloch calculations. Changing the scale a from 0.1 to 2.0 can tune the full frequency band gaps of mixed polarization modes. Using detailed calculations of dispersion relations of reticular geometric structures with scale a=0.1 to 2.0, Fig. 16 shows the band gap widths with the scale a from 0.1 to 2.0 when the vertical range ranges from 2400 to 5200 Hz. The horizontal axis ranges from 0 to 2.0, and the vertical axis represents frequency (Hz). No full frequency band gap exists when the scale a are 0.1, 0.2, 0.3, 0.4, 0.6, 0.7, 1.5, 1.6, and 1.7. These results clearly show that changing the scale a can increase or decrease the full frequency band gap.

It is noted that the unit cells with a=0.5 and 2.0 are the same in the Bloch calculations. However, the dispersion phenomena is similar except for the scalar of the eigenmode frequencies in the vertical axis of dispersion relations. In both cases, there is only one total band gap of the mixed polarization modes. The location of the band gap ranges from approximately 5009 to 5017.4 Hz with a=0.5, while that for a=2.0 ranges from approximately 2504.5 to 2508.7 Hz.

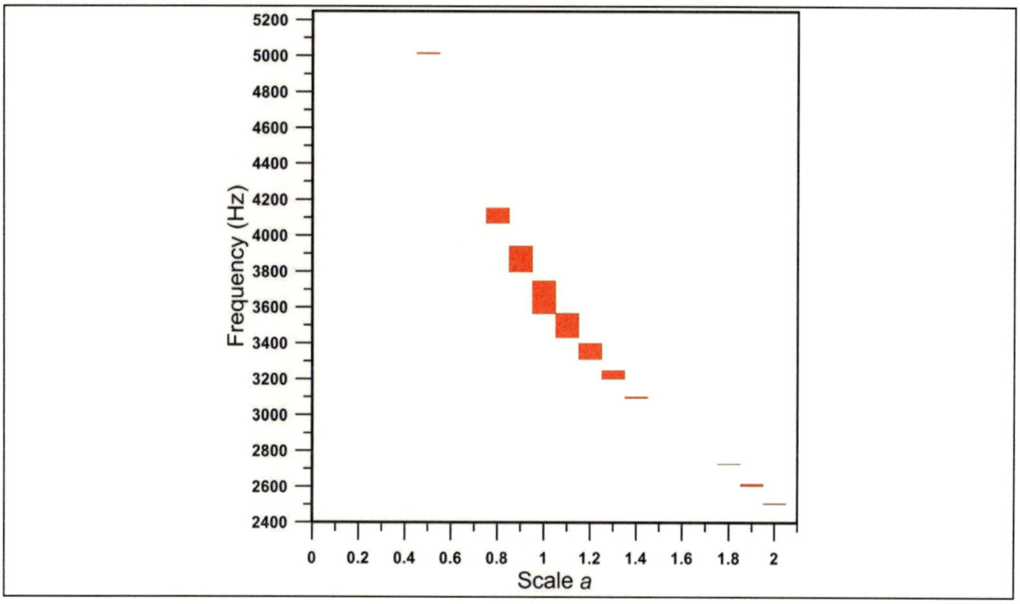

Fig. 16. The band gap widths with the scale a from 0.1 to 2.0

Finally, the rotating angles of reticular geometric structures were changed to analyze the distribution of total band gaps. Figure 17 shows the 2D diagrams of unit rectangular lattices in different rotating angles D=30 deg, 45 deg, 75 deg, and 90 deg. In these cases, the widths of aluminum remain constant, 0.14R, in the reticular geometric structures with different rotating angles in the calculations. Figure 18 shows the band gap widths of rectangular lattices with different rotating angles of reticular geometric structures. Based on the symmetry of the geometry, the different angles in the Bloch calculations were adopted from 15 deg ~ 90 deg. In the calculated results, no band gap is detected from D=5 deg to 65 deg.

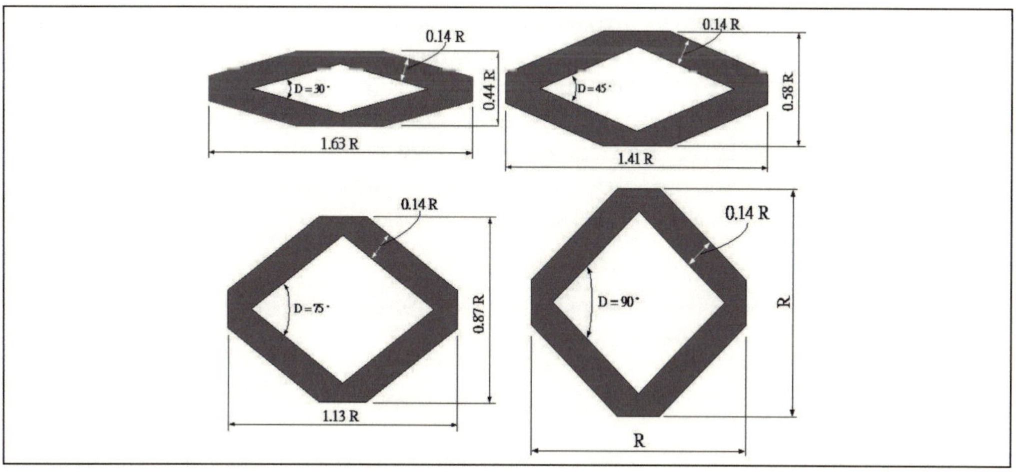

Fig. 17. 2D diagrams of unit rectangular lattices in different rotating angles D=30 deg, 45 deg, 75 deg, and 90 deg

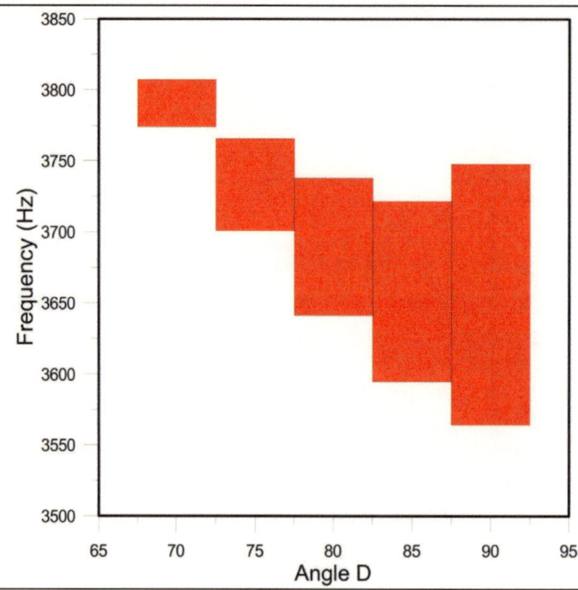

Fig. 18. The band gap widths of the rectangular lattices with different rotating angles of reticular geometric structures

5. Conclusion

This chapter examines and discusses the acoustic waves in homogeneous medium and inhomogeneous medium, periodic structures with two media and one medium with geometrical periodicity. The wave velocities of shear and longitudinal modes in an isotropic material and those of quasi-SV, quasi-SH, and quasi-L modes in an anisotropic material are obtained using the finite element method. This method also discusses the tunable frequency band gaps of bulk acoustic waves in two-dimensional phononic crystals with reticular geometric structures using the 2D and 3D finite element methods. This study adopts the finite element method to calculate dispersion relations, avoiding the numerical errors, Gibbs phenomenon, from the PWE method. Results show that changing the filling fraction, scale a, and the rotating angles of unit lattices in the reticular geometric structures can increase or decrease the elastic/acoustic band gaps. The effect discussed in this chapter can be utilized to enlarge the phononic band gap frequency and may enable the study of the frequency band gaps of elastic/acoustic modes in special phononic band structures.

6. Acknowledgment

The authors thank the National Science Council (NSC 97-2218-E-150-006, 98-2221-E-150-026, and 99-2628-E-150-001) of Taiwan for financial support.

7. References

Axmann, W. & Kuchment, P. (1999). An efficient finite element method for computing spectra of photonic and acoustic band-gap materials. J. Comput. Phys. Vol. 150, pp. 468, ISSN 0021-9991

Burger, M. S., Osher, J., & Yablonovitch, E. (2004). Inverse Problem Techniques for the Design of Photonic Crystals. IEICE Trans. Electron, E87C, 258-265.

Checoury, X. & Lourtioz, J. M. (2006). Wavelet method for computing band diagrams of 2D photonic crystals. Optics Communications Vol. 59, pp. 360, ISSN 0030-4018

Chiang, P. J., Yu, C. P., & Chang, H. C. (2007). Analysis of two-dimensional photonic crystals using a multidomain pseudospectral method. Phys. Rev. E Vol. 75, pp. 026703, ISSN 1539-3755

Dobson, D. C. (1999). An efficient method for band structure calculations in 2D photonic crystals. J. Comput. Phys. Vol. 149, pp. 363, ISSN 0021-9991

Garica-Pablos, D., Sigalas, M., Montero de Espinosa, F. R., Torres, M., Kafesaki, M., and Garcia, N. (2000). Theory and Experiments on Elastic Band gaps. Phys. Rev. Lett. Vol. 84, pp. 4349, ISSN 0031-9007

Huang, G. L. & Sun, C. T. (2010). Band Gaps in a Multiresonator Acoustic Metamaterial. ASME J. Vib. Acoust. Vol. 132, pp. 031003. ISSN 1048-9002

Huang, Z. G. & Chen, Z. Y. (2011). Acoustic Waves in Two-dimensional Phononic Crystals with Reticular Geometric Structures. ASME J. Vib. Acoust. Vol. 133(3), pp.031011, ISSN 1048-9002

Huang, Z. G. & Wu, T.-T. (2005). Temperature effects on bandgaps of surface and bulk acoustic waves in two-dimensional phononic crystals. IEEE Trans. Ultrason. Ferroelectr. Freq. Control Vol. 52, pp. 365, ISSN 0885-3010

Hussein, M. I. (2009). Reduced Bloch mode expansion for periodic media band structure calculations. Proceedings of the Royal Society A Vol. 465, pp. 2825-2848, ISSN 1364–5021

Joannopoulos, J. D., Meade, R. D. & Winn, J. N. (1995) . Photonic Crystals: Molding the flow of light, ISBN: 978-0691124568, Princeton University Press, Princeton, NJ.

Johnson, S. G. & Joannopoulos, J. D. (2001). Block-iterative frequency-domain methods for Maxwell's equations in a planewave basis. Optics Express Vol. 8, pp. 173, ISSN 1094-4087

Johnson, S. G. & Joannopoulos, J. D. (2003). PHOTONIC CRYSTALS: The road from theory to practice, ISBN 978-0792376095, Kluwer academic publishers, Boston.

Jun, S., Cho, Y. S., & Im, S. (2003). Moving least-square method for the band-structure calculation of 2D photonic crystals. Optics Express Vol. 11, pp. 541, ISSN 1094-4087

Kafesaki, M. & Economou, E. N. (1999). Multiple-scattering theory for three-dimensional periodic acoustic composites. Phys. Rev. B Vol. 60, pp. 11993, ISSN 1098-0121

Kittel, C. (1996). Introduction to Solid State Physics, ISBN 978-0-471-41526-8, 7th ed., John Wiley & Sons. Inc.

Kushwaha, M. S., Halevi, P., Dobrzynski, L. & Djafari-Rouhani, B. (1993). Acoustic Band Structure of Periodic Elastic Composites. Phys. Rev. Lett. Vol. 71, pp. 2022, ISSN 0031-9007

Laude, V., Wilm, M., Benchabane, S., Khelif, A. (2005). Full band gap for surface acoustic waves in a piezoelectric phononic crystal. Phys. Rev. E Vol. 71, pp. 036607, ISSN 1539-3755

Leung, K. M. & Liu, Y. F. (1990). Full vector wave calculation of photonic band structures in face-centered-cubic dielectric media. Phys. Rev. Lett. Vol. 65, pp. 2646, ISSN 0031-9007

Leung, K. M. & Qiu, Y. (1993). Multiple-scattering calculation of the two-dimensional photonic band structure. Phys. Rev. B Vol. 48, pp. 7767, ISSN 1098-0121

Moreno, E., Erni, D., & Hafner, C. (2002). Band structure computations of metallic photonic crystals with the multiple multipole method. Phys. Rev. B Vol. 65, pp. 155120, ISSN 1098-0121

Pendry, J. B. & MacKinnon, A. (1992). Calculation of photon dispersion relations. Phys. Rev. Lett. Vol. 69, pp. 2772, ISSN 0031-9007

Psarobas, I. E. & Stefanou, N. (2000). Scattering of elastic waves by periodic arrays of spherical bodies. Phys. Rev. B Vol. 62, pp. 278, ISSN 1098-0121

Salehian, A. & Inman, D. J. (2010). Micropolar Continuous Modeling and Frequency Response Validation of a Lattice Structure. ASME J. Vib. Acoust. Vol. 132, pp.011010, ISSN 1048-9002

Sun, J. H. & Wu, T.-T. (2005). Analyses of mode coupling in joined parallel phononic crystal waveguides. Phys. Rev. B Vol. 71, pp. 174303, ISSN 1098-0121

Tanaka, Y. & Tamura, S. (1998). Surface acoustic waves in two-dimensional periodic elastic structures. Phys. Rev. B Vol. 58, pp. 7958, ISSN 1098-0121

Tanaka, Y., Tomoyasu, Y., & Tamura, S. I. (2000). Band structure of acoustic waves in phononic lattices: Two-dimensional composites with large acoustic mismatch. Phys. Rev. B Vol.62, no. 11, 7387–7392.

Wang, X., Zhang, X. G., Yu, Q. & Harmon, B. N. (1993). Multiple-scattering theory for electromagnetic waves. Phys. Rev. B Vol. 47, pp. 4161, ISSN 1098-0121

Wu, T.-T. & Huang, Z. G. (2004). Level repulsion of bulk acoustic waves in composite materials. Phys. Rev. B Vol. 70, pp. 214304, ISSN 1098-0121

Wu, T.-T., Huang, Z. G., & Lin, S. (2004). Surface and bulk acoustic waves in two-dimensional phononic crystals consisting of materials with general anisotropy. Phys. Rev. B Vol. 69, pp. 094301, ISSN 1098-0121

Wu, T.-T., Huang, Z. G., Tsai, T. C. & Wu, T. C. (2008). Evidence of complete band gap and resonances in a plate with periodic stubbed surface. Applied Physics Letters Vol. 93, pp. 111902, ISSN 0003-6951

Yan, Z. Z. & Wang, Y. S. (2006). Wavelet-based method for calculating elastic band gaps of two-dimensional phononic crystals. Phys. Rev. B Vol. 74, pp. 224303, ISSN 1098-0121

Yang, H. Y. D. (1996). Finite difference analysis of 2-D photonic crystals. IEEE Trans. Microwave Theory Tech. Vol. 44, pp. 2688, ISSN 0018-9480

2

Topological Singularities in Acoustic Fields due to Absorption of a Crystal

V. I. Alshits[1,2], V. N. Lyubimov[1] and A. Radowicz[3]
[1]*A.V. Shubnikov Institute of Crystallography, Russian Academy of Sciences, Moscow,*
[2]*Polish-Japanese Institute of Information Technology, Warsaw,*
[3]*Kielce University of Technology, Kielce*
[1]*Russia*
[2,3]*Poland*

1. Introduction

The influence of energy dissipation on the properties of bulk elastic waves in crystals is not at all reduced to trivial decrease in their amplitudes along propagation. In anisotropic media the situation is much more complicated than it looks like at first glance, at least for such specific directions of propagation as acoustic axes. The latter are defined as directions $\mathbf{m}_0$ along which a degeneracy of the phase speeds of two isonormal waves occurs (Fedorov, 1968; Khatkevich, 1962a, 1964). The corresponding points of the contact of the degenerate sheets of the phase velocity surface P may be tangent or conical (Alshits & Lothe, 1979; Alshits, Sarychev & Shuvalov, 1985) (Fig.1).

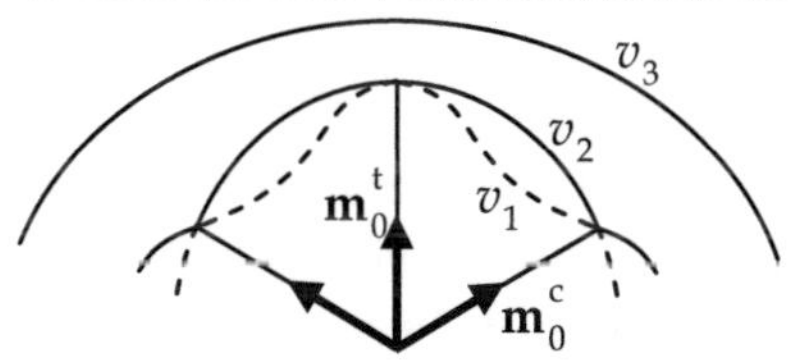

Fig. 1. Schematic plot of the section fragment of the three sheets of the phase velocity surface $v_\alpha(\mathbf{m})$ $(\alpha = 1, 2, 3)$ containing one tangent and two conical points of degeneracy

Taking into account that formally the wave attenuation may be described as an imaginary perturbation of the phase speed, one could expect due to the damping either a shift or a split of the acoustic axis, of course if it is not created by a symmetry. As we shall see below, for an acoustic axis of general position it is just splitting what is realized, and with quite a radical transformation of the local geometry of the phase velocity surface. The other possible reason for sensitivity of the wave properties to a small attenuation is related to a polarization aspect. Indeed, it is known (Alshits & Lothe, 1979; Alshits, Sarychev & Shuvalov, 1985) that the acoustic axes indicate on the unit sphere of propagation directions $\mathbf{m}^2 = 1$ the singular points in the vector fields of polarizations which are characterized by the definite vector rotation around these points on $\pm 2\pi$ or $\pm\pi$, i.e. by the Poincarè indices $n = \pm 1$ or $\pm 1/2$ (Fig.2). It is clear that a split of such singular points must be quite catastrophic for the corresponding polarization distribution. And that really occurs.

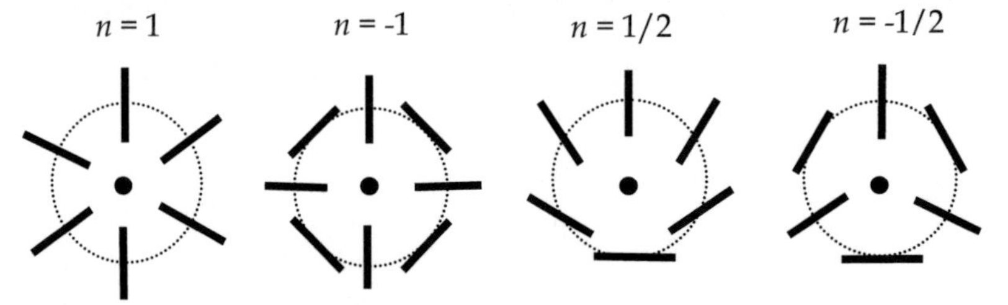

Fig. 2. Singular polarization distributions around the two types of tangent degeneracy points $n=\pm 1$ and the two types of conical degeneracies $n=\pm 1/2$

The above peculiarities are associated with space distribution of wave characteristic rather than with individual properties of bulk elastic waves. Meanwhile, as we shall see, in absorptive crystals the individual wave properties close to degeneracy directions also manifest quite unusual features, such as an almost circular polarization, in contrast to a quasi-linear one in the non-degenerate regions.

The theory of acoustic axes in non-absorptive anisotropic media is quite complete. For a review we address readers to the paper by Shuvalov (1998). The theory gives the general criteria of the degeneracy occurrence and describes all possible types of acoustic axes classifying them with respect to a local geometry of the degenerate velocity sheets and to specific features of the vector polarization fields around the degeneracy directions. This classification (Alshits & Lothe, 1979; Alshits, Sarychev & Shuvalov, 1985) includes more types than we presented in Figs. 1 and 2. However, apart from a line degeneracy known in hexagonal crystals, the rest additional types relate to the model media with accidentally coinciding or vanishing material constants (or some their combinations). Such media are beyond our interest in this paper. Note in addition that conical acoustic axes may exist in real crystals even in quite non-symmetric directions and always exist along the symmetry axis 3. In contrast, tangent degeneracies are realized in practice only due to a high symmetry of the crystals and are known only along symmetry axes ∞ and 4. As was shown in (Alshits & Lothe, 1979; Alshits, Sarychev & Shuvalov, 1985) , all the "model" acoustic axes together with any tangent or line degeneracies are unstable and must disappear, split or be transformed into other types under any small triclinic perturbation of the elastic moduli tensor $\hat{c}$. The only stable type of acoustic axes is the conical type. Under any real perturbation $\delta\hat{c}$ a conical degeneracy never split or disappear, but can only shift.

The wave attenuation can be interpreted as a perturbation of the tensor $\hat{c}$, however not real but imaginary. As was mentioned in (Alshits, Sarychev & Shuvalov, 1985), under such a non-hermitian perturbation even a conical degeneracy may lose its stability. Later Shuvalov & Chadwick (1997) rigorously investigated the stability of different acoustic axes with respect to a weak thermoelastic coupling. Their conclusion was: all types of degeneracies are unstable including a conical acoustic axis which splits into a pair. The same problem for viscoelastic and thermo-viscoelastic media has been studied by Shuvalov & Scott (1999, 2000) with similar conclusions.

It is evident that the considered physical mechanisms of the damping definitely do not disturb symmetry of the crystal and therefore cannot shift or split degeneracies along symmetry axes ∞, *4* and *3*. It means that any really existing tangent degeneracies and conical degeneracies along symmetry axes *3* must be stable under the damping perturbation. This statement was proved by Alshits & Lyubimov (1998) for viscoelastic media.

In this chapter we shall consider the attenuation in terms of viscoelasticity following to the approach of the papers (Alshits & Lyubimov, 1998, 2011). We shall analyse in detail the mentioned above geometrical peculiarities and polarization singularities related to a pair of the so-called singular acoustic axes representing a new type of stable degeneracy and arising as a result of the considered split of a conical acoustic axis. On this basis we shall develop an extension of the classical theory of internal conical refraction (Barry & Musgrave, 1979; De Klerk & Musgrave, 1955; Fedorov, 1968; Khatkevich, 1962b; Musgrave, 1957) for an absorptive crystal. As will be shown, the damping provides very radical and non-trivial modifications of fundamental features of the phenomenon.

2. Statement of the problem and general relations

Let us consider the viscoelastic medium characterized by the density ρ and the tensors of elastic moduli $\hat{c}$ and viscosity $\hat{\eta}$. The dynamic displacement field $\mathbf{u}(\mathbf{r},t)$ in such medium is described by the known equation (Landau & Lifshitz, 1986)

$$\rho \ddot{u}_i = c_{ijkl} u_{l,kj} - \eta_{ijkl} \dot{u}_{l,kj} , \tag{1}$$

where the vectors $\dot{\mathbf{u}}$ and $\ddot{\mathbf{u}}$ are the velocity and acceleration fields and the usual notation $..._{,k} \equiv \partial / \partial x_k ...$ is accepted. For the bulk wave

$$\mathbf{u}(\mathbf{r},t) = C\mathbf{A}\exp[ik(\mathbf{m}\cdot\mathbf{r} - vt)] \tag{2}$$

propagating along the wave vector $\mathbf{k} = k\mathbf{m}$ with the amplitude *C*, the frequency $\omega = kv$, the phase speed v and the polarization **A**, eqn. (1) is transformed into Christoffel's equation

$$(\hat{Q}' - i\hat{Q}'')\mathbf{A} = \rho v^2 \mathbf{A} \tag{3}$$

where $\hat{Q}'$ and $\hat{Q}''$ are the real symmetric matrices

$$Q'_{jk} = m_i c_{ijkl} m_l , \qquad Q''_{jk} = \omega m_i \eta_{ijkl} m_l . \tag{4}$$

Note, that the imaginary addition $-i\hat{Q}''$ to the usual acoustic tensor $\hat{Q}'$, in contrast to the latter, is dependent on the frequency. Eqn. (3) determines the three complex eigenvectors $\mathbf{A}_a$ and the three corresponding complex eigenvalues ρv_a^2 ($\alpha = 1, 2, 3$), i.e. the three phase speeds v_a as functions of the direction **m**:

$$\mathbf{A}_a = \mathbf{A}'_a + i\mathbf{A}''_a , \qquad v_a = v'_a - iv''_a . \tag{5}$$

Below the frequency will be supposed to be real. Hence, by (5), the value k_a should contain an imaginary addition determining the decay of the wave along its propagation:

$$k_a \equiv k'_a + ik''_a = \frac{\omega}{v'_a - iv''_a} \approx \frac{\omega}{v'_a}\left(1 + i\frac{v''_a}{v'_a}\right). \tag{6}$$

The complex phase speeds of eigenwaves are found from the equation

$$\rho v_a^2 = \frac{\mathbf{A}_a(\hat{Q}' - i\hat{Q}'')\mathbf{A}_a}{\mathbf{A}_a \cdot \mathbf{A}_a}. \tag{7}$$

The polarization vectors $\mathbf{A}_a$ as eigenvectors of the symmetric matrix $\hat{Q}'$ - $i\hat{Q}''$ for non-degenerate directions **m** of propagation must be mutually orthogonal

$$\mathbf{A}_a \cdot \mathbf{A}_\beta = \delta_{a\beta}, \qquad a \neq \beta. \tag{8}$$

As regards to their normalization, we cannot use the customary condition $\mathbf{A}_a^2 = 1$, bearing in mind the possibility of a circular polarization for which $\mathbf{A}_a^2 = 0$. Instead, the normalizing factor will be chosen so that

$$|\mathbf{A}_a|^2 = \mathbf{A}_a'^2 + \mathbf{A}_a''^2 = 1. \tag{9}$$

For a further development let us divide the basic eqn. (3) on the real and imaginary parts

$$\begin{aligned} \hat{Q}'\mathbf{A}'_a + \hat{Q}''\mathbf{A}''_a &= \rho(v_a'^2 - v_a''^2)\mathbf{A}'_a + 2\rho v'_a v''_a \mathbf{A}''_a, \\ \hat{Q}'\mathbf{A}''_a - \hat{Q}''\mathbf{A}'_a &= \rho(v_a'^2 - v_a''^2)\mathbf{A}''_a - 2\rho v'_a v''_a \mathbf{A}'_a. \end{aligned} \tag{10}$$

Multiplying these equations by $\mathbf{A}'_{a,\beta}$ or $\mathbf{A}''_{a,\beta}$ $(\alpha \neq \beta)$ and combining the results one obtains

$$2\rho v'_a v''_a = \mathbf{A}'_a \cdot \hat{Q}''\mathbf{A}'_a + \mathbf{A}''_a \cdot \hat{Q}''\mathbf{A}''_a, \tag{11}$$

$$\rho(v_a'^2 - v_a''^2) = \mathbf{A}'_a \cdot \hat{Q}'\mathbf{A}'_a + \mathbf{A}''_a \cdot \hat{Q}'\mathbf{A}''_a, \tag{12}$$

$$\rho(v_\beta'^2 - v_a'^2 + v_a''^2 - v_\beta''^2)\mathbf{A}''_a \cdot \mathbf{A}'_\beta = \mathbf{A}'_a \cdot \hat{Q}''\mathbf{A}'_\beta + \mathbf{A}''_a \cdot \hat{Q}''\mathbf{A}''_\beta - 2\rho(v'_a v''_a + v'_\beta v''_\beta)\mathbf{A}''_a \cdot \mathbf{A}''_\beta. \tag{13}$$

Eqns. (11)-(13) are exact. The first two of them show that the imaginary part of the phase speed v''_a being linear in small viscosity $\hat{\eta}$ is small compared to v'_a independently of the direction **m**. In accordance with Eqn. (13), one can also conclude that $A''_a << A'_a$ however not for any **m**, but only far enough from acoustic axes, when the difference $v'_\beta - v'_a$ is not small. In this case the value $A''_a = |\mathbf{A}''_a|$ is also linear in $\hat{\eta}$ and therefore small. Let us decompose the vector $\mathbf{A}''_a$ on the two components: $\mathbf{A}''_a = \mathbf{A}''^{\perp}_a + \mathbf{A}''^{\parallel}_a$, where $\mathbf{A}''^{\perp}_a \perp \mathbf{A}'_a$ and $\mathbf{A}''^{\parallel}_a \parallel \mathbf{A}'_a$. Thus, the ellipticity $\varepsilon = A''^{\perp}_a / A'_a$ of the wave polarization due to the damping is also small almost everywhere beyond small domains around acoustic axes. Let us estimate this ellipticity to the first order in $\hat{\eta}$.

Being perpendicular to $\mathbf{A}'_a$, the vector $\mathbf{A}''^{\perp}_a$ may be expressed in leading approximation as a superposition of two isonormal vectors $\mathbf{A}'_\beta$ and $\mathbf{A}'_\gamma$ which are almost orthogonal to $\mathbf{A}'_a$,

$$\mathbf{A}''^{\perp}_a \approx (\mathbf{A}''^{\perp}_a \cdot \mathbf{A}'_\beta)\mathbf{A}'_\beta + (\mathbf{A}''^{\perp}_a \cdot \mathbf{A}'_\beta)\mathbf{A}'_\beta \approx (\mathbf{A}''_a \cdot \mathbf{A}'_\beta)\mathbf{A}'_\beta + (\mathbf{A}''_a \cdot \mathbf{A}'_\beta)\mathbf{A}'_\beta. \tag{14}$$

In view of eqn. (13) this gives far from degeneracies

$$\mathbf{A}_a''^{\perp} = \frac{\mathbf{A}_a' \cdot \hat{Q}''\mathbf{A}_\beta'}{\rho(v_\beta'^2 - v_a'^2)}\mathbf{A}_\beta' + \frac{\mathbf{A}_a' \cdot \hat{Q}''\mathbf{A}_\gamma'}{\rho(v_\gamma'^2 - v_a'^2)}\mathbf{A}_\gamma' . \tag{15}$$

In fact, for considered non-singular directions the component $\mathbf{A}_a''^{\|}$ is physically unimportant. Indeed, the vector amplitude of the wave (2) in the accepted linear approximation is equal

$$C\mathbf{A}_a = C(\mathbf{A}_a' + i\mathbf{A}_a''^{\|} + i\mathbf{A}_a''^{\perp}) = C(1+i\varepsilon^{\|})\left(\mathbf{A}_a' + \frac{i\mathbf{A}_a''^{\perp}}{1+i\varepsilon^{\|}}\right) \approx C'(\mathbf{A}_a' + i\mathbf{A}_a''^{\perp}), \tag{16}$$

where the notations $\varepsilon^{\|} = A_a''^{\|} / A_a'$ and $C' = C(1+i\varepsilon^{\|})$ are introduced.
With (15) and (9), the wave ellipticity is readily estimated as $\varepsilon \approx A_a''^{\perp} = |\mathbf{A}_a''^{\perp}|$. For similar non-singular directions the speeds v_a'' and v_a' determined by eqns. (11) and (12) may be expressed in the same leading approximation as

$$v_a'' = \frac{\mathbf{A}_a' \cdot \hat{Q}''\mathbf{A}_a'}{2\rho v_a'}, \qquad \rho v_a'^2 = \mathbf{A}_a' \cdot \hat{Q}'\mathbf{A}_a' . \tag{17}$$

On the other hand, eqn. (15) demonstrates the tendency to increasing ellipticity ε when the wave normal $\mathbf{m}$ approaches the degeneracy direction ($v_a' = v_\beta'$ or $v_a' = v_\gamma'$) and one of the denominators in (15) decreases becoming singular. Of course, in the vicinity of the degeneracy it is necessary to replace eqns. (15) and (17) by some other relations.

3. General formalism for the neighbourhood of an acoustic axis

In fact, eqns. (15) and (17) quite hold for the description of the non-degenerate wave branch even along the direction where two other branches are degenerate. In the further development we shall choose for the non-degenerate wave characteristics the number $a = 3$. In this notation, by eqn. (15), the vector $\mathbf{A}_3''$ must be small addition to $\mathbf{A}_3'$. In view of the orthogonality condition (8) this allows us in the leading approximation to replace the complex polarization vectors $\mathbf{A}_{1,2}$ by their projections on the plane orthogonal to the vector $\mathbf{A}_3'$. This must work even close to acoustic axes where the imaginary components of $\mathbf{A}_{1,2}$ might be comparable in the length with their real counterparts. We are following here the ideology developed in the theory of acoustic axes for the case of zero damping (Alshits & Lothe, 1979; Alshits, Sarychev & Shuvalov, 1985).
Thus, let $\mathbf{m}_0$ is the direction of the acoustic axis in the crystal with the "switched off" attenuation. By definition, along $\mathbf{m}_0$ there must be $v_1 = v_2 \equiv v_0$ and, apart from the non-degenerate wave with the speed v_3 and the polarization $\mathbf{A}_{03}$, any polarization in the degeneracy plane $D \perp \mathbf{A}_{03}$ is permissible (Fig.3).
Let us choose in the D plane an arbitrary pair of unit orthogonal vectors $\mathbf{A}_{01}$ and $\mathbf{A}_{02}$ forming with $\mathbf{A}_{03}$ the orthonormal right-handed basis $\{\mathbf{A}_{01}, \mathbf{A}_{02}, \mathbf{A}_{03}\}$ (Fig. 3).
Now "switch on" the damping and consider eqn. (3) close to $\mathbf{m}_0$ at $\mathbf{m} = \mathbf{m}_0 + \delta\mathbf{m}$:

$$(\mathbf{m}_0+\delta\mathbf{m})(\hat{c}-i\omega\hat{\eta})(\mathbf{m}_0+\delta\mathbf{m})\mathbf{A}_a=\rho(v_{0a}+\delta v_a)^2\mathbf{A}_a\,. \qquad (18)$$

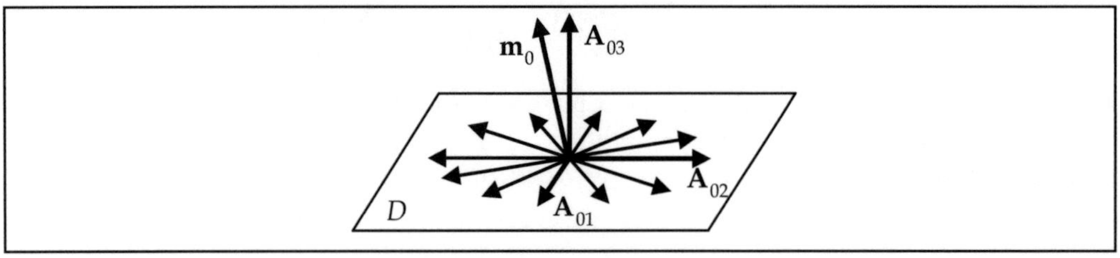

Fig. 3. Allowed polarizations along the acoustic axis $\mathbf{m}_0$ at "switched off" attenuation

In the linear approximation eqn. (18) is transformed to

$$(\hat{Q}_0+\delta\hat{Q})\mathbf{A}_a=\rho(v_{0a}^2+2v_{0a}\delta v_a)\mathbf{A}_a\,, \qquad (19)$$

where

$$\hat{Q}_0=\hat{Q}'(\mathbf{m}_0)\,,\qquad \delta\hat{Q}=\mathbf{m}_0\hat{c}\delta\mathbf{m}+\delta\mathbf{m}\hat{c}\mathbf{m}_0-i\hat{Q}_0''\,,\qquad \hat{Q}_0''=\hat{Q}''(\mathbf{m}_0)\,. \qquad (20)$$

The complex polarization vectors $\mathbf{A}_a$ may be decomposed in the basis $\{\mathbf{A}_{01},\mathbf{A}_{02},\mathbf{A}_{03}\}$ as

$$\mathbf{A}_a=a_{a\beta}\mathbf{A}_{0\beta}\,, \qquad (21)$$

where a, β = 1, 2, 3 and the summation over β is assumed. Substituting the linear superpositions (21) at a = 1 and a = 2 into eqn. (19) one obtains

$$\begin{aligned}\delta\hat{Q}\mathbf{A}_1&=2\rho v_0\delta v_1\mathbf{A}_1+\rho(v_0^2-v_{03}^2)a_{13}\mathbf{A}_{03},\\ \delta\hat{Q}\mathbf{A}_2&=2\rho v_0\delta v_2\mathbf{A}_2+\rho(v_0^2-v_{03}^2)a_{23}\mathbf{A}_{03}.\end{aligned} \qquad (22)$$

Eqns. (22) show that the coefficients a_{13} and a_{23} must be linearly small. So, indeed in the leading approximation one can replace the polarization vectors $\mathbf{A}_1$ and $\mathbf{A}_2$ by their projections on the D-plane

$$\mathbf{A}_1\approx a_{11}\mathbf{A}_{01}+a_{12}\mathbf{A}_{02}\,,\qquad \mathbf{A}_2\approx a_{21}\mathbf{A}_{01}+a_{22}\mathbf{A}_{02}\,. \qquad (23)$$

Multiplying eqns. (22) by $\mathbf{A}_1$ or $\mathbf{A}_2$ we obtain the two linear systems determining the coefficients $a_{a\beta}$ in (23):

$$\begin{cases}(\delta Q_{11}-2v_0\rho\delta v_1)a_{11}+\delta Q_{12}a_{12}=0,\\ \delta Q_{12}a_{11}+(\delta Q_{22}-2v_0\rho\delta v_1)a_{12}=0;\end{cases}\qquad \begin{cases}(\delta Q_{11}-2v_0\rho\delta v_2)a_{21}+\delta Q_{12}a_{22}=0,\\ \delta Q_{12}a_{21}+(\delta Q_{22}-2v_0\rho\delta v_2)a_{22}=0;\end{cases} \qquad (24)$$

where

$$\delta Q_{ij}=\mathbf{A}_{0i}\cdot\delta\hat{Q}\mathbf{A}_{0j}\,,\qquad i,j=1,2\,. \qquad (25)$$

The conditions for the existence of nontrivial solutions of the systems (24) give the common quadratic equation determining both δv_1 and δv_2

$$(\delta Q_{11} - 2v_0\rho\delta v)(\delta Q_{22} - 2v_0\rho\delta v) - \delta Q_{12}^2 = 0 \tag{26}$$

with the roots determining the unknown additions $\delta v_{1,2}$ to the degenerate speed v_0:

$$\delta v_{1,2} = \mathbf{s}_0 \cdot \delta\mathbf{m} - is'' \mp R\,. \tag{27}$$

Here the notations are introduced

$$R = \sqrt{(\mathbf{p}\cdot\delta\mathbf{m} - ip'')^2 + (\mathbf{q}\cdot\delta\mathbf{m} - iq'')^2}\,; \tag{28}$$

$$\left.\begin{array}{c}\mathbf{s}_0\\ \mathbf{p}\end{array}\right\} = \frac{1}{2\rho v_0}(\mathbf{A}_{01}\hat{c}\mathbf{A}_{01} \pm \mathbf{A}_{02}\hat{c}\mathbf{A}_{02})\mathbf{m}_0, \qquad \mathbf{q} = \frac{1}{2\rho v_0}(\mathbf{A}_{01}\hat{c}\mathbf{A}_{02} + \mathbf{A}_{02}\hat{c}\mathbf{A}_{01})\mathbf{m}_0\,; \tag{29}$$

$$\left.\begin{array}{c}s''\\ p''\end{array}\right\} = \frac{Q''_{11} \pm Q''_{22}}{4\rho v_0}, \qquad q'' = \frac{Q''_{12}}{2\rho v_0}\,; \tag{30}$$

$$Q''_{ij} = \mathbf{A}_{0i} \cdot \hat{Q}''_0 \mathbf{A}_{0j}\,. \tag{31}$$

The introduced vectors $\mathbf{s}_0$, $\mathbf{p}$ and $\mathbf{q}$ have the following projections on $\mathbf{m}_0$:

$$\mathbf{s}_0 \cdot \mathbf{m}_0 = v_0, \qquad \mathbf{p}\cdot\mathbf{m}_0 = \mathbf{q}\cdot\mathbf{m}_0 = 0\,. \tag{32}$$

Note, that the vectors $\mathbf{s}_0$, $\mathbf{p}$ and $\mathbf{q}$ were first introduced by Fedorov (1968) in his theory of internal conical refraction. Then the same vectors were used in the theory of acoustic axes (Alshits, Sarychev & Shuvalov, 1985). With (27), systems (24) are easily solved which allows us to find the polarization vectors $\mathbf{A}_{1,2}$ (23) (not normalized at this stage):

$$\mathbf{A}_{1,2} = -(\mathbf{q}\cdot\delta\mathbf{m} - iq'')\mathbf{A}_{01} + (\mathbf{p}\cdot\delta\mathbf{m} - ip'' \pm R)\mathbf{A}_{02}\,. \tag{33}$$

It is easily checked that $\mathbf{A}_1 \cdot \mathbf{A}_2 = 0$, i.e. the orthogonality property, eqn. (8), is fulfilled. Actually, eqns. (27) and (33) contain all necessary information for our further analysis. However, in the next section we shall have to make preliminary "step aside".

4. On the acoustic axes along directions of high symmetry

Note, that the above formalism linear in small parameters does not work for the case of tangent acoustic axes along which $\mathbf{p} = \mathbf{q} = 0$ (Alshits, Sarychev & Shuvalov, 1985) and one should keep the higher order terms in all expansions. The above criterion for a tangent degeneracy can be satisfied either because of an accidental vanishing of some combinations of material parameters (i.e. in model crystals) or due to a high symmetry of the direction $\mathbf{m}_0$. That is why tangent degeneracies are known in real crystals only along 4- and ∞-fold symmetry axes. In the first case the both Poincarè indices $n = \pm 1$ (Fig. 2) are possible, in the latter case only the index n =+1 can occur (Alshits, Sarychev & Shuvalov, 1985). We already mentioned that model media are beyond our interest in this paper. As to "symmetrical" tangent acoustic axes, their reaction to "switching on" the damping is predictable without any calculations. The answer is rather natural: existing due to a symmetry which is not disturbed by the attenuation, they keep their directions and linear polarizations of the elastic waves propagating along them also retain, though the phase speeds v_a of these waves certainly take small imaginary components.

Indeed, the tensors $\hat{c}$ and $\hat{\eta}$ have completely the same symmetrical structure. It is well known, that along the direction $\mathbf{m}_0$ of the symmetry axis ∞ or 4 the tensor $\hat{Q}_0 = \mathbf{m}_0\hat{c}\mathbf{m}_0$ has eigenvectors $\mathbf{A}_{01}, \mathbf{A}_{02}$ and $\mathbf{A}_{03}$ coinciding with the basis vectors of the crystallographic coordinate system. Clearly, the tensor $\hat{Q}''_0 = \omega\mathbf{m}_0\hat{\eta}\mathbf{m}_0$ (20) must have the same eigenvectors. Hence, the combined complex Christoffel tensor $\hat{Q}' - i\hat{Q}'' = \hat{Q}_0 - i\hat{Q}''_0$ along the direction $\mathbf{m}_0$ admits purely real polarizations of three isonormal eigenwaves: one longitudinal ($\mathbf{A}_{03}$ parallel to z) and two transverse ($\mathbf{A}_{01}$ parallel to x and $\mathbf{A}_{02}$ parallel to y). It is easy to check that the degeneracy along $\mathbf{m}_0$ also retains. In accordance with eqn. (7)

$$\begin{aligned} \rho v_1^2 &= c_{1331} - i\omega\eta_{1331} = c_{55} - i\omega\eta_{55}, \\ \rho v_2^2 &= c_{2332} - i\omega\eta_{2332} = c_{44} - i\omega\eta_{44}, \\ \rho v_3^2 &= c_{3333} - i\omega\eta_{3333} = c_{33} - i\omega\eta_{33}. \end{aligned} \tag{34}$$

But the symmetry axes ∞ or 4 present only in tetragonal, cubic and hexagonal crystals where

$$c_{44} = c_{55}, \qquad \eta_{44} = \eta_{55} \tag{35}$$

and therefore $v_1 = v_2$. Accordingly, the degenerate tensor $\hat{Q}_0 - i\hat{Q}''_0$ has the spectral representation

$$\hat{Q}_0 - i\hat{Q}''_0 = [c_{44} - c_{33} - i\omega(\eta_{44} - \eta_{33})]\hat{I} + (c_{33} - i\,\omega\eta_{33})\mathbf{A}_{03} \otimes \mathbf{A}_{03} \tag{36}$$

where $\hat{I}$ denotes the unit tensor and the symbol $\otimes$ means a dyadic product. So, one can see, that any linear combination $a\mathbf{A}_{01} + \beta\mathbf{A}_{02}$ is also an eigenvector of $\hat{Q}_0 - i\hat{Q}''_0$ and any transverse wave may propagate along $\mathbf{m}_0$ (Fig.3). Note in addition, that in the considered situation the nominator $\mathbf{A}_{01} \cdot \hat{Q}''\mathbf{A}_{02}$ of the singular term in eqn. (15) vanishes together with η_{45}. This explains why our qualitative expectations of increasing imaginary components $\mathbf{A}''_{1,2}$ close to $\mathbf{m}_0$ have not been realized.

The same arguments are equally applicable for the directions of 3-fold symmetry axes along which the conical acoustic axes necessarily occur (Alshits, Sarychev & Shuvalov, 1985) being characterized by the polarization singularity with the Poincarè index $n = -1/2$ (Fig. 2). However in this case $\mathbf{p} \neq \mathbf{q} \neq 0$ and the formalism developed in the previous section does work and may be used for the demonstration of the validity of the above considerations. For example, in the case of the 3-fold symmetry axis in trigonal crystal one has: $p'' \propto \eta_{1331} - \eta_{2332} = \eta_{55} - \eta_{44} = 0$ and $q'' \propto \eta_{1332} = \eta_{54} = 0$ but $s'' \propto \eta_{1331} + \eta_{2332} = 2\eta_{44} \neq 0$. As a result, the phase speed perturbations $\delta v_{1,2}$, eqn. (27), have the equal imaginary components and the difference between δv_1 and δv_2 remains purely real and vanishes at $\delta\mathbf{m} \to 0$. Thus, again there is neither a split nor a shift of the degeneracy. And in accordance with eqn. (33) the degenerate polarization fields $\mathbf{A}_{1,2}$ in the neighbourhood of the acoustic axis in the leading approximation remain linear, i.e. their imaginary components are small and vanish at $\delta\mathbf{m} \to 0$.

But such a trivial situation occurs only along the symmetry axes ∞, *4* and *3*. As we shall see, any other point of degeneracy, even in a symmetry plane, manifests instability with respect to attenuation and singular behaviour of basic wave parameters close to new acoustic axes.

Let us consider the other "symmetric" case known in real crystals: the line of degeneracy which occurs in some transversely isotropic media. According to (Alshits, Sarychev & Shuvalov, 1985), along such line: $\mathbf{p}\times\mathbf{q}=0$ i.e. the vectors **p** and **q** must be parallel or one of them should vanish (let $\mathbf{q}=\gamma\mathbf{p}$ say). Note that at $\mathbf{p}\times\mathbf{q}=0$ the point degeneracy is also possible (in model crystals, of course). In this case for its description one should keep in expansions the terms of the higher order. But for a line degeneracy the leading approximation used above is completely sufficient and eqns. (27) and (33) may be applied for an analysis. The condition of the degeneracy $\delta v_1 = \delta v_2$ is equivalent to the requirement of the vanishing square root in (27) which brings us to the following system

$$(\mathbf{p}\cdot\delta\mathbf{m})^2 + (\mathbf{q}\cdot\delta\mathbf{m})^2 - p''^2 - q''^2 = 0\,, \tag{37}$$

$$p''(\mathbf{p}\cdot\delta\mathbf{m}) + q''(q\cdot\delta\mathbf{m}) = 0\,. \tag{38}$$

At $\mathbf{q}=\gamma\mathbf{p}$ this system becomes clearly contradictive,

$$\mathbf{p}\cdot\delta\mathbf{m} = \sqrt{(p''^2+q''^2)/(1+\gamma^2)}\,,$$

$$\mathbf{p}\cdot\delta\mathbf{m} = 0\,,$$

and has no solutions unless the both parameters p'' and q'' simultaneously vanish which does not occur in hexagonal crystals. Thus the line degeneracy $\mathbf{p}\cdot\delta\mathbf{m}=0$ under the damping perturbation must completely disappear which coincide with the corresponding conclusion in (Shuvalov & Chadwick, 1997). But looking at eqns. (27) and (33) one can say more. It follows from (27) that at $\mathbf{p}\cdot\delta\mathbf{m}=0$ the perturbations $\delta v_{1,2}$ are purely imaginary. This means that real components of the phase speeds v_{01} and v_{02} coincide as before on the same line $\mathbf{p}\cdot\delta\mathbf{m}=0$. However, the imaginary components $\delta v''_{1,2}$ are different on this line which eliminates the degeneracy. As regards to polarizations, the only peculiarity of the polarization field on the line $\mathbf{p}\cdot\delta\mathbf{m}=0$ is the lack of even a symbolic ellipticity: by (33) it is purely linear.

5. Split of acoustic axes of general positions

At the switched off attenuation eqn. (27) transforms to the known equation (Alshits, Sarychev & Shuvalov, 1985) describing local geometry of sheets of the phase velocity surface *P*: $v_{1,2}(\mathbf{m}) = v_0 + \delta v_{1,2}(\mathbf{m}_0+\delta\mathbf{m})$ in the vicinity of the degeneracy point $v_1(\mathbf{m}_0)=v_2(\mathbf{m}_0)=v_0$:

$$\delta v_{1,2} = \mathbf{s}_0\cdot\delta\mathbf{m} \mp \sqrt{(\mathbf{p}\cdot\delta\mathbf{m})^2+(\mathbf{q}\cdot\delta\mathbf{m})^2}\,. \tag{39}$$

If the vectors $\mathbf{p}$ and $\mathbf{q}$ are neither vanishing nor parallel to each other, $\mathbf{p}\times\mathbf{q}\neq 0$, then eqn.(39) describes a conical contact of the sheets $v_{1,2}(\mathbf{m})$ and simultaneously of the sheets $1/v_{1,2}(\mathbf{m})$ of the slowness surface *S*. This is a conical degeneracy of a general type not related to symmetry of a crystal.

As we know, a "switching on" the attenuation causes a small imaginary addition to a phase speed of the wave: $v = v' - iv''$. As a result, apart from the wave surfaces *P* and *S* the new surface of attenuation, $v''(\mathbf{m})$, arises. And the real components of the phase speeds $v'_{1,2}(\mathbf{m})$ also manifest important changes providing a topological transformation of the wave surfaces *P* and *S*.

Let us return to eqns. (37), (38). At $\mathbf{p}\times\mathbf{q}\neq 0$ this system describes the split of conical acoustic axis due to a damping. The new positions of the degeneracies $\delta\mathbf{m}_1$ and $\delta\mathbf{m}_2$ are given by the intersection of the ellipse

$$\left(\frac{\delta\mathbf{m}\cdot\mathbf{p}}{r}\right)^2 + \left(\frac{\delta\mathbf{m}\cdot\mathbf{q}}{r}\right)^2 = 1, \qquad r = \sqrt{p''^2 + q''^2}, \tag{40}$$

with the straight line $\delta\mathbf{m}\cdot\mathbf{M} = 0$ passing through the end of the vector $\mathbf{m}_0$ perpendicularly to the vector (Fig. 4)

$$\mathbf{M} = p''\mathbf{p} + q''\mathbf{q} = r(\mathbf{p}\sin a + \mathbf{q}\cos a), \tag{41}$$

where the angle a is introduced by the expressions

$$\sin a = p''/r, \qquad \cos a = q''/r. \tag{41a}$$

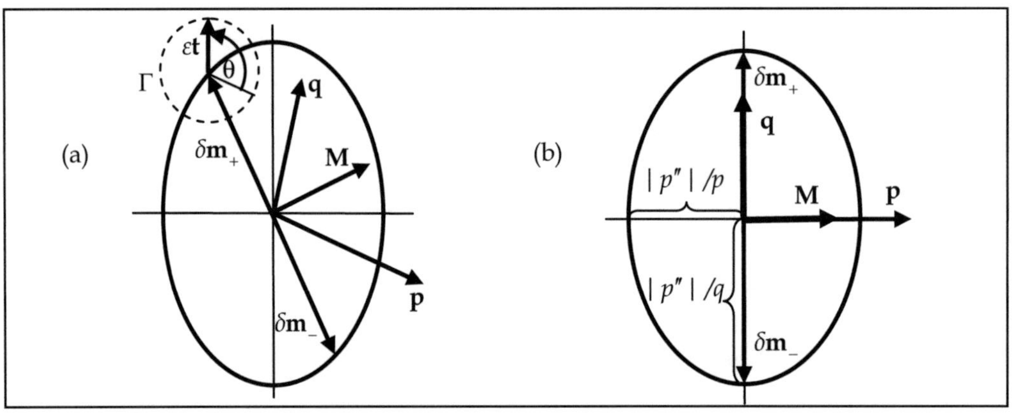

Fig. 4. Schematic plot of the ellipse, eqn. (40), in the general case (a), and for the case of the conical acoustic axis $\mathbf{m}_0$ splitting from the symmetry plane of a crystal (b)

In accordance with eqn. (32) the both vectors $\mathbf{p}$ and $\mathbf{q}$ are orthogonal to $\mathbf{m}_0$. Therefore the ellipse (40) (Fig.4) belongs to the plane tangent to the unit sphere $\mathbf{m}\cdot\mathbf{m}=1$ at the point $\mathbf{m}=\mathbf{m}_0$ which indicates the center of the ellipse. Thus, "switching on" the damping causes the split of the conical axis $\mathbf{m}_0$ into the two singular axes directed along the wave normals $\mathbf{m}_\pm = \mathbf{m}_0 + \delta\mathbf{m}_\pm$ where

$$\delta\mathbf{m}_{\pm} = \pm\frac{\mathbf{m}_0 \times (p''\mathbf{p} + q''\mathbf{q})}{\mathbf{m}_0 \cdot (\mathbf{p}\times\mathbf{q})}. \tag{42}$$

Note that the projections of $\delta\mathbf{m}_{\pm}$ (42) on $\mathbf{p}$ and $\mathbf{q}$ vectors look rather simple

$$\delta\mathbf{m}_{\pm} \cdot \mathbf{p} = \mp q'', \quad \delta\mathbf{m}_{\pm} \cdot \mathbf{q} = \pm p''. \tag{43}$$

Let us consider the example of splitting of a conical axis belonging to the symmetry plane S of the crystal. It is evident that in this case the polarization vector $\mathbf{A}_{03}$ also belongs to the plane S. The other vectors of our basis may be chosen so that, say, the vector $\mathbf{A}_{01}$ is directed along the normal to the plane S, and the vector $\mathbf{A}_{02}$ belongs to the same plane S together with the vectors $\mathbf{m}_0$ and $\mathbf{A}_{03}$ (Fig. 5a). It is easily checked that in the given case due to a crystal symmetry, which is not less than monoclinic, there must be

$$q'' = 0, \quad \mathbf{q} \parallel \mathbf{A}_{01}, \quad \mathbf{p} \parallel \mathbf{A}_{01} \times \mathbf{m}_0 \tag{44}$$

(Fig. 5b). By eqns. (43), (44), the split from the symmetry plane is determined by the vectors

$$\delta\mathbf{m}_{\pm} = \pm\frac{|p''|}{q}\mathbf{A}_{01}. \tag{45}$$

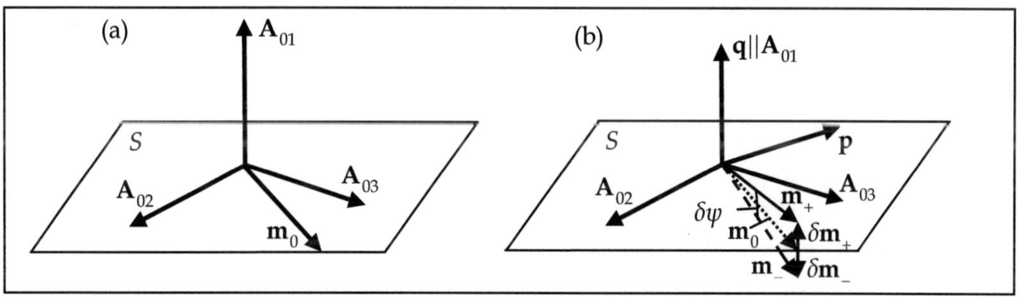

Fig. 5. Acoustic axis $\mathbf{m}_0$ in symmetry plane S (a), and its splitting due to the damping (b)

For the found mutual orthogonality of the vectors $\mathbf{p}$ and $\mathbf{q}$ ellipse (40) looks very symmetric (Fig. 4b). Thus, in the considered particular case the split of the acoustic axis occurs in the plane orthogonal to S and the angle $\delta\psi$ of splitting is proportional to the damping (Fig. 5b)

$$\delta\psi \approx 2|\delta m_{\pm}| = 2|p''|/q. \tag{46}$$

6. Local geometry of the velocity surfaces in the vicinity of split axes

Let us now return to eqns. (27), (28). We shall not divide eqn. (27) on the real and imaginary parts. It is more convenient to analyse this equation in its combined form. First of all, let us note that the expression under square root in eqn. (28) along the line $\delta\mathbf{m}\cdot\mathbf{M}=0$ is purely real, being negative between the degeneracy points (i.e. inside the ellipse, eqn. (40) and Fig.4) and positive beyond them (i.e. outside the ellipse). But this means that on the part of this line which is inside of the ellipse, the square root is purely imaginary. Accordingly, on this part of the line the real components of phase speed $v_1'(\mathbf{m})$ and $v_2'(\mathbf{m})$ must coincide which creates the lines of self-intersection of the wave surfaces $v_{1,2}'(\mathbf{m})$ and $1/v_{1,2}'(\mathbf{m})$. Quite similarly, we come to the conclusion that the corresponding sheets of the attenuation surface $v_{1,2}''(\mathbf{m})$ must intersect each other over the line $\delta\mathbf{m}\cdot\mathbf{M}=0$ outside the ellipse (40). Fig. 6 gives a schematic illustration of such self-intersection of the slowness surface.

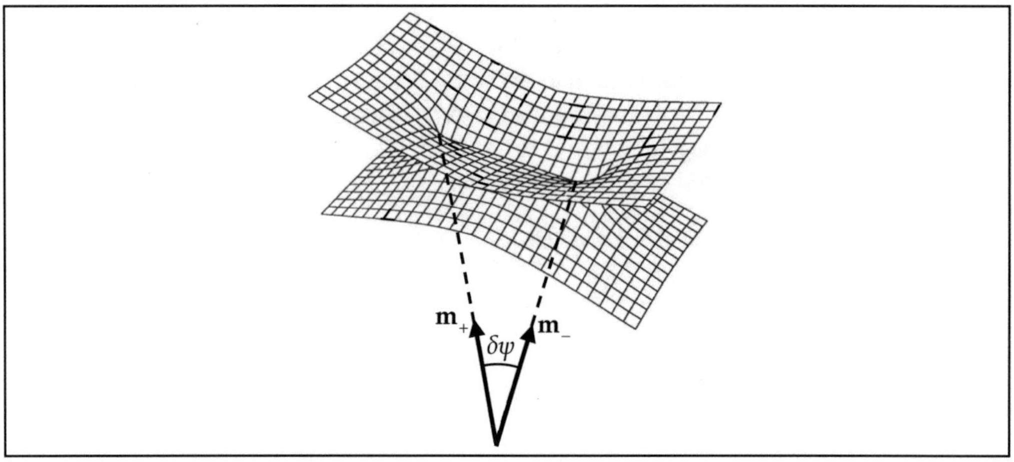

Fig. 6. Self-intersection of the slowness surface $1/v_{1,2}'(\mathbf{m})$ and split acoustic axes

Now let us study the above surfaces close to one of the new degeneracies, say, $\mathbf{m}_+=\mathbf{m}_0+\delta\mathbf{m}_+$. We are going to find the phase speeds of isonormal waves at the contour Γ (Fig. 4a): $\mathbf{m}=\mathbf{m}_+ +\delta\mathbf{m}_\varepsilon(\theta)$. The contour lies in the plane orthogonal to $\mathbf{m}_0$ and its radius is supposed to be small: $\varepsilon=|\delta\mathbf{m}_\varepsilon|<<|\delta\mathbf{m}_+|$. Denote

$$\delta\mathbf{m}_\varepsilon(\theta)=\varepsilon\mathbf{t}(\theta), \tag{47}$$

where $\mathbf{t}$ is the unit vector making the angle θ with the vector $\mathbf{p}$:

$$\mathbf{t}(\theta)=[\mathbf{p}\cos\theta+(\mathbf{m}_0\times\mathbf{p})\sin\theta]/p. \tag{48}$$

Thus, by changing θ from 0 to 2π, the vector $\delta\mathbf{m}_\varepsilon$ (47) path-traces the contour Γ around the degeneracy point $\delta\mathbf{m}_+$ (Fig. 4a).
With (47), (48), eqn. (27) gives in the leading approximation over ε at the contour Γ:

$$\delta v_{1,2}'(\varepsilon,\theta)=v_{1,2}'-v_{0+}'=\mp\sqrt{\varepsilon}\,\mathrm{Re}[f(\theta)], \tag{49}$$

where

$$f(\theta)=\sqrt{2\mathbf{t}\cdot(\mathbf{N}-i\mathbf{M})}, \qquad \mathbf{N}=-q''\mathbf{p}+p''\mathbf{q}. \tag{50}$$

As is seen from (49), the dependence $\delta v_{1,2}'(\varepsilon)\propto\sqrt{\varepsilon}$ at $\varepsilon\to 0$ is characterized by an infinite derivative over ε in any section $\theta\neq\theta_0$, where the angle θ_0 relates to a transition of the vector $\delta\mathbf{m}_\varepsilon$ through the self-intersection line, $\mathrm{Re}[f(\theta_0)]=0$. This singularity of the function

$\delta v'_{1,2}(\varepsilon)$ at the end of the wedge of self-intersection corresponds to a sharpening tip of the slowness surface $1 / v'_{1,2}(\mathbf{m})$ and to a plane fan of the normals to this surface at the contour Γ when $\varepsilon \to 0$ (Fig. 7).

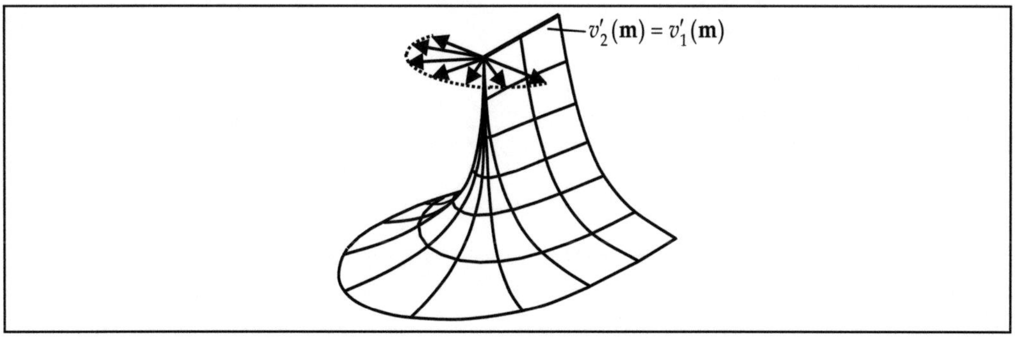

Fig. 7. The fragment of the internal degenerate sheet $1/v'_2(\mathbf{m})$ of the slowness surface close to the singular point at the end of the wedge of the self-intersection and the plane fan of normals to the surface at this point

7. Polarization field singularities around the split acoustic axes

At the same contour Γ (47) polarization vectors (33) up to normalizing factors are equal

$$\mathbf{A}_{1,2} = \mathbf{A}_{01} + i\left(1 \mp \frac{\sqrt{\varepsilon} f(\theta)}{q'' + ip''}\right)\mathbf{A}_{02} . \tag{51}$$

For the further analysis the function $f(\theta)$ here should be concretize to the form

$$f(\theta) = A\sqrt{\mathbf{p}\cdot\mathbf{q}\cos\theta + g\sin\theta - ip^2\cos\theta} , \tag{52}$$

where

$$A = \sqrt{2(p'' - iq'')/p} , \qquad g = \mathbf{m}_0 \cdot (\mathbf{p}\times\mathbf{q}) . \tag{53}$$

It is easily checked, that at the rotation of unit vector **t**, (48), over the whole circuit, i.e. at varying θ from 0 to 2π, the function $f(\theta)$ (52), changes its sign. Indeed, the phase of the complex function

$$f(\theta) / A \equiv R(\theta)\exp[i\Psi(\theta)] \tag{54}$$

must be twice less than the phase of its square

$$R^2\exp(2i\Psi) = \mathbf{p}\cdot\mathbf{q}\cos\theta + g\sin\theta - ip^2\cos\theta . \tag{55}$$

On the other hand, one can find from (55) the following relations

$$\Psi(\theta) = -\frac{1}{2}\left(\mathrm{Arctg}\frac{p^2}{\mathbf{p}\cdot\mathbf{q} + g\mathrm{tg}\theta}\right), \qquad \frac{\partial\Psi}{\partial\theta} = \frac{gp^2\cos^2 2\Psi}{(\mathbf{p}\cdot\mathbf{q}\cos\theta + g\sin\theta)^2} . \tag{56}$$

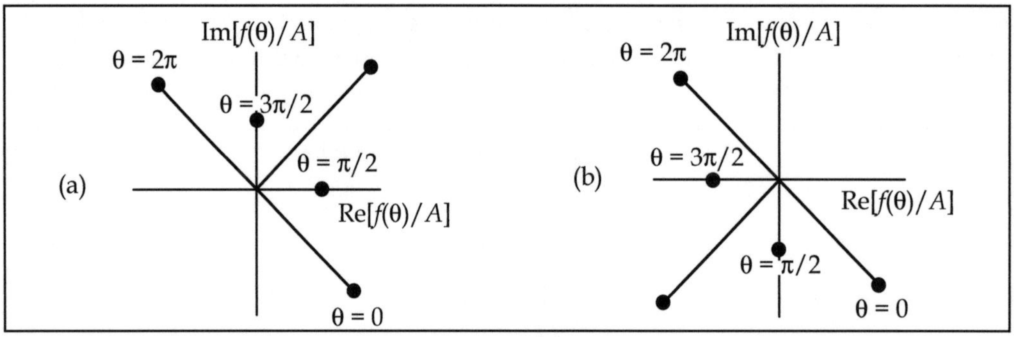

Fig. 8. The function $f(\theta)/A$ in the complex plane at $g > 0$ (a) and $g < 0$ (b)

This gives (see also Fig. 8)

$$\Psi(2\pi) - \Psi(0) = \pi \operatorname{sgn} g\,, \tag{57}$$

$$f(2\pi) = -f(0)\,. \tag{58}$$

Thus, after the whole turn over the contour Γ around the degeneracy point at $\delta\mathbf{m}_+$ (Fig. 4a) one has the identical transformation of the polarization field (51) in itself in the form

$$\mathbf{A}_1(2\pi) = \mathbf{A}_2(0)\,, \quad \mathbf{A}_2(2\pi) = \mathbf{A}_1(0)\,. \tag{59}$$

In other words, each of two orthogonal polarization ellipses rotates exactly on $\pi/2$ being transformed into the polarization of the isonormal wave (Fig. 9). And simultaneously the complex velocities $v_{1,2} = v'_{1,2} - iv''_{1,2}$ also are interchanging with their counterparts (Fig. 10).

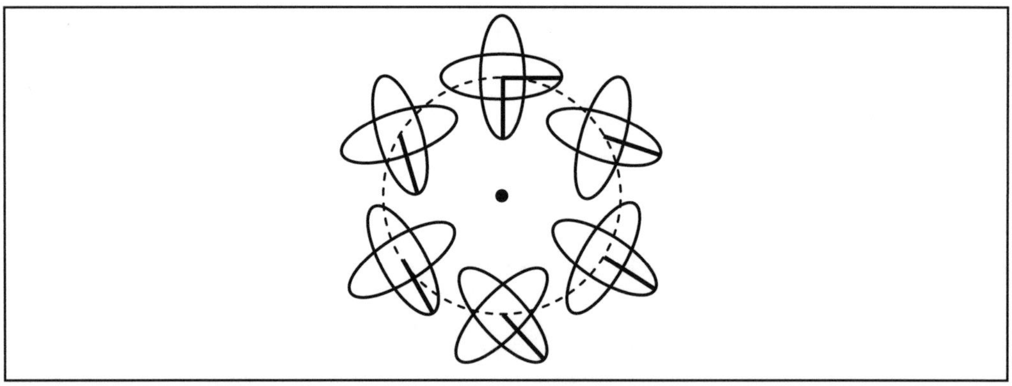

Fig. 9. The rotation of the polarization ellipses $\mathbf{A}_{1,2}$ in the degeneracy plane D when the wave normal $\mathbf{m}$ is scanning the contour Γ. The case $g > 0$ is shown when n = ¼.

The found singularity of the polarization field at the degeneracy point $\mathbf{m}_+$ (Fig. 9) may be characterized by the Poincarè index defined as the value of the total polarization rotation (in the 2π units) at a complete path-tracing over the contour Γ around this point. The found turn of the polarization ellipses is equal $\pi/2$, and the direction of the rotation, by eqn. (57), is determined by the sign of the parameter g (53). Hence, one has (Alshits & Lyubimov, 1998)

$$n = \frac{1}{4}\operatorname{sgn}[\mathbf{m}_0 \cdot (\mathbf{p} \times \mathbf{q})] \tag{60}$$

It is easily verified that the same relation is valid for the second degeneracy point $\mathbf{m}_-$.

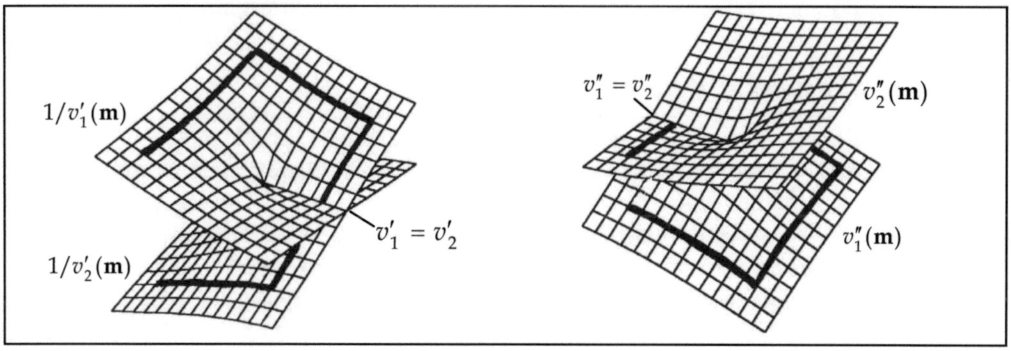

Fig. 10. The surfaces $1/v_{1,2}'(\mathbf{m})$ and $v_{1,2}''(\mathbf{m})$ in the vicinity of the acoustic axis $\mathbf{m}_+$. The transition between the sheets of the surfaces when $\mathbf{m}$ is scanning the contour around $\mathbf{m}_+$

Thus the physical equivalence of two pictures at $\theta = 0$ and $\theta = 2\pi$ is realized not by a coincidence of the wave characteristic inside each of the branches, as it occurs at zero damping, but by the identity of their superpositions. This becomes topologically possible due to such a new feature of the slowness surfaces as their self-intersections (Fig. 10). In the absence of damping, when the degenerate wave sheets locally have the only contact point, one of the branches along any direction is always "faster" than the other. And the related polarization cross, contained of isonormal linear non-directed vectors, has non-equivalent "differently coloured" crosspieces. Hence for a coincidence of such cross with itself it is required its turn on the minimum angle π, instead of $\pi/2$, as in the above case (Figs. 9, 10). The turn on $\pi/2$ is sufficient only when the change of "colours" of crosspieces occurs during the turn.

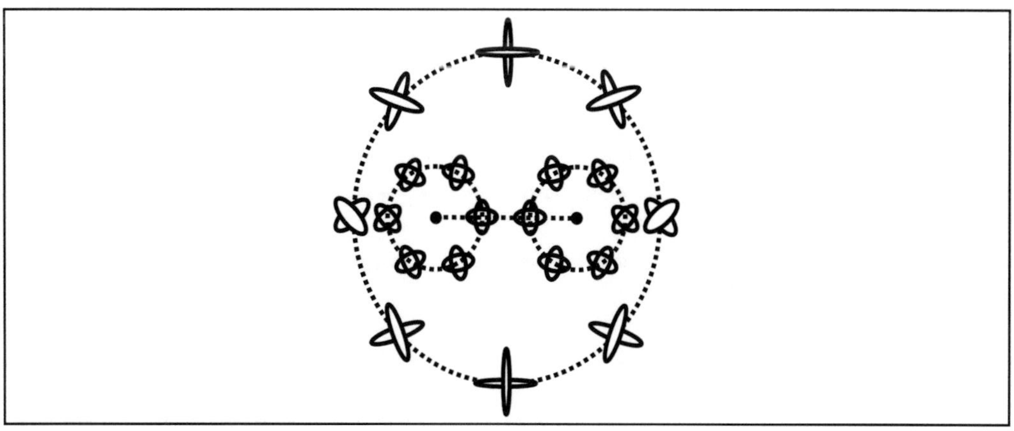

Fig. 11. The field of elliptic polarizations of degenerate branches in the vicinity of split axes of an absorptive crystal for $g < 0$. The Poincarè indices at small contours are $n = -1/4$, and the combined index at the external contour is $n = -1/2$

That is why (Alshits, Sarychev & Shuvalov, 1985) in the absence of the damping a conical axis along $\mathbf{m}_0$ is characterized by the Poincarè index $n = (1/2)\text{sgn}g$. This is the minimal index for a real polarization field. Its splitting into the two singularities (60) due to "switching on" attenuation satisfies the index conservation law. On the other hand, the same combined index ±1/2 arises at the path-tracing of the both points $\mathbf{m}_\pm$ (Fig. 11).

8. Conical refraction in absorptive crystals

Internal conical refraction of elastic waves in crystals is a good example of a non-trivial role of anisotropy, which may create new phenomena principally impossible in isotropic media. The energy flux **P** of the wave in crystal is, as a rule, non-parallel to its direction **m** of propagation. For any wave normal **m** the direction of the Poynting vector **P** is determined by the orientation of the normal **n** to the slowness surface. At the choice of the wave normal along a conical acoustic axis each polarization vector in the degeneracy plane *D* (Fig. 3) relates to the definite Poynting vector, i.e. to the definite normal to a cone. Rotation of the polarization in the plane *D* (e.g. in a circularly polarized wave) should create a precession of the energy flux **P**.

This phenomenon called the internal conical refraction was theoretically predicted and experimentally discovered by De Klerk & Musgrave (1955). They found a circular cone of refraction along the 3-fold symmetry axis in the cubic crystal Ni. Later on the more general cases of the refraction cones of elliptic section were theoretically studied (Barry & Musgrave, 1979; Khatkecich, 1962b; Musgrave, 1957) and experimentally found (Aleksandrov & Ryzhova, 1964). The complete theory of this phenomenon is presented in the monographs (Fedorov, 1968; Sirotin & Shaskolskaya, 1983). Below we shall develop an extension of this theory for absorptive crystals following to the recent paper (Alshits & Lyubimov, 2011).

8.1 Conical refraction in the absence of attenuation

As we have seen, in a crystal without damping along the acoustic axis $\mathbf{m}_0$, apart from the non-degenerate wave with the polarization vector $\mathbf{A}_{03}$, an infinite number of elastic waves may propagate with arbitrary polarization in the degeneracy plane *D* (Fig. 3). Thus in the basis $\{\mathbf{A}_{01}, \mathbf{A}_{02}\}$ belonging to the same plane, for any angle β the vector

$$\mathbf{A}(\beta) = \mathbf{A}_{01}\cos\beta + \mathbf{A}_{02}\sin\beta \tag{61}$$

determines polarization of the eigenwave propagating along $\mathbf{m}_0$ with the phase speed v_0. Certainly, the wave with a circular polarization $\mathbf{A} = \mathbf{A}_{01} + i\mathbf{A}_{02}$ can also propagate along the same direction.

Consider a monochromatic plane wave propagating along the acoustic axis $\mathbf{m}_0$ with the polarization **A** and the phase speed v_0:

$$\mathbf{u}(\mathbf{r},t) = C\mathbf{A}\exp(i\Phi_0), \qquad \Phi_0 = k(\mathbf{m}_0 \cdot \mathbf{r} - v_0 t). \tag{62}$$

The Poynting vector of such wave is equal (Fedorov, 1968)

$$\mathbf{P} = (\text{Re}\dot{\mathbf{u}})\hat{c}(\text{Re}\dot{\mathbf{u}})\mathbf{m}_0 / v_0 . \tag{63}$$

For linear and circular polarizations one has, respectively,

$$\mathrm{Re}\mathbf{u}_{\mathrm{lin}} = C\mathbf{A}(\beta)\cos\Phi_0 , \tag{64}$$

$$\mathrm{Re}\mathbf{u}_{\mathrm{cir}} = C[\mathbf{A}_{01}\cos\Phi_0 - \mathbf{A}_{02}\sin\Phi_0] . \tag{65}$$

In these two cases the Poynting vectors are given by different expressions:

$$\mathbf{P}_{\mathrm{lin}} = C^2\rho\omega^2(\mathbf{s}_0 + \mathbf{p}\cos 2\beta + \mathbf{q}\sin 2\beta)\sin^2\Phi_0 , \tag{66}$$

$$\mathbf{P}_{\mathrm{cir}} = C^2\rho\omega^2(\mathbf{s}_0 - \mathbf{p}\cos 2\Phi_0 + \mathbf{q}\sin 2\Phi_0) . \tag{67}$$

Quite similarly one finds the elastic energy densities $W = \rho(\mathrm{Re}\dot{\mathbf{u}})^2$:

$$W_{\mathrm{lin}} = C^2\rho\omega^2\sin^2\Phi_0 , \qquad W_{\mathrm{cir}} = C^2\rho\omega^2 . \tag{68}$$

With eqns. (66)-(68), the ray velocities of the considered waves are equal

$$\mathbf{s}_{\mathrm{lin}} = \mathbf{P}_{\mathrm{lin}}/W_{\mathrm{lin}} = \mathbf{s}_0 + \mathbf{p}\cos 2\beta + \mathbf{q}\sin 2\beta , \tag{69}$$

$$\mathbf{s}_{\mathrm{cir}} = \mathbf{P}_{\mathrm{cir}}/W_{\mathrm{cir}} = \mathbf{s}_0 - \mathbf{p}\cos 2\Phi_0 + \mathbf{q}\sin 2\Phi_0 . \tag{70}$$

During the period of the circularly polarized wave at a complete turn of the polarization vector in the degeneracy plane D, the ray velocity vector $\mathbf{s}_{\mathrm{cir}}$ (70) twice circumscribes a cone (Fig. 12). At that the end of the vector $\mathbf{s}_{\mathrm{cir}}$ twice path-traces the ellipse

$$\Delta\mathbf{s} = \mathbf{s} - \mathbf{s}_0 = -\mathbf{p}\cos 2\Phi_0 + \mathbf{q}\sin 2\Phi_0 . \tag{71}$$

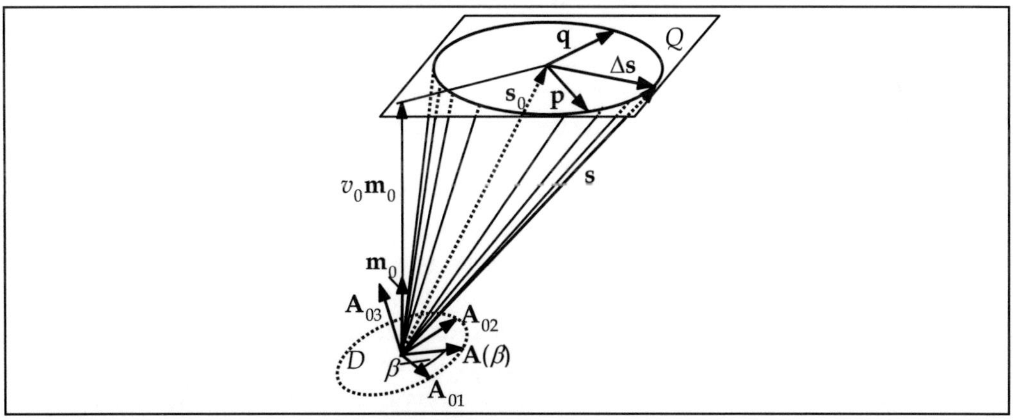

Fig. 12. The cone of the internal conical refraction

In view of (32), the plane Q of the ellipse is orthogonal to $\mathbf{m}_0$, and the directions of path-tracing of the vectors $\Delta\mathbf{s}$ and $\mathrm{Re}\mathbf{u}_{\mathrm{cir}}$ are the same when $g > 0$ and opposite when $g < 0$.
For a linearly polarized wave the same refraction cone is described by the vector $\mathbf{s}_{\mathrm{lin}}$ (69) when the angle β changes within the interval $0 \le \beta \le 2\pi$ (Fig. 12). This particular scheme was realized in the first experiments of De Klerk & Musgrave (1955).

8.2 The polarization ellipses at the ridge of the wedge of self-intersection

Consider now the wave characteristics of an absorptive crystal at the ridge of the wedge of self-intersection of the slowness surface. For a description of the set of wave normals related to the ridge between the two degeneracy points at the slowness surface let us introduce the vector $\delta\mathbf{m}_\xi = \delta\mathbf{m}_+ \sin\xi$. At changing ξ from $-\pi/2$ to $+\pi/2$ the vector $\delta\mathbf{m}_\xi$ moves through all the ridge from one degeneracy ($\delta\mathbf{m}_-$) to another one ($\delta\mathbf{m}_+$). Substitution $\delta\mathbf{m} = \delta\mathbf{m}_\xi$ into (33) gives the polarization vectors at any point of the line of self-intersection. Making use of relations (43) one obtains

$$\mathbf{A}_{1,2} = \frac{(\sin a \sin\xi - i\cos a)\mathbf{A}_{01} + [\cos a \sin\xi + i(\sin a \mp \cos\xi)]\mathbf{A}_{02}}{\sqrt{2(1 \mp \sin a \cos\xi)}}, \tag{72}$$

where the normalizing (9) is fulfilled and notations (41a) are used.

We remind that at the ridge of the wedge the real components of the phase speeds $v_{1,2}$ coincide: $v_1' = v_2' = v_\xi'$. The imaginary components $v_{1,2}''$ coincide only at the end points of the ridge, $\xi = \pm\pi/2$. In view of (6), the real components of the displacement vectors $\mathbf{u}_{1,2}$ take the form

$$\mathrm{Re}\mathbf{u}_{1,2}(\mathbf{r}, t) = C_{1,2}\exp(-k_{1,2}''\mathbf{m}_\xi \cdot \mathbf{r})\mathrm{Re}[\mathbf{A}_{1,2}\exp(i\Phi_\xi)] \equiv C_{1,2}\exp(-k_{1,2}''\mathbf{m}_\xi \cdot \mathbf{r})\mathbf{U}_{1,2}. \tag{73}$$

We introduced here the wave normal $\mathbf{m}_\xi$ and the real phase Φ_ξ at the ridge,

$$\mathbf{m}_\xi = \mathbf{m}_0 + \delta\mathbf{m}_\xi, \quad \Phi_\xi = k_\xi'\mathbf{m}_\xi \cdot \mathbf{r} - \omega t, \tag{74}$$

and the dimensionless displacement vectors

$$\mathbf{U}_{1,2} = \mathrm{Re}[\mathbf{A}_{1,2}\exp(i\Phi_\xi)]. \tag{75}$$

It is essential that in eqn. (73) a trivial damping of the wave $\propto \exp(-k_{1,2}''\mathbf{m}_\xi \cdot \mathbf{r})$ is separated from the vectors $\mathbf{U}_{1,2}$ describing much more important for us effects of attenuation.

In the considered stationary problem a choice of the time origin is certainly unessential and may be different for isonormal waves, independent from each other. Hence, the vectors $\mathbf{U}_{1,2}$ as well as the polarization vectors $\mathbf{A}_{1,2}$ are defined to the sign. Below this sign will be chosen so that our expression would be more compact.

Note, that at scanning the ridge by the wave normal $\mathbf{m}_\xi$, the elliptic polarization determined by eqns. (72), (75) is sharply changing. It is easily checked that this ellipticity provides rotations of the vectors $\mathbf{U}_{1,2}$ along the same directions corresponding to the right-hand screw along the propagation, until $\sin\xi > 0$, and to the left-hand screw, when $\sin\xi < 0$. At the ridge ends ($\xi = \pm\pi/2$) where the degeneracies occur, the isonormal waves, naturally, coincide: $\mathbf{U}_1 = \mathbf{U}_2 \equiv \mathbf{U}_0$. In both cases the polarization is circular however with different rotation "signs":

$$\begin{aligned} \mathbf{U}_0\big|_{\xi=\pi/2} &= \frac{1}{\sqrt{2}}(\mathbf{A}_{01}\cos\Phi_\xi - \mathbf{A}_{02}\sin\Phi_\xi), \\ \mathbf{U}_0\big|_{\xi=-\pi/2} &= \frac{1}{\sqrt{2}}(\mathbf{A}_{01}\cos\Phi_\xi + \mathbf{A}_{02}\sin\Phi_\xi). \end{aligned} \tag{76}$$

Here the angle α is excluded from the arguments by the choice of the time origin.
In any other points of the ridge the polarization ellipses of isonormal waves are different. In the middle point $\xi = 0$ the isonormal waves have linear polarizations orthogonal to each other:

$$\begin{aligned} \mathbf{U}_1\big|_{\xi=0} &= \left\{ \mathbf{A}_{01}\cos\left(\frac{\pi}{4} - \frac{a}{2}\right) + \mathbf{A}_{02}\sin\left(\frac{\pi}{4} - \frac{a}{2}\right) \right\} \sin\Phi_0, \\ \mathbf{U}_2\big|_{\xi=0} &= \left\{ \mathbf{A}_{01}\sin\left(\frac{\pi}{4} - \frac{a}{2}\right) - \mathbf{A}_{02}\cos\left(\frac{\pi}{4} - \frac{a}{2}\right) \right\} \sin\Phi_0. \end{aligned} \tag{77}$$

One can show that linear polarization retain on a whole line passing through the middle of the ridge ($\xi = 0$) perpendicular to it (at the unit sphere $\mathbf{m}^2 = 1$ this line passes through point $\mathbf{m}_0$ with local orientation along vector **M**).
Expressions for the polarization ellipses of isonormal waves at the ridge are remarkably simplified in the considered above particular case related to the unperturbed acoustic axis $\mathbf{m}_0$ situated in the symmetry plane of the crystal. In this case $q'' = 0$. Supposing for definiteness that $p'' > 0$, one can put $a = \pi/2$. Then, instead of (72), the polarization vectors of the isonormal waves are equal

$$\begin{aligned} \mathbf{A}_1 &= \mathbf{A}_{01}\cos(\xi/2) + i\mathbf{A}_{02}\sin(\xi/2), \\ \mathbf{A}_2 &= \mathbf{A}_{01}\sin(\xi/2) + i\mathbf{A}_{02}\cos(\xi/2). \end{aligned} \tag{78}$$

And the rotation of the displacement vectors $\mathbf{U}_{1,2}$ (75) over the ellipses is now described by

$$\begin{aligned} \mathbf{U}_1 &= \mathbf{A}_{01}\cos(\xi/2)\cos\Phi - \mathbf{A}_{02}\sin(\xi/2)\sin\Phi, \\ \mathbf{U}_2 &= \mathbf{A}_{01}\sin(\xi/2)\cos\Phi - \mathbf{A}_{02}\cos(\xi/2)\sin\Phi. \end{aligned} \tag{79}$$

These expressions represent ellipses in a parametric form. The lengths of the horizontal and vertical semi-axes of the first ellipse are equal $|\cos(\xi/2)|$ and $|\sin(\xi/2)|$, respectively. For the second ellipse the same length relate to the vertical and horizontal semi-axes. At the ridge ends $\xi = \pm\pi/2$ the above lengths of the semi-axes are equal to each other, and the polarization becomes circular. With a displacement of the "observation" point $\delta\mathbf{m}_\xi$ from the ridge ends to its middle the large semi-axes increase and the small semi-axes decrease to zero at $\xi = 0$.
Thus, both general expressions (76), (77) and the particular example (79) lead to the same picture of polarization distribution at the ridge of wedge of self-intersection. At passing along this line the isonormal waves, starting from a circular polarization of definite sign, monotonously decrease their ellipticity to zero in the middle of the ridge, where ellipses are transformed into non-directed vectors. At the second half of the ridge the ellipticity changes its sign and monotonously increases becoming circular at the other degeneracy point. Fig. 13 illustrates this behavior of polarization at the line of self-intersection of the slowness surface.
Consider now the kinematics of the motion of the displacement vectors of isonormal waves along the polarization ellipses. Express the radius-vectors $\mathbf{U}_a$ $(a = 1,2)$ at the ellipse in polar coordinates (U_a, φ_a):

$$\mathbf{U}_a = U_a(\mathbf{A}_{01}\cos\varphi_a + \mathbf{A}_{02}\sin\varphi_a). \tag{80}$$

Certainly, the lengths U_a of these radius-vectors at the ellipse depend on the azimuth φ_a. Comparing eqns. (72), (75) and (80), one has

$$U_{1,2}^2 = \frac{1}{2}\left(1 \pm \cos\xi \frac{(\sin a \mp \cos\xi)\cos 2\Phi_\xi + \sin\xi \cos a \sin 2\Phi_\xi}{1 \mp \sin a \cos\xi}\right), \tag{81}$$

$$\mathrm{tg}\varphi_{1,2} = \frac{\cos a \sin\xi - (\sin a \mp \cos\xi)\mathrm{tg}\Phi_\xi}{\sin a \sin\xi + \cos a \ \mathrm{tg}\Phi_\xi}. \tag{82}$$

Differentiating the latter expression with respect to time, it is easy to find the angular velocities $\dot{\varphi}_{1,2}$ of the radius-vectors $\mathbf{U}_{1,2}$ at their ellipses:

$$\dot{\varphi}_{1,2} = \frac{\omega \sin\xi}{2U_{1,2}^2}, \tag{83}$$

where we put $\dot{\Phi}_\xi = -\omega$. As is seen from (83), the angular velocities differently behave in time at different "observation" points at the ridge. Along the acoustic axes ($\xi = \pm\pi/2$), when the isonormal ellipses coincide into one circle (76), the denominator in (83) is equal to 1, and the circular motion has a constant angular velocity: $\dot{\varphi}_1 = \dot{\varphi}_2 = \pm\omega$. Here the upper and lower signs relate to different directions of the rotation at $\xi = \pm\pi/2$ (Fig. 13).

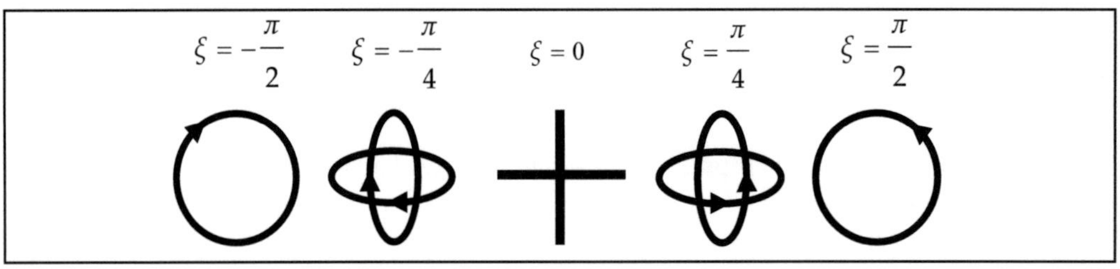

Fig. 13. Polarization distribution for isonormal waves at the line of self-intersection of the slowness surface

With decreasing $|\xi|$ a non-uniformity of the motion increases and at $|\xi| << 1$ acquires a singular character, when during the most part of the period the velocities $\dot{\varphi}_{1,2}$ are very small, and the azimuth angles $\varphi_{1,2}$ related to them are almost fixed. In this regime, the vectors $\mathbf{U}_{1,2}$ pass the most part of the ellipse in a short time with very high velocity. This is clearly seen from the analytical formulae related to the discussed above particular case of the acoustic axis splitting from the symmetry plane (Fig. 14):

$$\begin{aligned} \mathrm{tg}\varphi_1 &= -\mathrm{tg}(\xi/2)\mathrm{tg}\Phi_\xi, \\ \mathrm{tg}\varphi_2 &= -\mathrm{ctg}(\xi/2)\mathrm{tg}\Phi_\xi; \end{aligned} \tag{84}$$

$$\dot{\varphi}_{1,2} = \frac{\omega \sin\xi}{1 \pm \cos\xi \cos 2\Phi_\xi}. \tag{85}$$

Eqns. (81)-(85) and Fig. 14 show that the functions $\varphi_{1,2}(\Phi_\xi)$ and $\dot{\varphi}_{1,2}(\Phi_\xi)$ have the period twice less than the period of the wave. This means that the half-turn of the displacement vector over the polarization ellipse exhausts all its physically different orientations.

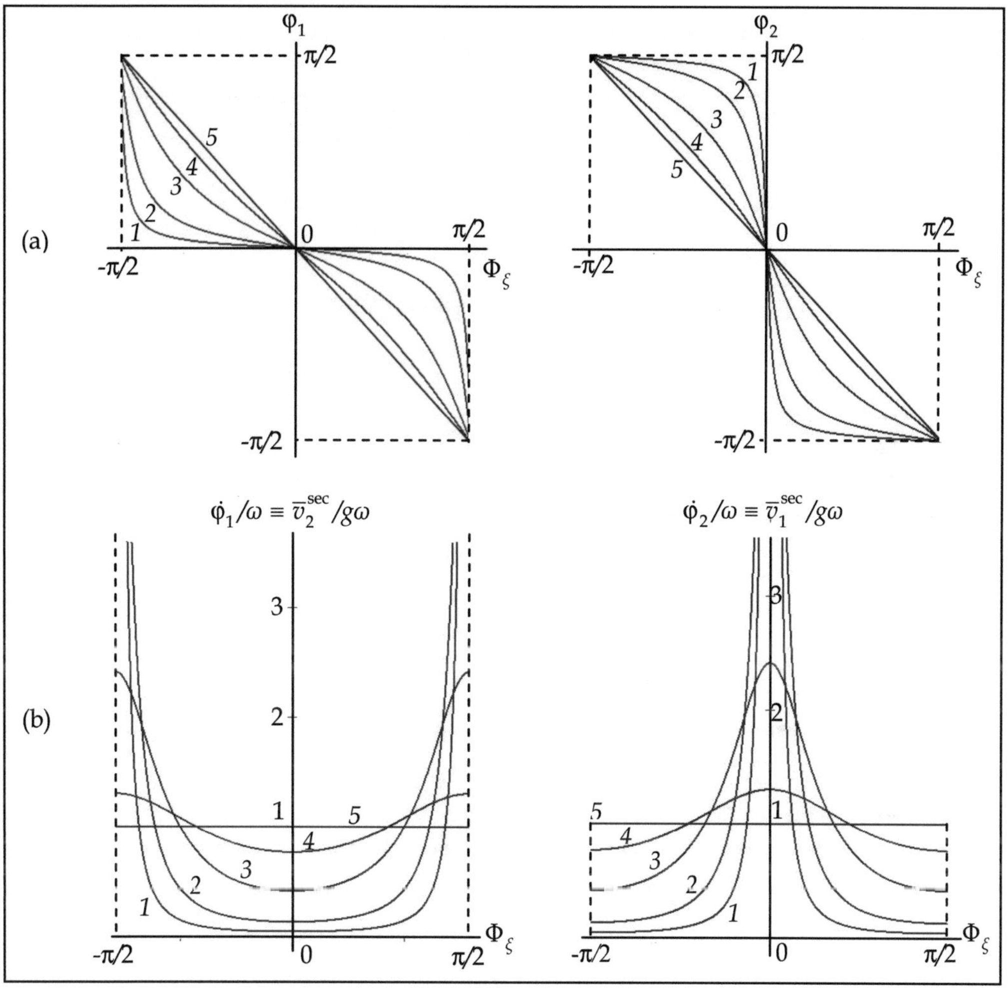

Fig. 14. Numerical plot of the azimuths $\varphi_{1,2}$ (84) (a) and the coinciding normalized speeds $\dot{\varphi}_{1,2} / \omega \equiv \overline{v}_{2,1}^{\rm sec} / g\omega$ (85), (106) (b) versus the phase Φ_ξ for the series of "observation" points at the line of self-intersection of the slowness surface for a particular case of the acoustic axis splitting from the symmetry plane. *1* - $\xi = 5^\circ$, *2* - 15°, *3* - 45°, *4* - 75°, *5* - 90°

The other kinematic characteristics of the considered motion could be the so-called sectorial velocities defined as area circumscribed by a rotating vectors $\mathbf{U}_{1,2}$ per unit time:

$$v_{1,2}^{\rm sec} = \frac{1}{2} U_{1,2}^2 \dot{\varphi}_{1,2} = \frac{1}{4} \omega \sin\xi \,. \tag{86}$$

The found expression is valid for unrestricted anisotropy. It is identical for the both isonormal waves being independent of the time. However velocity (86) strongly depends on the position of the "observation" point at the line of self-intersection. In particular, it vanishes in the center of the ridge ($\xi = 0$), where the polarization becomes linear.

8.3 Universal refraction cone at the line of self-intersection and kinematics of ray velocity precession on this cone

Let us find the ray velocities of isonormal waves (73) at the ridge of self-intersection of the slowness surface $1/v'_{1,2}(\mathbf{m})$. Substituting (73) into (63) one obtains the energy flux

$$\mathbf{P}_{1,2} = |C_{1,2}\exp(-k''_{1,2}\mathbf{m}_\xi\mathbf{r})|^2\,\rho\omega^2[\mathbf{s}_0(F^2+G_\mp^2)-\mathbf{p}(G_\mp^2-F^2)+2\mathbf{q}FG_\mp], \tag{87}$$

where

$$F = \sin a \sin\xi \sin\Phi_\xi - \cos a \cos\Phi_\xi \tag{88}$$

$$G_\mp = \cos a \sin\xi \sin\Phi_\xi + (\sin a \mp \cos\xi)\cos\Phi_\xi\,. \tag{89}$$

The energy density in the isonormal waves in the same terms is equal

$$W_{1,2} = \rho(\mathrm{Re}\dot{\mathbf{u}}_{1,2})^2 = |C_{1,2}\exp(-k''_{1,2}\mathbf{m}_\xi\mathbf{r})|^2\,\rho\omega^2(F^2+G_\mp^2)\,. \tag{90}$$

Accordingly, the ray velocities of these waves are given by

$$\mathbf{s}_{1,2} = \mathbf{P}_{1,2}/W_{1,2} = \mathbf{s}_0 - \mathbf{p}\cos2\Theta_{1,2} + \mathbf{q}\sin2\Theta_{1,2}\,, \tag{91}$$

where

$$\Theta_{1,2} = \Theta_{1,2}(a,\xi,\Phi_\xi), \qquad \mathrm{tg}\Theta_{1,2} = F/G_\mp\,. \tag{92}$$

Eqn. (91) is transformed from classic expression (70) for crystals without damping after the substitution in the latter $\Phi_0 \to \Theta_{1,2}$. This means that in an absorptive crystal at any point of the ridge of self-intersection the ends of the ray velocity vectors $\mathbf{s}_{1,2}$ move along the same trajectories, described by the universal ellipse

$$\Delta\mathbf{s}_{1,2} = -\mathbf{p}\cos2\Theta_{1,2} + \mathbf{q}\sin2\Theta_{1,2}\,. \tag{93}$$

The form of this ellipse is completely determined by the vectors **p** and **q**, and is independent of the parameters $\Theta_{1,2}$. In other words, it is insensitive neither to the phase Φ_ξ of the wave, nor to the angles α and ξ, related to parameters of damping and to a position of the "observation" point. The principal semi-axes of universal ellipse (93), coinciding with ellipse (71) for a non-attenuating medium, are equal

$$\lambda_{1,2} = \frac{1}{2}\left(\mathbf{p}^2+\mathbf{q}^2 \pm \sqrt{(\mathbf{p}^2+\mathbf{q}^2)^2 - 4(\mathbf{p}\times\mathbf{q})^2}\right). \tag{94}$$

Though the vectors **p** and **q** (29) do depend on a choice of the basis $\{\mathbf{A}_{01},\mathbf{A}_{02}\}$, one can easily check that their combinations $\mathbf{p}^2+\mathbf{q}^2$ and $\mathbf{p}\times\mathbf{q}$ determining semi-axes (94) are invariant with respect to orientation of this basis in the degeneracy plane D (Fig. 3).

With identical trajectories of the ray velocities precession at the whole ridge, the kinematics of their motion is very sensitive to the position ξ of the "observation" point. It may be shown that at the ridge ends $\xi = \pm\pi/2$ the values $\Theta_0(\pi/2)$ and $\Theta_0(-\pi/2)$ differ by only signs:

$$\Theta_0(\pm\pi/2) = \pm\Phi_\xi, \tag{95}$$

which gives, by (91),

$$\begin{aligned} \mathbf{s}(\pi/2) &= \mathbf{s}_0 - \mathbf{p}\cos2\Phi_\xi + \mathbf{q}\sin2\Phi_\xi, \\ \mathbf{s}(-\pi/2) &= \mathbf{s}_0 - \mathbf{p}\cos2\Phi_\xi - \mathbf{q}\sin2\Phi_\xi. \end{aligned} \tag{96}$$

This shows that the precession of the ray velocity vector along one of split acoustic axes is identical with the analogous process for a circularly polarized wave (70) propagating along the unsplit acoustic axis in the crystal without damping. Directions of rotation of the ray velocities $\mathbf{s}(\pm\pi/2)$ (96) have different signs. It is easily checked that at $g > 0$ they coincide with corresponding directions of circular polarization (76), and at $g < 0$ – are opposite to them.

In spite of the found identity of cones (70) and (96), there is an important difference between the related to them pictures of conical refraction. In the crystal without damping the ray velocities forming the refraction cone are directed along the appropriate normals to the slowness surface at the conical point of degeneracy. And the normals to the analogous surface in the vicinity of one of the split axes, as we have seen (Fig. 7), form a plane fan, which has nothing to do with a cone of ray velocities (96) (Fig. 12).

With passing of the "observation" point from the end of the ridge to its center, the motion of the ray velocity around universal cone (96) becomes less and less uniform. At the center point $\xi = 0$ the motion deceases at all: the isonormal vectors $\mathbf{s}_{1,2}$ are "frozen" at definite positions. Indeed, at $\xi = 0$ eqns. (88), (89) and (92) give the values $\Theta_{1,2}$ independent of time:

$$\Theta_{1,2} = \frac{a}{2} \pm \frac{\pi}{4}. \tag{97}$$

The corresponding fixed vectors of ray velocity are equal

$$\mathbf{s}_{1,2}(0) = \mathbf{s}_0 \pm (\mathbf{p}\sin a + \mathbf{q}\cos a) = \mathbf{s}_0 \pm \mathbf{M}/r. \tag{98}$$

As it would be expected, this result relates to expression (77) for a linear polarization of isonormal waves in the same way, as eqn. (69) to expression (64) from the refraction theory for crystals without damping. One can show that in this point ($\xi = 0$) of the ridge the two normals to the slowness surface are parallel not to vectors (98), but to their components belonging to the plane $\{\mathbf{m}_0, \mathbf{M}\}$ orthogonal to the ridge.

One should note that the fixed in time positions of the ray velocities (98) on ellipse (93) depend on the attenuation (the angle a, eqn. (41a)), whereas the universal ellipse does not "know" about a. This means that points (98), generally speaking, do not coincide with the ends of the principal semi-axes of the ellipse. Of course, in more symmetric situations the coincidence may occur, as it happens in the case of the splitting of the acoustic axis from a symmetry plane. What is more important that the vectors $\Delta\mathbf{s}_{1,2} = \mathbf{s}_{1,2}(0) - \mathbf{s}_0$, as it follows from eqns. (98), (41) and Fig. 4, remain universally orthogonal to the ridge of the wedge at any changes of the angle a.

It is evident that at any small deviation of ξ from zero the fixed vectors (98) acquire some increments dependent on the phase Φ_ξ. This will renew a motion of the ray velocities $\mathbf{s}_{1,2}$ over the cone. However, if not to pass far from the middle point $\xi = 0$, the most part of the period the vectors $\mathbf{s}_{1,2}$ will retain orientations close to directions (98). And the time-averaged vectors $\bar{\mathbf{s}}_{1,2}$ in these points will be close to directions (98). This means that in the middle domain of the self-intersection line of the slowness surface, the refraction will have rather a wedge than a cone character.

In the considered above particular case of the acoustic axis splitting from the symmetry plane, one can put $a = \pi / 2$ which remarkably simplifies expressions (88) and (89), and together with them also the formulae for angle parameters $\Theta_{1,2}$ (92):

$$\mathrm{tg}\Theta_1 = \mathrm{ctg}(\xi / 2)\mathrm{tg}\Phi_\xi, \qquad \mathrm{tg}\Theta_2 = \mathrm{tg}(\xi / 2)\mathrm{tg}\Phi_\xi . \tag{99}$$

The discussed problem of kinematics of the precession of ray velocities at the line of self-intersection of slowness surface may be quantitatively described. Introduce the polar coordinates ($S_{1,2}, \bar{\varphi}_{1,2}$) of the positions of the ends of the radius-vectors $\Delta\mathbf{s}_{1,2}$ at ellipses (93):

$$\Delta\mathbf{s}_{1,2} = S_{1,2}\left(\frac{\mathbf{p}}{p}\cos\bar{\varphi}_{1,2} + \frac{\mathbf{m}_0 \times \mathbf{p}}{p}\sin\bar{\varphi}_{1,2}\right). \tag{100}$$

Comparing (100) with (93) one obtains

$$S_{1,2}^2 = (\mathbf{q}\sin\Theta_{1,2} - \mathbf{p}\cos\Theta_{1,2})^2 = \frac{q^2F^2 - 2\mathbf{p}\cdot\mathbf{q}G_\mp + p^2G_\mp^2}{F^2 + G_\mp^2}, \tag{101}$$

$$\mathrm{ctg}\bar{\varphi}_{1,2} = (\mathbf{p}\cdot\mathbf{q} - p^2\mathrm{ctg}\Theta_{1,2}) / g . \tag{102}$$

Differentiating the latter equation gives the angular velocities

$$\dot{\bar{\varphi}}_{1,2} = -\frac{g}{S_{1,2}^2}\dot{\Theta}_{1,2} . \tag{103}$$

Here the derivatives $\dot{\Theta}_{1,2}$ are found from (92):

$$\dot{\Theta}_{1,2} = \frac{\dot{F}G_\mp - F\dot{G}_\mp}{F^2 + G_\mp^2} = -\frac{\omega\sin\xi}{U_{1,2}^2(\Phi_\xi + \pi / 2)} = -2\dot{\varphi}_{1,2}(\Phi_\xi + \pi / 2), \tag{104}$$

where $U_{1,2}^2$ and $\dot{\varphi}_{1,2}$ are given by functions (81) and (83) shifted in phase: $\Phi_\xi \to \Phi_\xi + \pi / 2$. The sectorial velocities $\bar{v}_{1,2}^{sec}$ of the motion of the vectors $\Delta\mathbf{s}_{1,2}$ over universal ellipse (93) are found in analogy with eqn. (86):

$$\bar{v}_{1,2}^{sec} = \frac{1}{2}S_{1,2}^2\dot{\bar{\varphi}}_{1,2} = -\frac{1}{2}g\dot{\Theta}_{1,2} = g\dot{\varphi}_{1,2}(\Phi_\xi + \pi / 2) . \tag{105}$$

Thus, this velocity differs from the angular velocity of the polarization, eqn. (83), only by the dimensional factor g and by the retardation $\pi/2$. Substituting into (105) the angular velocity $\dot{\varphi}_{1,2}$ (85) for the considered above symmetric example, one obtains the more compact form for the sectorial velocity:

$$\overline{v}_{1,2}^{sec}(\Phi_\xi) = \frac{g\omega\sin\xi}{1 \mp \cos\xi\cos2\Phi_\xi} \equiv g\dot{\varphi}_{2,1}(\Phi_\xi)\,. \quad (106)$$

Here it is bearing in mind that the phase shift of the velocity $\dot{\varphi}_{1,2}$ in simplified variant (85) is equivalent to the transition at the counterpart branch: $\dot{\varphi}_{1,2}(\Phi_\xi + \pi/2) = \dot{\varphi}_{2,1}(\Phi_\xi)$. The found relation (106) allowed us to use in Fig. 14 the same curves for a characterization of both angular velocities of polarization and sectorial velocities of ray speeds. The shown dependencies adequately reflect the discussed above properties of the ray velocity precession at the line of self-intersection of the slowness surface. Angle velocities (103) behave in a similar way, especially in the region of small ξ. With closing to acoustic axes $\xi = \pm\pi/2$, variations of angular velocities in time are smoothing, but retain finite until $p \neq q$, in contrast to the velocities $\overline{v}_{1,2}^{sec}$, which tend to constant at these limits.

9. Conclusions

Thus, we have found that specific features of the influence of attenuation on the basic wave properties are associated with two main qualities of the damping: i) it does not disturb the symmetry of a crystal, and ii) formally, it provides an imaginary, i.e. non-hermitian, perturbation of the acoustic tensor. Due to the first quality there is almost no influence of the damping on the acoustic axes which exist due to symmetry of the crystal (tangent degeneracies along ∞ and 4-fold symmetry axes and conical degeneracies along 3-fold axes). On the other hand, the conical acoustic axes of any other orientations manifest instability with respect to an imaginary perturbation of the acoustic tensor. They split into pairs of degeneracies of new type (the so-called singular acoustic axes), which never occurs without damping. In the neighbourhood of split acoustic axes the polarization of elastic waves proves to be strongly elliptical becoming almost circular close to the degeneracy points. A rotation of the polarization ellipses around those points is described by the Poincarѐ index $n = \pm1/4$. The slowness surface acquires lines of self-intersection connecting the split singular acoustic axes. Only the end points of these lines correspond to true degeneracies where the imaginary components of phase speeds of isonormal waves also coincide. The latter coincidence also occurs on the whole equi-damping lines at the attenuation surface. These self-intersection lines at the two different surfaces (Fig. 10) after their projection on the unit sphere $\mathbf{m}^2 = 1$ of propagation directions continue each other at the degeneracy points.

Topological transformations of wave surfaces and polarization fields create new features of the phenomenon of internal conical refraction. Still an extension of the theory may be done in terms of the same classic refraction cone bounded by the universal ellipse. As we have seen, in crystals without damping the classic picture of conical refraction automatically arises for a circularly polarized wave propagating along conical acoustic axis. In an absorptive crystal the same cone and universal ellipse as a trajectory of precession of the ray velocity vectors retain at the whole self-intersection line of the slowness surface between split degeneracy points.

Along singular axes the refraction does not differ from the classical picture: the isonormal waves degenerate into one circularly polarized wave with the ray precession of constant sectorial velocity $\overline{v}^{\text{sec}} = g\omega$ at the ellipse. A screen "illumination" related to such precession would look as a completely drawn ellipse (Fig. 15a). Some increase of intensity in the vicinity of large semi-axes ($S_{\max}$) is explained by a slower motion of the vector $\mathbf{s}_0$ in this region (its linear speed at the ellipse is equal $2\overline{v}^{\text{sec}} / S$). When the "observation" point passes along the ridge of the wedge to its middle, both the precession of the vectors $\mathbf{s}_{1,2}$ and the "illumination" pattern become less and less uniform (Fig. 15b,c). And in the center (ξ = 0) only two points (Fig. 15d) will turn to be "illuminated". They relate to the isonormal waves with linear polarization: the refraction becomes purely wedge-like. Thus, with scanning the ridge by the wave normal **m** the refraction continuously transforms from purely conic to purely wedge type.

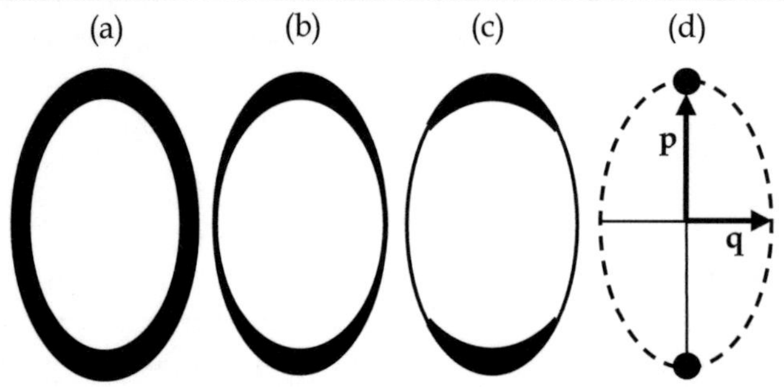

Fig. 15. Schematic plot of gradual transition from purely conical refraction (a) along a singular acoustic axis to purely wedge refraction in the middle of the ridge (d) for a particular case of the acoustic axis splitting from the symmetry plane

In conclusion, let us discuss the observability of the above beautiful and nontrivial physical effects. In principal, there is no threshold level of damping for the split of acoustic axes. Just the less damping, the less is the solid angle inside of which all the peculiarities manifest themselves. If this angle is less than the angle of the acoustic beam divergence, then we shall not observe neither splitting of acoustic axes, nor any accompanied effects. Thus, for the observability of our predictions the split angle $\delta\psi$ (46) must exceed the divergence of the beam. The best experimentally realizable collimation of sound beams is limited by the diffraction divergence, which is estimated as $\sim\lambda/d$, where λ is the wave length and d is the diameter of the beam. So, with increasing frequency ω the angle $\delta\psi$ increases and the beam divergence, on the contrary, decreases. Thus, we deal here with a frequency threshold from below. The order of the splitting angle is determined by the estimate $\delta\psi = 2\delta m_0 \sim \omega\eta / \mu$ (46), where μ is the shear modulus and η is the viscousity. Substituting this estimate to the inequality $\delta\psi > \lambda / d$, one obtains the following lower threshold for the frequency $\nu = \omega/2\pi$

$$\nu > \nu_{\text{th}} \sim \sqrt{\frac{c_s \mu}{2\pi\eta d}}, \tag{107}$$

where c_s is the sound speed. There are known a series of physical mechanisms of the sound attenuation η. Often phonons play in it an important role. The phonon viscosity at room (or high) temperature T may be estimated as

$$\eta_{\text{ph}} \sim \tau_{\text{ph}}(3k_{\text{B}}T / a^3)\,. \tag{108}$$

Here τ_{ph} is the phonon relaxation time, k_{B} is the Boltzmann constant, and a is the lattice parameter. Substituting into eqns. (107), (108) $c_s \sim 3\cdot10^5$ cm/s, $\mu \sim 10^{11}$ dyn / cm^2, $d \sim 0.5$ cm, $T \approx 300$ K, $a \sim 3\cdot10^{-8}$ cm, $\tau_{\text{ph}} \sim 10^{-9}$ s, we come to the estimate $\nu_{\text{th}} \sim 100$ MHz. Thus, at rather high-frequencies, which however belong to experimentally available ultrasound range, the properties and effects described in this chapter appear quite observable.

10. Acknowledgment

Authors are grateful to A.L. Shuvalov for helpful discussions and to W. Gerulski for the help in computations related to the illustrations. The support of the Polish Foundation MNiSW (grant No N N 501252334) is gratefully acknowledged. V.I.A. is also grateful to the Kielce University of Technology for a hospitality and support.

11. References

Aleksandrov, K.S. & Ryzhova, T.V. (1964). Internal conical refraction of elastic waves in ammonium dihydrogen phosphate. *Kristallografiya*, Vol. 9, No. 3 (June 1964) 373-376, ISSN 0023-4761 [*Sov. Phys. Crystallography*, Vol. 9, No. 3 (1964) 298-300, ISSN 1063-7745]

Alshits, V.I. & Lothe, J. (1979). Elastic waves in triclinic crystals. *Kristallografiya*, Vol. 24, No. 4, 6 (Aug., Dec. 1979) 972-993, 1122-1130, ISSN 0023-4761 [*Sov. Phys. Crystallography*, Vol. 24, No. 4, 6 (1979) 387-398, 644-648, ISSN 1063-7745]

Alshits, V.I. & Lyubimov, V.N. (1998). Elastic waves in absorptive media: peculiarities of wave surfaces and singularities in the polarization fields. In: *Dissipation in Physical Systems*, A. Radowicz (Ed.), pp. 15-43, Politechnika Swietokrzyska, ISSN 0239-4979, Kielce. *Proceedings of 2nd Workshop on Dissipation in Physical Systems*, Borkow, Poland, September 1-3, 1997

Alshits, V.I. & Lyubimov, V.N. (2011). Conical refraction of elastic waves in absorptive crystals. *Zh. Eksp. Teor. Fiz.*, Vol. 140, No. 2(8) (Aug. 2011) [JETP, Vol. 113, No. 2 (2011), ISSN 1063-7761]

Alshits, V.I.; Sarychev, A.V. & Shuvalov, A.L. (1985). Classification of degeneracies and analysis of their stability in the theory of elastic waves in crystals. *Zh. Eksp. Teor. Fiz.*, Vol. 89, No. 3(9) (Sept. 1985) 922-938, ISSN 0044-4510 [Sov. Phys. JETP, Vol. 62, No. 3 (1985) 531-539, ISSN 1063-7761]

Barry P.A. & Musgrave, M.J.P. (1979). On elliptical conical refraction of elastic waves in tetragonal crystals. *Q. J. Mech. & Appl. Math.*, Vol. 32, No. 3 (March 1979) 205-214, ISSN 0033-5614

De Klerk, J. & Musgrave, M.J.P. (1955). Internal conical refraction of transverse elastic waves in a cubic crystal. *Proc. Phys. Soc. Lond. B*, Vol. 68, No. 2 (Feb. 1955) 81-88, ISSN 1088-0370

Fedorov, F.I. (1968). *Theory of Elastic Waves in Crystals*, Plenum Press, ISBN, New York

Khatkievich, A.G. (1962a). The acoustic axes in crystals. *Kristallografiya*, Vol. 7, No. 5 (Oct. 1962) 742-747, ISSN 0023-4761 [*Sov. Phys. Crystallography*, Vol. 7, No. 5 (1963) 601-604, ISSN 1063-7745]

Khatkievich, A.G. (1962b). Internal conical refraction of elastic waves. *Kristallografiya*, Vol. 7, No. 6 (Dec. 1962) 916-920, ISSN 0023-4761 [*Sov. Phys. Crystallography*, Vol. 7, No. 6 (1963) 742-745, ISSN 1063-7745]

Khatkievich, A.G. (1964). Special directions for elastic waves in crystals. *Kristallografiya*, Vol. 9, No.5 (Oct. 1964) 690-695, ISSN 0023-4761 [*Sov. Phys. Crystallography*, Vol. 9, No. 5 (1964) 579-582, ISSN 1063-7745]

Landau L.D. & Lifshitz, E.M. (1986). *Theory of elasticity*. Pergamon Press, ISBN, London

Musgrave, M.J.P. (1957). On an elliptic cone of internal refraction for quasi-transverse waves in tetragonal crystals. *Acta Crystallogr.*, Vol. 10, No. 4 (Apr. 1957) 316-318, ISSN

Shuvalov, A.L. (1998). Topological features of polarization fields of plane acoustic waves in anisotropic media. *Proc. R. Soc. Lond. A*, Vol. 454, (Nov. 1998) 2911-2947, ISSN 1471-2946

Shuvalov, A.L. & Chadwick, P. (1997). Degeneracies in the theory of plane harmonic wave propagation in anisotropic heat-conducting elastic media. *Phil. Trans. R. Soc. Lond. A*, Vol. 355 (Jan. 1977) 156-188, ISSN 1471-2962

Shuvalov, A.L. & Scott, N.H. (1999). On the properties of homogeneous viscoelastic waves. *Q. J. Mech. Appl. Math.*, Vol. 52 (Sept. 1999) 405-417, ISSN 0033-5614

Shuvalov, A.L. & Scott, N.H. (2000). On singular features of acoustic wave propagation in weakly dissipative anisotropic thermoviscoelasticity. *Acta Mechanica*, Vol. 140, No 1-2 (March 2000) 1-15, ISSN

Sirotin Yu.I. & Shaskolskaya, M.P. (1979). *Fundamentals of Crystal Physics* (in Russian), Nauka, Moscow [(1982) translation into English, Mir, ISBN, Moscow]

3

An Operational Approach to the Acoustic Analogy Equations

Dorel Homentcovschi and Ronald Miles
Department of Mechanical Engineering
State University of New York at Binghamton
USA

1. Introduction

Great progress has been made in the last sixty years in the study of the important problem of noise generated by the interaction of flow with stationary or mobile bodies such as occurs in jets, rotating blade propulsion machinery (propellers, turbofans helicopter rotors) and last but not least in aircraft at all ranges of flight and speed. An important part of this progress was based on a rigorous theory known as the Acoustic Analogy initiated by Sir James Lighthill in (Lighthill, 1952) and (Lighthill, 1954). Lighthill considered a free flow, as for example with a jet engine, and the nonstationary fluctuations of the stream represented by a distribution of quadrupole sources in the same volume. The flow parameters such as the surface pressure and the Lighthill tensor T_{ij} are assumed known from solving the aerodynamic problem in the region of sound generation or furnished by measurements. For the first time, this revealed a clear distinction between Aerodynamic Theory, meant to determine mainly the aerodynamic parameters as the lift and damping on the moving object (and also supplying the data for the noise determination) and the Aeroacoustic Theory needed for studying the noise produced, generally at large distances, by the flying (or moving) objects. A primary aim of the following is to show that by using an operational calculus based on the multidimensional Fourier Transform all the theory involved in obtaining the Ffowcs Williams-Hawkings formula (Ffowcs Williams and Hawkings, 1969) can be performed using only classical mathematical analysis.

Curle's contribution (Curle, 1955) is a formal solution of the Acoustic Analogy which takes stationary hard surfaces into consideration. The theory developed by Ffowcs Williams and Hawkings (FW-H) (Ffowcs Williams and Hawkings, 1969) is valid for aeroacoustic sources in relative motion with respect to a hard surface, as is the case in many technical applications for example in the automotive industry or in air travel. The calculation involves quadrupole, dipole and monopole terms. An important point is that FW-H theory, developed in (Ffowcs Williams and Hawkings, 1969) assumed that the boundary surface coincides with the physical body surface and is impenetrable. In both Aerodynamic and Aeroacoustic theories the domain was the same: the infinite air domain external to the moving body.

When the Aeroacoustic Theory was developed by Lighthill, Curle, Ffowcs Williams and Hawkings there were not a lot of experimental or theoretical data to be used as input to their aeroacoustic theoretical work. For this reason, they derived mainly qualitative results which were quite useful in guiding many significant acoustic experiments and in designing low noise propulsion machinery.

The situation has changed dramatically in the last 20 years. The rapid growth in high speed digital computer technology, the availability of turbulent flow simulation codes as well as high quality measured fluid dynamic data and advances in the theory of partial differential equations, resulted in obtaining the needed data for many important problems of aeroacoustics. However, the development of reliable Computational Fluid Dynamics (CFD) methods made them also useful in the evaluation of the near-field aerodynamic parameters. Unfortunately, a fully CFD-based computational aeroacoustic methodology is so far too inefficient and beyond the capability of supercomputers of today.

To avoid these computational difficulties, the philosophy of approaching the Aeroacoustic problem has changed by introducing a surface S as a "permeable" control surface. The surface S is assumed to include inside, in the volume V, all the nonlinear flow effects and noise sources. This splitting of the problem into a linear problem for an infinite domain and a nonlinear setting for a bounded region allows the use of the most appropriate numerical methodology for each of them. In the bounded domain V the CFD methods or advanced measurement techniques will be used for obtaining the aerodynamic near-field and providing the data on the surface S needed for the external, infinite domain modelling. The analysis of the flow information inside V is, in general, expensive either using experiments or CFD. Therefore, it is advantageous to make the volume V as small as possible.

The FH-W equation involving a permeable surface is the proper model for determining the far-field pressure in the infinite domain. The case of permeable surfaces was analyzed by Ffowcs Williams in (Dowling and Ffowcs Williams, 1982) and (Crighton, et al.), by Francescantonio in (Francescantonio, 1997) who called it the KFWH (Kirchhoff FW-H) formula, by Pilon and Lyrintzis in (Pilon and Lyrintzis, 1997) calling it an improved Kirchhoff method and by Brentner and Farassat in (Brentner & Farassat, 1998). Besides the accessibility to the surface data the advantage of the methods using a permeable control surface is that the surface integrals and the first derivative needed can be evaluated more easily than the volume integrals and the second derivatives necessary for the calculation of the quadrupole terms when the traditional Acoustic Analogy is used.

The Acoustic Analogy approach and especially the theory based on the FW-H equation is the most widely used tool for deterministic noise sources. The beauty (and the power) of this model is that all the manipulations are completely rigorous without any *ad hoc* reasoning.

Besides the approach based on generalized functions, used in most papers approaching the FW-H formula, we note the work by Goldstein (Goldstein, 1976) where all the formulas are obtained starting with a generalized Green's formula. A similar approach was used in (Wu and Akay, 1992) . However, the algebra involved in their construction is substantial.

In the second of the two reports by Farassat (Farassat, 1994),(Farassat, 1996), which covers the details of the mathematics used for the wave equation with sources on a moving surface, the author correctly claims that the Ffowcs Williams-Hawkings famous paper published in 1969 used a level of mathematical sophistication including multidimensional generalized functions (distributions) and differential geometry unfamiliar to most researchers and designers working in the field. Many people use Dirac's δ (generalized) function starting with its integral definition. On the other hand, to learn about more complicated operations such as the derivative of a generalized function (distribution) we need a change of paradigm in the way we look at ordinary functions. For some people involved in practical applications this is not a simple task. The power of the theory of generalized functions stems from its operational properties. Thus, for example, discontinuous solutions of linear equations using the Green's function are easily obtained by posing the problem in generalized function spaces. In the following we show that the theory connected with the FW-H formula can be made much simpler, without manipulating multidimensional generalized functions, by using an operational calculus based on the multidimensional Fourier Transform. The approach based on using the Fourier Transform preserves all the good operational properties of the generalized functions without the need to introduce a new sophisticated mathematical tool. The method was used previously by us (Homentcovschi and Singler, 1999) for a direct introduction to the Boundary Element Method.

In the case of permeable surfaces an alternate method for solving the Aeroacoustic problem for the infinite external domain is based on the Kirchhoff formula for the wave equation. Due to its use in Aeroacoustic theory we included also in Section 6 the Kirchhoff formulation for the solution of the wave equation in the case of mobile surfaces. The proof is based again on 3-D Fourier Transform of discontinuous functions and the final formula includes the volume sources and the surface sources as well. As a comparison of the two approaches (that based on FW-K equation and that using Kirchhoff's equation) we notice that Kirchhoff's method requires less memory because fewer quantities on the control surface needed to be stored. On the other hand, Brentner and Farassat in (Brentner & Farassat, 1998) have shown that the FW-H equation is superior to the Kirchhoff formula for aeroacoustic problems because it is based on conservation laws of fluid mechanics rather than on the wave equation. Thus, the FW-H equation is valid even if the integration surface is in the nonlinear region being therefore more robust with the choice of control surface. Another advantage of the FW-H method is that it does not require computation of the normal derivatives on the permeable surface.

A comprehensive review of the use of Kirchhoff's method in computational aeroacoustics was given by Lyrintzis in (Lyrintzis, 1994). The same author reviewed the advances in the use of surface integral methods in aeroacoustics, including Kirchhoff's method and permeable Ffowcs-Williams Hawkings methods in (Lyrintzis, 2003). Morino in (Morino, 2003) addresses commonalities and differences between aeroacoustics and aerodynamics. A discussion about the acoustic analogy and alternative theories for jet noise prediction is the subject of (Morris and Farassat, 2002). Finally we note some interesting work about this subject included in the book edited by (Raman, 2009).

It is our hope that the elementary derivations included in this chapter will make the application of the FW-H equation more clear, avoiding in the future comments such as those generated by (Zinoviev and Bies, 2004) (See (Farassat, 2005), (Zinoviev and Bies, 2005), (Farassat and Myers, 2006), (Zinoviev and Bies, 2006)).

2. The equations of the acoustic analogy and their operational form

In order to apply the 3-D Fourier Transform to a certain physical variable this has to be defined in the whole space. Thus, to utilize the Fourier Transform to examine the sound field in a finite physical domain, it is necessary to imbed this domain within an infinite space. For example, in the case of studying the air motion around a finite body, the interior of the body is considered as an air-filled domain separated from the exterior, infinite domain by an impermeable surface, S_b enclosing the body. This introduces a first class of discontinuity surfaces of the motion between two air-filled regions. The second class contains the natural discontinuity surfaces S_s inside the flow domain as shock fronts, wakes, etc. Finally, to aid computations, it is often helpful to also define a virtual permeable surface, S_p, enclosing body along with the portion of the air domain where viscous effects and the nonlinear terms in the Navier-Stokes equations are important. It is convenient to evaluate the aerodynamic field in this region numerically using CFD, due to the difficulties imposed by viscosity and nonlinear effects. These calculations furnish data describing the hydrodynamic state on the virtual permeable surface, S_p. In the domain exterior to the permeable surface the acoustical analogy is applied to predict the sound field. In other words, the field inside the virtual permeable surface is assumed to be strongly influenced by hydrodynamic effects, while the field in the external, infinite domain can be modeled according to the usual assumptions of linear acoustics.

2.1 The equations of the acoustic analogy

In the case where the whole space $\mathbb{R}^3$ is filled by a compressible viscous fluid, containing several discontinuity surfaces, the flow inside each domain is governed by the following equations:
the continuity equation (mass conservation)

$$\frac{\partial \rho}{\partial t} + \frac{\partial}{\partial x_j}\left(\rho u_j\right) = 0, \tag{1}$$

and the conservation of momentum in the form written by Lighthill in Refs. (Lighthill, 1952) and (Lighthill, 1954)

$$\frac{\partial\left(\rho u_i\right)}{\partial t} + c^2 \frac{\partial \rho}{\partial x_i} = -\frac{\partial T_{ij}}{\partial x_j}. \tag{2}$$

Here ρ is the density, c is the velocity of sound in the uniform medium, u_j is the component of fluid velocity in the direction x_j $(j = 1, 2, 3)$, and a repeated index implies a summation over these values, and,

$$T_{ij} = P_{ij} + \rho u_i u_j - c^2 \rho\, \delta_{ij},$$

is Lighthill's stress tensor. Also, we denoted by

$$P_{ij} = p\delta_{ij} + \mu\left(-\frac{\partial u_j}{\partial x_j} - \frac{\partial u_j}{\partial x_i} + \frac{2}{3}\left(\frac{\partial u_\kappa}{\partial x_\kappa}\right)\delta_{ij}\right),$$

the compressive stress tensor, δ_{ij} is the Kronicker delta function, p is the pressure and μ is the viscosity.

Remark 1. *The acoustical analogy equations (1) and (2) are in fact the exact fluid flow equations in the form written by Lighthill. This means that if one solves these equations correctly for a problem satisfying the Lighthill assumptions, then one will get the correct answer to the aerodynamic and aeroacoustic problems simultaneously.*

In the case where inside the flow domain there are discontinuity surfaces of type S_s (as shock fronts) the conservation laws for mass and momentum yield the Rankine-Hugoniot type junction conditions

$$\left[\rho\left(u_n - v_n\right)\right] = 0 \tag{3}$$

$$\left[\rho u_\kappa\left(u_n - v_n\right) + P_{j\kappa} n_j\right] = 0. \tag{4}$$

We denote by square brackets the jump of its content across the discontinuity surface (see also formula (54) in Appendix A). u_n is the velocity projection on the normal to the surface, $\mathbf{n}$, and v_n is the normal velocity of the surface.

Finally, on solid (deformable) impermeable surfaces S_b inside the flow the obvious non-penetration condition proves true

$$u_n - v_n = 0. \tag{5}$$

2.2 The operational form of the acoustic analogy equations

Since the density ρ_0 and pressure p_0 at large distances (in the unperturbed fluid) are different from zero we introduce the perturbation of the density $\rho' = \rho - \rho_0$ and the perturbation of the stress tensor as $P'_{j\kappa} = P_{j\kappa} - p_0\delta_{j\kappa}$ (where $\delta_{j\kappa}$ is the Kronecker delta) and will write equations (1) and (2) as

$$\frac{\partial \rho'}{\partial t} + \frac{\partial}{\partial x_j}\left(\rho u_j\right) = 0 \tag{6}$$

$$\frac{\partial\left(\rho u_j\right)}{\partial t} + c^2\frac{\partial \rho'}{\partial x_j} = -\frac{\partial T'_{j\kappa}}{\partial x_\kappa}, \tag{7}$$

where

$$T'_{j\kappa} = P'_{j\kappa} + \rho u_j u_\kappa - c^2\rho'\delta_{j\kappa}$$

According to the previous discussion, these equations are valid within the flow domains in the whole space. Assuming, as in Appendix A, that there is a discontinuity surface S, separating the inner domain $D^{(i)}$ and the external domain $D^{(e)}$, by applying the Fourier transform to equations (6) and (7) and using the formulas (53) and (68) there results *the operational equations*

$$\frac{d\widetilde{\rho'}}{dt} + ik_j\widetilde{\rho u_j} = \int_S \left[\rho\left(u_n - v_n\right)\right] e^{-i\mathbf{k}\cdot\mathbf{y}} dS \tag{8}$$

$$\frac{d\,\widetilde{\rho u_j}}{dt} + ik_j c^2\widetilde{\rho'} = -ik_\kappa \widetilde{T'}_{j\kappa} + \int_S \left[\rho u_j\left(u_n - v_n\right) + P_{\kappa j}n_\kappa\right] e^{-i\mathbf{k}\cdot\mathbf{y}} dS \tag{9}$$

Here the overhead tilde denotes a Fourier Transform (see Appendix A), $\mathbf{k}(k_1, k_2, k_3)$ is the wave vector and $\mathbf{y}(y_1, y_2, y_3)$ is the position vector of the integration (source) point. The *operational* form of the conservation equations contains in the left-hand sides the Fourier transform of corresponding terms in (1) and (2) and in the right-hand sides the integrals accounting for the influence of the discontinuity surfaces. As was noted in Remark 1 in Appendix A, in the general case, in the right-hand side of equations (8) and (9) the sum of the contributions of all discontinuity surfaces will appear.

Remark 2. *The square brackets in equations (8) and (9) can also be written as* $\left[\rho_0 u_n + \rho'\left(u_n - v_n\right)\right]$ *and* $\left[\rho u_\kappa\left(u_n - v_n\right) + P'_{\kappa j}n_j\right]$ *respectively.*

2.3 Some particular cases

2.3.1 Shock-type discontinuity surfaces (S_s)

In the case of a discontinuity surface S_s (of shock-type) inside the fluid domain the junction conditions (3) and (4) will cancel out the integral terms in equations (8) and (9). Consequently the shock-type discontinuity surfaces are not introducing any supplementary terms in the operational equations.

2.3.2 Impermeable solid deformable surfaces (S_b)

In the case of a solid with a deformable and impermeable boundary surface the condition (5) will cancel out the integral term in the operational form of the continuity equation (8) and a part of the integral in the momentum equation (9). In this case, the Fourier transform of the traction $\mathbf{P}=(P_1, P_2, P_3)$ of the surface on the fluid enters in the integral term of the momentum equations as $P_j = P_{\kappa j} n_\kappa$. The operational momentum equation now contains in the right-hand side the action of the solid surface S_b on the fluid flow.

3. Solution of the operational form of the acoustical analogy equations.

3.1 The case of a permeable surface (S_p)

Let S_p be a discontinuity surface of the flow variables inside the external fluid flow such that in the domain $D^{(i)}$ we have $p^{(i)} = p_0, \rho^{(i)} = \rho_0, \mathbf{v}^{(i)} = \mathbf{0}$.
We write the system of equations (8), (9) as

$$\frac{d\widetilde{\mathbf{w}}}{dt} = \widetilde{\mathbb{A}}\widetilde{\mathbf{w}} + \widetilde{\mathbf{f}} \tag{10}$$

where

$$\widetilde{\mathbf{w}}^T = (\widetilde{\rho'}, \widetilde{\rho u_1}, \widetilde{\rho u_2}, \widetilde{\rho u_3}), \tag{11}$$

$$\widetilde{\mathbb{A}} = \begin{bmatrix} 0 & -ik_1 & -ik_2 & -ik_3 \\ -c^2 i k_1 & 0 & 0 & 0 \\ -c^2 i k_2 & 0 & 0 & 0 \\ -c^2 i k_3 & 0 & 0 & 0 \end{bmatrix}, \tag{12}$$

$$\begin{aligned} \widetilde{f}_1 &= \int_S \left[\rho_0 u_n + \rho' (u_n - v_n)\right] e^{-i\mathbf{k}\cdot\mathbf{y}} dS \\ \widetilde{f}_{j+1} &= -ik_\kappa \widetilde{T}_{j\kappa} + \int_S \left[\rho u_j (u_n - v_n) + P_j\right] e^{-i\mathbf{k}\cdot\mathbf{y}} dS, \quad j = 1,2,3. \end{aligned} \tag{13}$$

The solution of equation (10) can be obtained by using the exponential $\widetilde{\mathbb{H}}(t)$ of the matrix $\widetilde{\mathbb{A}}$ as

$$\widetilde{\mathbf{w}} = \int_{-\infty}^{t} \widetilde{\mathbb{H}}(t - t') \, \widetilde{\mathbf{f}}(t') \, dt' \tag{14}$$

The exponential of the matrix $\widetilde{\mathbb{A}}$ can be written as

$$\widetilde{\mathbb{H}} = \left[\widetilde{h}_{j\kappa}\right] \tag{15}$$

$$\begin{aligned} \widetilde{h}_{1,1} &= \cos(ckt), \quad \widetilde{h}_{1,\kappa+1} = -\frac{i k_\kappa \sin(ckt)}{ck}, \ \kappa = 1,2,3 \\ \widetilde{h}_{j+1,1} &= -\frac{i k_j c \sin(ckt)}{k}, \quad j = 1,2,3 \\ \widetilde{h}_{j+1,\kappa+1} &= \delta_{j\kappa} + \frac{k_j k_\kappa (\cos(ckt) - 1)}{k^2}, \quad j,\kappa = 1,2,3 \end{aligned} \tag{16}$$

where $k^2 = k_1^2 + k_2^2 + k_3^2 = |\mathbf{k}|^2$. By introducing the function

$$\begin{aligned} \widetilde{F}(\mathbf{k}, t) = & \frac{d}{dt} \int_S \{\rho_0 u_n + \rho' (u_n - v_n)\} e^{-i\mathbf{k}\cdot\mathbf{y}} dS \\ & - ik_j \int_S \left\{\rho u_j (u_n - v_n) + P_j'\right\} e^{-i\mathbf{k}\cdot\mathbf{y}} dS + ik_j ik_\kappa \widetilde{T}'_{j\kappa} \end{aligned} \tag{17}$$

the components of the vector $\widetilde{\mathbf{w}}$ can be written as

$$\widetilde{\rho'}(\mathbf{k},t) = \int_{-\infty}^{t} \widetilde{F}(\mathbf{k},\tau) \frac{\sin(ck(t-\tau))}{ck} d\tau \tag{18}$$

$$\begin{aligned}\widetilde{\rho u_j}(\mathbf{k},t) = &-\int_{-\infty}^{t} ik_\kappa \widetilde{T}_{j\kappa}(\mathbf{k},\tau) d\tau + \int_{-\infty}^{t} d\tau \int_{S(\tau)} \left\{ \rho u_j (u_n - v_n) + P'_j \right\} e^{-i\mathbf{k}\cdot\mathbf{y}} dS \\ &+ik_j \int_{-\infty}^{t} \widetilde{F}(\mathbf{k},\tau) \frac{\cos(ck(t-\tau)) - 1}{k^2} d\tau\end{aligned} \tag{19}$$

The method given in this section has the advantage of furnishing the integral representations for the operational density and operational velocity as well. A simpler deduction of the representation of the operational density is given in the next section.

3.2 The equation for the operational density

By eliminating the Fourier transform of the momentum $\widetilde{\rho u_j}$ between equations (8) and (9) the operational equation satisfied by the density perturbation becomes

$$\frac{d^2 \widetilde{\rho'}}{dt^2} + c^2 k^2 \widetilde{\rho'} = \widetilde{F}(\mathbf{k},t). \tag{20}$$

The relationship (20) is the operational form of the nonhomogeneous wave equation. The general solution of the homogeneous wave equation in operational form can be written as

$$\widetilde{\rho'}_h = A(k)\cos(c\,k\,t) + B(k)\sin(c\,k\,t))$$

and by using Lagrange's method of variation of parameters there results the same representation formula (18) for the operational density as the solution of equation (20).

3.3 The case of an impermeable surface (S_b)

In the case where the surface S_b is impermeable, the condition (5) cancels out some terms in formula (17). In this case the nonhomogeneous term is

$$\widetilde{F}_b(\mathbf{k},t) = \frac{d}{dt} \int_S \rho_0 u_n e^{-i\mathbf{k}\cdot\mathbf{y}} dS - ik_j \int_S P'_j e^{-i\mathbf{k}\cdot\mathbf{y}} dS + ik_j ik_\kappa \widetilde{T}'_{j\kappa} \tag{21}$$

which coincides with the nonhomogeneous term in the operational form of the FW-H equation.

By introducing the new variables suggested by Francescantonio (Francescantonio, 1997), in the form modified in (Brentner & Farassat, 1998)

$$U_j = \left(1 - \frac{\rho}{\rho_0}\right) v_j + \frac{\rho u_j}{\rho_0} \tag{22}$$

$$L_j = P'_j + \rho u_j (u_n - v_n) \tag{23}$$

the nonhomogeneous term of the operational wave equation in the case of a permeable surface coincides with that corresponding to an impermeable case

$$\widetilde{F}(\mathbf{k},t) = \frac{d}{dt} \int_S \rho_0 U_n e^{-i\mathbf{k}\cdot\mathbf{y}} dS - ik_j \int_S L_j e^{-i\mathbf{k}\cdot\mathbf{y}} dS + ik_j ik_\kappa \widetilde{T}_{j\kappa} \tag{24}$$

The terms U_j and L_j can be interpreted respectively as a modified velocity and a modified traction, which take into account the flow across S.

4. Determination of velocity

4.1 The lift component of velocity

The lift component of velocity is given by inverse Fourier transform of the term

$$\left(\widetilde{\rho u_j}\right)_L = ik_j ik_r \int_{-\infty}^{t} d\tau \int_{S_\tau} P'(\mathbf{y},\tau)\, n'_r(\mathbf{y},\tau)\, e^{-i\mathbf{k}\cdot\mathbf{y}} \frac{\cos\left(ck\left(t-\tau\right)\right)-1}{k^2} dS$$

Therefore, by using formula (73) we obtain

$$\left(\rho u_j\right)_L = \frac{\partial^2}{\partial x_j \partial x_r} \int_{-\infty}^{t} d\tau \int_{S_\tau} P' n'_r \frac{H\left(t-\tau-r/c\right)}{4\pi r} dS$$

where

$$r = |\mathbf{x}-\mathbf{y}| \equiv \sqrt{\left(x_1-y_1\right)^2+\left(x_2-y_2\right)^2+\left(x_3-y_3\right)^2}.$$

$\mathbf{x}(x_1, x_2, x_3)$ being the position vector of the observation point.

5. Determination of density

Since the general permeable case can be studied by means of an equivalent impermeable case we shall determine the density in the case where the nonhomogeneous term is (21).

5.1 The case of sources on a surface

Consider the simpler case when the noise sources are on a rigid surface having only translation and rotation motions. Then,

$$\widetilde{F}_S(\mathbf{k},t) = \int_S Q(\mathbf{y},t)\, e^{-i\mathbf{k}\cdot\mathbf{y}} dS \tag{25}$$

where $Q(\mathbf{y},t)$ is the surface intensity. In this case we have

$$\widetilde{\rho}'_S(\mathbf{k},t) = \int_{-\infty}^{t} d\tau \int_{S_\tau} Q(\mathbf{y},\tau)\, e^{-i\mathbf{k}\cdot\mathbf{y}} \frac{\sin\left(ck\left(t-\tau\right)\right)}{ck} dS \tag{26}$$

Hence,

$$\begin{aligned}\rho'_S(\mathbf{x},t) &= \frac{1}{(2\pi)^3}\int e^{i\mathbf{k}\cdot\mathbf{x}} d\mathbf{k} \int_{-\infty}^{t} d\tau \int_{S_\tau} Q(\mathbf{y},\tau)\, e^{-i\mathbf{k}\cdot\mathbf{y}} \frac{\sin\left(ck\left(t-\tau\right)\right)}{ck} dS \\ &= \int_{-\infty}^{t} d\tau \int_{S_\tau} Q(\mathbf{y},\tau)\, dS \int \frac{\sin\left(ck\left(t-\tau\right)\right)}{(2\pi)^3 ck} e^{i\mathbf{k}\cdot(\mathbf{x}-\mathbf{y})} d\mathbf{k} \\ &= \int_{-\infty}^{t} d\tau \int_{S_\tau} Q(\mathbf{y},\tau)\, \frac{\delta\left(\tau-\left(t-r/c\right)\right)}{4\pi c^2 r} dS\end{aligned} \tag{27}$$

Here the relationship (71) and the property $\delta(-t) = \delta(t)$ have been used.
The only contribution in the last integral in formula (27) comes from the time τ_e which is the solution of the equation

$$g(\tau,\mathbf{y},t,\mathbf{x}) = 0 \tag{28}$$

where

$$g(\tau,\mathbf{y},t,\mathbf{x}) = \tau - t + \frac{|\mathbf{x}-\mathbf{y}|}{c} \tag{29}$$

In other worlds, the value of the density at the observation point $\mathbf{x}$ at the moment t is determined by the noise sources at the *emission (radiating) time* τ_e on the *emission (radiating) surface* $S_e \equiv S_{\tau_e}$.

It is now necessay to consider the coordinate systems. Let a fixed point P_0 reside on a material surface such as an airplane wing or a blade etc. We consider two coordinate frames:

a) A frame η fixed relative to the moving material surface. This is the frame used by a designer to describe the structure geometry for purposes of fabrication. The variable η will be called the Lagrangian variable of the point P_0 and in the case supposed here (nondeformable material surface) is independent of time.

b) A coordinate frame fixed with respect to the undisturbed medium having the origin O (see Fig.1). The position of the observation point is given by the position vector $\mathbf{x}$. The position of the point P_0 is described by the position vector $\mathbf{y}(\eta, \tau)$. The position vectors $\mathbf{x}$ and $\mathbf{y}$ will give the Eulerian coordinates of the observation and source terms, respectively. The formula

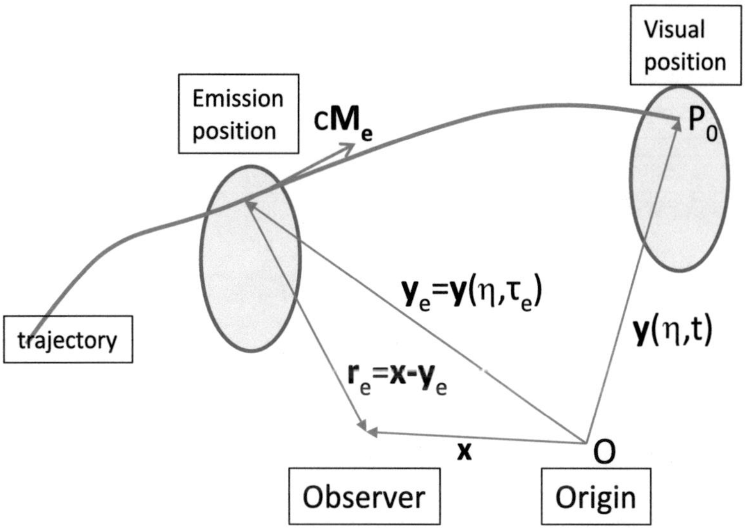

Fig. 1. Coordinate frame

$$\mathbf{y} = \eta + \int^{\tau} c\mathbf{M}(\eta, \tau')\, d\tau' \tag{30}$$

gives the connection between the Lagrangian and Eulerian coordinates of the point P_0. For η fixed the equation $\mathbf{y} = \mathbf{y}(\eta, \tau)$ gives the trajectory of the point P_0 and $c\mathbf{M} = d\mathbf{y}/d\tau$ is the velocity of the source point with respect to the undisturbed medium (source convection velocity). Formula (30) can be viewed as a transformation between the Lagrangian and Eulerian coordinates of the point P_0. The inverse transformation will be denoted by $\eta = \eta(\mathbf{y}, \tau)$. In the case where the transformation involves only translations and rotations we have $\det(\partial\mathbf{y}/\partial\eta) = \det(\partial\eta/\partial\mathbf{y}) = 1$. Fig.1 shows the observer's position $\mathbf{x}$ at time t, the emission (radiation) position of the material surface ($S_e \equiv S_{\tau_e}$), the position of the same surface at the observation time (S_t), the position vector $\mathbf{y}_e$ at the emission time and at the observation time $\mathbf{y}(\eta, t)$, the trajectory of the point P_0, the convection velocity of the source at the emission moment $c\mathbf{M}_e$, the radiation vector $\mathbf{r}_e = \mathbf{x} - \mathbf{y}_e$, and the emission distance $r_e = |\mathbf{r}_e|$.

To change the integration variable in the last integral in formula (27) from τ to g, we calculate

$$\frac{dg}{d\tau} = 1 - \frac{1}{c}\nabla r \cdot \frac{\partial \mathbf{y}}{\partial \tau} = 1 - \frac{\mathbf{r}}{r} \cdot \frac{\mathbf{v}}{c} = 1 - M_r \tag{31}$$

where M_r is the Mach number at the point η in the radiation direction at the time τ. The density perturbation can therefore be written as

$$\rho'_S(\mathbf{x},t) = \int_{S_e} \left[\frac{Q(\mathbf{y},\tau)}{4\pi c^2 r \left|1 - M_r\right|} \right]_{ret} dS_\eta \equiv \int_{S^*(\tau_e)} \frac{Q^*(\eta,\tau_e)}{4\pi c^2 r_e \left|1 - M_{r_e}\right|} dS_\eta \tag{32}$$

Here, $|1 - M_{r_e}|$ is the *Doppler factor.* The square brackets $[]_{ret}$ imply that the contents are to be evaluated at the retarded (emission or radiating time) τ_e given implicitly by $g(\tau) = 0$. The emission position is $\mathbf{y}_e = \mathbf{y}(\eta, \tau_e)$, the emission distance r_e of the source point η to the observer position $\mathbf{x}$ is $r_e = |\mathbf{x} - \mathbf{y}(\eta_e, \tau_e)|$, and $Q^*(\eta, \tau_e) = Q(\mathbf{y}_e, \tau_e)$.

5.1.1 The thickness noise

The thickness noise is given by the term

$$\widetilde{F}_{thickness}(\mathbf{k},t) = \frac{d}{dt}\int_S \rho_0 u_n e^{-i\mathbf{k}\cdot\mathbf{y}} dS$$

An analysis similar with that of the section 5.1 yields

$$\rho'_{thickness}(\mathbf{x},t) = \frac{\partial}{\partial t}\int_S \left[\frac{\rho_0 u_n}{4\pi c^2 r \left|1 - M_r\right|} \right]_{ret} dS_\eta \tag{33}$$

5.1.2 The loading noise

The last term in relationship (21) describes the loading noise

$$\widetilde{F}_{loading}(\mathbf{k},t) = -ik_j \int_S P'_j e^{-i\mathbf{k}\cdot\mathbf{y}} dS$$

Its contribution to the perturbed density ρ' is

$$\rho'_{loading}(\mathbf{x},t) = -\frac{\partial}{\partial x_j}\int_S \left[\frac{P'_j}{4\pi c^2 r \left|1 - M_r\right|} \right]_{ret} dS_\eta \tag{34}$$

We have given here a very short presentation of formulas for thickness noise and loading noise. A more complete presentation about these formulas and their implementation can be found in (Farassat, 2007).

5.2 The quadrupole noise term

The last term in formula (21) corresponds to a quadrupole noise source:

$$\begin{aligned}
\rho'_q(\mathbf{x},t) &= \mathcal{F}^{-1}\left\{ ik_j ik_\kappa \widetilde{T}_{j\kappa} \right\} \\
&= \int_{-\infty}^{t} d\tau \int ik_j ik_\kappa \widetilde{T}_{j\kappa} \frac{\sin\left(ck\left(t-\tau\right)\right)}{\left(2\pi\right)^3 ck} e^{i\mathbf{k}\cdot\mathbf{x}} d\mathbf{k} \\
&= \frac{\partial^2}{\partial x_j \partial x_\kappa}\int_{-\infty}^{t} d\tau \int_{D_\tau^{(e)}} T_{j\kappa}(\mathbf{y},\tau)\, d\mathbf{y} \int \frac{\sin\left(ck\left(t-\tau\right)\right)}{\left(2\pi\right)^3 ck} e^{i\mathbf{k}\cdot(\mathbf{x}-\mathbf{y})} d\mathbf{k},
\end{aligned} \tag{35}$$

where $D_\tau^{(e)}$ denotes the 3-D domain occupied by volume sources at the moment τ. The last integral in formula (35) was calculated in Appendix B. Introducing its expression given by formula (71) there results

$$\rho_q'(\mathbf{x},t) = \frac{\partial^2}{\partial x_j \partial x_\kappa} \int_{-\infty}^{t} d\tau \int_{D_\tau^{(e)}} T_{j\kappa}(\mathbf{y},\tau) \frac{\delta(\tau - (t - r/c))}{4\pi c^2 r} d\mathbf{y}, \tag{36}$$

the relationship (36) becomes

$$\rho_q'(\mathbf{x},t) = \frac{\partial^2}{\partial x_j \partial x_\kappa} \int_{D^{(e)}(\tau)} \left[\frac{T_{j\kappa}(\mathbf{y},\tau)}{4\pi c^2 r \left|1 - M_r\right|} \right]_{ret} d\eta. \tag{37}$$

where the effect of source convection is revealed by the Doppler factor; convection effectively increases the source strength by $\left|1 - M_r\right|^{-1}$. Further transformations of the formula (37) useful for its numerical implementation were made by Farassat and Brentner (Farassat and Brentner, 1988) and by Brentner in (Brentner, 1997). In the case where the discontinuity surface is permeable (of type S_p) this term is missing, the surface being usually chosen outside the space containing the quadruple sources. In the general case the sum of the solutions $\rho_q'(\mathbf{x},t)$, $\rho_{thickness}'$ and $\rho_{loading}'$ completely specify the density field.

6. The Kirchhoff method in Aeroacoustics

Besides the Acoustic Analogy approach for the solution of the Aeroacoustic noise, another widely used method is based on Kirchhoffs' solution of the wave equation. We start with the nonhomogeneous wave equation for the pressure perturbation

$$\left[\Delta - \frac{1}{c^2} \frac{\partial^2}{\partial t^2} \right] p'(\mathbf{x},t) = g(\mathbf{x},t) \tag{38}$$

where $g(\mathbf{x},t)$ represents the density of pressure sources. By applying the Fourier transform with respect to the spatial variables and using formulas (58) and (70) we obtain the operational form of the wave equation (38)

$$\frac{d^2 \widetilde{p'}}{d t^2} + c^2 k^2 \widetilde{p'} = -c^2 \widetilde{g}(\mathbf{k},t) + \widetilde{G}(\mathbf{k},t) \tag{39}$$

where

$$\widetilde{G}(\mathbf{k},t) = -\int_{S_t} \left(c^2 \left[\frac{\partial p'}{\partial n} \right] + v_n \left[\frac{\partial p'}{\partial t} \right] \right) e^{-i\mathbf{k}\cdot\mathbf{y}} dS \tag{40}$$
$$- \int_{S_t} c^2 i \mathbf{k} \cdot \mathbf{n} \left[p' \right] e^{-i\mathbf{k}\cdot\mathbf{y}} dS - \frac{d}{dt} \int_{S_t} v_n \left[p' \right] e^{-i\mathbf{k}\cdot\mathbf{y}} dS$$

Equation (39) is similar to equation (20). Consequently, its solution can be written as

$$\widetilde{p'}(\mathbf{k},t) = \int_{-\infty}^{t} \left(-c^2 \widetilde{g}(\mathbf{k},\tau) + \widetilde{G}(\mathbf{k},\tau) \right) \frac{\sin(ck(t-\tau))}{ck} d\tau \tag{41}$$

Hence the contribution of the nonhomogeneous term in equation (38) can be written as

$$\begin{aligned} p'_g(\mathbf{x},t) &= -c^2\int_{-\infty}^{t} d\tau \int \tilde{g}(\mathbf{k},\tau)\,\frac{\sin\left(ck\left(t-\tau\right)\right)}{\left(2\pi\right)^3 ck} e^{i\mathbf{k}\cdot\mathbf{x}} d\mathbf{k} \\ &= -\int_{-\infty}^{t} d\tau \int g(\mathbf{y},\tau)\,\frac{\delta\left(t-\tau-|\mathbf{x}-\mathbf{y}|/c\right)}{4\pi|\mathbf{x}-\mathbf{y}|} d\mathbf{y} \\ &= -\int g\left(\mathbf{y}, t-|\mathbf{x}-\mathbf{y}|/c\right)\frac{d\mathbf{y}}{4\pi|\mathbf{x}-\mathbf{y}|} \end{aligned} \tag{42}$$

Finally, the contribution of the terms corresponding to the boundary conditions on the mobile surface S can be written as

$$\begin{aligned} p'_G(\mathbf{x},t) &= -\frac{\partial}{\partial x_i}\int_{S_t}\left[\frac{p' n_i}{4\pi r\,|1-M_r|}\right]_{ret} dS \\ &-\frac{\partial}{\partial t}\int_{S_t}\left[\frac{p' M_n}{4\pi r\,|1-M_r|}\right]_{ret} dS \\ &-\int_{S_t}\left[\left(\frac{\partial p'}{\partial n}+M_n\frac{\partial p'}{\partial \tau}\right)\frac{1}{4\pi r\,|1-M_r|}\right]_{ret} dS \end{aligned} \tag{43}$$

which coincides with the relationship (5.3) given in (Ffowcs Williams and Hawkings, 1969).

7. Concluding remarks

Acoustic Analogy is one of the greatest contributions to the field of acoustics of the previous century. It is a major extension of acoustics made by Sir M. J. Lighthill (and other contributors) who formulated for the first time the science of how sound is created by fluid motion. This theory completes the previous work by famous researchers in the field of acoustics who had discovered how sound propagates through various media and across surrounding surfaces. In this chapter we have attempted to simplify the application of the Acoustic Analogy by showing how to apply it using only classical mathematical analysis tools.

8. Acknowledgment

This work has been supported by the National Institute on Deafness and Other Communication Disorders grant R01 DC009429 to RNM. The content is solely the responsibility of the authors and does not necessarily represent the official views of the National Institute on Deafness and Other Communication Disorders or the National Institutes of Health.

9. Appendix A: Fourier Transform of piecewise differentiable functions

Let $D^{(i)}(t)$ be a bounded mobile domain in $\mathbb{R}^3$ having a smooth boundary surface S_t. Denote by $D^{(e)}(t)$ the domain external to the surface S_t (See Fig.2): $D^{(i)} \cup S \cup D^{(e)} = \mathbb{R}^3$, $D^{(i)} \cap D^{(e)} = \phi$. Let also $\varphi^{(i)}(\mathbf{x},t)$ be a continuous differentiable function defined in the closed domain $\overline{D^{(i)}} \times \mathbb{R}$ and $\varphi^{(e)}(\mathbf{x},t)$ a continuous differentiable function defined in $\overline{D^{(e)}} \times \mathbb{R}$. We define also

$$\varphi(\mathbf{x},t) = \begin{cases} \varphi^{(i)}(\mathbf{x},t), \text{ in } D^{(i)} \\ \varphi^{(e)}(\mathbf{x},t), \text{ in } D^{(e)} \end{cases} \tag{44}$$

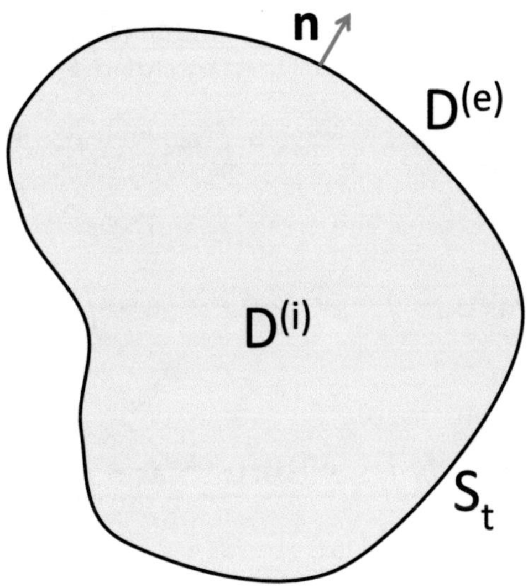

Fig. 2. Boundary Surface

Generally, the function $\varphi(\mathbf{x},t)$ is discontinuous across the surface S. We call it a piecewise differentiable function (pdf). Assuming that the function $\varphi^{(e)}(\mathbf{x},t)$ is decreasing sufficiently rapidly at infinity (for more precise conditions about the function $\varphi^{(e)}(\mathbf{x},t)$ see (Homentcovschi and Singler, 1999)) we can take the Fourier Transform of the function $\varphi(\mathbf{x},t)$ with respect to space variables

$$\widetilde{\varphi}(\mathbf{k},t) \equiv \mathcal{F}\{\varphi(\mathbf{x},t)\} = \int \varphi(\mathbf{x},t)\, e^{-i\mathbf{k}\cdot\mathbf{x}} d\mathbf{x} \tag{45}$$

Here $\mathbf{x} = (x_1,x_2,x_3)$, $\mathbf{k} = (k_1,k_2,k_3)$, $\mathbf{k}\cdot\mathbf{x} = k_1x_1 + k_2x_2 + k_3x_3$ is the inner product of the two vectors, $d\mathbf{x} = dx_1dx_2dx_3$ and the integral is extended over the whole $\mathbb{R}^3$ space. The inversion formula can be written as

$$\varphi(\mathbf{x},t) \equiv \mathcal{F}^{-1}\{\widetilde{\varphi}(\mathbf{k},t)\} = \frac{1}{(2\pi)^3}\int \widetilde{\varphi}(\mathbf{k},t)\, e^{i\mathbf{x}\cdot\mathbf{k}} d\mathbf{k} \tag{46}$$

where $d\mathbf{k} = dk_1dk_2dk_3$. Accounting for relationship (44) we can write

$$\mathcal{F}\{\varphi(\mathbf{x},t)\} = \int_{D^{(i)}} \varphi^{(i)}(\mathbf{x},t)\, e^{-i\mathbf{k}\cdot\mathbf{x}} d\mathbf{x} + \int_{D^{(e)}} \varphi^{(e)}(\mathbf{x},t)\, e^{-i\mathbf{k}\cdot\mathbf{x}} d\mathbf{x} \tag{47}$$

9.1 Fourier transform of the derivative with respect to a spatial variable of a piecewise differentiable function

9.1.1 The first basic formula

We write

$$\mathcal{F}\left\{\frac{\partial\varphi}{\partial x_1}\right\} = \int_{D^{(i)}} \frac{\partial\varphi^{(i)}(\mathbf{x},t)}{\partial x_1} e^{-i\mathbf{k}\cdot\mathbf{x}} d\mathbf{x} + \int_{D^{(e)}} \frac{\partial\varphi^{(e)}(\mathbf{x},t)}{\partial x_1} e^{-i\mathbf{k}\cdot\mathbf{x}} d\mathbf{x} \tag{48}$$

But

$$\int_{D^{(i)}} \frac{\partial\varphi^{(i)}}{\partial x_1} e^{-i\mathbf{k}\cdot\mathbf{x}} d\mathbf{x} = \int_{D^{(i)}} \frac{\partial}{\partial x_1}\left(\varphi^{(i)} e^{-i\mathbf{k}\cdot\mathbf{x}}\right) d\mathbf{x} + ik_1 \int_{D^{(i)}} \varphi^{(i)}(\mathbf{x},t)\, e^{-i\mathbf{k}\cdot\mathbf{x}} d\mathbf{x} \tag{49}$$

The first integral in the right-hand side of relationship (49) can be replaced, by using the divergence theorem by an integral over the boundary surface S_t.

$$\int_{D^{(i)}} \frac{\partial}{\partial x_1}\left(\varphi^{(i)} e^{-i\mathbf{k}\cdot\mathbf{x}}\right) d\mathbf{x} = \int_{S_t} n_1 \varphi^{(i)}(\mathbf{y},t)\, e^{-i\mathbf{k}\cdot\mathbf{y}} dS \tag{50}$$

where $\mathbf{n} = (n_1, n_2, n_3)$ is the external unit normal to S_t. Therefore,

$$\int_{D^{(i)}} \frac{\partial \varphi^{(i)}}{\partial x_1} e^{-i\mathbf{k}\cdot\mathbf{x}} d\mathbf{x} = ik_1 \int_{D^{(i)}} \varphi^{(i)}(\mathbf{x},t)\, e^{-i\mathbf{k}\cdot\mathbf{x}} d\mathbf{x} + \int_{S_t} n_1 \varphi^{(i)}(\mathbf{y},t)\, e^{-i\mathbf{k}\cdot\mathbf{y}} dS \tag{51}$$

Similarly,

$$\int_{D^{(e)}} \frac{\partial \varphi^{(e)}}{\partial x_1} e^{-i\mathbf{k}\cdot\mathbf{x}} d\mathbf{x} = ik_1 \int_{D^{(e)}} \varphi^{(e)}(\mathbf{x},t)\, e^{-i\mathbf{k}\cdot\mathbf{x}} d\mathbf{x} - \int_{S_t} n_1 \varphi^{(e)}(\mathbf{y},t)\, e^{-i\mathbf{k}\cdot\mathbf{y}} dS \tag{52}$$

Finally, the relationships (49), (51), and (52) give *the first basic formula*:

$$\mathcal{F}\left\{\frac{\partial \varphi}{\partial x_1}\right\} = ik_1\, \widetilde{\varphi}(\mathbf{k},t) - \int_S n_1 \left[\varphi(\mathbf{y},t)\right] e^{-i\mathbf{k}\cdot\mathbf{y}} dS \tag{53}$$

Here we denoted by square bracket the jump of the function $\varphi(\mathbf{x},t)$ across the surface S_t

$$\left[\varphi(\mathbf{y},t)\right] = \lim_{\mathbf{x}^{(e)}\to\mathbf{y}} \varphi^{(e)}\left(\mathbf{x}^{(e)},t\right) - \lim_{\mathbf{x}^{(i)}\to\mathbf{y}} \varphi^{(i)}\left(\mathbf{x}^{(i)},t\right) \tag{54}$$

for $\mathbf{y} \in S_t$, $\mathbf{x}^{(i)} \in D^{(i)}$ and $\mathbf{x}^{(e)} \in D^{(e)}$. Similar relationships to (53) can be proved for the derivatives in the directions x_2 and x_3.

Remark 3. *It is clear that the obtained relationships can be extended immediately to the case where there are more discontinuity surfaces of the given function. The resulting formulas will contain sums of integrals corresponding to each discontinuity surface.*

9.1.2 Other formulas

The relationship (53) gives also

$$\mathcal{F}\{\nabla\varphi\} \equiv \widetilde{\nabla\varphi} = i\mathbf{k}\, \widetilde{\varphi}(\mathbf{k},t) - \int_{S_t} \mathbf{n}\left[\varphi(\mathbf{y},t)\right] e^{-i\mathbf{k}\cdot\mathbf{y}} dS \tag{55}$$

In the case where we write $\mathbf{V}(\mathbf{x},t) = (V_1(\mathbf{x},t), V_2(\mathbf{x},t), V_3(\mathbf{x},t))$ where $V_j(\mathbf{x},t)$ is a piecewise differentiable function defined by a relationship similar to (44) we can write also the formulas

$$\mathcal{F}\{\nabla\cdot\mathbf{V}\} \equiv \widetilde{\nabla\cdot\mathbf{V}} = i\mathbf{k}\cdot\widetilde{\mathbf{V}}(\mathbf{k},t) - \int_{S_t} \mathbf{n}\cdot\left[\mathbf{V}(\mathbf{y},t)\right] e^{-i\mathbf{k}\cdot\mathbf{y}} dS \tag{56}$$

$$\mathcal{F}\{\nabla\times\mathbf{V}\} \equiv \widetilde{\nabla\times\mathbf{V}} = i\mathbf{k}\times\widetilde{\mathbf{V}}(\mathbf{k},t) - \int_{S_t} \mathbf{n}\times\left[\mathbf{V}(\mathbf{y},t)\right] e^{-i\mathbf{k}\cdot\mathbf{y}} dS \tag{57}$$

The formulas (55), (56) and (57) permit the calculation of the Fourier Transforms of a gradient of a scalar field of a divergence and a curl of a vector field in the case of piecewise differentiable scalar and vector fields.

Moreover, we can write

$$\mathcal{F}\{\Delta\varphi\} \equiv \mathcal{F}\{\nabla\cdot\nabla\varphi\} = i\mathbf{k}\cdot\widetilde{\nabla\varphi}(\mathbf{k},t) - \int_{S_t}\mathbf{n}\cdot[\nabla(\mathbf{y},t)\varphi]\,e^{-i\mathbf{k}\cdot\mathbf{y}}dS \tag{58}$$

$$= i\mathbf{k}\cdot\left(i\mathbf{k}\,\widetilde{\varphi}(\mathbf{k},t) - \int_{S_t}\mathbf{n}\,[\varphi(\mathbf{y},t)]\,e^{-i\mathbf{k}\cdot\mathbf{y}}dS\right) - \int_{S_t}\left[\frac{\partial\varphi}{\partial n}(\mathbf{y},t)\right]e^{-i\mathbf{k}\cdot\mathbf{y}}dS$$

Finally,

$$\widetilde{\Delta\varphi} = -k^2\widetilde{\varphi}(\mathbf{k},t) - \int_{S_t} i\mathbf{k}\cdot\mathbf{n}\,[\varphi(\mathbf{y},t)]\,e^{-i\mathbf{k}\cdot\mathbf{y}}dS - \int_{S_t}\left[\frac{\partial\varphi}{\partial n}(\mathbf{y},t)\right]e^{-i\mathbf{k}\cdot\mathbf{y}}dS \tag{59}$$

where $k^2 = k_1^2 + k_2^2 + k_3^2 = |\mathbf{k}|^2$.

9.2 Fourier transform of the time derivative of a piecewise differentiable function

9.2.1 The displacement velocity of a mobile surface

Let $S(y_1, y_2, y_3, t) = 0$ be the equation of the mobile surface S_t. Then, for $S(y_{01}, y_{02}, y_{03}, t_0)$ on S_{t_0} we can write

$$0 = S(y_1, y_2, y_3, t) - S(y_{01}, y_{02}, y_{03}, t_0) = \tag{60}$$
$$\left(\frac{\partial S}{\partial y_1}\frac{dy_1}{dt} + \frac{\partial S}{\partial y_2}\frac{dy_2}{dt} + \frac{\partial S}{\partial y_3}\frac{dy_3}{dt} + \frac{\partial S}{\partial t}\right)(t - t_0),$$

the partial derivatives of the function S being calculated at a certain point $\mathbf{x}'$ lying between the points $\mathbf{x}(t_0)$ and $\mathbf{x}(t)$. But,

$$n_1 = \frac{\partial S/\partial y_1}{|grad\,S|}, n_2 = \frac{\partial S/\partial y_2}{|grad\,S|}, n_3 = \frac{\partial S/\partial y_3}{|grad\,S|} \tag{61}$$

are the components of the external normal unit vector $\mathbf{n}$ and

$$\frac{dy_1}{dt} = v_1, \frac{dy_2}{dt} = v_2, \frac{dy_3}{dt} = v_3, \tag{62}$$

are the projections on the velocity vector of a point on the surface S_t on the three axes. The relation (60) yields

$$v_n = \frac{\partial S/\partial t}{|grad\,S|} \tag{63}$$

which is the displacement velocity of the geometric surface S_t. We mention that the displacement velocity of a surface has the direction of the normal vector to this surface.

9.2.2 Reynolds' transport theorem

For calculating the Fourier Transform of a time derivative of a certain function we write

$$\int_{D^{(i)}(t)}\varphi^{(i)}(\mathbf{x},t)\,e^{-i\mathbf{k}\cdot\mathbf{x}}d\mathbf{x} - \int_{D^{(i)}(t_0)}\varphi^{(i)}(\mathbf{x},t_0)\,e^{-i\mathbf{k}\cdot\mathbf{x}}d\mathbf{x} = \tag{64}$$
$$\int_{D^{(i)}(t)}\varphi^{(i)}(\mathbf{x},t)\,e^{-i\mathbf{k}\cdot\mathbf{x}}d\mathbf{x} - \int_{D^{(i)}(t)}\varphi^{(i)}(\mathbf{x},t_0)\,e^{-i\mathbf{k}\cdot\mathbf{x}}d\mathbf{x} +$$
$$\int_{D^{(i)}(t)}\varphi^{(i)}(\mathbf{x},t_0)\,e^{-i\mathbf{k}\cdot\mathbf{x}}d\mathbf{x} - \int_{D^{(i)}(t_0)}\varphi^{(i)}(\mathbf{x},t_0)\,e^{-i\mathbf{k}\cdot\mathbf{x}}d\mathbf{x} =$$
$$\int_{D^{(i)}(t)}\left\{\varphi^{(i)}(\mathbf{x},t) - \varphi^{(i)}(\mathbf{x},t_0)\right\}e^{-i\mathbf{k}\cdot\mathbf{x}}d\mathbf{x} + \int_{D^{(i)}(t)-D^{(i)}(t_0)}\varphi^{(i)}(\mathbf{x},t_0)\,e^{-i\mathbf{k}\cdot\mathbf{x}}d\mathbf{x}$$

Now, dividing by $(t - t_0)$ and taking the limit for $t \to t_0$ the first term gives the Fourier transform of the time derivative of the function $\varphi^{(i)}(\mathbf{x}, t)$ while in the second integral we can write $d\mathbf{x} = v_n (t - t_0)\, dS$ (see (Jacob, 1959), (Currie, 2003)). Finally, we obtain the following form of the Reynolds' transport theorem

$$\frac{d}{dt}\int_{D^{(i)}(t)} \varphi^{(i)}(\mathbf{x}, t)\, e^{-i\mathbf{k}\cdot\mathbf{x}} d\mathbf{x} = \int_{D^{(i)}(t)} \frac{\partial \varphi^{(i)}(\mathbf{x}, t)}{\partial t} e^{-i\mathbf{k}\cdot\mathbf{x}} d\mathbf{x} + \int_{S_t} v_n(\mathbf{y}, t)\, \varphi^{(i)}(\mathbf{y}, t)\, e^{-i\mathbf{k}\cdot\mathbf{y}} dS \tag{65}$$

v_n being the displacement velocity of the surface $S(t)$.

9.2.3 The second basic formula

We calculate now

$$\mathcal{F}\left\{\frac{\partial \varphi(\mathbf{x}, t)}{\partial t}\right\} = \int_{D^{(i)}(t)} \frac{\partial \varphi^{(i)}(\mathbf{x}, t)}{\partial t} e^{-i\mathbf{k}\cdot\mathbf{x}} d\mathbf{x} + \int_{D^{(e)}(t)} \frac{\partial \varphi^{(e)}(\mathbf{x}, t)}{\partial t} e^{-i\mathbf{k}\cdot\mathbf{x}} d\mathbf{x}$$

By using formula (65) we get

$$\int_{D^{(i)}(t)} \frac{\partial \varphi^{(i)}(\mathbf{x}, t)}{\partial t} e^{-i\mathbf{k}\cdot\mathbf{x}} d\mathbf{x} = \frac{d}{dt}\int_{D^{(i)}(t)} \varphi^{(i)} e^{-i\mathbf{k}\cdot\mathbf{x}} d\mathbf{x} - \int_{S_t} v_n\, \varphi^{(i)}(\mathbf{y}, t)\, e^{-i\mathbf{k}\cdot\mathbf{y}} dS \tag{66}$$

$$\int_{D^{(e)}(t)} \frac{\partial \varphi^{(e)}(\mathbf{x}, t)}{\partial t} e^{-i\mathbf{k}\cdot\mathbf{x}} d\mathbf{x} = \frac{d}{dt}\int_{D^{(e)}(t)} \varphi^{(e)} e^{-i\mathbf{k}\cdot\mathbf{x}} d\mathbf{x} + \int_{S_t} v_n \varphi^{(e)}(\mathbf{y}, t)\, e^{-i\mathbf{k}\cdot\mathbf{y}} dS \tag{67}$$

The sum of formulas (66) and (67) gives the *second basic formula*

$$\mathcal{F}\left\{\frac{\partial \varphi(\mathbf{x}, t)}{\partial t}\right\} = \frac{d\, \widetilde{\varphi}(\mathbf{k}, t)}{dt} + \int_{S_t} v_n(\mathbf{y}, t)\, [\varphi(\mathbf{y}, t)]\, e^{-i\mathbf{k}\cdot\mathbf{y}} dS. \tag{68}$$

Formula (68) permits us to calculate the Fourier transform of the time derivative of a piecewise differentiable function. For the second time derivative we can write

$$\mathcal{F}\left\{\frac{\partial^2 \varphi(\mathbf{x}, t)}{\partial t^2}\right\} = \frac{d}{dt}\frac{\widetilde{d\varphi}}{dt} + \int_{S_t} v_n(\mathbf{y}, t) \left[\frac{\partial \varphi(\mathbf{y}, t)}{\partial t}\right] e^{-i\mathbf{k}\cdot\mathbf{y}} dS \tag{69}$$

By using again formula (68) we obtain finally,

$$\widetilde{\frac{\partial^2 \varphi}{\partial t^2}} = \frac{d^2\, \widetilde{\varphi}}{dt^2} + \frac{d}{dt}\int_{S_t} v_n\, [\varphi]\, e^{-i\mathbf{k}\cdot\mathbf{y}} dS + \int_{S_t} v_n \left[\frac{\partial \varphi}{\partial t}\right] e^{-i\mathbf{k}\cdot\mathbf{y}} dS \tag{70}$$

such that in the Fourier transform of second time derivative of the piecewise differentiable function φ enters the jump of the function φ across the discontinuity and the jump of the first time derivative of φ as well.

10. Appendix B: Determination of the Greens' function for the wave equation

By using formula 11, given at page 364 in (Gelfand and Shilov, 1964) we can write

$$\mathcal{F}^{-1}\left\{\frac{\sin(ckt)}{k}\right\} = \frac{\delta(t - r/c)}{4\pi c r} \tag{71}$$

where $r = |\mathbf{x}|$.
By taking the derivative with respect to parameter t in the both sides of relationship (71) there results

$$\mathcal{F}^{-1}\{\cos(ckt)\} = \frac{\delta'(t - r/c)}{4\pi c^2 r} \tag{72}$$

Similarly, integrating the formula (71) over the interval $(0, t)$ we obtain

$$\mathcal{F}^{-1}\left\{\frac{1 - \cos(ckt)}{k^2}\right\} = \frac{H(t - r/c)}{4\pi r} \tag{73}$$

$H(x)$ being the Heaviside's function.

11. References

Brentner, K. S., An efficient and robust method for predicting helicopter high-speed impulsive noise, Journal of Sound and Vibration, 1997, 203(1), 87-100.

Brentner, K. S. & Farassat, F. 1998 Analytical comparison of the acoustic analogy and Kirchhoff formulation for moving surfaces. AIAA J. 36(8), 1998, 1379–1386.

Crighton, D. G., Dowling, A. P., Ffowcs Williams, J. E., Heckl, M., and Leppington, F. G., Modern Methods in Analytical Acoustics: Lecture Notes, Springer–Verlag, London, 1992. Chap.11, Sec. 10.

Curle, N., The influence of solid boundaries upon aerodynamic sound, Proceedings of the Royal Society A 231 (1955) 505–514.

Currie, I. G., Fundamental Mechanics of Fluids, 3rd edition, Marcel Dekker, 2003, pg.12.

Dowling, A. P., and Ffowcs Williams, J.E., Sound and Sources of Sound, Wiley &Sons, New York, 1982. Chap. 9, Sec. 2.

Farassat, F., Introduction to generalized functions with applications in aerodynamics and aeroacoustics, Corrected Copy (April 1996), NASA Technical Paper 3428, 1994,

Farassat, F., The Kirchhoff formulas for moving surfaces in aeroacoustics—the subsonic and supersonic cases, NASA Technical Memorandum 110285, 1996.

Farassat, F., Comments on the paper by Zinoviev and Bies "On acoustic radiation by a rigid object in a fluid flow", Journal of Sound and Vibration 281 (2005) 1217–1223.

Farassat, F., Derivation of Formulations 1 and 1A of Farassat, NASA Technical Memorandum 214853, 2007.

Farassat, F., and Brentner, K. S., The uses and abuses of the acoustic analogy in helicopter rotor noise prediction, Journal of the American Helicopter Society, 1988, 33, 29-36

Farassat, F., Myers, M.K., Further comments on the paper by Zinoviev and Bies, "On acoustic radiation by a rigid object in a fluid flow", Journal of Sound and Vibration 290 (2006), 538-547.

Ffowcs Williams, J.E., and Hawkings, D.L., Sound generation by turbulence and surfaces in arbitrary motion, Philosophical Transactions of the Royal Society A 264 (1969) 321–342.

Francescantonio, P. Di,A new boundary integral formulation for the prediction of sound radiation, J. Sound Vibr. 202 (1997) (4), pp. 491–509.

Gelfand, I. M., and Shilov, G. E., Generalized Functions, Volume 1, Academic Press, New York and London, 1964.

Goldstein, M. E., Aeroacoustics, McGraw-Hill Book Co., 1976.

Homentcovschi, D., and Singler, T., An introduction to BEM by integral transforms, Engineering Analysis with Boundary Elements, 23 (1999) 603-609.

Jacob, C., Introduction Mathematique a la Mechanique des Fluides, Gauthier-Villlars, Paris, 1959.

Lighthill, M.J., On sound generated aerodynamically, Proceedings of the Royal Society of London, A 211 (1952) 564–586.

Lighthill, M.J., On sound generated aerodynamically: Turbulence as a source of sound, Proceedings of the Royal Society of London, A 222 (1954), 1-32.

Lyrintzis, A. (1994). Review: The use of Kirchhoff's method in computational aerodynamics, *Transactions of the ASME, Journal of Fluid Engineering* 116(4): 665–676.

Lyrintzis, A.S., Surface integral methods in computational aeroacoustics.-From the (CFD) near field to the (Acoustic) far-field, Aeroacoustics, 2 (2003) pp. 95-128.

Morino, L. (2003). Is there a difference between aeroacoustics and aerodynamics? An aeroelastician's viewpoint, *AIAA JOURNAL* 41(7): 1209–1223.

Morris, P.J. and Farassat, F., Acoustic analogy and alternative theories for jet noise prediction, AIAA Journal, 40 (2002), pp. 671-680.

Pilon, A. R., and Lyrintzis, A. S., "Integral Methods for Computational Aeroacoustics," AIAA paper No. 97-0020, presented at the 35th Aerospace Science Meeting, Reno, NV, Jan. 1997.

Raman, G. (editor), Computational Aeroacoustics, Multiscience Publishing Co. Ltd., 2009, 507pp.

Wu, X-F, and Akay A., Sound radiation from vibrating bodies in motion, J. Acoust. Soc. Am., 91 (1992) pp.2544-2555.

Zinoviev, A., Bies, D.A., On acoustic radiation by a rigid object in a fluid flow, Journal of Sound and Vibration 269 (2004) 535–548.

Zinoviev, A., Bies, D.A., Author's Reply to: F. Farassat, Comments on the paper by Zinoviev and Bies "On acoustic radiation by arigid object in a fluid flow", Journal of Sound and Vibration 281 (2005) 1224–1237.

Zinoviev, A., Bies, D.A., Author's Reply, Journal of Sound and Vibration 290 (2006) 548-554.

4

Exact Solutions Expressible in Hyperbolic and Jacobi Elliptic Functions of Some Important Equations of Ion-Acoustic Waves

A. H. Khater[1] and M. M. Hassan[2]
[1] *Mathematics Department, Faculty of Science, Beni-Suef University, Beni-Suef*
[2] *Mathematics Department, Faculty of Science, Minia University, El-Minia*
Egypt

1. Introduction

Many phenomena in physics and other fields are often described by nonlinear partial differential equations (NLPDEs). The investigation of exact and numerical solutions, in particular, traveling wave solutions, for NLPDEs plays an important role in the study of nonlinear physical phenomena. These exact solutions when they exist can help one to well understand the mechanism of the complicated physical phenomena and dynamical processes modeled by these nonlinear evolution equations (NLEEs). The ion-acoustic solitary wave is one of the fundamental nonlinear wave phenomena appearing in fluid dynamics [1] and plasma physics [2, 3]. It has recently became more interesting to obtain exact analytical solutions to NLPDEs by using appropriate techniques and symbolical computer programs such as Maple or Mathematica. The capability and power of these software have increased dramatically over the past decade. Hence, direct search for exact solutions is now much more viable. Several important direct methods have been developed for obtaining traveling wave solutions to NLEEs such as the inverse scattering method [3], the tanh-function method [4], the extended tanh-function method [5] and the homogeneous balance method [6]. We assume that the exact solution is expressed by a simple expansion $u(x,t) = U(\xi) = \sum_{i=0}^{N} A_i F^i(\xi)$ where A_i are constants to be determined and the function $F(\xi)$ is defined by the solution of an auxiliary ordinary differential equation (ODE). The tanh-function method is the well known method as a direct selection of the function $F(\xi) = tanh(\xi)$. Recently, many exact solutions expressed by various Jacobi elliptic functions (JEFs) of many NLEEs have been obtained by Jacobi elliptic function expansion method [7-10], mapping method [11, 12], F-expansion method [13], extended F-expansion method [14], the generalized Jacobi elliptic function method [15] and other methods [16-20]. Various exact solutions were obtained by using these methods, including the solitary wave solutions, shock wave solutions and periodic wave solutions.

The main steps of the F-expansion method [13] are outlined as follows:

Step 1. Use the transformation $u(x,t) = u(\xi)$; $\xi = k(x - \omega t) + \xi_0$, ξ_0 is an arbitrary constant, and reduce a given NLPDE, say in two independent variables,

$$F(u, u_t, u_x, u_{tt}, u_{xx}, ...) = 0, \tag{1.1}$$

to the (ODE)

$$G(u,u',u'',...) = 0, \qquad u' = \frac{du}{d\xi}. \tag{1.2}$$

In general, the left hand side of Eq. (1.1) is a polynomial in u and its various derivatives.
Step 2. The F-expansion method gives the solution of (1.1) in the form

$$u(x,t) = u(\xi) = \sum_{i=0}^{N} a_i F^i(\xi), \qquad a_N \neq 0, \tag{1.3}$$

where a_i $(i = 0,1,2,...,N)$ are constants to be determined and $F(\xi)$ satisfies the first order nonlinear ODE in the form

$$(F'(\xi))^2 = q_0 + q_2F^2(\xi) + q_4F^4(\xi), \tag{1.4}$$

where q_0, q_2 and q_4 are constants and N in Eq. (1.3) is a positive integer that can be determined by balancing the nonlinear term(s) and the highest order derivatives in Eq. (1.1).
Step 3. Substituting the F-expansion (1.3) into (1.2) and using (1.4); setting each coefficient of the polynomial to zero yields a system of algebraic equations involving a_0, a_1,...a_N, k and ω.
Step 4. Solving these equations, probably with the aid of Mathematica or Maple, then a_0, a_1,...a_N, k and ω can be expressed by q_0, q_2, q_4.
Step 5. Substituting these results into F-expansion (1.3), then a general form of traveling wave solution of the NLPDE (1.1) can be obtained. Many solutions of equation (1.4) have been reported in [13, 14]. Substituting the values of q_0, q_2, q_4 and the corresponding JEF solution $F(\xi)$ into the general form of solution, we may get several classes of exact solutions of equations (1.1) involving JEFs.
Also, we give a brief description of the mapping method to seek the traveling wave solutions of (1.1) in the form $u(x,t) = u(\eta)$, $\eta = kx - \omega t + \eta_0$, η_0 is an arbitrary constant. Thus, Eq. (1.1) reduces to Eq. (1.2), whose solution can be express in the form

$$u(\eta) = \sum_{i=0}^{n} A_i f^i(\eta), \tag{1.5}$$

where n is a balancing number, A_i are constants to be determined and $f(\eta)$ satisfies the nonlinear ODE

$$f'^2(\eta) = 2p\,f(\eta) + q\,f^2(\eta) + \frac{2}{3}r\,f^3(\eta). \tag{1.6}$$

Here p, q and r are constants. After substituting Eq. (1.5) into the ODE (1.2) and using Eq. (1.6), the constants A_i, k and ω may be determined. By using the solutions of auxiliary nonlinear equation (1.6), many JEF solutions of NLEEs have been obtained [19, 20].
The JEFs $\mathrm{sn}(\xi) = \mathrm{sn}(\xi,m)$, $\mathrm{cn}(\xi) = \mathrm{cn}(\xi,m)$ and $\mathrm{dn}(\xi) = \mathrm{dn}(\xi,m)$ are double periodic and have the following properties:

$$\mathrm{sn}^2(\xi) + \mathrm{cn}^2(\xi) = 1, \quad \mathrm{dn}^2(\xi) + m^2\,\mathrm{sn}^2(\xi) = 1.$$

In the limit $m \longrightarrow 1$, the JEFs degenerate to the hyperbolic functions, i.e.,

$$\mathrm{sn}(\xi,1) \longrightarrow \tanh(\xi), \;\; \mathrm{cn}(\xi,1) \longrightarrow \mathrm{sech}(\xi), \;\; \mathrm{dn}(\xi,1) \longrightarrow \mathrm{sech}(\xi).$$

Detailed explanations about JEFs can be found in [21].
Some of the nonlinear models in fluids, plasma and dust plasma are described by canonical models and include the Korteweg-de Vries (KdV) and the modified KdV equations [22-25].

The evolution of small but finite-amplitude solitary waves, studied by means of the Korteweg-de Vries (KdV) equation, is of considerable interest in plasma dynamics. In the study of multidimensional version two type of nonlinear waves are well known, the so called Kadomtsev-Petviashvilli (KP) equation and Zakharov - Kuzentsov (ZK) equation. Employing the reductive perturbation technique on the system of equations for hydrodynamics and the dynamics of plasma waves to derive such equation.
We construct several classes of exact JEF solutions of some nonlinear evolution equations of plasma physics by using the mapping method and the F-expansion method. The rest of this chapter is organized as follows: in section 2, we present the JEF solutions to the KdV equation, combined KdV - modified KdV equation. In section 3, we apply the F-expansion method to the Schamel- KdV equation. Moreover, using the ansatz solution (1.5) and the solutions of nonlinear ODE (1.6), many exact solutions of Schamel equation, ZK equation and modified fifth order KdV equation are given in sections 4, 5, 6.

2. The KdV and modified KdV equations

The Korteweg de-Vries (KdV) equation

$$u_t + \alpha u u_x + u_{xxx} = 0,$$

models a variety of nonlinear phenomena, including ion acoustic waves in plasmas, dust acoustic solitary structures in magnetized dusty plasmas, and shallow water waves. On the other hand, the modified KdV equation (mKdV)

$$u_t + b u^2 u_x + u_{xxx} = 0,$$

models the dust-ion acoustic waves, electromagnetic waves in size-quantized films, ion acoustic solitons, traffic flow problems, and in other applications. The KdV equation and the modified KdV equation are completely integrable equations that have multiple-soliton solutions and possess infinite conservation quantities. The KdV equation is the earliest soliton equation that was firstly derived by Korteweg and de Vries to model the evolution of shallow water wave in 1895. In the study of the KdV equation, traveling wave solution leads to periodic solution which is called cnoidal wave solution [22, 23]. Exact solutions of KdV equation have been studied extensively since they were first found. Solitary wave solutions and periodic wave solutions were obtained for the KdV and modified KdV equations [3, 7, 22]. The JEF solutions to two kinds of KdV equations with variables coefficients have been constructed by using the method of the auxiliary equation [19]. The reductive perturbation method [24] has been employed to derive the KdV equation for small but finite amplitude electrostatic ion-acoustic waves [23, 25, 26]. The basic equations describing the system in dimensionless variables is studied by El-Labany [26] and the KdV equation for the first-order perturbed potential has been obtained using the reductive perturbation method.
We consider the combined KdV and mKdV equation [22, 27, 28]

$$u_t + \alpha u u_x + \beta u^2 u_x + \delta u_{xxx} = 0, \quad \beta \neq 0. \tag{2.1}$$

where α, β and δ are constants. Equation (2.1) is widely used in various fields such as quantum field theory, dust-acoustic waves, ion acoustic waves in plasmas with a negative ion, solid-state physics and fluid dynamics.
Let $u = u(\xi)$, equation (2.1) transformed to the reduced equation

$$-\omega u' + \alpha u u' + \beta u^2 u' + \delta k^2 u''' = 0. \tag{2.2}$$

Balancing u''' with u^2u' yields $N = 1$, so the F-expansion method gives

$$u(x,t) = a_0 + a_1F(\xi). \tag{2.3}$$

Substituting (2.3) into (2.2) and equating the coefficients of like powers of $F(\xi)$ to zero, we obtain a set of algebraic equations. Solving these algebraic equations, we obtain the exact solutions of (2.1) as follows:
When $q_0 = 1,\ q_2 = -1 - m^2,\ q_4 = m^2$, solutions of Eq. (1.4) is $F(\xi) = \mathrm{sn}\xi$, we have

$$u = -\frac{\alpha}{2\beta} \pm k\sqrt{\frac{-6m^2\delta}{\beta}}\,\mathrm{sn}\Big(k(x + (\frac{\alpha^2 + 4\beta\delta k^2(m^2+1)}{4\beta})t) + \xi_0\Big), \tag{2.4}$$

If $q_0 = m^2 - 1,\ q_2 = 2 - m^2,\ q_4 = -1$, the solution of Eq (1.4) is $F(\xi) = \mathrm{dn}\xi$. Thus, we obtain the periodic wave solutions of Eq. (2.1)

$$u = -\frac{\alpha}{2\beta} \pm k\sqrt{\frac{6\delta}{\beta}}\,\mathrm{dn}(k(x + \frac{\alpha^2 - 4\beta\delta k^2(2-m^2)}{4\beta}t) + \xi_0), \tag{2.5}$$

Selecting the values of the q_0, q_2 and q_4 of equation (1.4) and the corresponding function F, we can construct various JEF solutions of (2.1). Other JEF solutions are omitted here for simplicity. If we put $\alpha = 0$ in (2.4), we get the periodic solution of the modified KdV equation which coincides with that given by Liu et al. [7]. Moreover, the solutions (2.5) to equation (2.1) given in [28] are recovered. With $m \longrightarrow 1$ in (2.4) , (2.5), the solitary wave solutions to (2.1) given in [7, 27, 28] are also recovered.
We notice that the solutions of the KdV equation cannot obtain from (2.4) and (2.5) as $\beta = 0$. In this case, the general form of cnoidal wave solutions of the KdV equation are given by

$$u(x,t) = -\frac{3\omega q_4}{\alpha\, q_2}F^2(\xi), \quad \xi = \sqrt{\frac{\omega}{4\delta q_2}}\,(x - \omega t) + \xi_0. \tag{2.6}$$

Thus we can obtain abundant cnoidal wave solutions of the KdV equation in terms of JEFs. Some periodic wave solutions of the KdV equation and modified KdV equation have been studied in [7,23, 28]. As $m \longrightarrow 1$, these solutions will degenerate into the corresponding solitary wave solutions.

3. The JEF solutions of Schamel- KdV equation

We consider the Schamel- KdV equation [29, 30]

$$u_t + (\alpha u^{1/2} + \beta u)u_x + \delta u_{xxx} = 0, \quad \beta \neq 0 \tag{3.1}$$

where α, β and δ are constants and u is the wave potential.
In order to find the periodic wave solution of (3.1), we use the transformations $u = v^2$, $v(x,t) = V(\xi)$; $\xi = k(x - \omega t) + \xi_0$, then (2.7) becomes

$$-\omega V V' + (\alpha V^2 + \beta V^3)V' + \delta k^2[VV''' + 3V'V''] = 0. \tag{3.2}$$

The balancing procedure implies that $N = 1$. Therefore, the F-expansion method gives the solution

$$V(x,t) = V(\xi) = a_0 + a_1F(\xi), \tag{3.3}$$

where a_0 and a_1 are constants to be determined and $F(\xi)$ is a solution of Eq. (1.4). Substituting Eq. (3.3) into Eq. (3.2) and equating the coefficients of the like powers of F to zero, yields a set of algebraic equations for a_0, a_1, k and ω:

$$\begin{aligned} [\beta a_1^2 + 12\delta k^2 q_4]a_1 &= 0, \\ [\alpha a_1^2 + 3\beta a_0 a_1^2 + 6\delta k^2 a_0 q_4]a_1 &= 0, \\ [-\omega + 2\alpha a_0 + 3\beta a_0^2 + 4\delta k^2 q_2]a_1 &= 0, \\ [-\omega + \alpha a_0 + \beta a_0^2 + \delta k^2 q_2]a_0 &= 0. \end{aligned} \tag{3.4}$$

Solving these algebraic equations, we gave a general form of traveling wave solutions of Eq. (3.1)

$$u = \frac{4\alpha^2}{25\beta^2}\left[1 \pm \sqrt{\frac{-2q_4}{q_2}}\,F(\xi)\right]^2. \tag{3.5}$$

Therefore, we obtained in [30] the JEF solutions of Eq. (3.1) as follows:
When $q_0 = 1,\ q_2 = -1 - m^2,\ q_4 = m^2$, solutions of Eq. (1.4) is $F(\xi) = \mathrm{sn}\xi$, we have

$$u_1 = \frac{4\alpha^2}{25\beta^2}\left[1 \pm \sqrt{\frac{2m^2}{m^2+1}}\mathrm{sn}\left(\frac{2\alpha}{5\sqrt{-6\delta\beta(m^2+1)}}(x + \frac{16\alpha^2}{75\beta}t) + \xi_0\right)\right]^2, \quad \beta\delta < 0, \tag{3.6}$$

If $q_0 = 1 - m^2,\ q_2 = 2m^2 - 1,\ q_4 = -m^2,\ F(\xi) = \mathrm{cn}\xi$, thus yields the exact solutions of Eq. (3.1)

$$u_2 = \frac{4\alpha^2}{25\beta^2}\left[1 \pm \sqrt{\frac{2m^2}{2m^2-1}}\,\mathrm{cn}\left(\frac{2\alpha}{5\sqrt{6\delta\beta(2m^2-1)}}(x + \frac{16\alpha^2}{75\beta}t) + \xi_0\right)\right]^2, \quad \beta\delta > 0, \tag{3.7}$$

If $q_0 = m^2 - 1,\ q_2 = 2 - m^2,\ q_4 = -1$, the solution of Eq (1.4) is $F(\xi) = \mathrm{dn}\xi$. So, we obtained the exact solutions of Eq. (3.1) in the form

$$u_3 = \frac{4\alpha^2}{25\beta^2}\left[1 \pm \sqrt{\frac{2}{2-m^2}}\,\mathrm{dn}\left(\frac{2\alpha}{5\sqrt{6\delta\beta(2-m^2)}}(x + \frac{16\alpha^2}{75\beta}t) + \xi_0\right)\right]^2, \quad \beta\delta > 0, \tag{3.8}$$

Many types of JEF solutions of Eq. (3.1) are given [30]. As $m \longrightarrow 1$, Eqs. (3.6)-(3.8) degenerate to

$$\begin{aligned} u_4 &= \frac{4\alpha^2}{25\beta^2}\left[1 \pm \tanh\left(\frac{\alpha}{5\sqrt{-3\delta\beta}}(x + \frac{16\alpha^2}{75\beta}t) + \xi_0\right)\right]^2, \quad \beta\delta < 0, \\ u_5 &= \frac{4\alpha^2}{25\beta^2}\left[1 \pm \sqrt{2}\,\mathrm{sech}\left(\frac{2\alpha}{5\sqrt{6\delta\beta}}(x + \frac{16\alpha^2}{75\beta}t) + \xi_0\right)\right]^2, \quad \beta\delta > 0, \end{aligned} \tag{3.9}$$

The solitary wave solutions (3.9) in terms of tanh are equivalent to the solutions given in [31]. The JEF solutions of (3.1) may be describe various features of waves and may be helpful in understanding the problems in ion acoustic waves.

4. Schamel equation and modified KP equation

The equation describing ion-acoustic waves in a cold-ion plasma where electrons do not behave isothermally during their passage of the wave is

$$u_t + u^{1/2}\, u_x + \delta\, u_{xxx} = 0. \tag{4.1}$$

Schamel [29] derived this equation and a simple solitary wave solution having a sech^4 profile was obtained. Therefore the Schamel equation (4.1) containing a square root nonlinearity is very attractive model for the study of ion-acoustic waves in plasmas and dusty plasmas.
In order to find the periodic wave solution of (4.1), we use the transformations
$u = v^2,\ v(x,t) = V(\eta);\ \eta = kx - \omega\, t + \eta_0,$, then (4.1) becomes

$$-\,\omega V\, V' + k\, V^2 V' + \delta k^3 [V V''' + 3V' V''] = 0. \tag{4.2}$$

According to the mapping method, we assume that Eq. (4.2) has the following solution:

$$V(\eta) \;=\; A_0 + A_1 f(\eta), \tag{4.3}$$

where A_0 and A_1 are constants to be determined and $f(\eta)$ satisfies Eq. (1.6).
Substitution of Eq. (4.3) into Eq. (4.2) and selecting the values of p, q and r, we have the solutions of Eq. (4.1) which was given in [20] as follows:
Case 1. $p = 2,\ q = -4\,(1 + m^2),\ r = 6\,m^2$. In this case, we have $f(\eta) = \mathrm{sn}^2\eta$. Thus the periodic wave solutions of Eq. (4.1) are

$$\begin{aligned} u_1(x,t) &= 100\delta^2\, k^4 \left[1 + m^2 \pm \sqrt{1 - m^2 + m^4} - 3\, m^2\, \mathrm{sn}^2\, \eta\right]^2, \\ \eta &= kx \mp 16\,\delta k^3 \sqrt{1 - m^2 + m^4}\; t + \eta_0. \end{aligned} \tag{4.4}$$

Case 2. $p = \frac{-(1-m^2)^2}{2},\ q = 2(1 + m^2),\ r = \frac{-3}{2}$. The solutions of Eq. (1.6) are $f(\eta) = (m\,\mathrm{cn}\eta \pm \mathrm{dn}\eta)^2$. Thus the exact solutions of Eq. (4.1) are

$$\begin{aligned} u_2(x,t) &= \tfrac{25\delta^2 k^4}{4} \left[-2(1 + m^2) \pm \sqrt{1 + 14m^2 + m^4} + 3\,(m\,\mathrm{cn}\eta \pm \mathrm{dn}\eta)^2\right]^2, \\ \eta &= kx \mp 4\,\delta k^3 \sqrt{1 + 14m^2 + m^4}\; t + \eta_0. \end{aligned} \tag{4.5}$$

Case 3. $p = \frac{m^2}{2},\ q = 2(m^2 - 2),\ r = \frac{3\,m^2}{2}$. The solutions of Eq. (1.6) are
$f(\eta) = \left(\frac{m\,\mathrm{sn}\eta}{1 \pm \mathrm{dn}\,\eta}\right)^2$. So, we obtained the exact solutions of Eq. (4.1)in the form

$$\begin{aligned} u_3(x,t) &= \tfrac{25\delta^2 k^4}{4} \left[2(2 - m^2) \pm \sqrt{16 - 16m^2 + m^4} - 3\, m^4 \left(\frac{\mathrm{sn}\,\eta}{1 \pm \mathrm{dn}\,\eta}\right)^2\right]^2, \\ \eta &= kx \mp 4\,\delta k^3 \sqrt{16 - 16m^2 + m^4}\; t + \eta_0. \end{aligned} \tag{4.6}$$

There are several exact solutions for the Eq. (4.1) which are omitted here for simplicity. As $m \to 1$, these solutions reduce to the solitary wave solutions

$$\begin{aligned} u_4(x,t) &= 900\delta^2\, k^4\, \mathrm{sech}^4(kx - 16\,\delta k^3\, t + \eta_0), \\ u_5(x,t) &= 100\delta^2\, k^4\, [2 - 3\,\mathrm{sech}^2(kx + 16\,\delta k^3\, t + \eta_0)]^2. \end{aligned} \tag{4.7}$$

$$u_6(x,t) = \frac{225\delta^2 k^4}{4}\left[1 - \left(\frac{\tanh(kx-4\delta k^3 t+\eta_0)}{1+\operatorname{sech}(kx-4\delta k^3 t+\eta_0)}\right)^2\right]^2. \tag{4.8}$$

The KdV equation in two dimensions, known as Kadomtsev Petviashivili (KP) equation [32], was derived for ion-acoustic waves in a non magnetized plasma by Kako and Rowlands [33]. Therefore the modified KP equation containing a square root nonlinearity is very attractive model for the study of ion-acoustic waves in plasma and dusty plasma [34- 36]. Extensive work has been devoted to the study of nonlinear waves associated with the dust ion-acoustic waves, particularly the dust ion-acoustic solitary and shock waves in dusty plasmas in which dust particles are stationary and provide only the neutrality [37]. The KP equation is derived [38] for the propagation of nonlinear waves in warm dusty plasmas with variable dust charge, two-temperature ions and nonthermal electrons by using the reductive perturbation theory. Consider the modified KP equation

$$(u_t + \alpha u^{1/2}u_x + \beta u_{xxx})_x + \delta u_{yy} = 0, \tag{4.9}$$

where α and β are constants. The modified KP equation (4.9) for ion-acoustic waves in a multi species plasma consisting of non-isothermal electrons have been derived by Chakraborty and Das [34]. We applied the mapping method with the ansatz solution (4.3) and the solutions of auxiliary equation (1.6) to find the solutions of equation (4.9) (see [39]).

5. The ZK equation and modified ZK equation

The equation

$$u_t + \beta u^2 u_x + u_{xxx} + u_{yyx} = 0, \tag{5.1}$$

is the modified ZK in (2+1) dimensions which is a model for acoustic plasma waves [40, 41]. The ZK equation was first derived for describing weakly nonlinear ion- acoustic waves in a strongly magnetized lossless plasma in two dimension [41]. The ZK equation and modified ZK equation possess traveling wave structures [28, 42]. Peng [42] studied the exact solutions of ZK equation by using extended mapping method. Various types of solutions of Schamel-KdV equation and modified ZK equation arising in plasma and dust plasma are presented in [43].

We apply the F-expansion method to the modified ZK equation. Thus, Eq. (5.1) has a solution in the form

$$u(\xi) = a_0 + a_1 F(\zeta), \quad \zeta = k(x + ly - \omega t) + \xi_0.$$

Substituting this equation into Eq. (5.1), we obtain the following classes of exact solutions of the modified ZK equation:

$$\begin{aligned} u &= m\sqrt{\frac{6\omega}{(m^2+1)\beta}}\,\mathrm{sn}\Big(\sqrt{\frac{-\omega}{(m^2+1)(1+l^2)}}\,(x+ly-\omega t+\xi_0)\Big),\\ u &= m\sqrt{\frac{6\omega}{(2-m^2)\beta}}\,\mathrm{dn}\Big(\sqrt{\frac{\omega}{(2-m^2)(1+l^2)}}\,(x+ly-\omega t+\xi_0)\Big). \end{aligned} \tag{5.2}$$

In the following we apply the mapping method to the ZK equation

$$u_t + \alpha u\, u_x + u_{xxx} + u_{yyx} = 0. \tag{5.3}$$

In this case, we have $n = 1$. Thus Eq. (5.3) has a solution in the form

$$u(\eta) = A_0 + A_1 f(\eta), \quad \eta = kx + ly - \omega t + \eta_0.$$

Substituting this equation into Eq. (5.3) to determine A_0, A_1, k, ω and using the solutions of auxiliary equation (1.6), we obtained the following classes of exact solutions of the ZK equation [39]:

$$\begin{aligned} u_1(x,y,t) &= \frac{\omega}{k\alpha} + \frac{4(1+m^2)(l^2+k^2)}{\alpha} - \frac{12m^2(l^2+k^2)}{\alpha}\,\text{sn}^2(kx+ly-\omega t+\eta_0), \\ u_2(x,y,t) &= \frac{\omega}{k\alpha} + \frac{4(m^2-2)(l^2+k^2)}{\alpha} + \frac{12(l^2+k^2)}{\alpha}\,\text{dn}^2(kx+ly-\omega t+\eta_0), \end{aligned} \tag{5.4}$$

$$u_3 = \frac{\omega}{k\alpha} - \frac{2(1+m^2)(l^2+k^2)}{\alpha} + \frac{3(l^2+k^2)}{\alpha}\,[m\,\text{cn}(\eta) \pm \text{dn}(\eta)]^2, \tag{5.5}$$

$$u_4(x,y,t) = \frac{\omega}{k\alpha} - \frac{2(1+m^2)(l^2+k^2)}{\alpha} - \frac{3(1-m^2)(l^2+k^2)}{\alpha}\left(\frac{\text{cn}(kx+ly-\omega t+\eta_0)}{1\pm\text{sn}(kx+ly-\omega t+\eta_0)}\right)^2. \tag{5.6}$$

When $m \longrightarrow 1$, some of these solutions degenerate as solitary wave solutions of ZK equation. The solutions (5.3) are coincide with the solutions given in [44].

Recently, some properties of the quantum ion-acoustic waves were also investigated in dense quantum plasmas by studying the quantum hydrodynamical equations in different conditions, which includes the quantum Zakharov Kuznetsov equation, the extended quantum Zakharov Kuznetsov equation, and the quantum Zakharov system [45]. The three-dimensional extended quantum Zakharov Kuznetsov (QZK) equation [46] was investigated in dense quantum plasmas which arises from the dimensionless hydrodynamics equations describing the nonlinear propagation of the quantum ion-acoustic waves. The three-dimensional extended QZK equation was given in [46]

$$\Phi_t + (A\Phi + B\,\Phi^2)\Phi_x + C\,\Phi_{zzz} + D\,(\Phi_{xxz} + \Phi_{yyz}) = 0, \tag{5.7}$$

where A, B, C and D are constants. This equation has the following JEF solutions (see [45, 46]):

$$\begin{aligned} \Phi_1 &= -\frac{A}{2B} + m\,k\sqrt{\frac{-6E}{B}}\,\text{sn}(k(x+ly+\gamma z-\omega t+\eta_0)), \quad \omega = -\frac{4BEk^2(1+m^2)+A^2}{4B}, \quad BE<0, \\ \Phi_2 &= -\frac{A}{2B} + mk\sqrt{\frac{6E}{B}}\,\text{cn}(k(x+ly+\gamma z-\omega t+\eta_0)), \quad \omega = -\frac{4BEk^2(1-2m^2)+A^2}{4B}, \quad BE>0, \end{aligned} \tag{5.8}$$

with $E = C\gamma^2 + D(1+l^2)$. Moreover, many types of analytical solutions of the extended QZK equation are constructed in terms of some powerful ansatze, which include doubly periodic wave solutions, solitary wave solutions, kink-shaped wave solutions, rational wave solutions and singular solutions [46].

6. The modified fifth order KdV equation

Higher order KdV equations have many applications in different fields of mathematical physics. For example the fifth-order KdV equations can be derived in fluid dynamics and in magneto-acoustic waves in plasma and its exact solutions was given in [47-51]. The higher-order KdV equation can be derived for magnetized plasmas by using the reductive perturbation technique. Traveling wave solutions of Kawahara equation and modified Kawahara equation have been studied [9, 48, 49]. Moreover, the solitary wave solutions of nonlinear equations with arbitrary odd-order derivatives were studied by many authors [47, 51].

Consider the modified fifth order KdV equation

$$u_t + \beta u^2 u_x + c_3 u_{xxx} + c_5 u_{xxxxx} = 0, \tag{6.1}$$

where β, c_3 and c_5 are constants. Here, we review the exact traveling wave solutions of equation (6.1) using exact solutions of the auxiliary equation (1.5) and applied the mapping method. Thus, Eq. (6.1) has the solutions in the form

$$u(\eta) = A_0 + A_1 f(\eta), \quad \eta = kx - \omega t + \eta_0, \tag{6.2}$$

Substituting equation (6.2) into (6.1) and equating the coefficients of like powers of f to zero, yields a system of algebraic equations for A_0, A_1, k and ω and then solve it. Therefore, the solutions of the modified fifth order KdV equation (6.1) was given in [39] as follows:

$$\begin{aligned} u_1 &= \pm\frac{(10k^2c_5(1+m^2)+c_3)}{\sqrt{-10\beta c_5}} \mp k^2\sqrt{\frac{-90\,c_5}{4\beta}}\left(m\,\mathrm{cn}\,\eta \pm \mathrm{dn}\,\eta\right)^2, \\ \eta &= k\left[x + \frac{(15c_5^2k^4(m^4+14m^2+1)+c_3^2)}{10c_5}\,t\right] + \eta_0. \end{aligned} \tag{6.3}$$

If we choose $A_0 = 0$, equation (6.3) takes the form

$$\begin{aligned} u_2 &= \pm\frac{3c_3}{2(1+m^2)\sqrt{-10\,\beta\,c_5}}\left(m\,\mathrm{cn}\,\eta \pm \mathrm{dn}\,\eta\right)^2, \\ \eta &= \pm\sqrt{\frac{-c_3}{10c_5(1+m^2)}}\left[x + \frac{(23m^4+82m^2+23)\,c_3^2}{200c_5(1+m^2)^2}\,t\right] + \eta_0. \end{aligned} \tag{6.4}$$

Moreover, we have obtained the exact solutions

$$\begin{aligned} u_3 &= \pm\frac{3m^2c_3}{2(m^2-2)\sqrt{-10\,\beta\,c_5}}\left(\frac{m\,\mathrm{sn}\,\eta}{1\pm\mathrm{dn}\,\eta}\right)^2, \\ \eta &= \pm\sqrt{\frac{c_3}{10c_5(2-m^2)}}\left[x + \frac{(23m^4-128m^2+128)\,c_3^2}{200c_5(m^2-2)^2}\,t\right] + \eta_0, \end{aligned} \tag{6.5}$$

$$\begin{aligned} u_4 &= \pm\frac{3c_3}{2(1-2m^2)\sqrt{-10\,\beta\,c_5}}\left(\frac{\mathrm{sn}\,\eta}{1\pm\mathrm{cn}\,\eta}\right)^2, \\ \eta &= \pm\sqrt{\frac{-c_3}{10c_5(1-2m^2)}}\left[x + \frac{(128m^4-128m^2+23)\,c_3^2}{200c_5(1-2m^2)^2}\,t\right] + \eta_0. \end{aligned} \tag{6.6}$$

There are several other JEFs of Eq. (6.1) which are omitted here for simplicity. When $m \longrightarrow 1$, then (6.4)-(6.6) become the solitary wave solutions

$$u_5 = \pm\frac{3c_3}{\sqrt{-10\beta c_5}}\,\mathrm{sech}^2\Big(\frac{1}{2}\sqrt{\frac{-c_3}{5c_5}}\,(x + \frac{4\,c_3^2}{25c_5}\,t) + \eta_0\Big), \tag{6.7}$$

$$u_6 = \mp\frac{3c_3}{2\sqrt{-10\,\beta\,c_5}}\left(\frac{\tanh\eta}{1\pm\mathrm{sech}\,\eta}\right)^2, \quad \eta = \pm\sqrt{\frac{c_3}{10c_5}}\left[x + \frac{23\,c_3^2}{200c_5}\,t\right] + \eta_0. \tag{6.8}$$

We notice that Eq. (6.7) is the solution given by Example 2 in Ref. [47].
Finally, we can construct various types of exact and explicit solutions of the generalized ZK equation

$$u_t + (\alpha + \beta\,u^p)\,u^p u_x + u_{xxx} + \delta u_{yyx} = 0, \tag{6.9}$$

by using suitable method and using an appropriate transformation. Also, we can study the exact solution of the generalized KdV equation ($\delta = 0$) which studied by many authors [22, 23, 31]. The generalized ZK equation was first derived for describing weakly nonlinear ion-acoustic waves in strongly magnetized lossless plasma in two dimensions and governs the behavior of weakly nonlinear ion-acoustic waves in plasma comprising cold ions and hot isothermal electrons in the presence of a uniform magnetic field. Eq. (6.9) includes considerable interesting equations, such as KdV equation, mKdV equation, ZK equation and mZK equation. Exact traveling wave solutions for the generalized ZK equation with higher-order nonlinear terms have obtained in [52-54]. Moreover, we can use the symbolic computations and apply the mapping method with the ansatz solution (1.5) to find the several classes of traveling wave solutions of the fifth order KdV equation

$$u_t + c_1 u\, u_x + c_2\, u_{xxx} + \delta u_{xxxxx} = 0.$$

This equation appears in the theory of shallow water waves with surface tension and the theory of magneto-acoustic waves in plasmas [9]. Wazwaz [55] studied soliton solutions of fifth-order KdV equation. We can use a suitable method to construct the exact solutions of some special types of nonlinear evolution equations aries in plasma physics such as Liouville, sine-Gordon and sinh-Poisson equations.

7. References

[1] G. Whitham, Linear and Nonlinear Waves, New York, Wiley (1974).

[2] R. Davidson, Methods in Nonlinear Plasma Theory, New York, Academic Press (1972).

[3] M. J. Ablowitz and P. A. Clarkson, Solitons, Nonlinear Evolution Equations and Inverse Scattering Transform, *Cambridge, Cambridge University Press* (1991).

[4] W. Malfliet, Solitary wave solutions of nonlinear wave equations, *Am. J. Phys.* 60 (1992) 650-654;
W. Malfliet, The tanh method: a tool for solving certain classes of nonlinear evolution and wave equations, *J. Comput. Appl. Math.* 164-156 (2004) 529-541.

[5] E.G. Fan, Extended tanh-function method and its applications to nonlinear equations, *Phys. Lett.* A 277 (2000) 212-218;
E.G. Fan and Y.C. Hong, Generalized tanh method to special types of nonlinear equations, *Z. Naturforsch.* A 57 (2002) 692-700.

[6] M. Wang, Exact solutions for a compound KdV - Burgers equation, *Phys. Lett.* A 213 (1996) 279-287.

[7] S. K. Liu, Z. T. Fu, S. D. Liu, and Q. Zhao, Jacobi elliptic function expansion method and periodic wave solutions of nonlinear wave equations, *Phys. Lett. A* 289 (2001) 69-74;

[8] Z.T.Fu, S. K. Liu, S. D. Liu, and Q. Zhao, New Jacobi elliptic function expansion method and new periodic solutions of nonlinear wave equations, *Phys. Lett. A* 290 (2001) 72-76.

[9] E.J. Parkes, B.R. Duffy and P.C. Abbott, The Jacobi elliptic-function method for finding periodic wave solutions to nonlinear evolution equations, *Phys. Lett.* A 295 (2002) 280-286.

[10] H. T. Chen and H. Q. Zhang, Improved Jacobin elliptic method and its applications, *Chaos, Solitons and Fractals* 15 (2003) 585-591.

[11] Y. Peng, Exact periodic wave solutions to a new Hamiltonian amplitude equation, *J. Phys. Soc. Japan* 72 (2003) 1356-1359;
Y. Peng, New exact solutions to a new Hamiltonian amplitude equation II, *J. Phys. Soc. Japan* 73 (2004) 1156-1158.

[12] Y. Peng, Exact periodic wave solutions to the Melnikov equation, *Z. Naturforsch A* 60 (2005) 321-327.

[13] Y. B. Zhao, M. L. Wang and Y. M. Wang, Periodic wave solutions to a coupled KdV equations with variable coefficients, *Phys. Lett. A* 308 (2003) 31-36.

[14] J. Liu and K. Yang, The extended F-expansion method and exact solutions of nonlinear PDEs, *Chaos, Solitons and Fractals* 22 (2004) 111-121.

[15] H. T. Chen and H. Q. Zhang, New double periodic and multiple soliton solutions of the generalized (2+1)-dimensional Boussinesq equation, *Chaos, Solitons and Fractals* 20 (2004) 765-769.
[16] S. A. Elwakil, S. K. El-labany, M. A. Zahran and R. Sabry, Modified extended tanh-function method for solving nonlinear partial differential equations, *Phys. Lett. A.* 299 (2002) 179-188.
[17] M. A. Abdou and S. Zhang, New periodic wave solutions via extended mapping method, *Commun. Nonlinear Sci. Numer. Simul.* 14 (2009) 2-11.
[18] D. Baldwin, Ü. Göktas, W. Hereman et al., Symbolic computions of exact solutions expressible in hyperbolic and elliptic functions for nonlinear PDEs, *J. Symb. Comput.* 37 (2004) 669-705.
[19] Taogetusang and Sirendaoerji, The Jacobi elliptic function-like exact solutions to two kinds of KdV equations with variable coefficients and KdV equation with forcible term, *Chinese Phys.* 15 (2006) 2809-2818.
[20] A. H. Khater, M. M. Hassan, E. V. Krishnan and Y.Z. Peng, Applications of elliptic functions to ion-acoustic plasma waves, *Eur. Phys. J. D* 50 (2008) 177-184.
[21] M. Abramowitz and I. A. Stegun, Handbook of Mathematical Functions, *Dover, New York* (1965).
[22] P. G. Drazin and R. S. Johnson, Solitons,: An Introduction, *Cambridge University press, Cambridge* (1989).
[23] A. Jeffrey and T. Kakutani, Weak nonlinear dispersive waves: A discussion centered around the Korteweg de Vries equation, *SIAM Rev.* 14 (1972) 582-643.
[24] H. Washimi and T. Taniuti, Propagation of ion acoustic solitary waves of small amplitude, J Phys Rev Lett. 17 (1966) 996 -998.
[25] E.K. El-Shewy, H.G. Abdelwahed and H.M. Abd-El-Hamid, Computational solutions for the KortewegŰdeVries equation in warm Plasma, Computat. Methods in Sci. Technology 16 (2010) 13-18.
[26] S. K. El-Labany, Contribution of higher-order nonlinearity to nonlinear ion acoustic waves in a weakly relativistic warm plasma. Part1. isothermal case, J. Plasma Phys. 50 (1993) 495 .
[27] J. F. Zhang, New solitary wave solution of the combined KdV and m KdV equation, *Int. J. Theoret. Phys.* 37 (1998) 1541-1546.
[28] A. H. Khater and M. M. Hassan, Travelling and periodic wave solutions of some nonlinear wave equations, *Z. Naturforsch.* 59 a (2004) 389-396.
[29] H. Schamel, A modified Korteweg de Vries equation for ion acoustic waves due to resonant electrons, *J. Plasma Phys.* 9 (1973) 377-387.
[30] A. H. Khater, M. M. Hassan and R. S. Temsah, Exact solutions with Jacobi elliptic functions of two nonlinear models for ion-acoustic plasma waves, *J. Phys. Soc. Japan* 74 (2005) 1431-1435.
[31] M. M. Hassan, Exact solitary wave solutions for a generalized KdV - Burgers equation, *Chaos, Solitons and Fractals* 19 (2004) 1201-1206.
[32] B. B. Kadomtsev and V. I. Petviashvili, On the stability of solitary in weakly dispersive media, *Soviet Phys. Dokl.* 15 (1970) 539-541.
[33] M. Kako and G. Rowlands, Two-dimensional stability of ion acoustic solitons, *Plasma Phys.* 18 (1976) 165-170.
[34] D. Chakraborty and K. P. Das, Stability of ion acoustic solitons in a multispecies plasma consisting of non-isothermal electrons, *J. Plasma Phys.* 60 (1998) 151-158.
[35] A. H. Khater, A. A. Abdallah, O. H. El-Kalaaway and D. K. Callebaut, Backlund transformations, a simple transformation and exact solutions for dust-acoustic solitary

waves in dusty plasma consisting of cold dust particles and two-temperature isothermal ions, *Phys. Plasmas* 6 (1999) 4542-4547.

[36] A. H. Khater and M. M. Hassan, Exact Jacobi elliptic function solutions for some special types of nonlinear evolution equations, *Il Nuovo Cimento* 121 B (2006) 613- 622.

[37] A. A. Mamun and P.K. Shukla, Cylindrical and spherical dust ion-acoustic solitary waves, *Phys. Plasmas 9* (2002) 1468.

[38] H.R. Pakzad, Soliton energy of the Kadomtsev Petviashvili equation in warm dusty plasma with variable dust charge, two-temperature ions, and nonthermal electrons, *Astrophys Space Sci.* 326 (2010) 69 -75.

[39] A. H. Khater, M. M. Hassan and D. K. Callebaut, Travelling wave solutions to some important equations of mathematical physics, *Reports on Math. Phys.* 66 (2010) 1-19.

[40] K. P. Das and F. Verheest, Ion acoustic solitons in magnetized multi-component plasmas including negative ions, *J. Plasma. Phys.* 41 (1989) 139-155.

[41] V. E. Zakharov and E. A. Kuznetsov, On three-dimensional solitons, *Sov. Phys. JETP* 39 (1974) 285-286.

[42] Y.Z. Peng, Exact travelling wave solutions of the Zakharov Kuznetsov equation, *Appl. Math. Comput.* 199 (2008) 397- 405.

[43] M. M. Hassan, New exact solutions of two nonlinear physical models, *Commun. Theor. Phys.* 53 (2010) 596-604.

[44] M. M. Hassan, Exact solutions for some models of nonlinear evolution equations, *J. Egypt Math. Soc.* 12 (2004) 31- 43.

[45] R. Sabry, W.M. Moslem, F. Haas, S. Ali, P.K. Shukla, Nonlinear structures: Explosive, soliton and shock in a quantum electron- positron-ion magentoplasma, Phys. Plasmas 15 (2008) 122308.

[46] Z. Yan, Periodic, Solitary and rational wave solutions of the 3D extended quantum Zakharov Kuznetsov equation in dense quantum plasmas, *Phys. Let.* A 373 (2009) 2432-2437.

[47] E.J. Parkes, Z. Zhu, B.R. Duffy and H. C. Hang, Sech-polynomail traveling solitary-wave solutions of odd-order generalized KdV equations, *Phys. Lett.* A 248 (1998) 219-224.

[48] Z.J. Yang, Exact solitary wave solutions to a class of generalized odd-order KdV equations, *Int. J. Theor. Phys.* 34 (1995) 641-647.

[49] D. Zhang, Doubly periodic solution of modified Kawahara equation, *Chaos, Solitons and Fractals* 25 (2005) 1155-1160.

[50] A. H. Khater, M. M. Hassan and R. S. Temsah, Cnoidal wave solutions for a class of fifth-order KdV equations, *Math. Comput. Simulation* 70 (2005) 221-226.

[51] J. Sarma, Solitary wave solution of higher-order Korteweg de Vries equation, *Chaos, Solitons and Fractals* 39 (2009) 277 - 281.

[52] L.- H. Zhang, Travelling wave solutions for the generalized Zakharov-Kuznetsov equation with higher-order nonlinear terms, *Appl. Math. Comput.* 208 (2009) 144-155.

[53] C. Deng, New exact solutions to the Zakharov -Kuznetsov equation and its generalized form, *Commun. Nonlinear Sci. Numer. Simulat.* 15 (2010) 857 - 868.

[54] SUN Yu-Huai, MA Zhi-Min and LI Yan, Explicit solutions for generalized (2+1)-dimensional nonlinear Zakharov- Kuznetsov equation, *Commun. Theor. Phys.* 54 (2010) 397-400.

[55] A.-M., Wazwaz, Soliton solutions for the fifth-order KdV equation and the Kawahara equation with time-dependent coefficients, *Phys. Scr.* 82 (2010) 035009.

5

Acoustic Wave

P. K. Karmakar
Department of Physics, Tezpur University, Napaam, Tezpur, Assam India

1. Introduction

An acoustic wave basically is a mechanical oscillation of pressure that travels through a medium like solid, liquid, gas, or plasma in a periodic wave pattern transmitting energy from one point to another in the medium [1-2]. It transmits sound by vibrating organs in the ear that produce the sensation of hearing and hence, it is also called acoustic signal. This is well-known that air is a fluid. Mechanical waves in air can only be longitudinal in nature; and therefore, all sound waves traveling through air must be longitudinal waves originating in the transmission form of compression and rarefaction from vibrating matter in the medium. The propagation of sound in absence of any material medium is always impossible. Therefore, sound does not travel through the vacuum of outer space, since there is nothing to carry the vibrations from a source to a receiver. The nature of the molecules making up a substance determines how well or how rapidly the substance will carry sound waves. The two characteristic variables affecting the propagation of acoustic waves are (1) the inertia of the constituent molecules and (2) the strength of molecular interaction. Thus, hydrogen gas, with the least massive molecules, will carry a sound wave at 1,284.00 ms^{-1} when the gas temperature is 0^0 C [1]. More massive helium gas molecules have more inertia and carry a sound wave at only 965.00 ms^{-1} at the same temperature. A solid, however, has molecules that are strongly attached, so acoustic vibrations are passed rapidly from molecule to molecule. Steel, for an instant example, is highly elastic, and sound will move rapidly through a steel rail at 5,940.00 ms^{-1} at the same temperature. The temperature of a medium influences the phase speed of sound through it. The gas molecules in warmer air thus have a greater kinetic energy than those of cooler air. The molecules of warmer air therefore transmit an acoustic impulse from molecule to molecule more rapidly. More precisely, the speed of a sound wave increases by 0.60 ms^{-1} for each Celcius degree rise in temperature above 0^0 C.

Acoustic waves, or sound waves, are defined generally and specified mainly by three characteristics: wavelength, frequency, and amplitude. The wavelength is the distance from the top of one wave's crest to the next (or, from the top of one trough to the next). The frequency of a sound wave is the number of waves that pass a point each second [1]. Sound waves with higher frequencies have higher pitches than sound waves with lower frequencies and vice versa. Amplitude is the measure of energy in a sound wave and affects volume. The greater the amplitude of an acoustic wave, the louder the sound and vice versa. An acoustic wave is what makes humans and other animals able to hear. A person's ear perceives the vibrations of an acoustic wave and interprets it as sound [1]. The outer ear, the visible part, is shaped like a funnel that collects sound waves and sends them into the ear canal where they hit the ear drum, which is a tightly stretched piece of skin that vibrates in time with the wave. The ear drum starts a chain reaction and sends the vibration through three little bones in the middle ear that amplify sound. Those bones are called the hammer, the anvil, and the stirrup.

Furthermore, acoustic waves from a purely hydrodynamic point of view are small-amplitude disturbances that propagate in a compressible medium (like a fluid) through the interplay between fluid inertia, and the restoring force of fluid pressure. The propagation of small-amplitude disturbances in homogeneous medium is observed as acoustic waves such as water waves, and in self-gravitationally stratified medium like stellar atmosphere [36-37, 41-44], acoustic-gravity waves such as p-modes, g-modes, f-modes, etc., as found by helio- and astero-seismological studies. Acoustic waves propagating through a dispersive medium may get dynamically converted into *solitons* or *shocks* depending on the physical mechanisms responsible for their saturation. When fluid nonlinearity (convective effect) is balanced by dispersion (geometrical effect), solitons usually result [4]. Conversely, shocks are formed if fluid nonlinearity is balanced by dissipation (damping effect). The nonlinear hydrodynamic equations of various forms (like KdV equation, Burger equation, NLS equation, BO equation, etc.) in the context of the generation, structure, propagation, self-organization and dissipation of solitons or shocks have long been developed applying the hydrodynamic views of the usual conservation laws of flux, momentum and energy [4]. Similar outlook is needed to understand the formation of other nonlinear localized structures of low frequency acoustic waves like double layers, vortices, etc. They are important in a wide variety of space, astrophysical and laboratory problems for the investigation of dynamical stability against perturbation [3-4]. In addition, these equations have wide applications to study a nonlinear, radial, energetic, and steady-flow problem that provides a first rough approximation to the physics of stellar winds and associated acoustic wave kinetics, which are responsible for stellar mass-loss phenomena via supersonic flow into interstellar space [2].

Acoustic mode in plasmas of all types [2-44], similarly, is actually a pressure driven longitudinal wave like the ordinary sound mode in neutral gas. In normal two-component plasmas, the electron thermal pressure drives the collective ion oscillations to propagate as the ion sound (acoustic) wave. Here the electron thermal pressure provides the restoring force to allow the collective ion dynamics in the form of ionic compression and rarefaction to propagate in the plasma background and ionic mass provides the corresponding inertial force. Thermal plasma species (like electrons) are free to carry out thermal screening of the electrostatic potential. In absence of any dissipative mechanism, the ion sound wave moves with constant amplitude. For mathematical description of the ion sound kinetics, the plasma electrons are normally treated as inertialess species and the plasma ions, with full inertial dynamics. However, recent finding of ion sound wave excitation in *transonic plasma* condition of hydrodynamic equilibrium offers a new physical scope of acoustic turbulence due to weak but finite electron inertial delay effect [5-12]. Qualitative and quantitative modifications are introduced into its nonlinear counterpart as well, under the same transonic plasma equilibrium configuration [12]. The *transonic transition* of the plasma flow motion quite naturally occurs in the neighborhood of boundary wall surface of laboratory plasmas, self-similar expansion of plasmas into vacuum, in solar wind plasmas and different astrophysical plasmas, etc. The self-similar plasma expansion model predicts supersonic motion of plasma flow into vacuum. This model is widely used to describe the motion of intense ion plasma jets produced by short time pulse laser interaction with solid target [17-

23]. Recently, the self-similar plasma expansion into vacuum is modeled by an appropriate consideration of space charge separation effect on the expanding front [13].

According to the recently proposed inertia-induced ion acoustic excitation theory [5-8], the large-scale plasma flow motion feeds the energy to the short scale fluctuations near the pre-sheath termination at sonic point. This is a kind of energy transfer process from large-scale flow energy to wave energy through short scale instability of cascading type. In order to maintain the turbulence type of hydrodynamic equilibrium, there must be some source to feed large-scale flow and sink to arrest the infinite growth of the excited short waves. The growing wave energy could be used to re-modify the global transonic equilibrium such that the transonic transition becomes a natural equilibrium with smooth change in flow motion from subsonic to supersonic regime. Of course, this is a quite involved problem to handle the self-consistent turbulence theory of transonic plasma in terms of anomalous transport [5]. Now one may ask how to produce such boundary layer with sufficient size of the transonic plasma layer for laboratory experimentations?

This, in fact, is an experimental challenge to design and set up such experiments to produce extended length of the transonic zone to sufficient extent to resolve the desired unstable wave spectral components. Creation of a thick boundary layer of transonic flow dynamics is, no doubt, an important task. This zone lies between subsonic and supersonic domains, and is naturally bounded by low supersonic and high subsonic speeds. It should be mentioned here that the sonic velocity corresponds to the phase velocity of the bulk plasma mode of the dispersionless ion acoustic wave. In case of sheath edge boundary, transonic layer could be probed by high-resolving diagnosis of the Debye length order. The desired experiments of spectral analysis of the unstable ion acoustic waves in *transonic plasma condition* may be quite useful to resolve the mystery of sheath edge singularity. Using *de-Lavel nozzle mechanism* of hydrodynamic flow motion, experiments could be designed to produce transonic transition layer of desired length and characteristics [6-8].

Study of the ambient acoustic spectrum associated with plasma flow motion can be termed as the *acoustic spectroscopy* of equilibrium homogeneous plasma flows [6, 26]. This may be useful for expanding background plasmas [13], solar wind plasmas and also in space plasmas through which the space vehicles' motion and aerodynamic motion occur [3, 25, 28]. Basic principles of the acoustic spectroscopy have concern to the linear and non-linear ion acoustic wave turbulence theory and properties of the transonic plasma equilibrium [5-12, 26]. These properties may be used to develop the required diagnostic tools to study and describe the hydrodynamic equilibrium states of plasma flows by suitable observations and analysis of the waves and instabilities they exhibit. In fact, the ambient turbulence-driven plasma flow is quite natural to occur in toroidal and poloidal directions of the magnetic confinement of tokamak device. Similar physical mechanism is supposed to be operative in the transonic transition behavior of equilibrium plasma flow motion [5-12]. Thorough investigations of acoustic wave turbulence theory in transonic plasma condition will be needed to explore transonic flow dynamics on a concrete footing.

Recently, there has been an outburst of interest in plasma states where the assumption of static equilibrium practically is violated [28-30]. Great deals of research activities are now going on in transonic and supersonic magnetohydrodynamic (MHD) flows in laboratory and astrophysical plasmas. Similar activities are also important for understanding the designing of supersonic aerodynamics having relevance in spacecraft-based laboratory experimentations of space plasma research as well [8, 30]. This is also argued that future tokamak reactors need the consideration of rotation of fusion plasma with high speeds that do not permit the assumption of static equilibrium to hold good. This may be brought about due to neutral beam heating and pumped divertor action for the extraction of heat and exhaust.

In astrophysics [3, 28-32, 35-44], the primary importance of plasma flows is revealed in such diverse situations as coronal flux tubes, stellar winds, rotating accretion disks, torsional modes, and jets emitted from radio galaxies. This is to argue that the basic understanding of the acoustic wave dynamics in transonic plasma system constitutes an important subject of future interdisciplinary research [5-12, 26-30]. This may be useful for development of the appropriate diagnostics for acoustic spectroscopy to measure and characterize the hydrodynamic equilibrium of flowing transonic plasmas [8-10]. Such concepts of acoustic wave dynamics in a wider horizon may also applied to understand some helio- and astero-seismic observations in astrophysical contexts.

Most of the plasma devices of industrial applications like dense plasma focus machine, plasma torches, etc. depend on the plasma flows that violate the static equilibrium [26-30]. In fusion plasmas of future generation too, the static approximation of the equilibrium plasma description may not be suitable to describe the acoustic wave behavior. In future course of fusion research, rotational motions of fusion plasmas in poloidal and toroidal directions may decide the equilibrium. This is important to state that in toroidal plasmas, the geodesic acoustic mode becomes of fundamental importance in comparison to the ordinary sound modes [30]. This may be more important when these rotational motions are in the defined range of the transonic limit. Simplicity is correlated to the local mode approximation of the acoustic wave description in transonic limit of uniform and unidirectional plasma flow motion without magnetic field.

The lowest order nonlinear wave theory of the ion acoustic wave dynamics predicts that the usual KdV equation is not suitable to describe the kinetics of the nonlinear traveling ion acoustic waves in transonic plasma condition [8-9, 12, 26]. A self-consistent linear source driven KdV equation, termed as *d*-KdV equation, is prescribed as a more suitable nonlinear differential equation to describe the nonlinear traveling ion acoustic wave dynamics in transonic plasma condition. By mathematical structure of the derived *d*-KdV equation, it looks analytically non-integrable and physically non-conservative dynamical system [8-9]. Due to linear source term, an additional class of nonlinear traveling wave solution of oscillatory shock-like nature is obtained. This is more prominent in the shorter scale domain of the unstable ion acoustic wave spectrum, but within the validity limit of weak nonlinearity and weak dispersion.

If there is multispecies ionic composition in a plasma system, varieties of plasma sound waves are likely to exist depending on, in principle, the number of inertial ionic species. In plasmas containing two varieties of dust or fine suspended particles, two distinct kinds of natural plasma sound modes are possible [15-16]. Such plasmas, termed as the *colloidal plasmas* [16], have become the subject of intensive study in various fields of physics and engineering such as in space, astrophysics, plasma physics, plasma-aided manufacturing technique, and lastly, fusion technology [14-23]. The dust grains or the solid fine particles suspended in low temperature gaseous plasmas are usually negatively charged. It is also observed that plasmas including micro scale-sized and nano scale-sized suspended particles exist in many natural conditions of technological values. Such plasmas have been generated in laboratories with a view to investigate the dust grain charging physics, plasma wave physics as well as some acoustic instability phenomena.

The two distinct sound modes, however, in bi-ion colloidal plasma are well-separated in space and time scales due to wide range variations of mass scaling of the normal ions and the charged dust grains and free electrons' populations. The charged dust grains are termed as the *Dust Grain Like Impurity Ions* (DGLIIs) [15] to distinguish from the normal impurity ions. The present contributory chapter, additionally, applies the inertia-induced acoustic excitation theory to nonlinear description of plasma sound modes in colloidal plasma [15-16] under different configurations. Two separate cases of ion flow motion and dust grain motions are considered. It is indeed found that the *modified Ion Acoustic Wave* (m-IAW) or *Dust Ion Acoustic* (DIA) *wave* and the *so-called (Ion) Acoustic Wave* (s-IAW) or *Dust Acoustic* (DA) *wave* both become nonlinearly unstable due to an active role of weak but finite inertial correction of the respective plasma thermal species [15, 26]. Proper mass domain scaling of the dust grains for acoustic instability to occur is estimated to be equal to that of the asymptotic mass ratio of plasma electron to ion as the lowest order inertial correction of background plasma thermal species. This contributory chapter is thus a review organized to aim at some illustrative examples of linear and nonlinear acoustic wave propagation dynamics through transonic plasma fluid, particularly, under the light of current scenario. Some important reported findings on nonlinear acoustic modes found in space and astrophysical situation [31-44], like in solar plasma system [10-11, 31-44], will also be presented in concise to understand space phenomena. Incipient future scopes of the presented contribution on transonic flow dynamics in different astrophysical situations will also be briefly pointed out.

2. Physical model description

A simple two-component non-isothermal, field-free and collisionless plasma system under fluid limit approximation is assumed. The plasma ions are supposed to be drifting with uniform velocity at around the sonic phase speed under field-free approximation. Global plasma equilibrium flow motion over transonic plasma scale length at hydrodynamic equilibrium is assumed to satisfy the global quasi-neutrality. Such situations are realizable in the transonic region of the plasma sheath system as well as in solar and other stellar wind plasmas [3, 10-11, 35-41]. Its importance has previously been discussed [1, 5-12], where the ion-beam driven wave phenomena are supposed to be involved in *Q*-machine or in unipolar/ bipolar ion rich sheath formed around an electrode wall or grid in Double Plasma Device (DPD) experiments of plasma sheath driven low frequency instabilities of relaxation type [5]. The unstable situation is equally likely to occur on both the sides of the sheath structures with plasma ion streamers [12].

3. Linear normal acoustic mode analyses

3.1 Basic governing equations

The basic set of governing dynamical evolution equations for the linear normal mode behavior of fluid acoustic wave consist of electron continuity equation, electron momentum equation, ion continuity equation, and ion momentum equation [5-6]. The set is closed by coupling the plasma thermal electron dynamics with that of plasma inertial ion dynamics through a single Poisson's equation for electrostatic potential distribution due to localized ambipolar effects. Applying Fourier's wave analysis for linear normal mode behavior of ion acoustic wave over the basic set of governing dynamical evolution equations [5-6], the linear dispersion relation is derived as follows

$$(\Omega + k.v_0)^2 = \frac{k^2 v^2_{te}\left(1 + k^2\lambda^2_{De}\right)}{\Omega^2}\left(\Omega_a^2 - \Omega^2\right). \tag{1}$$

All the notations in the equation (1) are usual and conventional. Here Ω is the Doppler-shifted frequency of the ion acoustic wave, Ω_a is the ion acoustic wave frequency in laboratory frame of reference, k is the angular wave number of the ion acoustic wave such that $k\lambda_{De}$ is a measure of the acoustic wave dispersion scaling and v_{te} is the electron thermal velocity. Now the kinematics of any mode can be analyzed in two different ways: one in lab-frame and the other, in Doppler-shifted frame of reference. This is to note that the obtained dispersion relation differs from those of the other known normal modes of low frequency relaxation type of instability, ion plasma oscillations and waves. This is due to the weak but finite electron inertial delay effect in the dispersion relation of the wave fluctuations. This is mathematically incorporated by a weak inertial perturbation over electron inertial dynamics over the leading order solution obtained by virtue of electron fluid equations neglecting electron inertial term.

It is thus obvious from the mathematical construct of equation (1) that the LHS is a non-resonant term whereas RHS is a resonant term. The RHS gets artificially transformed into a resonant term if and only if $k.v_{i0} < 0$. Now, it can be inferred that equation (1) represents a resonantly unstable situation at Doppler shifted resonance frequency of $\Omega \approx |k.v_0| \geq \Omega_a$, if and only if $k.v_0$ <0. This means that only the mode counter moving with respect to the plasma beam mode gets resonantly unstable. The resonance growth rate for this resonant instability [5-6] is found to be of the following form

$$\gamma = \sqrt{\frac{m_i}{m_e} 2\Omega_a \left(1 + k^2\lambda_{De}^2\right)}\left|\left(\Omega - |k.v_0|\right)^{1/2}\right|. \tag{2}$$

This is important to add that the resonance condition required by equation (1) dictates the propagation direction of the unstable ion acoustic wave (counter moving with respect to plasma ion streams) at reduced frequencies. It is clear from equation (2) that there is the physical appearance of two distinct classes of eigen mode frequencies of the resonantly coupled mode-mode system of linearly growing ion acoustic oscillations in lab-frame: near-zero frequency (standing mode pattern) and non-zero frequency (propagating mode pattern). These two distinct eigen modes are generated by the process of repeated Doppler-shifting of the ion acoustic wave frequency under the unique mathematical compulsion of the hydrodynamic tailoring of the electron fluid density perturbation over ion acoustic time scale. The unstable condition decides the resonant acoustic excitation threshold value for the onset of the instability in terms of normalized value of the eigen mode frequency of the acoustic fluctuations.

3.2 Graphical analysis

It is well-known that the graphical method is a more informative, simple and quick tool for analyzing the stability behavior of a plasma-beam system even without solving dispersion relation. To depict the clear-cut picture of the poles, relation (1) is rewritten as,

$$F(\Omega,k) = \frac{1}{\left(1 + k^2\lambda_{De}^2\right)} = k^2 v_{te}^2\left[\frac{\Omega_a^2}{\Omega^2\left(\Omega + k.v_0\right)^2} - \frac{1}{\left(\Omega + k.v_0\right)^2}\right]. \tag{3}$$

It is clear from the equation (3) that two poles are possible to exist in Ω -space at $\Omega=0$ and $\Omega=|k.v_0|$ for $k.v_0<0$. According to graphical method, the beam-plasma system will exhibit instability only when the curve of $F(\Omega,k)$ versus Ω has multiple singular values in Ω - space having finite minima in between the two successive singularities, which do not intersect with the line $F(\Omega,k)=1/(1+k^2\lambda_{De}^2)$. The required condition for minimization of $F(\Omega,k)$ in Ω -space can be obtained by equating $dF/d\Omega=0$. Now this condition, when applied to equation (3), results into the following equality to derive the value of Ω where dispersion function is supposed to be minimum

$$\Omega_a^2(\Omega+k.v_0)+\Omega(\Omega_a^2-\Omega^2)=0. \tag{4}$$

In principle, equation (4) is to be solved to determine the value of Ω. This is obvious to note that this equality is satisfied at resonance value of $\Omega\sim|k.v_0|\sim\Omega_a$ for $k.v_0<0$. Now to indemnify the complex nature of Ω, the functional value of $F(\Omega,k)>1/(1+k^2\lambda_{De}^2)$. This can, however, be further simplified to yield the following inequality to determine the threshold value for the onset of the inertia-induced instability

$$k^2v_{te}^2(\Omega_a^2-\Omega^2)>(\Omega-|k.v_0|)^2/(1+k^2\lambda_{De}^2). \tag{5}$$

The threshold condition for the instability is satisfied for equality sign at resonance frequency $\Omega\sim|k.v_0|\sim\Omega_a$ that characterizes the case of a marginal instability. A few typical plots of the function $F(\Omega,k)$ in Ω -space for shorter and longer acoustic wavelengths (perturbation scale lengths) are represented in Fig. 1.

3.3 Numerical analysis

Numerical techniques for solving polynomials over years have developed to a vast extent for solving polynomials even with complex coefficients and complex variables. For the present case, the Laguerre's algebraic root-finding method [6] to solve the normalized form of polynomial equation has been used. The polynomial $P(\Omega')$ in the normalized form of the dispersion relation (1) in ion-beam frame is given below

$$P(\Omega')=a_0+a_1\Omega'+a_2\Omega'^2+a_3\Omega'^3+a_4\Omega'^4=0. \tag{6}$$

Here all the normalized notations used are usual, generic and defined by $\Omega'=\Omega/\omega_{pi}$, $\Omega_a'=\Omega_a/\omega_{pi}$ $k'=k\lambda_{De}$, $v'_{te}=v_{te}/c_s=\sqrt{m_i/m_e}$ and $M=v_0/c_s$. The expressions for the various coefficients in the polynomial $P(\Omega')$ are defined as follows

$$a_0=-(1+k'^2)(k'^2v_{te}'^2\Omega_a'^2),$$

$$a_1=0,$$

$$a_2=(|k'.M|)^2+(1+k'^2)(k'^2v_{te}'^2),$$

$$a_3=-2|k'.M|, \text{ and}$$

$$a_4 = 1.$$

It is found that out of four possible roots of $P(\Omega')$, only two roots are complex and these are the complex conjugates as a pair. For all the complex conjugated roots, only the complex root with positive imaginary part is useful, since this determines the growth rate of the instability. Real and imaginary parts of the corresponding complex roots are then plotted as shown in Figs. 2 and 3, respectively. Numerical characterization of the unstable mode of the instability clearly depicts the resonant character of the electron inertia-induced resonant acoustic instability [5].

3.4 Evaluation of wave energy

This is important to evaluate the wave energy in order to have a more complete picture of the basic source mechanism of the discussed instability. In presence of the beam, it is expected that one of the modes involved, has positive energy and the other has negative energy. The dispersion relation (1) can be put in the laboratory frame for a more clear identification and characterization of the positive and negative energy modes in the form of dispersion function $\varepsilon(\omega,k)$ as follows

$$\varepsilon(\omega,k) = 1 + \frac{1}{k^2\lambda_{De}^2}\left(1 + \frac{\omega_{pe}{}^2}{k^2 v_{te}^2}\right) - \frac{\omega_{pi}^2}{(\omega - k.v_0)^2} = 0. \tag{7}$$

The average electric field energy stored in a propagating electrostatic (created by ambipolar effect) wave in a medium is given by the following relation [6]

$$W_\omega(\omega,k) = \frac{1}{2}\varepsilon_0 \left\langle \left|\delta E(\omega,k)\right|^2 \right\rangle \frac{\partial}{\partial\omega}\left[\omega\varepsilon(\omega,k)\right]. \tag{8}$$

Here ε_0 is the dielectric constant of free space, $\delta E(\omega,k)$ is the electric field amplitude of the ion acoustic fluctuations and $W_E = 1/2\left|\delta E(\omega,k)\right|^2$ is the corresponding counterpart of electric energy of the acoustic fluctuations through free space. Applying the equations (7) and (8), the following can explicitly be derived

$$\frac{W_\omega}{W_E} = \omega\frac{\partial\varepsilon(\omega,k)}{\partial\omega} = \left[\frac{2\omega^2}{k^2\lambda_{De}^2 k^2 v_{te}^2} + \frac{2\omega\omega_{pi}^2}{(\omega - k.v_0)^3}\right]. \tag{9}$$

Now, clearly, it is evident that the second term of equation (9) contributes negative energy value to the defined wave-plasma system. This occurs as because the sign of this term becomes negative for the values of $\omega < k.v_0$, which is the case for the reported instability. From a few typical plots in Fig.4, one can notice that the total wave energy suffers a sharp transition from negative to positive values at resonance frequency point of zero energy value. The resonance point lies in the domain of near-zero and non-zero frequencies in lab-frame. According to conventional definition and understanding, the wave energy expression in equation (9) classifies the near-zero frequency mode as the negative energy mode. Then immediately the non-zero frequency mode may be classified as the positive energy mode.

This is important to clarify that the theoretical concept of near-zero frequency mode is an outcome of the mathematical construct of weak but finite electron inertial response to the ion acoustic wave fluctuations. The blowing up character, as shown in Fig. 4, of the total wave energy in opposite directions suggests referring the discussed instability to as an 'explosive instability' in accordance with the law of conservation of energy. It signifies the transonic plasma condition with the resonant mode-mode coupling of the positive and negative energy modes. The time average of the hydrodynamic and wave potential energies of the considered wave-plasma system over the growth time scale is conserved during the energy exchange process between the unstable resonant eigen modes and the main source of ion flow dynamics. These two modes are clearly identified from equation (9) as the natural resonant modes of the defined plasma system that undergo linear resonant mode-mode coupling to produce the defined wave instability.

3.5 Estimation of quenching time

Under the cold ion approximation, even the small electrostatic potential will be able to distort the ion particle motion and associated trajectories, affecting the driving source flow velocity of the resonant instability under consideration. In wave frame, the streaming ion energy (E_i) can be expressed by the following relation

$$E_i = \frac{1}{2} m_i \left(v_0 - \frac{\omega}{k} \right)^2 . \tag{10}$$

For $v_o >> \omega/k \sim \left|(c_s - v_0)\right|$, which is a valid case for the considered instability [5], the condition for ion orbit distortion becomes of the following form,

$$W_w \geq \frac{1}{2} m_i n_0 v_0^2 . \tag{11}$$

From this condition, the quenching time is estimated under the assumption that the wave amplitude grows sufficiently from thermal noise level to physically measurable level such that

$$W_E(t) = W_i e^{\gamma t} . \tag{12}$$

Here W_i is the initial energy of the acoustic wave amplitude, which is of the order of the thermal fluctuations, i.e., $W_i \sim T_e / \lambda_{De}^3$ and is the unnormalized linear growth rate. Using the resonance values of $\omega = k\left|c_s - v_0\right|$ and $\omega - k.v_0 \sim kc_s$ as derived in [5] for long wavelength case of resonant mode, equation (12) for the quenching time τ with the help of (9) can be rewritten as follows

$$\tau = \sqrt{\frac{m_e}{2m_i}} \frac{1}{k\lambda_{De}} \left(\left|1-M\right|\right)^{-1/2} \ln \left[\frac{1}{4} \left(n_0 \lambda_{De}^3\right)\left(k^2 \lambda_{De}^2\right) \frac{M^2}{\left|1-M\right|} \right] . \tag{13}$$

For some typical plasma parameters in hydrogen plasma, $\ln\left(n_0 \lambda_{De}^3\right) \sim 15-30$. For $k\lambda_{De} \sim 0.3, 0.1, 0.05$ near resonant M as in Figs. 2 and 3, equation (13) gives $\tau > 1$, i.e., $\tau_q > \tau_{pi}$. This physically means that the resonant growth time scale is greater than that of the plasma ion oscillation time scale. Thus the resonant nature of the instability is observable in the present analysis.

3.6 Physical consequences

Wave energy analyses are carried out to depict the graphical appearance of poles (Fig. 1), nature of real parts of the roots (Fig. 2), nature of imaginary parts of roots (Fig. 3) and positive-negative energy modes (Fig. 4).

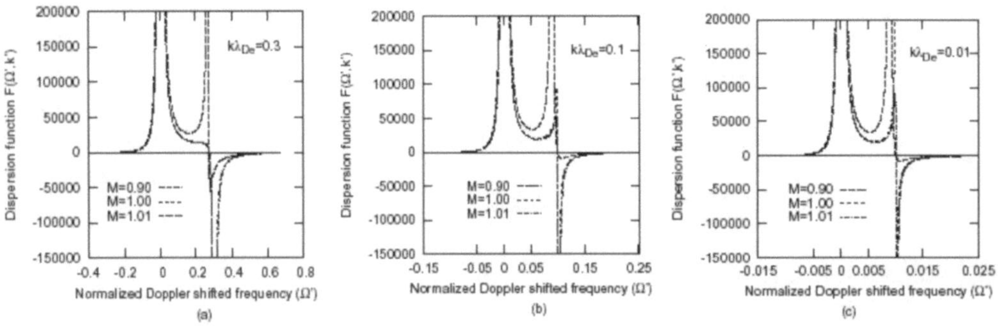

Fig. 1. Graphical appearance of resonance poles as a variation of the dispersion function $F(\Omega,k)$ with normalized Doppler-shifted frequency for dispersion scaling (a) $k\lambda_{De}=0.3$, (b) $k\lambda_{De}=0.1$, and (c) $k\lambda_{De}=0.01$

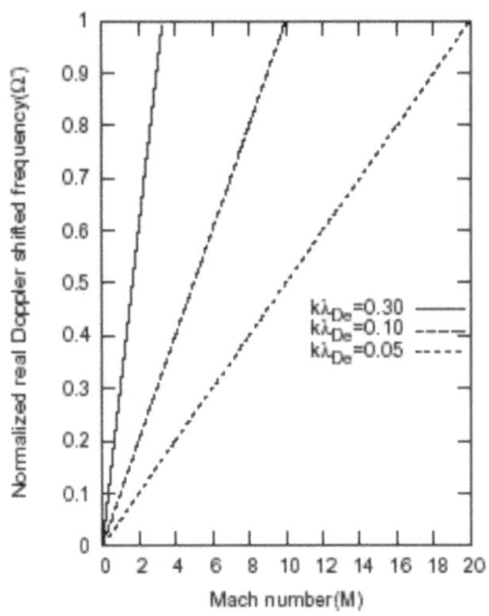

Fig. 2. Variation of the real part of the normalized Doppler shifted eigen mode frequency (Ω') with respect to Mach number (M) for different values of $k\lambda_{De}=0.30,0.10,0.05$

It is found that the instability arises out of linear resonance mode-mode intermixed coupling between the negative and positive energy modes. The total energy of the coupled mode-mode system comprising of hydrodynamical potential energy and wave kinetic energy, however, is in accordance with the law of conservation of energy in the observation time scale on the order of ion acoustic wave time scale. Identification and characterization of the resonance nature of the said instability through *transonic plasma* is presented in order to explore the acoustic richness in terms of collective waves, oscillations and fluctuations. This is an important point to be mentioned here that the same type of instability features are expected to happen in plasma-wall interaction process and sheath-induced instability phenomena in other similar situations as well.

There are different sorts of analytical and numerical tools for studying the linear instabilities in a given plasma system. *Energy method,* based on energy minimization principle and the *normal mode analysis,* based on equilibrium perturbations are the two basic mathematical tools for analyzing the stability behavior of the given plasma systems. However, the latter is most popular and simple for common use in analyzing the threshold conditions of the instabilities and their growth rates. In the normal mode analysis, a linear dispersion relation is derived which can be put in the form of a polynomial with real or imaginary coefficients. The limitation of the analytical method depends upon the degree of the polynomial.

Computational technique broadly takes into account two ways of investigating instability. First, an unstable mode can be deduced by the derived dispersion relation. The obtained polynomial is then solved to delineate the complex roots having concern to the desired instability. Second, a more comprehensive computational method involves solving for the time dependent solution. Simulation technique used to solve the basic set of equations is supposed to give more complete picture of the space and time evolution of the wave phenomena. However, there is another very informative and simple method for analyzing the derived dispersion relation to predict for the unstable behavior of the plasma system under consideration. This is the graphical method in which the dispersion relation is graphically represented for different values of resonance characterization parameters. Source perturbation scale length $(k\lambda_{De})$ and deviation from sonic point $(1-M)$ are the characterization parameters for the defined acoustic resonance.

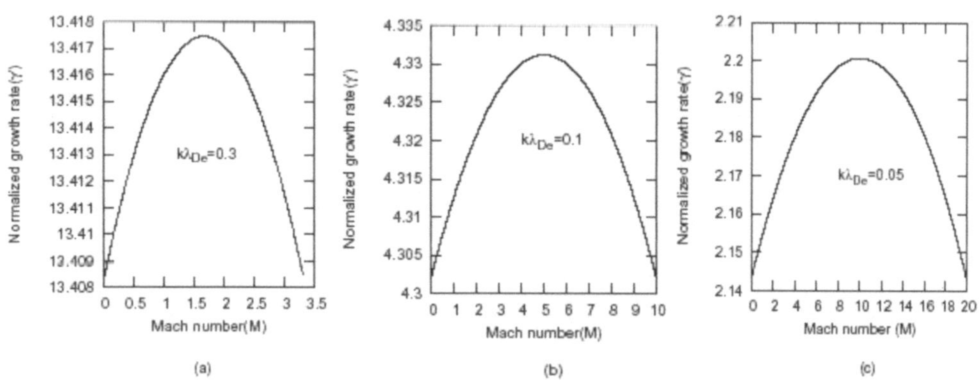

Fig. 3. Variation of the normalized growth rate of the electron inertia-induced resonant acoustic instability with Mach number for (a) $k\lambda_{De} = 0.3$, (b) $k\lambda_{De} = 0.1$ and (c) $k\lambda_{De} = 0.05$ showing that transonic plasma is rich in wide range acoustic spectral components and hence, an unstable zone

This is quite natural and interesting to argue that the transonic plasma condition offers a unique example where the physical situation of localized hydrodynamic equilibrium of quasi-neutral plasma flow dynamics exists. Previous publication reports that the transonic plasma layer, assumed to have finite extension, can be considered as a good physical situation to study the acoustic instability, wave and turbulence driven by electron inertia-induced ion acoustic excitation physics.

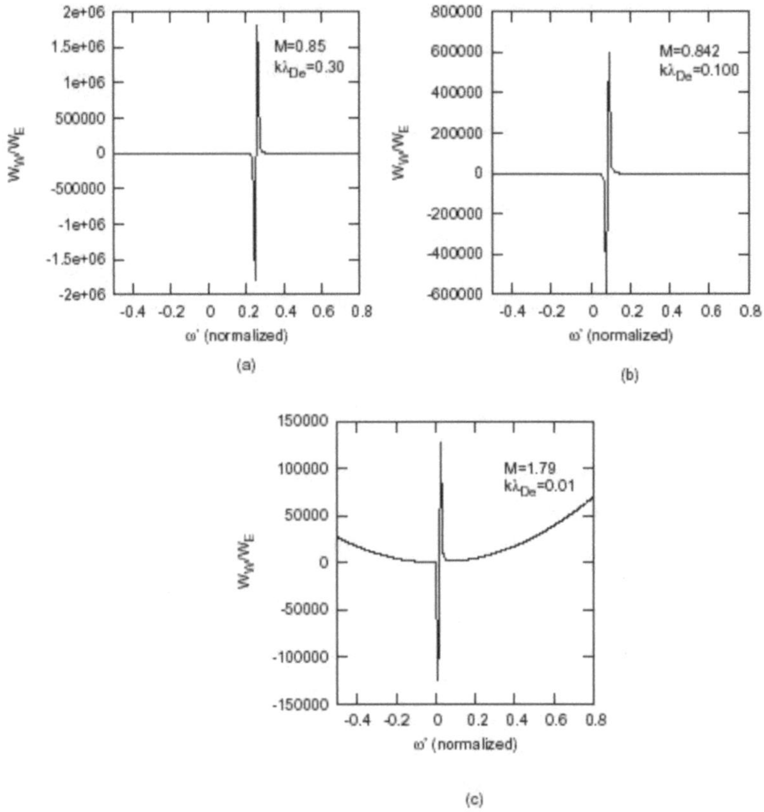

Fig. 4. Explosive nature of the electron inertia-induced ion acoustic wave instability as an outcome of an interplay for the linear resonant mode-mode coupling of positive and negative energy eigen modes. It shows how the normalized wave energy varies with normalized frequency under a set of fixed values of $M$and $k\lambda_{De}$ as (a) $M = 0.85, k\lambda_{De} = 0.30$; (b) $M = 0.842, k\lambda_{De} = 0.100$; and (c) $M = 1.79, k\lambda_{De} = 0.01$

In the present sections of the chapter, many features about the electron inertia-induced ion acoustic wave instability are observed. For example, we physically identify and demonstrate the following features of the instability obtained by theoretical and numerical means of analysis of the desired dispersion relation:

1. The transonic plasma layer is an unstable zone of hydrodynamic equilibrium of quasineutral plasma gas flow motion,
2. The instability is an outcome of the linear resonant mode-mode coupling of positive and negative energy modes,
3. The quenching time of the instability is estimated for some typical values of plasma and wave parameters as mentioned in the previous section. It is found to moderately exceed the ion plasma oscillation time scale, and
4. Lastly, this indicates that in lab frame observation the unstable mode of ion acoustic wave fluctuations at reduced frequencies may look like a purely growing mode. This is very likely to occur for almost entire unstable frequency domain of the frequency transformed ion acoustic waves.

In fact, the electron inertial responses naturally appear only at electron oscillation frequency. However, the *transonic plasma* condition creates a natural physical situation for the same to occur even at the ion acoustic wave frequency of the transformed reduced values. The linear process of resonant mode-mode coupling produces this and makes the coupled system of wave modes unstable.

We have identified and demonstrated the following features of the instability obtained by theoretical and numerical analysis of the dispersion relation: (i) The transonic plasma layer is indeed an unstable zone of hydrodynamic equilibrium of quasi-neutral plasma gas flow motion. (ii) The instability is an outcome of linear resonant mode-mode coupling of positive and negative energy modes. (iii) The normalized values of Doppler-shifted resonant frequencies of the unstable ion acoustic wave fluctuations in ion beam frame come out to be almost equal to 0.5. (iv) The estimated quenching time of the instability exceeds the ion plasma oscillation time scale moderately and hence, (v) In the lab-frame, the unstable modes of ion acoustic wave fluctuations at reduced frequencies may look physically like a purely growing mode.

This is further argued that the physical insights as listed above can be useful as theoretical, graphical and numerical recipes to (1) formulate and solve the problems of saturation mechanisms of the unstable ion acoustic wave fluctuations, (2) formulate and solve the problems of the ion acoustic wave turbulence, and (3) design and setup experiments to study the basic physics of linear and nonlinear ion acoustic wave activities in unique transonic plasma system. These investigations may be useful to improve the existing conceptual framework of physical and mathematical methods of two-scale theory of plasma sheath research to resolve the long-term mystery of the sheath edge singularity. These, in brief, are added to judge the didactic *vis-à-vis* the scientific qualities of the current research work too much specialized in the subject of ion acoustic wave physics.

3.7 Comments

The main conclusive comment here is that the graphical method successfully explains the unstable behavior of the fluid acoustic mode of the ion acoustic wave fluctuations in drifting plasmas with cold ions and hot electrons. A more vivid picture of linear resonant mode-mode coupling of positive and negative energy waves is obtained. This is important to note that simple formulae for wave energy and quenching time calculations [6] are used. This calculation further confirms the earlier results of stability analysis of drifting plasmas against the acoustic wave perturbations [5]. It is, therefore, reasonable to think of logical hypothesis of wave turbulence model approach to solve the sheath edge singularity problem [1, 4]. Actually, the local normal mode theory of the discussed instability implies that the entire transonic plasma zone should be rich in wide frequency range spectrum of the ion acoustic wave fluctuations. This leads to develop the conceptual framework of *situational definition* of the Debye sheath edge to behave as a turbulent zone with finite extension [12]. This hypothetical scenario of the transonic plasma condition can be examined by appropriate experiments of measuring wide range spectral components of the ion acoustic wave fluctuations.

This is a nontrivial problem to explicitly characterize the turbulent properties of the transonic region. The more realistic problem of wave turbulence analysis demands the self-consistent consideration of flow induced quasi-neutral plasma with inhomogeneity in equilibrium plasma background. Similar situations are likely to occur in stellar wind plasmas, where, the transonic behavior is brought about by *deLaval nozzle mechanism* [6-10] of gas flow through a tube of varying cross section. Recent experimental observation [12] in double plasma device (DPD) reports an instability even in a condition of symmetric bipotential ion-rich sheath case. Its frequency falls within zero frequency range and its source is believed to lie in presheath.

Finally, in a nutshell, it is concluded that the graphical method of analyzing the dispersion relation of the inertia-induced instability offers a simple and more informative method of practical importance in transonic plasma equilibrium. Moreover, the plasma environment of Debye sheath edge locality offers a realistic situation for self-excitation of the ion acoustic wave turbulence through resonant ion acoustic wave instability. This is induced by hydrodynamic tailoring of the ion acoustic wave-induced electron density fluctuations. Of course, no experimental observation of instability in transonic plasma has yet been reported to directly compare with the theoretical results. However it cannot be undermined in understanding wave turbulence phenomenon of flowing plasmas. This is informative to add that the frequency and amplitude transformation of the normal ion acoustic wave into unstable ion plasma wave at higher frequency is reported in high intense laser-plasma interaction processes [6-7] through the nonlinear ponderomotive action. This leads to the formation of soliton, double layers, etc. through the saturation mechanism of strong laser-plasma interaction processes due to non-zero average value of the spatially varying electric field associated with laser pulse.

4. Nonlinear normal acoustic mode analyses

4.1 Basic governing equations

A large amount of literature of theoretical and experimental investigations has been produced on the solitary wave propagation in plasmas since the theoretical discovery of ion acoustic soliton [4, 11-12, their references]. Varieties of physical situations of drifting ions of high energy with [5-12] and without [13-33] electron inertial correction have been considered in the ion acoustic wave dynamics. It is shown that the electron inertial motion becomes more important than the ion relativistic effect. Such situations exist in Earth's magnetosphere, stellar atmosphere and in Van Allen radiation belts [3]. Similar studies have been carried out in plasmas with additional ion beam fluid with full electron inertial response in motion [12 and references].

A number of experiments were performed in the unstable condition of beam plasma system in laboratory in order to observe soliton amplification [12]. There are many theoretical calculations and experiments on linear [7-8] and nonlinear [9-11] wave propagation properties of acoustic waves to see their behavior near the transonic point. For an assumed transonic region, it has been theoretically shown that the small amplitude acoustic wave fluctuations exhibit linear resonant growth of relaxation type under the consideration of weak but finite electron inertial delay effect [12-13]. In contrast to earlier claim [3] that the complex nature of coefficients in KdV equation prevents the soliton formation, we argue that their interpretation seems to be physically inappropriate. Instead, by global phase modification technique [12], we show that the usual soliton solution exists (even under the unstable condition), but only for infinitely long wavelength source perturbations. Otherwise, oscillatory shock-like solutions are more likely to exist.

Under fluid approximation, the self-consistently closed set of basic dynamical equations for transonic plasma system with all usual notations in normalized form is given as follows

Electron continuity equation:

$$\frac{\partial \phi}{\partial t}+v_e.\frac{\partial \phi}{\partial x}+\frac{\partial v_e}{\partial x}=0 \text{, and} \tag{14}$$

Electron momentum equation:

$$\frac{m_e}{m_i}\left(\frac{\partial v_e}{\partial t}+v_e.\frac{\partial v_e}{\partial x}\right)=\frac{\partial \phi}{\partial x}-\frac{1}{n_e}\frac{\partial n_e}{\partial x}. \tag{15}$$

This is to remind the readers that equation (15) is obtained by substituting zero-order solution of Boltzmann electron density distribution into the normal electron continuity equation. In fact, in the asymptotic limit of $m_e/m_i \to 0$, electron continuity equation as such is redundant as because the left hand side (electron inertial effect) of (15) is ignorable. Equation (14) basically offers a scope to introduce the weak but finite role of electron to ion inertial mass ratio on the normal mode behavior of acoustic wave.

Ion continuity equation:

$$\frac{\partial n_i}{\partial t}+\frac{\partial}{\partial x}(n_i v_i)=0, \tag{16}$$

Ion momentum equation:

$$\frac{\partial v_i}{\partial t}+v_i\frac{\partial v_i}{\partial x}=-\frac{\partial \phi}{\partial x} \text{, and} \tag{17}$$

Poisson equation:

$$\frac{\partial^2 \phi}{\partial x^2}=n_e-n_i. \tag{18}$$

Following form of the derived *d*-KdV equation obtained from the above equations by the standard methodology of reductive perturbation [12] describes the nonlinear ion acoustic wave dynamics under transient limit (~soliton transit time scale) in a new space defined by the stretched coordinates (ξ,τ). This is to mention that $\phi(x,t)=\phi(\xi,\tau)e^{-\gamma\tau}$ and $\gamma\tau\to 0$ under the transient time action of the propagating ion acoustic soliton through transonic plasma

$$K_0\frac{\partial \phi}{\partial t}+M_0\phi\frac{\partial \phi}{\partial x}+\frac{1}{2}\frac{\partial^3 \phi}{\partial x^3}=\gamma K_0\phi. \tag{19}$$

Here the notations K_0 and M_0 termed as *complex response coefficients* [11-12, 26] and the linear resonant growth rate (γ) of the ion acoustic wave with complex Doppler-shifted Mach number $M_D=M_{Dr}+iM_{Di}$ and lab-frame Mach number $M=M_r+iM_i$ in transonic equilibrium appearing in equation (19) are as follows,

$$K_0=\left[A^2+B^2\right]^{1/2} \text{ where,}$$

$$A=\left(\frac{M_r}{\varepsilon_m}+\frac{M^3{}_{Dr}-3M_{Dr}M_i^2}{\left(M_{Dr}^2+M_i^2\right)^3}\right), \text{ and}$$

$$B=\left(\frac{M_i}{\varepsilon_m}+\frac{M_i^3-3M_{Dr}^2M_i}{\left(M_{Dr}^2+M_i^2\right)^3}\right),$$

$$M_0=\left[C^2+D^2\right]^{1/2} \text{ where,}$$

$$C=\frac{1}{2}\left[\frac{3\left\{\left(M_{Dr}^2-M_i^2\right)^2-4M_{Dr}^2M_i^2\right\}}{\left(M_{Dr}^2+M_i^2\right)^4}-\frac{\left(M_r^2-M_i^2\right)^2-4M_r^2M_i^2}{\varepsilon_m^2}-1\right], \text{ and}$$

$$D=-\frac{1}{2}\left\{\frac{12\left(M_{Dr}^2-M_i^2\right)M_{Dr}M_i}{\left(M_{Dr}^2+M_i^2\right)^4}+\frac{4\left(M_r^2-M_i^2\right)M_rM_i}{\varepsilon_m^2}\right\},$$

$$\gamma=\sqrt{2\left(\frac{m_i}{m_e}\right)}k\lambda_{De}\left|(1-v_{i0})^{1/2}\right|.$$

The notations are usual and generic as discussed earlier [12]. In the system, plasma ions are self consistently drifting or streaming through a negative neutralizing background of hot electrons having relatively zero inertia. The time response of the electron fluid here is normally ignored. As a result, the unique role of weak but finite electron inertia to destabilize the plasma ion sound wave in transonic plasma equilibrium even within fluid model approach of normal mode description is masked.

4.2 Physical consequences

Now equation (19) after being transformed into an equivalent stationary ODE form by the Galilean transformations is numerically solved as an initial value problem. Some very small simultaneous values of ϕ, $\partial\phi/\partial\xi$ and $\partial^2\phi/\partial\xi^2$ are required for the numerical programme to proceed. A few numerical plots for the desired nonlinear evolutions are shown in Figs. 5-6. This is to note that the calculated amplitudes (as shown in Figs. 5a-6a) are the solutions of the present *d*-KdV equation (19) with bounded and unbounded phase potraits (as shown in Figs. 5b-6b). Now, the actual amplitudes of the resulting solutions can be deduced by multiplying the numerically obtained values with $\varepsilon\sim(k\lambda_{DE})^2\approx10^{-2}$ [12]. In principle, the parameter ε is an arbitrary smallness parameter proportional to the dispersion strength or the amplitude of the weakly dispersive and weakly nonlinear plasma wave.
The unique motivation here is to characterize the possible nonlinear normal mode structure of ion acoustic fluctuations under unstable condition of the ion drifts [8-9,12]. By this very specific example, we show that the complex nature of the coefficients of the derived KdV equation in the unstable zone of transonic plasma doesn't prevent the existence of localized nonlinear solutions including usual soliton solution, too. The concept of global phase

modification technique (DPMT) [11-12, 26] results into a *d*-KdV equation [8-9, 12] with variable nonlinear and dispersion coefficients.
Two distinct classes of solutions are obtained: *soliton* and *oscillatory shock-like* structures. Amplification and damping of the driven KdV soliton over the usual KdV soliton is noted for extremely large wavelength (dc) acoustic driving in source term as shown in Figs. 5-6. The amplification near resonance is associated with considerable reduction in nonlinear coefficient than unity as confirmed by numerical calculation. In other cases of shorter acoustic driving in source term as shown in Figs. 5-6, nonlinear solutions of oscillatory shock-like nature are obtained depending on the small deviation from resonant values. It is clearly seen that the peaks of oscillatory shock-like solutions are of either sinusoidal or non-sinusoidal nature with continuous elevation of the initial values of the successive peaks beyond the main nonlinear acoustic peak.
Most of the experimental results in Double Plasma Device (DPD) are reported to show that the obtained theoretical results may have practical relevance to understand the basic physics of ion acoustic wave activities in the transonic region [12] as in Fig. 7. The experiment is performed in a DPD of 90 *cm* in length and 50 *cm* in diameter equipped with multi-dipole magnets for surface plasma confinement [12]. The chamber is divided into source and target by a mesh grid of 85% transparency kept electrically floating. It is evacuated down to a pressure of $(5-6)\times10^{-5}$ Pa with a turbomolecular pump backed by a rotary pump. *Ar*-gas is bled into the system at a pressure $(3-5)\times10^{-2}$ Pa under continuous pumping condition. The source and target plasmas are produced by dc discharge between the tungsten filament of 0.1 *mm* diameter and magnetic cages.

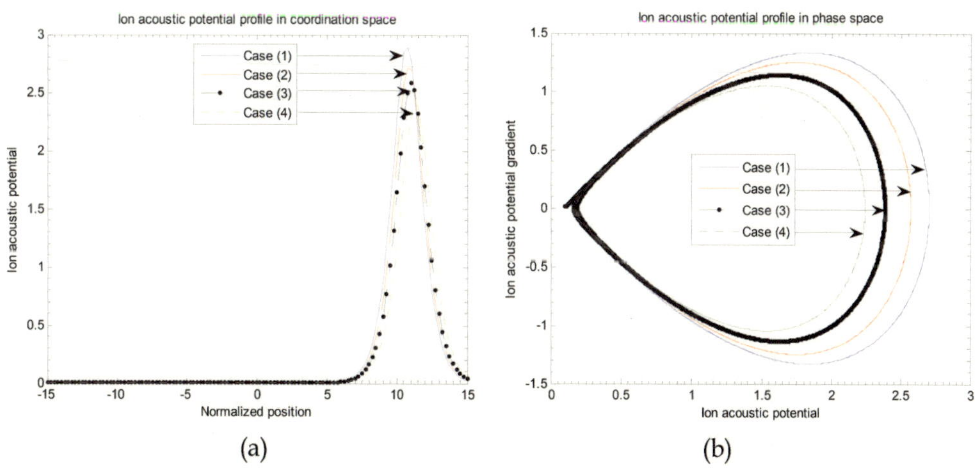

Fig. 5. Profile of (a) ion acoustic potential (ϕ) with normalized space variable (ξ), and (b) phase space geometry of ion acoustic potential in a phase space described by ϕ and $(\phi)_\xi$ with $k\lambda_{De}=2.5\times10^{-8}$ (fixed) for Case (1): $\delta=1.0\times10^{-7}$, Case (2): $\delta=2.5\times10^{-7}$, Case (3): $\delta=5.0\times10^{-7}$, and Case (4): $\delta=7.5\times10^{-7}$

The plasma parameters are measured with the help of a plane Langmuir probe of 5 *mm* diameter and Retarding Potential Analyzer (RPA) of 2.2 *cm* in diameter. The probe and the analyzer are movable axially by a motor driving system so as to take data at any desired position. The plasma parameters are: electron density $n_0 = 10^8 - 10^9 cm^3$, electron temperature $T_e = 1.0 - 1.5eV$ and ion temperature $T_i = 0.1eV$. An ion-acoustic wave is excited with a positive ramp voltage of which the rise time is controllable and is applied to the source anode of the system. Propagating signals are detected by an axially movable Langmuir probe which is biased to $+4V$ with respect to the plasma potential in order to detect the perturbation in the electron current saturation region. The current is then converted into voltage by a resistance of 100Ω and the resultant signals are fed to the oscilloscope. The probe surface is repeatedly cleaned with ion bombardment by applying $-100V$ to it for a short time scale.

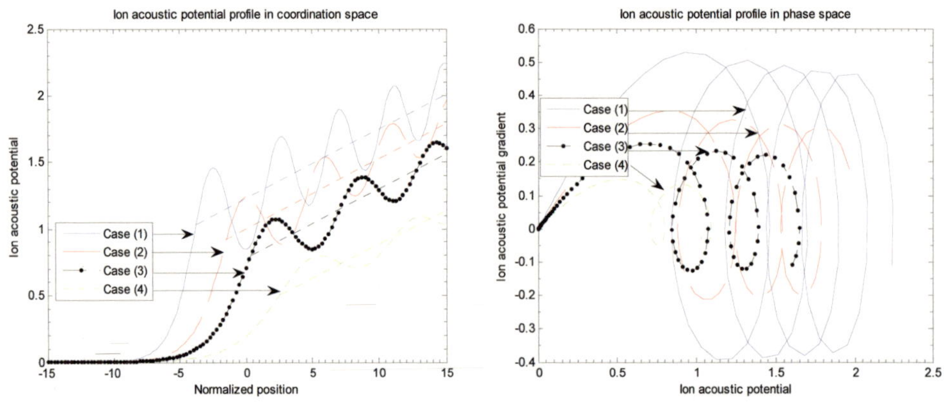

Fig. 6. Same as Fig. 5 but with $k\lambda_{De} = 1.0\times10^{-1}$ (fixed) for Case (1): $\delta = 1.0\times10^{-5}$, Case (2): $\delta = 2.0\times10^{-5}$, Case (3): $\delta = 3.0\times10^{-5}$, and Case (4): $\delta = 5.0\times10^{-5}$

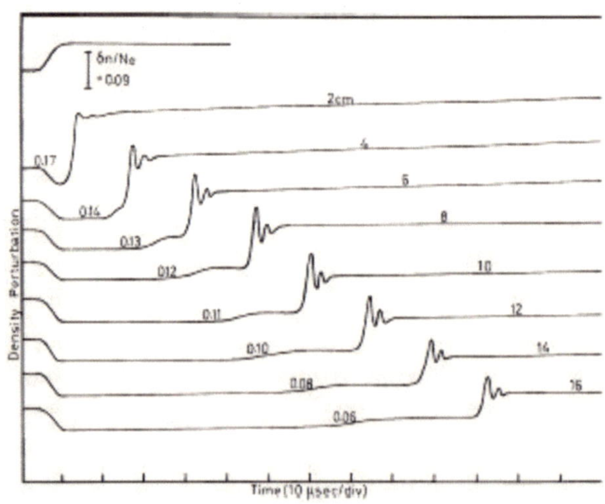

Fig. 7. Experimental profiles of variation of plasma density perturbation (δn) against time (t) at different position of the probe from the grid is shown. Along the *x*-axis, each division represents $10\mu s$ and along the *y*-axis, the density perturbation scale is given as $\delta n/n_e = 0.09$

It is also seen that the value of $\delta = 1 - M_{Dr}$ where resonance occurs remains invariant to spectral variation in source term even by orders of $1.0 \times 10^{-1} - 1.0 \times 10^{-10}$. The nonlinear and dispersive coefficients exhibit very sensitive role on even slight variation of δ from its resonance. It is noted that as the value of $\delta = 0$, the usual KdV soliton is recovered, irrespective of any wave number value in the source term. The source term plays an effective role only when finite $k\lambda_{De}$ and δ-values are assigned simultaneously

4.3 Comments

As per experimental observations, oscillatory shocks are reported to emerge from the transonic zone in the target plasma as shown in Fig. 7. One can qualitatively argue that as soon as the solitary wave passes through the unstable transonic zone, it may experience the transient phase modifications leading thereby to the formation of oscillatory shock. The observed damping of the oscillatory shock may be correlated to the non-resonant type of dissipation through phase incoherence among ion acoustic spectral components of the usual solitary wave. It seems to be more plausible to argue that the input energy to the usual soliton due to transonic plasma equilibrium may be shared among different spectral components through adiabatic energy exchange processes. This is concluded here that the complex coefficients of the KdV equation should, in principle, not become the criterion for the non-existence of localized nonlinear solutions including usual solilton, too. But the usual soliton solution exists only for infinitely long wavelength source perturbation. This conclusion is derived subject to the validity condition of our arguments of global temporal phase modification of usual soliton amplitude under unstable condition of the plasma medium. The unstable condition of the medium may cause structural deformation of the non-driven KdV solution. Such deformations may result into sinusoidal (linear) or non-sinusoidal (nonlinear) peaks of oscillatory shock-like solution depending on the wavelength of the source perturbations [8-9, 12].

Applying the wave packet model for a moving soliton leaving behind an acoustic tail of dispersive waves known as precursor or acoustic wind (in soliton frame), the asymmetry can be associated with elevation of the bottom potential by a finite dc value superposed with periodic repetition of linear or nonlinear peaks. The amplification or suppression of a single soliton can be possible only for infinitely long wavelength (dc) source. For shorter wavelength source driving, the transition from usual soliton solution to oscillatory shock-like solutions is more likely to occur. It is, in brief, concluded that the present mathematical study of *d*-KdV equation offers a significant contribution of analytical supports to our numerical prediction of structural transformation of the traveling nonlinear ion acoustic waves in transonic plasma equilibrium of desired quality. It clearly shows that the actual solution of *d*-KdV equation is a resultant of linear mixing (superposition) of soliton and shock both.

Dominating features of the individual nonlinear modes is decided by an appropriate choice of the specific values of unstable wave number (or wavelength) for a given value of the ion flow Mach number. It is obvious to note that in zero growth limit of *d*-KdV equation, the shock-term disappears and only soliton remains. This limit is correlated with dc range of the chosen unstable wave number of quite weaker dispersion strength. As the dispersion strength becomes significant to influence the original soliton strength of weak nonleariy and weak dispersion in the defined transonic plasma of finite extension, structural modification of the usual KdV soliton profile occurs.

We further argue that the linear and nonlinear normal mode behaviors of the ion acoustic waves in transonic plasma condition differ qualitatively from those derived for static and

dynamic equilibriums without electron inertial correction. The finite but weak hydrodynamic tailoring of the electron fluid motion on ion acoustic space and time scales brings about this difference. It is then argued that the plasma flows in transonic equilibrium should exhibit rich spectrum of linear and nonlinear ion acoustic waves and oscillations. Of course, under Vlasov model the hot electrons with streaming velocity comparable to the phase speed of the ion sound wave may destabilize the ion sound mode through wave-particle resonance effect [8 and references] too. However, our excitation mechanism of ion sound wave differs from the other known mechanisms [8] to excite the same ion sound wave on many grounds [8]. This kind of theoretical scenario of *transonic plasmas* offers a unique scope of *acoustic spectroscopy* to describe the internal state of transonic equilibrium of plasma flows [28].
These calculations have potential applications [26] extensively to understand plasma acoustic dynamics in colloidal plasmas too, but under transonic equilibrium configuration. A generalized statement thereby is reported that all possible sound modes in multi-species colloidal plasmas with drift motions (of inertial ionic species) could be destabilized by the inertial delay effect of the corresponding plasma thermal species that carry out thermal screening of acoustic potential developed due to respective inertial ionic species. Of course, threshold values may differ depending on the choice of the plasma sound mode under consideration. In technological application point of view, one may argue that the proposed theoretical model for inertia-induced acoustic instability mechanism may be utilized to make a plasma-based micro device for *acoustic amplifier* [26]. The amplified acoustic signals (developed due to respective inertial ionic species) from the amplifier could be detected, received and analyzed for the diagnosis and characterization of hydrodynamic flow of plasmas with embedded inertial dust contaminations. These analyses may have potential applications in different ion acoustic wave turbulence-related situations like aerodynamics, solar wind and space plasmas, fusion plasmas, industrial plasmas and plasma flows in astrophysical context, etc.

5. Astrophysical normal acoustics

A plasma-based Gravito-Electorstatic Sheath (GES) model is proposed to discuss the fundamental issues of the solar interior plasma (SIP) and solar wind plasma (SWP). Basic concepts of plasma-wall interaction physics are invoked. Here the wall is defined by a continuous variation of gravity associated with the SIP mass. The neutral gas approximation of the inertially confined SIP is relaxed, and as such the scope of quasi-neutral plasma sheath formation is allowed to arise near the self-consistently defined solar surface boundary (SSB). Analytical and numerical results are obtained to define the SSB and discuss the physics of the surface properties of the Sun, and hence, those of the SWP.

5.1 Physical model description

The SIP system can be idealized as a self-gravitationally bounded quasi-neutral plasma with a spherically symmetric surface boundary of nonrigid and nonphysical nature. The self-gravitational potential barrier of the solar plasma mass distribution acts as an enclosure to confine this quasi-neutral plasma. An estimated typical value $\sim 10^{-20}$ of the ratio of the solar plasma Debye length and Jeans length of the total solar mass justifies the quasi-neutral behavior of the solar plasma on both the bounded and unbounded scales. Here the zeroth-order boundary surface can be defined by the exact hydrostatic condition of gravito-electrostatic force balancing of the enclosed plasma mass at some arbitrary radial position from the center of the mean solar gravitational mass. With this much background in mind, let us now formulate the problem of the physical and mathematical descriptions GES formation around the SSB. For simplicity, we consider spherical symmetry of the inertially confined SIP mass, which helps to reduce the three dimensional problem of describing the GES into a simplified one dimensional problem in the radial direction. Thus, only a single radial degree of freedom is required for description of the dynamical behavior of the SWP under the assumed spherically symmetric self-gravitating solar plasma mass distribution.

The idea of the GES formation can be appreciated with quantitative estimates of the gravito-thermal coupling constants for the SIP electrons and hydrogen ions. Henceforth, "ions" and "hydrogen ions" will be used in the sense of the same ionic species. These parameters [10] can be defined and estimated as follows: The gravitothermal coupling constant for electrons can be estimated as $\Gamma_e = k_B T_e / m_e g_\Theta R_\Theta \approx 10$, for a mean electron temperature of $T_e \sim 10^5$ K and as $\Gamma_e \approx 800$ for mean $T_e = 10^6$ K. The notation k_B (=1.3806×10^{-23} JK^{-1}) denotes the Boltzmann constant. Similarly, the gravito-thermal coupling constant for ions can be estimated as $\Gamma_i = (T_i m_e / T_e m_i)\Gamma_e << 1$ for mean $T_e \sim 10^5$ K, and $\Gamma_i \approx 1$ for mean $T_e \sim 10^6 K$. Here $g_\Theta = GM_\Theta / R_\Theta^2$ denotes the value of the solar surface gravity. The values of the other constant quantities are taken to be $G = 6.6726 \times 10^{-8}$ dyn cm^2 g^{-2}, $M_\Theta = 1.90 \times 10^{33}$ g, and $R_\Theta = 6.97 \times 10^{10}$ cm.

These estimates are based on the condition of an isothermal SIP, where T_e and T_i respectively denote the electron and ion temperatures. It is now easy to see that the electrons can very well overcome the gravitational potential barrier at the SSB in the standard solar model, whereas the ions cannot. This is the reason why a surface polarization-induced space charge (electrostatic) field is likely to appear, due to thermal leakage of the electrons from the SSB in the radially outward direction. Moreover, the neutral gas approximation for the SIP may not be a good one for describing the properties of the SSB. Similar realizations have already occurred to previous authors [5, 9, 11, 14] for the SWP as well. We take the SIP to be an ideal nonisothermal plasma gas with relatively cold ions. The mean electron temperature $T_e > 10^6$ K for the SIP emerges as a more suitable choice for our theoretical consideration.

According to our GES-model analyses, the GES divides into two scales: one bounded, and the other unbounded. The former includes the steady state equilibrium description of the SIP dynamics bounded by the solar self-gravity. This extends from the solar center to the self-consistently defined and specified SSB. The unbounded scale encompasses the SWP dynamics extending from the SSB to infinity. The SIP electrons can easily escape from the defined SSB. On the other hand, the SIP ions cannot cross the gravitational potential barrier of the solar mass on their thermal energy alone. However, surface leakage of the SIP electrons is bound to produce an electrostatic field by virtue of surface charge polarization. This, in turn, provides an additional source to act on the SIP ions to further energize and encourage them cross over the solar self-gravitational potential barrier.

5.2 Basic governing equations

In order to describe the plasma-based GES physics of our model system, we adopt a collisionless unmagnetized plasma fluid for simplicity in mathematical development to obtain some physical insight into the solar wind physics. The role of magnetic field is also ignored (just for mathematical simplicity) in discussing the collisionless SIP and SWP dynamics. Applying the spherical capacitor charging model [3], the coulomb charge on the SSB comes out to be $Q_{SSB} \sim 120C$. The mean rotational frequency of the SSB about the centre of the SIP system is is determined to be $f_{SSB} \sim 1.59 \times 10^{-12}\, Hz$ [42]. Applying the electrical model [42] of the Sun, the mean value of the strength of the solar magnetic field at the SSB in our model analysis is estimated as $\langle |B_{SSB}| \rangle = 4\pi^2 Q_{SSB}\, f_{SSB} \sim 7.53 \times 10^{-11}\, T$, which is negligibly small for producing any significant effects on the dynamics of the solar plasma particles. Thus the effects of the magnetic field are not realized by the solar plasma particles due to the weak Lorentz force, which is now estimated to be $F_L = e(v \times B) \approx 3.61 \times 10^{-33}\, N$ corresponding to a subsonic flow speed $v \sim 3.00\, cm\, s^{-1}$ with the input data available [2, 42] with us and hence, neglected. Therefore our unmagnetized plasma approximation is well justified in our model configuration. In addition, the effects of solar rotation, viscosity, non-thermal energy. For further simplification, the electrons are assumed to obey a Maxwellian velocity distribution. Although these approximations may not be realistic, but they may be considered working hypotheses to begin with an ideal situation. Deviations indeed exist from a Maxwellian velocity distribution. We however use it as a working hypothesis for our model considerations. As a result, the usual form of the Boltzmann density distribution for plasma thermal electrons with all usual notations is given as

$$N_e = e^{\theta}. \tag{20}$$

Here $N_e = n_e / n_0$ denotes the normalized electron density. The generic notation $\theta = e\phi / T_e$ denotes the normalized value of the plasma potential associated with the GES on the bounded scale and with the SWP on unbounded scale. The general notation n_e stands for the nonnormalized electron density and $n_0 = \rho_\Theta / m_i$ defines the average bulk density of the equilibrium SIP. The notation $\rho_\Theta = 1.43$ g cm^{-3} stands for the average but constant solar plasma mass density and $m_i = 1.67 \times 10^{-24}$ g for the ionic (protonic) mass. Again e represents the electronic charge unit and ϕ, the nonnormalized plasma potential associated with both the GES and SWP.

The hydrogen ions are described by their full inertial response dynamics. This includes the ion momentum equation as well as the ion continuity equation. The first describes the change in ion momentum under the action of central gravito-electrostatic fields of potential gradient and forces induced by thermal gas pressure gradients. The latter equation is considered a gas dynamic analog of plasma flowing through a spherical chamber of radially varying surface area. In normalized forms, the ion momentum equation is

$$M \frac{dM}{d\xi} = -\frac{d\theta}{d\xi} - \varepsilon_T \frac{1}{N_i} \frac{dN_i}{d\xi} - \frac{d\eta}{d\xi}. \tag{21}$$

Here the minus sign in the gravitational potential term indicates the radially inward direction of the solar self-gravity. The deviation from the conventional neutral gas treatment of the SIP is introduced through the electric space charge-induced force (first term on right-hand side) effect. The normalized expression for conservation of ion flux density is

$$\frac{1}{N_i} \frac{dN_i}{d\xi} + \frac{1}{M} \frac{dM}{d\xi} + \frac{2}{\xi} = 0. \tag{22}$$

The normalizations are defined as follows:

$$\theta = \frac{e\phi}{T_e},\ \eta = \frac{\psi}{C_s^2},\ N_e = \frac{n_e}{n_0},\ N_i = \frac{n_i}{n_0},$$

$$M = \frac{v_i}{c_s},\ \xi = \frac{r}{\lambda_J},\ \lambda_J = \frac{c_s}{\omega_J},\ c_s = \left(\frac{T_e}{m_i}\right)^{1/2},$$

$$\omega_J = \left(4\pi\rho_\Theta G\right)^{1/2},\ \varepsilon_T = \frac{T_i}{T_e}.$$

The notations ϕ and ψ respectively stand for the dimensional (*unnormalized*) values of the plasma electrostatic potential and the self-solar gravitational potential as variables associated with the GES. The dimensional values of the electron and ion population density variables are respectively denoted by n_e and n_i. Likewise, the dimensional ion fluid velocity variable is represented by the symbol v_i. The notation η stands for the normalized variable of the self-solar gravitation potential. The notation N_i denotes the normalized value of the ion particle population density variable. Notation M stands for the ion flow Mach number.
The notations r and ξ stand for the nonnormalized and normalized radial distance respectively from the heliocenter in spherical co-ordinates. The other notations λ_J, c_s and ω_J defined as above stand for the Jeans length, sound speed and Jeans frequency respectively. Finally, the notation ε_T as defined above stands for the ratio of ion to electron temperature. The ion flux density conservation (eq. 22) contains a term that includes the effect of geometry on the ion flow dynamics of the SIP mass, self-gravitationally confined in a spherical region, whose size is to be determined from our own model calculations. Equations (21) and (22) can be combined to yield a single expression representing the well-known steady state hydrodynamic flow,

$$\left(M^2 - \varepsilon_T\right)\frac{1}{M}\frac{dM}{d\xi} = -\frac{d\theta}{d\xi} + \varepsilon_T \frac{2}{\xi} - \frac{d\eta}{d\xi}. \tag{23}$$

There is an obvious difference in the above equation from the corresponding momentum equation under the neutral gas approximation for the SIP. The difference appears, as discussed above, in the form of a space charge effect originating from the Coulomb force on a collective scale (first term on the right-hand side of eq. (21)).
The gravito-electrostatic Poisson equations complement the steady dynamical equation (23) for a complete description of the gravito-electrostatic sheath structure, which is formed inside the non-rigid SSB. This is important to emphasize that in the case of a real physical boundary, the plasma sheath is always formed both inside and outside the boundary surface in its close vicinity [12]. The normalized forms of the gravitational and electrostatic Poisson equations for the SWP description are respectively given by

$$\frac{d^2\eta}{d\xi^2} + \frac{2}{\xi}\frac{d\eta}{d\xi} = N_i, \text{ and} \tag{24}$$

$$\left(\frac{\lambda_{De}}{\lambda_J}\right)^2\left[\frac{d^2\theta}{d\xi^2}+\frac{2}{\xi}\frac{d\theta}{d\xi}\right]=N_e-N_i\,. \tag{25}$$

Here $\lambda_{De}=\left(T_e/4\pi n_0 e^2\right)^{1/2}$ denotes the plasma electron Debye length of the defined SIP system. The other quantities are as defined above as usual. Equations (21)-(25) constitute a completely closed set of basic governing equations with which to discuss the basic physics of the GES-potential distribution on the bounded scale. Of course, the discussion also includes the associated ambipolar radial flow variation of the SIP towards an unknown SSB which we have to determine self-consistently in this problem with GES-based theory. For a typical value $T_e=10^6$ K, one can estimate that $\lambda_{De}/\lambda_J\approx 10^{-20}$ which implies that the Debye length is quite a bit smaller than the Jeans scale length of the solar plasma mass. Thus, on the typical gravitational scale length of the inertially bounded plasma, the limit $\lambda_{De}/\lambda_J\rightarrow 0$ represents a realistic (physical) approximation. By virtue of this limiting condition, the entire SIP extending up to the solar boundary and beyond obeys the plasma approximation. Thus, the quasi-neutrality condition as given below holds good

$$N_e=N_i=N=e^{\theta}. \tag{26}$$

This is to note that equation (26) does not mean that the plasma ions are Boltzmannian in thermal character, but inertial species. Equation (26) can be differentiated once in space and further rewritten as,

$$\frac{1}{N}\frac{dN}{d\xi}=\frac{d\theta}{d\xi}. \tag{27}$$

By virtue of the plasma approximation, one can justify that the GES of the SIP origin should behave as a quasi-neutral space charge sheath on the Jeans scale size order. The formation mechanism of the defined GES, however, is the same as in the case of plasma-wall interaction process in laboratory confined plasmas. From equations (26)-(27), it is clear that for the electrostatic potential and its gradient being negative, causes the exponential decrease of the plasma density. Finally, the reduced form of the basic set of autonomous closed system of coupled nonlinear dynamical evolution equations under quasi-neutral plasma approximation is enlisted as follows

$$\left(M^2-\varepsilon_T\right)\frac{1}{M}\frac{dM}{d\xi}=-\frac{d\theta}{d\xi}+\varepsilon_T\frac{2}{\xi}-\frac{d\eta}{d\xi}, \tag{28}$$

$$\frac{d\theta}{d\xi}+\frac{1}{M}\frac{dM}{d\xi}+\frac{2}{\xi}=0, \text{ and} \tag{29}$$

$$\frac{d^2\eta}{d\xi^2}+\frac{2}{\xi}\frac{d\eta}{d\xi}=e^{\theta}. \tag{30}$$

This set of differential evolution equations constitutes a closed dynamical system of governing hydrodynamic equations that will be used to determine the existence of a bounded GES structure on the order of the Jeans scale length (λ_J) in our GES-model of the subsonic origin of the SWP of current interest. Thus the solar parameters $M(\xi)$, $g_s(\xi)$ and $\theta(\xi)$ representing the equilibrium Mach number, solar self-gravity and electrostatic potential, respectively, will characterize the gravito-electrostatic acoustics in our approach.

5.3 Theoretical analysis of solar surface boundary

5.3.1 Analytical calculations

We first wish to specify the overall condition for the existence of the SSB. Such existence demands the possibility of a self-consistent bounded solution for the solar self-gravity. The boundary will correspond to a maximum value of the solar self-gravity at some radial distance from the heliocenter. This defines a self-consistent location of the SSB. Before we proceed further, let us argue that the radially outward pulling bulk force effect of the GES-associated potential term in equation (28) demands a negative electrostatic potential gradient, that is, $d\theta/d\xi < 0$. This makes some physical sense because the ion fluid has to overcome the gravitational barrier to create a global-scale flow of the SIP in a quasi-hydrostatic way.

Now, if we invoke the concept of exact hydrostatic formation under gravito-electrostatic force balancing $(d\theta/d\xi \approx d\eta/d\xi)$, the surface potential can be solved to get $\theta - \theta_\Theta \approx \eta - \eta_\Theta$. Here the unknown boundary values of $\theta = \theta_\Theta$, $\eta = \eta_\Theta$ and $M = M_{SSB}$ are to be self-consistently specified numerically. The notation (M_{SSB}) stands for the Mach value associated with the SIP flow at the SSB. Now, by the exact hydrostatic equilibrium condition in the set of equations (28)-(30), one can get the following set of equations for the SSB description:

$$\left(M^2 - \varepsilon_T\right)\frac{1}{M}\frac{dM}{d\xi} = \varepsilon_T \frac{2}{\xi}, \tag{31}$$

$$-\frac{d\eta}{d\xi} + \frac{1}{M}\frac{dM}{d\xi} + \frac{2}{\xi} = 0, \text{ and} \tag{32}$$

$$\frac{d^2\eta}{d\xi^2} + \frac{2}{\xi}\frac{d\eta}{d\xi} = e^{\theta}. \tag{33}$$

For purpose of the GES analysis, we define the solar self-gravitational acceleration as $g_s = d\eta/d\xi$. Equation (34) thus reads

$$\frac{dg_s}{d\xi} + \frac{2}{\xi}g_s = e^{\theta}. \tag{34}$$

Finally, the SIP and hence, the SSB are described and specified in terms of the relevant solar plasma parameters $M(\xi)$, $g_s(\xi)$ and $\theta(\xi)$ representing respectively the equilibrium Mach number, solar self-gravity and electrostatic potential as a coupled dynamical system of the closed set of equations recast as the following

Solar self-gravity equation:

$$\frac{dg_s}{d\xi} + \frac{2}{\xi}g_s = e^{\theta}, \tag{34a}$$

Ion continuity equation:

$$\frac{d\theta}{d\xi}+\frac{1}{M}\frac{dM}{d\xi}+\frac{2}{\xi}=0, \text{ and} \tag{34b}$$

Ion momentum equation:

$$\left(M^2-\alpha\right)\frac{1}{M}\frac{dM}{d\xi}=\alpha\frac{2}{\xi}-g_s, \tag{34c}$$

where $\alpha=1+\in_T=1+\left(T_i/T_e\right)$, T_e is the electron temperature and T_i is the inertial ion temperature for the bounded solar plasma on the SIP-scale as already mentioned.
Let us now denote the maximum value $\left(g_\Theta\right)$ of solar gravity at some radial position $\xi=\xi_\Theta$ where $\theta=\theta_\Theta$ and apply the necessary condition for the maximization of g_s at a spatial coordinate $\xi=\xi_\Theta$. This condition $\left(dg_s/d\xi\right)_{\xi=\xi_\Theta}=0$ when used in equation (34) yields $\xi_\Theta=2g_\Theta e^{-\theta_\Theta}$. However, it is not sufficient to justify the occurrence of the maximum value of g_s until and unless the second derivative of g_s is shown to have negative value. To derive the sufficient condition for the maximum value of g_s at $\xi=\xi_\Theta$, let us once spatially differentiate equation (34) to yield

$$\frac{d^2g_s}{d\xi^2}-\frac{2}{\xi^2}g_s+\frac{2}{\xi}\frac{dg_s}{d\xi}=e^{\theta}\frac{d\theta}{d\xi}. \tag{35}$$

Now the condition for the maximization of g_s at the location $\xi=\xi_\Theta$ can be discussed by considering $d\theta/d\xi<0$ in equation (35) under the exact hydrostatic equilibrium approximation $\left(\left|d\theta/d\xi\right|\approx\left|d\eta/d\xi\right|=g_\Theta\right)$ near the solar surface to yield the following inequality

$$\left.\frac{d^2g_s}{d\xi^2}\right|_{\xi=\xi_\Theta}=\frac{2}{\xi_\Theta^2}g_\Theta-e^{\theta_\Theta}g_\Theta=g_\Theta\left(\frac{2}{\xi_\Theta^2}-e^{\theta_\Theta}\right)<0. \tag{36}$$

From these analytical arguments one can infer that the maximization of g_s indeed occurs at some arbitrary radial position that satisfies the inequality: $\xi_\Theta>\sqrt{2}e^{-\theta_\Theta/2}(=2.33)$ for $\theta_\Theta\sim-1$ (Figs. 8b, 9b, and 10b). Numerically the location of the SSB is found to lie at $\xi_\Theta\sim3.5$ that matches with $\xi_\Theta=2g_\Theta e^{-\theta_\Theta}$ for $\theta_\Theta=1.07$ and $g_\Theta=0.6$. It satisfies the analytically derived inequality (36) too. Now the other two equations (32)-(33) can be simultaneously satisfied in the SSB only for a subsonic solar plasma ion flow speed if Mach number gradient acquires some appropriate negative minimum near zero $\left(M\sim\left|\left(dM/d\xi\right)\right|\sim10^{-6}\right)$.
It is indeed seen numerically that near the maximum solar self-gravity of the SIP mass, the first and third terms in equation (32) are almost equal and hence the Mach number gradient term, which is negative in the close vicinity of the SSB, should be smaller than the other two terms so as to satisfy equations (31) and (32), simultaneously. Actually, the three equations (31)-(32) and (34) are solved numerically to describe the SSB of the maximum self-gravitational potential barrier properly where g_s associated with the self SIP mass is maximized.

5.3.2 Numerical calculations

Determination of the autonomous set of the initial values of the defined physical variables is a prerequisite to solve the nonlinear dynamical evolution equations (34a)-(34c) in general as an initial value problem. The initial values of the physical variables like $M(\xi), g_s(\xi)$ and $\theta(\xi)$ are defined inside the solar interior and are determined on the basis of extreme condition of the *nonlinear stability analysis* [4]. The self-consistent choice of the initial values is obtained by putting $dM/d\xi|_{\xi_i} = -e^{\theta_i/2}$, $dg_s/d\xi|_{\xi_i} = 0$ and $d\theta/d\xi|_{\xi_i} = 0$ in these three equations (34a)-(34c). But the realistic SWP model demands that $d\theta/d\xi|_{\xi_i} \neq 0$. Finally, we determine the expressions for a physically valid set of the initial values of the given physical variables as follows,

$$M_i = \frac{1}{2}\xi_i e^{\theta_i/2} \tag{37}$$

$$g_{si} = \frac{1}{2}\xi_i e^{\theta_i} \tag{38}$$

This is to note that the initial values of θ_i and ξ_i are chosen arbitrarily. As discussed later, we find that the SSB acquires a negative potential bias ($\theta_s \sim -1$) of about -1 kV. It also acquires the maximum value of solar interior gravity ($g_\Theta \sim 0.6$) and minimum value of non-zero SIP flow speed ($M_{SSB} \sim 10^{-7}$) at the SSB. The value of the electrostatic potential gradient at the surface comes out to be ~ -0.6. This means that the strength of the GES-associated solar surface gravity and electrostatic potential gradient is almost equal. As a result the SSB is defined by some constant values of the physical variables (g_s, θ, M). The SSB values of these parameters are determined through spatial evolution of the coupled system of equations (34a)-(34c) from the given initial values (37)-(38) inside the SIP zone.

We have used the well-known fourth-order Runge-Kutta method (RK 4) for numerical solutions of equations (34a)-(34c). By numerical analysis (Figs 8-11), we find that the solar radius is equal to thrice of the Jeans length (λ_J) for mean solar mass density $(\rho_\Theta = 1.41 g.cm^{-3})$. From this observation one can easily estimate that $\lambda_j = R_\Theta / 3.5$. Now comparing our own theoretical value of the solar mass self-gravity with that of the standard value, we arrive at the following relationship between the solar plasma sound speed (c_s) and the Jeans length (λ_J)

$$0.6\left(c_s^2/\lambda_J\right) = 2.74\times10^4 \; cm/sec^2. \tag{39}$$

By substituting the value of the Jeans length expressed in terms of the solar radius, one can determine and specify the mean value of the electron temperature, which is $T_e \sim 10^7$ K. The sound speed in the SIP under the cold ion model approximation is thus obtained as $c_s \sim 3\times10^7$ cm s^{-1}. Note that the SWP speed at 1 AU is fixed by the sound speed of the SWP medium, which is determined and specified by the requirement that a transonic transition solution occurs on the unbounded scale of the SWP dynamics description.

The gravito-acoustic coupling coefficient could be estimated as $\Gamma_{g-a} = m_i g_\Theta R_\Theta / T_e \sim 2.0$. In absence of the gravitational force, the bulk SIP ion fluid will acquire the flow speed corresponding to $M \sim 1.41$ for a negative potential drop of the order of $(-T_e/e)$ over a distance equal to that of the solar radius. If one estimates the value of gravito-acoustic coupling coefficient at this velocity defined by $M \sim 1.41$, it comes out to be unity. Thus the quasi-hydrostatic type of equilibrium gravitational surface confinement of the SIP is ensured. The GES-potential induced outward flow of the SIP is also ensured. Due to comparable strength of the solar surface gravitational effect of deceleration, the net SIP ion fluid flow is highly suppressed to some minimum value (~1.0-3.0 *cm/sec*) corresponding to $M \sim 10^{-7}$ at the SSB.

An interesting point to note here is that near the defined SSB, the electrostatic potential gradient terminates into an almost linear type of profile. The value of its gradient value will provide an estimate of the second order derivative's contribution into the electrostatic potential which measures the level of local charge imbalance near the solar surface. From our computational plots (Figs. 8b, 9b and 10b), this local charge imbalance comes out to be of the order -0.17, which is equivalent to 17% ion excess charge distributed over a region of size on the order of the plasma Debye sheath scale length. However, the same level of the electrostatic local charge imbalance on the Jeans scale length does not require the inclusion of the role of the Poisson term for the evolution of the electrostatic potential's profile under the GES-model. Hence, in this sense the GES is practically equivalent to a quasi-neutral plasma sheath with its potential profile tailored and shaped by the potential barrier of the self-gravity of the SIP mass distribution.

5.3.3 Properties of solar surface boundary

Table I lists the defined initial values of the physical variables (g_s, θ, M) as already discussed and their corresponding boundary values numerically obtained for the description of the desired SSB. The initial values of g_s, θ, and M are associated with the normalized mean SIP mass density, enclosed within a tiny spherical globule having normalized radius equal to an arbitrarily chosen value of ξ_i.

Parameter	At the Initial Radial Point (ξ_i)	At the Solar Surface Boundary (ξ_Θ)	Initial Values
Potential θ	$\frac{d\theta}{d\xi}=0$	$\frac{d\theta}{d\xi} \sim -0.62$, $\theta_\Theta \sim -1.00$	θ_i, arbitrarily chosen
Gravity g_s	$\frac{dg_s}{d\xi}=0$	$\frac{dg_s}{d\xi}=0$, $g_\Theta \sim 0.60$	$g_{si}=\frac{1}{2}\xi_i e^{\theta_i}$, derived
Mach number M	$\frac{dM}{d\xi}=-e^{\theta_i/2}$	$\frac{dM}{d\xi}=0$, $M_{SWP} \sim 10^{-7}$	$M_i=\frac{1}{2}\xi_i e^{\theta_i/2}$, derived

Table 1. Initial and Boundary Values of Relevant Solar Parameters

From the numerical plots shown in Figs. 8-10, we find that the minimum Mach number (M_{SSB}) at the specifically defined SSB comes out to be of order 10^{-7}. For this value of Mach number, equation (31) can be simplified to show that near the boundary, $dM/d\xi \approx -M/\xi_\Theta = -3\times10^{-8} \sim 0$. This corresponds to a quasi-hydrostatic type of the SSB equilibrium. It arises from gravito-electrostatic balancing with an outward flow of the SIP having a minimum speed of about 1-3 cm s^{-1}. With these inferences one can argue that the

SWP originates by virtue of the interaction of the SIP with the SSB. Hence an interconnection between the Sun and the SWP can be observed by applying the GES model. Here the boundary is not sharp but distributed over the entire region of the solar interior volume. The basic principles of the GES coupling govern the solar surface emission process of the subsonic SWP.

As depicted in table I, the time-independent solar g_s-profile associated with the SIP mass distribution terminates into a diffuse surface boundary. This is characterized and defined by the quasi-hydrostatic equilibrium $g_s = g_\Theta \sim |d\theta/d\xi|$, which occurs at about $\xi = \xi_\Theta \sim 3.5$ (see Figs. 8-10). As such, the basic physics of the subsonic origin of the SWP from the SSB is correlated with the bulk SIP dynamics. We note that the precise definition of the SSB influences the SWP velocity at 1 AU. Other models report similar observations too [3, 31-41]. The dependence on the ion to electron temperature ratio is quite visible in Fig. 11a. Let us now discuss the numerical results in the figures individually.

Figure 8 depicts the time-independent profiles of (g_s, θ, M) and their variations with the ion-to-electron temperature ratio ε_t for fixed values of the initial point (ξ_i=0.01) and plasma sheath potential ($\theta_i = -0.001$). As shown in Fig. 8a, the location of the SSB remains the same but its maximum value changes, and a most suitable choice of ε_t = 0.4 is identified for which the quasi-hydrostatic condition is fulfilled. The *E*-field profile is invariant for all chosen values ε_t =0-0.5. Again, as shown in Fig. 8b, the electrostatic potential corresponding to ε_t~ 0.4 comes out to be $\theta = \theta_\Theta \sim -1$ (i.e. ~1 kV). Similarly, Fig. 8c depicts the minimum Mach value of M_{SSB}~10^{-7} for ε_t ~0.4 varying by a factor of 2 for other values of ε_t.

Figure 9 depicts the time-independent profiles of g_s, θ, and M and their variations with initial position for fixed values of ε_t =0-0.4 and $\theta_i = -0.001$. It can be seen that the most suitable choice of the initial position for our fixed values θ_i and ε_T comes out to be $\xi_i \sim 0.01$, which is consistent with the earlier value shown in Fig. 8a. Moreover, the minimum value of M~10^{-7} (Fig. 9c) is also consistent with the earlier value shown in Fig. 8c.

Figure 10 depicts the time-independent profiles of g_s, θ, and M and their variations with electrostatic potential for fixed values of ξ_i=0.01 and $\varepsilon_T = 0.4$. It is observed fascinatingly that the most suitable choice of the initial value of the normalized electrostatic potential for our fixed values of ξ_i and ε_T comes out to be $\theta_i = -0.001$.

It is notable that high initial drop of *M*-profile occurs as shown in Fig.8c, Fig. 9c and Fig. 10c. This indicates the over dominance of the solar interior gravity up to about $\xi \sim 1.5$, and thereafter, the *E*-field becomes comparable, balancing at about $\xi \sim 3.5$. Thus the normalized width of the *gravito-electrostatic sheath* could be estimated and denoted by $\xi_{G-E} \sim 2$. This is a quasi-neutral space charge region with positive charge (ion) excess near the defined SSB wall. Thus a self-consistent bounded solution of nonlinearly coupled gravito-electrostatic potential profiles exists. It forms a quasi-hydrostatic equilibrium at the SSB for the choice of the appropriate set of the initial parameter values $\theta_i = -0.001$ & $\xi_i = 0.01$ for $\varepsilon_T = 0.4$. This is not a rigid boundary at all. As a result the SSB is capable to exhibit many kinds of global oscillation dynamics governed by the nonlinear coupling of the gravito-electrostatic sheath potentials.

For a laboratory hydrogen plasma, the normalized floating potential can be estimated as $\theta_f = -3.76$ under the flat surface approximation. Now, if we consider the numerically calculated minimum value of M for our solar surface characterization, the estimated value of

θ_f is about -20 using the flat surface approximation. This is a crude estimate because the solar surface potential drop occurs over a distance of the order of the Jeans scale where the effect of curvature should not be ignored. The numerically obtained solar surface potential is quite a bit smaller than the floating potential. Simply put, this means that the defined SSB of the GES-model draws a finite amount of electron-dominated electric current that flows toward the heliocenter.

Let us invoke a generalized concept of the plasma sheath, which is traditionally associated with a localized electrostatic potential only in the plasma physics community. We argue that any localized nonneutral space charge layer (in our case, on the order of the Jeans length) is the result of a self-consistent nonlinear coupling of gravito-electrostatic force field variations. This is what we mean by the GES, which of course, obeys the global quasi-neutrality condition because of the smallness of the ratio λ_{De}/λ_J for the SIP parameters.

5.4 Acceleration of solar interior plasma

5.4.1 Basic equations for SWP descriptions

We have already argued that the subsonic origin of the SWP from the SSB is an outcome of the condition of quasi-hydrostatic equilibrium at the boundary. This is a result of the comparable, but competiting strengths of the gravitational deceleration and the electrostatic acceleration of the SIP near the SSB. Now we will try to look at the problem of solar wind acceleration from subsonic to supersonic speed. This is referred to as the *transonic transition* behavior of the outward-moving SIP in the form of the SWP. Let us now argue that the radial variation of Mach number and electrostatic potential beyond the defined SSB should be described by the following autonomous set of coupled nonlinear differential equations

$$\left(M^2-\varepsilon_T\right)\frac{1}{M}\frac{dM}{d\xi}=-\frac{d\theta}{d\xi}+\varepsilon_T\frac{2}{\xi}-\frac{1}{\xi^2}\frac{GM_\Theta}{C_s^2\lambda_J}, \text{ and} \tag{40}$$

$$\frac{d\theta}{d\xi}+\frac{1}{M}\frac{dM}{d\xi}+\frac{2}{\xi}=0\,. \tag{41}$$

Let us note that the constant SIP mass acts as an external object to offer a source of gravity for tailoring and monitoring the outgoing SIP flow with the initially subsonic speed specified at the defined SSB. The Poisson equation for gravity is now redundant. It is important to comment that the electrostatic force field is not imposed from outside to control the solar wind's motion. In fact, the required electric field for the SWP acceleration is of internal origin. Equations (40) and (41) can be combined to yield a single coupled form as given below

$$\left[M^2-\left(1+\varepsilon_T\right)\right]\frac{1}{M}\frac{dM}{d\xi}=\left(1+\varepsilon_T\right)\frac{2}{\xi}-\frac{1}{\xi^2}\frac{GM_\Theta}{C_s^2\lambda_J}. \tag{42}$$

The quantity $a_0=GM_\Theta/c_s^2\lambda_J$ (*normalization coefficient*) is treated as a free parameter, which eventually provides a way to estimate the SWP electron temperature. The value of this parameter is determined by the condition that a transonic solution for the SWP exists for a given set of initial values of the required physical variables. The above equation can now be rewritten as

$$\left[M^2-(1+\varepsilon_T)\right]\frac{1}{M}\frac{dM}{d\xi}=(1+\varepsilon_T)\frac{2}{\xi}-\frac{a_0}{\xi^2}. \tag{43}$$

5.4.2 Numerical results

Equations (41) and (43) can be solved by numerical methods (by Runge-Kutta IV method) to determine the time independent $M-$ and $\theta-$profiles associated with the SWP for some arbitrary values of a_0. However, we choose the minimum value of a_0 that yields transonic transition solutions. It is obvious from equation (43) that the critical distance will exist at $\xi=\xi_c=a_0/2(1+\varepsilon_T)$. This critical distance corresponds to $\sim 14R_\Theta$ from the defined solar surface. As shown in Fig. 10, the critical point for transonic transition, indeed, exists for a_0 = 95 for narrow range variation of $\varepsilon_T=0.0-0.1$, for the already derived solar surface values of $M_{SSB}=10^{-7}$ & $\theta_\Theta=-1.0$ as a set of initial values. The *M*-values at a distance of 1AU from the defined SSB, i.e., at $\xi\sim 750$ are about 3.3 and 3.5 for $\varepsilon_T=0.0$ & 0.1, respectively, as shown in Fig. 11a. The corresponding values of the electrostatic potential at the same distance are found to be $\theta=-31$ & -30 for $\varepsilon_T=0.0$ & 0.1, respectively as shown in Fig. 11b. This is to note that for higher values of ε_T, solar breeze solutions are obtained.
Substitution of $\lambda_J=R_\Theta/3.5$ in the defined expression of a_0 = 95, we can estimate c_s ~100 *km/sec* and T_e ~100 *eV* for the SWP. The critical distance for transition behavior to occur for M_{SSB} ~10^{-7} (Fig. 11a) as an initial value for Mach number exists at about $14R_\Theta$ distance apart from the defined solar surface. This is to note that if we consider M_{SSB} ~10^{-6} as an initial value for the numerical solution of equations (41) and (43), the transonic transition occurs for a_0 = 71 that yields almost the same values of c_s ~100 *km/sec* and T_e ~100 *eV* for the SWP. But the critical transition location point exists at about $10R_\Theta$ distance apart from the defined solar surface. This implies the initial value of M_{SSB} plays an important role in the proper fixation of a_0 that determines the exact location of transonic transition point and the SWP-property. Accordingly, the speed of the SWP at 1 AU comes out to be 330-350 km s $^{-1}$, as shown in Fig. 11a (dotted vertical line).
Let us now look at Fig. 11b which the electrostatic potential's profile for a predetermined set of initial values of $M_{SSB}=10^{-7}$, and $\theta_\Theta\sim -1$ at the SSB, as in the case of Fig. 11a. It can be seen that the normalized value of the SWP-associated electrostatic potential at 1 AU is about -30 to -31 for $\varepsilon_T=0.0-0.1$. With some simple calculations, as illustrated in the next subsection, we can argue that beyond the transonic transition, the SWP seems to obey the zero-electric current approximation, but not before. This is inferred from the floating surface condition, which is defined by the equalization of escaping flux of the SWP particles in accordance with the law of conservation of particle flux.

5.4.3 Theoretical estimation of floating potential

In absence of any particle source and/or sink of a stellar origin under spherical geometry approximation, we get an expression for the steady state mass conservation of the SWP flow

$$r^2 n_{iT} v_i = r_{SSB}^2 n_0 v_{SSB}. \tag{44}$$

In normalized form the above expression (44) for $N_i=n_i/n_0=1$, can be written as

$$M = \left(\xi_{SSB}^2 / \xi^2 \right) M_{SSB}. \tag{45}$$

Now using normal practice for floating potential estimation under net zero-electric current approximation, i.e. $J_e = J_i$,, one gets

$$e^{\theta_f} \sqrt{(m_i/m_e)} = M = \left(\xi_{SSB}^2 / \xi^2 \right) M_{SSB}. \tag{46}$$

Now, from equation (47) the normalized floating potential at any normalized radial position from the SSB can be expressed as

$$\theta_f = \log \left[\sqrt{\frac{m_e}{m_i}} \left(\frac{\xi_{SSB}}{\xi} \right)^2 M_{SSB} \right]. \tag{47}$$

By simple calculations, one can generate the following comparative data of theoretical estimation of the SWP floating potential (using above expression (47)) at different distances from the obtained SSB as follows.

ξ	θ_f
3.50 (at ξ_Θ)	-19.57
47.50 (at ξ_c)	-24.78
100	-26.27
200	-27.66
300	-28.47
400	-29.04
500	-29.49
600	-29.86
700	-30.16
750 (1 AU)	-30.30

Table 2. Values of the Floating Potential

It looks as if the SSB was in non-floating condition as because it does not acquire floating potential during evolution of the GES-potential distribution of the SIP. However, beyond the critical distance and up to a distance of 1 *AU*, the calculated values of the floating potential almost match with those of the SWP obtained numerically (Fig. 11b). This implies that a finite divergence-free electric current exists at the SSB up to the transonic transition region! Beyond the transonic point, zero electric current approximation seems to hold good.

It is commented that the zero-electric current approximation at the SSB assumed in previous model calculations [3, 11, 31-41] for the qualitative description of the SWP properties seems to be physically unjustified. Furthermore, our model calculation does not require outside imposition of the electric field to ensure the validity of the zero-electric current approximation at the SSB. Probably the imposition of the zero-electric current approximation is not suitable for proper description of the SWP properties. Now the natural question may arise, "What happens to the SWP current after the transonic transition?" It seems the current dissipates mainly through a channel of inertial resistance of the plasma ions due to solar gravity.

5.5 Physical consequences

5.5.1 Description of numerical results

The proposed GES-model predicts that the GES formation (of the SIP origin) drives the subsonic SWP at the solar surface. The quasi-hydrostatic equilibrium defines the solar boundary and ensures the GES formation. Numerically $\theta_\Theta \sim -1$, M_{SSB} ~10^{-7}, and $g_\Theta \approx |d\theta/d\xi| \sim 0.60$ prescribe the defined solar boundary (Table I). It requires specific initial values $\theta_i = -0.001 \,\&\, \xi_i = 0.01$ in the solar interior for $\varepsilon_T = 0.4$.

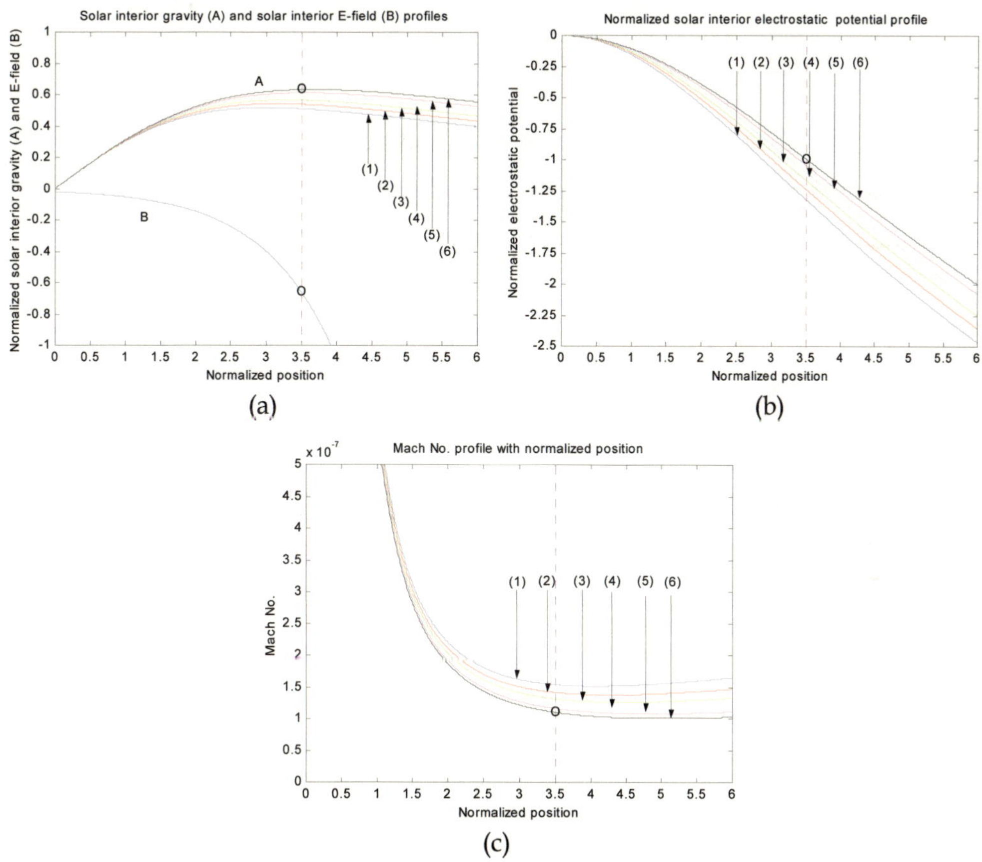

Fig. 8. Variation of normalized values of (a) solar interior gravity $d\eta/d\xi$ (upper group of curves) and electric field $d\theta/d\xi$ (lower curve), (b) electrostatic potential θ, and (c) speed M associated with solar interior plasma flow dynamics with normalized position (ξ) from the heliocenter $(\xi = 0)$. The values of initial position $\xi_i = 0.01$ and initial electrostatic potential $\theta_i = -0.001$ are held fixed. The lines correspond to the cases ε_T = 0.0 (graph 1), 0.1 (graph 2), 0.2 (graph 3), 0.3 (graph 4), 0.4 (graph 5), and 0.5 (graph 6) respectively. The defined solar surface boundary lies at a radial position $\xi_\Theta \sim 3.5$ (implying $R_\Theta \sim 3.5\lambda_J$) with circled points corresponding to the solar surface values

Item	Parker's model	GES Model
1.	Deals with an unbounded solution of steady state hydrodynamic equilibrium of the solar wind (SW)	Deals with bounded (SIP) and unbounded (SWP) solutions of a continuum steady state hydrodynamic equilibrium
2.	Considers a single neutral fluid (gas) model approximation for the SW gas flow dynamics	Considers a two-fluid ideal plasma (gas) model for the SIP and SWP gas flow
3.	Predicts an unbounded solution of supersonic expansion of the SW provided that a sub-sonic flow pre-exists at the SSB	Predicts a bounded solution of the SIP mass distribution with its subsonic outflow at the SSB
4.	Genesis of the subsonic solar surface origin of the SW is not precisely known: discusses the acceleration of the SW by analogy with the de Laval nozzle	Discusses the genesis of the subsonic SSB origin of the SIP in terms of the basic principles of the GES acceleration of ions: the transonic acceleration mechanism of the SWP is the same as Parker's
5.	Does not specify precisely the SW-base definition and prescription for the self-consistent SSB	Offers a precise definition of and prescription for a self-consistent SSB
6.	Standard solar surface is electrically uncharged and unbiased	SSB acquires a negative electrostatic potential (~1 kV) at the cost of thermal loss of the electrons
7.	Does not consider plasma-boundary wall interaction, plasma sheath formation and spontaneous thermal leakage through squeezing mechanism	Considers it
8.	Concept of floating surface (at which no net electric current) is not involved	It is involved
9.	Considers one-scale (SW) theoretical description	Considers two-scale (SIP and SWP) theoretical description
10.	Extensive research has already been done on the SW acceleration and heating	Opens a new chapter of the GES-based theory for interior (bounded) and exterior (unbounded) solar plasma flow dynamics

Table 3. Parker versus GES Model

The GES-formation occurs due to solar surface leakage of the thermal electrons of solar interior plasma outgoing radially outwards. It causes an appreciable space charge polarization effect near the boundary. The depth of the electrostatic potential well for the plasma ions, so formed, is such as to allow the incoming ions from the solar interior bulk plasma to acquire the kinetic energy of ion motion to overcome the maximum gravitational potential barrier height near the boundary. The SIP ions come out of the solar gravitational barrier with a minimum speed M_{SSB} ~10^{-7}. From the floating potential calculation with no net current flow, we infer that the solar surface boundary drives out some finite electric current in the outward flow. That is, it seems a finite electric current loss of the SIP occurs through its leakage process near the SSB! It can be shown that the total surface charge in the boundary, however, comes out to be about 10^{20} times the electronic charge. Table III gives a glimpse of distinction between Parker's model and GES-model of the subsonic SWP origin and its acceleration from subsonic to supersonic flow speed.

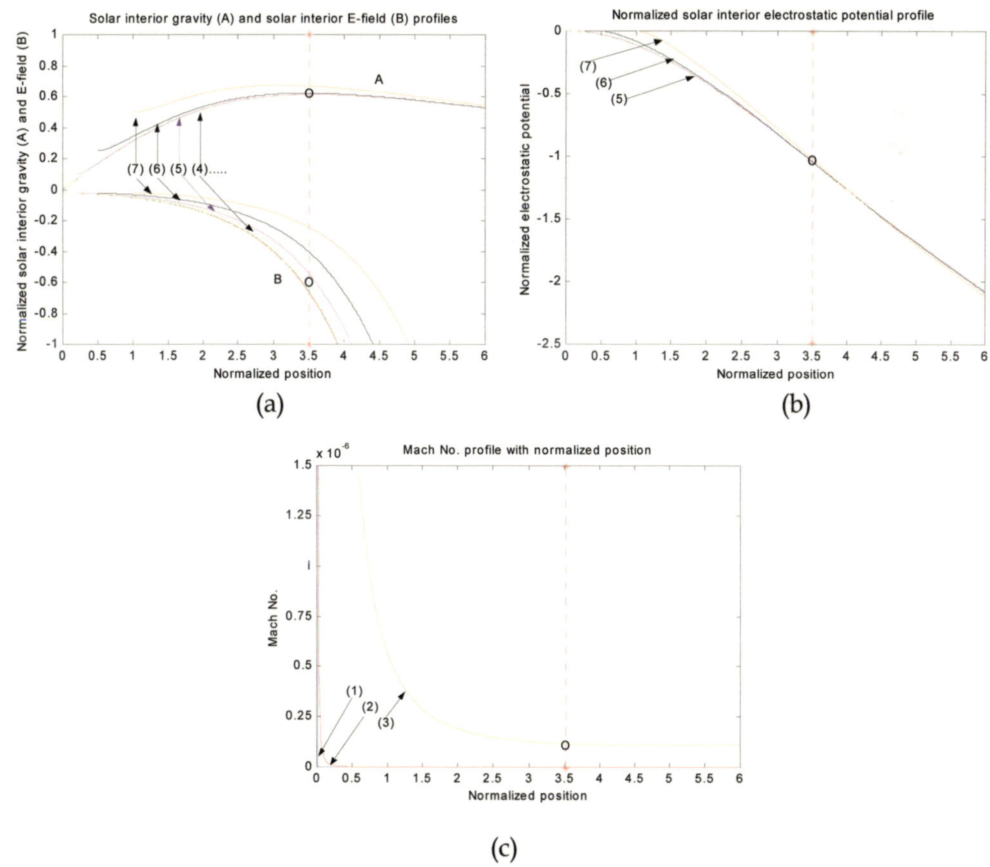

Fig. 9. Same as Fig. 8 but with the ion-to-electron temperature ratio $\varepsilon_T = 0.40$ and electrostatic potential $\theta_i = -0.001$ held fixed. The lines correspond to the cases with initial positions $\xi_i = 10^{-4}$ (graph 1), 10^{-3} (graph 2), 10^{-2} (graph 3), 10^{-1} (graph 4), 0.2 (graph 5), 0.5 (graph 6), and 1.0 (graph 7), respectively. The circled points indicate the most suitable choice of the solar surface values

This is to clarify that the GES-model is a quite simplified one in the sense that it does not include any role of magnetic forces, interplanetary medium or any other complications like rotations, viscosity, etc. It opens a new chapter for further study on the coupled system of the solar interior and exterior plasma flow dynamics.

According to GES-model, the normal acoustic modes of the global solar surface oscillations can be analyzed in terms of the local and global gravito-electrostatic plasma sheath oscillations governed by the basic principles of linear and nonlinear nonlocal theory of the Jeans collapse model [24-25] of charged dust clouds in plasma environment.

The magnified view of the Mach number variation in *transonic transition zone* of the SWP (Fig. 11c) indicates the existence of an extended region having almost uniform sonic flow speed. It can be deduced from Fig. 11c that the transonic point does not always coincide with the critical point. We define the critical point as a radial point (away from that defined solar surface) where the net force on the SWP ions, due to GES-induced *E*-field and external gravity due to total solar interior plasma mass, becomes almost zero.

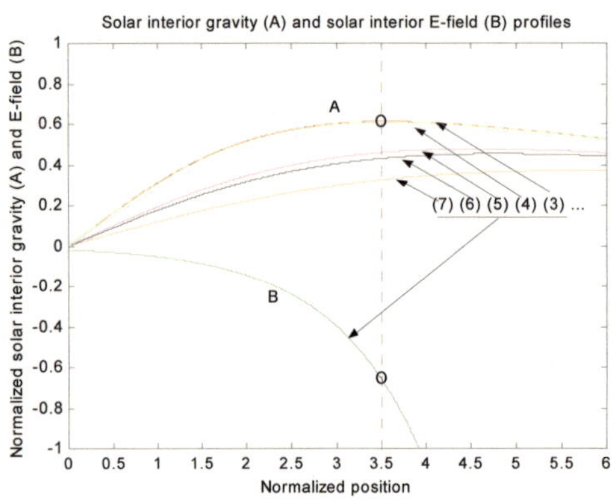

(a)

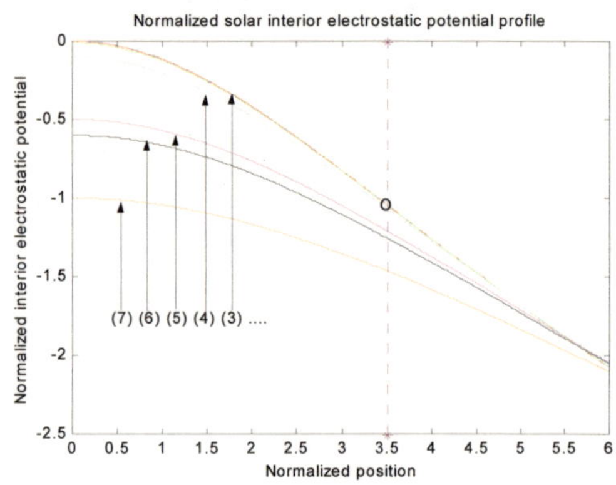

(b)

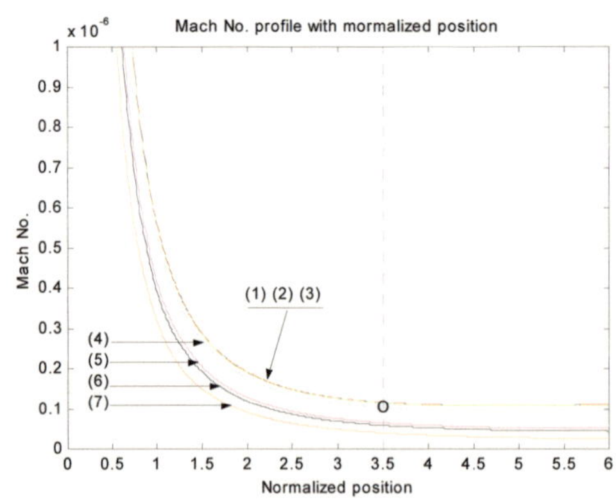

(c)

Fig. 10. Same as Fig. 8 but with the initial position $\xi_i = 0.01$ and ion-to-electron temperatue ratio $\varepsilon_T = 0.40$ held fixed. The lines correspond to the cases of θ_i = 0.0 (graph 1), -0.001 (graph 2), -0.01 (graph 3), -0.1 (graph 4), -0.5 (graph 5), -0.6 (graph 6), and –1.0 (graph 7), respectively. The circled points indicate the most suitable choice of the solar surface values

Numerical solution in the GES-model reproduces the Parker model values of the SWP speed at 1 AU (Fig. 11a) for the numerically predetermined set of initial values of M_{SSB} ~10^{-7} and SWP ion-to-electron temperature ratio $\varepsilon_T = 0.0 - 0.1$. The estimated critical point for the transonic transition to occur $(r_c \sim 14R_\Theta)$ differs from that $(r_c \sim 10R_\Theta)$ in Parker's model. We find that the latter can be obtained with a choice of M_{SSB} ~10^{-6} (or larger than this by orders of magnitude) as an initial Mach value at the solar surface.

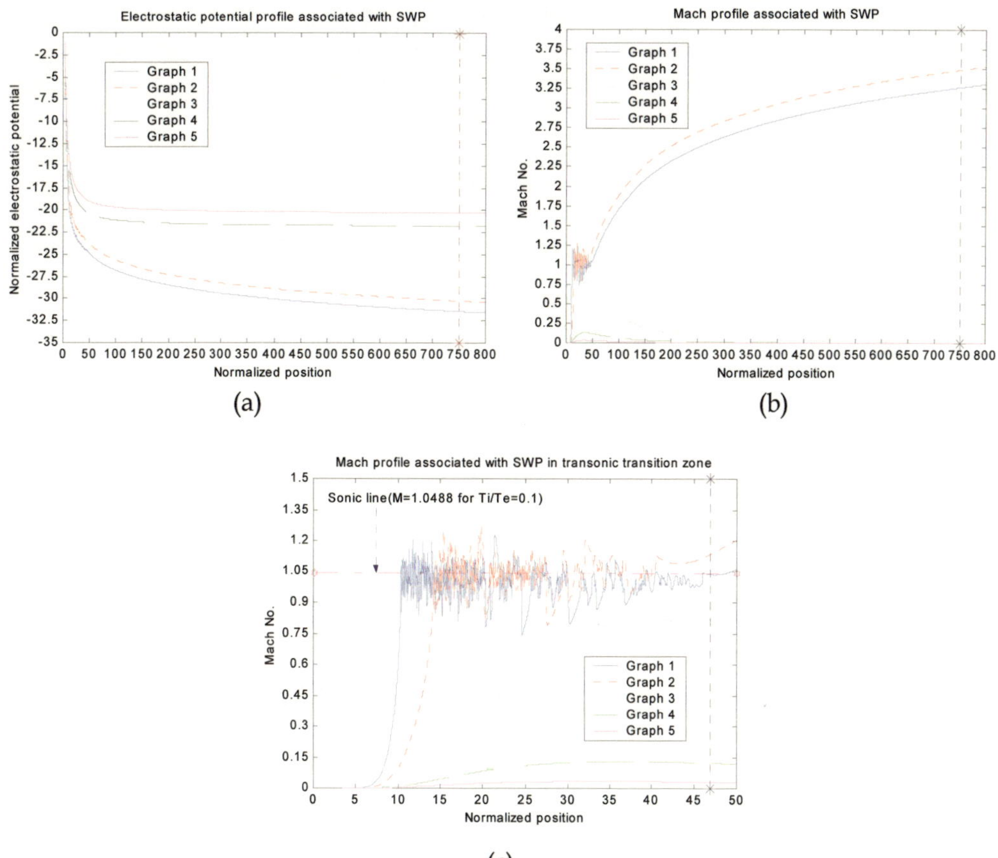

Fig. 11. Variation of normalized values of (a) speed M, (b) electrostatic potential θ, and (c) speed Min the transonic transition zone associated with SWP flow dynamics with respect to normalized position (ξ) from the solar surface boundary $(\xi_\Theta = 3.5)$ in magnified form. The predetermined solar surface boundary parameter values of $M_{SSB} \sim 10^{-7}$, $\theta_\Theta \sim -1.0$ and $a_0 = GM_\Theta / c_s^2 \lambda_J = 95$ are considered as the set of initial values. The lines correspond to the cases of ε_T = 0.0 (graph 1), 0.1 (graph 2), 0.2 (graph 3), 0.3 (graph 4), and 0.4 (graph 5), respectively. The critical distance lies at $\xi_c \cong 47.5$, which corresponds to a radial position of $r \sim 14R_\Theta$ from the solar surface boundary

This is to clarify that the GES-model is a quite simplified one in the sense that it does not include any role of magnetic forces, interplanetary medium or any other complications like rotations, viscosity, etc. It opens a new chapter for further study on the coupled system of the solar interior and exterior plasma flow dynamics.

According to GES-model, the normal acoustic modes of the global solar surface oscillations can be analyzed in terms of the local and global gravito-electrostatic plasma sheath oscillations governed by the basic principles of linear and nonlinear nonlocal theory of the Jeans collapse model [24-25] of charged dust clouds in plasma environment.

The magnified view of the Mach number variation in *transonic transition zone* of the SWP (Fig. 11c) indicates the existence of an extended region having almost uniform sonic flow speed. It can be deduced from Fig. 11c that the transonic point does not always coincide with the critical point. We define the critical point as a radial point (away from that defined solar surface) where the net force on the SWP ions, due to GES-induced *E*-field and external gravity due to the total SIP mass, becomes almost zero.

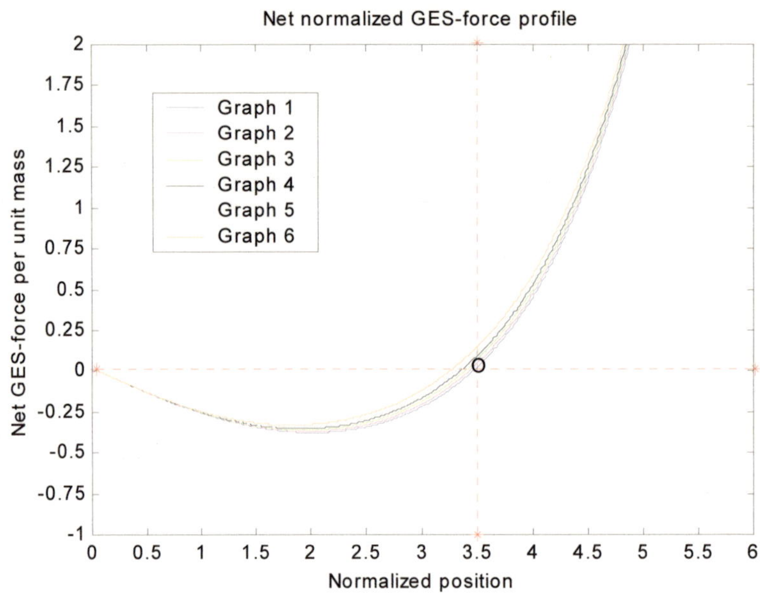

Fig. 12. Variation of net normalized GES force per unit mass associated with solar interior plasma flow dynamics with respect to normalized position (ξ) from the heliocentre $(\xi = 0)$. The initial position $\xi_i = 0.01$ and initial electrostatic potential $\theta_i = -0.001$ are held fixed. The lines correspond to the cases of ε_T = 0.5 (graph 1), 0.4 (graph 2), 0.3 (graph 3), 0.2 (graph 4), 0.1 (graph 5), and 0.0 (graph 6), respectively. The defined solar surface boundary is found to lie at a radial position $\xi_\Theta \sim 3.5$ for a more suitable choice of ε_T = 0.4-0.5 variations

Now, one can see that the sonic point for $\varepsilon_T = 0.0$ (graph 1 in Fig. 11c) falls around $\xi \cong 10$ whereas the transonic point for the same value of $\varepsilon_T = 0.0$ falls around the critical point $\xi_c \cong 47.5$. Similarly, one can see that the sonic point for $\varepsilon_T = 0.1$ (graph 2 in Fig. 11c) falls around $\xi \cong 15$ whereas the transonic point for the same value of $\varepsilon_T = 0.1$ falls around the non-critical point $\xi \cong 50$. From these numerical observations of the transonic transition region, one can clearly notice that an extended zone of about $40\lambda_J$ exists having almost a uniform sonic flow speed of the SWP between sonic and transonic points.

In this region, the inertia-induced acoustic excitation theory [5-12] may have potential applications provided it is improved with a proper inclusion of the solar gravity under nonlocal normal-mode analysis. Similar situations are likely to arise in laboratory plasmas of gravitationally sensitive multi-ion colloidal plasma systems [14-15, 24-25].
This is to point out that the intense acoustic fluctuations appearing in the Mach-profile (Fig. 11c) are merely the results of a numerical instability arising due to the mathematically indeterminate situations localized mainly near the sonic speed. These fluctuations, however, are found to disappear beyond the critical distance. Physically, however, the physical appearance of such indeterminate structures in the graphical plots is because of some chaotic interference and intermittency of acoustic background fluctuations in the emitted SWP. It may produce some dissipative effects in course of the electrodynamical process of the electrodynamical interaction of the SWP particles with background particles of ambient stellar atmosphere.

5.6 Comparison with exospheric model

Exospheric model [38-39] is a simple kinetic model for the solar coronal plasma expansion. This model assumes that beyond a given altitude termed as the *'exobase'* from the SSB, binary collisions between the SWP-particles are negligible. The coronal plasma expansion is believed to occur due to thermal evaporation of the hot protons that have velocity exceeding the escape velocity so as to cross over the barrier of the external solar gravitational field. The generalized model [19-23] considers the non-Maxwellian velocity distribution function for the coronal plasma electrons.
Of course, this model has succeeded to explain the observation of the high speed SWP. This model indeed explains the high speed SWP without requiring any additional source to heat the coronal plasma electrons. In this model, the exobase is defined by the condition that the density scale length equals the mean free path of the SWP particles. Due to the complexities of coronal based physics and multiplicities of plasma species different exobases are likely to coexist. Moreover, the appropriate electrostatic potential is determined by applying the approximations of both local quasi-neutrality and zero current. Of course, this model has succeeded in explaining the observation of the high-speed SWP. But according to our model calculations, the zero-current approximation of the exospheric model seems to be valid only on large scales but not near and above the SSB.
By our GES-model analyses, a finite electron-dominated [10-11] current with a positive finite divergence exists on the solar interior scale for $d\theta/d\xi < 0$. Immediately after the SSB, i.e., on the unbounded scale of the SWP, a divergence-free current exists. This seems to exhibit a discontinuous behavior. How to resolve this? In reality electron temperature has variable profiles on both the bounded and unbounded scales. Probably a self-consistent profile of two distinct electron temperatures on two regions of bounded and unbounded scales separately may resolve the interfacial transition problem of the proposed two-scale theory of the GES-associated solar plasma current system.
According to our GES theory and model calculations, the zero-electric current approximation of exospheric model calculations [38-39] requires further review. The appropriate electrostatic potential estimate from numerical analysis emphasizes that the zero electric current approximation is an outcome of the GES model on the large scale of the SWP. Now the question may naturally arise, "What happens to the SWP current after the transonic transition?" It seems that the electron-dominated electric current dissipates mainly through a channel of inertial resistance of the plasma ions due to solar gravity as a barrier.

The other dissipation channels of the electric current may be through the SWP heating generation of fluctuations in thermal noise level, etc. The uniform flow region of the SWP is, in addition, found to have a large number of conservation rules [11] under the lowest order inertial correction of the thermal electrons in the solar plasma system approximately from applied mathematical point of view. The details of the associated physical mechanism and fluctuations will be communicated to somewhere else.

5.7 Comments

Before we conclude with any physical comment, we must admit that the neglect of collisional dissipation and deviation from a Maxwellian velocity distribution of the plasma particles is not quite realistic. But our GES model under these simple approximations may provide quite interesting results. For example, it provides deep physical insight into the interconnection between the Sun and the SWP. The violation of the zero-current approximation is indeed noted in the neighborhood of the SSB and above. Of course, the zero current approximation seems to be satisfied beyond the transonic region. This conclusion is based on the well-known condition of the floating surface boundary in basic plasma sheath physics.

An estimated value $\lambda_{De}/\lambda_J \sim 10^{-20}$ of the ratio of the solar plasma Debye length and the Jeans length of the total solar mass justifies the quasi-neutral behavior of the solar plasma on both the bounded and unbounded scales.

Applying the spherical capacitor charging model, the coulomb charge on the SWP at a distance of ~ 1 AU comes out to be . For rotation frequency of the solar plasma system corresponding to the mean angular frequency about the centre of the system (Gunn 1931), the mean value of the strength of the solar magnetic field associated with the SWP in our model analysis is estimated as $\langle |B_{SWP}| \rangle = 4\pi^2 Q_{SWP} f_{SWP} \sim 3.01 \times 10^{11} T$. This is obviously considerably higher for producing any significant effects on the dynamics of the SWP particles. Thus the effects of the magnetic field are not ignorable for the SWP particles due to the significantly strong Lorentz force, which is now estimated to be $F_L = e(v_0 \times B_{SWP}) \approx 1.64 \times 10^{-2} N$ corresponding to a supersonic flow speed $v_0 \sim 340.00\ km\, s^{-1}$. Thus the Lorentz force may have some remarkable effect on the SWP particles and hence, may not justifiably neglected for the unbounded scale description. It justifies the convective and circulation dynamics to be considered in that context. Therefore our unmagnetized plasma approximation may not prove well justified in our GES model configuration for the SWP flow dynamics description. Although collision processes are dominant in the realistic solar interior [2, 39-44], collisionless models [2, 39] are also equally useful for the solar plasma description. Thus our collisionless model approximation for mathematical simplicity may be justified here. In our GES model, the calculated values of the mean free paths for the solar plasma electrons, $\lambda_e \sim 1.50 \times 10^{198} m$ and for ions, $\lambda_i \sim 3.05 \times 10^{132} m$ justify the collisionless model approximation. This approximation holds good justifiably under the fulfillment of the validity condition $\lambda_e, \lambda_i >> \lambda_J$.

One can note that the SIP electron temperature, specified by T_{e1}, differs from (exceeds) the SWP electron temperature specified T_{e2} by one order of magnitude. This is discussed already discussed above. It simply means the SWP has been relatively cooled. It is quite natural for expanding plasma gas to be cool. This is to further comment that these two different electron temperatures are considered constant over their respective scales.

Actually, a discontinuity exists at the interface of the bounded and unbounded scales. This is an open problem to resolve.
Let us clarify once again that equalizing the maximum value of the numerically determined solar self-gravity with the standard value specifies the SIP electron temperature. Similarly, the appropriate choice of the defined constant a_0 specifies the SWP electron temperature, which ensures that a transonic solution of the SWP dynamics exists. Now, with this simplified treatment our theoretical model calculations yield the following conclusions.

1. Contrary to the general belief that the SWP emerges from the SSB, our theory provokes us to argue that the genesis of the subsonic origin of the SWP at the SSB in fact lies in the SIP dynamics. It is governed by the basic principles of the GES formation near the SSB and beyond. The surface boundary is located at a radial distance defined by $\xi \sim 3.5$ (Figs. 8–10) from the heliocentric origin. This specific location in the plots (Figs. 8a, 9a, and 10a) is marked by a vertical line with small circles.
2. Thereafter, the outward moving SIP forms the SWP with a highly subsonic speed at the SSB. Initially the outward acceleration of the SWP is quite rapid allowing a *transonic transition* solution to exist for a specific choice of $\varepsilon_T = 0.0 - 0.1$ (Fig. 11a). This occurs as a consequence of the predominantly self-consistent electric field associated with the SWP (Fig. 11b). It produces a *transonic transition region* of sufficient length scale with the critical point lying at about $14R_{\Theta}$ (Fig. 11c) from the SSB.
3. It is noted that initially the gravitational potential barrier decelerates the SIP dynamics rapidly. As soon as the *E*-field of the SIP origin gathers sufficient strength, an outward flow occurs with a reduced minimum speed of $M_{SSB} \sim 10^{-7}$ (Figs. 8c, 9c, 10c) at the SSB defined by the quasi-hydrostatic equilibrium condition at a point of the maximum solar gravity, as clearly depicted in Fig. 12. This figure clearly shows the strong solar self-gravity up to the solar boundary and relatively weaker strength of the solar external gravity beyond the boundary.
4. According to our model calculations, the SSB behaves as a negatively biased grid with a bias potential of about 1 kV. The surface draws a finite current dominated by the thermal electrons and flows towards the surface. As a result, the solar surface oscillations may naturally be attributed to the resulting consequences of the GES oscillations. Under the neutral ideal gas approximation of the SIP, this property cannot be deduced.
5. We therefore conclude that our GES-based model may be useful to study the properties of the SSB and the properties of the slow speed SWP. Of course, the properties of the high speed SWP description under our model will require a kinetic treatment as already reported by previous workers in the case of the exospheric model.

A few more reminders are in order:

1. The exact location of the SSB and that of 1 AU distance as specified in Figs. 8-12 on the normalized scale are estimated for the normalization factor, which is, decided by the SIP parameters.
2. In the absence of magnetic field in our model approach, the Lorenz force term is absent, but it will be needed for further improvements under the fluid and/or kinetic regime to see the realistic dynamics of the solar plasma system. However, the estimated mean value of the solar magnetic field $\langle |B_{SIP}| \rangle \sim 7.53 \times 10^{-11}\, T$ in the SIP justifies and supports our unmagnetized plasma approximation in the present context.

3. The GES-model is a useful theoretical construct with which to study SWP dynamics in terms of solar interior dynamical behavior (generator of the SWP) through active dynamical coupling processes of solar exterior regions in the light of localized electric space charge effects.

Finally, it is important to comment that the further improvements and modifications to the model will be needed to make it more realistic for actual SWP conditions. These form the basic problems of future research on the GES model. The genesis of the SWP is now found to be associated with the coupling of the SIP potential and self-gravitational potential of the SIP mass. We finally argue that the lines of communications should be kept open between theorists and observers and solar and stellar physicists, and more importantly also between the solar and plasma physics communities, in order to further the study of stellar wind plasmas. Ours is a first step, albeit very simplified and external-field free and ideal, in this particular direction. We have tried to provide an integrated theoretical outlook on the SIP dynamics on the bounded scale, and SWP on the unbounded scale. This model could further be useful to study the properties of the helio-seismic dynamics of the Sun and other stars [36-37] too.

5.8 Overall summary

The presented chapter reviews the latest findings of normal acoustic mode analyses through different types of transonic plasma equilibrium models [5-12] under the lowest order inertial correction of plasma thermal species. Different types of acoustic resonances are observed in transonic plasma equilibria depending on different plasma inertial ions. The linear analyses show the graphical nature of the associated resonance poles. This implies that transonic plasma is an unstable zone, which is rich in wide range spectra of acoustic wave fluctuations. The acoustic wave kinetics in the nonlinear normal mode analyses in different types of plasmas [8-9, 12, 26] is describable by a linear source driven KdV (*d*-KdV) equation. After integration, it shows two distinct classes of soluations, i. e., solitons and oscillatory shocks. The fundamental condition to observe inertia-induced (ion) acoustic wave resonant excitation is that the ion flow speed must be uniform. Accordingly, the same applies to the solar wind dynamics [10-11, 35-41] in self-gravitating plasma systems as well. A large number of conservation laws of applied mathematical significance associated with the *d*-KdV flow dynamics are also pointed out [9] in transonic plasma domain in different situations including solar plasma. Of course, convective and circulation dynamics which are the primary sources of magnetic field [41], are neglected throughout for simplicity. Similar observations of acoustic kinetics of the formation of soliton-type structures are also found in self-gravitating dust molecular clouds in presence of partially ionized dust grains through the active mechanism of gravito-acoustic coupling processes [27]. Some future scopes including realistic sources of acoustic perturbation of the presented analyses are also pointed out in brief.

Very similar to Geoseismology dealing with the Earth's interior through the various seismic (acoustic) waves produced during the earthquakes, Helioseismology is the study of the various linear and nonlinear surface waves and oscillations of solar origin (like *p-mode*, and *f-mode*) to measure the internal structure and dynamics of the Sun [36-44]. The acoustic dynamics in the Sun (or Star) is understandable by considering it to have a resonant cavity like an organ pipe in which acoustic waves are trapped (by reflections or refractions) [41]. One of the earliest studies of solar oscillations and fluctuations established that the power spectrum of the Sun's full disk contained a multitude of Doppler shift peaks between 2.5 mHz - 4.5 mHz [36-37 and references therein]. The Global Oscillation Network Group

(GONG), Stellar Observations Network Group (SONG), Helio- and Asteroseismology (HELAS) Network, and Birmingham Solar Oscillations Network (BiSON) are examples of recent studies being undertaken to measure these surface oscillations through space and ground based remote-sensing observations [36-37, 44]. Michelson Doppler Imager (MDI) onboard Solar and Heliospheric Observatory (SOHO) and recently, Helioseismic Magnetic Imager (HMI) onboard Solar Dynamics Observatory (SDO) also measure these oscillations from space [36-37, 41, 44]. Significant power has been observed at frequencies ranging from 1.4 mHz to 5.6 mHz, corresponding to periods of 3 to 12 minutes. They are called *'5 minute oscillations'* due to their dominant mean period [44]. Besides, the behavior of the solar intermediate-degree modes (during extended minimum) is also investigated to explore the time-varying solar interior dynamics with the help of contemporaneous helioseismic GONG and MDI data [44]. The basic physics behind these helioseismic and helioacoustic observations (*in situ*) reported in the literature, however, needs to be more clearly understood in a broader horizon. Moreover, there are many more experimental observations [3, 35-44] on seismic activities that will require self-gravitating plasma wave theory for further development of our stability analyses and seismic diagnostics. In conclusion, we strongly believe that the presented mathematical strategies and techniques of linear and nonlinear acoustic mode analyses amidst more realistic plasma-boundary interaction processes may have some potential applications in such future helio- and astero-seismic directions.

6. Acknowledgements

I am thankful to Ms. Sandra Bakic, Publishing Process Manager, InTech - Open Access Publisher, for the invitation, chapter proposal and continuous cooperation. I am also thankful to each and everybody of the InTech family whoever involved for extending cooperation. Lastly, I gratefully recognize the Intech Editorial Board for giving me this rare opportunity to publish this chapter in the book "Acoustic Waves – From Microdevices to Helioseismology" without any article processing charge.

7. References

[1] B. W. Tillery, *Physical Sciences* (8th Edition, McGraw-Hill 2009), Ch. 5, pp. 115-134.
[2] F. F. Chen, *Introduction to Plasma Physics and Controlled Fusion* (Plenum Press, New York and London, 1988), Ch.1, pp. 1-17.
[3] S. R. Cranmer, *Space. Sci. Rev.* 101, 229 (2002).
[4] M. Lakshamanan and S. Rajasekhar, *Nonlinear Dynamics: Integrability, Chaos, and Patterns* (Springer-Verlag Heidelberg 2003), Ch. 11-14, pp. 341-454.
[5] C. B. Dwivedi and R. Prakash, *J. Appl. Phys.* 90, 3200 (2001).
[6] P. K. Karmakar, U. Deka and C. B. Dwivedi, *Phys. Plasmas* 12, 032105 (2005).
[7] P. K. Karmakar, U. Deka and C. B. Dwivedi, *Phys. Plasmas* 13, 104702 (1) (2006).
[8] P. K. Karmakar and C.B. Dwivedi, *J. Math. Phys.* 47, 032901(1) (2006).
[9] P. K. Karmakar, *J. Phys.: Conf. Ser.* 208, 012059(1) (2010).
[10] C. B. Dwivedi, P.K. Karmakar and S.C. Tripathy, *Astrophys. J.* 663 (2), 1340 (2007).
[11] P. K. Karmakar, *J. Phys.: Conf. Ser.* 208, 012072(1) (2010).
[12] U. Deka, A. Sarma, R. Prakash, P. K. Karmakar and C. B. Dwivedi, *Phys. Scr.* 69, 303 (2004).
[13] P. Mora, *Phys. Rev. Lett.*, 90, 185002-1 (2003).

[14] C. B. Dwivedi, *Pramana-J. Phys.* 55, 843 (2000).
[15] C. B. Dwivedi, *Phys. Plasmas* 6, 31 (1999).
[16] M. S. Sodha and S. Guha, *Advances in Plasma Physics*, edited by A. Simon and W. B. Thompson (Wiley, New York, 1971) vol. 4.
[17] C. K. Goertz, *Rev. Geophys.* 27, 271 (1989).
[18] F. Verheest, *Space Sci. Rev.* 77, 267 (1996).
[19] H. Thomas, G. E. Morfill, V. Demmel, J. Goerce, B. Feuerbacher and D. Mohlmann, *Phys. Rev. Lett.* 73, 652 (1994).
[20] G. S. Selwyn, J. E. Heidenreich and K. L. Haller, *Appl. Phys. Lett.* 57, 1876 (1990).
[21] L. Boufendi, A. Bouchoule, P. K. Porteous, J. Ph. Blondeau, A. Plain and C. Laure, *J. Appl. Phys.* 73, 2160 (1993).
[22] A. Gondhalekar, P. C. Stangeby and J. D. Elder, *Nucl. Fusion* 34, 247 (1994).
[23] B. N. Kolbasov, A. B. Kukushkin, V. A. Rantsev Kartinov, and P. V. Romanov, *Phys. Lett. A* 269, 363 (2000).
[24] N. N. Rao, P. K. Shukla and M. Y. Yu, *Planet. Space Sci.* 38, 345 (1990).
[25] P. K. Shukla, *Waves in dusty, solar wind and space plasmas,* (AIP Conference proceedings, Leuven, Belgium) 537, 3 (2000).
[26] P. K. Karmakar, *Pramana- J. Phys.* 68, 631 (2007).
[27] P. K. Karmakar, *Pramana- J. Phys.* 76 (6), 945 (2011).
[28] J. P. Goedbloed, R. Keppens and S. Poedts, *Space Sci. Rev.* 107, 63 (2003).
[29] S. P. Kuo and D. Bivolaru, *Phys. Plasmas* 8 (7), 3258 (2001).
[30] K. Itoh, K. Hallatschek and S-I Itoh, *Plasma Phys. Control. Fusion* 47, 451 (2005).
[31] I. Ballai, *PADEU* 15, 73 (2005).
[32] I. Ballai, E. Forgacs and A. Marcu, *Astron. Nachr.* 328 (8), 734 (2007).
[33] P. K. Shukla and A. A. Mamun, *New J. Phys.* 5, 17.1 (2003).
[34] M. Khan, S. Ghosh, S. Sarkar and M. R. Gupta, *Phys. Scr.* T116, 53 (2005).
[35] M. S. Ruderman, *Phi. Trans. R. Soc. A*, 364, 485 (2006).
[36] W. J. Chaplin, *Astron. Nachr.* 331 (9), 1090 (2010).
[37] J. C. Dalsgaard, *Astron. Nachr.* 331 (9), 866 (2010).
[38] J. Lemaire and V. Pierrard, *Astrophys. Space Sci.* 277, 169 (2001).
[39] E. Marsch, *Liv. Rev. Solar Phys.* 3, 1 (2006) (http: //livingreviews.org/lrsp-2006-1).
[40] U. Narain and P. Ulmschneider, *Space Sci. Rev.* 75, 453 (1996).
[41] A. Nordlund, R. F. Stein and M. Asplund, *Living Rev. Solar Phys.* 6, 2 (2009) (http: //livingreviews.org/lrsp-2009-2).
[42] R. Gunn, *Phys. Rev.* 37, 983 (1931).
[43] V. M. Nakariakov and E. Verwichte, *Living Rev. Solar Phys.* 2, 3 (2005) (http: //livingreviews.org/lrsp-2005-3).
[44] S. C. Tripathy, F. Hill, K. Jain and J. W. Leibacher, *Astrophys. J.* 711, L84 (2010).

Part 2

Acoustic Waves as Investigative Tools

6

Acoustic Waves: A Probe for the Elastic Properties of Films

Marco G. Beghi
Politecnico di Milano, Energy Department and NEMAS Center, Milano Italy

1. Introduction

Films and thin films are exploited by an ever increasing number of technologies. The properties of films can be different from those of the same material in bulk form, and can depend on the preparation process, and on thickness. Specific techniques are needed for their measurement. Whenever films or thin layers have structural functions, as in micro electro-mechanical systems (MEMS), a precise characterization of their stiffness is crucial for the design of devices. The same can be said for the design of devices which exploit thin layers to support surface acoustic waves (surface acoustic wave filters). More generally, knowledge of the elastic properties is interesting because such properties depend on the structural properties.

The most widespread technique for the mechanical characterization of films, instrumented indentation, induces both elastic and inelastic strains. It also characterizes irreversible deformation, but the extraction of the information concerning the elastic behaviour is non trivial. To overcome this difficulty, several measurement methods have been developed, which exploit vibrations as a probe of the material behaviour. These methods intrinsically involve only elastic strains, and are non destructive. This is true at any length scale, and is peculiarly useful at micrometric and sub-micrometric scales, where the exploitation of other types of probes can become critical.

Both propagating waves and standing waves can be exploited, with excitation which can be either monochromatic (e.g. resonance techniques) or impulsive, therefore broadband, requiring an analysis of the response either in the time domain (the so called picoseconds ultrasonics) or in the frequency domain (the so called laser ultrasonics). The propagation velocities of vibrational modes are obtained, from which the stiffness is derived if an independent value of the mass density (the inertia) is available.

Older resonance techniques have been developed to be operated with thin slabs, also exploiting optical measurements of displacement. In the measurement of films and small structures, the advantages of light, a contact-less and inertia-less probe, are substantial, and are increasingly exploited. The laser ultrasonics technique, commercially available since some years ago, measures waves travelling along the film surface. The so called picosecond ultrasonics technique measures waves travelling across the film thickness; it is a relatively sophisticated optical technique, which exploits femtosecond laser pulses in a pump-and-probe measurement scheme. For best performance it needs the deposition of

an interaction layer, possibly combined with microlithography techniques to obtain a patterned layer.
Techniques for optical detection of acoustic vibrations include inelastic scattering of light: Brillouin scattering. Brillouin spectroscopy does not excite vibrations, but relies on thermal excitation, which has the broadest band but small amplitude, and measures the spectrum of inelastically scattered laser light. Brillouin Spectroscopy and Surface Brillouin Spectroscopy are relatively simple optical techniques, which do not require a specific specimen preparation. They operate at sub-micrometric acoustic wavelengths, which are selected by the scattering geometry, and have been exploited to characterize the elastic properties of bulk materials and of films.
An overview of this variety of techniques is presented here, underlining analogies and differences. The increasing demand of precise characterization raised the point of precision and accuracy achievable by vibration based techniques, and specifically by the optical techniques. In the overview, attention is drawn to the steps or the parameters which are the limiting factor for the achievable precision, and to the way of characterizing them. The effects of inaccuracies of the mass density are common to all the techniques based on vibrations, while other sources of uncertainty are more specific to each technique.

2. Acoustic modes in elastic solids

In the continuum description the instantaneous configuration of a solid undergoing deformation can be represented by the displacement vector field $\mathbf{u}(\mathbf{r},t)$, where $\mathbf{u}=(u_1,u_2,u_3)$, $\mathbf{r}=(x_1,x_2,x_3)$ and t is time. The local state of the solid being represented by the strain and stress tensors, the linear elastic behaviour is characterized by the tensor of the elastic constants C_{ijmn}, which can be conveniently represented by the matrix of the elastic constants C_{ij}. When other phenomena, like e.g. viscoelasticity or elasticity of higher orders, can be neglected, the tensor of the elastic constants fully characterizes the stiffness. Inertia is characterized by the mass density ρ. Within a homogeneous linear elastic solid, in the absence of body forces, the equations of motion for the displacement vector field are homogeneous, and read (Auld, 1990; Every, 2001; Kundu, 2004)

$$\rho\frac{\partial^2 u_i}{\partial t^2}=\sum_{j,m,n} C_{ijmn}\frac{\partial^2 u_m}{\partial x_j \partial x_n}, \quad i=1,2,3 \ . \tag{1}$$

These equations describe vibrational elastic excitations, which are typically called acoustic also in the ultrasonic frequency range. The basic solutions of Eqs.(1), and the most important ones when boundary effects are irrelevant, are the plane acoustic waves, or modes (Auld, 1990; Kundu, 2004), of the form

$$\mathbf{u}=\mathbf{e}\Re\left\{\tilde{A}\exp\left[i(\mathbf{k}\cdot\mathbf{r}-\omega t)\right]\right\} \ , \tag{2}$$

where $\mathbf{k}$ is the wavevector, $\omega = 2\pi f$ the circular frequency, f the frequency, $\tilde{A}$ an arbitrary complex amplitude, and $\mathbf{e}$ the polarization vector, which is normalized. The continuum description, underlying Eq. (1), is appropriate until the wavelength $\lambda=2\pi/|\mathbf{k}|$ is much larger than the interatomic distances. The three translational degrees of freedom of each infinitesimal volume element correspond, for each wavevector $\mathbf{k}$, to three independent

modes, having different polarization vectors and frequencies. In general the phase velocity $v = \omega / |\mathbf{k}| = \lambda f$ depends on both the direction of **k** and the polarization vector **e**. In an infinite homogeneous medium, travelling waves of the type given by Eq. (2) exist for any frequency *f* compatible with the mentioned lower limit for wavelength.
In the simplest case, the isotropic solid, the matrix of the elastic constants is fully determined by only two independent quantities; the only non null matrix elements are $C_{11} = C_{22} = C_{33}$, $C_{44} = C_{55} = C_{66}$, $C_{12} = C_{13} = C_{23} = C_{11} - 2C_{44}$. In this case the shear modulus *G* coincides with C_{44}, while Young modulus *E*, Poisson's ratio ν and bulk modulus *B* are respectively given by (Every, 2001; Kundu, 2004)

$$E = \frac{C_{44}(3C_{12} + 2C_{44})}{C_{12} + C_{44}} = \frac{C_{44}(3C_{11} - 4C_{44})}{C_{11} - C_{44}}, \tag{3}$$

$$\nu = \frac{C_{12}}{C_{11} + C_{12}} = \frac{C_{11} - 2C_{44}}{2(C_{11} - C_{44})} = \frac{E}{2G} - 1, \tag{4}$$

$$B = \frac{C_{11} + 2C_{12}}{3} = C_{11} - \frac{4}{3}C_{44}, \tag{5}$$

In the isotropic case the phase velocities are independent from the direction of **k**, only depending on the relative orientation of **e** with respect to **k**; one of the three modes is longitudinal ($\mathbf{e} \parallel \mathbf{k}$) and has velocity v_l, the other two are transversal ($\mathbf{e} \perp \mathbf{k}$), are independent (the two polarization vectors are orthogonal) and degenerate: they have the same velocity v_t (Auld, 1990; Kundu, 2004). The two velocities are

$$v_l = \sqrt{C_{11}/\rho} \quad \text{and} \quad v_t = \sqrt{C_{44}/\rho} \ . \tag{6}$$

In the non isotropic case more than two independent quantities are needed to determine the matrix of the elastic constants, and the phase velocities, beside depending on the direction of **k**, have a more complex dependence on the C_{ij} values.
In a finite geometry the search for standing waves having the harmonic time dependence of the type $e^{-i\omega t}$ transforms Eq. (1) into an eigenfunction / eigenvalue equation of the Helmholtz type (Auld, 1990); an appropriate set of basis functions allows to transform this equation into a matrix eigenvalue problem (Nakamura et al., 2004). The eigenvalues are proportional to ω^2, the square of the frequencies of the acoustic modes of the structure, or natural frequencies of the structure. In other words, the finiteness of the geometry converts the continuum spectrum of frequencies of the modes of the infinite medium, given by Eq. (2), into the discrete spectrum of the natural frequencies. These frequencies depend on the (C_{ij}/ρ) values and on the geometry.
In a schematic way: also in non isotropic media the acoustic velocities depend on stiffness and inertia as in Eqs. (6): $v^2 = C/\rho$, indicating generically by *C* the relevant combination of elastic constants and, possibly, direction cosines of **k**. In the simplest case, the one dimensional geometry of length *L*, the standing waves are identified by the constructive self interference condition $L = n\lambda/2 = (n/2)v/f$ (*n* is an integer number), such that $f = (n/2)\sqrt{C/\rho}/L$. Therefore, a measurement of the frequencies of the acoustic modes

allows to derive $C=\rho(fL)^2/(n/2)^2$. Also in more complex geometries, the dependence is of the same type

$$C=\rho(fL)^2 N \tag{7}$$

where L is now a characteristic length of the structure (for a slender rod, essentially one dimensional, the length), and N is a dimensionless numerical factor which, beside the mode order n, can depend on dimensionless quantities like geometrical aspect ratios or Poisson's ratio. The factor N also depends on the character of the mode whose frequency f is being measured, and therefore on the specific modulus C which is involved.

Structures can be finite in one or two dimensions and practically infinite in others, as it happens e.g. in a slab or a long cylinder. The free surface of an otherwise homogeneous solid is a case of semi-infiniteness along a single dimension. The translational symmetry is broken in the direction perpendicular to the surface, and new phenomena, absent in the infinite medium, are found: the reflection of bulk waves, and the existence of surface acoustic waves. Namely, at a stress free surface Eqs. (1) admit a further solution: the Rayleigh wave, the paradigm of the surface acoustic waves (SAWs). Such waves have peculiar characters (Farnell & Adler, 1972): a displacement field confined in the neighborhood of the surface, with the amplitude which declines with depth, a wavevector parallel to the surface, and a velocity lower than that of any bulk wave, such that the surface wave cannot couple to bulk waves, and does not lose its energy irradiating it towards the bulk. Pseudo surface acoustic waves can also exist, which violate this last condition. The velocity v_R of the Rayleigh wave cannot be given in closed form; in the isotropic case a good approximation is (Farnell & Adler, 1972)

$$v_R \cong v_t \frac{0.862+1.14\nu}{1+\nu} \quad . \tag{8}$$

The continuum model of a homogeneous solid does not contain any intrinsic length scale. Accordingly, all the solutions for this model are non dispersive, meaning that the velocities (Eqs. (6) and (8)) are independent from wavelength (or frequency).

More complex modes occur in non homogeneous media. Layered media are a particularly relevant case, in which new types of acoustic modes can occur; namely, modes confined around the interfaces and modes which are essentially guided by one layer or another, like the Sezawa waves. In this case, also in the continuum model the physical system has an intrinsic length scale, identified by the layer thicknesses. For wavelengths much smaller than the thicknesses wave propagation occurs within each layer as if it was infinite, with reflections and refractions at the surfaces. Instead, for wavelengths comparable to, or larger than, the thicknesses, the acoustic modes extend over several layers, and are modes of the whole structure. Such modes are dispersive: their velocities depend on wavelength, or, more precisely, on the wavelength to thickness ratio(s). Also the simplest surface wave, the Rayleigh wave of a bare homogeneous substrate, is modified by a layer deposited on it, and becomes dispersive: the propagation velocity depends on wavelength, therefore on frequency. The velocities of the acoustic modes in layered structures can be numerically computed, as non trivial functions of the properties of the substrate and the layer(s), and of the wavelength to thickness ratio. The dispersion relations $\omega(k)$ or $v(f)$ are thus obtained.

3. Stiffness measurements

3.1 Vibration based methods

It has always been recognized that since the phase velocities of acoustic waves and the natural frequencies of the acoustic modes depend on stiffness and inertia, their measurement gives access, by Eq.(6) or Eq. (7), to the elastic constants C_{ij}, if the mass density ρ, and possibly the geometry, are known. Many experimental methods have been devised, which exploit vibrations to measure the elastic properties of solids. These methods measure the dynamic, or adiabatic, elastic moduli; these moduli do not coincide with the isothermal moduli which are measured in monotonic tests (if strain rate is not too high), but in elastic solids the difference between adiabatic and isothermal moduli seldom exceeds 1% (Every, 2001). Furthermore, when the elastic constants are needed to design a device which operates dynamically, like most microdevices, the dynamic moduli are exactly the ones which are needed in the design process.

Some methods measure the wave propagation velocity by measuring the transit time over a finite, macroscopic distance, other methods measure the frequency of standing modes defined by the sample geometry, or the frequency of propagating waves of well defined wavelength. The excitation can be either monochromatic, at a frequency which typically should be adjustable until resonance conditions are achieved, or broadband. The latter is typically obtained by an impulsive excitation, which can be provided by a mechanical percussion or by a laser pulse. Generally, the response to a broadband excitation is spectrally analyzed. The availability of ultrafast lasers (femtosecond laser pulses) also allows an analysis in the time domain, by pump-and-probe techniques.

In homogeneous specimens each propagation velocity, or the frequency of each standing wave, has a single value, from which the corresponding elastic modulus can be derived. In non homogeneous specimens, typically in supported films, each propagation velocity can depend on wavelength or frequency. Dispersion relations $\omega(k)$ or $v(\lambda)$ can be measured over a finite interval of frequency or wavelength, and the film properties can be obtained fitting the computed dispersion relations to the measured ones.

3.2 Vibration excitation and detection techniques

The various experimental methods operate in different frequency ranges. The range is determined by both the excitation and the detection techniques, and is strictly correlated to the spatial resolution. It is worth remembering that the acoustic velocities in typical elastic solids like metals and ceramics are of the order of a few km/s = mm×MHz, and that a phase velocity v links the frequency f to a characteristic length L, which can be a characteristic dimension of a structure supporting standing waves, or the wavelength of a travelling wave.

Characteristic lengths of centimetres imply frequencies in the tens of kHz range, which are easily excited by a mechanical percussion and measured by a microphone; Nieves et al. (2000) estimate at around 0.1 MHz the upper limit of the frequencies excited by the mechanical percussion, with a steel ball of a very few millimetres. Piezoelectric actuators, and sensors are also available.

Characteristic lengths of several micrometers correspond to frequencies in the tens of MHz range. Structures of this size can be built by micromachining techniques, and their vibration can be excited and detected by capacitive actuators and sensors. Measurement techniques of this type are essentially a miniaturization of the vibrating reed technique (Kubisztal, 2008). Czaplewski et al. (2005) built flexural and torsional resonators of tetrahedral amorphous carbon (also known as amorphous diamond), by standard techniques for the production of micro-electromechanical systems. They exploited piezo-electric actuation, and an interferometric technique to measure the oscillation. They were able to perform measurements at variable temperatures, determining the elastic moduli of this material as function of temperature. They analyzed the uncertainty sources, finding that the leading contribution to the uncertainty comes, for the flexural oscillator, from the value of the mass density of this material, while for the torsional oscillator it comes from the exact dimensions of the thin member undergoing torsion.

In larger structures, waves at frequencies in the tens of MHz range can be excited and detected by piezoelectric elements, possibly operating simultaneously as actuators and sensors. Excitation can also be performed by a laser pulse; if the pulse is short enough, the upper limit of the measurable frequency range can be set by the piezoelectric sensor. Optical detection techniques are also available. Specific devices like interdigitated transducers (IDTs) can be built by lithographic techniques on, or within, an appropriate layer stack, which must include a piezoelectric layer. Such devices emit and receive waves at the wavelength which resonates with the periodicity of the transducer, typically at micrometric scale. This configuration was exploited to measure the material properties (Bi et al. 2002; Kim et al., 2000), but it is seldom adopted, because it requires the production of a dedicated micro device.

Micrometric and sub-micrometric wavelengths correspond to frequencies in the GHz to tens of GHz range. Detection of such frequencies requires optical techniques; excitation of such frequencies can be obtained by laser pulses of short enough duration.

The variety of vibration based methods to measure the stiffness of solids the can be classified according to various criteria. In this chapter methods are reviewed grouping them by the main vibration excitation techniques: mechanical excitation, either periodic or by percussion, laser pulse excitation, and inelastic light scattering (Brillouin spectroscopy). Similarly to Raman spectroscopy, Brillouin spectroscopy does not excite vibrations at all, but relies on the naturally occurring thermal motion. This gives access to the broadest band, but with small vibration amplitudes, which require time consuming measurements.

3.3 Precision and accuracy

In all the techniques based on vibrations the elastic constants themselves are not the direct outcome of the measurement, but are derived from direct measurements of a primary quantity like frequency or velocity, and 'auxiliary' quantities like thickness, or mass density. The uncertainty to be associated to the resulting value of each elastic constant must be evaluated considering the uncertainties associated to each of the raw measurements. For a quantity q which is derived from directly measured quantities a, b and c, the uncertainty σ_q depends on the 'primary' uncertainties σ_a, σ_b and σ_c. For a functional dependence of the type $f = Aa^\alpha b^\beta c^\gamma$, where A is a numerical constant, the usual error propagation formula can be written in terms of the relative uncertainties (σ_a / a), (σ_b / b), (σ_c / c) as

$$\left(\frac{\sigma_q}{q}\right)^2 = \alpha^2\left(\frac{\sigma_a}{a}\right)^2 + \beta^2\left(\frac{\sigma_b}{b}\right)^2 + \gamma^2\left(\frac{\sigma_c}{c}\right)^2 \quad (8)$$

However, the various uncertainties can have different meanings and consequences. The frequency is typically measured either identifying the frequency of a periodic signal which achieves resonance, or by the spectral analysis of the response to a broadband excitation. In both cases each frequency reading is associated to a finite degree of uncertainty, mainly due to random errors. In a set of repeated measurements such errors are uncorrelated, and tend to be cancelled by an averaging process; the error of each measurement affects the dispersion of results around the average, but not the average itself. In other words, these errors affect precision, but not accuracy. The accuracy of results is at most affected by the finite accuracy in the calibration of the frequency meter, of whichever nature it be. When then deriving the elastic moduli, the frequency reading can be exploited as such (see e.g. Eq.(7)), or via the determination of a propagation velocity. In both cases, the obtained moduli also depend on further 'auxiliary' parameters. In a very simple case, from Eq. (6) we have $C_{11} = \rho v_l^2$, and the resulting value of C_{11} is also affected by the finite uncertainty of the best available value of ρ, exploited in the derivation. However, in a set of repeated measurements the uncertainty of ρ does not contribute to the dispersion of results around the average, but it affects the average itself. This means that it affects accuracy, but not precision. The same can be said for the sample geometry (see Eq. (7)).

4. Mechanical excitation

Mechanical excitation can be either impulsive and broadband, as obtained by a percussion, or narrow band, as provided by a periodic excitation. Most methods exploiting mechanical excitation rely on the identification of the natural frequencies, or resonances, of a structure. With a periodic excitation, such frequencies are identified scanning the excitation frequency until resonance conditions (maximum oscillation amplitude for given excitation force) are detected. With a broadband excitation the response (measured amplitude) is frequency analyzed to identify the resonant frequencies.

Among the methods adopting harmonic excitation, acoustic microscopy (Zinin, 2001) exploits a piezoelectric actuator, typically in the form of an acoustic lens, and often operating also as a transducer. The acoustic lens is mechanically coupled to the sample by a liquid drop. Acoustic microscopy can be operated with imaging purposes; in the quantitative acoustic microscopy version (Zinin, 2001) it aims at measuring the acoustic properties of the sample.

Beside acoustic microscopy, two main types of methods have been developed. The first one measures the bulk properties. It adopts macroscopic homogeneous samples, and can exploit either broadband or narrow band excitation. These methods have also been ruled by norms (ASTM, 2008, 2009). A second group of methods, collectively called Resonance Ultrasound Spectroscopy, aims at measuring the properties of thin supported films. It almost invariably exploits a periodic excitation, whose frequency is swept in order to achieve resonance conditions.

4.1 Measurement of bulk properties

Macroscopic homogeneous samples are self supporting. They can be tested as free standing samples, provided the disturbances to free oscillations are minimized. Such a minimization includes sample suspension by thin threads, or specimen support by adequate material (cork, rubber), the supports having contact of minimum size, and positions at the nodes of the fundamental vibrational mode of interest, either flexural or torsional. Also the sensor contact, if oscillation is detected by a contact device, must be devised aiming at the minimization of the disturbance induced by the contact. Non contact detection techniques are available, including all optical techniques, and acoustic techniques: in the proper frequency range, oscillations can be detected through the air, by a microphone, and even excited, with harmonic excitation, by a similar technique, exploiting an audio oscillator and an audio amplifier. The optical techniques, intrinsically contact-less and inertia-less, have the broadest band, only limited by the light detection and analysis apparatus.

Since the full characterization of the elasticity of an isotropic medium requires two independent parameters, it typically requires excitation of at least two modes of different nature. For a specific simple geometry, the slender rod of length L, test methods have been regulated by norms (American Society for Testing and Materials [ASTM], 2008, 2009, and other ASTM norms cited by these two). Mainly flexural and torsional modes are considered, for a slender rod of mass m, and either rectangular section of width b and thickness t, or circular section of diameter D. Mass density being measured as $\rho = m/(tbL)$, Eq.(7) takes in this case, for the rectangular section, the forms (ASTM, 2008)

$$E = \left[m/(tbL)\right]\left(f_f L\right)^2 \times 0.9465 (L/t)^2 T_E(\nu, L/t) \tag{9a}$$

$$G = \left[m/(tbL)\right]\left(f_t L\right)^2 \times 4 T_G(b/t, b/L) \tag{9b}$$

where f_f and f_t are the frequencies of respectively the fundamental flexural and torsional modes, and T_E and T_G are numerical factors, functions of the indicated dimensionless quantities. Similar formulas hold for rods of circular section. The fundamental frequencies can be identified by either sweeping the frequency of a periodic excitation (ASTM, 2008), or by the broadband excitation by a mechanical percussion (ASTM, 2009), followed by the frequency analysis of the response. In both cases the displacement can be sensed by a contact transducer or by a microphone. The estimates for E and G being coupled by the value of Poisson's ratio (Eqs. 4 and 9a), an iterative procedure is indicated, to obtain consistent estimates. The algorithm of Eq.(8) applied to Eq. (9a) gives

$$\left(\frac{\sigma_E}{E}\right)^2 = \left(\frac{\sigma_m}{m}\right)^2 + 3^2\left(\frac{\sigma_t}{t}\right)^2 + \left(\frac{\sigma_b}{b}\right)^2 + 3^2\left(\frac{\sigma_L}{L}\right)^2 + 2^2\left(\frac{\sigma_f}{f}\right) \tag{10}$$

similar expressions being obtained for (σ_G/G) and for rods of circular section; such equations allow to derive the uncertainties to be associated to the obtained moduli, from the intrinsic uncertainties of the primary quantities. It was estimated (ASTM, 2008) that the major sources of uncertainty come from the fundamental frequency f and from the smallest specimen dimension (thickness or diameter). Uncertainties of the moduli in the 1% range are achievable.

Other free standing sample geometries were considered, and analyzed by detailed finite elements computations, to identify the appropriate values of the numerical factor N (see Eq. (7)). Nieves et al. (2000) consider a cylinder with length equal to diameter (L = D). They excite vibrations by a longitudinal percussion, and detect the displacement by an optical technique; several modes are typically observed. Alfano & Pagnotta (2006) consider instead a thin square plate, and, analyzing the displacement distributions of the first modes, they identify the best positions for the plate supports and for perpendicular percussion. They measure the response by a microphone. D'Evelyn & Taniguchi (1999) similarly exploited thin disks, exciting different modes by different impact points of a hollow zirconia bead, and measuring the response by a microphone. They exploit computations of the resonant frequencies of thin disks performed by others, and they estimate the accuracies of these computations to 1 % or better.

In all these cases, from the numerical computations the mode frequencies are tabulated or interpolated; Nieves et al. exploit the scaling parameter $\sqrt{G/\rho}/(\pi L)$. The ratios of the mode frequencies depend essentially on Poisson's ratio: the ratios of the observed frequencies allow therefore to identify the modes and to evaluate Poisson's ratio. The frequency values allow then, by the scaling parameter, to derive the shear modulus.

Both Nieves et al. and Alfano & Pagnotta perform detailed analyses of the measurement uncertainties. They both find that, also with their experimental set-up, the frequency measurement has a crucial role in the precision of the obtained moduli. Since Alfano & Pagnotta consider thin plates, they find that the precision of the thickness t is also crucial, simply because, thickness t being much smaller than the plate size a, a small value of the relative uncertainty σ_t/t is more difficultly achievable than for σ_a/a.

For comparison purposes, Nieves et al. (2000), beside considering the axial modes, also excite, by tangential percussion, the torsional modes, whose frequencies can be computed in closed form. Once the torsional modes are discriminated from the bending modes, the results agree to better than 1%, indicating a precision of this order. They also perform measurements by the pulse-echo method, finding instead discrepancies of 2% or more; they suggest that this method, which measures the propagation velocity of travelling waves, might become intrinsically less accurate when performed in a confined geometry of small size.

4.2 Resonance ultrasound spectroscopy

The Resonance Ultrasound Spectroscopy (RUS) methods have been developed (Migliori et al., 1993; Ohno, 1976; Schwarz et al., 2005; So et al., 2003) aiming in particular at the measurement of the properties of supported thin films. A recent implementation (Nakamura et al., 2004, 2010) exploits, as film support, a thin plate which, to be measured, is located on a tripod. One of the three legs is rigid, and contains a thermocouple which monitors the sample temperature; the second leg is a piezoelectric actuator, feeding a harmonic excitation whose frequency is swept, the third one is a piezoelectric sensor, which detects the oscillation amplitude. The resonance spectrum, i.e. the oscillation amplitude as function of the frequency, is measured sweeping the excitation frequency. Several peaks are found, sometimes with partial overlaps; their amplitudes, but not their frequencies, depend on the position of the piezoelectric sensor. The measurement is precise enough to clearly detect the difference between measurements in air and in vacuum, and the reproducibility of the resonance frequencies is at the 0.1 % level (Nakamura et al., 2010). The elastic constants are found fitting the computed frequencies to the measured ones; to this end, mode identification is crucial. Identification is performed keeping the excitation frequency at resonance, and scanning the specimen surface by a laser-Doppler interferometer. The map of the out-of-plane displacement of the vibrating specimen is thus obtained, which allows an unambiguous mode identification.

The elastic constants of the substrate are previously found by performing the same type of measurement on a bare substrate, and the elastic constants of the film are derived from the measured modifications of the resonance spectrum of the substrate. Since the vibration amplitude of the standing waves in a plate is maximum at the surface, the sensitivity of the method is better than the ratio of film thickness to the support thickness. In particular, the sensitivity of each resonance frequency to each elastic constant is assessed evaluating numerically the derivatives $\partial f / \partial C_{ij}$, and from these values the uncertainties ΔC_{ij} are derived from the estimated uncertainties Δf as

$$\Delta C_{ij} = \frac{1}{\partial f / \partial C_{ij}} \Delta f \ . \quad (11)$$

Deposited thin films often have a significant texture, with one crystalline direction preferentially oriented perpendicularly to the substrate surface, and random in-plane orientations, resulting, at a scale larger than that of the single crystallite, in in-plane isotropy, with different out-of-plane properties. This type of symmetry corresponds to the hexagonal symmetry, in which the tensor of the elastic constants is determined by five independent quantities. Among these, the resonance frequencies turn out to be almost insensitive to C_{44}, which therefore remains not determined, while the highest sensitivity is to C_{11} (Nakamura et al., 2010).

5. Laser pulse excitation

A laser pulse which is absorbed induces a sudden local heating which, by thermal expansion, produces an impulsive mechanical loading. Such a mechanical impulse excites waves in a broad frequency range, which are then detected. The accessible frequency band can be limited by either the excitation bandwidth or the detection device. According to the nature of the material being investigated, the deposition of an interaction layer can be needed. In particular, a short absorption depth is required, in order to excite a pulse which has small spatial and temporal duration, and therefore a broad band. The power density threshold for ablation must also be considered: measurements are typically conducted with high repetition rate pulses, and if ablation occurs the specimen undergoes a continuous modification during the measurement.

Two main configurations have been developed up to maturity; they differ for the propagation geometry and for the technique to detect vibrations. They are respectively called laser ultrasonics and picoseconds ultrasonics.

5.1 Laser ultrasonics

The so called laser ultrasonics technique is mainly exploited to characterize thin supported films. Oscillations are excited by a focused laser pulse, and propagation along the surface is detected, measuring the surface displacement at a distance from excitation pulse ranging from millimetres to centimetres. It is mainly the Rayleigh wave, modified by the presence of the film, which is excited and detected. The laser pulse, typically of nanosecond duration, is focused by a cylindrical lens on a line. The sudden expansion of this line source launches surface waves of limited divergence, propagating along the surface, perpendicularly to the focusing line. The component of the surface displacement normal to the surface itself can be measured, at various distances from the line source, by optical interferometry (Neubrand & Hess, 1992; Withfield et al., 2000), or, in a simpler and more robust way, by a piezoelectric sensor (Lehmann et al., 2002; Schneider et al., 1997, 1998, 2000).

The recorded displacement is frequency analyzed, yielding the dispersion relation $v(f)$ for a frequency interval that can extend over a full frequency decade (e.g. 20 to 200 MHz).

Fitting the computed dispersion relation to the measured one allows to derive the film properties. The width of the measured frequency interval can allow to to derive the Young modulus and also the film thickness (Schneider et al., 1997, 2000). The uncertainties of the results is evaluated numerically by the fitting procedure. Since the observed propagation distance is of the order of millimetres, the obtained properties are representative of an average over the propagation distance. The performance of the method could be pushed to the measurement of the properties of diamond-like carbon films having thickness down to 5 nm (Schneider et al., 2000). By stretching the observed propagation path to 20 mm it was possible to reduce the uncertainty of the measured propagation velocity of the Rayleigh wave to below 0.25 m/s. It was thus possible to detect the small variation of the Rayleigh velocity (5081 m/s for the bare (001) silicon substrate) induced by the presence of the film.
In a different configuration, the laser pulse is focused on a point instead of a line, with consequent excitation of waves which expand in all the radial directions, and a common path interferometer is adopted, whose light collection point scans the specimen surface (Sugawara et al., 2002). It is thus possible to visualize the wavefronts, circular for isotropic samples and non circular for anisotropic ones, with time resolution of the order of picoseconds and spatial resolution of a few micrometres.

5.2 Picosecond ultrasonics

The so called picosecond ultrasonics technique owes its name to the picosecond laser pulses which were available at the time it was introduced (Thomsen et al., 1984, 1986). It is nowadays implemented by femtosecond laser pulses, and is intrinsically suited to characterize thin supported films and multilayers. It follows the optical pump-and-probe scheme (Belliard et al., 2009; Bienville et al., 2006; Bryner et al. 2006; Vollmann et al. 2002). The pump beam, a femtosecond laser pulse, is focused, by a spherical lens, at the specimen surface and, at least partially, absorbed. The focusing spot, a few to tens of micrometers wide, is orders of magnitude larger than the characteristic lengths of the involved phenomena: mainly the optical absorption length, and also the thermal diffusion length and the acoustic wave propagation length, both for a femtosecond time scale. Bryner et al. (2006) estimate that with an aluminium surface, a near infrared laser (800 nm), and a pulse width of 70 fs, the absorption depth and therefore the dominant acoustic wavelength are of the order of 10 nm. Quite consistently, for a similar laser pulse but for Pt and Fe ultra thin films, Ogi et al. (2007) estimate at several THz the upper bound of the frequencies excited by the laser pulse.
The thermal and mechanical fields are thus (almost perfectly) laterally uniform, and essentially one dimensional: the absorbed pulse has a depth of the order of nanometres, and propagates like an acoustic wave with a plane wavefront which travels perpendicular to the surface, towards the specimen depth. At each interface this wave is partly transmitted and partly reflected, according to the acoustical impedances of the layers, and gives rise to echoes, which return to the surface, where they again are reflected back.
The surface is then probed by the probe beam, much weaker than the pump beam, which reaches the surface with a variable delay, controlled by a delay line. The probe beam, similarly to the acoustic wave, is partly transmitted and partly reflected at each interface, leading, in each layer, to a forward and a backward electromagnetic field, whose complex amplitudes can be calculated using a transfer-matrix formalism. The external reflectivity of the surface is the ratio of the backward to the forward complex amplitudes in the outer

space. This reflectivity is modified by the acoustic strain, by two mechanisms. Firstly, each interface is displaced by the acoustic wave, and, secondly, the refractive index in each layer is modified by the elastic strain, by the acousto-optic (also called photoelastic) coupling. In particular, it must be remembered that the traveling acoustic pulse extends over a depth of a few nanometers, inducing a localized perturbation of the refractive index, which partially scatters the optical beam. Interference can occur between the beam reflected at the outer surface and that reflected by the traveling acoustic pulse.

Interferometric techniques allow to measure the variation of both amplitude and phase of the reflected beam, thus measuring the variation of the complex reflection coefficient. With probe pulses in the femtosecond range, and varying the probe pulse delay, the time evolution of the surface reflectivity can be monitored with high temporal resolution. This time evolution typically shows several features. The diffusion, towards the sample depth, of the heat deposited by the laser pulse gives a slowly varying reflectivity background. The echoes of the acoustic pulse which, after partial reflection at the film/substrate interface, are again reflected at the outer surface are generally visible. The so called Brillouin oscillations, due to the interference between the beam reflected at the outer surface and that reflected by the traveling acoustic pulse, can then be found.

The analysis of the various features allows to characterize the waves which cross the layers travelling perpendicularly to the surface. In the derivation of the film properties, the knowledge of film thickness, typically obtained by X-ray reflectivity, has a crucial role; the uncertainty about thickness is one of the leading terms in the uncertainty of the final results. The achievable resolution depends on the excited wavelength. In copper the absorption depth is larger than the value cited above for aluminium. Since the smallness of the absorption depth determines the localization of the acoustic pulse and the achievable resolution, the deposition of an aluminium interaction layer, which guarantees a very small absorption depth, is a common practice. The interaction layer, typically a few tens of nanometres thick, then participates to the vibrational behaviour of the structure being investigated. Accurate measurements of stiffness therefore require consideration of the effects of the interaction layer, e.g. by measurements with layers of different thicknesses, followed by an extrapolation to null thickness (Mante et al., 2008). Obviously this deconvolution of the effects of the interaction layer contributes to the uncertainty of the final results.

Near infrared lasers are a common choice, because at shorter wavelength more complex phenomena can occur, which were attributed to electronic interband transitions (Devos & Cote, 2004); obviously, if one is interested in elastic properties, electronic transitions are a spurious effect to be avoided.

By picoseconds ultrasonics it was possible to characterize a layer stack, including a buried layer of about 20 nm thickness (Bryner et al., 2006). The uncertainty for the elastic constants of this layer is estimated at 20-25%, which is however remarkable for a layer of this type. The lowest limit for layer detection is also estimated at about 10 nm thickness. In a different configuration, namely a single Pt or Fe layer on a silicon substrate or a borosilicate glass substrate, Ogi et al. (2007) were able to characterize metallic films of thickness down to 5 nm. They found, at nanometric thicknesses, a dependence of the elastic moduli on thickness. This was explained by the impossibility of plastic flow at such low thicknesses: the elastic strains can thus reach levels which are non reachable in thicker samples, such that higher order elastic constants are no longer negligible.

Periodic Mo/Si multilayers (superlattices) were investigated by Belliard et al. (2009), exploring various periodicities in the nanometric range. They detect bulk waves crossing back and forth the multilayer; such waves show pulses of the order of 10 ps, which correspond to propagation lengths much larger than the superlattice period. These waves therefore see the whole multilayer as an effective medium. The theoretical prediction for the properties of the effective medium and the reflection coefficient at the superlattice / Si substrate interface are confirmed by the experimental findings. They also detect higher frequency oscillations, which correspond to localized waves. The periodic multilayer acts as a Bragg reflector, and opens forbidden gaps in the spectrum. It can confine a mode in the neighbourhood of the outer surface (acoustic-phonon surface modes), but only if the outer layer is the lower acoustic impedance layer (in this case Si), which therefore acts as a perfect reflector. The properties of such modes could be correctly predicted only taking into account the nanometric top silicon oxide layer, which spontaneously forms at he silicon surface. The presence of this additional layer is also consistent with the X-ray reflectivity measurements. The behaviour of a Mo cavity sandwiched between Mo/Si mirrors was also analyzed.

The picoseconds ultrasonics technique was also exploited to investigate non laterally homogeneous specimens (Bienville et al., 2006; Mante et al., 2008). One limitation of this technique in the measurement of the elastic constants is that it involves only plane waves travelling perpendicular to the surface, thus allowing only the out-of-plane elastic characterization of the film. To overcome this limitation, Mante et al. and Robillard et al. (2008) proposed a technique by which the film to be characterized, and the aluminium interaction layer deposited on it, are cut by lithographic techniques to obtain a periodic square lattice of square (200 nm × 200 nm) pillars. As confirmed by the measurements performed on square lattices of different lattice constants, the pump pulse also excites, in this nanostructured film, acoustic collective modes of the pillars, which propagate along the surface in various directions. Various branches are measured, from which also the in-plane properties of the film can be measured, achieving a complete elastic characterization.

6. Brillouin spectroscopy

Aggregates of atoms, from molecules to clusters, to nanoparticles and nanocrystals, up to mesoscopic and macroscopic aggregates, can interact with electromagnetic waves either elastically or inelastically. Inelastic interactions include emission/absorption phenomena, and inelastic scattering. We consider here inelastic scattering by vibrational excitations. At the molecular scale the atomic structure of matter has a crucial role, and quantum phenomena are relevant. At this level, vibrational excitations are the vibrations of molecules, or, in a crystal, the vibrations of the internal degrees of freedom of each unit cell, which form the so called optical branches of the dispersion relation, or optical phonons. Broadly speaking, inelastic scattering by these excitations is called Raman scattering.

Aggregates above the nanometric scale also support collective vibrational excitations which begin to resemble to acoustic waves, and can be described by a continuum model. In a crystal, the vibrations of the degrees of freedom of the centre of mass of each unit cell form the so called acoustic branches of the dispersion relation, or acoustic phonons. In the long wavelength limit they are the are acoustic waves, accurately described by the continuum model (Eq. (1)). Broadly speaking, inelastic scattering by these excitations is called Brillouin scattering.

Visible light has sub micrometric wavelength. In media, either crystalline or amorphous, which are homogeneous, and therefore translationally invariant, over at least a few micrometres, vibrational excitations of sub-micrometric wavelength have a well defined wavevector (see Eq. (2)). Due to translational invariance, the kinematics of scattering selects the excitations whose wavevector is close to the Brillouin zone center (**k** = 0). The acoustic and optical phonon branches have very different behaviours close to the zone center (**k** → 0). The acoustic branch frequency ω_a goes to zero, with phase velocity ω_a/k and group velocity $\partial\omega_a/\partial k$ which tend to coincide with the sound velocity (which depends on polarization and, in an anisotropic crystal, on the wavevector direction), while the optical branch frequency ω_o typically goes to a maximum, with group velocity $\partial\omega_o/\partial k$ which goes to zero.

Correspondingly, with visible light and with typical properties of solids, the two types of branches produce inelastic scattering with frequency shifts ranging from a fraction of cm^{-1} to a few cm^{-1} (i.e from a few GHz to tens of GHz) for Brillouin scattering, and from hundreds to thousands of cm^{-1} (i.e from THz to tens of THz) for Raman scattering. The spectral analysis of so widely different frequency ranges requires different types of spectrometer. However, for both types of scattering the experiments are performed without exciting the vibrations, but relying on the naturally occurring thermal motion.

Brillouin spectrometry thus offers a fully optical, and therefore contact-less, method to measure the dispersion relations of bulk and surface acoustic waves, whose wavelength is determined by the scattering geometry and the optical wavelength, and is typically sub-micrometric. The frequency results from the medium properties, and typically falls in the GHz to tens of GHz range. Measurements are performed illuminating the sample by a focused laser beam, and analyzing the spectrum of scattered light, which is dominated by the elastically scattered light, but also contains weak Stokes/anti-Stokes doublets due to inelastic scattering by thermally excited vibrations (Beghi et al., 2004; Comins, 2001; Every, 2002; Grimsditch, 2001; Sandercock, 1982).

In sufficiently transparent materials scattering can occur in the bulk, by bulk acoustic waves. The coupling mechanism is the elasto-optic (or acousto-optic) effect: the periodic modulation of the refractive index by the periodic strain of the acoustic wave. In both transparent and opaque materials scattering can also occur by surface acoustic waves, by the ripple mechanism: the periodic corrugation of the surface due to the surface wave.

In more detail: the incident beam, of angular frequency Ω_i and wavelength λ_0, impinges on a sufficiently transparent sample with wavevector $\mathbf{q}_i$ and is refracted into the wavevector $\mathbf{q}'_i$. Scattered light, of wavevector $\mathbf{q}'_s$, emerges with wavevector $\mathbf{q}_s$. The probed wavevector, $\mathbf{k} = \pm(\mathbf{q}'_s - \mathbf{q}'_i)$, is determined by λ_0, the directions of $\mathbf{q}_i$ and $\mathbf{q}_s$, and the refractive index n. Light inelastically scattered by a vibrational excitation of angular frequency $\omega(\mathbf{k})$ gives a Stokes/anti-Stokes doublet at frequencies $\Omega_s = \Omega_i \pm \omega$. Detection, in the spectrum of scattered light, of such a doublet allows to measure $\omega = |\Omega_s - \Omega_i|$ and to derive the excitation velocity $v = \omega/|\mathbf{k}|$. In both transparent and opaque samples scattering occurring by surface waves only depends on the components of wavevectors parallel to the surface: the probed wavevector is $\mathbf{k}_{\parallel} = \pm(\mathbf{q}'_s - \mathbf{q}'_i)_{\parallel}$, and the surface wave velocity is $v = \omega/|\mathbf{k}_{\parallel}|$. In other words, the spontaneous thermal motion can be viewed as spatially Fourier transformed into an incoherent superposition of harmonic waves having all the possible wavevectors; the scattering geometry (the directions of wavevectors $\mathbf{q}_i$ and $\mathbf{q}_s$) selects a specific wavevector $\mathbf{k}$ or $\mathbf{k}_{\parallel}$ which is probed by the inelastic light scattering event.

Although also other geometries have been exploited (Beghi et al. 2011), in Brillouin spectroscopy the most frequently adopted scattering geometry is backscattering: $\mathbf{q}_s = -\mathbf{q}_i$. For bulk scattering it corresponds to $\mathbf{k} = \pm 2\mathbf{q}'_i$, such that

$$|\mathbf{k}| = 2\frac{2\pi}{\lambda_0}n \,, \tag{12}$$

which depends on the refractive index, but depends on geometry only when the sample is anisotropic, while for surface scattering it means $\mathbf{k}_{\|} = \pm 2\mathbf{q}_{i\|}$, i.e.

$$|\mathbf{k}_{\|}| = 2\frac{2\pi}{\lambda_0}\sin\theta \,, \tag{13}$$

where θ is the incidence angle (the angle between the incident beam and the surface normal). In this case the probed wavevector depends the incidence angle, but not on the refractive index, because Snell's law implies that upon refraction the optical parallel components $\mathbf{q}_{\|}$ remain unchanged.

The data analysis for Brillouin spectroscopy results is common to all the methods, like laser ultrasonics, which measure the velocity of travelling waves. In the simplest cases the velocity is a function of the elastic constants which can be given in closed form. For instance, if scattering by the longitudinal bulk wave, of velocity $v_l = \sqrt{C_{11}/\rho}$ (Eq. (6)) is detected, C_{11} is directly obtained as $C_{11} = \rho v_l^2 = \rho\omega^2/k^2$, and its uncertainty $\sigma_{C_{11}}$ is evaluated by Eq. (8). In other cases, and namely in the case of supported films, the mode velocity can be computed as function of the elastic constants only numerically. In that case it must be remembered that the stiffness of an elastic solid is determined by as many independent parameters as are needed to completely identify the tensor of the elastic constants. In other words the stiffness is identified by a point in a multidimensional space, the dimensionality being 2 in the simplest case of the isotropic medium, and being higher for lower symmetry media.

Focusing here on the isotropic case, the stiffness can be represented, among other possible choices, by the (E,ν) couple (see Eqs. (3) and (4)). In Brillouin spectroscopy, for each mode the velocity is measured for various wavevectors k (in laser ultrasonics it is measured for various frequencies) as $v_m(k)$ with uncertainty $\sigma_{v(k)}$; since it can also be computed as $v_c(E,\nu,k)$, the stiffness, represented by the (E,ν) couple, can be determined by a standard least squares minimization procedure. The sum of squares is computed as

$$\chi^2(E,\nu) = \sum_k \left(\frac{v_c(E,\nu,k) - v_m(k)}{\sigma_{v(k)}} \right)^2 \,, \tag{14}$$

where, for each wavevector k, the sum is further extended to all the detected acoustic modes. Following standard estimation theory, the minimum of $\chi^2(E,\nu)$ identifies the most probable value $(\overline{E,\nu})$ of the (E,ν) couple, and the isolevel curves of the normalized estimator $\left(\chi^2(E,\nu) - \chi^2(\overline{E,\nu})\right)/\chi^2(\overline{E,\nu})$ identify the confidence region at any predetermined confidence level (Beghi et al., 2001, 2004, 2011; Lefeuvre et al., 1999). In some cases a well defined minimum of $\chi^2(E,\nu)$ is found, allowing a good identification of the parameters (Beghi et al., 2001; Comins et al., 2000; Zhang et al., 2001a), while in other cases a broad, valley-shaped minimum is found. In such cases a good identification of the parameters is not possible (Beghi et al., 2002; Zhang et al., 1998), although sometimes some combination of the parameters can be identified with better precision than individual

parameters (Comins et al., 2000; Zhang et al., 1998, 2001a). When several acoustic modes are measured, the larger amount of available information allows a precise and complete elastic characterization, as obtained e.g. for SiC films of micrometric thickness (Djemia et al, 2004).
In Eq. (14) each value $v_m(k)$, being obtained as ω/k, has an uncertainty $\sigma_{v(k)}$ which depends in turn on the uncertainty of the frequency of each spectral doublet, and on the precision of the incidence angle (see Eq. (13)) or of the refractive index (see Eq. (12)). It can be noted that the uncertainties of frequency and angle are of the random type, which affects precision but not accuracy, while the uncertainty of the refractive index affects accuracy but not precision (see Section 3.3). As with other techniques, also the uncertainties concerning the mass density and the layer thickness(es) affect accuracy but not precision. These uncertainties were the object of detailed investigations (Beghi et al., 2011; Stoddart et al., 1998). The effects of the uncertainties of the quantities which are directly measured ('primary uncertainties') on the values of the elastic constants which are finally obtained were evaluated. It was found that with appropriate sets of measurement uncertainties at the 1% level are reachable (Beghi et al., 2011).
As already noted, Brillouin spectroscopy measures the acoustic modes at frequencies of the order of GHz to tens of GHz, therefore at wavelengths much shorter than those corresponding to frequencies of tens to hundreds of MHz, typically observed with piezoelectric excitation and/or detection. This gives Brillouin spectroscopy an intrinsically higher sensitivity to the properties of films, or to the perturbation induced by the presence of very thin films. Brillouin spectroscopy was exploited to characterize tetrahedral amorphous carbon films of thicknesses of hundreds of nanometres (Chirita et al., 1999), tens of nanometres (Ferrari et al., 1999), down to a few nanometres (Beghi et al., 2002). It was also shown that inelastic light scattering can be sensitive to nanometric thickness differences (Lou et al., 2010). By Brillouin spectroscopy it was also possible to characterize buried layers in silicon-on-insulator structures (Ghislotti & Bottani, 1994).
On the other hand, the techniques which excite vibrations operate with oscillation amplitudes significantly larger than those measured by Brillouin spectroscopy; this allows more precise measurements of frequencies, which at least partially compensates for the lower intrinsic sensitivity due to the larger wavelengths. Combinations of techniques were also exploited: thicker tetrahedral amorphous carbon films films (3 micron) were characterized combining Brillouin spectroscopy and laser ultrasonics. A wide range of frequencies was thus accessible, allowing a detailed characterization of the elastic properties of the film (Berezina et al., 2004). A combination of Brillouin spectroscopy and picosecond ultrasonics was instead exploited to characterize superlattices formed by periodic multilayers of permalloy/alumina, with various periodicities at the nanometric scale (Rossignol et al., 2004). Picosecond ultrasonics characterizes the out-of-plane properties by waves travelling normal to the surface, while Brillouin spectroscopy characterizes the in-plane properties by waves travelling along the surface. The combination of techniques elucidated the effects of the interfaces.
Another algorithm for data analysis, different from that outlined above, has also been recently proposed (Every et al., 2010). Both algorithms refer to the types of waves most frequently measured in Brillouin spectroscopy of films or layered structures: surface acoustic wave, or pseudo surface acoustic waves, which essentially travel parallel to the surface, or however have a significant wavevector component parallel to the surface. It can also be mentioned that it was also possible, by Brillouin spectroscopy, to detect standing

acoustic waves trapped within a film, which are reflected back and forth, crossing the film perpendicularly to its surface (Zhang, 2001b).
Brillouin spectroscopy lends itself to the characterization of structures other than films or layers. In particular, single-walled carbon nanotubes were characterized, measuring Brillouin scattering by a free-standing film of pure, partially aligned, single-walled nanotubes, and analyzing the results in terms of continuum models (Bottani et al., 2003). The dependence of the measured spectra on the angle between the exchanged wavevector and the preferential direction of the tubes shows that the tube-tube interactions are weak: the tubes are vibrationally almost independent. The tubes are modelled as continuous membranes, at two different levels: at the first one the membrane is infinitely flexible, only able to transmit in-plane forces, while at the second level of approximation the tube wall is treated as also able to transmit shear forces and torques not belonging to the shell surface. In both cases scattering was essentially due to longitudinal waves travelling along the tubes. Taking into account that AFM images suggest that the tube segments contributing to scattering are not in the infinite tube length approximation, it was possible to derive the 2D Young modulus for the tube wall, achieving the first dynamic estimation of the stiffness of the tube wall. Scattering from carbon nanotubes was observed also in a different geometry, with an ordered array of tubes, clamped at one end (Polomska et al., 2007).
Due to its intrinsic contact-less nature, Brillouin spectroscopy is the natural choice for the measurement of elastic properties in conditions, like high temperature and/or high pressure, in which physical contact with the specimen is difficult, if possible at all. Brillouin spectroscopy only requires optical access, which can be obtained by an appropriate window, and even in the extreme conditions achievable in a diamond anvil cell, optical access is guaranteed by the transparency of the same diamond anvils.
Measurements were performed at high temperatures (Pang, 1997; Stoddart, 1995; Zhang et al., 2001a), as well as at low temperatures, which were crucial to single out a particular mechanism of hypersound propagation in alkali-borate glasses (Carini et al., 2008). After pioneering experiments at high pressures (Crowhurst et al., 1999; Whitfield et al., 1976,), in recent years dedicated Brillouin spectrometers were built at synchrotron facilities, allowing simultaneous performance of high resolution x-ray diffraction and Brillouin spectroscopy (and possibly other optical investigations like Raman spectroscopy, fluorescence, absorption) on specimens subjected to extreme pressures in a diamond anvil cell, with possible heating (Murakami et al., 2009; Prakapenka et al., 2010; Sinogeikin et al., 2006). This set-up, of particular interest for geophysicists since it allows to characterize the behaviour of minerals in the conditions which are found in the Earth's interior, was exploited to perform measurements on SiO_2 glass (Murakami & Bass, 2010) and other minerals, but also on polymers (Stevens et al., 2007) and liquid methane (Li et al., 2010).

7. Conclusion

The stiffness of films, characterized by the elastic constants, depends on the film microstructure, and its precise characterization is crucial when thin layers have structural functions. The interest in the measurement of the elastic constants is witnessed by the number of new techniques, or of improvements of existing techniques, being proposed.
The techniques which exploit either propagating acoustic waves or standing oscillations involve exclusively elastic strains: they therefore offer the most direct and clean access to the elastic properties, and potentially the most accurate measurements. Among the methods

based on vibrations, those which exploit, for excitation and/or detection, the contact-less and inertia-less nature of light, have an important role.
An overview of the variety of existing methods was presented here, trying to present a unified picture, and underlining the peculiarities of each of them, in particular for what concerns the experimental uncertainties. It turns out that, under appropriate conditions and experimental procedures, several techniques can achieve significant precision and accuracy.

8. References

Alfano, M. & Pagnotta, L., (2006). Measurement of the dynamic elastic properties of a thin coating, *Review of Scientific Instruments,* Vol.77, Paper No. 056107

ASTM E1875-08 (2008). *Standard Test Method for Dynamic Young's Modulus,Shear Modulus, and Poisson's Ratio by Sonic Resonance,* ASTM International, West Conshohocken, PA.

ASTM E1876-09 (2009). *Standard Test Method for Dynamic Young's Modulus, Shear Modulus, and Poisson's Ratio by Impulse Excitation of Vibration,* ASTM International, West Conshohocken, PA.

Auld, B. A. (1990). *Acoustic fields and Waves in Solids,* Robert E. Krieger Publishing Company, Malabar, Florida

Beghi, M. G., Bottani, C. E. & Pastorelli, R. (2001). High accuracy measurement of elastic constants of thin films by surface Brillouin scattering. In *Mechanical properties of structural films,* C. Muhlstein, C. & Brown, S.B. (Eds.), pp. 109-126. ASTM STP 1413, American Society for Testing and Materials, Conshohoken, PA

Beghi, M.G., Ferrari, A. C., Teo, K.B.K., Robertson, J., Bottani, C.E., Libassi, A. & Tanner, B.K. (2002). Bonding and mechanical properties of ultrathin diamond-like carbon films, *Applied Physics Letters,* Vol.81, pp. 3804-3806

Beghi, M.G., Every, A.G. & Zinin, P.V. (2004). Brillouin scattering measurement of SAW velocities for determining near-surface elastic properties, in *Ultrasonic nondestructive evaluation,* T. Kundu (Ed.), pp. 581-651, CRC Press, Boca Raton, FL. Revised edition in press (2011)

Beghi, M. G., Di Fonzo, F., Pietralunga, S., Ubaldi, C. & Bottani, C. E. (2011). Precision and accuracy in film stiffness measurement by Brillouin spectroscopy, *Review of Scientific Instruments,* Vol.102, Paper No. 053107

Belliard, L., Huynh, A., Perrin, B., Michel, A., Abadias, G. & Jaouen, C. (2009). Elastic properties and phonon generation in Mo/Si superlattices, *Physical Review B* Vol.80, Paper No. 155424

Berezina, S., Zinin, P. V., Schneider, D., Fei, D. & Rebinsky, D. A. (2004). Combining Brillouin spectroscopy and laser-SAW technique for elastic property characterization of thick DLC films, *Ultrasonics,* Vol.43, pp. 87 - 93

Bi, B., Huang, W.-S., Asmussen, J. & Bolding, B. (2002). Surface acoustic waves on nanocrystalline diamond, *Diamond and Related Materials,* Vol.11, pp. 677--680.

Bienville, T., Robillard, J.F., Belliard, L., Roch_Jeune, I, Devos, A. & Perrin, B. (2006). Individual and collective vibrational modes of nanostructures studied by picosecond ultrasonics, *Ultrasonics,* Vol.44, pp. e1289-e1294

Bottani, C.E., Li Bassi, A., Beghi, M.G., Podesta, A., Milani, P., Zakhidov, A., Baughman, R., Walters, D.A. & Smalley, R.E. (2003). Dynamic light scattering from acoustic modes in single-walled carbon nanotubes, *Physical. Review. B* , Vol.67, paper n. 155407

Bryner, J., Profunser, D.M., Vollmann, J., Mueller, E. & Dual, J. (2006). Characterization of Ta and TaN diffusion barriers beneath Cu layers using picosecond ultrasonics, *Ultrasonics*, Vol.44, pp. e1269-e1275

Carini, G., Tripodo, G. & Borjesson, L. (2008). Thermally activated relaxations and vibrational anharmonicity in alkali-borate glasses: Brillouin scattering study. *Physical Review B*, Vol.78, paper n. 024104

Chirita, M., Sooryakumar, R., Xia, H., Monteiro, O. R. & Brown, I. G. (1999). Observation of guided longitudinal acoustic modes in hard supported layers. *Physical Review B*, Vol.60, pp. 5153-5156.

Comins, J. D., Every, A. G., Stoddart, P. R., Zhang, X., Crowhurst, J. C. & Hearne, G. R. (2000). Surface Brillouin scattering of opaque solids and thin supported films. *Ultrasonics*, Vol.38, pp. 450-458.

Comins J.D. (2001). Surface Brillouin scattering, in *Handbook of Elastic Properties of Solids, Liquids, and Gases*, M. Levy, H. E. Bass and R. R. Stern & V. Keppens (Eds.), Volume I: *Dynamic Methods for Measuring the Elastic Properties of Solids*, pp. 349-378, Academic Press, New York

Crowhurst, J.C., Hearne, G.R., Comins, J.D., Every, A.G. & Stoddart, P.R.. (1999). Surface Brillouin scattering at high pressure: Application to a thin supported gold film. *Physical Review B*, Vol.60, pp. R14990-R14993

Czaplewski, D.A., Sullivan, J.P., Friedmann, T.A. & Wendt, J.R. (2005). Temperature dependence of the mechanical properties of tetrahedrally coordinated amorphous carbon thin films, *Applied Physics Letters*, Vol.87, paper n. 161915

D'Evelyn M.P. & Taniguchi, T. (1999). Elastic properties of translucent polycrystalline cubic boron nitride as characterized by the dynamic resonance method, *Diamond and Related Materials*, Vol.8, pp. 1522-1526

Devos, A. & Côte, R. (2004). Strong oscillations detected by picoseconds ultrasonics in silicon: evidence for an electronic structure effect, *Physical Review B*, Vol.70, paper n. 125208

Djemia, P., Roussigné, Y., Dirras, G.,F. & Jackson, K.M. (2004). Elastic properties of SiC films by Brillouin light scattering, *Journal of Applied Physics*, Vol. 95, pp. 2324-2330

Every, A.G. (2001). The Elastic Properties of Solids: Static and Dynamic Principles, In: *Handbook of Elastic Properties of Solids, Liquids, and Gases*, M. Levy, H. Bass, R. Stern & V. Keppens (Eds.), Volume I: *Dynamic Methods for Measuring the Elastic Properties of Solids*, pp. 3-36, Academic Press, New York

Every, A.G. (2002). Measurement of the near surface elastic properties of solids and thin supported films, *Measurement Science and Technology*, Vol.13, pp. R21-R39

Every, A.G., Kotane, L.M. & Comins, J. D. (2010). Characteristic wave speeds in the surface Brillouin scattering measurement of elastic constants of crystals. *Physical Review B*, Vol.81, paper n. 224303

Farnell, G.W. & Adler, E.L. (1972). Elastic wave propagation in thin layers. In: *Physical Acoustics*, W.P. Mason & R.N. Thurston (Eds.), Vol. 9, pp. 35-127, Academic, New York

Ferrari, A. C., Robertson, J., Beghi, M. G., Bottani, C. E., Ferulano, R. & Pastorelli, R. (1999). Elastic constants of tetrahedral amorphous carbon films by surface Brillouin scattering, *Applied Physics Letters*, Vol.75, pp. 1893-1895.

Ghislotti, G. & Bottani, C. E. (1994). Brillouin scattering from shear horizontal surface phonons in silicon-on-insulator structures - Theory and experiment, *Physical Review B,* Vol.50, pp. 12131-12137.

Grimsditch, M., (2001). Brillouin scattering, in *Handbook of Elastic Properties of Solids, Liquids, and Gases,* M. Levy, H. E. Bass and R. R. Stern & V. Keppens (Eds.), Volume I: *Dynamic Methods for Measuring the Elastic Properties of Solids,* pp. 331-347, Academic Press, New York

Kim, J. Y., Chung, H. J., Kim, H. J., Cho, H. M., Yang, H. K. & Park, J. C. (2000). Surface acoustic wave propagation properties of nitrogenated diamond-like carbon films, *Journal of Vacuum Science and Technology A,* Vol.18, pp. 1993—1997

Kubisztal, M., Kubisztal, J., Chrobak, A., Haneczok, B., Budniok, A. & Rasek, J. (2008). Elastic properties of Ni and Ni + Mo coatings electrodeposited on stainless steel substrate, *Surface and Coatings Technology,* Vol.202, pp. 2292–2296

Kundu, T. (2004). Mechanics of elastic waves and ultrasonics non-destructive evaluation, in *Ultrasonic nondestructive evaluation,* T. Kundu (Ed.), pp. 1-142, CRC Press, Boca Raton, FL. Revised edition in press (2011)

Lefeuvre, O., Pang, W., Zinin, P., Comins, J.D., Every, A.G., Briggs, G.A.D., Zeller, B.D. & Thompson, G.E. (1999). Determination of the elastic properties of a barrier film on aluminium by Brillouin spectroscopy. *Thin Solid Films,* Vol.350, pp. 53-58.

Lehmann, G., Hess, P., Weissmantel, S., Reisse, G., Scheible, P. & Lunk, A. (2002). Young's modulus and density of nanocrystalline cubic boron nitride films determined by dispersion of surface acoustic waves, *Applied Physics A,* Vol.74, pp. 41—45

Li, M., Li, F.F., Gao, W., Ma, C.L., Huang, L.Y., Zhou, Q.A. & Cui, Q.L. (2010). Brillouin scattering study of liquid methane under high pressures and high temperatures. *Journal of Chemical Physics,* Vol.133, paper n. 044503

Lou, N., Groenen, J., Benassayag, G. & Zwick, A. (2010). Acoustics at nanoscale: Raman-Brillouin scattering from thin silicon-on-insulator layers, *Applied Physics Letters,* Vol.97, article n. 141908

Mante, P.A., Robillard, J.F. & Devos, A. (2008). Complete thin film mechanical characterization using picoseconds ultrasonics and nanostructured transducers: experimental demonstration on SiO_2, *Applied Physics Letters,* Vol.93, paper n. 071909

Migliori, A., Sarrao, J.L., Visschera, W.M., Bella, T.M., Leia, M., Fisk, Z. & Leisure, R.G. (1993). Resonant ultrasound spectroscopic techniques for measurement of the elastic moduli of solids, *Physica B,* Vol.183, pp. 1–24.

Murakami, M., Asahara, Y., Ohishi, Y., Hirao N. & Hirose, K. (2009). Development of in situ Brillouin spectroscopy at high pressure and high temperature with synchrotron radiation and infrared laser heating system: Application to the Earth's deep interior. *Physics of the Earth and Planetary Interiors,* Vol.174, pp. 282-291

Murakami, M. & Bass, J.D. (2010). Spectroscopic Evidence for Ultrahigh-Pressure Polymorphism in SiO_2 Glass. *Physical Review Letters,* Vol.104, pp. 025504.

Nakamura, N., Ogi, H. & Hirao, M. (2004). Resonance ultrasound spectroscopy with laser-Doppler interferometry for studying elastic properties of thin films, *Ultrasonics,* Vol.42, pp. 491–494

Nakamura, N., Nakashima, T., Oura, S., Ogi, H. & Hirao, M. (2010). Resonant-ultrasound spectroscopy for studying annealing effect on elastic constant of thin film, *Ultrasonics,* Vol.50, pp. 150–154

Neubrand, A. & Hess, P. (1992). Laser generation and detection of surface acoustic waves: Elastic properties of surface layers, *Journal of applied physics,* Vol.71, pp. 227-238

Nieves, F.J., Gascòn, F. & Bayòn, A. (2000). Precise and direct determination of the elastic constants of a cylinder with a length equal to its diameter, *Review of Scientific Instruments,* Vol.71, pp. 2433-2439

Ogi, H., Fujii, M., Nakamura, N., Shagawa, T. & M. Hirao, M. (2007). Resonance acoustic-phonon spectroscopy for studying elasticity of ultrathin films, *Applied Physics Letters,* Vol. 90, paper n. 191906.

Ohno, I. (1976). Free vibration of a rectangular parallelepiped crystal and its application to determination of elastic constants of orthorhombic crystal, *Journal of Physics of the Earth,* Vol.24, pp. 355–379.

Pang, W., Stoddart, P.R., Comins, J.D., Every, A.G., Pietersen, D. & Marais P.J. (1997). Elastic properties of TiN hard films at room and high temperatures using Brillouin scattering, *International Journal of Refractory Metals and Hard Materials,* Vol.15, pp. 179-185.

Polomska, A.M., Young, C.K., Andrews, G.T., Clouter, M.J., Yin, A. & Xu, J. M. (2007). Inelastic laser light scattering study of an ordered array of carbon nanotubes. *Applied Physics Letters,* Vol.90, paper n. 201918

Prakapenka, V. (2010). On-line Brillouin Spectroscopy at GSECARS: Basic Principles and Application for High Pressure Research, *Synchrotron Radiation News* Vol.23, pp. 14-15.

Robillard, J.-F., Devos, A., Roch-Jeune, I. & Mante, P. A. (2008). Collective acoustic modes in various two-dimensional crystals by ultrafast acoustics: Theory and experiment, *Physical Review B,* Vol.78, paper n. 064302

Rossignol, C., Perrin, B., Bonello, B., Djemia, P., Moch, P., & Hurdequint, H. (2004). Elastic properties of ultrathin permalloy/alumina multilayer films using picosecond ultrasonics and Brillouin light scattering, *Physical Review.B,* Vol.70, paper n. 094102

Sandercock J.R. (1982). Trends in Brillouin scattering - Studies of opaque materials, supported films, and central modes, in *Light Scattering in solids III,* M. Cardona and G. Güntherodt (Eds.), pp. 173-206, Springer, Berlin

Schneider, D., Schwarz, T., Scheibe, H.-J. & Panzner, M. (1997) Non destructive evaluation of diamond-like carbon films by laser induced surface acoustic waves, *Thin Solid Films, Vol.*295, pp. 107–116

Schneider, D., Schultrich, B., Scheibe, H.-J., Ziegele, H. & Griepentrog, M. (1998). A laser acoustic method for testing and classifying hard surface layers, *Thin Solid Films,* Vol. 332, pp. 157–163.

Schneider, D., Witke, T.H., Schwarz, T.H., Schoneich, B. & Schultrich, B. (2000). Testing ultra-thin films by laser-acoustics, Surface and Coatings Technology, Vol.126, pp.136-141

Schwarz, R.B., Hooks, D.E., Dick, J.J. & Archuleta, J.I. (2005). Resonant ultrasound spectroscopy measurement of the elastic constants of cyclotrimethylene trinitramine, *Journal of Applied Physics,* Vol.98, paper N. 056106.

Sinogeikin, S., Bass, J., Prakapenka, V., Lakshtanov, D., Shen, G.Y., Sanchez-Valle, C & Rivers, M. (2006). Brillouin spectrometer interfaced with synchrotron radiation for simultaneous X-ray density and acoustic velocity measurements. *Review of Scientific Instruments,* Vol.77, article n. 103905

So, J.H., Gladden, J.R., Hu, Y.F., Maynard, J.D. & Qi Li (2003). Measurements of elastic constants in thin films of colossal magnetoresistance material, *Physical Review Letters*, Vol.90, paper N. 036103.

Stevens, L.L., Orler, E.B., Dattelbaum, D.M., Ahart, M. & Hemley, R. J. (2007). Brillouin scattering determination of the acoustic properties and their pressure dependence for three polymeric elastomers. *Journal of Chemical Physics*, Vol.127, article n. 104905

Stoddart, P.R., Comins, J.D. & Every, A.G. (1995). Brillouin-scattering measurements of surface-acoustic-wave velocities in silicon at high-temperatures. *Physical Review B*, Vol.51, pp. 17574-17578

Stoddart, P.R., Crowhurst, J.C., Every, A.G. & Comins, J. D. (1998). Measurement precision in surface Brillouin scattering, *Journal of the Optical Society of America B*, Vol.15, pp. 2481-2489.

Sugawara, Y., Wright, O.B., Matsuda, O. & Gusev, V.E. (2002). Spatiotemporal mapping of surface acoustic waves in isotropic and anisotropic materials, *Ultrasonics*, Vol.40, pp. 55-59

Thomsen, C., Strait, J., Vardeny, Z., Maris, H.J., Tauc, J. & Hauser, J.J. (1984). Coherent phonon generation and detection by picoseconds light pulses, *Physical Review Letters*, Vol. 53, pp. 989-992

Thomsen, C., Grahn, H.T., Maris, H.J. & Tauc, J. (1986). Surface generation and detection of phonons by picoseconds light pulses, *Physical Review B*, Vol. 34, pp. 4129-4138

Vollmann, J., Profunser, D.M. & Dual, J., (2002). Sensitivity improvement of a pump-probe set-up for thin film and microstructure metrology, *Ultrasonics*, Vol.40, pp. 757–763

Whitfield, C.H., Brody, E.M. & Bassett, W.A. (1976). Elastic moduli of NaCl by Brillouin scattering at high pressure in a diamond anvil cell. *Review of Scientific Instruments*, Vol.47, pp. 942-947

Whitfield, M. D., Audic, B., Flannery, C. M., Kehoe, L. P., Crean, G. M. & Jackman, R. B. (2000). Charactrization of acoustic Lamb wave propagation in polycrystalline diamond film by laser ultrasonics, *Journal of Applied Physics*, Vol.88, pp. 2984—2993.

Zhang, X., Comins, J.D., Every, A.G. & Stoddart, P.R. (1998). Surface Brillouin scattering studies on vanadium carbide, *International Journal of Refractory Metals and Hard Materials*, Vol.16, pp. 303-308

Zhang, X., Stoddart, P.R., Comins, J.D. & Every, A. G. (2001a). High-temperature elastic properties of a nickel-based superalloy studied by surface Brillouin scattering, *Journal of Physics: Condensed. Matter*, Vol.13, pp. 2281-2294

Zhang, X., Sooryakumar, R., Every, A.G. & Manghnani, M.H. (2001b). Observation of organpipe acoustic excitations in supported thin films. Physical Review B, Vol.64, paper n. 081402.

Zinin, P.V (2001). Quantitative acoustic microscopy of solids, in *Handbook of Elastic Properties of Solids, Liquids, and Gases,* M. Levy, H. E. Bass and R. R. Stern & V. Keppens (Eds.), Volume I; *Dynamic Methods for Measuring the Elastic Properties of Solids*, pp. 187-226, Academic Press, New York

7

Evaluation Method for Anisotropic Drilling Characteristics of the Formation by Using Acoustic Wave Information

Deli Gao and Qifeng Pan
China University of Petroleum at Beijing, Beijing
China

1. Introduction

In drilling engineering, we must have solid understanding of the underground geological environment which is not only complicated and diversified but also someway concealed. Thus, in order to find an effective method of predicting it, a long-term research and practice must be required. In drilling engineering for oil & gas, problems including borehole deviation & its control, wellbore instability & its control, influence directly drilling quality & efficiency of a deep or complicated well for exploration and production of oil & gas fields. For instance, because of the complicated surface and underground conditions as well as the depth (over 5000 m) of oil & gas reservoirs in western China, such unstable factors such as hole deviation and instability often encountered with each other in deep drilling engineering. Because we did not have access to the geological parameters of the formations to be drilled including the rock drillability anisotropy and so on, huge economic loss had been caused and the steps to explore and produce oil & gas in western China had been seriously restricted. Hence, there are many researches and development programs to do for the right cognition and scientific evaluation of the geological environments, and for the further study of mechanism of the drilling process instability, and so on. The solution of these problems is the key to improve the performance of drilling & HSSE (health, safety, security, environment) and lower well construction cost.

The factors influencing the instability can be sorted into subjective category and objective category. In the objective category, there are the types of geological structure and in-situ stress, rock anisotropy, porosity, permeability, lithology, pressures, and mineral components, as well as rock strength and weak layer of the formation to be drilled, and so on. In the subjective category, there are the performance of down hole drilling system, the drilling parameters (weight and torque on bit, etc.), the drilling fluid performances (water loss, viscosity, rheological property and density) and its hydration on shale, the direction and open time of wellbore, the erosion and surge pressure of drilling fluid on the hole wall, the interaction between drillstring and hole wall. Thereby, in researches on the instability, the factors from the both categories should be taken into comprehensive consideration.

Whether in vertical drilling or in directional drilling, it is always a complicated academic and technological problem how to control the well trajectory exactly along the designed track to reach the underground targets. In rotary drilling, the forming of wellbore & its trajectory is the result of the rock-bit interaction. In this interaction, the drill bit anisotropy and its mechanical behavior (i.e. the drill bit force and tilt angle) are important factors that can directly affect the well trajectory. The mechanical behavior depends on by the bottom hole assembly (BHA) analysis. Accordingly, principal factors influencing the well trajectory generally contain BHA, drill bit, operating parameters in drilling, drilled wellbore configuration and the formations to be drilled. Of which the BHA, drill bit and operating parameters in drilling are the factors that can be artificially controlled, and the formation property (such as rock drillability and its anisotropy) is the objective factor which can not be changed by us. The trajectory can be predicted before drilling and also can be determined after drilling through surveys and calculations. Besides, the drilled wellbore will not only generate a strong reaction on the drill bit force and the drillstring deflection, but also will exert an influence on the anisotropic drilling characteristics of the formation. Due to the above-complicated factors, the hole deviation is always inevitable, which may seriously influence the wellbore quality and the drilling performance.

The well trajectory control is the process which forces drill bit to break through formations along the designed track forward by applying reasonable techniques. The anisotropic drilling characteristics of the drill bit & the formation and their interaction effects are the factors which will cause a direct influence on the well trajectory control. Thereby, it is a complicated scientific and technological problem for us how to make the cognition, evaluation and utilization of anisotropic drilling characteristics of the formation, as well as the prediction & control of mechanical action of the drill bit on the formations.

Rock drillability anisotropy of the formation to be drilled has significant effects on the well trajectory control so that it is very important to evaluate it. Definitions of rock drillability anisotropy and acoustic wave anisotropy of the formation to be drilled are presented in this chapter. The acoustic velocities and the drillability parameters of some rock samples from Chinese Continental Scientific Drilling (CCSD) are respectively measured with the testing device of rock drillability and the ultrasonic testing system in laboratory. Thus, their drillability anisotropy and acoustic wave anisotropy are respectively calculated and discussed in detail by using the experimental data. Based on the experiments and calculations, the correlations between drillability anisotropy and acoustic wave anisotropy of the rock samples are illustrated through regression analysis. What's more, the correlation of rock drillability in directions perpendicular to and parallel to the bedding plane of core samples is studied by means of mathematical statistics. Thus, a mathematic model is established for predicting rock drillability in direction parallel to the formation bedding plane by using rock drillability in direction perpendicular to the formation bedding plane with the well logging or seismic data. The inversion method for rock anisotropy parameters (ε,δ) is presented by using well logging information and the acoustic wave velocity in direction perpendicular to the bedding plane of the formation is calculated by using acoustic wave velocity in any direction of the bedding plane. Then, rock drillability in direction perpendicular to the bedding plane of the formation can be calculated by using acoustic wave velocity in the same direction. Thus, rock drillability anisotropy and anisotropic drilling characteristics of the formation can be evaluated by using the acoustic wave information based on well logging data. The evaluation method has been examined by case study based on oilfield data in west China.

2. Anisotropic drilling characteristics of the formation

Although many theories have been proposed to explain the hole deviation since the 1950s (Gao et al, 1994), it is only the rock drillability anisotropy theory (Lubinski & Woods, 1953) that was recognized by petroleum engineers and widely applied to petroleum engineering because it can be used to quantify the anisotropic drilling characteristics of the formation and to explain properly the actual cases of hole deviation encountered in drilling engineering. The theory suggested that since values of rock drillability are not always the same in the directions perpendicular and parallel to the bedding plane of the formation, the formation will bring the bit a considerable force, which may likely cause changes on the original drilling direction and hole deviation.

The orthotropic or the transversely isotropic formations are the typical formations encountered frequently in drilling engineering. The anisotropic effects of the formations (rock drillability) on the well trajectory must be considered in hole deviation control and directional drilling. Based on the rock-bit interaction model, the formation force is defined and modeled in this section to describe quantitatively anisotropic drilling characteristics of the formations to be drilled.

2.1 Definition of rock drillability anisotropy

Because of rock drillability anisotropy, the real drilling direction does not coincide with the resultant force direction of the drill bit (supposed that it is isotropic) on bottom hole. Besides calculating the drill bit force by BHA (bottom hole assembly) analysis, rock drillability anisotropy of the formation must be considered in hole deviation control.

The formation studied here is typical orthotropic one, and the transversely isotropic formation discussed previously is regarded as its particular case. Let $\mathbf{e}_d$, $\mathbf{e}_u$ and $\mathbf{e}_s$ represent unit vectors in the directions of inner normal, up-dip and strike of the formation respectively, as shown in Fig.1.There are different physical properties along different directions of them. γ in Fig.1 represents dip angle of the formation to be drilled. Rock drillability anisotropy of the formation can be expressed by rock drillability anisotropy index. If the components of penetration rate of the drill bit (isotropic) along inner normal, up-dip and strike of the orthotropic formation are noted as R_{dip}, R_{str} and R_{n} respectively, correspondingly the net applied forces are F_{dip}, F_{str} and F_{n} respectively, the rock drillability can be defined as:

$$D_{\mathrm{n}} = \frac{R_{\mathrm{n}}}{F_{\mathrm{n}}} \;,\; D_{\mathrm{dip}} = \frac{R_{\mathrm{dip}}}{F_{\mathrm{dip}}} \;,\; D_{\mathrm{str}} = \frac{R_{\mathrm{str}}}{F_{\mathrm{str}}} \tag{1}$$

Rock drillability anisotropy of the orthotropic formation may be represented by two indexes (I_{r1} and I_{r2}) which are defined as:

$$I_{\mathrm{r1}} = \frac{D_{\mathrm{dip}}}{D_{\mathrm{n}}} \;,\; I_{\mathrm{r2}} = \frac{D_{\mathrm{str}}}{D_{\mathrm{n}}} \tag{2}$$

Dip angle and strike of the formation can be obtained from the analysis of well logging and geological structure survey. The values of I_{r1} and I_{r2} for the orthotropic formation can be evaluated by the experimental analysis or using the acoustic wave information.

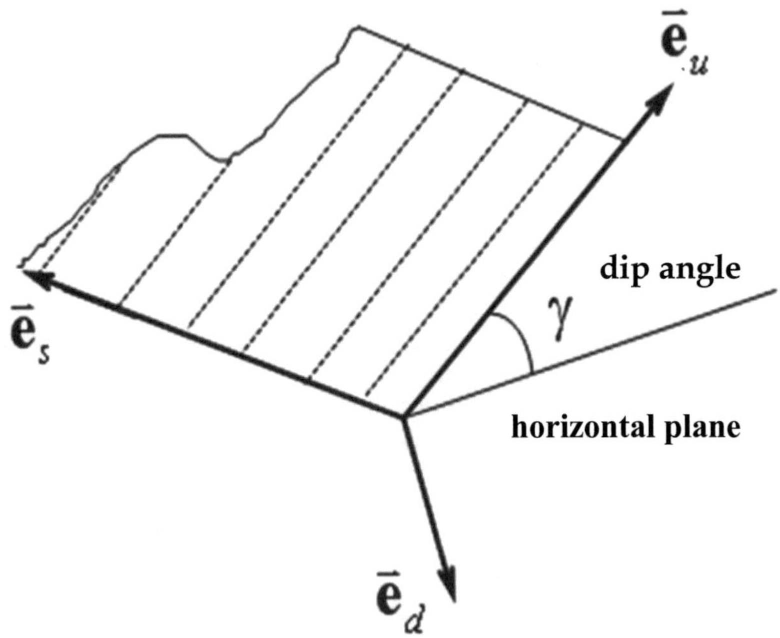

Fig. 1. Descartes coordinates for the formation geometry

2.2 The formation force

Assumed that the drill bit is isotropic for eliminating the effects of its tilt angle on hole deviation, the effects of the orthotropic formation on hole deviation can be presented by the formation force analysis. The two parameter equations related to the formation forces can be derived from the rock-bit interaction model (Gao & Liu, 1989):

$$\begin{cases} G_\alpha = \dfrac{t_{22}t_{13} - t_{12}t_{23}}{t_{11}t_{22} - t_{12}t_{21}} \\ G_\phi = \dfrac{t_{11}t_{23} - t_{21}t_{13}}{t_{11}t_{22} - t_{12}t_{21}} \end{cases} \tag{3}$$

Where G_α and G_ϕ are called as the building angle parameter (positive for building up the inclination of well trajectory) and the drifting azimuth parameter (positive for left walking of well trajectory) of the formation respectively , and the t_{ij} (i, j=1,2,3) can be expressed as (Gao & Liu, 1990):

$$\begin{cases} t_{ij} = I_{r1}\delta_{ij} + (1 - I_{r1})a_{ij} + (I_{r2} - I_{r1})c_{ij} \\ \delta_{ij} = \begin{cases} 0, i \neq j \\ 1, i = j \end{cases} \end{cases} \tag{4}$$

where $a_{ij} = a_{ji}$, $c_{ij} = c_{ji}$ (i, j=1, 2, 3) can be calculated by the following equations:

$$\left.\begin{aligned}
a_{11} &= (\sin\alpha\cos\gamma - \cos\alpha\sin\gamma\cos\Delta\varphi)^2 \\
a_{12} &= (\cos\alpha\sin\gamma\cos\Delta\varphi - \sin\alpha\cos\gamma)\sin\gamma\sin\Delta\varphi \\
a_{13} &= (\cos\alpha\sin\gamma\cos\Delta\varphi - \sin\alpha\cos\gamma)(\sin\alpha\sin\gamma\cos\Delta\varphi + \cos\alpha\cos\gamma) \\
a_{21} &= (\cos\alpha\sin\gamma\cos\Delta\varphi - \sin\alpha\cos\gamma)\sin\gamma\sin\Delta\varphi \\
a_{12} &= a_{21} \\
a_{22} &= (\sin\gamma\sin\Delta\varphi)^2 \\
a_{23} &= \sin\gamma\sin\Delta\varphi(\sin\alpha\sin\gamma\cos\Delta\varphi + \cos\alpha\cos\gamma) \\
a_{31} &= a_{13} \\
a_{32} &= a_{23} \\
a_{33} &= (\sin\alpha\sin\gamma\cos\Delta\varphi + \cos\alpha\cos\gamma)^2
\end{aligned}\right\} \quad (5)$$

$$\left.\begin{aligned}
c_{11} &= (\sin\Delta\varphi\cos\alpha)^2 \\
c_{12} &= -\cos\Delta\varphi\sin\Delta\varphi\cos\alpha \\
c_{13} &= (\sin\Delta\varphi)^2\sin\alpha\cos\alpha \\
c_{21} &= c_{12} \\
c_{22} &= (\cos\Delta\varphi)^2 \\
c_{23} &= -\cos\Delta\varphi\sin\Delta\varphi\sin\alpha \\
c_{31} &= c_{13} \\
c_{32} &= c_{23} \\
c_{33} &= (\sin\Delta\varphi\sin\alpha)^2
\end{aligned}\right\} \quad (6)$$

Where $\Delta\varphi = \varphi - \psi$; φ and α are respectively azimuth and inclination of well trajectory on the bottom hole; γ and ψ are respectively dip angle and up dip azimuth of the formation to be drilled.

It is obviously that the values of G_α and G_ϕ are not only controlled by rock drillability anisotropy of the formation, but also affected by the formation geometry and the well trajectory. Therefore, G_α and G_ϕ can be used to describe the anisotropic drilling characteristics of the formation to be drilled. Thus, the formation force can be mathematically defined as:

$$\begin{cases} GF_\alpha = G_\alpha W_{ob} \\ GF_\phi = G_\phi W_{ob} \end{cases} \quad (7)$$

Where GF_α and GF_ϕ are called as the inclination force (positive for building up the inclination) and the azimuth force (positive for decreasing the azimuth) of the formation respectively, and W_{ob} is weight on bit. It should be pointed out that both GF_α and GF_ϕ are only an equivalent expression of anisotropic drilling characteristics of the formation and they are completely different from the mechanical action forces of the drill bit on the formation. Rock drillability anisotropy of the formation is the internal cause of the generations of GF_α and GF_ϕ , while weight on bit is the its external cause.

2.3 G_α and G_φ of the transversely isotropic formation

By using equations (5) and (6) and making $I_{r1} = I_{r2} = I_r$, equation (3) can be simplified as the following expressions of G_α and G_ϕ for the transversely isotropic formation:

$$G_\alpha = \frac{(1-I_r)(\cos\alpha\sin\gamma\cos\Delta\varphi - \sin\alpha\cos\gamma)(\cos\alpha\cos\gamma + \sin\alpha\sin\gamma\cos\Delta\varphi)}{I_r + (1-I_r)\left[(\sin\gamma\sin\Delta\varphi)^2 + (\cos\alpha\sin\gamma\cos\Delta\varphi - \sin\alpha\cos\gamma)^2\right]} \tag{8}$$

$$G_\varphi = \frac{(1-I_r)\sin\gamma\sin\Delta\varphi(\cos\alpha\cos\gamma + \sin\alpha\sin\gamma\cos\Delta\varphi)}{I_r + (1-I_r)\left[(\sin\gamma\sin\Delta\varphi)^2 + (\cos\alpha\sin\gamma\cos\Delta\varphi - \sin\alpha\cos\gamma)^2\right]} \tag{9}$$

Where all the symbols here express the same meanings as the previous ones.

3. Experiments on rock anisotropy

Evaluation of rock drillability anisotropy is necessary for hole deviation control in drilling engineering. Many efforts have been made to evaluate rock drillability of the formation through the core testing, the inverse calculation and the acoustic wave. Proposed in this section is an alternative solution by using the acoustic wave to evaluate rock drillability anisotropy of the formation. First, a correlation between the P-wave velocity anisotropy coefficient and the rock drillability anisotropy index of the formation which are calculated according to the core testing data in laboratory, is established by means of mathematical statistics. Then, a mathematical model is obtained for predicting the rock drillability anisotropy index by using the P-wave velocity anisotropy coefficient. Thus, rock drillability anisotropy of the formation can be evaluated conveniently by using the well logging or seismic data (Gao & Pan, 2006).

3.1 Rock drillability anisotropy

3.1.1 Definition

The transversely isotropic formation is a typical anisotropic formation, whose anisotropy can be expressed by a rock drillability anisotropy index:

$$I_r = \frac{D_h}{D_v} \tag{10}$$

where $D_v = V_v/F_v$ and $D_h = V_h/F_h$ are respectively rock drillability parameters in the directions perpendicular and parallel to the bedding plane of the transversely isotropic formation; V_v & F_v and V_h & F_h are the corresponding components of the penetration rate & the net applied force of the isotropic bit to the formation.

When the rock drillability is tested in laboratory using the core samples, the weight on the bit and the rotary speed are constant so that rock drillability anisotropy index of the transversely isotropic formation can also be expressed as:

$$I_r = \frac{T_v}{T_h} \tag{11}$$

where T_v and T_h are two parameters representing the drilling time (seconds) in directions perpendicular and parallel to bedding plane of the core samples respectively. The standard definition of rock drillability can be expressed by the following equation (Yin, 1989):

$$K_d = \log_2 T \tag{12}$$

where K_d is the rock drillability and T the drilling time. Taking two sides of equation (11) into logarithm to the base 2, we can obtain the following equations:

$$\log_2 I_r = \log_2 T_v - \log_2 T_h = K_{dv} - K_{dh} = -\Delta K_d \tag{13}$$

$$I_r = 2^{-\Delta K_d} \tag{14}$$

3.1.2 Rock samples

Fourteen core samples used in laboratory came from the measured depth interval of 48m~1027 m of the well KZ-1 for scientific drilling in China, which were supplied by the Engineering Center for Chinese Continental Scientific Drilling (CCSD). In the directions perpendicular and parallel to the bedding plane, these core samples were cut into shapes of cube or cuboid and their surfaces of both ends were polished and kept parallel to each other, with an error of less than 0.2 mm. Then, the machined samples were put into an oven with a temperature of 105-110°C and roasted for 24 h. Finally, all of the samples can be used for the testing of rock drillability after cooling down to room temperature.

3.1.3 Testing method

The rock drillability can be measured with a device for testing the rock drillability (shown in Fig.2). During the measurement, some weight is applied on the micro-bit by the function of a hydraulic pressure tank with the fixed poises, so that the weight on the micro-bit is kept at a constant value. The measured depth to be drilled to is set with the standard indicator, and the drilling time is logged with a stopwatch. Both the roller bit (bit of this kind has three rotating cones and each cone will rotate on its own axis during drilling) drillability and the PDC (the acronym of Polycrystalline Diamond Compact) bit drillability can be tested with the above-mentioned instrument, which is of the following standard data.

The diameter of the micro-bit is 31.75 mm.

Weight is 90±20 N on the roller bit and 500±20 N on the PDC bit.

The rotary speed is 55±1 r/min.

The total depth to be drilled to is 2.6 mm for the roller bit with a pre-drilled depth of 0.2 mm and 4 mm for the PDC bit with a pre-drilled depth of 1.0 mm.

During testing the rock drillability, the micro-bit is often checked so that each of the worn micro-bits should be replaced in time to ensure the testing accuracy. The testing points of drilling time for each tested side of a rock sample should be gained as many as possible and their average value is taken as the test value of the side. The grade value of each side drillability of the rock sample can be calculated by equation (16) with the test data of drilling time for each side of the rock sample.

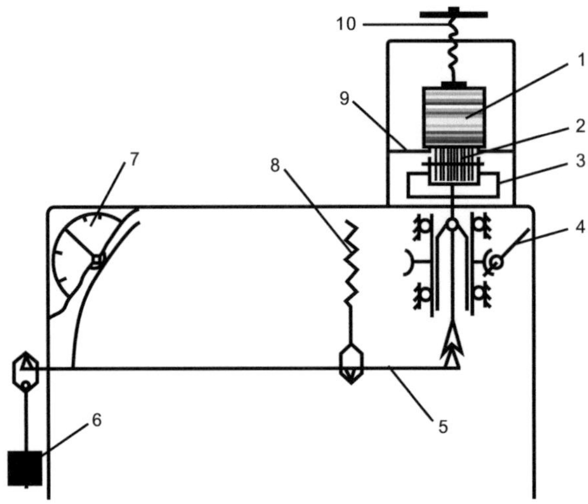

Fig. 2. Testing device for rock drillability(Note: 1. Rock sample; 2. micro-bit; 3. cutting tray; 4. turbine rod; 5. lever; 6. weight; 7. meter for measuring depth; 8. bar with thread for adjusting lever; 9. worktable; 10. compaction bar with thread)

3.1.4 Experimental result and analysis

Some testing results of rock drillability for the 14 core samples from CCSD are obtained in laboratory and shown in Table 1 and Table 2.

No. of the cores from CCSD	Measured depth, m	Rock drillability with the roller bit (K_{dRB})		Rock drillability anisotropy index
		Perpendicular to the bedding plane	Parallel to the bedding plane	
9	48	6.03	6.12	0.94
38	145	9.18	9.86	0.62
57	197	10.29	10.79	0.71
104	305	11.11	11.39	0.82
143	400	8.21	8.17	1.03
179	504	8.70	8.99	0.82
218	607	9.25	10.29	0.49
252	698	10.64	10.78	0.91
281	775	8.78	9.21	0.74
288	795	10.17	8.22	3.86
304	834	7.92	8.57	0.64
340	925	8.89	9.08	0.88
363	998	9.15	10.20	0.48
373	1027	8.12	10.21	0.23

Table 1. Experimental results of rock drillability with the roller bit

No. of the cores from CCSD	Measured depth, m	Rock drillability with the PDC bit (K_{dPDC})		Rock drillability anisotropy index
		Perpendicular to the bedding plane	Parallel to the bedding plane	
9	48	4.55	4.34	1.16
38	145	8.57	10.65	0.24
57	197	9.89	10.06	0.89
104	305	10.78	10.90	0.92
143	400	7.82	7.40	1.34
179	504	8.48	8.47	1.01
218	607	8.61	9.14	0.69
252	698	9.52	9.86	0.79
281	775	8.22	9.03	0.57
288	795	9.55	7.31	4.72
304	834	6.06	7.71	0.32
340	925	8.14	8.95	0.57
363	998	8.18	8.64	0.73
373	1027	7.93	8.77	0.56

Table 2. Experimental results of rock drillability with the PDC bit

It is observed clearly from Table 1 and Table 2 that the rock samples from CCSD have the anisotropic characteristics in the rock drillability. The rock drillability perpendicular to the bedding plan is different from that parallel to the bedding plane, whether it is for the roller bit or for the PDC bit. For the roller bit, indices of drillability anisotropy of the rock samples are ranged from 0.23 to 0.94, except the anisotropy indices of rock samples of 143# and 288#, which are 1.03 and 3.86 respectively. The case is similar to the PDC bit; indices of drillability anisotropy of the rock samples are between 0.24 and 0.92, except the anisotropy indices of rock samples of 9#, 143#, 179# and 288#, corresponding to 1.16, 1.34, 1.01 and 4.72, respectively. Generally, the rock drillability perpendicular to the bedding plan is less than that parallel to the bedding plane, so that the formation can be penetrated more easily in the direction perpendicular to the bedding plane.

3.2 Acoustic anisotropy of rock sample

3.2.1 Definition

It is supposed that the formation is the transversely isotropic, and thus the acoustic anisotropy of the formation rock can be expressed by an acoustic anisotropy index (I_v):

$$I_v = V_{av} / V_{ah} \tag{15}$$

where V_{av} and V_{ah} are the acoustic velocities in rock along the directions perpendicular and parallel to the bedding plane of the formation respectively.

3.2.2 Testing method

With the method of making the ultrasonic pulse penetrating through a rock sample, the acoustic velocities V_{av} and V_{ah} can be measured in laboratory. The ultrasonic testing system used in laboratory is shown in Fig. 3, in which the ultrasonic transducers can provide a frequency of 0.5 MHz and the butter and honey can be used as its coupling media. The pulse generator can generate electric pulses with a strength range of 1-300 V. The width and iteration frequency of the electric pulse can be adjusted and controlled. During testing, the signal generator makes an electric pulse signal which will touch off the emission end of the energy exchanger to generate ultrasonic pulses. The ultrasonic pulses (acoustic waves) propagating through the rock sample are incepted by the reception end of the energy exchanger. Finally, the propagation time and the signal strength of the ultrasonic pulses (acoustic waves) through the rock sample are logged by a digital memory oscillograph.

In order to reduce the errors from the artificial operations, the emission end of the energy exchanger is aimed at its reception end as accurately as possible during testing. Before each test, the ultrasonic testing system should be calibrated using the aluminum rod to ensure the accuracy of the test results. Testing for each point of a rock sample is conducted for three times in the actual testing. The average value of the test data of three times for each point is taken as a final test result for the point of a rock sample. With the test data, the acoustic velocity may be calculated by the following equation:

$$V = \frac{l}{t - t_0} \tag{16}$$

where V is the acoustic velocity; l is length of the rock sample, mm; t is propagation time of the acoustic wave, μs; and t_0 is delayed time of the testing system, μs.

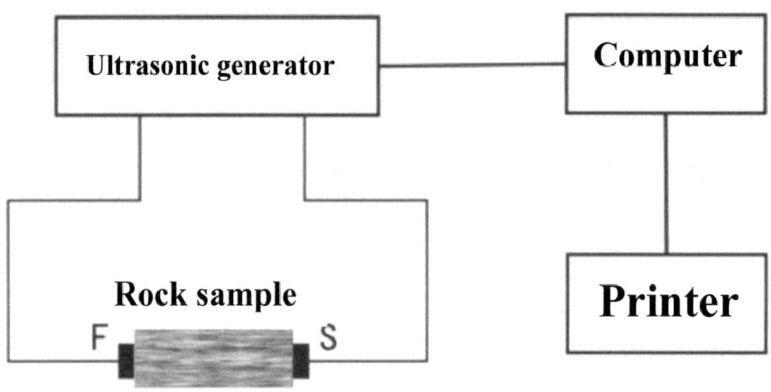

Fig. 3. The ultrasonic testing system

3.2.3 Experimental result and analysis

Some ultrasonic test results of the 14 core samples from CCSD are logged with the above test method and with the ultrasonic testing system in laboratory, and the rock acoustic velocities shown in Table 3 can be calculated by equation (16).

It can be obviously observed from Table 3 that the rock samples from CCSD are of the rock acoustic anisotropy. The rock acoustic velocity perpendicular to the bedding plan is different from that parallel to the bedding plane. Based on the acoustic velocity data in Table 3, the acoustic anisotropy of the rock samples can be calculated by equation (15). The acoustic anisotropy indices of the rock samples are between 0.85 and 0.98, except the 363# and 373#, which are 0.77 and 0.76 respectively. For the test of the rock samples from CCSD, the rock acoustic velocity perpendicular to the bedding plan is less than that parallel to the bedding plane, as shown in Table 3. The main reason for this difference is that there are many fractures with different scales in the rock sample. When the acoustic wave penetrates through the fractures, the fractures cause a loss of the pulse energy so as to make the acoustic velocity reduce more quickly, on the other hand, the pulse energy is dissipated in the process of propagation. According to some progress in geophysics (Patrick & Richard, 1984), the fractures can play a role in guiding the wave when the elastic wave has propagated in the direction parallel to the bedding plane of the rock sample, and play a role in obstructing the wave when the elastic wave has propagated in the direction perpendicular to the bedding plane. Therefore, the propagation of the acoustic wave penetrating through the rock sample is probably controlled by such a kind of geophysical mechanism.

No. of the core from CCSD	Measured depth, m	P-wave velocities of the rock samples, m/s		Acoustic anisotropy index of the rock sample
		Perpendicular to bedding plane (V_{av})	Parallel to bedding plane (V_{ah})	
9	48	4387	4457	0.98
38	145	4442	5120	0.87
57	197	6365	6826	0.93
104	305	5431	5714	0.95
143	400	4410	4928	0.89
179	504	4568	4744	0.96
218	607	4177	4671	0.89
252	698	5805	5990	0.97
281	775	4776	5516	0.87
288	795	5142	5453	0.94
304	834	4392	5129	0.86
340	925	4325	5096	0.85
363	998	3816	4928	0.77
373	1027	3761	4930	0.76

Table 3. Experimental results of acoustic velocities of the rock samples

3.3 Correlations between I_r and I_v

With the experimental data in table 1 to table 3 and the corresponding calculations, it can be found that the rock drillability anisotropy is inherently related to the acoustic anisotropy of the rock samples. Therefore, exponential function, logarithmic function, polynomial function, and linear function, are used to make a regression analysis of the data obtained by experiments. With the matching & extrapolating effects of these regression functions comprehensively considered, exponential function is finally selected as the regression model of correlation between I_r and I_v . The results of regression calculations for the correlations are listed in Table 4.

In Table 4, I_{rRB} is denoted as the drillability anisotropy index of the rock sample with a roller bit, I_{vp} as the acoustic anisotropy index of P-wave through the rock sample, ΔK_{dRB} as the rock drillability difference between both directions perpendicular and parallel to the bedding plane of the rock sample with a roller bit, calculated by equation (13), and I_{rPDC} as the drillability anisotropy index of the rock sample with a PDC bit.

Regression functions		$I_{rRB} = e^{(a+bI_{vp})}$	$\Delta K_{dRB} = e^{(a+bI_{vp})}$	$I_{rPDC} = e^{(a+bI_{vp})}$
a	Value	−3.418	8.129	−4.286
	Standard error	0.840	1.893	1.491
	t-ratio	−4.0679	4.295	−2.875
	Prob(t)	0.00226	0.00157	0.02068
b	Value	3.401	-9.993	4.366
	Standard error	0.914	2.361	1.661
	t-ratio	3.721	−4.232	2.628
	Prob.(t)	0.00397	0.00174	0.03026
Correlation coefficient R		0.793	0.832	0.710
F-ratio		16.92	22.41	8.09
Prob.(F)		0.0021	0.0008	0.0217

Table 4. Results of the regression calculations

4. Evaluation method based on acoustic wave information

Many studies have been made to evaluate the rock drillability anisotropy with the core testing method (Gao & Pan, 2006) and the inversion method (Gao et al, 1994). However, as for the core testing method, its result may not reflect the actual rock drillability anisotropy since the experimental conditions are different from the downhole conditions. Moreover, the profile of rock drillability anisotropy along the hole depth can not be established because of the limitation of the core samples. The inversion method needs to work with a bottom hole assembly (BHA) analysis program and some parameters in the inversion model are not easy to obtain so that its applications are limited to some extent. Thus, the evaluation method will be presented in this section so as to predict rock drillability anisotropy of the formation by using the acoustic wave information (Gao et al, 2008).

4.1 Acoustic wave velocity of the formation

The formation studied here is the transversely isotropic formation which is frequently encountered in drilling for oil & gas. Experimental investigation shows that layered rock has the transversely isotropic characteristics.

4.1.1 Phase velocity in the transversely isotropic formation

For the transversely isotropic media, Hooke's law can be written as

$$\begin{pmatrix}\sigma_{xx}\\ \sigma_{yy}\\ \sigma_{zz}\\ \sigma_{yz}\\ \sigma_{xz}\\ \sigma_{xy}\end{pmatrix}=\begin{pmatrix}C_{11} & C_{11}-2C_{66} & C_{13} & 0 & 0 & 0\\ C_{11}-2C_{66} & C_{11} & C_{13} & 0 & 0 & 0\\ C_{13} & C_{13} & C_{33} & 0 & 0 & 0\\ 0 & 0 & 0 & C_{44} & 0 & 0\\ 0 & 0 & 0 & 0 & C_{44} & 0\\ 0 & 0 & 0 & 0 & 0 & C_{66}\end{pmatrix}\begin{pmatrix}\varepsilon_{xx}\\ \varepsilon_{yy}\\ \varepsilon_{zz}\\ \varepsilon_{yz}\\ \varepsilon_{xz}\\ \varepsilon_{xy}\end{pmatrix} \tag{17}$$

Elastodynamic equation of the elastic media can be obtained from the textbook and expressed as

$$\begin{cases}\dfrac{\partial\sigma_{xx}}{\partial x}+\dfrac{\partial\sigma_{yx}}{\partial y}+\dfrac{\partial\sigma_{zx}}{\partial z}+X-\rho\dfrac{\partial^2 u}{\partial t^2}=0\\ \dfrac{\partial\sigma_{yy}}{\partial y}+\dfrac{\partial\sigma_{zy}}{\partial z}+\dfrac{\partial\sigma_{xy}}{\partial x}+Y-\rho\dfrac{\partial^2 v}{\partial t^2}=0\\ \dfrac{\partial\sigma_{zz}}{\partial z}+\dfrac{\partial\sigma_{xz}}{\partial x}+\dfrac{\partial\sigma_{yz}}{\partial y}+Z-\rho\dfrac{\partial^2 w}{\partial t^2}=0\end{cases} \tag{18}$$

where X, Y, and Z are respectively the body force in directions of x, y and z (Xu, 2011). u, v and w are the corresponding displacements. ρ is the density of the elastic media, $\mathrm{g/cm^3}$. Substituting equation (17) into equation (18) and solving with geometric equations without considering body force, we can get the following wave equation :

$$\begin{cases}\rho\dfrac{\partial^2 u}{\partial t^2}=C_{11}\dfrac{\partial^2 u}{\partial x^2}+C_{66}\dfrac{\partial^2 u}{\partial y^2}+C_{44}\dfrac{\partial^2 u}{\partial z^2}+(C_{11}-C_{66})\dfrac{\partial^2 v}{\partial x\partial y}+(C_{13}+C_{44})\dfrac{\partial^2 w}{\partial x\partial z}\\ \rho\dfrac{\partial^2 v}{\partial t^2}=C_{66}\dfrac{\partial^2 v}{\partial x^2}+C_{22}\dfrac{\partial^2 v}{\partial y^2}+C_{44}\dfrac{\partial^2 v}{\partial z^2}+(C_{11}-C_{66})\dfrac{\partial^2 u}{\partial x\partial y}+(C_{13}+C_{44})\dfrac{\partial^2 w}{\partial y\partial z}\\ \rho\dfrac{\partial^2 w}{\partial t^2}=C_{44}\dfrac{\partial^2 w}{\partial x^2}+C_{44}\dfrac{\partial^2 w}{\partial y^2}+C_{33}\dfrac{\partial^2 w}{\partial z^2}+(C_{13}+C_{44})\dfrac{\partial^2 u}{\partial x\partial z}+(C_{13}+C_{44})\dfrac{\partial^2 v}{\partial y\partial z}\end{cases} \tag{19}$$

Because of the symmetry of the stress and strain in the direction normal to z direction, the wave equation can be simplified to two dimensions without any loss of generality. In the plane of y=0(that is the xz plane), the wave equation (19) can be written as

$$\begin{cases}\rho\dfrac{\partial^2 u}{\partial t^2}=C_{11}\dfrac{\partial^2 u}{\partial x^2}+C_{44}\dfrac{\partial^2 u}{\partial z^2}+(C_{13}+C_{44})\dfrac{\partial^2 w}{\partial x\partial z}\\ \rho\dfrac{\partial^2 v}{\partial t^2}=C_{66}\dfrac{\partial^2 v}{\partial x^2}+C_{44}\dfrac{\partial^2 v}{\partial z^2}\\ \rho\dfrac{\partial^2 w}{\partial t^2}=C_{44}\dfrac{\partial^2 w}{\partial x^2}+C_{33}\dfrac{\partial^2 w}{\partial z^2}+(C_{13}+C_{44})\dfrac{\partial^2 u}{\partial x\partial z}\end{cases} \tag{20}$$

The solutions of equation (20) are

$$\rho v_{Pa}^2(\theta) = \frac{1}{2}[C_{33} + C_{44} + (C_{11} - C_{33})\sin^2\theta + D(\theta)] \quad (21)$$

$$\rho v_{SVa}^2(\theta) = \frac{1}{2}[C_{33} + C_{44} + (C_{11} - C_{33})\sin^2\theta - D(\theta)] \quad (22)$$

$$\rho v_{SHa}^2(\theta) = C_{66}\sin^2\theta + C_{44}\cos^2\theta \quad (23)$$

Where

$$D(\theta) = \{(C_{33} - C_{44})^2 + 2[2(C_{13} + C_{44})^2 - (C_{33} - C_{44})(C_{11} + C_{33} - 2C_{44})]\sin^2\theta + \\ + [(C_{11} + C_{33} - 2C_{44})^2 - 4(C_{13} + C_{44})^2]\sin^4\theta\}^{1/2},$$

where v_{Pa} is phase velocity of the P-wave; v_{SVa} is phase velocity of P-SV wave; v_{SHa} is phase velocity of SH-wave; θ is phase angle which is the angle between the wave front normal and the unique (vertical) axis as shown in Fig.4.

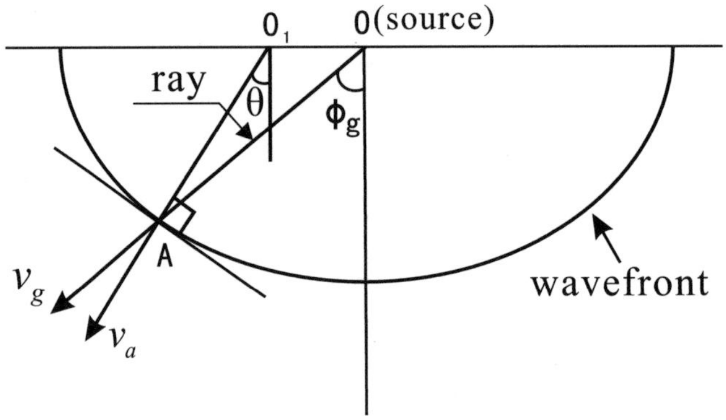

Fig. 4. Phase angle and group angle

It is defined that

$$\varepsilon = \frac{C_{11} - C_{33}}{2C_{33}} \quad (24)$$

$$\gamma = \frac{C_{66} - C_{44}}{2C_{44}} \quad (25)$$

$$\delta^* = \frac{1}{2C_{33}^2}[2(C_{13} + C_{44})^2 - (C_{33} - C_{44})(C_{11} + C_{33} - 2C_{44})] \quad (26)$$

and

$$\alpha_0 = \sqrt{C_{33}/\rho}, \quad \beta_0 = \sqrt{C_{44}/\rho} \quad (27)$$

where α_0 is the vertical P-wave velocity; β_0 is the vertical SV-wave velocity; ρ is rock density. ε , γ and δ^* are rock anisotropy parameters of the formation.
Substituting equation (24), (25), (26) and (27) into equation (21), (22) and (23), we can get

$$v_{Pa}^2(\theta) = \alpha_0^2[1+\varepsilon\sin^2\theta + D^*(\theta)] \, , \tag{28}$$

$$v_{SVa}^2(\theta) = \beta_0^2[1+\frac{\alpha_0^2}{\beta_0^2}\varepsilon\sin^2\theta - \frac{\alpha_0^2}{\beta_0^2}D^*(\theta)] \tag{29}$$

$$v_{SHa}^2(\theta) = \beta_0^2[1+2\gamma\sin^2\theta] \tag{30}$$

where

$$D^*(\theta) = \frac{1}{2}(1-\frac{\beta_0^2}{\alpha_0^2})\{[1+\frac{4\delta^*}{(1-\beta_0^2/\alpha_0^2)^2}\sin^2\theta\cos^2\theta + \frac{4(1-\beta_0^2/\alpha_0^2+\varepsilon)\varepsilon}{(1-\beta_0^2/\alpha_0^2)^2}\sin^4\theta]^{1/2} - 1\} \tag{31}$$

Letting $\theta = \frac{\pi}{2}$ and substituting it into equations (28), (29), (30) and (31), we can get

$$\begin{cases} v_{Pa,90} = \alpha_0(1+2\varepsilon)^{0.5} \\ v_{SVa,90} = \beta_0 \\ v_{SHa,90} = \beta_0(1+2\gamma)^{0.5} \end{cases} \tag{32}$$

For the case of weak rock anisotropy (i.e. the quantity of ε , γ, θ and δ^* is small), expanding equation (31) in the Taylor series at fixed θ and neglecting the second order of small quantity can be simplified as

$$D^* \approx \frac{\delta^*}{(1-\beta_0^2/\alpha_0^2)}\sin^2\theta\cos^2\theta + \varepsilon\sin^4\theta \tag{33}$$

Using equations (33) in equation (28) and (29), expanding v_{Pa} , v_{SVa} and v_{SHa} in a Taylor series at the fixed θ and neglecting the second order of small quantity can be expressed as

$$v_{Pa}(\theta) = \alpha_0(1+\delta\sin^2\theta\cos^2\theta + \varepsilon\sin^4\theta) \tag{34}$$

$$v_{SVa}(\theta) = \beta_0[1+\frac{\alpha_0^2}{\beta_0^2}(\varepsilon-\delta)\sin^2\theta\cos^2\theta] \tag{35}$$

$$v_{SHa}(\theta) = \beta_0(1+\gamma\sin^2\theta) \tag{36}$$

Where

$$\delta = \frac{1}{2}\left[\varepsilon + \frac{\delta^*}{(1-\beta_0^2/\alpha_0^2)}\right].$$

4.1.2 Phase velocity and group velocity

The phase velocity is the velocity in the direction of the phase propagation vector, normal to the surface of constant phase, which is also called the wave front velocity since it is the propagation velocity of the wave front along the phase vector. The phase angle is formed between the direction of phase vector and the vertical axis. In contrast, the ray vector points always from the source to the considered point on the wave front. The energy propagates along the ray vector with the group velocity, while the group angle is formed between the propagation direction and the vertical axis. The difference between the phase angle and the ray angle is illustrated in Fig.4.

The relationship between the phase angle and the group angle can be expressed by the following equation:

$$\tan(\phi_g - \theta) = \frac{\frac{dv_a}{d\theta}}{v_a} \tag{37}$$

Where ϕ_g is the group angle; v_a is the phase velocity; θ is the phase angle. Expanding equation (37) leads to

$$\tan(\phi_g(\theta)) = (\tan\theta + \frac{1}{v_a}\frac{dv_a}{d\theta}) / (1 - \frac{\tan\theta}{v_a}\frac{dv_a}{d\theta}) \tag{38}$$

The group velocity (v_g) is related to the phase velocity (v_a) as shown by the following formula

$$v_g^2(\phi_g(\theta)) = v_a^2(\theta) + \left(\frac{dv_a}{d\theta}\right)^2 \tag{39}$$

Where v_g is the group velocity.

The following section makes the solution for the relationship between the group velocity and the phase velocity, the group angle and the phase angle of the P-wave at any angle.

Another rock anisotropy parameter δ is introduced and expressed as

$$\delta = \frac{1}{2}\left[\varepsilon + \frac{\delta^*}{(1-\beta_0^2/\alpha_0^2)}\right] = \frac{(C_{13}+C_{44})^2 - (C_{33}-C_{44})^2}{2C_{33}(C_{33}-C_{44})} \tag{40}$$

Letting $t = 1 - \beta_0^2/\alpha_0^2$, equation (38) can be rescaled as

$$\frac{v_{Pa}^2(\theta)}{\alpha_0^2} = 1 + \varepsilon\sin^2\theta + D(\theta) \tag{41}$$

Where

$$D(\theta) = \frac{1}{2}\sqrt{4(\varepsilon^2 + 2t\varepsilon - 2t\delta)\sin^4\theta + 4t(2\delta - \varepsilon)\sin^2\theta + t^2} - \frac{1}{2}t\,.$$

Making derivation to both sides of equation (41), we can get

$$\frac{dv_{Pa}(\theta)}{d\theta}=\frac{\alpha_0^2\sin\theta\cos\theta}{v_{Pa}(\theta)R(\theta)}\left[2(\varepsilon^2+2\varepsilon t-2t\delta)\sin^2\theta+2t\delta-\varepsilon t+\varepsilon R(\theta)\right] \tag{42}$$

Where

$$R(\theta)=\left(4(\varepsilon^2+2t\varepsilon-2t\delta)\sin^4\theta+4t(2\delta-\varepsilon)\sin^2\theta+t^2\right)^{\frac{1}{2}}=2D(\theta)+t\,.$$

Substituting equations (41) and (42) into equation (38), we can get

$$\tan\phi_g=\frac{\left\{2\left[M_3(\theta)-M_2(\theta)-2M_1\right]\sin^2\theta-M_4(\theta)-2M_3(\theta)\right\}\tan\theta}{2\left[M_3(\theta)-M_2(\theta)\right]\sin^2\theta-M_4(\theta)} \tag{43}$$

Where

$$M_1=\varepsilon^2-2t\delta+2t\varepsilon\,;$$

$$M_2(\theta)=4t\delta+\varepsilon R(\theta)-2t\varepsilon\,;$$

$$M_3(\theta)=2t\delta+\varepsilon R(\theta)-t\varepsilon\,;$$

$$M_4(\theta)=t^2-tR(\theta)+2R(\theta)\,.$$

Substituting equation (41) and (42) into equation (39), we can get the following equation:

$$v_{Pg}\left(\phi_g(\theta)\right)=\frac{1}{v_{Pa}(\theta)R(\theta)}\left\{v_{Pa}^4(\theta)R^2(\theta)+\alpha_0^4\sin^2\theta\cos^2\theta\left[2M_1\sin^2\theta+M_3(\theta)\right]^2\right\}^{\frac{1}{2}} \tag{44}$$

Where v_{Pg} is the group velocity of the P-wave; v_{Pa} is the phase velocity of the P-wave.
In the case of $\theta=0^\circ$ and $\theta=90^\circ$, $\frac{dv_{Pa}(\theta)}{d\theta}=0$, the phase velocity is equal to the group velocity.Thus, when the group angle of the P-wave at the considered point of the formation is given, its phase angle can be calculated by using equation (43). Phase velocity and group velocity of the point can also be made out by using equation (41) and (44).

4.1.3 Methodology for determining rock anisotropy parameters

From the above discussion, we know that if rock anisotropy parameters δ, ε and γ, are known, the wave velocity at any direction can be calculated by using acoustic wave velocity perpendicular to the bedding of the formation. In other words, the acoustic wave velocity perpendicular to the bedding of the formation can be make out if the wave velocity at any direction and rock anisotropy parameters, δ, ε and γ, are known.

It is assumed that the formation to be drilled is transversely isotropic with symmetry axis perpendicular to the bedding of the formation and the formation properties do not change significantly from one well to another. Acoustic wave logging provides a way to measure the velocity of P-wave or S-wave in the formation (or slowness time). The schematic figure for measuring the velocity of P-wave or S-wave is illustrated in Fig.5.

In the figure 5, S_1 and S_2 are monopole sonic transducers. R_1, R_2, R_3 and R_4 are sonic receivers. When the sonic is transmitted from S_1, time difference between R_2 and R_4 is recorded. In the same way, when the sonic is transmitted from S_2, time difference between R_1 and R_3 is recorded. The average of time difference between R_2 and R_4 and time difference between R_1 and R_3 is the velocity in the formation measured.

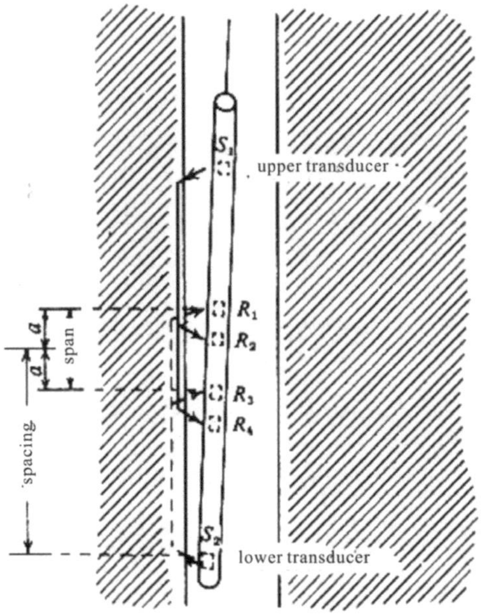

Fig. 5. The principle of acoustic wave logging

Since available S-wave velocity is limited in logging data, we restrict ourselves to take consideration of the P-wave only. The frequency of the acoustic wave logging is about 20kHz~25kHz， which has long wave length. Since a monopole sonic transducer has a mini-bulk, a monopole borehole sonic tool may be approximated by a point source in line with an array of point receivers. The group velocity surface is the response from a point source and so the monopole sonic tool response is approximated as a point source coupled with a series of point receivers in an infinite media (neglecting borehole effects). Therefore, we measure group velocity with borehole sonic tools.

Three parameters, the vertical P-wave velocity (α_0) and the anisotropy parameters ε and δ can be recovered using borehole sonic measurement at different angles relative to the axis of symmetry by following objective function:

$$\Delta v_P = \frac{1}{n}\sum_{i=1}^{n}\left[v_{Pmi}(\theta) - v_{Pci}(\theta)\right]^2 \tag{45}$$

where $v_{Pmi}(\theta)$ is the measured P-wave velocity; $v_{Pci}(\theta)$ is the P-wave velocity calculated by equation (32); n is the total number of the measured signals. The goal of the inversion is to find the optimization value of C_{11}, C_{13}, C_{33} and C_{44}, to minimize value of Δv_P, by which α_0, δ and ε can be calculated, as shown in Fig.6 (Gao et al, 2008).

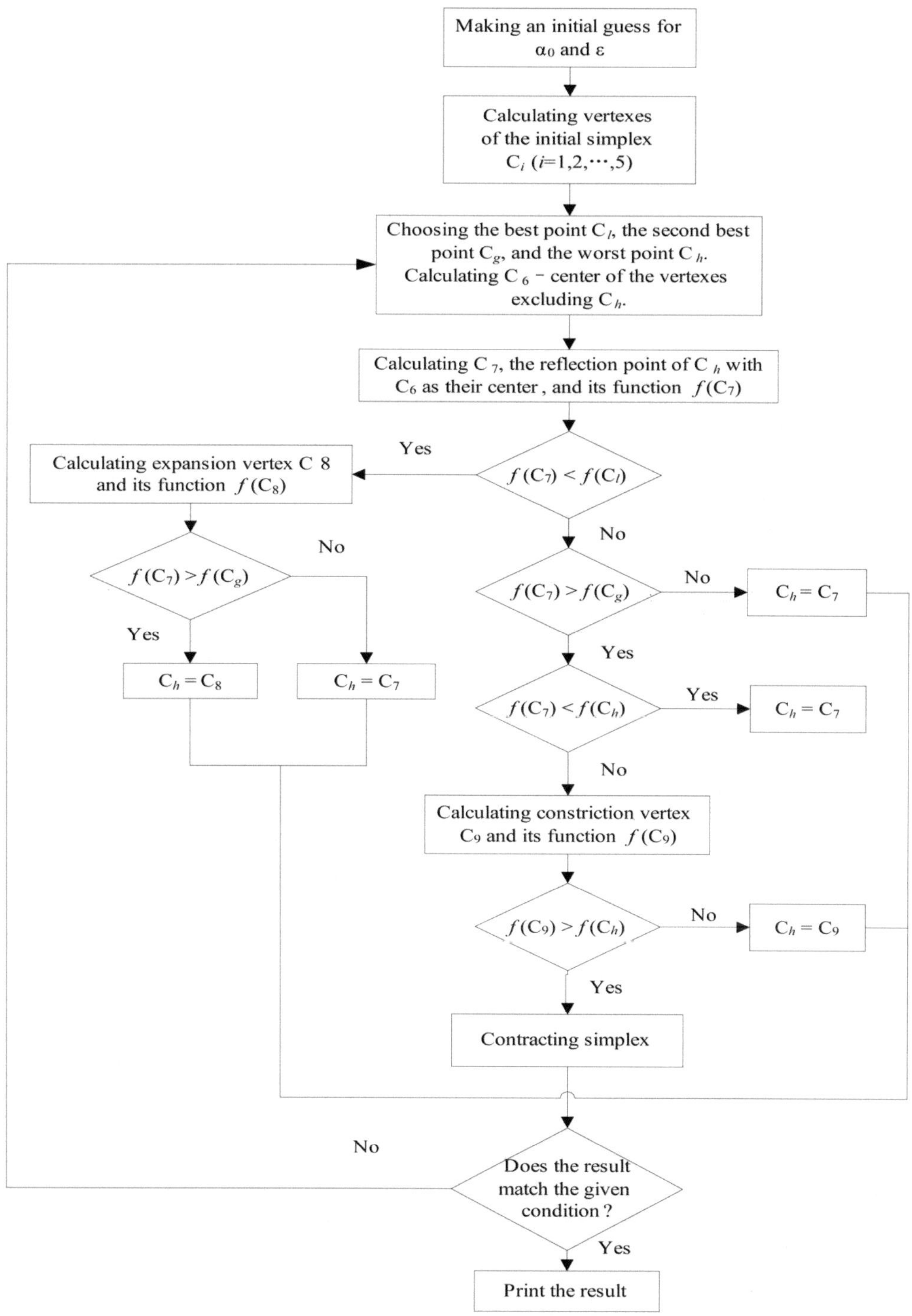

Fig. 6. Flow chart for the inversion calculations of rock anisotropy parameters

4.2 Prediction model of rock drillability anisotropy

Based on the previous section 3, a calculation model has been established to predict rock drillability anisotropy of the formation:

$$I_r = 2^{C_1 K_{dv} + C_2} \tag{46}$$

$$K_{dv} = C_3 + C_4 \ln(\Delta t) \tag{47}$$

Where K_{dv} is the rock drillability perpendicular to the bedding plane of the formation; Δt is the time interval of acoustic wave in the same direction, us/m; C_j (j=1,2,3,4) are the regression coefficients based on the experimental data and the survey data in drilling engineering. For example, by the regression analysis based on some oilfield data in west China, we can get such coefficients as C_1=0.05246, C_2=-0.76732, C_3=32.977, C_4=-4.950.

4.3 Evaluation method of rock drillability anisotropy based on acoustic wave

From equations (46) and (47), it is shown that the key point for the evaluation of rock drillability anisotropy is how to obtain the rock drillability perpendicular to the bedding plane of the formation which depends on the time interval of acoustic wave in the same direction. Thus, the evaluation of rock drillability anisotropy comes down to determine the time interval of acoustic wave perpendicular to the bedding plane of the formation.

Provided that the formation is of the transversely isotropy and has the symmetry axis perpendicular to the bedding plane of the formation, the angle between hole axis and the formation normal can be calculated by the following formula which is derived from transformation of the formation coordinates to the bottom hole coordinate.

$$\omega = \arccos\left[\cos\alpha\cos\beta - \sin\alpha\sin\beta\cos(\phi - \phi_f)\right] \tag{48}$$

where ω is the angle between hole axis and normal of the formation; α is hole inclination, degree or radian; ϕ is azimuth, degree or radian; β is stratigraphic dip, degree or radian; ϕ_f is azimuth of the formation tendency, degree or radian.

When rock anisotropy parameters of a hole section is known, its acoustic wave velocity perpendicular to the bedding plane of the formation can be calculated by the following procedures:

1. Calculating group angle according to stratigraphic dip angle & up dip direction, and inclination & azimuth of hole.
2. Making an initial guess for the acoustic wave velocity $v_{P,0}$.
3. Reading shear wave velocity from shear wave logging or calculating it by equation (46).
4. Calculating phase angle by equation (43).
5. Calculating phase velocity and group velocity of the P-wave by equation (41) and equation (44), respectively.
6. Comparing the P-wave group velocity with the measured velocity.
7. If group velocity of the P-wave matches the measured velocity, $v_{P,0}$ is what we find. Otherwise, we should repeat step 2 to step 7 until they are matched.

The flow chart for inversion of the acoustic wave velocity perpendicular to the bedding plane of the formation is shown in figure 7.

The rock drillability can be calculated by equation (47) after obtaining the time interval of the acoustic wave perpendicular to the bedding plane of the formation. Thus, the profile of rock drillability anisotropy index can be established by using equation (46).

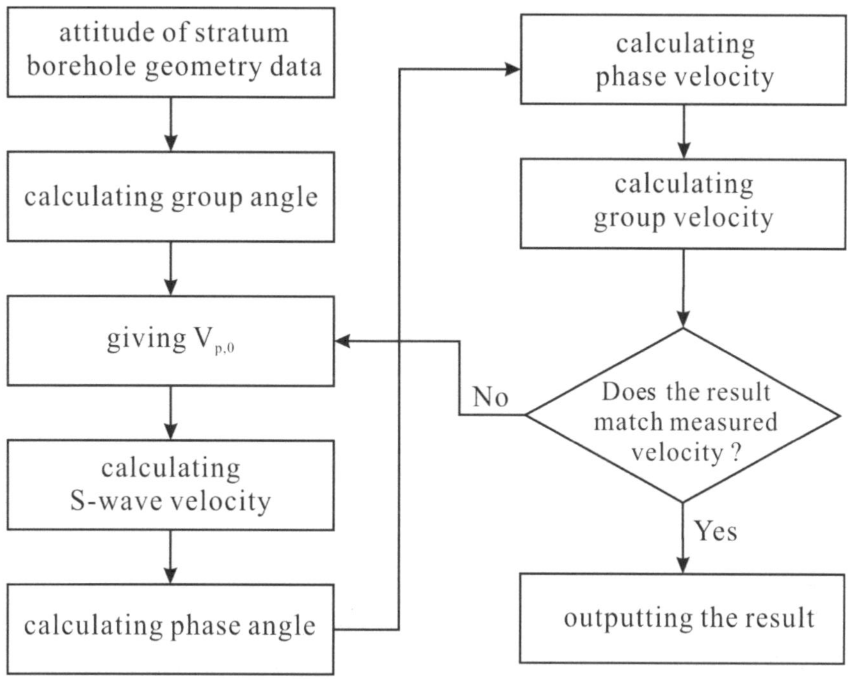

Fig. 7. Inversion method of the acoustic wave velocity perpendicular to the bedding plane

5. Case study

Based on some well logging data and drilling information from Qinghai oilfield in west China, the case study is presented in this section to verify the evaluation method for the anisotropic drilling characteristics of the formation to a certain extent.

Based on these data in table 5, rock drillability anisotropy of the fromation and its anisotropic drilling characteristics can be calculated by using the evaluation method described above. The inversion result of shale anisotropy parameters is shown in table 6.

Well Number	Well logging information
Well 5	gamma-ray, compensated acoustic wave and compensated density, inclinometer data and geologic stratification data
Well 6	gamma-ray, compensated acoustic wave and compensated density, inclinometer data and geologic stratification data
Well 7	gamma-ray, compensated acoustic wave and compensated density, inclinometer data and geologic stratification data, and some other records

Table 5. Well drilling & logging information from some completed wells at the Honggouzi conformation in Qinghai oilfield

$v_{P,0}$ (m/s)	ε	δ	root-mean-square (m/s)
4029.1	1.6833	1.6098	98.6153

Table 6. The inversion result of shale anisotropy parameters for the Honggouzi conformation

From the data in the table 6, we can see that the shale is of strong rock anisotropy. The acoustic wave front of the shale section is shown in figure 8.

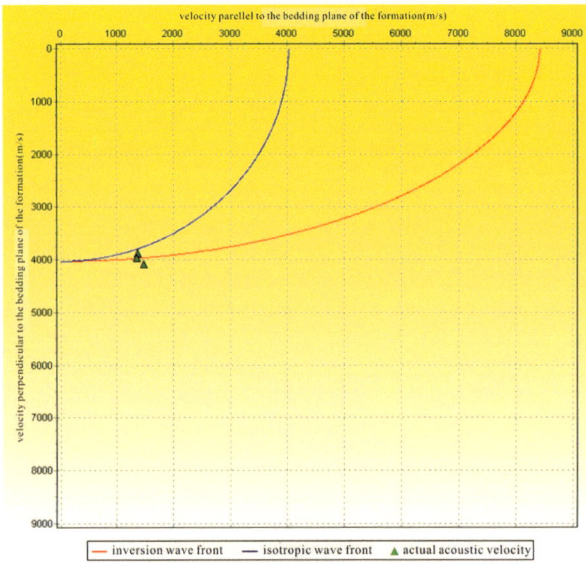

Fig. 8. Acoustic wave front of the shale section of the Honggouzi conformation

Based on the wave front of the shale shown in Fig.8, the rock anisotropy parameters of the formation at any measured depth, ε and δ, can be approximately calculated by the following equations:

$$\left.\begin{aligned} &\varepsilon = V_{sh}\varepsilon_c \\ &\delta = V_{sh}\delta_c \\ &V_{sh} = (2^{\Delta GR \cdot G_{cur}} - 1)/(2^{G_{cur}} - 1) \\ &\Delta GR = (GR - GR_{min})/(GR_{max} - GR_{min}) \end{aligned}\right\} \tag{49}$$

where ε and δ are rock anisotropy parameters of the formation at any measured depth; ε_c and δ_c are rock anisotropy parameters of the shale section; V_{sh} is the shale content, %; GR is gamma ray value; GR_{max} is the maximum value of gamma ray; GR_{min} is the minimum value of gamma ray; G_{cur} is the Hilchie index whose value is 3.7 for the Neogene Stratigraphy and 2 for old strata.

After obtaining the rock anisotropy parameters(ε and δ), we can calculate the acoustic wave perpendicular to the bedding plane of the formation by using the inversion method shown in Fig.7 and the well logging data of acoustic wave. Thus, the rock drillability anisotropy of the formation can be calculated by equations (46) & (47), and the corresponding anisotropic drilling characteristics can be evaluated by equations (8) & (9), as shown in figure 9.

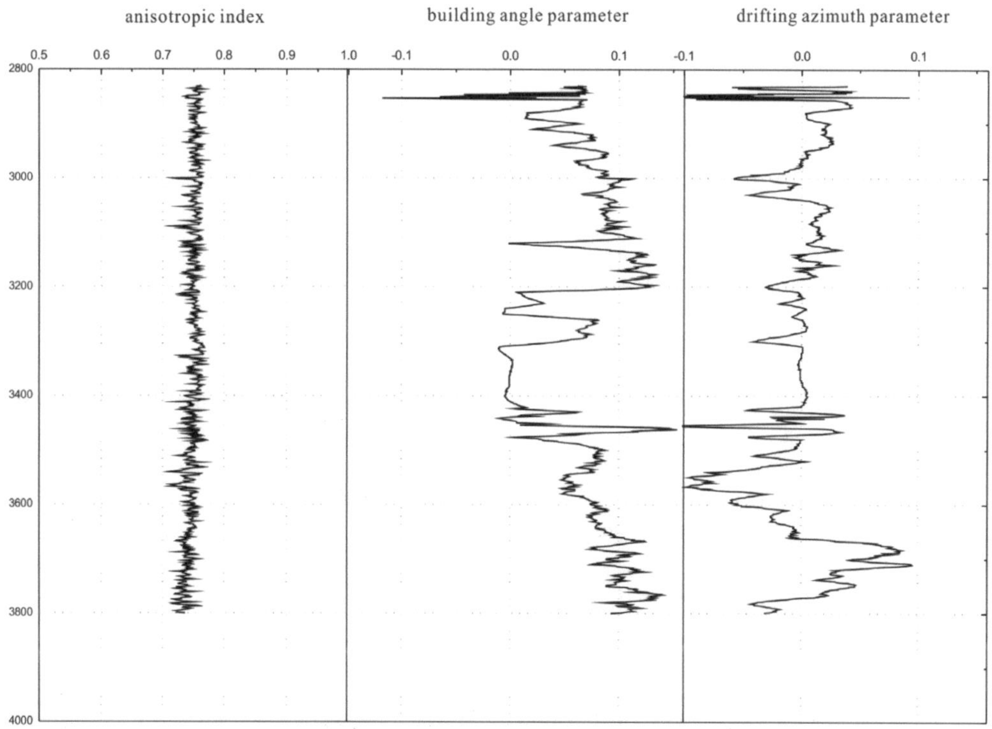

Fig. 9. The evaluation results of anisotropic drilling characteristics of the formation at the Honggouzi conformation in Qinghai oilfield

6. Conclusion

The orthotropic formation and the transversely isotropic formation are the typical formations encountered frequently in drilling engineering. Based on rock-bit interaction model, the two parameter equations have been derived for us to calculate the anisotropic drilling characteristics of them as soon as rock drillability anisotropy of the formations is evaluated quantificationally by using the oilfield data.

The correlation between rock drillability anisotropy and acoustic wave anisotropy of the formation can be matched to each other by an exponential function which is of the best extrapolative performance and relativity. Coefficients in the model are various for different formations to be drilled.

To a certain extent, the research results presented here have shown a new way for us to evaluate conveniently rock drillability anisotropy of the formation by using well logging or seismic data. Case study shows that this evaluation method is better for applications of rock drillability anisotropy of the formation in drilling engineering.

7. Acknowledgment

The authors are grateful for the financial support from the national research project (Grant No. 2010CB226703) and the supply of many core samples by CCSD.

8. References

Gao, D.; Dai, D. & Pan, Q. (2008). Evaluation of deflecting characteristics of anisotropy formation by using well-log information. *ACTA PETROLEI SINICA*, Vol.29, No.6, (December 2008), pp. 927-932, ISSN 0253-2697

Gao, D. & Pan, Q.(2006). Experimental study of rock drill-ability anisotropy by acoustic velocity. *Petroleum Science*, Vol. 3, No.1, (March 2006), pp.50-55, ISSN 1672-5107

Gao, D. (1995). Predicting and scanning of wellbore trajectory in horizontal well using advanced model. *Proceedings of the Fifth International Conference on Petroleum Engineering Held in Beijing*, China, SPE 29982, pp.297-308, 14-17 Nov. 1995

Gao, D.; Liu, X. & Xu B.(Dec. 1994). *Prediction and Control of Well Trajectory*, Petroleum University Press, ISBN 7-5636-0584-3/TE·95, Dongying, China

Gao, D. & Liu, X.(1990). Anisotropic drilling characteristics of the typical formations. *Journal of the University of petroleum, China*, Vol.14, No.5, (Oct. 1990), pp.1-8, ISSN 1000-7393

Gao, D. & Liu, X.(1989). A new model of rock-bit interaction. *Oil Drilling and Production Technology*, Vol.11, No.5, (Oct. 1989), pp.23-28, ISSN 1000-7393

Lubinski, A. & Woods,H.(1953). Factors affecting the angle of inclination and dog-legging in rotary bore holes. DPP, 1953: 222-242

Patrick, J. & Richard L.(1984). An experimental test of P-wave anisotropy in stratified media. *Geophysics*, Vol.49,No.4, (1984),pp. 374-378

Xu, B.(Jan. 2011). *Concise Elasticity and Plasticity*, Higher Education Press, ISBN 978-7-04-030725-2, Beijing, China

Yin H.(1989). Study of formation anisotropy-rock drillability. *Oil Drilling and Production Technology*, Vol.11, No.1, (Feb. 1989),pp.15-22, ISSN 1000-7393

8

Machinery Faults Detection Using Acoustic Emission Signal

Dong Sik Gu and Byeong Keun Choi
Gyeongsang National University
Republic of Korea

1. Introduction

Application of the high-frequency acoustic emission (AE) technique in condition monitoring of rotating machinery has been growing over recent years. This is particularly true for bearing defect diagnosis and seal rubbing (Mba et al., 1999, 2003, 2005; Kim et al., 2007; Siores & Negro, 1997). The main drawback with the application of the AE technique is the attenuation of the signal and as such the AE sensor has to be close to its source. However, it is often practical to place the AE sensor on the non-rotating member of the machine, such as the bearing or gear casing. Therefore, the AE signal originating from the defective component will suffer severe attenuation before reaching the sensor. Typical frequencies associated with AE activity range from 20 kHz to 1 MHz.

While vibration analysis on gear fault diagnosis is well established, the application of AE to this field is still in its infancy. In addition, there are limited publications on application of AE to gear fault diagnosis. Siores explored several AE analysis techniques in an attempt to correlate all possible failure modes of a gearbox during its useful life. Failures such as excessive backlash, shaft misalignment, tooth breakage, scuffing, and a worn tooth were seeded during tests. Siores correlated the various seeded failure modes of the gearbox with the AE amplitude, root mean square, standard deviation and duration. It was concluded that the AE results could be correlated to various defect conditions (Siores et al., 1997). Sentoku correlated tooth surface damage such as pitting to AE activity. An AE sensor was mounted on the gear wheel and the AE signature was transmitted from the sensor to data acquisition card across a mercury slip ring. It was concluded that AE amplitude and energy increased with increased pitting (Sentoku, 1998). In a separated study, Singh studied the feasibility of AE for gear fault diagnosis. In one test, a simulated pit was introduced on the pitch line of a gear tooth using an electrical discharge machining (EDM) process. An AE sensor and an accelerometer for comparative purposes were employed in both test cases. It was important to note that both the accelerometer and AE sensor were placed on the gearbox casing, it was observed that the AE amplitude increased with increased rotational speed and increased AE activity was observed with increased pitting. In a second test, periodically occurring peaks were observed when natural pitting started to appear after half an hour of operation. These AE activities increased as the pitting spread over more teeth. Singh concluded that AE could provide earlier detection over vibration monitoring for pitting of gears, but noted it could not be applicable to extremely high speeds or for unloaded gear conditions. (Singh et al., 1996) Tan offered that AE RMS (Root Mean Square) levels from the pinion were linearly correlated to pitting rates; AE showed better sensitivity than vibration at higher toque level (220Nm) due to fatigue gear testing using spur gears. He made sure that the linear relationship between AE, gearbox running time and pit progression implied that the AE technique offers good potential in prognostic capabilities for monitoring the health of rotating machines. (Tan et al., 2005, 2007)

On the other hand, the signal processing method for AE signal was studied using bearing and gearbox. In the results of the research (Sheen, 2008, 2010; Yang et al., 2007), the envelope analysis was found to be useful to detect fault in rolling element bearing. The fault detection frequency of bearing can be presented in the power spectrum. Wavelet transform was used for the signal processing method for the gearboxes (Wu et al., 2006, 2009), but wavelet transforms can give the different results with the envelope analysis. It can be shown the defect frequency, but the efficiency is lower than that of envelope analysis. Thus, the signal processing method for AE signal has not been completed until now, and it must be developed in the future.

Therefore, in this paper, a signal processing method for AE signal by envelope analysis with discrete wavelet transforms is proposed. For the detection of faults generated by gear systems and a cracked rotor using the suggested signal processing, these were installed in each test rig system. In gearbox, misalignment was created by a twisted case caused by arc-welding to fix the base and bearing inner race fault was generated by severe misalignment. Through the 15 days test using AE sensor, misalignment was observed and bearing faults were also detected in the early fault stage. To identify the sensing ability of the AE, vibration signal was acquired through an accelerometer and compared with the AE signal. Also, to find the advantage of the proposed signal processing method, it was compared to traditional envelope analysis. The detection results of the test were shown by the power spectrum and comparison of the harmonics level of the rotating speed. Modal test and zooming by a microscope were performed to prove the reason of the other faults. And the crack was seeded by wire cutting with 0.5 mm depth. The shaft was coupled with motor and non-drive-end was left 6.5 mm by lifting tool. During rotating the shaft, AE signals were acquired by AE sensor with 5MHz sampling frequency and 0.5 seconds storing time. The AE signals were transformed by FFT to create the power spectrums, and in the spectrums several peaks were occurred by the crack growth. Along the growth of the crack, the characteristic of the power spectrum was changed and displayed different frequencies.

2. Signal processing method

Envelope analysis typically refers to the following sequence of procedure: (1) band-pass filtering (BPF), (2) wave rectification, (3) Hilbert transform or low-pass filtering (LPF) and (4) power spectrum. The purpose of the band-pass filtering is to reject the low-frequency high-amplitude signals associated with the *i*th mechanical vibration components and to eliminate random noise outside the pass-band. Theoretically, in HFRT (High Frequency Resonance Technique) analysis, the best band-pass range includes the resonance of the bearing components. This frequency can be found through impact tests or theoretical calculations involving the dimensions and material properties of the bearing. However, it is very difficult to predict or specify which resonant modes of neighboring structures will be excited. It will be costly and unrealistic in practice to find the resonant modes through experiments on rotating machinery that may also alter under the different operational conditions. In addition, it is also difficult to estimate how these resonant modes are affected in the assembly of a complete bearing and mounting in a specific housing, even if the resonant frequencies of individual bearing elements can be tested or calculated theoretically (Misiti et al., 2009). Therefore, most researchers decide on the band-pass range as on option. To recover the disadvantage of this option, wavelet analysis is included in the process of traditional envelope analysis in this paper.

Wavelet theory (Burrus et al., 1997) is introduced that is a tool for the analysis of transient, non-stationary, or time-varying phenomena. Wavelet analysis is also called wavelet transform. There are two kinds of wavelet transform: continuous wavelet transform (CWT) and discrete wavelet transform (DWT). CWT is defined as the sum over all time of the signal multiplied by scaled, shifted versions of the wavelet function. To use CWT, one signal can be decomposed into a series of "small" waves belonging to a wavelet family. The wavelet family is composed of scaling functions, $\phi(t)$ deduced by father wavelet and wavelet functions, $\psi(t)$ deduced by mother wavelet. The scaling function can be represented by the following mathematical expression:

$$\phi_{j,k}(k) = 2^{j/2}\phi\left(2^j t - k\right)$$

where j represents the scale coefficient and k represents shift coefficient. Scaling a wavelet simply means stretching (or compressing) it. Shifting a wavelet simply means delaying (or hastening) its onset. Mathematically, delaying a function $f(t)$ by k is represented by $f(t + k)$. Similarly, the associated wavelet function can be generated using the same coefficients as the scaling function.

$$\psi_{j,k}(t) = 2^{j/2}\psi\left(2^j t - k\right)$$

The scaling functions are orthogonal to each other as well as the wavelet functions as shown in the following equations:

$$\int_{-\infty}^{\infty}\phi(2t-k)\cdot\phi(2t-l)\,dt = 0 \text{ for all } k \neq l$$

$$\int_{-\infty}^{\infty}\psi(t)\cdot\phi(t)\,dt = 0$$

Using an iterative method, the scaling function and associated wavelet function can be computed if the coefficients j and k are known.

For many signals, the low-frequency content is the most important part. It is what gives the signal its identity. The high-frequency content, on the other hand, imparts flavour or nuance that is often useful for singular signal detection. In wavelet analysis, we often speak of approximations and details. The approximations are the high-scale, low frequency components of the signal. The details are the low-scale, high-frequency components. A signal can be decomposed into approximate coefficients $a_{j,k}$, through the inner product of the original signal at scale j and the scaling function.

$$\alpha_{j,k} = \int_{-\infty}^{\infty} f_j(t)\cdot\phi_{j,k}(t)\,dt$$

Similarly the detail coefficients $d_{j,k}$ can be obtained through the inner product of the signal and the complex conjugate of the wavelet function.

$$d_{j,k} = \int_{-\infty}^{\infty} f_j(t) \cdot \psi_{j,k}(t)\, dt$$

However, CWT takes a long time due to calculating the wavelet coefficient at all scales and it produces a lot of data. To overcome such a disadvantage, we can choose scales and positions based on powers of two - the so-called dyadic scales and positions - then wavelet analysis will be much more efficient and just as accurate. Such an analysis is obtained from the discrete wavelet transform (DWT). The approximate coefficients and detail coefficients decomposed from a discredited signal can be expressed as

$$\alpha_{(j+1),k} = \sum_{k=0}^{N} \alpha_{j,k} \int \phi_{j,k}(t) \cdot \phi_{(j+1),k}(t)\, dt = \sum_k \alpha_{j,k} \cdot g[k]$$
$$\alpha_{(j+1),k} = \sum_{k=0}^{N} \alpha_{j,k} \int \phi_{j,k}(t) \cdot \psi_{(j+1),k}(t)\, dt = \sum_k \alpha_{j,k} \cdot h[k]$$

The decomposition coefficients can therefore be determined through convolution and implemented by using a filter. The filter, $g[k]$, is a low-pass filter and $h[k]$ is a high-pass filter. The decomposition process can be iterated, with successive approximations being decomposed in turn, so that one signal is broken down into many lower resolution components. This is called the wavelet decomposition tree as shown in Fig. 1.

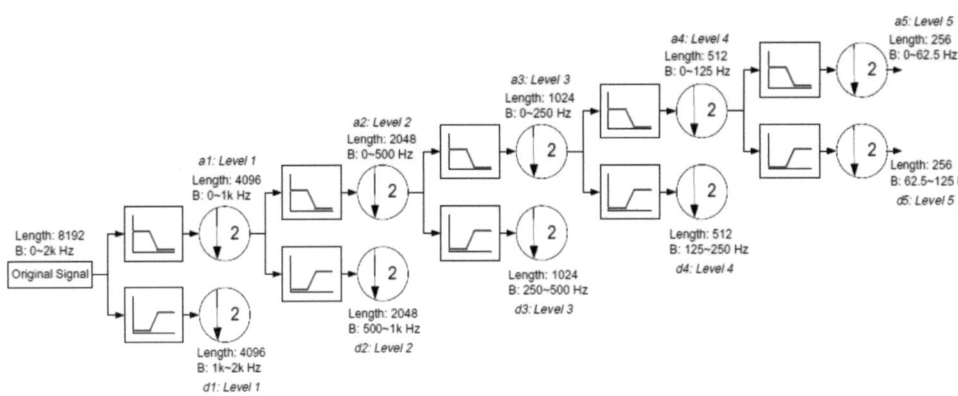

Fig. 1. Wavelet decomposition tree

DWT has a de-noise function and a filter effect focused on impact signal. To make up the weak point of BPF of the envelope analysis, DWT was intercalated on typical envelope analysis, between BPF and wave rectification exactly. The signal by DWT will be separated to different band widths by decomposition level and adapted to the signal with impact.

For more complicated signals which are expressible as a sum of many sinusoids, a filter can be constructed which shifts each sinusoidal component by a quarter cycle. This is called a Hilbert transform filter. Let $H_t\{x\}$ denotes the output at time t of the Hilbert-transform filter applied to the signal x. Ideally, this filter has magnitude 1 at all frequencies and introduces a phase shift of $-\pi/2$ at each positive frequency and $+\pi/2$ at each negative frequency. When a real signal $x(t)$ and its Hilbert transform $y(t) = H_t\{t\}$ are used to form a new complex signal $z(t)= x(t)+jy(t)$, the signal $z(t)$ is the (complex) analytic signal corresponding to the real signal $x(t)$. In other words, for any real signal $x(t)$, the corresponding analytic signal $z(t) = x(t)+jH_t\{x\}$ has the property that all 'negative frequencies' of $x(t)$ have been 'filtered out' (Douglas & Pillay, 2005). Hence, the coefficients of complex term in the corresponding analytic signal were used for FFT.

Fig. 2 shows an analytic signal of the Hilbert transform for envelope analysis. The solid line is a time signal and the dash is its envelope curve. A high frequency signal modified by wavelet transform is modulated to a low frequency signal with no loss of the fault information due to envelope effect. According to that, the fault signals in the low frequency region can be detected using the analytic signal. That is an important fact for the proposed signal processing method. Therefore, the proposed signal processing method in this paper is an envelope analysis with DWT and using the coefficients of the complex term in Hilbert transform.

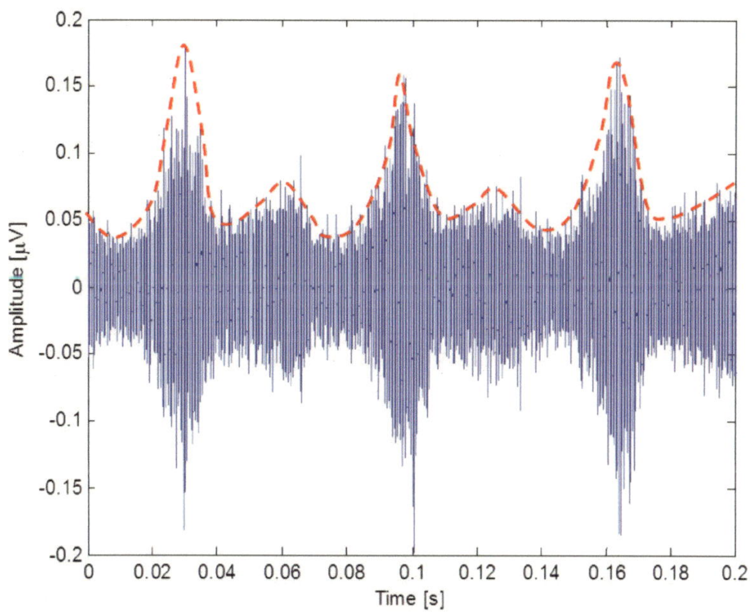

Fig. 2. Analytic signal (dash) of the envelope effect

Furthermore, to reduce the noise level in the power spectrum, the spectrum values were presented as the mean value of each day. Fig. 3 shows the power spectrums of the two different signal processing method. Fig. 3(a) is from envelope analysis, and Fig. 3(b) shows the envelope analysis intercalated DWT using Daubechies mother function between BPF and wave rectification. In Fig. 3, the DWT has an effect the amplifying sidebands peaks, especially about gear mesh frequencies, so the peaks of the harmonics of the rotating speed (f_r) and gear mesh frequencies (f_m) are bigger than another, and we can check up them easily. Therefore, in the following result, the power spectrum through envelope analysis with DWT will be shown.

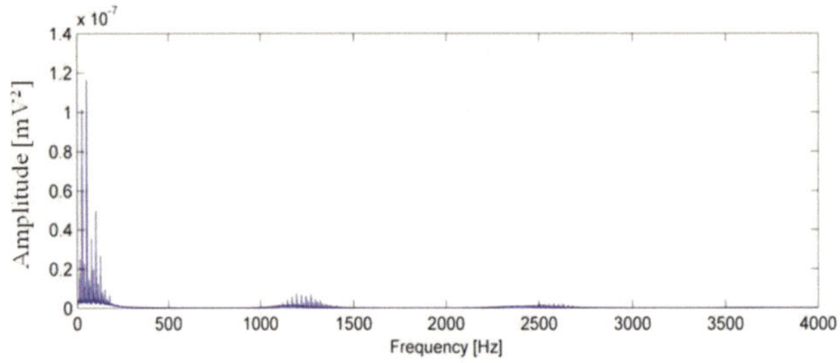

(a) Envelope analysis

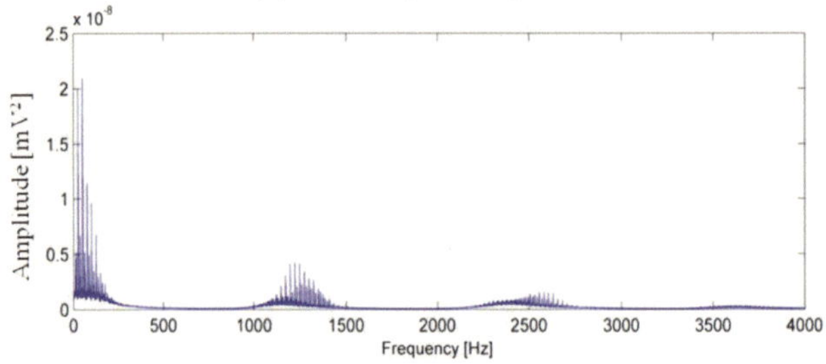

(b) Envelope analysis with DWT

Fig. 3. The comparison of Power spectrums in Envelope analysis with/without DWT

3. Gearbox

3.1 Test-rig

The test-rig employed for this investigation consists of one identical oil-bath lubricated gearbox, 3 HP-motor, rigid coupling, tapper-roller bearing, pinion, gear, control panel and break system, as seen Fig. 4. The pinion was made from steel with heat treatment, the number of teeth is 70, and diameter is 140 mm. The gear was made from steel, but it was produced without any heat treatment process during manufacturing. The number of gear teeth is 50 and diameter is 100 mm, and module is 2 mm for the gear and pinion, respectively.

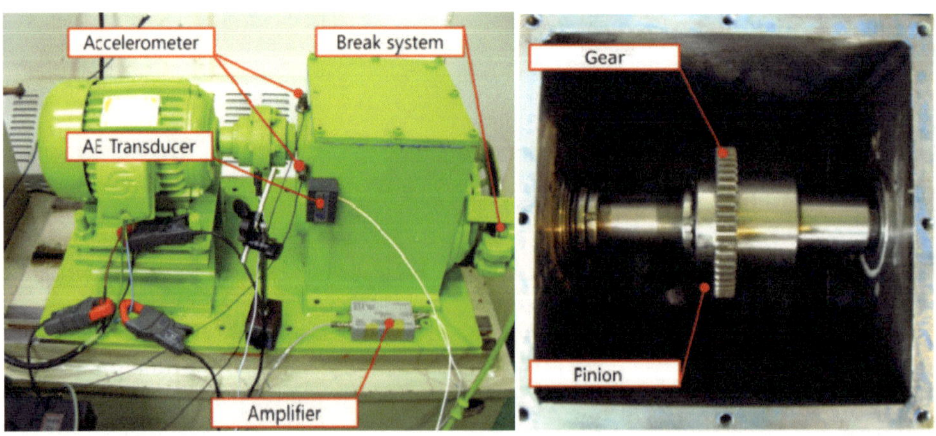

Fig. 4. Test-rig

A simple mechanism that permitted a break of disk-pad type to be rotated relative to each other was employed to apply torque to the gear. Contact ratio (Pinion/Gear) of the gears was 1.4. The motor used to drive the gearbox was a 3-phase induction motor with a maximum running speed of 1800 rpm respectively and was operated for 15 days with 1500rpm. The torque on the output shaft was 1.2 kN·m while the motor was in operation, and other specifications of the gearbox are given as in Table 1.

	Gear		Pinion	
No. of teeth	50		70	
Speed of shaft	25.01 rev/s			
Meshing frequency	1250 Hz		1750 Hz	
Bearing (NSK HR 32206J)				
No. of rolling element	17	Type	Defect Freq. (f_d)	Fault Freq. (f_d X f_r)
Diameter of outer race	62 mm	BPFO BPFI FTF BSF	8.76 Hz 11.24 Hz 3.84 Hz 0.44 Hz	219.3 Hz 281.38 Hz 96.13 Hz 11.01 Hz
Diameter of inner race	30 mm			
BPFO : ball pass frequency of outer race BPFI : ball pass frequency of inner race FTF : fundamental train frequency BSF : ball spin frequency				

Table 1. Specification of gearbox and bearing

3.2 Acquisition system and test procedures

AE sensors used in this paper are a broadband type with a relative flat response in the range frequency from 10 kHz to 1 MHz. They are placed on the right side of the gearbox cases near the coupling in the horizontal direction at the same height with the shaft center (Fig. 4).

AE signals are pre-amplified by 60 dB and the output from the amplifier is collected by a commercial data acquisition card with 10 MHz sampling rate during the test. Prior to the analog-to-digital converter (ADC), anti-aliasing filter is employed that can be controlled DAQ software. And Table 2 is shown the detail specifications of the data acquisition system. Before the test, attenuation test on the gearbox components was taken in order to understand the characteristics of the test-rig. The gearbox was run for 30 minutes prior to acquiring AE data for the unload condition. Based on the sampling rate of 10 MHz, the available recording acquisition time was 2 sec.

2 Channel AE system on PCI-Board	18-bit A/D conversion 10M samples/s rate (on one channel, 5M samples/s on 2 AE channels)
AE Sensor (Wideband type)	Peak sensitivity [V/μbar] : -62 dB Operating frequency range : 100-1,000 kHz Resonant freq. [V/ μbar] : 650 kHz Directionality : ±1.5 dB
Preamplifier Gain	Wide dynamic range < 90 dB Single power/signal BNC or optional separate power/signal BNC 20/40/60 dB selectable gain

Table 2. Specifications of data acquisition system

3.3 Experiment result and discussion

In general, the misaligned gear which almost always excites higher order f_m harmonics is shown as in Fig. 5. Often, only small amplitudes will be at the fundamental f_m, but much higher levels will be at $2f_m$ and/or $3f_m$. The sideband spacing about f_m might be $2f_r$ or even $3f_r$ when gear misalignment problems are involved. When significant tooth wear occurs, not only will sidebands appear about f_m, but also about the gear natural frequencies. In the case of those around f_m, the amplitude of the sidebands themselves is a better indicator for wear than the amplitude of f_m.

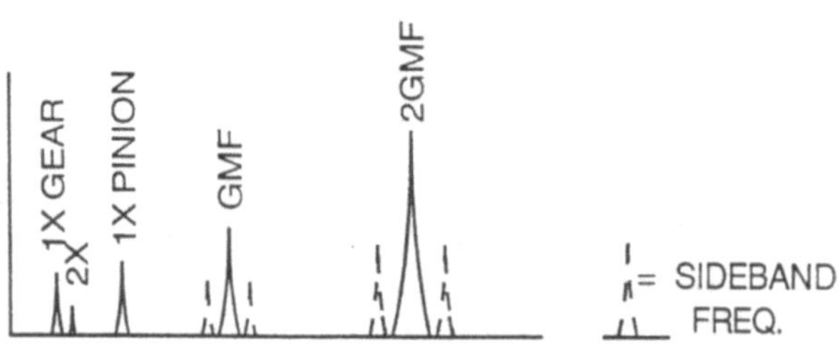

Fig. 5. Spectrum indicating misalignment of gear

As for significant gear eccentricity and/or backlash, these problems display the following characteristics:

- Both eccentricity and backlash excite the gear natural frequencies as well as f_m. They also may generate a number of sidebands about both the natural and gear mesh frequencies.
- If a gear is eccentric; it will modulate the natural frequency and gear mesh frequencies, both of which will be sidebanded around the f_r of the eccentric gear. An eccentric gear can generate significant forces, stresses and vibration if it is forced to bottom out with the meshing gears. (James & Bery, 1994)

In the results of the envelop analysis with DWT, the high harmonics of f_m occurred by strong wearing phenomena caused by misaligned teeth. In the power spectrum (Fig. 6), 25Hz (f_r) and its harmonics are generated and 11.32Hz was the ball pass frequency of inner race (BPFI [f_d]). In Fig. 6(c) and (d), the center dash line is shown for f_m and $2f_m$, and their side lines are the sidebands with difference 25Hz (f_r).

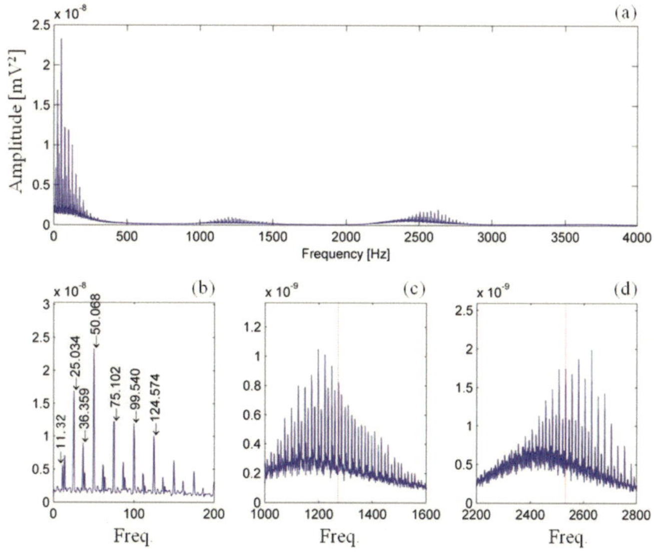

Fig. 6. Power spectrum of the second day

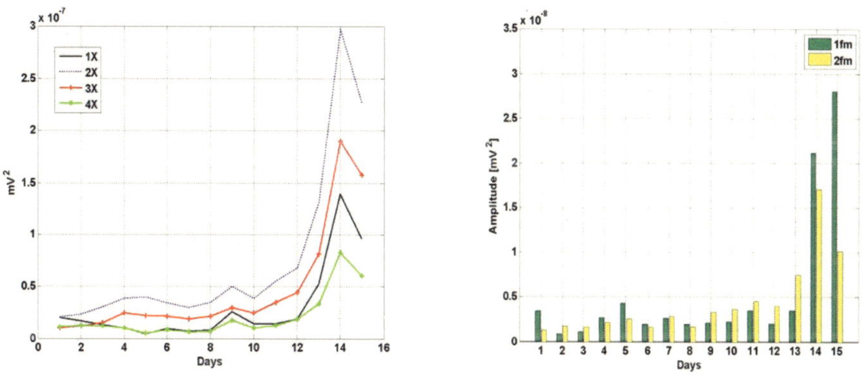

(a) Harmonics of f_r

(b) Harmonics of f_m

Fig. 7. Peak level trend among days

Fig. 8. Gear tooth weaned by misalignment

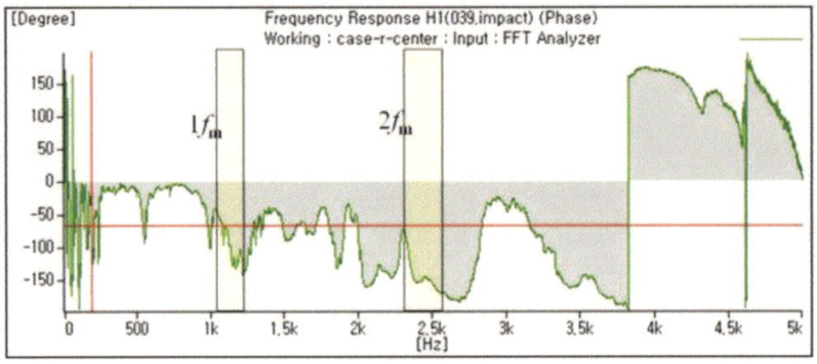

(a) Phase

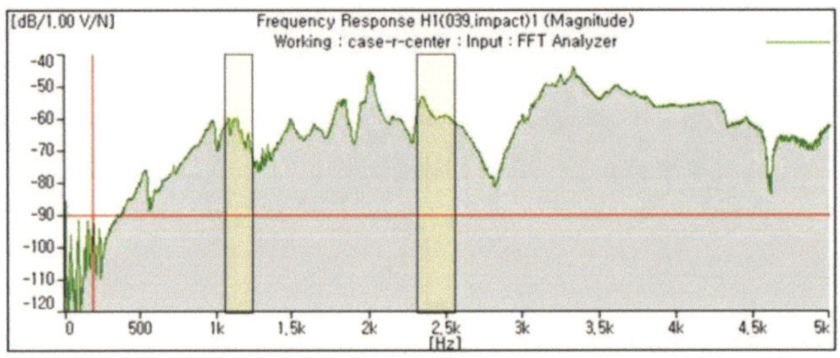

(b) Frequency response function

Fig. 9. Modal test result

In condition monitoring for general rotating machinery, the harmonics ($2f_r$, $3f_r$, $4f_r$ ...) of f_r are occurred higher than f_r when the misalignment was happened. According to the phenomena of misalignment as shown in Fig.5, high level harmonics of f_r were generated such as in Fig. 6(b), and $2f_r$ was always bigger than f_r as shown in Fig. 7(a). The level of $2f_m$ from second to thirteenth day was higher than or similar to f_r as shown in Fig. 7(b). Thus, it is easily catching up to the misalignment that occurred in this test rig. However, it might be that faults of this system are not only misalignment but also resonance trouble, looseness, bearing fault, etc.

Wearing effect by misalignment pollutes the lubrication oil. In Fig. 8, it could be found by the worn teeth and the spots near the pieces of gear teeth. The dripped pieces from the unloading surface raised the wearing effect on the loading surface, and then the gap between gear and pinion was increased. In addition, we could know that the impact marks on the unloading surface (Fig. 8(b)) were generated by misalignment; the impacting force was strong in the initial condition. In this way, the gear teeth were seriously damaged as in Fi g. 8. In Fig. 6(c) and (d), the sidebands are created on wide-spread frequency range near f_m and $2f_m$. That is similar to a state excited by impact force. To confirm the natural frequencies of the test-rig, modal test was fulfilled. The result of the modal test as in Fig. 9 show that f_m and $2f_m$ exist on the exiting frequency range. On the other hand, partial frequency bands close to f_m and $2f_m$ were excited by the impact force, but it is not an exact natural frequency because the phase did not shift enough. Therefore, the peaks near f_m and $2f_m$ were amplified and have many sidebands of f_r and 11.32Hz (BPFI [f_d]). Therefore, it is considered that excessive backlash occurred. Moreover, Fig. 10(a) shows the zooming power spectrum of Fig. 6(a) focused on f_r harmonics. We could clearly know that if the sidebands were caused by BPFI [f_d], then the inner race had some kind of fault. To find out the fault, the surface of the bearing inner race was carried out and viewed by a microscope with 100X zoom as shown in Fig. 11. Small spots were found on the surface, and small cracks were found out on the spots. However, this trouble was not seeded and existed from the initial condition. Thus, it is as assumed that the problem happened in assembly and/or was caused by misalignment.

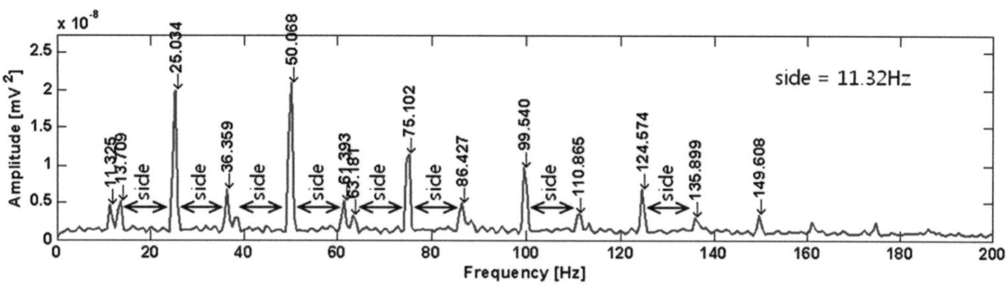

(a) Power spectrum of the AE signal using envelope analysis with DWT(suggested)

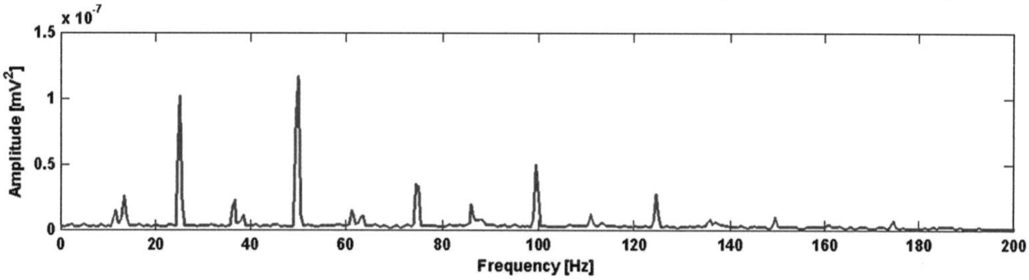

(b) Power spectrum of the AE signal using traditional envelope analysis

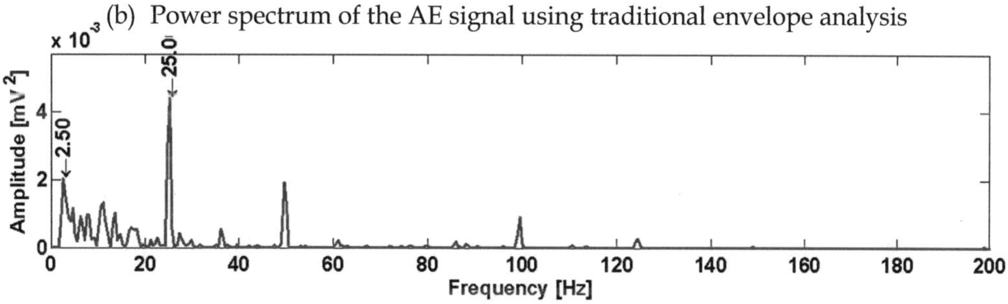

(c) Power spectrum of the signal from accelerometer

Fig. 10. Sidebands BPFI in the first day power spectrum

To identify the sensing ability of the AE, vibration signal was acquired through accelerometer and compared with the AE signal. Also, to find the advantage of the proposed signal processing method, it was compared to traditional envelope analysis.

The power spectrum of the AE signal using traditional envelope analysis is shown in Fig. 10(b), and the power spectrum using the vibration signal by accelerometer is displayed on Fig. 10(c). The vibration signal was treated by the same method with AE signal. The harmonics of f_r are generated, and 2 f_r for detecting the misalignment is created and can be found in all spectrums (Fig. 10) but the power spectrum of the AE signal, Fig. 10(a) and (b), can explicitly display the defect frequencies as compared to the accelerometer signal (Fig. 10(c)). For example, in Fig. 10(c), the sidebands of BPFI are not easily found because of the higher level of noise in the low frequency range below f_r than in the AE signal with or without DWT.

Fig. 11. Zooming of the inner race surface of the fault bearing

Frequency [Hz]		Traditional Method	Proposed method
13.709	1X-BPFI	0.2146	0.2459
25.034	1X	0.8727	0.9525
36.360	1X+BPFI	0.1936	0.3196
38.750	2X- BPFI	0.0952	0.1435
50.070	2X	1.0000	1.0000
61.393	2X+ BPFI	0.1277	0.2470
63.181	3X- BPFI	0.0970	0.1560
75.102	3X	0.3044	0.5465
86.427	3X+ BPFI	0.1699	0.2352
99.540	4X	0.4269	0.4616
110.866	4X+ BPFI	0.1062	0.1675
124.574	5X	0.2312	0.3253
135.899	5X+ BPFI	0.0714	0.1443

Table 3. Ratio of peaks versus the maximum peak in respective spectrum

According to the above results, we can understand that the AE signal can detect the fault more easily than accelerometers and can be used in the condition monitoring system for early detection fault. Moreover, as shown in Table 3 which is the ratio of peaks versus the maximum peak in the respective spectrum, the peak levels of the harmonics of f_r and sidebands caused by BPFI are highly generated in the proposed signal processing method (Fig. 10(a)) than the traditional method. This can lead good feature values to evaluate the condition of the machinery. Therefore, the power spectrum of the proposed envelope analysis using AE signal can be shown the clean result with harmonics and sidebands and is a better technique for condition monitoring system.

4. Cracked rotor

4.1 Experiment system

Test rig consisted of a motor, a flexible coupling, rolling element bearings (NSK6200), three steel bearing housings, a lifting tool and a cracked shaft. The transverse crack was seeded by

wire-cutting with 0.5 mm depth on the shaft made from SM45C. As shown in Fig. 12, the crack was positioned at 5mm near to the second drive-end bearing, and the non-driven end of the shaft was left 6.5 mm with bearing housing by the lifting tool.
AE signal was acquired by an AE sensor and transferred to amplifier, analog-filter, DAQ board and HDD of a desktop. AE sensor is a wideband type with a relative flat response in the range frequency range from 100 kHz to 1 MHz. AE signals were pre-amplified with 60 dB and the output from the amplifier was collected by a commercial data acquisition card with 5 MHz sampling rate during the test. The signals were stored 0.5sec by every 30sec until the shaft was fractured, and the rotating speed of the motor was 600rpm (10Hz).

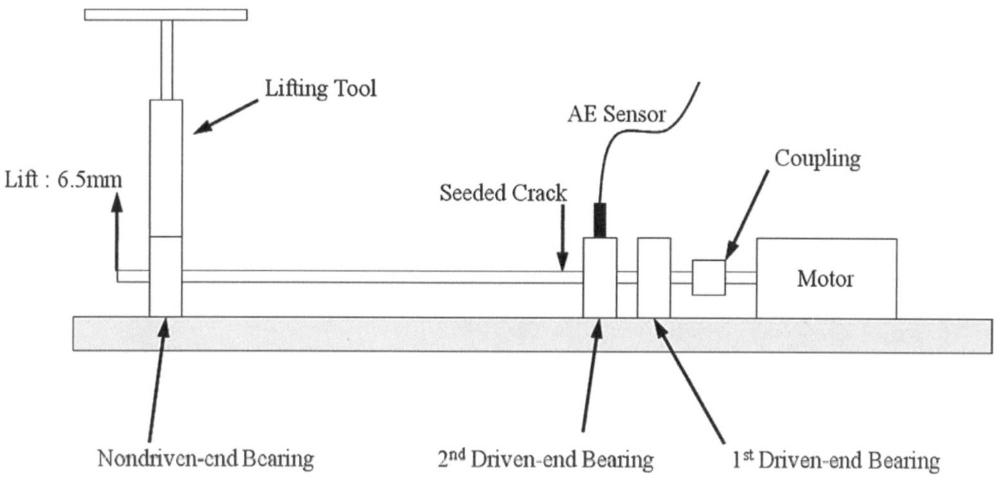

Fig. 12. Experiment system

4.2 Test result and discussion

The operating speed was 600rpm, and the initial radial load for 160N was employed. The radial load was a variable parameter because it was applied by keeping the lifting distance with 6.5mm of the non-drive end of the shaft, and the test terminated on a fracture of the shaft. Fig. 13 shows the observations of continuous feature values as mean value, RMS, peak value and entropy estimation. In information theory, uncertainty can be measured by entropy. The entropy of a distribution is the amount of a randomness of that distribution. Entropy estimation is two stage processes; first a histogram is estimated and thereafter the entropy is calculated. Here, we estimate the entropy of AE signal with using unbiased estimated approach. Fig. 13(a), relatively high level of AE activity was noted from 18 minutes, and it was increased until 60 minutes. But in RMS, Peak and Entropy estimation, the levels were kept to 18minutes beside a peak in around 9 minutes, since that these were continuously decreased to 70 minutes and were increased with hunting with several minute intervals until the fracture.
Interestingly observations of the AE waveform, sampled at 5 MHz showed changing characteristics as a function of time. Fig. 14 shows a contour map of the peaks level of each frequency with time. Rotating speed (9.5Hz) and 3rd harmonic of rotating speed (28.6Hz, 3X) dominated while the test as shown in Fig. 14(b). It is normally known that 3X is caused by misalignment of the bearings created by the loading system for this test (Hatch & Bently,

2002). However, the harmonic component (3X) was kept the level in the wavelet level 6(Fig. 14(b)), but it was increased from 30 minutes in the wavelet level 4(Fig. 14(a)). Additionally, 1X started increasing earlier than 3X as shown in Fig. 14(a).

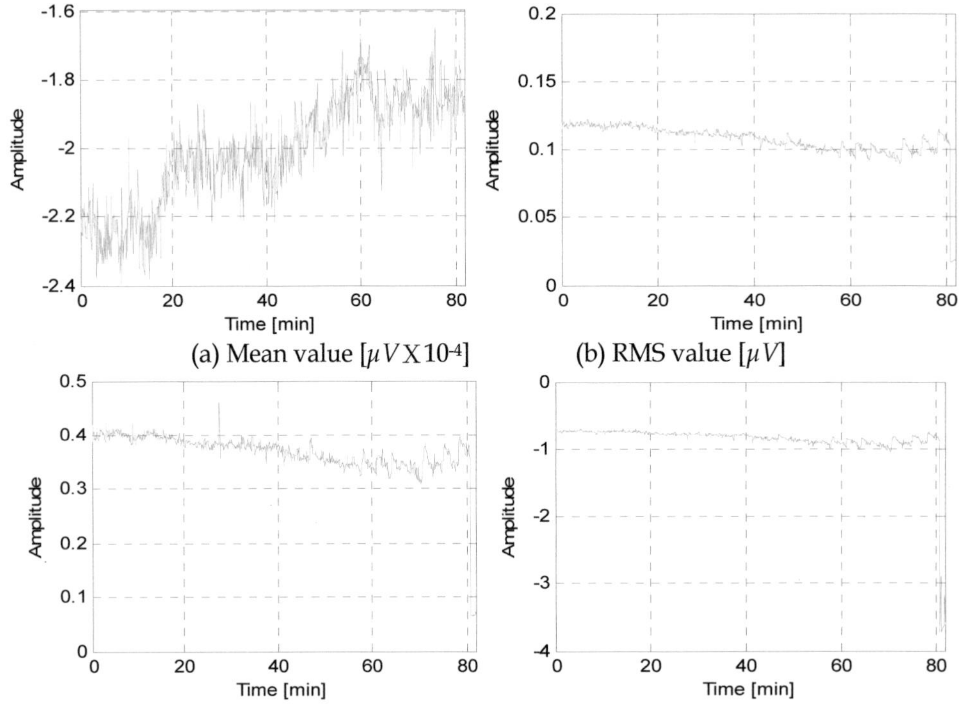

(a) Mean value [μV X 10^{-4}] (b) RMS value [μV]

(c) Peak value [μV] (d) Entropy estimation value

Fig. 13. Shaft test results; run-to-failure

31Hz and 62Hz that were the harmonics of the fundamental train frequency (FTF or cage noise) of the bearing were continuously occurred. Cage noise can be generated in any type of bearing and the magnitude of it is usually not very high. Characteristics of this noise include: (1) it occurs with pressed steel cages, machined cages and plastic cages. (2) It occurs with grease and oil lubrication. (3) It tends to occur if a moment load is applied to the outer ring of a bearing. (4) It tends to occur more often with greater radial clearance. In Fig. 14, 62Hz was continuously detected; 31Hz was detected with 3X. In a general bearing system, the amplitude of the bearing fault frequency is depended on the load grade and is increased along the growing load grade. However, 31Hz of the case noise of this test was not followed the load scale because loading force for this test was decreased with the crack growth. So, it was shown that 31Hz was related with the crack growth.

According to this result, we could know that the reason of 3X (28.6Hz) was the moment load by the loading system; 31Hz was connected with the crack growth. Therefore, the peak levels around 3X and 31Hz was excited by the two frequencies and was increased with the crack growth.

To clear more the characteristic of the crack growth, in addition, PAC energy value was observed. In acoustic emission technology, PAC-Energy is a 2-byte parameter derived from the integral of the rectified voltage signal over the duration of the AE hit (or waveform), hence the voltage-time units ($\mu V^2 \cdot$ sec). So, PAC energy value was determined by an integral of the square sum total of the transferred time signal in each wavelet level. Fig. 15 shown the energy level of every wavelet levels along time, and its value was transferred to logarithmic value because of too low resolution in linear scale.

Fig. 15 shows the energy trend of wavelet level 1 to 8. Many peaks were created while the test in wavelet level 1(Fig. 15(a)), a high peak was created around 9 minutes existed in wavelet level 1 to 4. Wavelet level 2, 4, 5 and 6 shows a similar trend after 10minutes. The energy level was slowly increased with time until about 30 minutes, and then it was increased fast until 35 minutes (additionally, it was considered that this increasing was related with the growth of the 3X and 31Hz in Fig. 14(a)), after that it was decreased a little for 10 minutes. And it was hunted with every several minutes about 4 minutes until close fracture. In this trend, we had considered of two phenomena, the high peak and the period hunting.

We could mind that the high peak was related with initial crack growth. Because it was shown as follows,

- The increasing ratio of the energy was changed after it was happened.
- The mean value was began fast change (Fig. 13(a)).
- Generated in high frequency range (Fig. 15(a) ~ (d)).

The period hunting was clearly occurred and displayed in wavelet level 2, 4 ~ 8. It was considered that it could indicate a state of the final stage of the fracture in the rotating shaft because of follows,

- It had a period about 4 minutes.
- It was displayed in lower frequency range than the high peak (Fig. 15(e) ~ (h)).
- Its level was increased along near to the fracture.

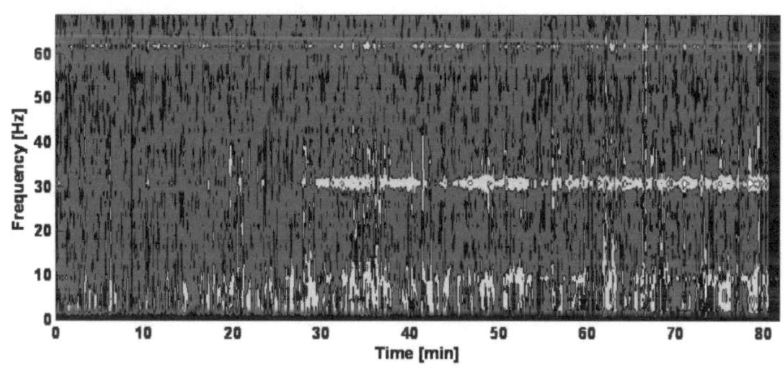

(a) Wavelet level 4

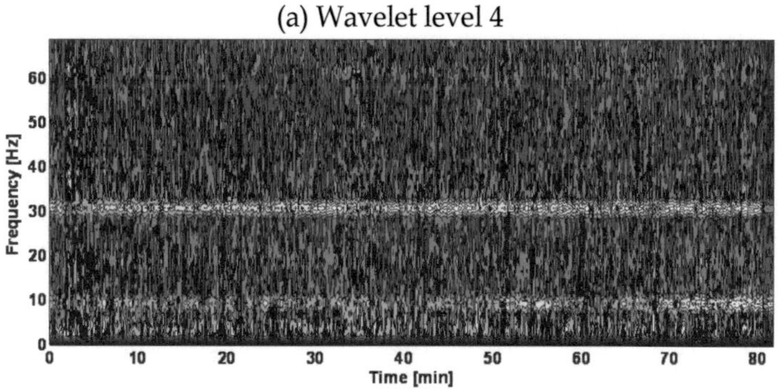

(b) Wavelet level 6

Fig. 14. Peak level trend by frequency

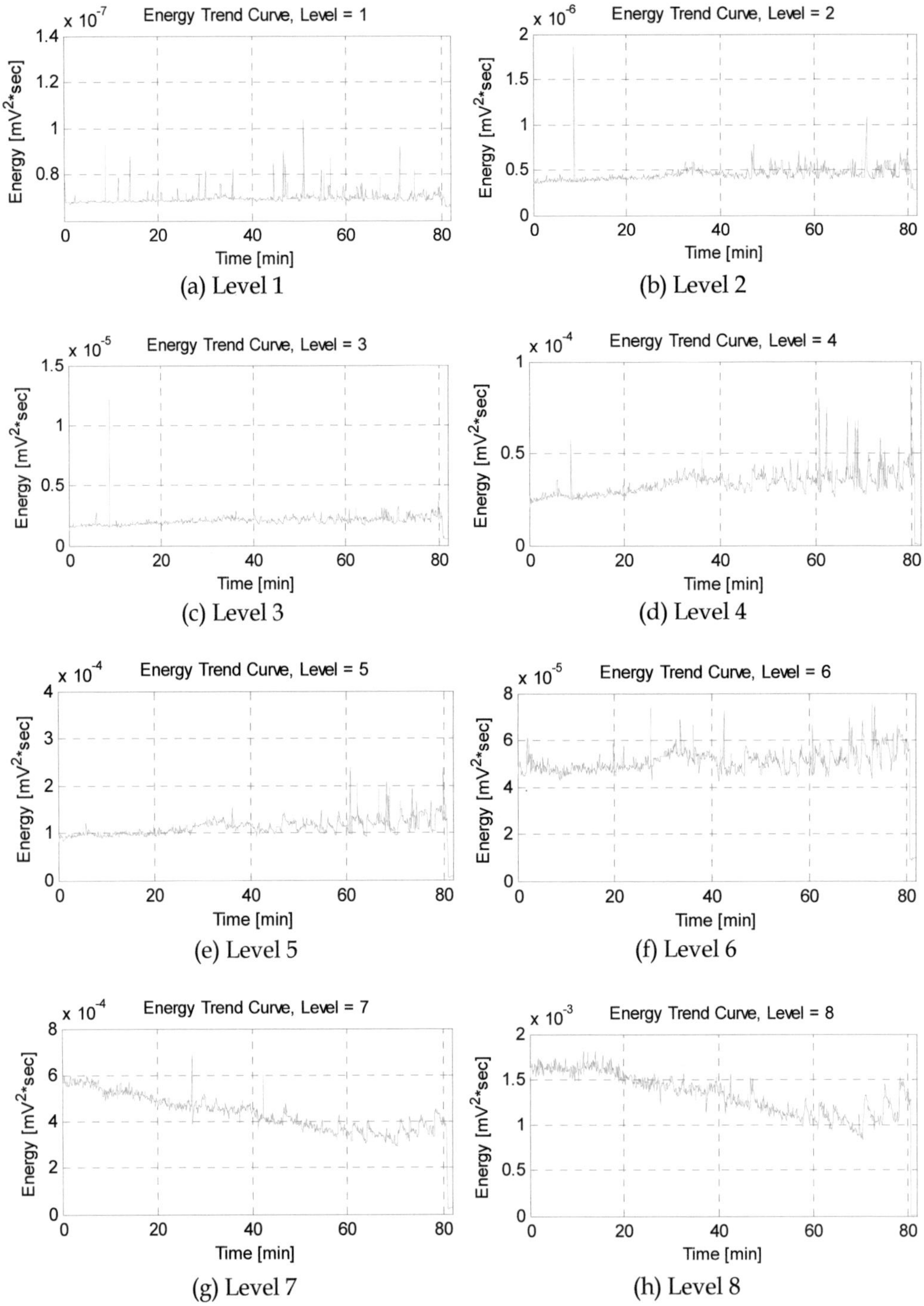

Fig. 15. PAC energy trend of each wavelet level

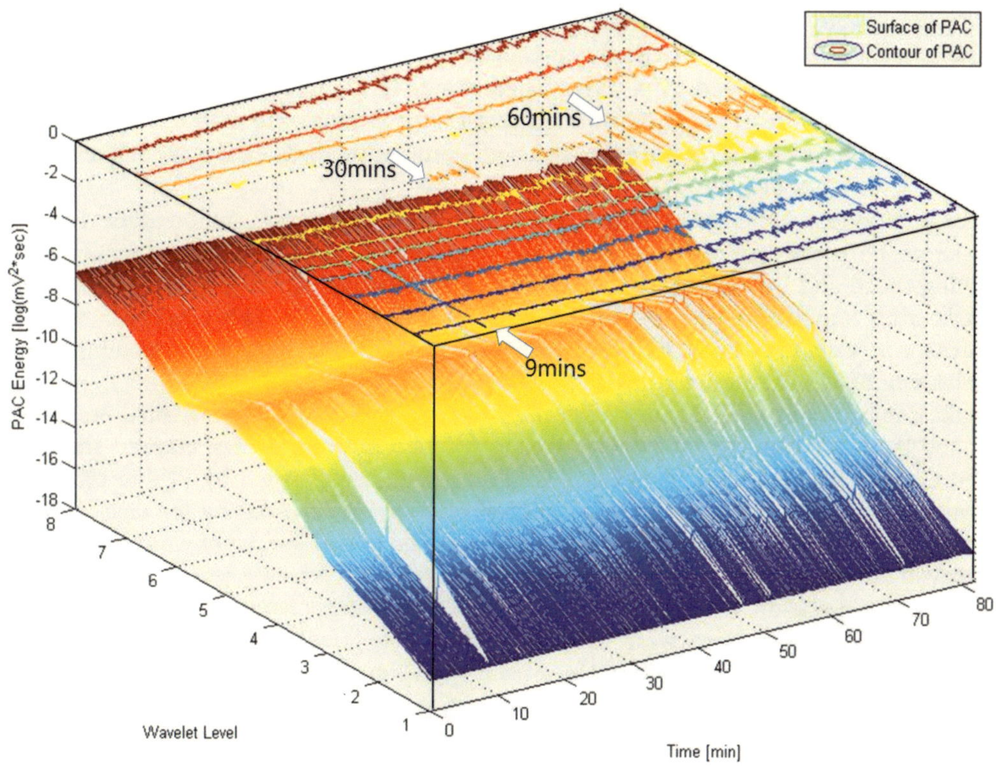

Fig. 16. PAC-Energy level of total wavelet level

To compare absolutely the energy of each wavelet level, all of PAC energy was displayed on a 3D graph with time and wavelet levels as shown in Fig. 16. In wavelet level 5, after approximately 30 minutes, a large transient rise in PAC energy level was observed and this AE activity gradually observed frequently after 60 minutes until the shaft was fractured. In addition, a peak created at 9 minutes was indicated. In here, we could consider that the frequency range of wavelet level 5 could be shown a good relationship between the PAC energy and the crack growth of the middle and final stage. Even so, the trend of the wavelet level 7 and 8 was not clearly connected with the others because the frequency range of wavelet level 7 and 8 was lower than the useful frequency range (100kHz to 1MHz) of AE sensor for this research.

Therefore, the AE signal caused by the crack growth was generated on the whole ultra-sound frequency range; the initial crack could be detected using the PAC energy on wavelet level 1 to 4. In addition, it could be presented on wavelet level 5 until the fracture of the shaft. In the frequency domain, it was shown that the harmonic components of the rotating speed and bearing cage frequency were excited by the crack growth, especially on the 3X (28.6Hz) and 31Hz.

5. Conclusion

In this paper, a signal processing method for AE signal by envelope analysis with discrete wavelet transforms is proposed. For the detection of faults generated from a gear system using the suggested signal processing, a gearbox was installed in the test rig system. Misalignment was created by twisted case caused by arc-welding to fix the base and bearing inner race fault is generated by severe misalignment. To identify the sensing ability of the AE, vibration signal was acquired through accelerometer and compared to the AE signal. Also, to find the advantage of the proposed signal processing method, it was compared with traditional envelope analysis.

According to the experiment result, AE sensor can detect the fault earlier than an accelerometer because of high sensitivity and in the power spectrum, the harmonics of the rotating speed and the gear mesh frequency clearly occurred. Misalignment was observed and bearing faults were also detected in the early fault stage. The proposed envelope analysis is worked to evaluate the faults and indicated the faults frequencies, rotating speed, sideband of BPFI, gear mesh frequency and harmonics, explicitly.

For the detection of the crack growth on the shaft, a cracked shaft was installed on the test rig, and the crack was seeded by wire-cutting with 0.5 mm depth. The cracked shaft was lifted 6.5 mm by the lifting tool. The AE signals were transformed by FFT to create the power spectrums, and in the spectrums several peaks were occurred by the crack growth. Along the growth of the crack, the characteristic of the power spectrum was changed and displayed different frequencies.

In the power spectrum, it was shown that the harmonic components of the rotating speed and bearing cage frequency were excited by the crack growth as shown in the Fig. 6, especially on the 3X (28.6Hz) and 31Hz. And the AE signal caused by the crack growth is generated on the whole ultrasonic frequency range; the initial crack could be detected using the PAC-Energy on wavelet level 1 to 4, and after that, it could be presented on wavelet level 5 until the fracture of the shaft. Therefore, in this paper, it could be shown that the crack growth in rotating machinery is able to be considered and to be detected; in addition, PAE-Energy can be used to detect the early detection of the crack.

Therefore, the proposed signal processing method that is the envelope analysis intercalated DWT using Daubechies mother function between BPF and wave rectification can be shown to provide better result than traditional envelope analysis.

6. Acknowledgment

This work has been supported by Basic Science Research Program through the National Research Foundation of Korea (NRF) funded by the Ministry of Education, Science and Technology (2011-0013652) and the 2nd Phase of Brain Korea 21.

7. References

Burrus, C.S., Gopinath, R.A. & Guo, H. (1997). *Introduction to wavelet and wavelet transforms: A Primer*, Prentice-Hall, ISBN 0134896006, Upper Saddle River, NJ.

Douglas, H. & Pillay, P. (2005). The impact of wavelet selection on transient motor current signature analysis, *Proceedings of 2005 IEEE International Conference on Electric Machines and Drives*, ISBN 0780389875; 978-078038987-8, San Antonio, TX, May 2005.

Hatch, C.T. & Bently, D.E. (2002). *Fundamentals of Rotating Machinery Diagnostics*, Bently Pressurized Bearing Press, ISBN 0-9714081-0-6, Minden, NV.

James, E. & Bery, P.E. (1994). *IRD Advancement Training Analysis II: Concentrated Vibration Signature Analysis and Related Condition Monitoring Techniques*, IRD Mechanalysis Inc., Columbus, Ohio.

Kim, Y.H., Tan, Andy. C.C., Mathew, J., Kosse, V. & Yang, B.S. (2007). A comparative study on the application of acoustic emission technique and acceleration measurements for low speed condition monitoring, *12th Asia-Pacific Vibration Conference*, Hokkaido Univ. Japan, August 2007.

Kim, Y.H., Tan, Andy. C.C., Mathew, J. & Yang, B.S. (2005). Experimental study on incipient fault detection of low speed rolling element bearings: time domain statistical parameters, *12th Asia-Pacific Vibration Conference*, Hokkaido Univ. Japan, August 2007.

Li, H., Zhang, Y. & Zheng, H. (2009). Gear fault detection and diagnosis under speed-up condition based on order cepstrum and radial basis function neural network, *Journal of Mechanical Science and Technology*, Vol. 23, No. 10, (October 2009), pp.2780-2789, ISSN 1738494X.

Mba, D. & Bannister, R. H. (1999). Condition monitoring of low-speed rotating machinery using stress waves: Part 1 and Part 2, *Journal of Process Mechanical Engineering*. Vol. 213, No. 3, (1999), pp.153-185, ISSN 0954-4089(Print), 2041-3009(Online)

Mba, D., Cooke, A., Roby, D. & Hewitt, G. (2003). Opportunities offered by acoustic emission for shaft-seal rubbing in power generation turbines; a case study. *Conference sponsored by the British Institute of NDT. International Conference on Condition Monitoring*, ISBN 1901892174, Oxford, UK, July 2003.

Mba, D., Cooke, A., Roby, D. & Hewitt, G. (2004). The detection of shaft-seal rubbing in large-scale power generation turbines with acoustic emission; Case study, *Proceedings of the Institution of Mechanical Engineers, Part A: Journal of Power and Energy*, Vol. 218, No. 2, (March 2004), pp.71-81, ISSN 0957-6509.

Misiti, M., Misiti, Y., Oppenheim, G. & Poggi, J.M. (2009) *Wavelet Toolbox TM 4 user's guide*, The MathWorks. Inc., Retrieved from <http://www.mathworks.com >

Ronnie, K.M. & V. K. Eric, (2005). *Nondestructive testing handbook Volume 6; Acoustic Emission Testing*, American Society for Nondestructive Testing, ISBN 978-1-57117-137-5, Columbus, Ohio.

Sato, I. (1990). Rotating machinery diagnosis with acoustic emission techniques, *Electrical Engineering of Japanese*, Vol. 110, No. 2, (1990), pp.115-127, ISSN 04247760.

Sentoku, H. (1998). AE in tooth surface failure process of spur gear, *Journal of Acoustic Emission*, Vol. 16 No. 1-4, (August 1998) pp.S19-S24, ISSN 0730-0050.

Sheen, Y.T. (2008). An envelope detection method based on the first-vibration-mode of bearing vibration, *Measurement: Journal of the International Measurement Confederation*, Vol. 41, No. 7, (August 2008), pp.797-809, ISSN 0263-2231.

Sheen, Y.T. (2010). An envelope analysis based on the resonance modes of the mechanical system for the bearing defect diagnosis, *Measurement: Journal of the International Measurement Confederation*, Vol. 43, No. 7, (August 2010), pp.912-934, ISSN 0263-2241.

Shiroishi, J., Li, Y., Lian, S., Danyluk, S. & Kurfess, T. (1999). Vibration analysis for bearing outer race condition diagnostics, *Journal of Brazilian Society of Mechanical Science*, Vol. 21, No. 3, (September 1999), pp.484-492, ISSN 0100-7386.

Singh, A., Houser, D. R. & Vijayakar, S. (1996). Early detection of gear pitting, *American Society of Mechanical Engineers*, Vol. 88, (1996) pp.673-678, ISSN 15214613.

Siores, E. & Negro, A.A. (1997). Condition monitoring of a gear box using acoustic emission testing, *Material Evaluation*, Vol. 55, No. 2, (February 1997), pp.183-187, ISSN 00255327

Tan, C.K. & Mba, D. (2005). Limitation of acoustic emission for identifying seeded defects in gearboxes, *Journal of Non-Destructive Evaluation*, Vol. 24, No. 1, (March 2005), pp.11-28, ISSN 0195-9298.

Tan, C.K., Irving, P. & Mba, D. (2007). A comparative experimental study on the diagnostic and prognostic capabilities of acoustics emission, vibration and spectrometric oil analysis for spur gears, *Mechanical Systems and Signal Processing*, Vol. 21, No. 1, (January 2007), pp.208-233, ISSN 0888-3270

Wu, J.D., Hsu, C.C. & Wu, G.Z. (2009). Fault gear identification and classification using discrete wavelet transform and adaptive neuro-fuzzy inference, *Expert Systems with Applications*, Vol. 36, No. 3, (April 2009), pp.6244-6255, ISSN 0957-4174.

Wu, J.D. & Chen, J.C. (2006). Continuous wavelet transform technique for fault signal diagnosis of internal combustions engines, *NDT&E International*, Vol. 39, No. 4, (June 2006), pp.304-311, ISSN 0963-8695.

Yang, Y., Yu, D. & Cheng, J. (2007). A fault diagnosis approach for roller bearing based on IMF envelope spectrum and SVM, *Measurement: Journal of the International Measurement Confederation*, Vol. 40, No. 9-10, (November 2007), pp.943-950, ISSN 0263-2231.

9

Compensation of Ultrasound Attenuation in Photoacoustic Imaging

P. Burgholzer[1,2], H. Roitner[1,2], J. Bauer-Marschallinger[1,2], H. Grün[2], T. Berer[1,2] and G. Paltauf[3]

[1]*Christian Doppler Laboratory for Photoacoustic Imaging and Laser Ultrasonics,*
[2]*Research Center for Non Destructive Testing (RECENDT),*
[3]*Institute of Physics, Karl-Franzens-University Graz*
Austria

1. Introduction

Photoacoustic imaging is a non-destructive method to obtain information about the distribution of optically absorbing structures inside a semitransparent medium. It is based on thermoelastic generation of ultrasonic waves by the absorption of a short laser pulse inside the sample. From the ultrasonic waves measured outside the object, the interior distribution of absorbed energy is reconstructed. The ultrasonic waves, which transport information from the interior to the surface of the sample, are scattered or absorbed to a certain extent by dissipative processes. The scope of this work is to quantify the information loss which is equal to the entropy production during these dissipative processes and thereby to give a principle limit for the spatial resolution which can be gained in photoacoustic imaging. This theoretical limit is compared to experimental data. In this book chapter state-of-the-art methods for modeling ultrasonic wave propagation in the case of attenuating media are described. From these models strategies for compensating ultrasound attenuation are derived which may be combined with well-known reconstruction algorithms from the non-attenuating case for photoacoustic imaging.

Section 2 gives a short description of photoacoustic imaging, especially photoacoustic tomography, and the available image reconstruction algorithms to reconstruct the interior structure from the detected ultrasound signal at the sample surface. Beside small point-like detectors also large detectors, so called integrating detectors are used for photoacoustic tomography. The latter ones require different image reconstruction algorithms. Spatial resolution is an essential issue for any imaging method. Therefore we describe the influencing factors of the resolution in photoacoustic tomography.

Section 3 is dedicated to acoustic attenuation. The spatial resolution in photoacoustic imaging is limited by the acoustic bandwidth. To resolve small objects shorter wavelengths with higher frequencies are necessary. For such high frequencies, however, the acoustic attenuation increases. This effect is usually ignored in photoacoustic image reconstruction but as small objects or structures generate high frequency components it limits the minimum detectable size, hence the resolution. Several models for acoustic attenuation, especially used for ultrasound propagation in biological tissue, are compared with experimental data.

Section 4 describes two different attempts to compensate this acoustic attenuation: either to include the compensation directly in the image reconstruction, e. g. in a modified time reversal method, or to calculate first the acoustic signal without attenuation from the measured attenuated signal and then perform the conventional photoacoustic image reconstruction. As any compensation of acoustic attenuation is mathematically an ill-posed problem both methods need regularization to prevent small measurement fluctuations from growing infinitely high in the reconstructed image. The possible degree of compensation depends on the size of these fluctuations. On the other hand acoustic attenuation is a dissipative process that causes entropy production equal to a loss of information, which cannot be compensated by any compensation algorithm. Therefore one can use the entropy production caused by acoustic attenuation to determine the minimal fluctuations in the measurement data, which turn out to be equal to thermal fluctuations. In statistical physics this fact is well known as the fluctuation-dissipation theorem, but the information theoretical background as a starting point to derive this theorem was not mentioned before in the literature.

Section 5 uses stochastic processes to understand theoretically how information can be lost and its connection to entropy production. Therefore, the measured pressure signal is treated as a random variable with a certain mean value as a function of time and certain fluctuations around that value. First for the simple model of a damped harmonic oscillator, it is shown how information is lost during a dissipative process and to what extent we can reconstruct the original information after some time. Then attenuated acoustic waves can be treated in a similar way: the spatial Fourier transform of the pressure wave can be described by a similar stochastic process as the damped harmonic oscillator - only in a higher dimension. Each wave vector is represented by a damped oscillator of different frequency.

Thinking about acoustic attenuation as a stochastic process helps to understand how entropy production and loss of information "work" on a microscopic scale. Beside a better theoretical insight the stochastic view on the acoustic wave answers a very important question: which is the best compensation method and the corresponding practicable spatial resolution in photoacoustic imaging? This question can be answered without taking fluctuations on a microscopic scale into account: the entropy production, which can be calculated from macroscopic mean values, is set equal to the information loss.

2. Photoacoustic imaging

In 1880, Alexander Graham Bell discovered that pulsed light striking a solid substrate can produce a sound wave, a phenomenon called the photoacoustic effect (Bell, 1880). Practical imaging methods based on this effect have been developed and reported the last decade (Xu & Wang, 2006). Today, photoacoustic imaging, which is also referred to as optoacoustic imaging or, when using microwaves instead of light for excitation, as thermoacoustic imaging, is attracting intense interest for cross-sectional or three-dimensional imaging in biomedicine.

In photoacoustic imaging, short laser pulses are fired at a sample and the absorbed energy causes local heating (Fig. 1). This heating causes thermoelastic expansion and thereby generation of broadband elastic pressure waves (ultrasound) which can be detected outside the sample, for example by a piezoelectric device or by an optical detector. Two methods are used for photoacoustic imaging: photoacoustic microscopy uses focused ultrasonic detectors and the sample is imaged by scanning the focus through the sample. In photoacoustic tomography (PAT) an unfocused detector is used which detects the pressure from the ultrasound wave arriving from all different locations of the source. A map or "image" of the photo-generated pressure distribution in the sample can be made by collecting the ultrasound at many different locations and processing it using a suitable algorithm e.g. by a filtered backprojection algorithm or by a time reversal algorithm.

Only if the pulse is short enough, thermal expansion causes a pressure rise proportional to the locally absorbed energy density. Short enough means that the pressure wave does not "run out" of the smallest structure which should be resolved in the photoacoustic image during the pulse time. This so called "stress confinement" is therefore fulfilled if the sound velocity multiplied by the pulse time of the laser is small compared to the spatial resolution one wants to achieve in imaging. Another constraint is the "thermal confinement" which is fulfilled if the heat induced in a structure by the absorbed laser pulse does not diffuse out of this structure during the time of the laser pulse. As heat diffusion is usually slower than the propagation of sound the thermal confinement is fulfilled if stress confinement is fulfilled.

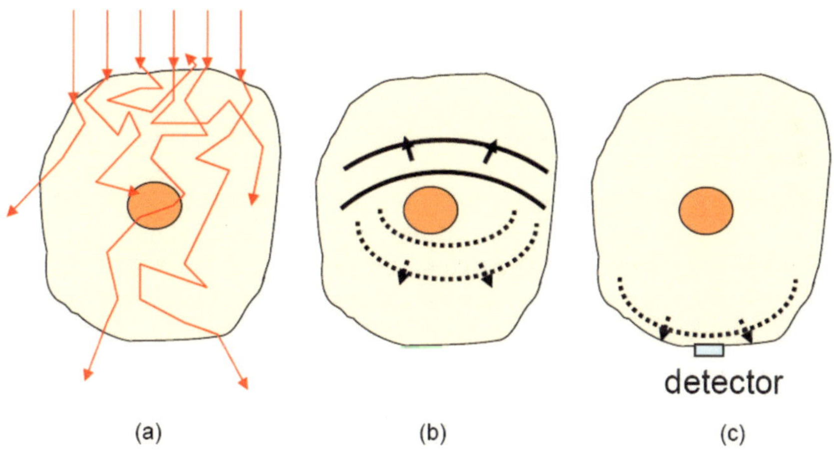

Fig. 1. Photoacoustic Imaging - the spatial resolution is determined by excitation, propagation, and detection of the acoustic wave. (a) Thermoelastic generation of acoustic wave by laser light (arrows indicate scattered photons): excited pressure is proportional to the absorbed optical energy density, if laser pulse is short enough to satisfy thermal and stress confinement. (b) Propagation of ultrasonic wave to sample surface: frequency dependent acoustic attenuation causes entropy production and therefore a loss of information. (c) Detection of ultrasonic wave: bandwidth and size of detector limits spatial resolution

Any photons, either unscattered or scattered (see arrows in Fig. 1), contribute to the absorbed energy as long as the photon excitation is relaxed thermally. Therefore PAT visualizes the product of the optical absorption distribution and the local light fluence.

Using a Nd:YAG laser and an optical parametric oscillator (OPO) light pulses from the infrared to the visible regime can be selected with a repetition rate from 10 Hz up to 100 Hz. Some high speed PAT systems can even go up to 1000 Hz. The pulse duration in the nanosecond range enables a theoretical resolution of several microns in tissue (sound velocity similar to water at approx. 1500 m/s). For biomedical applications the light energy should not exceed 20 mJ/cm^2 in the visible spectral range and 100 mJ/cm^2 in the near infrared light.

If acoustic attenuation and shear waves (in liquid and soft tissue) are neglected, the acoustic pressure p as a function of time and space obeys the equation

$$\Delta p(\mathbf{r},t)-\frac{1}{c^2}\frac{\partial^2}{\partial t^2}p(\mathbf{r},t)=-\frac{\beta}{C_p}\frac{\partial}{\partial t}\mathrm{H}(\mathbf{r},t) \tag{1}$$

where Δ is the three-dimensional Laplace operator, c the sound velocity, β the thermal expansion coefficient, C_p the specific heat capacity and $H(\mathbf{r},t)$ the deposited energy per time and volume ("heating function") caused by the absorption of the electromagnetic radiation in the sample.

For short electromagnetic pulses $H(\mathbf{r},t)=A(\mathbf{r})\cdot\delta(t)$, where $A(\mathbf{r})$ is the energy density of the absorbed electromagnetic radiation and $\delta(t)$ the Dirac delta function. Then the acoustic pressure $p(\mathbf{r},t)$ solves the homogeneous scalar wave equation with the initial conditions $p(\mathbf{r},0)=p_0(\mathbf{r})=\beta c^2/C_p\cdot A(\mathbf{r})\equiv\Gamma\cdot A(\mathbf{r})$ and $\partial/\partial t\, p(\mathbf{r},0)=0$. The initial pressure p_0 at time $t=0$ is therefore directly proportional to the absorbed energy density A with the dimensionless constant Γ, the Grüneisen coefficient.

As shown in Fig. 1 (c) bandwidth and size of the detectors for collecting the ultrasound signals are important for the resolution of this imaging modality. A photoacoustically generated ultrasound signal is a broadband signal and contains frequencies in the range from kilohertz up to a few megahertz. Conventional point like piezo elements such as known from arrays for medical ultrasonic imaging have their maximum sensitivity close to their center frequency and can detect frequencies only within a certain bandwidth around this frequency. Therefore high frequencies are not detected which correspond to small structures and are necessary for image reconstruction with a high spatial resolution. Other approaches are necessary for high resolution photoacoustic imaging. A hydrophone could be one solution (Wang, 2008), or the utilization of optical point like detection as demonstrated e.g. by (Zhang et al., 2009) or (Berer et al., 2010).

Point like detectors show a limit in achievable resolution by their size. The smaller the point detector the better is the spatial resolution. Unfortunately thermal and other fluctuations increase for a smaller detector, which results in a reduction of resolution. A totally different approach is the use of so called integrating detectors which are at least in one dimension larger than the object. This way the drawback of finite dimensions of point like detectors can be overcome. Such an integrating detector for photoacoustic imaging was introduced by (Haltmeier et al, 2004). They showed the mathematical proof of integrating area and line detectors and introduced new reconstruction methods which are necessary when using such a detector. First measurements using an integrating detector were shown by (Burgholzer et al., 2005). The first integrating detector was an area detector which was bigger than the object in two dimensions. For sufficient data for 3D image reconstruction the area detector had to be scanned around the object tangential to the surface of a sphere. This detector movement is difficult to realize. Hence the idea of the integrating line detector was developed. A fragmentation of the area detector into an array of line detectors results in an easier setup with only one rotation axis for the object and a linear motion of the integrating line detector (Fig. 2).

An integrating line detector is a line which has at least a length $\sqrt{8}*D$, where D is the diameter of a circle enclosing the sample and tangentially touching the line detector (Haltmeier et al., 2004). The line detector integrates the pressure along the line on a cylindrical surface with the radius $c \cdot t$ where c is the speed of sound in the medium and t the time. Thus, integrating line detectors arranged in an array around the sample, e.g. in a circle, measure projection images of the object in a first measurement and reconstruction step. By rotating the sample and measuring such projection images from different angles it is possible to reconstruct a 3D image (Fig. 2). Three dimensional image reconstruction from a set of projection images requires only the application of the inverse Radon transform. Therefore 3D imaging using integrating line detectors is not computationally intensive compared with other algorithms which reconstruct a 3D image from a set of signals acquired from point like detectors.

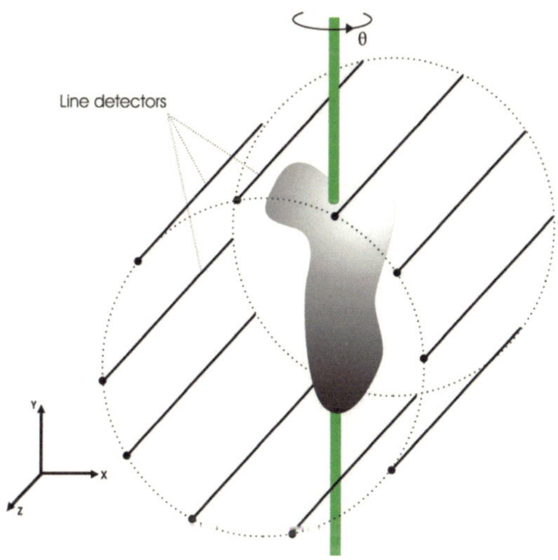

Fig. 2. Line detectors around a sample rotated on one axis. Either one line detector is scanning around the object or a detector array is used

Since the first measurement results from (Burgholzer et al, 2005) the integrating line detector was further improved in sensitivity and spatial resolution. Several types of such line detectors were implemented. One premature approach was a line made of PVDF (piezo foil) which provides high sensitivity but with the drawback of directivity. The next consequential step was an optical line detector realized by an interferometer. A laser beam that is part of an interferometer measures variations of the refractive index induced by the acoustic pressure (elasto-optic effect) (Paltauf et al., 2006). Optical detectors offer a broadband characteristic and due to the circular shape of a laser beam or light guiding fiber there is no such directivity like when using a film of piezo foil. Two main approaches can be distinguished: free-beam implementations and the use of fiber-based interferometers. Independent of the realization different types of the interferometer can be used, e.g. a Mach-Zehnder or a Fabry-Perot interferometer which is in general more sensitive than the first one.

(Paltauf et al., 2006) presented measurements using a free-beam Mach-Zehnder interferometer. They used a focused laser beam as line detector. When placing the object next to the beam waist of the focused laser beam the best spatial resolution due to the smallest beam diameter could be achieved (Paltauf et al., 2008).

(Grün et al, 2010) implemented fiber-based line detectors. The advantages of fiber-based line detectors are the easy handling and the small and constant beam diameter in the fiber. A small diameter of the laser beam is necessary for a good spatial resolution. The smaller the diameter of the detecting part the better is the spatial resolution. A typical single mode fiber for near infrared has a core diameter of 9 microns; single mode fibers for the visible range of detection wavelength have typically about 6 microns. Due to the constant diameter along the whole line detector this type of integrating line detector is dedicated for the imaging of big samples. After first implementations of a Mach-Zehnder and a Fabry-Perot interferometer in glass fibers now polymer fibers are used. Due to the much better impedance matching of polymer fibers to the surrounding water, their sensitivity is higher than for glass fibers, where approximately 2/3 of the incoming signal is reflected before reaching the core (Grün et al., 2010). Furthermore the Young's modulus is much lower in polymer fibers than in glass fibers for which reason the deformation of a polymer fiber is bigger than of a glass fiber applying the same pressure wave. As the strain optic coefficients are in the same order, this results in an enhanced change of refractive index and thus to higher signal amplitudes in the polymer fiber (Kiesel et al., 2007).

(Nuster et al, 2009) did a comparison of the different implementations of an integrating line detector. At this stage of development the free-beam Mach-Zehnder interferometer was the most sensitive integrating line detector. But these measurements showed some new approaches how the fiber-based line detector could be made more sensitive, e.g. by building up a Fabry-Perot interferometer in a single mode polymer fiber.

The next step after developing a sensitive line detector, no matter of which approach is the most sensitive one, is the creation of an array of many integrating line detectors, e.g. 200 detectors arranged in a curve around the sample. This way one could acquire all data for a projection image within one excitation pulse of the laser.

3. Acoustic attenuation

The imaging resolution in photoacoustic imaging is limited by the acoustic bandwidth and therefore by the laser pulse duration as mentioned above, but also by the attenuation of the acoustic wave on its way to the sample surface, and finally by the bandwidth and size of the ultrasonic transducer. The acoustic attenuation can be substantial for high frequencies. This effect is usually ignored in reconstruction algorithms but can have a strong impact on the resolution of small objects or structures within objects. Stokes could already show in 1845 that for liquids with low viscosity, such as water, the acoustical absorption increases by the square of the frequency (Stokes, 1845). If we do not take acoustic attenuation for photoacoustic image reconstruction into account, especially small structures (corresponding to shorter wavelengths and therefore higher frequencies) appear blurred (La Riviere et al., 2006). To what extent this blurring can be compensated by regularization methods (as performed by (La Riviere et al., 2006)) and how much information is lost due the irreversibility of attenuation is investigated in this chapter.

For thin layers (1D), small cylinders (2D), and small spherical inclusions (3D) the effect of attenuation is simulated and experimental results for several types of tissue are given. For photoacoustic tomography a new description of attenuation seems to be useful: like for a standing wave in a resonator the wave number is real but the frequency is complex. The complex part of the frequency is the damping in time. The resulting pressure wave as the solution of the wave equation is of course the same as by decomposing into plane waves with complex wave number. But with the complex frequency description acoustic attenuation can be included in all "k-space" methods well known in photoacoustic tomography just by introducing a factor describing the exponential decay in time (Roitner & Burgholzer, 2011).

Acoustic attenuation is an irreversible process and therefore the wave equation is not invariant to time reversal. Several important reasons for acoustic attenuation have been reported: viscosity, heat conduction, relaxation processes and chemical reactions. Stokes derived the scalar wave equation

$$\Delta p - \frac{1}{c^2}\frac{\partial^2 p}{\partial t^2} + \tau\,\Delta\frac{\partial p}{\partial t} = 0 \tag{2}$$

under the assumption of adiabatic conditions (thus neglecting the loss due to heat conduction) (Stokes, 1845). This equation can also be found in (Shutilov, 1988) and can generally be derived from a relaxation behavior of pressure and density (Royer & Dieulesaint, 2000), where the density change follows the pressure change after a relaxation time τ. If τ is further expressed in terms of viscosity and specific heat, this equation is also known as the thermoviscous equation. Eq. (2) describes acoustic attenuation which is approximately proportional to the square of the frequency. Other wave equations can describe a more general power law frequency dependence of the attenuation of the form $\alpha(\omega) = \alpha_0 \cdot |\omega|^y$ ($0 < y < 3$). (Szabo, 1994) has suggested adding the loss term $L(t) * p(\mathbf{r},t)$ to the wave equation in order to account for such attenuation behavior:

$$\Delta p(\mathbf{r},t) - \frac{1}{c_0^2}\frac{\partial^2}{\partial t^2} p(\mathbf{r},t) + L(t) * p(\mathbf{r},t) = 0 \tag{3}$$

Other models for acoustic wave propagation in acoustic media have been proposed also by (Nachman et al., 1990) and (Treeby & Cox, 2010).

Measurement and simulation of broadband acoustic attenuation:

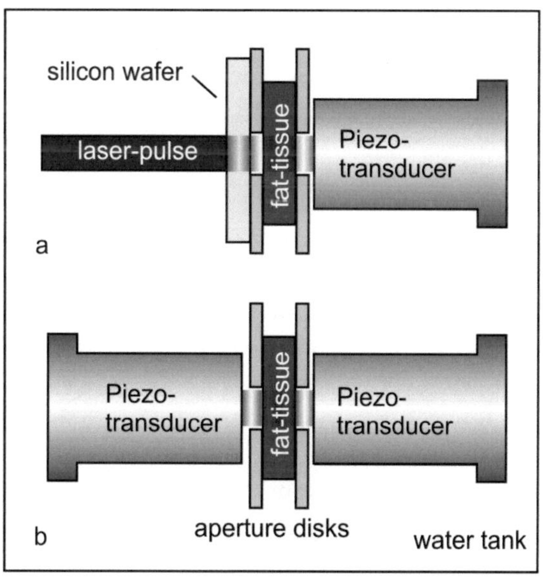

Fig. 3. Experimental set-up to measure broadband high frequency acoustic attenuation in tissue with ultrasound generated by short laser pulses (a) and by piezoelectric transducers (b). In both cases a piezoelectric transducer receives the ultrasonic signals

Fig. 3 shows the set up to determine frequency dependent acoustic attenuation in tissue by a single transmission experiment. High frequency ultrasonic pulses are either generated by a pulse laser (pulse duration 6 ns) heating up the surface of a silicon wafer (Fig. 3a) or by a piezoelectric transducer (Fig. 3b). The resulting planar-like ultrasound waves propagate through a fat tissue with varying thickness and through distilled water as a coupling medium to a piezoelectric transducer. The ultrasonic attenuation α is determined by comparing transmission measurements of the investigated samples and distilled water. In Fig. 4 Fourier transformed attenuation results for subcutaneous fat of pig, human blood and olive oil are shown. A power law $\alpha(f) = \alpha_0 \cdot f^y$ can be applied to those substances, where f denotes the frequency (Bauer-Marschallinger et al., 2011).

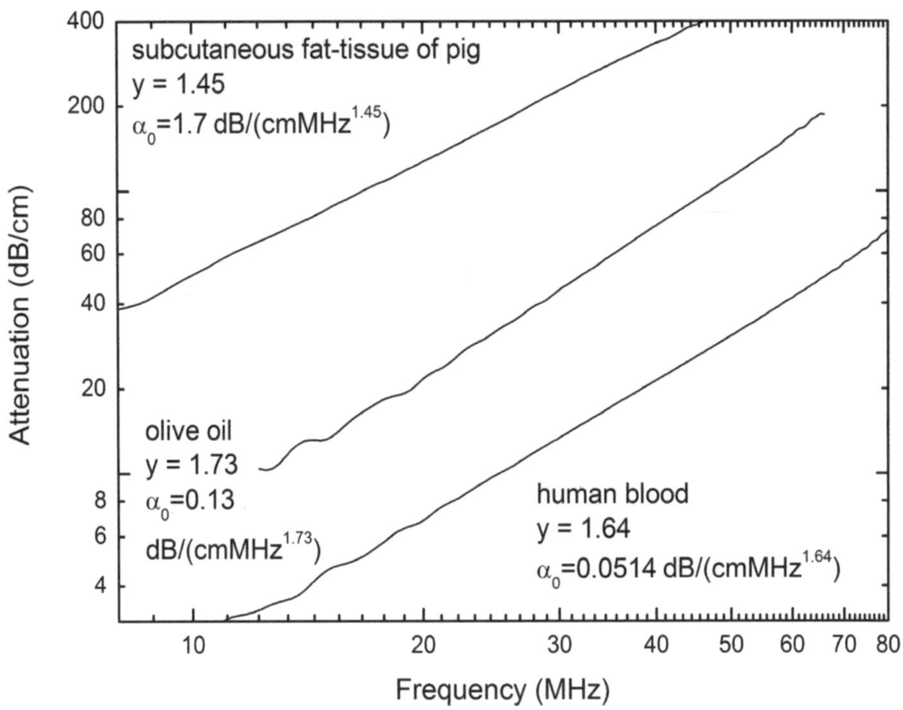

Fig. 4. Attenuation as a function of frequency for three biological substances (double logarithmic scale)

The attenuated planar ultrasound waveform can be calculated by a number of available simulation methods. In a comprehensive study (Roitner et al., 2011) we compared measured waveforms for two fat thicknesses (3.2 mm and 6.2 mm) to simulated waveforms obtained with three current simulation methods. The methods using the Matlab toolbox by (Treeby & Cox, 2010b) or the relation (4) by (LaRiviere et al., 2006) rely on a frequency power law absorption and are described in detail below. A third method by (Nachman et al., 1990) assumes multiple molecular relaxation processes as the cause of attenuation. These processes are characterized by their contributions $\kappa_n (n = 1 \ldots N)$ to the isothermal compressibility of the tissue and the relaxation times $\tau_n (n = 1 \ldots N)$ of their vibrational energies. With these parameters a complex wave number is obtained involving a generalized compressibility $\kappa(t)=\kappa_{\infty}\delta(t)+\sum_{n=1}^{N}\frac{\kappa_n}{\tau_n}\exp(-t/\tau_n)H(t)$ where $\kappa_{\infty}=\chi-\sum_{n=1}^{N}\kappa_n$ and $H(t)$ is the Heaviside step function. χ denotes the usual compressibility calculated with the formula $\chi=(c^2\rho)^{-1}$ which holds in liquids as well as in soft biological tissues of density ρ. Stokes' equation (2) is seen to be the special case of a single relaxation mechanism $N=1$, $\kappa_1=\chi$.

All three simulation methods produce waveforms in good agreement to the measured waveform. Of course, the approximation will become less accurate with longer propagation distance in the absorbing medium.

Simulation results may also be validated if we recall that the frequency domain transfer function of the pressure signal propagating through a layer with complex wavenumber $K(\omega)$ and thickness d equals $\exp(\mathrm{i}K(\omega)d)$. So the simulated 'water-fat-water' waveform may be obtained from the measured 'water-only' waveform by inverting water attenuation over the fat thickness and then applying fat attenuation over the same fat thickness.

4. Compensation of acoustic attenuation

When taking acoustic attenuation into account, the wave equation, e.g. (2) or (3), is not invariant to time reversal any more. First order terms in time t or higher order odd terms change sign if time is reversed. The equations then do not behave "well" any more, noise is amplified exponentially and regularization methods have to be used for solving these time reversed equations. Using such regularized methods the spatial resolution, that is limited by the frequency-dependent damping, is improved (Burgholzer et al., 2007). To prevent high-frequency noise from growing exponentially, Fourier spectral methods (Trefethen, 2000) in space have been used. They utilize the spatial Fourier transform to calculate the Laplacian and therefore allow for incorporating a damping of higher frequencies when calculating the time reversal. A similar algorithm to compensate acoustic attenuation step by step was proposed by (Treeby et al., 2010c).

In the second strategy, attenuation is compensated in one step where an approximation to the 'un-attenuated' measured signal is calculated from the attenuated measured signal by solving an appropriate integral equation (La Riviere et al., 2006); then an ordinary reconstruction algorithm for photoacoustic tomography is applied as in the absence of attenuation.

The Matlab Toolbox by Treeby et al. implements a state-of-the-art algorithm to simulate ultrasound wave propagation ('the direct problem') and image reconstruction ('the inverse problem') for PAI. For the direct and inverse problems arbitrary source distributions and detector geometries, variable medium densities and sound speeds and an arbitrary constant medium absorption are supported in 1D-3D. The algorithm is first-order finite difference in time and pseudospectral in space (working in spatial Fourier space = 'k-space'). The linearized Euler equations in a fluid $\frac{\partial \mathbf{u}}{\partial t}=-\frac{1}{\rho^*}\nabla p$, $\frac{\partial \rho}{\partial t}=-\rho^*\,\nabla\cdot\mathbf{u}$ together with an adiabatic equation of state $p=c_0^2\rho$ are solved up to time T for the sound velocity vector $\mathbf{u}$, pressure p and density ρ. In the case of absorbing media the equation of state is extended to

$p(\mathbf{r},t) = c_0^2 \rho(\mathbf{r},t) + \mathrm{F}^{-1}(\tau k^{y-2} \frac{\partial \hat{\rho}(\mathbf{k},t)}{\partial t} + \eta k^{y-1} \hat{\rho}(\mathbf{k},t))$, where τ is related to attenuation and η to dispersion (Treeby & Cox, 2010) and the inverse Fourier transform (IFT) in d spatial dimensions is defined by $f(\mathbf{r}) = \mathrm{F}^{-1}(\hat{f}(\mathbf{k})) = (2\pi)^{-d} \int_{R^d} \hat{f}(\mathbf{k}) \exp(i\mathbf{k}\cdot\mathbf{r}) d\mathbf{k}$. With this extension of the equation of state frequency-domain ultrasound absorption power laws of the form $\alpha(\omega) = \alpha_0 \cdot |\omega|^y$ can be supported.

For the inverse problem, the same algorithm as for the direct problem is applied, but with zero initial conditions $p(\mathbf{r},0) = 0$, $\mathbf{u}(\mathbf{r},0) = \mathbf{0}$ and time-varying Dirichlet boundary conditions on detector surface points $\mathbf{r}_S$ in the form of the time-reversed sensor data $p(\mathbf{r}_S,t) = p_{meas}(\mathbf{r}_S, T-t)$. Absorption is compensated by inverting the sign of τ but leaving η unchanged. Additionally, a regularization filter is applied in k-space that suppresses the higher frequencies.

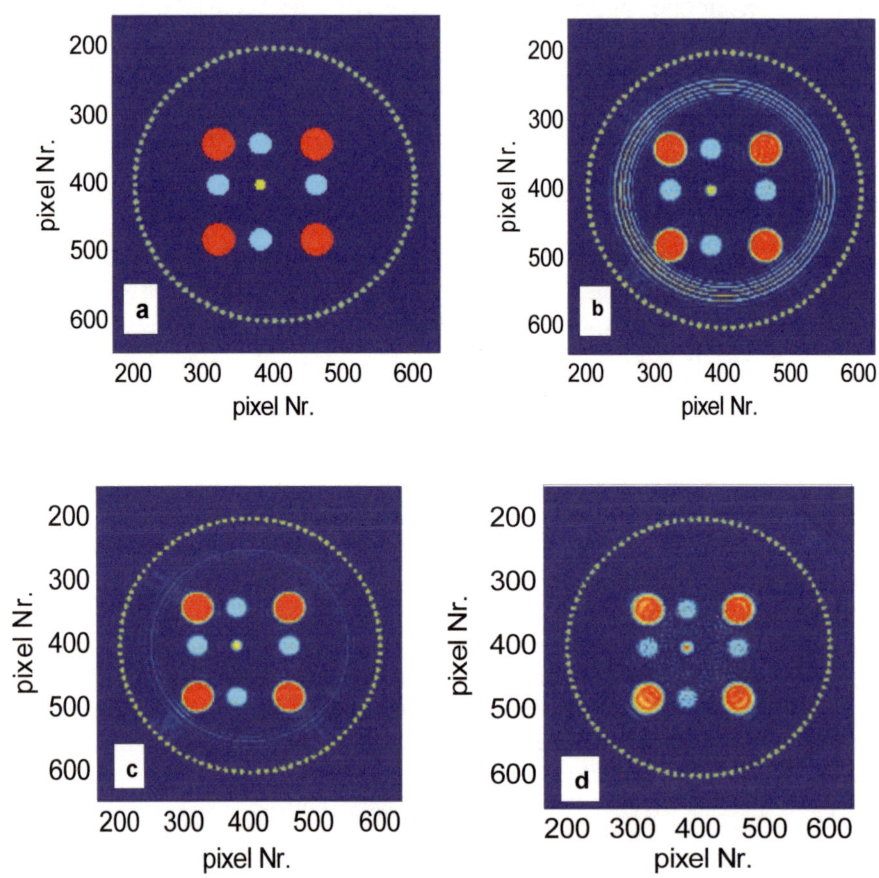

Fig. 5. Reconstruction of nine-circle phantom: (a) initial pressure distribution, (b) reconstruction Matlab toolbox with taper ratio r = 0.5, (c) reconstruction Matlab toolbox with taper ratio r = 0.75, (d) reconstruction one-step compensation plus fast series algorithm

A Tukey (tapered cosine) window is chosen for this filter. This window is parameterized by its cutoff-frequency in k-space (here $2\pi \cdot 10^4 rad/m$, corresponding to $2\pi \cdot 15MHz$ in ω-space) and its taper ratio r. The ratio controls the filter roll-off: raising r suppresses more high-frequency content in the measurement noise. In Fig. 5a we display a complex phantom consisting of nine circles of different sizes and source strengths, surrounded by a circular array of detectors (which could represent line detectors in 3D). This phantom was reconstructed first with taper ratio $r = 0.5$ (Fig. 5b) and then with taper ratio $r = 0.75$ (Fig. 5c). The absorption parameters were an exponent of $y = 1.5$ and $\alpha_0 = 3dB/(cm \cdot MHz^{1.5})$, and the 'measurement noise' consisted only of numerical noise.

Note that in Fig. 5b ring-like artifacts are produced during reconstruction, which are much reduced in Fig. 5c because there the regularization filter is more restrictive. With taper ratio $r = 1$ they disappear altogether. If a simpler phantom is used, e.g. just a single circle, taper ratio $r = 0.5$ does not produce any artifacts. To sum up, complex, fine-structured objects exhibit more signal content and hence more measurement noise at higher frequencies. Then the regularization filter must be parameterized more restrictively than for compact objects.

The formula used for calculating and compensating attenuation in one step was first presented by (LaRiviere et al., 2006) and reads

$$\tilde{p}_{att}(\mathbf{r}_S, \omega) = \frac{\omega}{c_0 \cdot K(\omega)} \int_{-\infty}^{+\infty} p_{ideal}(\mathbf{r}_S, t) \exp(ic_0 K(\omega) \cdot t)\, dt \tag{4}$$

This relation yields the temporal Fourier transform of the attenuated pressure at a given point $\mathbf{r} = \mathbf{r}_S$ (and, via an IFT, the attenuated pressure itself) if the ideal (i.e. un-attenuated) pressure is known over time at that point. This can be exploited for the direct problem where the evaluation involves only two nested numerical integrations. For the inverse problem, the left-hand side of (4) is known and the relation represents an integral equation to be solved for $p_{ideal}(\mathbf{r}_S, t)$. Since this solution is very sensitive to measurement noise, i.e. noise in the spectrum of $p_{att}(\mathbf{r}_S, t)$, regularization is necessary, for instance in the form of truncated singular value decomposition (SVD, see e.g. La Riviere et al., 2006).

If the detector geometry is regular, fast reconstruction algorithms using series expansions in eigenfunctions of the Laplacian on the regular domain are available for the lossless case making the whole reconstruction with compensation much faster than reconstruction with the mentioned toolbox algorithm. In Fig. 5d we display a reconstruction of the nine-circle phantom using this procedure. To apply the SVD, a certain noise level in temporal Fourier space must be assumed. It is desirable to base this assumption on physical reasons - this will be discussed below. Here we assume a white Gaussian noise with a standard deviation of $0{,}1\%$ of the largest amplitude in the spectrum.

Fig. 5d shows that the reconstruction with the one-step compensation is almost of the same quality as the reconstruction with the Matlab toolbox algorithm. However, computation time is reduced by a factor of 10 in comparison to the toolbox algorithm.

We already observed that the quality of reconstructions depends crucially on a good analysis of the amplitude and spectral distribution of the measurement noise. On the source side, advance knowledge of the type of object to be imaged is helpful (more bulky and compact, or more fine-structured). On the detector side, noise is created by fundamental physical processes as well as by technological limitations.

First, for any material volume V at temperature T pressure fluctuations of variance $Var(p) = k_B T / \chi V$ (Landau&Lifshitz, 1980) will occur where χ is the adiabatic compressibility and k_B is the Boltzmann constant. So if detectors are e.g. thin foils with small volume, this may play a role since noise amplitudes grow like $V^{-1/2}$.
Second, the fluctuation-dissipation theorem states that sound absorption processes in media create pressure fluctuations. If the absorption is described by Stokes' equation (2), a straightforward calculation yields a power spectral density of these pressure fluctuations $S_P(\omega) = \frac{2k_B T}{\pi} \cdot \chi^{-2} \frac{\tau}{(1+\omega^2\tau^2)^2}$.
Third, noise is created by the analog-to-digital conversion process during electronic acquisition of the pressure signal. If this quantization noise is assumed to be white, its power spectral density at sampling frequency f_s will be constant and equal $q^2 / 12 f_s$ (Widrow & Kollar, 2008) where q is the quantization interval.
We have given only an outline of the considerations that have to be taken into account for a good estimation of the measurement noise. In a concrete application the relative importance of the mentioned contributions will depend on sizes and shapes of objects and detectors, the physics of the ultrasound propagation medium (compressibility, temperature) and the properties of the measurement devices.

5. Stochastic processes for modeling acoustic attenuation

As for any other dissipative process the energy of the attenuated acoustic wave is not lost but is transferred to heat, which can be described in thermodynamics by an entropy increase. This increase in entropy is equal to a loss of information, as defined by Shannon, and no compensation algorithm can compensate this loss of information. This is a limit given by the second law of thermodynamics. But can this limit be found in the algorithms for compensation of acoustic attenuation given in the previous section? Fluctuations of the measured pressure are "amplified" exponentially during the compensation and therefore we suspect a relation between the entropy production caused by acoustic attenuation and the fluctuations. Indeed such a relation is known in statistical physics as the fluctuation-dissipation theorem and is due to (Callen & Welton, 1951), (Callen & Green, 1952), and (Greene & Callen, 1952). It represents in fact a generalization of the famous (Johnson, 1928) (Nyquist, 1928) formula in the theory of electric noise.
In this section we use the theory of non-equilibrium thermodynamics presented by S.R. de Groot and P. Mazur (De Groot & Mazur). More about random variables and stochastic processes can be found e.g. in the book about Statistical Physics from (Honerkamp, 1998). An introduction to stochastic processes on an elementary level has been published by. (Lemons, 2002), also containing "On the Theory of Brownian Motion" by (Langevin, 1908). An introduction to Markov Processes is given by (Gillespie, 1992).
In this section the measured pressure signal is treated as a time-dependent random variable with a mean value and a variance as a function of time. To be able to use the results of some "model" stochastic processes given in literature (Ornstein-Uhlenbeck process or damped harmonic oscillator) for a model of photoacoustic reconstruction we have changed the initial conditions: instead of a defined initial value (with zero variance) we have taken the stochastic process at equilibrium before time zero and at a time zero a certain perturbation has been applied to the process (e. g. a rapid change in momentum for the damped

harmonic oscillator - called kicked damped oscillator in this chapter). Reconstruction of the size of this perturbation at time t=0 from the measurement after a time t shows how the information about the size of this kick at t=0 gets lost with increasing time if dissipative processes occur. This change of the initial conditions has a significant advantage: it turns out that the variance stays constant in time, while the mean value is a function of time. This facilitates the calculations of the entropy production and of the information loss due to the stochastic process.

Gauss-Markov processes and entropy

In this section we shall assume that the time varying stochastic processes will have Gauss-Markov character. In doing so we do not wish to assume that all dissipative macroscopic processes considered belong to this specific class of Gauss-Markov processes. It may, however, be surmised that a number of real phenomena may, with a certain approximation, be adequately described by such processes (De Groot & Mazur). The advantage of specifying more precisely the nature of the processes considered is that it enables us to discuss, on the level of the theory of random processes, the behavior of entropy production and of information loss.
Following the theory of random fluctuations given e.g. by (De Groot & Mazur), we take as a starting point equations analogous to the Langevin equation used to describe the velocity of a Brownian particle:

$$\frac{d\boldsymbol{\alpha}}{dt} = -\mathbf{M}\cdot\boldsymbol{\alpha} + \boldsymbol{\varepsilon}(t) \tag{5}$$

The components of the vector $\boldsymbol{\alpha}$ are the random variables $a_i (i = 1,2,...,n)$, having zero mean value at equilibrium. The matrix $\mathbf{M}$ of real phenomenological coefficients is independent of time. The vector $\boldsymbol{\varepsilon}(t)$ represents white noise, which is uncorrelated at different times. The distribution density of $\boldsymbol{\alpha}$ turns out to be an n-dimensional Gaussian distribution:

$$f(\boldsymbol{\alpha},t) = \frac{1}{\sqrt{(2\pi)^n |\Sigma|}} e^{-\frac{1}{2}(\boldsymbol{\alpha}-\bar{\boldsymbol{\alpha}}(t))^T \Sigma^{-1} (\boldsymbol{\alpha}-\bar{\boldsymbol{\alpha}}(t))} \tag{6}$$

with the mean value $\boldsymbol{\alpha}(t)$ and the covariance matrix Σ, which is usually also a function of time but for the initial conditions used later on the covariance matrix will be constant in time. $|\Sigma|$ is the determinant of the covariance matrix. If the initial value of $\boldsymbol{\alpha}$ is given by a given $\bar{\boldsymbol{\alpha}}_0$, the mean value at a later time t will be:

$$\bar{\boldsymbol{\alpha}}(t) = e^{-\mathbf{M}t}\cdot\bar{\boldsymbol{\alpha}}_0 \tag{7}$$

By an adequate coordinate transformation of $\boldsymbol{\alpha}$ the matrix $\mathbf{M}$ can be diagonalized: the eigenvalues of $\mathbf{M}$ are the elements of the diagonal matrix.
According to the second law of thermodynamics the entropy of an adiabatically insulated system must increase monotonously until thermodynamic equilibrium is established within the system. Then the entropy will be set to zero and the entropy at a time t is:

$$S(t) = -k_B \int f(\boldsymbol{\alpha},t)\ln\frac{f(\boldsymbol{\alpha},t)}{f(\boldsymbol{\alpha},t\to\infty)}\, d\boldsymbol{\alpha} \tag{8}$$

with the Boltzmann constant k_B. For a constant covariance matrix the above integration results in:

$$S(t) = -\frac{1}{2} k_B \overline{\boldsymbol{\alpha}}(t)^T \Sigma^{-1} \overline{\boldsymbol{\alpha}}(t) \tag{9}$$

Before modeling the attenuated acoustic wave as a Gauss-Markov process we give two simple examples: the Orstein-Uhlenbeck process with only one component of $\boldsymbol{\alpha}$ as a model for the velocity of a Brownian particle and the damped harmonic oscillator with two components of $\boldsymbol{\alpha}$.

Example: kicked Ornstein-Uhlenbeck process

If the random vector $\boldsymbol{\alpha}$ in eq. (5) has only one component we get the Langevin equation

$$\frac{dv(t)}{dt} = -\gamma \cdot v(t) + \sigma\, \eta(t) \tag{10}$$

which was used to describe Brownian motion of a particle. The random variable v is the particle velocity, $-\gamma \cdot v$ is the viscous drag, and σ is the amplitude of the random fluctuations. The Langevin equation governs an Ornstein-Uhlenbeck process, after (Uhlenbeck & Ornstein, 1930), who formalized the properties of this continuous Markov process. Now we assume that initially we have thermal equilibrium with zero mean velocity. At time zero the particle is kicked which causes an immediate change in velocity of v_0. Following eq. (7) the mean value $\overline{v}(t)$ shows an exponential decay:

$$\overline{v}(t) = e^{-\gamma t} \cdot \overline{v}_0 \tag{11}$$

The variance of the velocity $Var(v)$ is $\sigma^2/2\gamma$ and is constant in time. In Fig. 6 time and velocity are scaled to be dimensionless and the standard deviation (square root of the variance) of the velocity is normalized.
From eq. (9) the information loss equal to the entropy production till the time t after the kick is:

$$\Delta S(t) = k_B \frac{\gamma}{\sigma^2} \overline{v}(t)^2 \tag{12}$$

On the other hand the entropy production known from thermodynamics is the dissipated energy ΔQ, which is the kinetic energy of the Brownian particle of mass m, divided by the temperature T:

$$\Delta S(t) = \frac{\Delta Q}{T} = \frac{m\overline{v}(t)^2}{2T} \tag{13}$$

The thermodynamic entropy production in eq. (12) has to be equal to the loss of information in eq. (13), and therefore we get for the variance of the velocity:

$$\frac{\sigma^2}{2\gamma} = \frac{kT}{m} \tag{14}$$

Eq. (14) has been derived previously by the equipartition theorem: the equilibrium energy associated with fluctuations in each degree of freedom is $k_B T/2$. We have used the equity of entropy production and information loss. Eq. (14) states a connection between the strength of the fluctuations, given by σ^2, and the strength of the dissipation γ. This is the fluctuation-dissipation theorem in its simplest form for uncorrelated white noise.

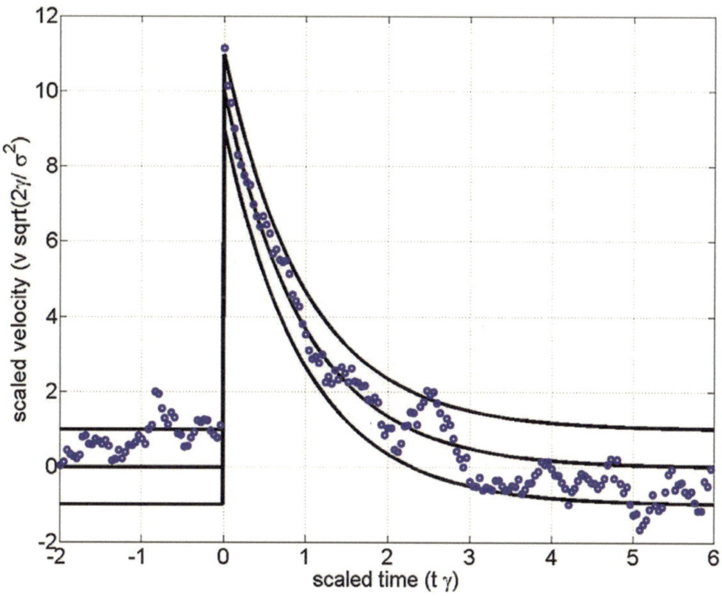

Fig. 6. Points on a sample path of the normalized kicked Ornstein-Uhlenbeck process defined by the Langevin eq. (10). The solid lines represent the mean, and mean ± standard deviation of the scaled velocity coordinate. At the time t=0 a value of v_0=10 has been added to the scaled velocity. After some time the information of the amplitude gets more and more lost due to the fluctuations

It is instructive to determine the least square estimator (Honerkamp, 1998) for the initial velocity v_0. If we write for the estimated initial velocity v_r

$$v_r = R(t) \cdot v(t) \tag{15}$$

we calculate $R(t)$ by minimizing the mean error $\left\langle (v_0 - v_r)^2 \right\rangle$:

$$R(t) = \frac{e^{-\gamma t}}{e^{-2\gamma t} + \sigma^2 / 2\gamma {v_0}^2} \tag{16}$$

This gives the Tikhonov regularization with $\sigma^2/2\gamma {v_0}^2 = Var(v)/{v_0}^2$ as regularization parameter. The inverse square root of the regularization parameter $v_0/\sqrt{Var(v)}$ can be interpreted as signal-to-noise-ration (SNR) of $v(t)$.

Example: kicked harmonic oscillator

For modeling of acoustic waves one needs in addition to the dissipation also an oscillating term. For pure oscillation without damping we have no loss of information (Fig. 7).

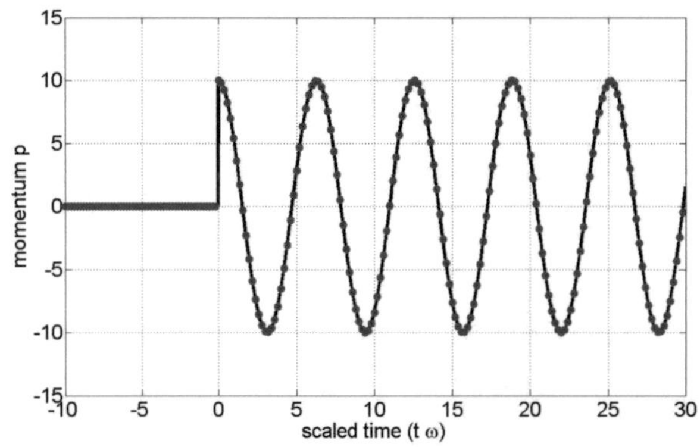

Fig. 7. Points on a sample path of the kicked harmonic oscillator without damping. The momentum used to kick the oscillator can be reconstructed without loss of information at any time after the kick if ω is known

The stochastically damped harmonic oscillator combines the oscillatory and the diffusive behavior and therefore it is a good starting point to model attenuated acoustic waves. The equations of motion are (using the momentum p instead of the velocity v):

$$\frac{dx(t)}{dt} = \frac{1}{m} p(t) \tag{17}$$

$$\frac{dp(t)}{dt} = -\frac{\gamma}{m} \cdot p(t) - m\omega_0^2 x + \sigma\,\eta(t) \tag{18}$$

These equations can be combined using a two dimensional random vector $\boldsymbol{\alpha} = (x, p)$ and were solved already by (Chandrasekhar, 1943) for definite initial conditions $x(0)$ and $p(0)$. Again we have changed the initial conditions to an oscillator with zero mean values kicked by an initial momentum p_0. In Fig. 8 the damping is chosen to be $\gamma = m\omega_0/3$.
Using the fluctuation-dissipation theorem $\sigma^2/2\gamma = kT$ one obtains for the distribution function

$$f(x,p,t) = \frac{1}{2\pi \frac{kT}{\omega_0}} e^{-\frac{1}{2mkT}(p-\overline{p}(t))^2 - \frac{1}{2kT} m\omega_0^2 (x-\overline{x}(t))^2}, \tag{19}$$

where $\overline{x}(t)$ and $\overline{p}(t)$ are the solutions of the ordinary (non-stochastic) damped harmonic oscillator. Then one gets for the information loss from eq. (9)

$$S(t) = -\frac{1}{T}\left(\frac{1}{2} m\omega_0^2 \overline{x}(t)^2 + \frac{\overline{p}(t)^2}{2m}\right) = -\frac{1}{T}(E_{pot} + E_{kin}), \tag{20}$$

which is equal to the entropy from thermodynamics, where E_{pot}+E_{kin} is the total energy of the harmonic oscillator (sum of the potential and kinetic energy). This fact confirms again that the fluctuation-dissipation theorem for the damped harmonic oscillator $\sigma^2/2\gamma = k_B T$ can be derived from the equity of entropy production and information loss.

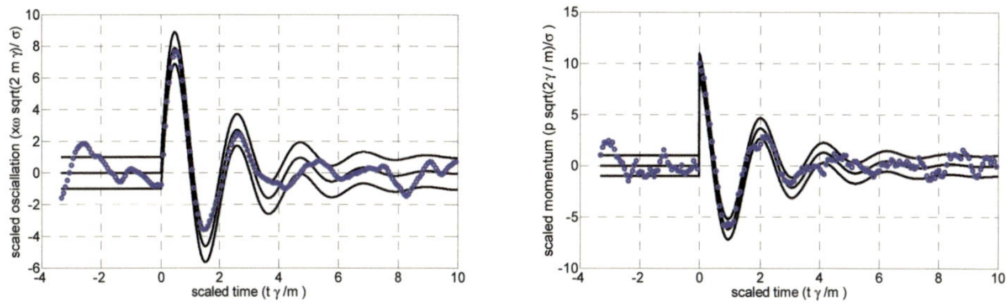

Fig. 8. Points on a sample path of the normalized kicked damped harmonic oscillator (eq. (17) and (18)). The solid lines represent the mean, and mean ± standard deviation of the scaled oscillation (left) and momentum (right). At the time t=0 a value of p_0=10 has been added to the scaled momentum. After some time the information about the value of p_0 gets more and more lost due to the fluctuations

For the mean value of the momentum $\overline{p}(t)$ one gets from eq. (7):

$$\overline{p}(t) = p_0[\cos(\omega t) - \frac{\gamma}{2m}\frac{\sin(\omega t)}{\omega}]e^{-\frac{\gamma t}{2m}} \tag{21}$$

Compared to the Ornstein-Uhlenbeck process (eq. (11)) the mean value has not only an exponential decay in time but also an oscillation with a frequency $\omega = \sqrt{\omega_0^2 - \gamma^2 / (4m^2)}$. As mentioned above (Fig. 7) the oscillating term does not change the entropy and no information is lost. In the average only the exponential decay causes production of entropy and the information about the value of p_0 gets more and more lost due to the fluctuations. The entropy production which is proportional to the total energy is shown in Fig. 9.

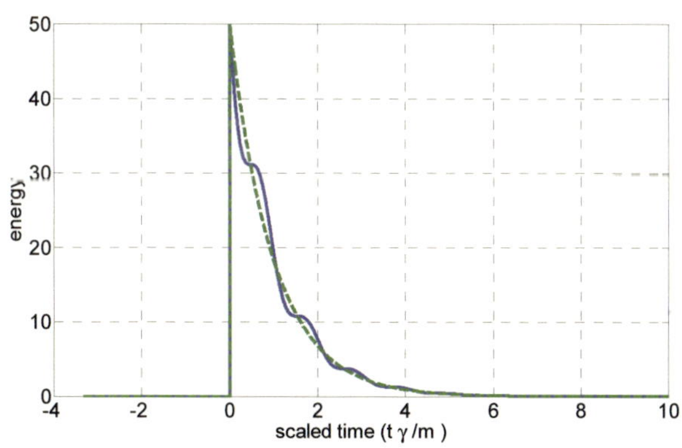

Fig. 9. Total energy of the damped harmonic oscillator (solid line) shows in the average an exponential decay with the time constant γ / m (dashed line)

Attenuated acoustic wave as a stochastic process

In photoacoustic imaging, the laser pulse at a time $t = 0$ generates an initial pressure distribution $p_0(\mathbf{r})$ (see section 2). For numerical calculations we use a discrete space $\mathbf{r} = \mathbf{r}_j$ (*j=1,..,N*), where $\mathbf{r}_j$ are N points on a cubic lattice with a spacing of Δr within the sample volume V. At a time t the pressure distribution $p_j(t)$ can be represented by a Fourier series (Barret et al. 1995), including the time $t = 0$ with the initial pressure distribution:

$$p_j(t) = p(\mathbf{r}_j, t) = \sum_{k=1}^{N} \hat{p}_k(t)\varphi_k(\mathbf{r}_j) \quad with \;\; \varphi_k(\mathbf{r}) = e^{2\pi i \boldsymbol{\rho}_k \cdot \mathbf{r}} \cdot D(\mathbf{r}) \tag{22}$$

$D(\mathbf{r})$ is a support function which is one within the sample volume V and zero outside. $\boldsymbol{\rho}_k$ are integer points on an infinite 3D lattices. From eq. (1) we get (if acoustic attenuation can be neglected):

$$\hat{p}_k(t) = \cos(\omega_k t)\hat{p}_k(0) \quad with \;\; \omega_k^{\,2} = 4\pi^2 c^2 \boldsymbol{\rho}_k^{\,2}. \tag{23}$$

Therefore we have only an oscillating term as shown in Fig. 7, but in higher dimensions, and no information is lost.

For an attenuated acoustic wave instead of the wave equation (1) we have used the Stokes equation (2) or the wave equation (3), giving for the spatial Fourier components $\hat{p}_k(t)$ not only an oscillating term but also an exponential decay:

$$\hat{p}_k(t) = [A_k \sin(\omega_k t) + B_k \cos(\omega_k t)]\, e^{-\lambda_k t} \tag{24}$$

where the phase factors A_k and B_k can be derived from the initial conditions and ω_k and λ_k are functions of $|\boldsymbol{\rho}_k|$. Like for the damped harmonic oscillator this exponential decay causes the loss of information and can be modeled by a Gauss-Markov process. The information content in Fourier space is the same as in real space (Fig. 10). Therefore we can describe the information loss in the attenuated acoustic wave by the same model as for the damped harmonic oscillator, only in higher dimensions, as for each wave-vector-index k an oscillator is needed.

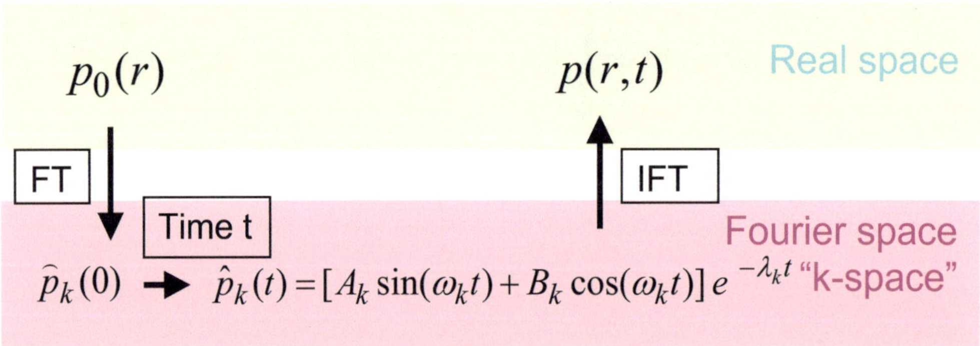

Fig. 10. The initial pressure distribution just after the laser pulse is Fourier transformed (FT). The time evolution of the Fourier series coefficients can be described similar to the mean value of a stochatsically damped harmonic oscillator. The pressure distribution after a time t is then calculated by an inverse Fourier transform (IFT)

One dimensional (1D) example: photoacoustic signal of a 0.2 mm thin layer in glycerin:

The effect of acoustic attenuation for the reconstructed image is similar in 1D, 2D, and 3D (Burgholzer et al., 2010a and 2010b). As in 1D the reconstructed image is just the shifted measured signal, the effect of attenuation and of its compensation can be directly seen. The photoacoustic signal of a 0.2 mm thin absorbing layer in glycerin is calculated by using the scheme of Fig. 10 after a time of 4.5 microseconds. In glycerin the acoustic pressure can be described well by the Stokes' equation with a relaxation time of 244 picoseconds (Shutilov, 1988).

For an attenuated acoustic wave instead of the wave equation (1) we have used the Stokes equation (2) or the wave equation (3), giving for the spatial Fourier components $\hat{p}_k(t)$ not only an oscillating term but also an exponential decay:

Putting eq. (24) into the Stokes equation (2) one gets:

$$\omega_k^2 = 4\pi^2 c^2 \rho_k^2 (1 - \frac{1}{4} c^2 \rho_k^2 \tau^2) \text{ and } \lambda_k = \pi c^2 \rho_k^2 \tau \quad (25)$$

The initial pressure distribution is a one dimensional square pulse corresponding to an absorbing layer of thickness 2a with a = 0.1 mm. In Fourier space we get:

$$\hat{p}_k(0) = \frac{1}{\pi k} e^{-ikr} \sin(ka) \, with \, k = 2\pi \rho_k \quad (26)$$

The calculated signal after a time of 4.5 microseconds is shown in Fig. 11 (dashed line). The dashed dotted line shows the signal without attenuation (no relaxation time) and therefore represents the ideal reconstructed image.

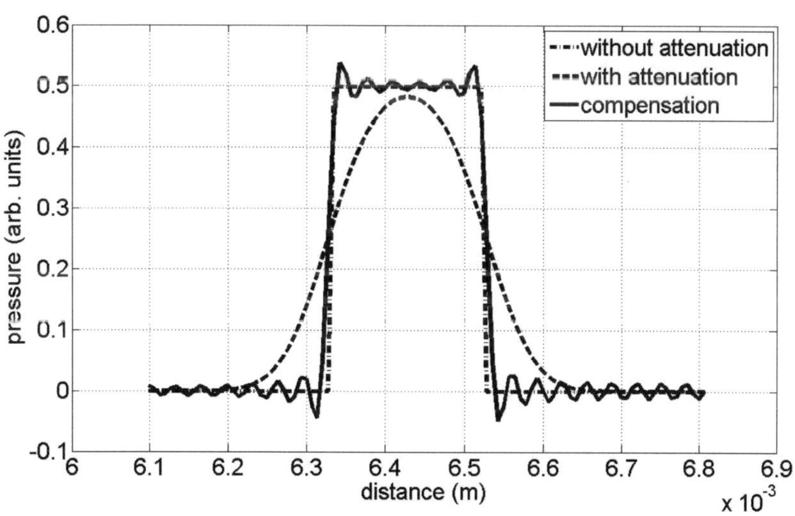

Fig. 11. Simulation result as an example for Stoke's equation: it is shown how an initial square pulse would look after a time of 4.5 microseconds in glycerine (τ=244ps) (dashed line). The dashed dotted line shows the pulse without attenuation (τ=0), which correspond to the reconstruction of the initial pressure pulse, and the solid line is the SVD reconstruction of the initial pressure pulse (see text)

For compensation of the acoustic attenuation we use the time reversed process of the Gauss-Markov process from Fig. 10. Similar to the stochastically damped oscillator the entropy production for the acoustic wave is set equal to the loss of information from the Gauss-

Markov process with standard deviation s_k and is approximated by an exponential decay in time (Fig. 9):

$$\Delta S(t) \approx \frac{1}{2} k_B \sum_k \frac{1}{s_k^2} \exp(-2\lambda_k t) \hat{p}_k(0)^2 \tag{27}$$

$$\Delta S(t) = \frac{\Delta Q}{T} \approx \frac{1}{2 c_0^2 \rho_0 T} \sum_k \frac{1}{s_k^2} \exp(-2\lambda_k t) \hat{p}_k(0)^2 \tag{28}$$

which gives a variance of $\hat{p}_k(t)$ in the spatial Fourier space ("k-space") independent from k:

$$s_k^2 = k_B c_0^2 \rho_0 T \tag{29}$$

In real space this gives the thermodynamic fluctuations. Or the other way around, the thermodynamic fluctuations have a quantity such that the information loss of the Gauss-Markov process is equal to the dissipated heat divided by the temperature. Using the thermodynamic fluctuations for a detector size of 1 cm^2 and the SVD regularization from section 4 the compensated $\hat{p}_k(t)$ is calculated. By applying a subsequent inverse Fourier transform (IFT) one gets the compensated pressure profile in real space (solid line in Fig. 11).

6. Summary and conclusions

Acoustic attenuation is modeled as a stochastic process: this helps to understand how thermodynamic entropy production and the decrease of information, which is "transported" in the acoustic wave, are closely connected on a microscopic scale. This theoretical insight enables to answer an important question in photoacoustic imaging: what is the highest possible compensation of attenuation and therefore the best spatial resolution one can achieve?

We could show that for thermal fluctuations the information loss of the reconstructed image is equal to the entropy production due to attenuation of the acoustic wave. Therefore it is sufficient to calculate the entropy production from the macroscopic mean values and it is not necessary to take the fluctuations of the pressure into account.

The size and locations of detectors in photoacoustic imaging should be optimized to get the best resolution and sensitivity. Up to now in such models it is not assumed that the pressure is a random variable, which favors small point like detectors. Taking thermal fluctuations and other noise into account will help to get a more realistic model for detectors and the reconstructed images from measured signals with these detectors.

7. Acknowledgments

This work has been supported by the Christian Doppler Research Association, by the Federal Ministry of Economy, Family and Youth, by the Austrian Science Fund (FWF) project numbers S10503-N20 and TRP102-N20, by the European Regional Development Fund (EFRE) in the framework of the EU-program Regio 13, and the federal state Upper Austria.

8. References

Barret, H. H., Denny, J. L., Wagner, R. F. & Myers, K. J. (1995). Objective assessment of image quality II, *J. Opt. Soc. Am.* A 12(5) 834-852

Bauer-Marschallinger, J., Berer, T., Roitner, H., Grün, H., Reitinger, B. & Burgholzer, P. (2011). Ultrasonic attenuation of biomaterials for compensation in photoacoustic imaging, *Proc. SPIE* 7899, 789931

Bell, A. G. (1880). On the production and reproduction of sound by light: the photophone, *American Journal of Science* 20, 305–324

Berer, T., Hochreiner, A., Zamiri, S. and Burgholzer, P. (2010) Remote photoacoustic imaging on solid material using a two-wave mixing interferometer, *Opt. Lett.* 35, 4151-4153

Burgholzer, P., Hofer, C., Paltauf, G., Haltmeier, M. & Scherzer, O. (2005). Thermoacoustic tomography with integrating area and line detectors, *IEEE Trans. Ultrason. Ferroelectr. Freq. Control* 52, 1577–1583

Burgholzer, P., Grün, H., Haltmeier, M., Nuster, R. & Paltauf, G. (2007). Compensation of acoustic attenuation for high resolution photoacoustic imaging with line detectors, *Proc. of SPIE* Vol. 6437 643724-1

Burgholzer, P., Roitner, H., Bauer-Marschallinger, J. & Paltauf, G. (2010a). Image Reconstruction in Photoacoustic Tomography using Integrating Detectors accounting for Frequency-Dependent Attenuation, *Proc. of SPIE* Vol. 7564 756423-1

Burgholzer, P., Berer, T., Gruen, H., Roitner, H., Bauer-Marschallinger, J., Nuster, R. & Paltauf, G. (2010b). Photoacoustic Tomography using Integrating Line Detectors Invited, *Journal of Physics: Conference Series*, 214(1).

Callen, H. B. & Welton, T. A. (1951), *Phys. Rev.* 83, 34

Callen, H. B. & Greene, R. F. (1952). *Phys. Rev.* 86, 702

Chandrasekhar, S. (1943). Stochastic Problems in Physics and Astronomy, *Reviews of Modern Physics* 15, 1-89

De Groot, S.R. & Mazur, P. (1984) *Non-Equilibrium Thermodynamics*, Dover Publications, Inc., New York

Gillespie, D. T. (1992). *Markov Processes*, Academic Press, New York

Greene, R. F. & Callen, H. B. (1952). *Phys. Rev.* 88, 1387

Grün, H., Berer, T., Nuster, R., Paltauf, G. & Burgholzer, P. (2010). Three-dimensional photoacoustic imaging using fiber-based line detectors, *Journal of Biomedical Optics*, 15(2), 021306-1 - 021306-8

Haltmeier, M., Scherzer, O., Burgholzer, P. & Paltauf, G. (2004). Thermoacoustic computed tomography with large planar receivers, *InverseProbl.* 20, 1663–1673

Hansen, P. C. (1987). *Rank-deficient and discrete ill-posed problems: Numerical aspects of linear inversion*, SIAM, Philadelphia

Honerkamp, J. (1998). *Statistical Physics*, Springer-Verlag

Johnson, J. B. (1928). Thermal Agitation of Electricity in Conductors, *Phy. Rev.* 32, 97

Kiesel, S., Peters, K., Hassan, T. & Kowalsky, M. (2007). Behavior of intrinsic polymer optical fibre sensor for large-strain, *Meas. Sci.Technol.* 18, 3144–3154

Landau, L. D. & Lifshitz, E. M. (1980). *Statistical physics, Part 1*, Pergamon Press, Oxford

Langevin, P. (1908). Sur la théorie du mouvement brownien, Compets rendus Académie des Sciences (Paris) 146, 530-533. Translation by Anthony Gythiel, published in American Journal of Physics 65, 1079-1081 (1997)

La Rivière, P. J., Zhang, J. & Anastasio, M. A. (2006). Image reconstruction in optoacoustic tomography for dispersive acoustic media, *Optics Letters*. 31(6), 781-783

Lemons, D. S. (2002). An Introduction to Stochatic Processes in Physics, The Johns Hopkins University Press

Nachman, A. I., Smith III, J. F. & Waag, R. C. (1990). An equation for acoustic propagation in inhomogeneous media with relaxation losses, *J.Acoust.Soc.Am.* 88, 1584-1595

Nuster, R., Gratt, S., Passler, K., Grün, H., Berer, T., Burgholzer, P. & Paltauf, G. (2009). Comparison of optical and piezoelectric integration line detectors, in *Biomedical Optics: Photons Plus Ultrasound: Imagingand Sensing 2009*, edited by A. A. Oraevsky and L. H. Wang, *Proc. SPIE* 7177, 71770T

Nyquist, H. (1928). *Phys. Rev.* 32, 110

Paltauf, G., Nuster, R., Haltmeier, M. & Burgholzer, P. (2006). Photoacoustic tomography using a Mach-Zehnder interferometer as acoustic line detector, *Appl. Opt.* 46, 3352–3358

Paltauf, G., Nuster, R., Passler, K., Haltmeier, M. & Burgholzer, P. (2008). Optimizing Image Resolution in Three-Dimensional Photoacoustic Tomography With Line Detectors, in *Biomedical Optics: Photons Plus Ultrasound: Imaging and Sensing 2008, Proc SPIE,* 6856

Roitner, H. & Burgholzer, P. (2011). Efficient modeling and compensation of ultrasound attenuation losses in photoacoustic imaging, *Inverse Problems* 27, 015003

Roitner, H., Bauer-Marschallinger, J., Berer, T. & Burgholzer, P. (2011). Experimental Evaluation of ultrasound attenuation losses in photoacoustic imaging, submitted to *J.Acoust.Soc.Am.*

Royer, D. & Dieulesaint, E. (2000). *Elastic Waves in Solids I*, Springer

Shutilov, V. A. (1988). *Fundamental Physics of Ultrasound*, Gordon and Breach Science Publishers

Stokes, G. G. (1845). On the theories of the internal friction of fluids in motion, and of the equilibrium and motion of elastic solids, *Trans. Cambridge Philos. Soc.* 8, 287-319

Szabo, T. L. (1994). Time domain wave equations for lossy media obeying a frequency power law, *J.Acoust. Soc. Am.* 96, 491–500

Treeby, B. E. & Cox, B. T. (2010). Modeling power law absorption and dispersion for acoustic propagation using the fractional Laplacian, *J.Acoust. Soc. Am.* 127 , 2741-2748

Treeby, B.E. & Cox, B. T. (2010b). k-wave: Matlab toolbox for the simulation and reconstruction of photoacoustic wave fields, *J. Biomed. Optics* 15(2), 021314-1 - 021314-12

Treeby, B.E., Zhang, E. Z. & Cox, B. T. (2010c). Photoacoustic tomography in absorbing acoustic media using time reversal, *Inverse Problems* 26, 115003

Trefethen, L. M. (2000). *Spectral Methods in Matlab*, SIAM, Philadelphia

Uhlenbeck, G. E. & Ornstein, L. S. (1930). On the Theory of Brownian Motion, *Phys. Rev.* 36, 823-41

Wang, L. V. (2008). Prospects of photoacoustic tomography, *Medical Physics* 35, 5758–5767

Widrow, B. & Kollar, I. (2008). *Quantization noise*, Cambridge Books, Cambridge

Xu, M. & Wang, L. V. (2006). Photoacoustic imaging in biomedicine, *Review of Scientific Instruments*. 77(4), 041101-041122

Zhang, E. Z., Laufer, J. G., Pedley, R. B. & Beard, P. C. (2009). In vivo high-resolution 3D photoacoustic imaging of superficial vascular anatomy, *Phys. Med. Biol.* 54, 1035–1046

10

Low Frequency Acoustic Devices for Viscoelastic Complex Media Characterization

Georges Nassar
Université de Lille –Nord de France
France

1. Introduction

The evolution in consumer expectations in terms of quality and safety in the agro-industry has led to the need to develop new methods of investigating product quality and the processes involved. Many fields of production still rely too much on the know-how of the operators who, with their experience acquired over time, have become key players in the company. In addition, the manufacturing of quality food products frequently relies on artisanal know-how that is difficult to industrialise and often synonymous of high production losses, therefore prohibitive costs. In contrast, so as to limit costs, the industrial production process is often associated with poorer quality. The objective evaluation of product quality involves the development of methods and sensors adapted to the product or the manufacturing process.

Indeed, beneath an apparent simplicity, agro-industry products have complex physical properties linked to elasticity, viscosity and plasticity. One of the major difficulties lies in the complexity of the processes which depend on numerous physical parameters. The matter is subjected to numerous mechanical, thermal or chemical treatments thus migrating towards viscoelastic or even plastic properties that are more difficult to quantify.

The originality of the approach adopted consists in the study and set up of an ultrasonic measuring device associated with its electronic environment in order to reply to a specific need due to the complexity of the physico-chemical phenomena involved. A global approach to this problem is very tricky as the physical properties of the media evolve significantly throughout processing.

We thus focused on the development of sensors and methods of characterisation dedicated to different phases of the industrial processes. Two very closely linked aspects were therefore studied targeting product characterisation and process control.

Work has been carried out to develop acoustic and ultrasonic instrumentation designed to monitor the change in state of the matter (liquid-gel transition and product cohesion), then to monitor the evolution of its elastic properties. The process control applications concern the development of a very low frequency, non-destructive monitoring method to reply to the specificities of the physical properties of the matter.

In this document, we report the scientific approach highlighting the design of the ultrasonic sources which dispenses with classic design through the choice of specific resonance modes for the sensors. Their design aims at promoting low frequency resonance in a relatively small scale composite structure. This sensor technology was adapted according to the

frequency chosen to study the change in physical state of the media and to monitor the evolution of the acoustic properties of products that are often heterogeneous.
Several approaches were used to optimise this technology: an analytical approach to determine the sensor's first vibratory mode which was consolidated by a numerical study, then confirmed and validated experimentally.
The application is based on two points:

- The physical or physico-chemical phenomenon linked to the transition phase and the sol-gel transition in the dairy field; the opaqueness and the fragility of this type of gel justifies the importance of quantifying the metrological parameters such as measurement accuracy, stability over time and mechanics of the sensor implemented.
- The interaction of the sensor with its environment in a process. A bivariate study (sensor/propagation medium) was carried out in order to select the required geometry to ensure that the sensor is adapted to its environment: loaded metallic plates subjected to mechanical stress and heated to a temperature of around 100°C (plate heat exchanger) or strongly absorbent media (fermenting bread dough...). This part of the study consisted in finding a good compromise between the geometry of the sensor, its location in the overall system and the required sensitivity.

2. Context

Many biochemical industrial activities involve very complex physicochemical phenomena which enable products to be processed. These products often go through a variety of states during processing.
Processing uses energy from chemical reactions (e.g. enzymatic), thermal or mechanical energy, or even a combination of these different forms of energy. Physical modifications can also occur (incorporation of air in the matter…). All these forms of energy are often combined within the same process and it is difficult to quantify the contribution of each in the product processing phenomena.
Due to the complexity of the processes (several processing stages, multiplicity of forms of energy…) and the products (viscoelastic matter, visco-elasto-plastic…), associated with the legitimate concern of not interfering with the process, few measurements have been carried out during the various stages of processing. The temperature, pressure and flow are often monitored during processing even though they are not necessarily correlated to the desired properties of the product under development.
This is why we chose to develop acoustic sensors adapted to the constraints imposed by either the product or the process.
Among the essential parameters sought-after, rheological measurements are often determining in terms of the consumers' perception of the qualities of the end product such as texture, viscosity, elasticity.... Several techniques of investigation exist but the majority are laboratory applications and are difficult to adapt for in-line controls. The difficulty thus arises of using multiple techniques to obtain a more complete characterization of the process and the interpretation of the data obtained with these analysis techniques.
Furthermore, the quality control of the processed products also involves evaluating the performance of the process. The temperature, pressure and flow are of course part of the characteristics measured for the control and/or closed-loop control of the various stages of product processing. However, in the case of complex processes they are difficult to correlate to the final properties of the product.

The processes can also evolve over time. This is the case with heat exchangers for which the performance varies over time due to fouling. Only preventive maintenance leading to additional production costs can ensure stable performances of the process over time. The development of sensors integrated in the process to provide information on the evolution of the performance remains essential.

This work presents a selection of studies which have led to the development of low frequency acoustic sensors specifically adapted to monitor changes in the physical state of complex media and the process: fragile gel, highly heterogeneous or highly absorbent media, media with complex rheological behaviour...

Several cases were studied:

- A low frequency acoustic sensor adapted to the characterization of complex products using an omni-directional source in the case of media undergoing a change in physical state;
    - continuous homogeneous medium: sol-gel transition,
    - complex heterogeneous medium: transition from a suspension of particles in a liquid to a cohesive visco-elasto-plastic solid.
- A very low frequency acoustic sensor used to monitor the response of a medium subjected to mechanical vibrations. Such technology is designed to study the processing phenomena of a highly absorbent product such as bread dough during fermentation.

Finally, the identification of the needs and constraints imposed by certain environments (temperature, hygiene, attenuation...) have led to the combination of these types of technology to monitor a process (e.g. fouling of plate heat exchangers, search for an optimum point in the kneading phase...). By taking into account the coupling of the sensor with its environment this technique can, in certain cases, exploit the noise emitted by the process itself, as in kneading for example.

In this work, we chose to illustrate the potential of low frequency acoustic methods on applications from the agri-foodstuffs sector. These same states can also be found in the pharmaceutical and cosmetics industries, the aviation industry, the medical field as well as in material chemistry.

The methodology implemented can be divided into several phases:

- Analysis of the product and/or process
- Definition and/or optimisation of the appropriate method and a sensor meeting the various constraints :
    - Analytical study
    - Numerical modelling
- Experimental validation of the sensor
- Validation of the application

3. Sensors suitable for studying media of which the physical properties evolve over time

3.1 Monitoring changes in the physical state of the matter

The analysis of the different stages in the formation of macromolecular networks is of major importance, since understanding the structure and properties (physical or chemical) of gels requires the understanding of the process of organization. In many physical, chemical or biological processes, the union of small separate elements to form aggregates of different sizes and further macroscopic phases makes connectivity an essential characteristic of this type of process. Many models have been proposed to explain the phenomenon of aggregation. The most important ones are those of Flory (Flory, 1953), Stockmayer (Stockmayer, 1943), Case (Case, 1960), Gupta (Gupta et al., 1979), Eichinger (Eichinger, 1981), Allsopp (Allsopp, 1981) and San Biagio (San Biagio et al., 1990). In most cases the phenomenon is described by the classical theory as a "particular case of percolation" and the two-dimensional growth of the network according to Caylay's tree. Other studies including those of De Gennes (De Gennes, 1989) and Stauffer (Stauffer, 1981, 1985) describe the phenomenon of random aggregation and the problems of percolation and gelation. However, the different characteristics of the macromolecular chain-making system can be evaluated according to Clerc (Clerc et al., 1983), using for example a Monte-Carlo simulation, predicting the influence of the characteristics of the starting solution and the gelation conditions on the structure and the arrangement of the masses.

In fact, the gelation process is a transition from an entirely soluble system to a heterogeneous two-phase system: composed of an insoluble entity (infinite-size macromolecule) and a soluble phase. This transition is accompanied by radical changes in some physical properties of the medium. Below the gelation "point", the viscosity of the medium increases and the medium ceases to flow by developing an elasticity.

To study this phenomenon, several physical measurement techniques exist i.e. optical, thermal, rheological and acoustic (Nassar, 1997). However, sampling and sensitivity to a limited range of physical properties are often drawbacks. Consequently, different techniques are required to explore an entire process with the difficulty of bringing together the heterogeneous data provided by these techniques. This is, for example, the case of optical methods which are penalized by the opacity of the substances analyzed as well as the size of the molecules formed in relation to the wavelength. Thermal methods are insensitive to the mechanical characteristics of the medium. The fragility of some gels (milk gel) limits rheological techniques. In many cases, several analytical techniques exist, but they are only used in the laboratory.

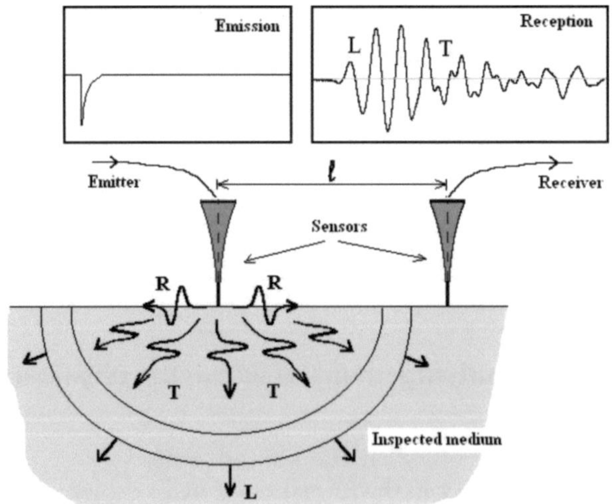

Fig. 1. Basic principle

To develop further instrumentation in order to understand and to quantify the modification process of media in real conditions, a low-frequency ultrasonic technique using sensors with

extremely pointed ends that act as point sources was examined. The application of this technology using two near-field coupled sensors to explore the relationship between the physical properties measured during the evolution of the time of flight of the wave and the structural changes during matter formation (Figure 1) was investigated.
The sensors were near-field coupled through the medium to be characterised. Such disposition privileges the Signal/Noise ratio and avoids the loss of acoustic pressure which is inversely proportional to the ray of the spherical wave. In a metal or ceramic solid all the waves are generated simultaneously, but in the media we are concerned with, the dominating longitudinal wave is the fastest and is relatively simple to exploit.
The advantage of these sensors is that they can be adapted to the measurement configuration envisaged according to the nature of the wave and the appropriate resonance mode.

3.2 Study of a low-frequency ultrasonic device

The aim of the study was to define and develop optimal ultrasonic instrumentation to understand the phenomenon and quantify the viscoelastic properties of changing media.
The usual ultrasonic characterization techniques are generally based on the use of a resonant piezoelectric transducer in thickness mode. As the resonant frequency of a transducer is inversely proportional to its size, it is greater for low frequencies around 100 KHz. Some researchers like Degertekin (Degertekin & Khury-Yakub, 1996a, 1996b, 1996c), Shuyu (Shuyu, 1996, 1997) and Nikolovski (Nikolovski & Royer, 1997) used this physical principle, but associated a tapered volume with the ceramic components to concentrate the mechanical energy.
The aim of this part of the work was to obtain a low frequency acoustic point source to generate a spherical wave in the medium. To do this a different procedure from that traditionally used in classic sensor design was implemented. A new technique was used which consisted in setting in resonance the entire mechanical structure of a reduced-size unit through the contact of an extremely pointed end with the material to be analyzed. In order to behave like an acoustic point source, the size of the point was smaller than the wavelength in the medium.
The first part presents a theoretical analysis of low-frequency ultrasonic resonators, beaming a spherical wave in the medium. The choice of a triangular shaped resonator and its mechanical behavior will be assessed and the study completed by a numerical approach based on the application of the finite elements method to characterize all the resonator vibration modes and visualize the corresponding distortions when the structure is excited. As the analytical results were in good agreement with the numerical results, they were applied to the whole triangular-shaped sensor to validate the findings experimentally. The resonance mode frequencies determined by the numerical calculation were then correlated with the electrical impedance measurements.

3.3 Study and design

3.3.1 Analytical approach

For a possible analytical analysis, the structure of a standard ultrasonic sensor is based on a simple triangle shape (Figure 2).
The propagation of longitudinal waves in the triangular part of the sensor was studied to determine the resonance frequency of the elongation mode and the velocity amplification

ratio between the ends. The analysis is based on an extension of Ensminger's (Ensminger, 1960) theory.
According to figure 2, the x section is written:

$$S = e \cdot \ell(x) \text{ Then } S = e \cdot \ell_1 (x_1 + x) / x_1 \tag{1}$$

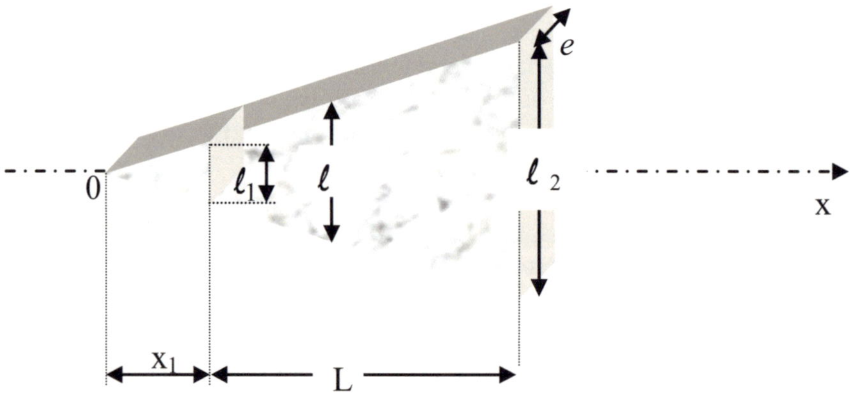

Fig. 2. Basic analytical shape

Ensminger studied the propagation of a wave in extensional mode in a cone with no loss of which the lateral dimensions were short in comparison with the length. In the case of a triangular shape, this equation takes the following form:

$$\frac{\partial^2 v}{\partial x^2} + \frac{1}{(x_1 + x)}\frac{\partial v}{\partial x} + \frac{\omega^2}{c^2} v = 0 \tag{2}$$

Where:
v is the velocity of the particles, ω is the pulsation and c is the velocity of the longitudinal wave in the material making up the vibratory element.
On the basis of the dimensions given in figure 3, the solution to this differential equation leads to an approximate velocity amplification ratio between the two extremities (Nassar, 1997); $|v(0)/v(L)| = 1/0.46 = 2.16$ for a resonance frequency: f = 60 KHz.

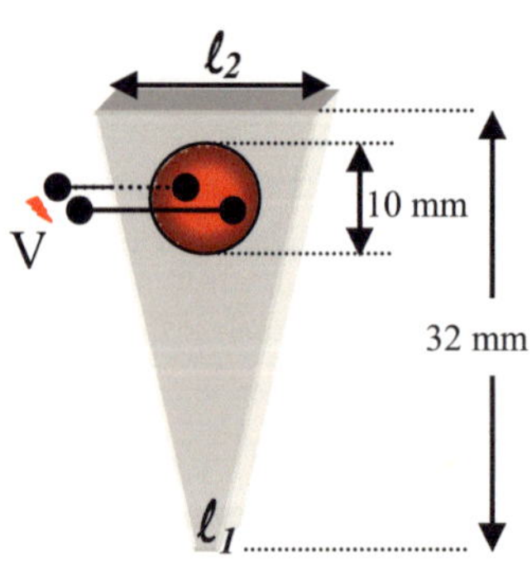

Fig. 3. Basic shape. Triangular sensor of thickness e = 1 mm, L = 32 mm, ℓ_1 = 2 mm and ℓ_2 = 16 mm

3.3.2 Numerical approach

While an analytical study can only take into consideration one particular mode of vibration of the triangular part of the sensor, a numerical study based on the finite elements method can determine all the vibrating modes of these parts as well as those of the realised sensor.
For a real structure; whole sensor included a binding rod and a triangular truncated part (Figure 4), the displacement differential equations were solved with a continuous regime, taking into account the boundary conditions at the surfaces. The materials were defined by Young's modulus E, Poisson's coefficient ν and density ρ. The results presented below were applied without loss and they were compared to the characteristics of the longitudinal mode determined by the analytical calculations. This comparison was also made for the triangular sensor which was studied as a whole.
For our study, the ANSYS analysis software was used. The sensors used were made essentially of piezoelectric material. A source of excitation was engraved in the general structure of the vibrating element (triangle part), providing mechanical continuity without any break (Figure 4a). This type of engraving was considered as it has been demonstrated (Nassar, 1997) that for the same longitudinal mode, the amplification ratio at the ends is **71** times bigger when there is one engraved source providing mechanical continuity with the vibrating element so that one source can impact the structure by gluing (Figure 3b).

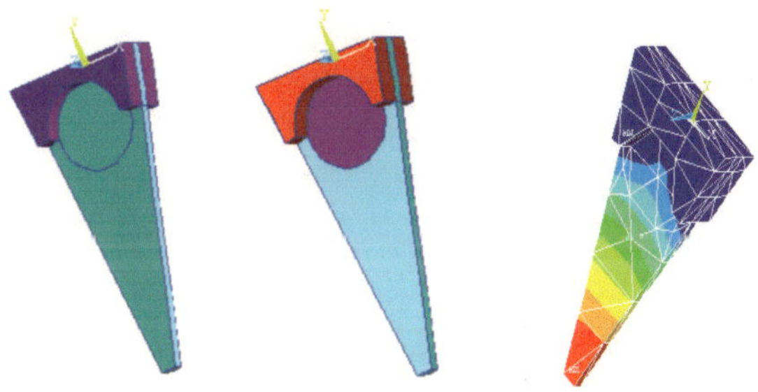

Fig. 4. From left to right: (a) engraved source; (b) embedded source; (c) elongation mode

Table 1 show the resonance frequencies and the vibration velocity transformation ratio (|v(0) /v(L)|) and table 2 present the difference rate of this ratio according to the position of the excitation source

	Longitudinal resonance frequency Hz	Velocity ratio v(0)/v(L) at the extremities
Analytical calculation	60132	2.16
Numerical calculation	**57611**	2.32
Electrical impedance measurements	59 kHz	

Table 1. Comparative study. Analytical and numerical approach

sensors	Embedded source	Engraved source
Frequency (Hz)	57729	**57611**
Standardised velocity ratio	$1.4\ 10^{-2}$	1.00

Table 2. Impact of the nature and the location of the source

The results show a good correlation between the frequencies determined by the calculations and those determined numerically or using impedance measurements.
A significant increase in the amplitude of vibration was observed resulting from the design of the electrode on an active element.

3.4 Application for monitoring changes in state

3.4.1 Pointed sensors for sol-gel transition

The milk gelation can be considered as an aggregation of different sized molecules (Walstra& Vliet, 1986; Fox, 1989; Dalgleish, 1993). This model was explored for several reasons: the available knowledge, the experimental conditions that are known and relatively easy to conduct, the complex medium with the physical properties of liquid and gel states in close contact.
As the reaction progresses, the average mass of each aggregate increases and the number of molecules in the medium tested decreases. The aggregation process results in a giant macromolecule defining the gel.
This process was examined using two identical ultrasonic sensors near-field coupled through the medium to be characterized. The working frequency was 60 kHz.
Figure 5 presents a schematic diagram of the measuring device. The emitter is driven by a sharp electrical pulse lasting 15 µs. These conditions provide a longitudinal vibration mode at the end of the sensor which behaves like a point source. This phenomenon generates a divergent ultrasonic wave in the medium, one part of which was measured using a receiver located at a constant distance from the transmitter by the first the zero-crossing of the wave.
The propagation of the wave in the medium is more or less a compressional wave, as suggested by the time of flight corresponding to a velocity of 1600 m/s in reconstituted milk samples at 25 °C.

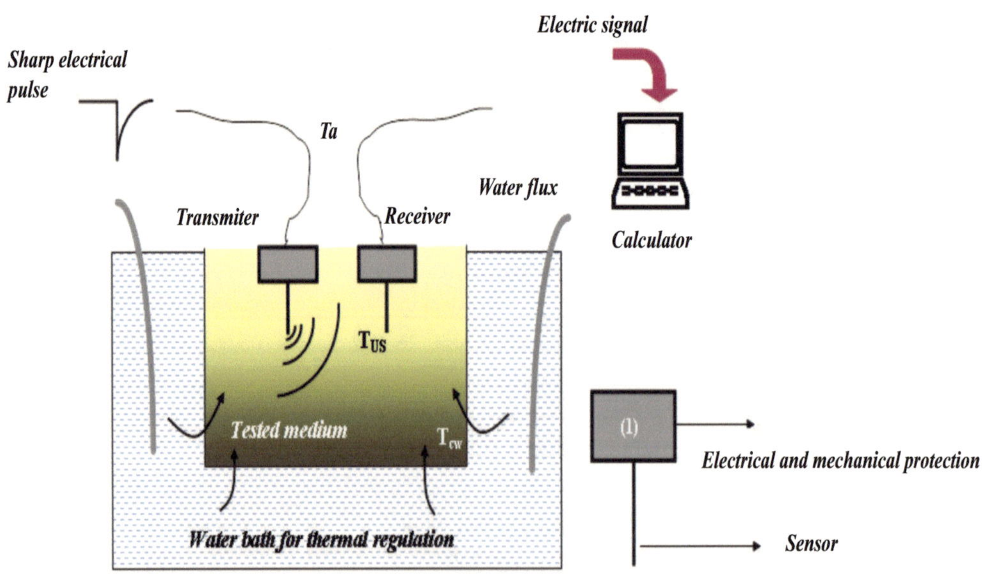

Fig. 5. Diagram of the measuring cell. Tus is the temperature of the product at "sensor level", Tcw is the temperature of the container walls and Ta is the ambient temperature

3.4.2 Ultrasonic monitoring of gelation: measurement of the variation in the time-of-flight of the wave

3.4.2.1 Measurement stability

The time of flight dt(ns) of the signal, measured in distilled temperature-controlled water at 30±0.1°C (reference medium) remains stable. The precision obtained was 1ns over a global reply time of 10μs, given a relative precision of 10^{-4} per measurement at the zero-crossing point

3.4.2.2 Measurement reliability in gelation process

The reliability of a measurement system resides in its reproducibility and its faculty to follow all the stages of the gelation process. In the standard conditions using 12 grams of skimmed milk powder dissolved in 100 ml of distilled water and in accordance with the literature (Noël & al., 1989), the sol-gel transition clotted between 15 and 16 min. At a regulated ambient temperature at 30 °C similar to the one of the medium under test, as shown in figure 6, the time of flight of the signal decreased indicating an increase in the mechanical resistance of the product. This variation had not reached a plateau value, indicating that the medium was still changing.

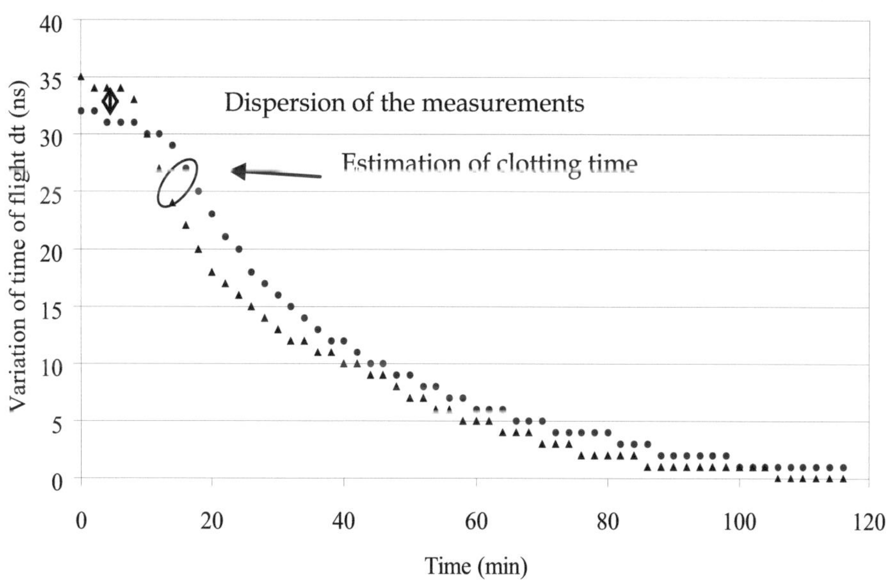

Fig. 6. Typical curves of the gelation process of two media prepared in the same conditions at a reaction temperature of 30.1 °C observed using the ultrasonic technique. The difference between the two curves provides a quantitative estimation of the global dispersion of the ultrasonic measurement due to the electronics and sample preparation

Figure 6 also provides a qualitative estimation of the reproducibility of the ultrasonic measurements. The curves show the progress of the action of the rennet in two media prepared in the same conditions. The maximum dispersion of the measurements was 5 ns due to the electronic parts and the milk reconstitution process.

3.4.2.3 Evolution of the molecular network

Figure 7 presents the variations in the time of flight resulting from the variation of the ultrasonic wave velocity during the milk gelation process at different ambient temperatures.

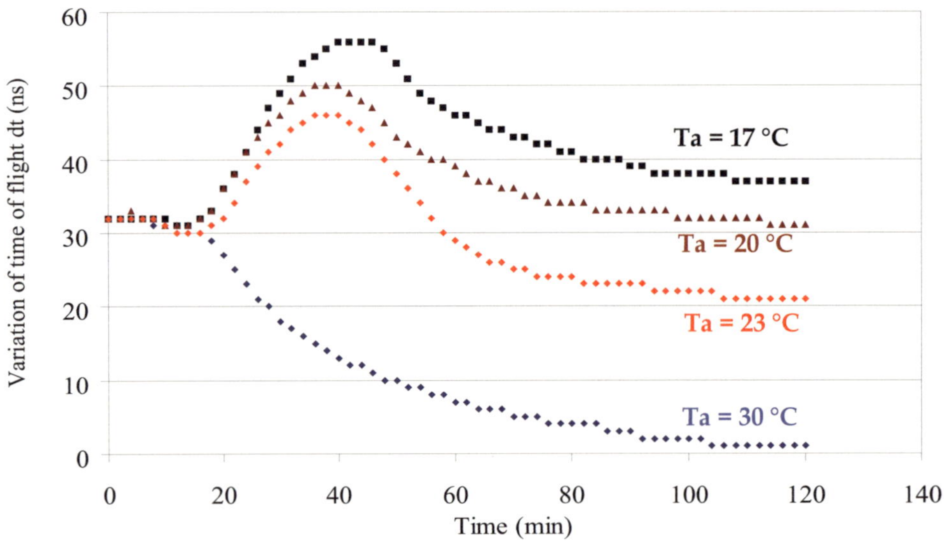

Fig. 7. Ultrasonic sensor responses during milk gelation at 30.1 °C, but at different ambient temperatures (from 17 °C to 30 °C)

The curve observed at an ambient temperature of 30 °C, similar to the temperature of the milk, was similar to those obtained by measurements using other physical methods (McMahon & Brown, 1984). When the ambient temperature was significantly different from the product temperature, ultrasonic curves showed a specific pattern.

According to Dalgleish (Dalgleish, 1982), gelation process was defined like a transition phenomenon in which a soluble suspension made of macromolecules (liquid phase) becomes insoluble when a giant mass forms. We can assume that basic macromolecules are synthesized by linking monomers (building units) via covalent bonds. This chemical reaction is due to the presence of functional groups of the monomers that are able to form chemical bonds with other functional groups of the monomers (Mercier & Marechal, 1993). The network formation occurs if the functionality of the units is greater than two (Mercier & Marechal, 1993). As the reaction progresses, the conversion status of the system is characterized by the connectivity rate p (Flory, 1953; Stockmayer, 1943). Below the gelation threshold, the viscosity of the medium increases as the connectivity rate p approaches the critical advancement rate pc. This phenomenon is known as a critical connectivity transition. Above the threshold, the medium ceases to flow and the gel phase develops some elasticity. This phenomenon introduces structural changes in the physical properties and more particularly in the mechanical behavior of the medium, thus resulting in the transition from a liquid state to a viscoelastic solid state.

To illustrate this process schematically, let us consider an initial solution containing units that can link together. At the beginning, the medium behaves like a viscous solution in a sol phase due to the presence of a single type of finite-size masses: in this case, p is low (Figure

8b). When p increases, bigger and bigger masses are formed (Figure 8c). For a certain critical value of p, pc, a giant chain appears (continuous connectivity of the space from one side to the other $A \leftrightarrow B$; Figure 8d) defining the gel point. Above the threshold pc, the medium in the gel phase has the macroscopic behavior of a viscoelastic solid. For $p = 1$ all the units belong to the giant mass (Figure 8e).

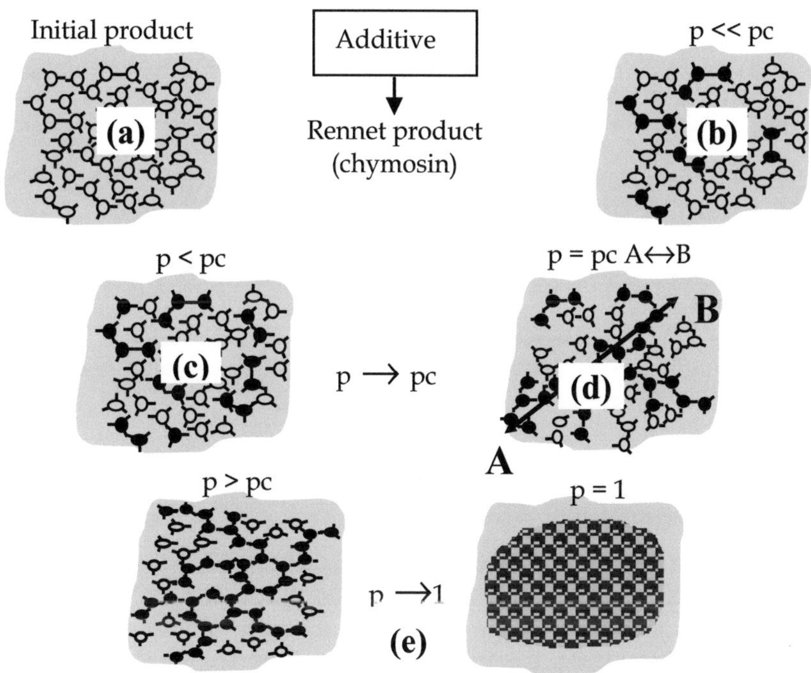

Fig. 8. Schematic representation of the sol-gel transition. a) Initial phase, b) Suspension of molecules of finite sizes, c) Agglomeration and formation of macromolecules of large masses, d) Critical connection phase: p = pc, e) Network continuity connection to give a single giant macromolecule, the gel

As the properties of a gelling medium are proportional to the reaction progress, it is possible to represent the behavior of the viscoelasticity in terms of the connectivity rate p according to the following cases (Figure 8):

- For p < pc, the system is a liquid whose viscosity increases as the gel point approaches.
- For p = pc, elastic behavior appears.
- For p > pc, the medium becomes a solid gel whose elasticity increases with p.

In order to relate the theoretical aspect to the experimental results, the following curve (Figure 9) has been divided into five different stages. The phenomenon describes the gelation process when the ambient temperature was different from the product temperature. It was undetectable when these temperatures were the same.

According to figure 9, "stage (a)" could be interpreted as a proteolytic phase characterized by the appearance of two polypeptides resulting from the effect of the rennet product. This stage is followed by the formation of aggregates of finite size (Figure 8b). This molecular reorganization might be related to the change of slope of the curve (stage b) reflecting a decrease in the time of flight, which means an increase in the velocity in the sol medium.

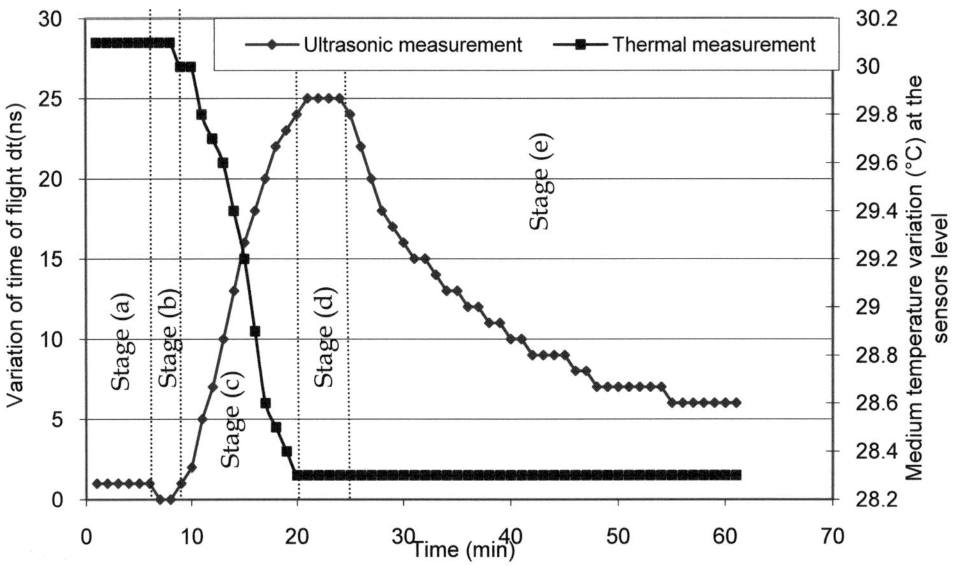

Fig. 9. The specific ultrasonic response during milk gelation when the ambient temperature was different from the temperature of the milk

The temperature inside the medium remained constant during this stage and resulted from the propagation of heat by pure and free convection. The formation of more or less voluminous masses in the medium (stage (c)) induced the transition from a viscous state to a viscoelastic state, slowing down the free convection. This led to a slight temperature decrease in the medium at the level of the sensors (T_{US} in Fig. 4) to reach a new equilibrium where heat was mainly transmitted through the container walls (regulated temperature: T_{cw}°C), by conduction. The changes in the medium during this stage could be interpreted in the following manner:

1. Due to a thermal conduction phenomenon, a slight temperature gradient appears in the medium, between the container walls (T_{cw} = 30.1°C) and the center of the vat, at the location of the measuring point; T_{US}°C (following on the ambient temperature). The temperature decrease induced an increase of the time of flight.
2. The time of flight decreased as the reaction progressed. This decrease can be attributed to the development of an elastic modulus resulting from the formation of macromolecules, changing the medium from a viscous liquid state to a viscoelastic solid state. The phenomenon was expressed physically by the evolution of the connectivity rate p towards its critical value pc .

These two phenomena make "stage (c)" a competition between :

1. An increase of the time of flight resulting from a decrease in temperature.
2. A decrease in the time of flight resulting from the appearance of an elastic component in the changing medium.

The connectivity rate p, reached a critical value pc in "stage (d)", at the maximum of the curve when the existence can be assumed of a giant macromolecular chain linking the two extreme sides of the considered space (Figure 8d). During this stage, the mechanical aspect

of the medium could dominate the remaining part of the reaction, due to the weak thermal variation (≈ 2 °C) resulting from the difference in temperature between the medium and the environment (ambient temperature).
During "stage (e), the gel strengthened. The gel was stronger when the ambient temperature was very close to the temperature medium (Figure 7). This phenomenon can be shown experimentally by an increase in Δdt(ns) resulting from a decrease in time of flight, whereas theoretically it was explained by the evolution of the connectivity rate p towards 1 following the establishment of continuous connections of finite size masses on the giant molecule, the gel.

3.4.3 Monitoring the formation dynamics of the cohesion forces in a fractionated medium: measurement of the variation in the wave amplitude

A key step often met in agro-industry processes is the formation of the matrix of the final product. In cheese-making, this phase involves the cohesion of the elements making up the medium. Generally, it is the conversion of the matter from a heterogeneous state (made up of overlapping grains) to a homogenous state. In this particular case, the cohesion of the curd grains, essential step in the process, varies according to the process conditions as well as the enzymatic and bacterial activities in the medium. It is thus necessary to take these into account in the description of the cohesion.

Analysis of the medium during the cohesion process

During draining, very different physical states are involved in the conversion of the curd grains from a heterogeneous medium to a more homogenous medium. It is therefore difficult to describe the interaction between the ultrasonic wave and the curd grains during draining using just one physical model. So, for a better evaluation of the different phases in the processing of the medium:

1. From moulding and for a very short period of time the grains are touching and cohesive links begin to form between the contact surfaces, thus forming a skeleton containing connected porosities through which the whey continues to drain (Figure 10a).
2. When the whey evacuation channels become blocked (Figure 10b), this phase is described by the multilayer model by Brekovskikh (Brekovskikh, 1980). It is equivalent to a material made up of layers of grains and whey of which the thickness is equivalent to the size of the grains as well as whey evacuation channels. The layers of whey become thinner and thinner until they disappear (Figure 10c), producing a homogenous medium (final phase). This approximation is valid insofar as the main signal beam is confined to a narrow area of the medium, which is the case in our measurements. Indeed, the zone of interest is comparable to the size of the grains (Figure 11).

The evolution of the medium throughout the entire draining phase in the mould was described using these two models: the outflow of the whey and the cohesion of the grains.

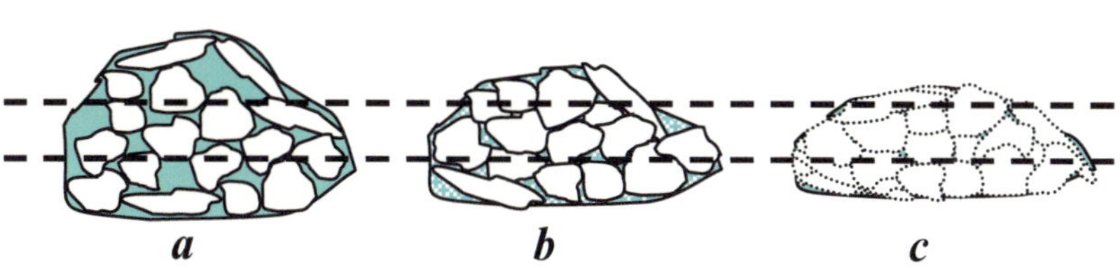

Fig. 10. Evolution of the medium over time

Fig. 11. Estimation of the propagation zone of the main signal beam

Heated to a constant temperature of 35 °C, the experimental mould was instrumented with a transmitter and two ultrasonic receivers spread out so as to integrate the signals transmitted through two paths presenting enough interfaces between the grains. This thus reduced measurement dispersion linked to the random number of interfaces (Figure 12).

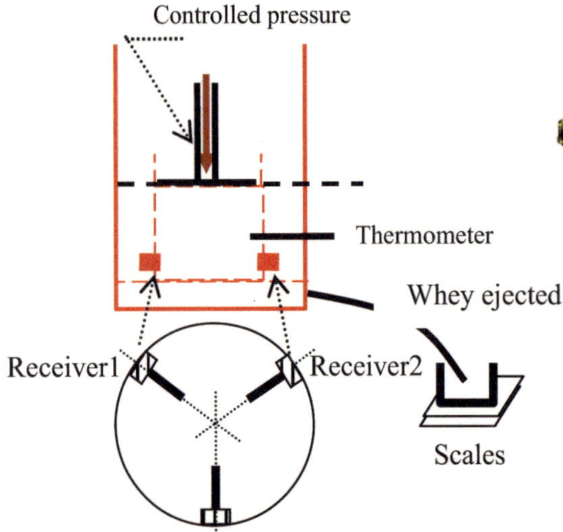

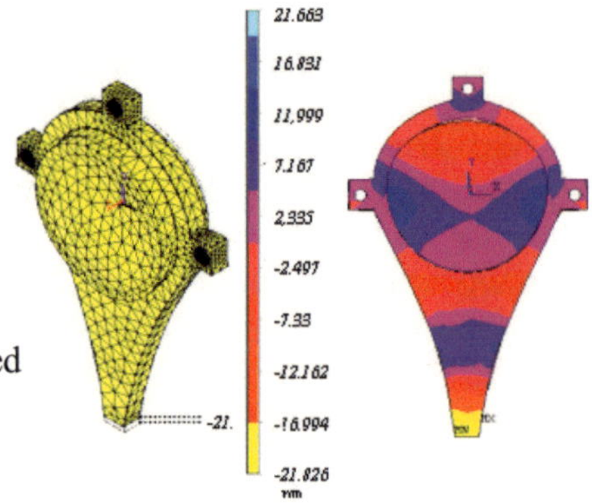

Fig. 12. On the left: The experimental measuring device with a horizontal cross-section of the mould at sensor level and on the right: omni-directional, optimised sensor in elongation mode at 246 kHz

Figure 13 shows that the amplitude of the ultrasonic signal is a parameter that is sensitive to variations in the properties of the medium under investigation.

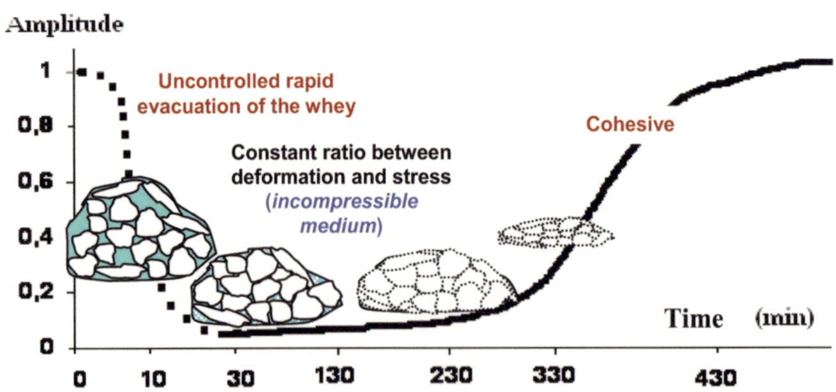

Fig. 13. Typical curve reflecting the cohesion phenomenon as seen by the variation in the ultrasonic amplitude

3.5 Case of highly absorbent matter

The characterisation of media using ultrasounds is often limited by the heterogeneous nature of the matrix which can, in the case of cosmetic, pharmaceutical and agro-food products, be viscoelastic and heterogeneous (foam or emulsion for example). Wave attenuation in such media is mainly due to viscous absorption and scattering from heterogeneities. The higher the frequency the greater the attenuation, hence the necessity to analyse the media using low frequencies in order to characterise the evolving matter.
The search for a compromise between the analysis frequency and the volume of the medium to be characterised led us to propose specific sensor geometries associated with specific excitation conditions.
In order to manage this constraint, a very low frequency acoustic technique was adapted so as to communicate sufficient energy to a particularly absorbent sample. This was achieved by mechanical excitation caused by a shock. An electrical image of this excitation is obtained using a second identical sensor used as the synchronisation reference (Figure 14).

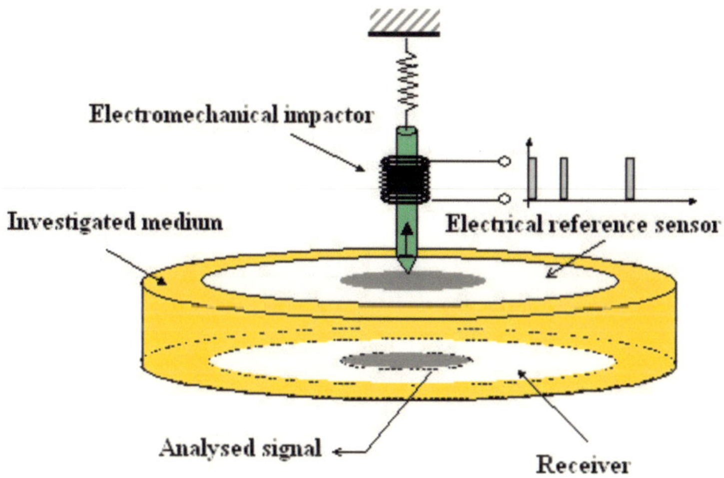

Fig. 14. Schematic diagram

3.5.1 Principle of the sensor proposed

The sensor proposed is illustrated in figure 15. In the shape of a thin disc, its structure is made up of a ring with an embedded piezoelectric disc.
A structure like this offers the advantage of being able to work at resonances lower than those of the piezoelectric disc and thus several resonance modes can be used of which the main ones are flexion and radial modes. This type of sensor also offers a large area of contact with the medium studied, which, in the case of soft and aerated materials, can be advantageous.
The resonance modes of a circular structure, notably those of a disc or a thin ring, have been studied for many years by several authors (Aggarwal, 1952a, 1952b; Moseley, 1960; Vogel & Skinner, 1965; Leissa, 1969; Blevins, 1979; Irie & al., 1984; Lee & Singh, 1994) . The main modes of resonance of a disc are radial modes in the disc plane and flexion modes outside the disc plane (Tables 3, 4 & 5). The tables show a good correlation between the theoretical, numerical and experimental analyses.

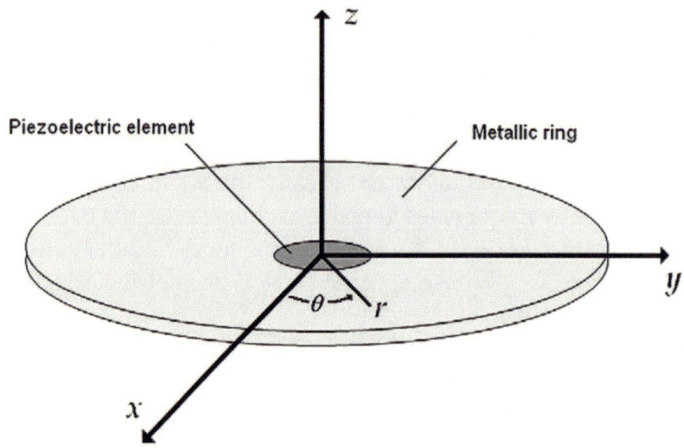

Fig. 15. Set-up of the sensor proposed

Flexion modes	(s, n)	Theoretical (Hz)	FEM (Hz)
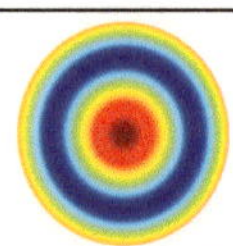	(1.0)	1798	1794
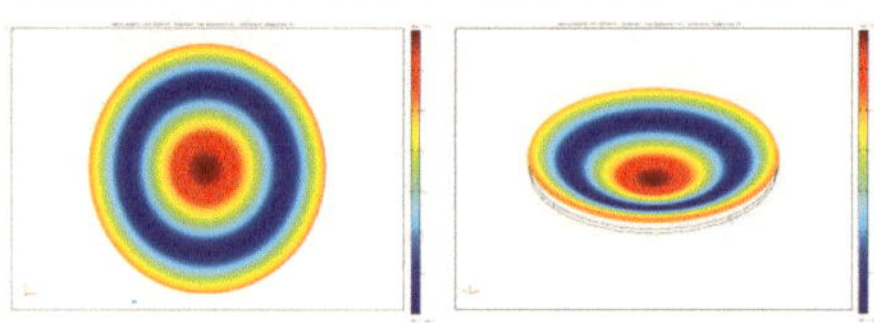			

Table 3. Illustration of the deformations. Amplitude perpendicular to the disc plane Flexion mode (1.0) for a frequency of 1.8 kHz (theoretical and using finite elements)

Radial modes	Analytical calculation (kHz)	FEM (kHz)
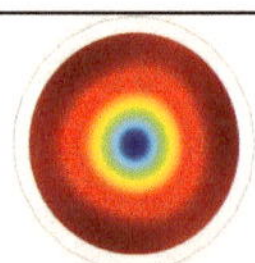	35.5	35.5
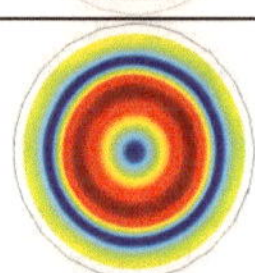	92.6	92.7

Table 4. The resonance frequencies of the first two radial modes for a free aluminium disc (R=5cm; h=2mm) obtained analytically and numerically using finite elements

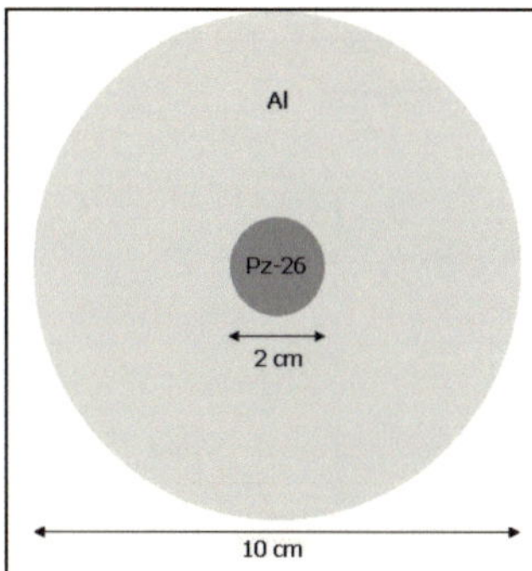 Radial modes	Frequency (kHz)
1	35.9
2	90
Flexion modes	
(1.0)	1.6
(2.0)	6.5

Table 5. Resonance frequencies of the first radial and flexion modes for the composite sensor

3.5.2 Application for monitoring fermenting bread dough

The objective of this application was to establish the links between the product evolution kinetics and the acoustic characteristics measured.
From a practical point of view, impulse excitation was used in this system. The excitation was obtained by a controlled mechanical impact (rod of an electromagnet), thus exciting the disc used for the synchronisation. The vibration induced in the dough is received by a receiver disc identical to that of the synchronisation disc (Figure 16).

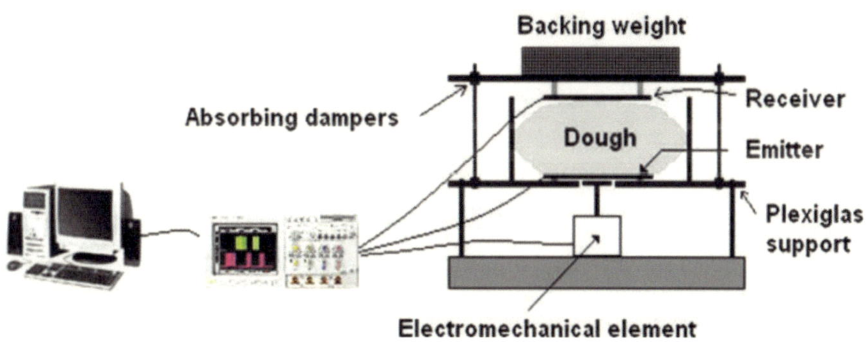

Fig. 16. Experimental measuring device

A metrological study of the measuring device carried out using standard samples (for example a pocket of water at 25°C) showed that the standard deviation of the amplitude and the velocity was approximately 2%. Signal acquisition was carried out over 3 hrs.

3.5.3 Dynamic monitoring of the fermentation process of bread dough

After controlled kneading of the dough, the measurement chamber was placed in an enclosure in order to control the temperature and humidity. The acoustic values studied were the variation of the time-of-flight and the wave amplitude on reception.

Figure 17 shows the variations in these two values. It can be noted that the critical points and phases appear simultaneously on the two curves.

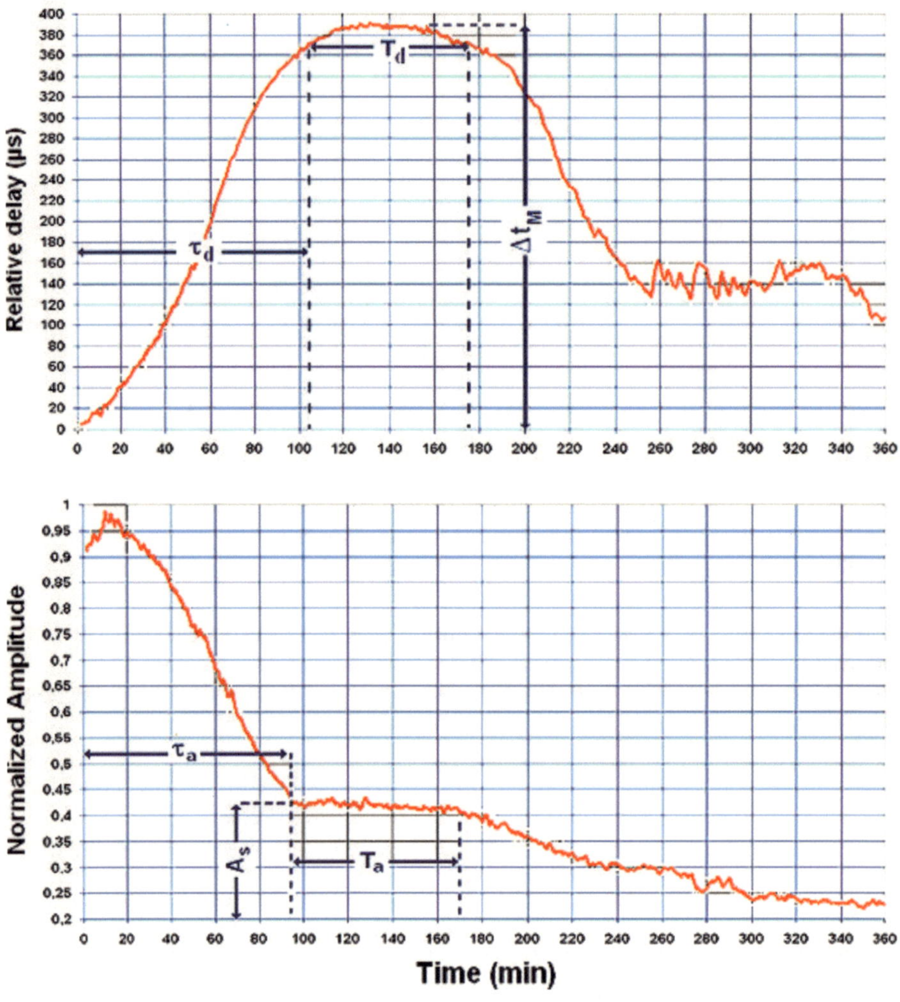

Fig. 17. Evolution of the standardised amplitude and the relative signal delay on reception during the fermentation phase

Where:

- τ_r is the time necessary to reach a relatively stable zone,
- T_r reflects the period of stability during which the relative delay reaches its maximum and remains relatively constant,
- Δt_M is the maximum relative delay. It is linked to the gas fraction contained in the dough and therefore the extensibility of the latter.
- τ_a is the period during which the amplitude of the signal decreases before reaching a plateau,
- T_a is the period of stability of the amplitude,
- A_S is defined as being the amplitude of the signal during the period of stability.

A repeatability study was carried out to estimate the dispersion of the parameters (delay and amplitude). Several tests were performed under the same operating conditions. The standard deviation of the measurements of these parameters was around 3%.
Table 6 summarises the variations in the characteristic parameters observed on the curves according to the evolution in the temperature

	20°	27°	34°
τ_r *(min)*	165	105	60
T_r *(min)*	145	70	55
Δt_M *(μs)*	380	385	374
τ_a *(min)*	160	95	55
T_a *(min)*	130	75	50
A_S *(%)*	40	43	44

Table 6. Parameters relating to the variation in temperature

It can be noted that the maximum relative delay is relatively constant (approximately 380μs) for the three products made under the same operating conditions. This parameter seems to be independent of the temperature, which is in agreement with the hypothesis that it varies according to the gas fraction contained in the matter and the elastic properties of the matrix.

4. Acoustic sensor for in-line monitoring of a manufacturing process

In certain industrial processes it is often difficult to access useful information in real-time due to the conditions imposed on the mechanical and thermal parameters, pressure, hygiene..., conditions which require a specific installation of the sensor with regard to its environment. The difficulty thus arises of an integration taking into account both the process constraints and the acoustic constraints. This is the case of a plate heat exchanger which can be considered as a typical example in this category (Figure 18).

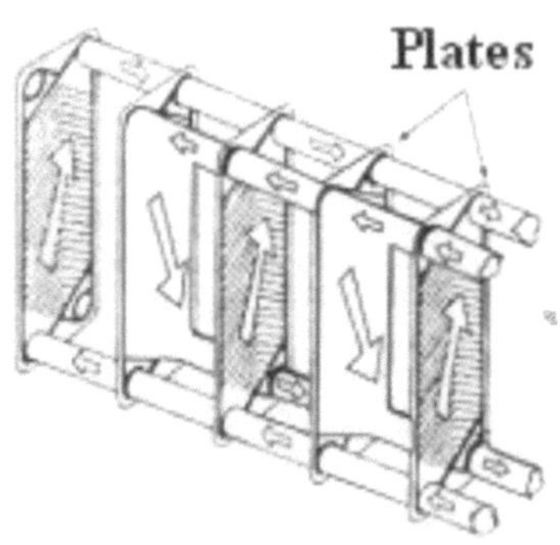

Fig. 18. Standard plate heat exchanger

4.1 Sensor selection criteria

For the exchanger, the sensor selected is not cumbersome and is sensitive over a temperature range reaching over 100°C (Figure 19). The excitation and synchronisation

modes remain the same as the previous case (disc sensor). The principle of the measurement is to excite a vibration mode in one or several plate exchangers and to analyse the evolution under the effect of fouling by measuring the response of the plates using a receiver.
A bivariate system-sensor study enabled the geometry of the latter to be defined over the same vibration frequency range as the system (exchanger).

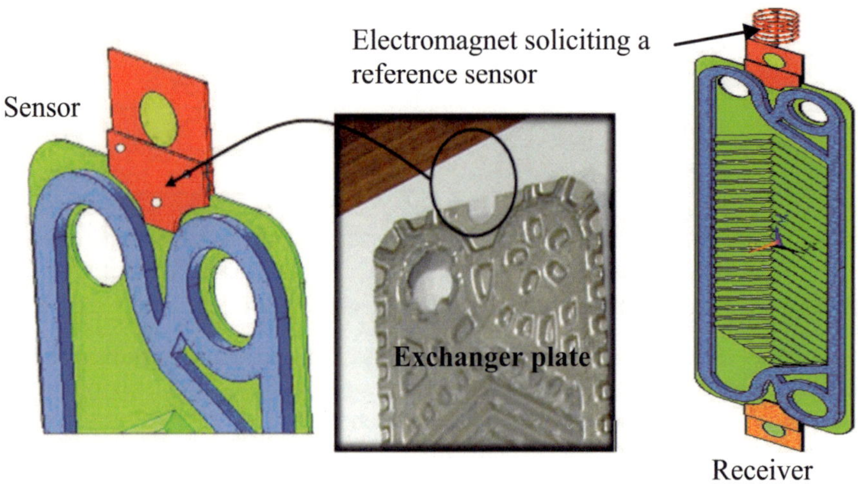

Fig. 19. Positioning of the sensors on an exchanger plate

4.1.1 Sensor excitation mode

In order to monitor the evolution of the damping of the plate modes due to fouling of the exchanger, it is necessary to excite these modes with enough energy to preserve the signal-noise ratio (of the signal received) after going through the exchanger.
A mechanical shock is the only way of producing enough energy for local excitation.
The frequency response obtained by modal analysis in the absence of structural constraints is given in the first column in table 7. This column gathers the different modes specific to the structure studied. Some correspond to simple, longitudinal or transversal displacements, others to more complex displacements (flexions, torsions...).

Mode	Frequency (Hz) - numerical	Frequency (Hz) - experimental
1	1683	1586
2	2387	2894
3	3557	3639
4	5422	5330
5	5734	6639
6	7417	7390

Table 7. The first 6 modes specific to the sensor

The second column shows the modal frequencies obtained from the analysis of the impedance of the sensor mounted on a heat exchanger.
The mean standard deviation between the frequencies obtained by modal analysis and those obtained experimentally is 5 %. The good correlation between these results indicates that the numerical modelling provides a good estimation of the resonance frequencies of the sensor.

4.1.2 Excitation by mechanical shock: estimation of the frequency range

The mechanical excitation in question is ensured via the core of an electromagnet.
As an indication, figures 20a and 20b show the temporal and frequency responses of the sensor.

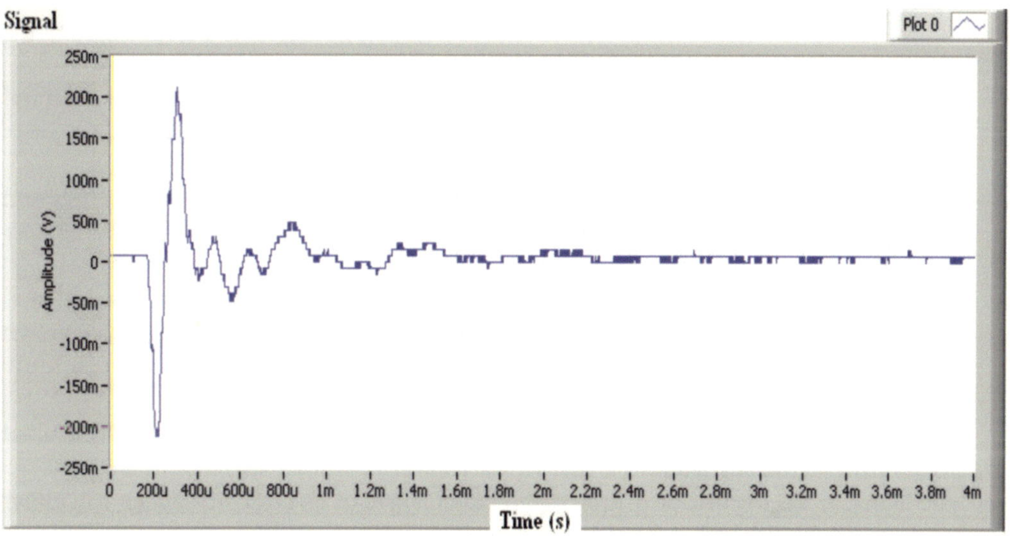

Fig. 20a. Temporal response of a mechanical shock

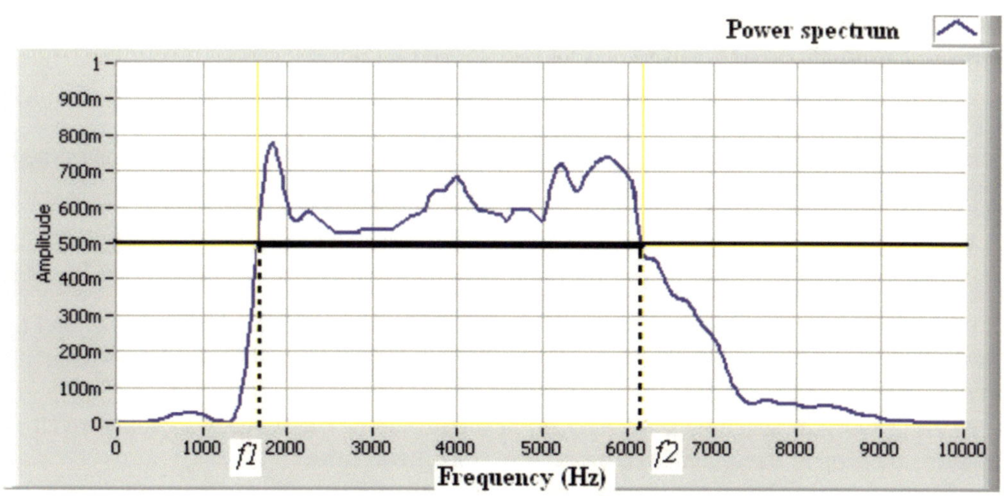

Fig. 20b. Spectral response associated with the shock

The curves show the temporal response and the frequency range of the sensor following a stress induced by a mechanical shock of short duration. The experiments carried out on the overall system (sensor & exchanger) in real configuration show that the temporal response is maximum 4 ms and its frequency response is around a central frequency of approximately 4 kHz.

4.2 Application

4.2.1 Fouling mechanism

Heat exchanger fouling is a dynamic process. The phenomenon continues to evolve, generally until equilibrium is reached or cleaning is required. The period of fouling can vary from a few hours to several months.

Müller (Müller-Steinhagen & Middis, 1989) looked at five stages in the process of the appearance and development of particulate fouling:

- The initiation, which corresponds to the time necessary before fouling, can be observed on a clean surface. The duration depends on the nature of the deposit, the initial state of the surface (material, roughness) and the temperature of the wall.
- The denaturing of the product (protein, organic matter...) under the effect of heat and the surrounding parameters (pH...), their aggregation and transport within the vicinity of the wall.
- The adhesion of the particles transported to the wall, controlled by surface adhesion forces (Van der Waals, electrostatic...) and cohesion of the deposit. It has been shown that the particles can adhere to a clean surface or adhere to other particles already deposited.
- The dislodging of deposited particles, caused by hydrodynamic forces which exert shear stress on the deposit.
- The aging of the deposit over time results in changes in its structure which can either weaken or consolidate it.

Generally, the initiation phase is rarely taken into account in particulate fouling models. The mechanisms that govern the deposit of particles are generally presented as being the transport of the particles to the surface, then the "adhesion" to the wall and finally the possible dislodging of the particles.

4.2.2 Results

Before studying the phenomenon of fouling, the metrological variation of the measurement system was taken into account according to the main technological parameters:

- Variation in temperature at constant flow.
- Variation in flow at constant temperature.
- Variation in viscosity at constant temperature and flow.

This phase is essential in order to separate the interferences of acoustic values generated by the fouling phenomenon from those linked to the technological conditions of the exchanger and its environment.

The curves in figure 21 show the evolution of the energy of the acoustic signals as well as the pressure drop in the system as a function of the process time.

The "Power" curve shows the damping effect linked to the load on the plate caused by fouling.

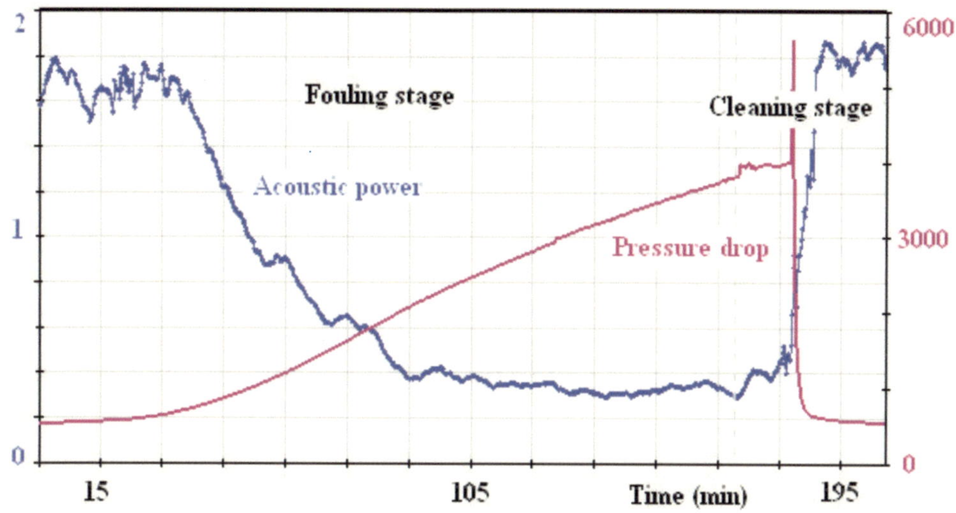

Fig. 21. Evolution of the power of the acoustic signal received during the fouling test and cleaning

In conclusion, this work concerned the monitoring of fouling using acoustics. By adopting a multi-stage experimental protocol we have been able to show that the variation in the acoustic signal can be used to predict variations in the pressure drop as well as the state of fouling in the plate heat exchanger under very specific operating conditions.
Finally, this study illustrates an example of a non-intrusive acoustic technique for the local monitoring in real time of the fouling of plate heat exchangers. The results show that it is possible to follow the relative kinetics of the state of fouling in each zone of the exchanger with the right choice and positioning of the sensors.

5. Conclusion

This chapter has proposed a synopsis of all the work that has led to the development of novel low frequency sensors. By using structural resonance modes excited by a transducer, these sensors present the advantage of having small sized sources with regard to the acoustic wavelength generated. These sensors are omni-directional but can nevertheless present significant contact areas with the medium to be characterised. This is the case for sensors developed for the characterisation of gels. The close contact of the elements set in resonance with the medium enables phenomena linked to changes in state to be monitored easily. Various applications have led us to develop sensors with very different geometries and which are optimised with the application in mind.
Indeed, for each need expressed, the approach consisted in optimising not only the geometry of the sensors but also their optimum position according to the problem posed. Three different cases were thus studied:

- identical near-field coupled sensors, through the medium to be characterised. They were used for monitoring the evolution of the ultrasonic values to characterise a sol-gel transition or the cohesion kinetics of a medium. For certain applications, the sensors are immersed in the medium. This direct immersion is essential for characterising fragile media.

- a low frequency receiver associated with an excitation of the medium via a mechanical shock in the case of very absorbent and scattering media. A second identical sensor is used for the synchronisation of the acquisitions thus reducing, by standardisation, the scattering of the values measured. The mechanical shock produces significant vibratory energy over a broad frequency range.
- finally, the sensors were coupled to heat exchanger plates in order to characterise fouling. This work has shown the interest of using acoustic sensors to monitor processes, providing an often local and dynamic response to the evolution of the performances of the process.

The work carried out provides a solid base of knowledge on ultrasound-complex media interactions. This knowledge could be put to good use in the development of sensors and integrated ultrasonic methods and their applications in the analysis and monitoring of local properties.

6. References

Aggarwal R. R., (1952a). Axially Symmetric Vibrations of a Finite Isotropic Disk. I, Journal of acoustical society of America, Vol. 24, N0. 5, pp. 463-467

Aggarwal R. R., (1952b). Axially Symmetric Vibrations of a Finite Isotropic Disk. II, Journal of acoustical society of America, Vol. 24, N°. 6, pp. 663-666

Allsopp, M. W. (1981). The developement and importance of suspension PVC morphology, Pure an applied chemistry, Vol. 53, pp. 449-465.

Blevins R. D., (1979). Formulas for natural frequency and mode shape, Van Nostrand Reinhold Co., ISBN 0-4422-0710-7, New York, USA

Brekhovskikh, L.M., (1980). Waves in layered media, Academic Press, ISBN 0-12-130560-0, New York, USA

Case, L. C. (1960). Molecular distributions in polycondensations involving unlike reactants. VII. Treatment of reactants involving nonindependent groups, Journal of polymer science, Vol. 48, pp. 27-35

Clerc, J. P. ; Giraud, G. ; Roussenq, J. ; Blanc, R. ; Carton, J.P. ; Guyon, E. ; Ottavi, H. & Stauffer, D. (1983). La Percolation: modèles, simulation analogiques et numériques, Annales de Physique, Vol. 8, Masson, Paris, France.

Dalgleish, D. G. (1982). Developments in Dairy Chemistry, edited by P. F. Fox (Applied Science, London,), Vol. 1, Chap. 5, ISBN 0-8533-4142-7, London, United kingdom

Dalgleish, D.G. (1993). Cheese: Chemistry, Physics and Microbiology, General Aspect, 2nd ed., Vol. 1, p. 69, Fox, P.F., Chapman & Hall, ISBN 0-1226-3652-X, London, United kingdom.

De Gennes, P. G. (1989). Scaling Concepts in Polymer Physics, Cornell University Press, Ithaca, ISBN 0-8014-1203-X, New York, USA

Degertekin, F. L. & Khury-Yakub, B.T. (1996). Hertzian contact transducers for non-destructive evaluation, Journal of acoustical society of America, Vol. 99, pp. 299-308

Degertekin, F. L. & Khury-Yakub, B.T. (1996). Lamb wave excitation by Hertzian contacts with applications in NDE. IEEE Transactions on Ultrasonics Ferroelectrics and Frequency, Vol. 44, N°. 4, pp. 769-778

Degertekin, F. L. & Khury-Yakub, B.T. (1996). Single mode lamb wave excitation in thin plates by Hertzian contacts, Applied physics letters, Vol. 69, N°. 2, pp. 146-148

Eichinger, B. E. (1981). Random elastic networks. I. Computer simulation of linked stars, Journal of chemical physics, Vol. 75, pp. 1964-1979

Ensminger, D. E. (1960). Solid cone in longitudinal half-wave resonance, Journal of acoustical society of America, Vol. 32, pp. 194-196

Flory, P. J. (1953). Principles of polymer chemistry, Cornell University Press, Ithaca & London, ISBN 0-8014-0134-8, New York, USA

Fox, P. F. (1989). Proteolysis during cheese manufacture and ripening. A review, Journal of dairy science, Vol. 72, pp. 1379-1385

Gupta, S. K.; Kumar A. & Bhargava, A. (1979). Molecular weight distribution and moments for condensation polymerization of monomers having reactivity different from their homologues, Polymer, Vol. 20, pp. 305-310

Irie T., Yamada G. & Muramoto Y., (1984). Natural frequencies of in-plane vibration of annular plates, Journal of sound and Vibration, Vol. 97, N°. 1, pp. 171-175

Lee M. & Singh R., (1994). Analytical formulations for annular disk sound radiation using structural modes, Journal of acoustical society of America, Vol. 95, N°. 6, pp. 3311-3323

Leissa A. W., (1969). Vibration of plates, NASA SP-160, U.S. Government Printing Office, Washington, D.C.

McMahon D. J. & Brown R. J., (1984). Enzymic coagulation of caseine micelles: a review, Journal of dairy science, Vol. 67, pp. 919-929

Mercier, J. P. & Marechal, E. (1993). Chimie des Polymères 1st ed. Presses polytechniques et universitaires romandes, Lausanne, Chap. 1, 3, 8. Lausanne, Swiss.

Moseley D. S., (1960). Contribution to the Theory of Radial Extensional Vibrations in Thin Disks, Journal of acoustical society of America,Vol. 32, N°. 8, pp. 991-995

Müller-Steinhagen H. & Middis J., (1989). Particulate fouling in plate heat exchangers, Heat Transfer Engineering, Vol. 10, N°. 4, pp. 30-36

Nassar, G. (1997). Etude et Optimisation d'un Dispositif Ultrasonore De Suivi en Ligne des propriétés viscoélastiques, Doctoral dissertation, Valenciennes University-France.

Nikolovski, J. P. & Royer, D. (1997). Local and selective detection of acoustic waves at the surface of a material", IEEE Ultrasonics Symposium, pp. 699-703, ISBN 0-7803-4153-8, Toronto, Ontario, Canada, October 5-8, 1997

Noël, Y. ; Flaud, P. & Quemada, D. (1989). Traitement Industriel des Fluides Alimentaires Non Newtoniens, Tome II, Actes du 2ème Colloque la Baule, La Baule, France, September 11-13, 1989, pp. 215-224.

San Biagio, P. L.; Bulone, D.; Emanuele, A.; Madonia, F.; Di Stefano, L.; Giacomazza, D.; Trapanese, M.; Palma-Vittorelli, M. B.; & Palma, M.U. (1990). Spinodal demixing, percolation and gelation of biosttural polymers, IUPAC 10th Int. Symp. on Polymer Networks, Vol. 40, pp. 33-44, Jerusalem, Israel, December, 1990

Shuyu, L. (1996). Study on the longitudinal-torsional composite mode exponential ultrasonic horns, Ultrasonics, Vol. 34, pp. 757-762

Shuyu, L. (1997). Study on the longitudinal-torsional composite vibration of a sectional exponential horn, Journal of acoustical society of America, Vol. 102, pp.1388-1393

Stauffer, D. (1981). Can percolation theory be applied to critical phenomena at gel point?, Pure an applied chemistry, Vol. 53, pp. 1479-1487

Stauffer, D. (1985). Introduction to Percolation Theory, Taylor & Francis Ltd., ISBN 0-7484-0253-5, London, United Kingdom

Stockmayer, W. H. (1943). Theory of molecular size distribution and gel formation in branched -chain polymers, Journal of chemical physics, Vol. 11, pp. 45-55

Vogel S. M. & Skinner D. W., (1965). Natural frequencies of transversely vibrating uniform annular plates, Journal of applied mechanics, Vol. 32, pp. 926-931

Walstra, P. & Vliet, V. (1986). The physical chemistry of curd making, Netherlands. milk dairy journal, Vol. 40, pp. 241-259

11

Modeling of Biological Interfacial Processes Using Thickness–Shear Mode Sensors

Ertan Ergezen et al.*
School of Biomedical Engineering, Health and Sciences, Drexel University, Philadelphia USA

1. Introduction

Biological interfaces and accompanying interfacial processes constitute one of the most dynamic and expanding fields in science and technology such as biomaterials, tissue engineering, and biosensors. For example, in biomaterials, the bio-interfacial processes between biomaterials and surrounding tissue plays a crucial role in the biocompatibility of the layer (Werner, 2008). In tissue engineering, cellular adhesion plays an important role in the regulation of cell behavior, such as the control of growth and differentiation during development and the modulation of cell migration in wound healing, metastasis, and angiogenesis (Hong et al., 2006). Performance of a biosensor is highly dependent on interfacial processes involving the sensor sensing interface and a target analyte. Therefore, quantitative information on the novel and robust immobilization of detector molecules is one the most important aspects of the biosensor field (Kroger et al., 1998).

Thickness shear mode (TSM) sensors have been used in a variety of studies including interfacial biological processes, cells, tissue and properties of various proteins and their reaction (Cote et al., 2003). Phenomena such as cell adhesion (Soonjin et al., 2006.), superhydrophobicity (Sun et al., 2006, Roach et al., 2007), particle-surface interactions (Zhang et al.,2005), organic and inorganic particle manipulation (Desa et al., 2010) and rheological and interfacial properties of blood coagulation (Ergezen et al. 2007) were studied using TSM sensors. Due to the high interfacial sensitivity of TSM sensors, it has been shown that cell motility can be monitored by analyzing the noise of the TSM sensor response (Sapper et al., 2006). It has also been demonstrated that the number of motile sperm in a semen sample can be assessed in real-time using a flow-chamber integrated with a thickness shear mode sensor (Newton et al., 2007).

1.1 Quantification of Thickness Shear Mode (TSM) sensor response

The TSM sensor response is affected by the complex nature of the interface. Its response is influenced by the geometrical and material properties of the interacting surfaces such as surface roughness (Cho et al., 2007), hydrophobicity (Ayad and Torad, 2009), interfacial

* Johann Desa, Matias Hochman, Robert Weisbein Hart, Qiliang Zhang, Sun Kwoun, Piyush Shah and Ryszard Lec
School of Biomedical Engineering, Health and Sciences, Drexel University, Philadelphia USA

slippage (Zhuang et al., 2008), coverage area (Johanssmann et al., 2008), sensitivity profile (Edvardsson et al., 2005) and penetration depth of the shear acoustic wave (Kunze et al., 2006).
Various theoretical models have been developed for quantitative characterization of the TSM sensor response to interfacial interactions. Nunalee et al (2006) developed model to predict of the TSM sensor response to a generalized viscoelastic material spreading at the sensor surface in a liquid medium. Cho et al (2007) created a model system to study the viscoelastic properties of two distinct layers, a layer of soft vesicles and a rigid bilayer. Urbakh and Daikhin (2007) developed a model to characterize the effect of surface morphology of non-uniform surface films on TSM sensor response in contact with liquid. Hovgaard et al (2007) have modeled TSM sensor data using an extension to Kevin-Voigt viscoelastic model for studying glucagon fibrillation at the solid-liquid interface. Kanazawa and Cho (2009) discussed the measurement methodologies and analytical models for characterizing macromolecular assembly dynamics.
The physical description based on a wave propagation concept in a one-dimensional approximation has been proven as the best model of thickness shear mode (TSM) sensors. The fundamentals have been published in several books (Rosenbaum, 1998). Martin et al. have (1994) applied this background to sensors by using Mason's equivalent circuit to describe the thickness shear mode sensor itself and transmission lines as well as lumped elements for viscoelastic coatings, semi-infinite liquids etc.. Follow-up papers have introduced a more straightforward definition of the elements of the BVD-model (Behling et al, 1998) as well as several additional approximations, e.g. based on perturbation theory, to derive less complex equations, have suggested a simplified notation to separate the mass from so-called nongravimetric effects, or have applied the transmission line model to several subsystems (Voinova et al, 2002) for demonstration of specific situations just to call some examples. More recent papers deal with deviations from the one-dimensional approximations, e.g. by introducing generalized parameters by deriving specific solutions e.g. for surface roughness or with discontinuity at boundaries.
TSM sensors combined with the theoretical models mentioned above were used to determine the properties of liquids (Lin et al., 1993), high protein concentration solutions (Saluja et al., 2005), and thin polymer films (Katz et al., 1996).
For viscoelastic layers, their mechanical impedance depends upon the density, thickness, and the complex shear modulus of the loading. Identification of the all the system parameters from the impedance measurements has been very challenging and uncertain without a priori knowledge of the thicknesses and/or some of the material properties (Lucklum et al. 1997).
Furthermore, Kwoun (2006) showed the beneficial features of the multi-resonance operation of the TSM (called as "multi-resonance thickness shear mode) sensor to study the formation of biological samples, specifically collagen and albumin, on the sensor surface. In this work, it was demonstrated that the different harmonic frequency clearly showed the different characteristics of mechanical properties, especially shear modulus, of the biological sample. Although this work was one of the pioneer studies to demonstrate the strengths of the MTSM measurement technique, it is limited as it is a semi-quantitative method. Exact values of mechanical properties of anisotropic collagen and albumin samples were not able to be defined due to complexity of the non-linear simultaneous equations of the model. An improved MTSM technique combined with an advanced data analysis technique was proposed by Ergezen et al (2010). A new approach merging the multi-harmonic thickness

shear mode (MTSM) measurement technique and genetic algorithm-based data analysis technique has been used. This novel method was utilized to solve two unmet needs:

1. Identification of all four parameter by using the MTSM sensor's single harmonic response results in an under-determined problem. The MTSM sensor response enables the identification of two parameters by providing imaginary and real components of the mechanical impedance. In other words, there are fewer equations than the material/geometrical parameters of the interface, therefore, the stochastic method is the only approach that can address this problem mathematically. In this project it was shown that combination of the MTSM measurement technique and the genetic algorithm-based data analysis technique (called as MTSM/GA technique) was used to solve this under-determined problem. *It was reported for the first time, a novel approach that enables determining all four parameters, which define the response of the MTSM technique.*
2. Most of the biological interfaces constitute multi-layer structures. Multi-layer modeling of biological interfacial processes was proposed by several researchers and by us (Wegener et al., 1999, Ergezen et al., 2007). In contrast, there has been very limited (Lucklum et al., 2001) theoretical study and no experimental studies based on the MTSM sensor for quantitative characterization of multi-layer biological processes. *It was reported, for the first time, the most comprehensive theoretical and experimental study for quantitative characterization of multi-layer biological interfacial processes.*

A new approach merging the multi-harmonic thickness shear mode (MTSM) sensor and a data extraction technique based on stochastic global optimization procedure has been proposed. For this purpose, the MTSM/GA technique is being developed and calibrated with a polymer layer (having known properties). This was then used to estimate the properties of a protein layer with unknown properties adsorbed to the MTSM sensor surface. It was demonstrated that this new method has the potential to be a novel tool for quantitatively characterization of interfacial biological layers.

2. Theory

2.1 Multi-Harmonic Thickness Shear Mode (MTSM) sensor

Piezoelectric MTSM sensors transmit acoustic shear waves into a medium under test, and the waves interact with the medium. Shear waves monitor local properties of a medium in the vicinity of the sensor and of the medium/sensor interface (on the order of nm - µm); thus, they provide a very attractive technique to study interfacial processes. Measured parameters of acoustic waves are correlated with medium properties such as interfacial mass/density, viscosity, or elasticity changes taking place during chemical or biological processes.

The shear acoustic wave penetrates the medium over a very short distance. The square of the depth of penetration of an acoustic shear wave in MTSM sensor is related to medium viscosity, elasticity, density and the frequency of the wave (please see Appendix IA.) (Kwoun et al. 2006). Figure 1a shows the acoustic wave penetrating the adjacent medium and Figure 1b shows that the depth of penetration decreases at higher harmonic frequencies in a semi-infinite medium.

Therefore, by changing the frequency, one can control the distance at which the wave probes the medium. Multi-harmonic operation of MTSM sensor will enable to control the interrogating depth into the biological processes. Therefore it will provide a more in depth characterization of the biological interfacial processes. For example, it was suggested that cell adhesion on extra cellular matrix should be modeled as a multi-layered structure (Wegener et al. 2000). Therefore MTSM sensors can provide information about mechanical and structural properties of the biological processes from different depths (slicing the medium).

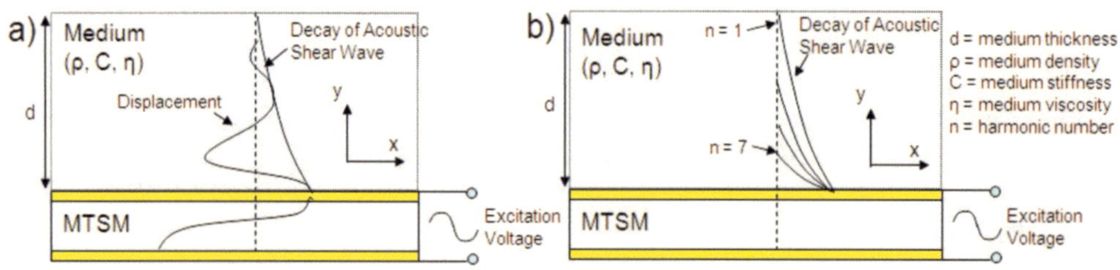

Fig. 1. a) Acoustic wave penetrating into the medium b) depth of penetration decreases at higher harmonic frequencies

It should be noted that it was assumed that the medium is semi-infinite and the mechanical properties are not frequency dependent in fig. 1.

2.2 Electrical response of MTSM sensor

The MTSM sensor is a piezoelectric-based sensor which has the property that an applied alternating voltage (AC) induces mechanical shear strain and vice versa. By exciting the sensor with AC voltage, standing acoustic waves are produced within the sensor, and the sensor behaves as a resonator. The electrical response of the MTSM sensor in air over a wide frequency range is shown in figure 2, where S_{21} is the magnitude response of the MTSM sensor ($|S_{21}|=20\log(100/(100+Z_t))$, Z_t=total electromechanical impedance of the MTSM sensor (Rosenbaum 1998). As an example, the magnitude and phase responses of MTSM sensor are presented at the first (5 MHz), third (15 MHz), fifth (25 MHz) and seventh (35 MHz) harmonics in air.

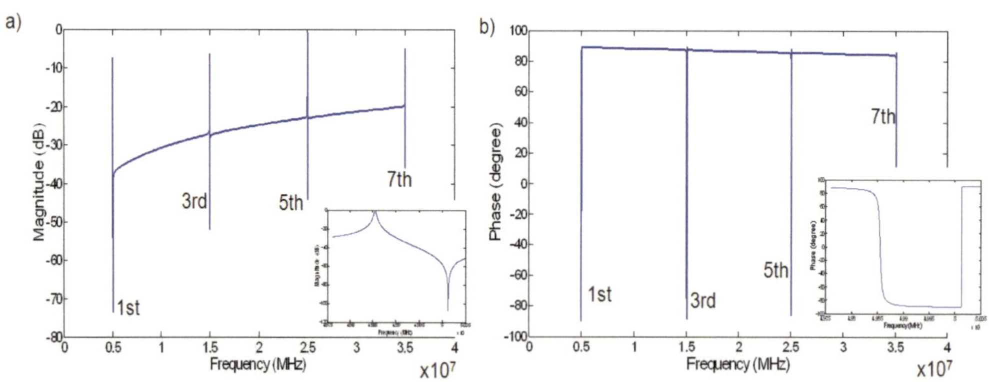

Fig. 2. A typical a) frequency vs. magnitude response and b) frequency vs. phase response characteristic and the associated resonance harmonics for the MTSM sensor, spanning a wide frequency range (5 MHz to 35 MHz). (Insets) Magnified view of magnitude and phase response at 5 MHz

An example of the MTSM's magnitude response in the vicinity of the fundamental resonant frequency is given below (figure 3a). When the TSM sensor is loaded with a biological media, there will be a shift in resonant frequency and a decrease in the magnitude. These changes can be correlated with changes in the mechanical and geometrical properties of the medium such as thickness, viscosity, density and stiffness. Depending on the changes at the interface of the sensor surface-medium interface, a positive and/or negative shift can be seen in the frequency response (Figure 3b).

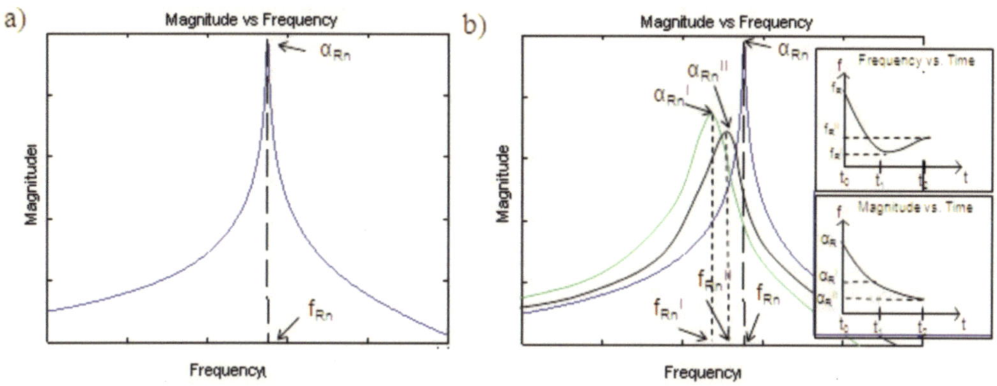

Fig. 3. (a)Demonstration of a typical qualitative frequency-dependent response curve for the MTSM sensor in the vicinity of the resonant frequency; n = harmonic number, α_{Rn}=Initial maximum magnitude, f_{Rn}=Initial resonant frequency, (b) In the case of both positive and negative frequency shifts throughout the experiment, α_{Rn}^{I}, α_{Rn}^{II} =Instantaneous maximum magnitudes of loaded MTSM sensor at time t_1 and t_2 respectively, f_{Rn}^{I}, f_{Rn}^{II} =Instantaneous resonant frequencies of the loaded MTSM sensor at time t_1 and t_2 respectively (Inlet) resonant frequency and magnitude are monitored as a function of time

2.3 MTSM/GA data processing technique

This section will be structured in the following manner; first, the general structure of a genetic algorithm will be explained. Second, advantages of genetic algorithm over other techniques will be discussed. Finally, implementation of MTSM-GA technique for determination of material parameters will be explained.

Principles of operation of a genetic algorithm (GA)

Basic definitions of GA terms are defined in Appendix IB. Genetic algorithm (GA) is based on the genetic processes of biological organisms (figure 4). GA works with a population of individuals, each representing a possible solution to a given problem. Each individual is assigned a fitness score according to how good a solution to the problem it is. The highly-fit individuals are given opportunities to reproduce, by cross breeding with other individuals in the population. This produces new individuals as offspring, which share some features taken from each parent.

Comparison of GA to other data processing techniques

Complex models are ubiquitous in many applications in the fields of engineering and science. Their solution often requires a global search approach. Therefore the objective of optimization techniques is to find the globally best solution of models, in the possible presence of multiple local optima. Conventional optimization and search techniques include; (1) gradient-based local optimization method, (2) random search, (3) stochastic hill climbing, (4) simulated annealing, (5) symbolic artificial intelligence and (6) genetic algorithms. The detailed information on each technique and comparisons to Genetic Algorithms (GA) are already explained by Depa and Sivanandam (2008). Here, the aim is not to analyze these techniques in detail but to show the suitability of GA as a parameter estimation algorithm. As discussed by Depa and Sivanandam, some of the advantages of GA over other techniques are: (1) it is good for multi-mode problems, (2) it is resistant to becoming trapped in local optima, (3) it performs well in large-scale optimization problems, (4) it handles large, poorly understood search spaces easily. These advantages match with the requirements for an optimization technique to be applied in this application. Therefore GA was chosen as an optimization technique and successfully combined with the MTSM technique.

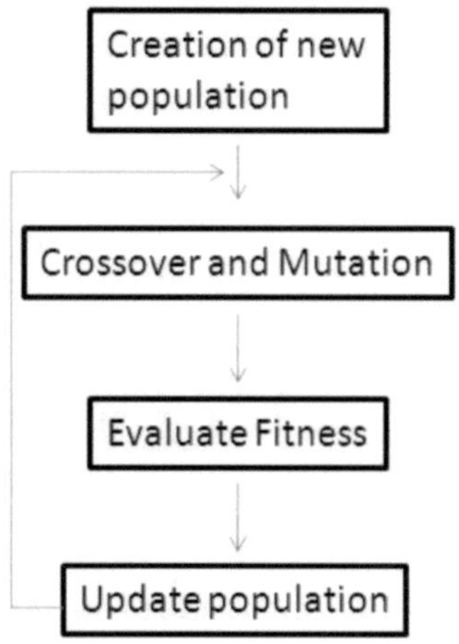

Fig. 4. Flow chart of a genetic algorithm

Structure of the MTSM/GA technique

The structure of MTSM-GA technique is presented in figure 5. As seen from the figure, there are two inputs to the GA, namely; range of variables and MTSM sensor response. GA outputs the determined values of the variables by using GA functions such as crossover, mutation and fitness evaluation. In the following sections, initially, the inputs to the GA will be explained. Then the structure of GA and its internal functions will be presented.

MTSM sensor response

The first input to the GA is the MTSM sensor response. Both magnitude and phase responses were continuously monitored during the experiments (see materials and methods section). Then the specific points on these responses such as resonant frequency, maximum magnitude, minimum phase, frequency at minimum phase, and phase at maximum magnitude were input to GA for calculating the fitness score for each individual. The changes in these target points were calibrated with the diwater/glycerin changes.

Selection of the ranges for variables

The next step of the technique is to set the ranges for the variables (chromosomes). These ranges represent the bounded space within which the GA will search for solutions. The ranges should be reasonable for each parameter in order to determine accurate solutions. For example, for a Newtonian liquid the stiffness is 0, therefore one should not set the range to be between 1e5 N/m^2 and 1e7 N/m^2. If this were done the algorithm will not converge to a solution because of the inappropriate choice of ranges.

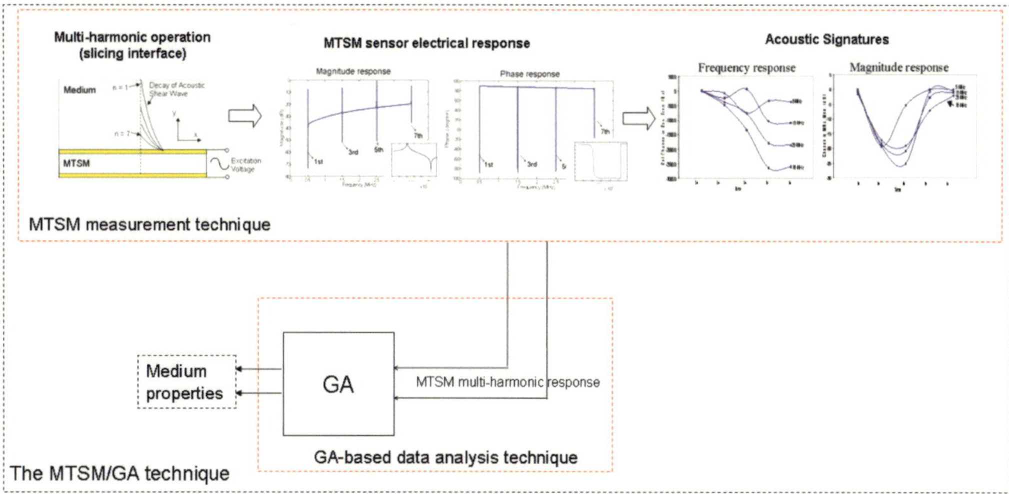

Fig. 5. Basic structure of MTSM/GA technique

As shown by Kwoun (2006), the viscoelastic materials can be divided in to four regimes, namely; liquid like, soft rubber, hard rubber and solid like. As seen from table 1, the viscosity values might change between 0.001 and 0.1 kg/m.s and stiffness value changes between 0 - 1e9 N/m^2 . Typical range of density values for a polymer was determined to be between 1000 - 1400 kg/m^3.

Phase	η (kg/m.s)	C (N/m^2)
Liquid like	0.001 - 0.01	0-1e5
Soft Rubber	0.01 - 0.1	0-1e5
Hard Rubber	0.01 - 0.1	1e5 - 1e7
Solid Like	- 0.1	1e7 - 1e9

Table 1. Four regimes of a viscoelastic system

Genetic Algorithm and its internal functions

This section will be divided into three sections. First, the GA's main parameters such as number of populations, crossovers, mutation rates and genes per chromosome will be analyzed. Then the fitness function of the GA will be explained. Finally, the technique combination of sub-spacing and zooming to determine the values for four variables will be presented.

Selection of GA parameters

Different combinations of the GA parameters were evaluated. Here, the combination that gives the best result is presented. Each variable was represented by a binary chromosome that contains 16 genes. A random population of 100 individuals was generated. Tournament selection was implemented for selection of individuals for mutation and crossover. In order to carry out the crossovers the entire population is divided into groups of 5 individuals each, these groups are randomly selected. From each group, the individual with the highest fitness together with another individual of this group are selected for crossover. The two selected individuals are the parents and yield two offspring. Both the parents and the offspring pass to the next generation. This idea was implemented in order to reduce the selection pressure.

The crossover between the parents is a simple one meaning that a random crossover point is selected and two kids' genome are formed with the left and right genes of the crossover point of each parent. A relatively high mutation probability (0.5) is present in order to avoid local minimum, otherwise all the individuals might end up having the same genome and this genome corresponding to a not optimal solution. Also elitism was implemented to assure that the best individual of a generation survives to the next generation. This ensures that the algorithm keeps the best solution until a better one is found.

Fitness function

One of the most important parts of a genetic algorithm is the fitness function. The fitness function must reflect the relevant measures to be optimized. This function evaluates the function being searched for the set of parameters of each member of the population. The output of the fitness function is a vector that contains the fitness for each member of the population. This vector helps in the selection of individual for generating new offspring or individuals that will be included in the new generated population.

The approach used, in this study to model biolayers on a MTSM sensor, is Mason's transmission line model (please see Appendix C). This model is a one-dimensional model that describes the electrical characteristics of an acoustic structure wherein, each layer of load can be represented as a T-network of impedances.

Once the initial population is created the algorithm randomly generates a population (includes 100 individuals) chosen from the ranges of the variables (the section titled "selection of the ranges for variables"). Then each individual was input to fitness function (transmission line model). The error between the model (transmission line model) and the experimental results were compared by using the following equation:

$$fit\_func = \frac{100}{1+(\sqrt{(\alpha_{Re}-\alpha_{Rt})^2}+\sqrt{(f_{Re}-f_{Rt})^2}+\sqrt{(P_{Me}-P_{Mt})^2}+\sqrt{(f_{Me}-f_{Mt})^2}+\sqrt{(\alpha_{ARe}-\alpha_{ARt})^2}+\sqrt{(f_{ARe}-f_{ARt})^2})}$$

The denominator of this function represents the difference between the model and the experimental data (we use the plus one in order to avoid the eventual division by zero). In this project, rather than fitting the whole magnitude and phase curve, certain points such as α_R = maximum magnitude, f_R = resonant frequency, P_M = minimum phase, f_M = resonant frequency at minimum phase, α_{AR} = minimum magnitude, f_{AR} = anti-resonant frequency has been compared between the model and the experimental results. Subscript "e" indicates experimental results and subscript "t" stands for theoretical model. This function is monotonously increasing with the kindness of the solution provided by the genetic algorithm. The algorithm was terminated at after 500 generations.

Set-up of the Genetic Algorithm

Acoustic impedance seen at the sensor/film interface is derived from transmission line theory (Martin and Frye 1991). Surface mechanical impedance is related to density and thickness of the film, and complex modulus (= $G^I + jG^{II}$). Therefore there are four independent variables to define the surface acoustic impedance. The MTSM sensor response contributes two parameters by providing real and imaginary part of mechanical impedance. Hence using single harmonic response results in an under-determined problem. Genetic optimization technique has been applied to under-determined problems to obtain approximate solutions with satisfactory accuracy (Wang and Dhawan, 2008). Here genetic algorithm has been improved by combining sub-space and zooming techniques. It was shown that this combination provides very good approximation with less than 1% error.

First, sub-spacing method was applied. This method gives a quick idea of where the solution can be and also it decreases algorithm running time dramatically (Garaia and Chaudhurib, 2007). Therefore the solution space was divided in 10 sub-spaces. Genetic algorithm was run 5 times in each subspace. Each subspace's convergence performance was evaluated. The sub-space with the best fitness score was considered to be the candidate solution space. It was observed that the candidate sub-space had a distinct convergence performance compared to the others. This method dramatically increased the efficiency of GA by eliminating the irrelevant solution spaces.

Secondly, GA was run 100 times (this number was chosen to have 95% confidence level and 10% confidence interval statistically). The termination criterion for each run was 500 generations. After 100 runs, it was observed that, for two out of four variables, observed points having a uniform distribution (skewness < 0.5) were accumulating around one number in a narrow range (in ±20% of candidate solution point). The average value of the observed points was also equal or very close (<5%) to solution (theoretically shown). Therefore GA was always able to converge to "the most likely" values for two out of four variables after these two steps (from our observations, mostly stiffness and thickness, and sometimes, viscosity and thickness). It was shown theoretically that one can always put these numbers, and calculate the other two variables with the error of less than <15% at this step.

Then zooming method was applied to reduce the search space around the candidate optimum solution point. Several zooming methods have been developed for different applications (Ndiritu and Daniel, 2001, Kwon et al. 2003). In this project, the GA was run 30 times, and then the new range was set to be between maximum and minimum numbers of the 30 points. This zooming continued until the error was less than 1% for all variables. This error was achieved after 6 zooming.

These results showed that the MTMS/GA technique combined with sub-spacing and zooming methods can be applied successfully to approximate the solution with good accuracy for this under-determined problem.

3. Materials and methods

The MTSM/GA technique first experimentally tested with the polymer SU8-2002 layer spin coated on sensor surface. The determined properties of the layer were compared with the values obtained from literature. The technique was then applied to obtain the mechanical and geometrical properties of a protein layer adsorbed on gold layer. The methods and chemicals used in the experiments are described below.

a. Deposition of the thin polymer film

The SU 8-2002 (MicroChem) polymer solution was spin coated on MTSM sensor by using the following procedure. First, the gold electrode surface of TSM sensors was cleaned using

Piranha solution (one part of 30% H_2O_2 in three parts H_2SO_4). After 2 min exposure time, the sensors were rinsed with distilled water. The surface was dried in a stream of nitrogen gas. The SU 8 - 2002 sample was dispensed on MTSM sensor surface and sensors were spin coated for 40 seconds. The sensors were then soft baked for 1 min at 95 °C. The SU 8-2002 films were exposed to UV light for 4 seconds under 25 mJ/cm^2. This was followed by 1 min hard baking on hot plate at 95 °C.

b. Antibody adsorption on MTSM sensor surface

The reference measurements were taken for air and phosphate buffer saline (PBS). Next, the sensors were exposed to rabit-immunoglobulin G (IgG) (50 µg/ml) suspended in diwater (Fisher Scientific, pH: 5.34, Cat No: 25—555-CM) for 50 minutes to allow IgG coating of the sensor surface by adsorption.

c. Characterization of geometrical properties of the thin film

The thicknesses of the SU 8 - 2002 films were determined by using optical profilometer (Zygo Inc. Model #: NV6200). For the thickness measurements, a very small portion of MTSM sensor surface was not exposed to UV light. After the films were developed, the SU 8-2002 layer was removed from this portion. To obtained different thicknesses of film layer, 1:1 solution of SU8-2002 and cyclopentanone (Acros Organics) was prepared.
The surface topography of the film layer was measured using atomic force microscopy (AFM). The prepared samples were placed on a glass slide installed on the atomic force microscope (Bioscope; Veeco), that was mounted on the inverted fluorescence microscope (TE2000; Nikon, Melville, N.Y.). Measurements were made using contact mode with a scan rate of 2 Hz.

d. Measurement system and MTSM sensor data analysis technique

A 14 mm diameter, 0.33 mm thick, 5 MHz quartz crystal with deposited 7 mm gold electrodes was placed in a custom fabricated brass sensor holder (ICM). The sensor holder was connected to a Network Analyzer (NA) (HP4395A). A LabView program on a personal computer was used to control the network analyzer and collect the data at 5, 15, 25 and 35 MHz. The experiments were done in room temperature (24°C±1°C). Magnitude and phase responses of MTSM sensor were monitored during the experiments (figure 6). The sampling rate was 30 seconds. Each experiment was repeated three times.

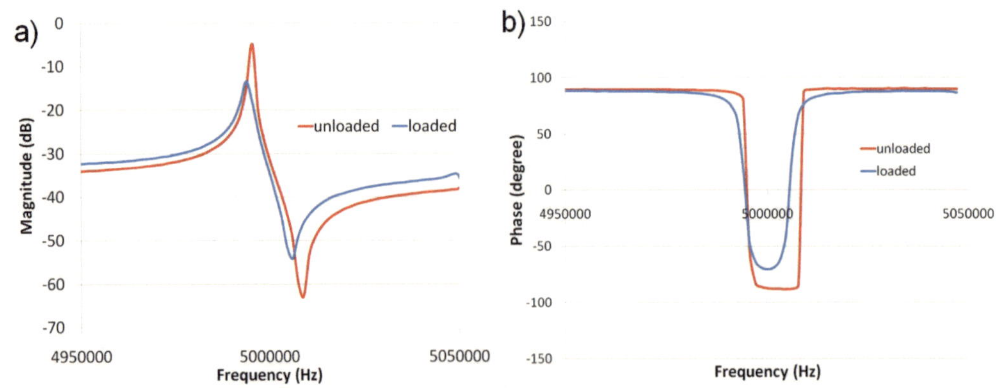

Fig. 6. a) Magnitude and b) phase responses of MTSM sensor

4. Results and discussions

Initially, two different thicknesses of SU8 2002 layers were spin coated on sensor surface and changes in the frequency and magnitude responses were monitored at 5, 15, 25 and 35 MHz. The thicknesses of the layers were measured by using optical profilometer (fig. 7a). The average thicknesses of the layers were 1920±25 nm and 770±50 nm respectively. Surface topography of the SU8 - 2002 layers was measured by using AFM (fig. 7b). The average roughness of the layer was 20 nm and no cracks on the surface were observed.

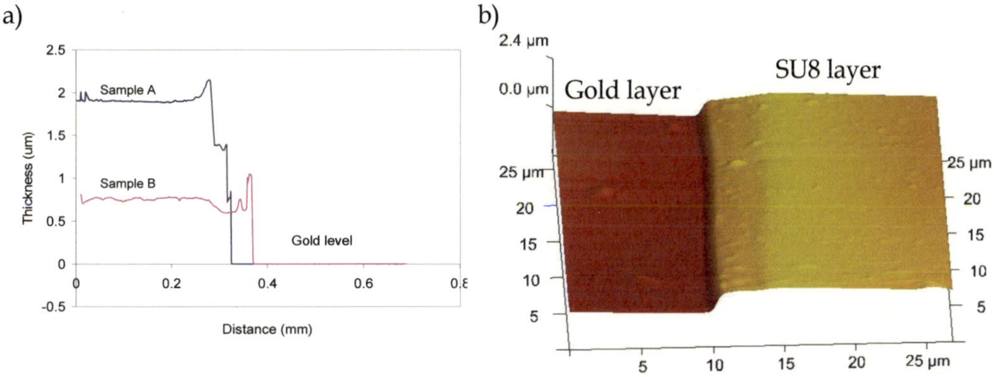

Fig. 7. A) Thickness measurements from optical profilometer sample a. SU8-2000 solution sample b. 1:1 dilution of SU8-2002 and cyclopentanone B) Surface topography of SU 8 layer

a. Determination of mechanical and geometrical properties of SU8 layer of 1.92 µm thickness

First set of experiments were performed by spin coating 2 µm thick SU 8 - 2002 layer on sensor surface. The MTSM/GA determined properties are presented in table 2. The average thickness of the polymer layer determined to range from 2080 nm to 2140 nm among the harmonics. Although these values are slightly higher than the value (1920±25 nm) obtained in control experiments, they are still in less than 10% experimental errors. The variation between the frequencies for density value was also very small, ranging from 1240 to 1253 kg/m³. These numbers correlate well with the literature value of 1200 kg/m³ (Jiang et al., 2003) for SU8.

MTSM Frequency	MTSM/GA Results		Profilometer	Jiang et al. [38]
(MHz)	d(nm)	ρ (kg/m³)	d (nm)	ρ (kg/m³)
5	2120±60	1253±10	1920±25	1200
15	2140±50	1246±11		
25	2080±110	1240±50		
35	2080±60	1240±28		

Table 2. Comparison density and thickness values of SU 8-2002 layer determined using MTSM/GA sensor at 5, 15, 25 and 35 MHz with profilometer and Jiang et al. (Jiang et al. 2003)

The frequency dependent shear modulus of SU 8-2002 layer obtained using the MTSM/GA is presented in table 3. Both loss and storage modulus varies with the operating frequency. These extracted values were compared with the values obtained by Jiang et al (2003) (table 3). Jiang et al calculated the shear modulus of SU8 layer by using the impedance-admittance characteristics of the equivalent circuit models of loaded and unperturbed TSM sensors operating at 9 MHz.

MTSM Frequency	MTSM/GA Results		Jiang et al.[38] (at 9 MHz)	
(MHz)	G^I (N/m²)	G^{II} (N/m²)	G^I (N/m²)	G^{II} (N/m²)
5	(4.55±2.12) x10⁷	(1.89±0.26) x10⁵	7.80e7	2.00e5
15	(2.33±0.18)x10⁸	(1.00±0.03) x10⁶		
25	(3.82±0.52)x10⁸	(4.69±0.52) x10⁶		
35	(5.81±0.71) x10⁸	(6.49±0.18) x10⁶		

Table 3. Comparison of determined G^I and G^{II} values of SU8 layer using MTSM/GA at 5, 15, 25 and 35 MHz and Jiang et al (2003)

As seen in table 3, the values obtained by Jiang et al. fall between the values obtained using the MTSM/GA method for 5 and 15 MHz. The small variation in the G^I and G^{II} may be due to difference in the film preparations. Alig et al. (1996) has shown that variations in film preparation methods can affect the mechanical properties of the polymer layers.

b. Determination of mechanical and geometrical properties of SU8 layer of 0.770 µm thickness

The second set of experiments was done with the ~770 nm thick SU 8-2002 layer on MTSM sensor. As seen from the table 4, the thickness of the layer determined using the MTSM/GA method correlates well with the expected thickness for each harmonic (less than 10% error). Furthermore the results vary only 10 nm between the harmonics. Similarly, determined values for density were consistent between the harmonics, which is around ~1200 kg/m³.

MTSM Frequency	MTSM/GA Results		Profilometer	Jiang et al.[39]
(MHz)	d(nm)	ρ (kg/m³)	d (nm)	ρ (kg/m³)
5	820±45	1180±40	770±50	1200
15	820±20	1190±30		
25	810±35	1190±50		
35	810±52	1213±35		

Table 4. Determined density and thickness values by MTSM/GA or 770 nm thick SU8 layer at 5, 15, 25 and 35 MHz

The initial losses before coating were -0.53 dB and -2.5 dB for 5 and 35 MHz respectively. The losses increase to -0.59 dB for 5 MHz and -4.18 dB for 35 MHz. As seen from these results, the losses remain relatively low when the thickness of the layer was decreased to 770 nm in contrast to the phenomenon observed when the film thickness was 2 µm. For 2 µm

film thickness, the losses increase to -1.9 dB and -11.5 dB at 5 MHz and 35 MHz respectively, while initial loses were similar to what observed for 770 nm film thickness.
The shear modulus values determined via the MTSM/GA technique are presented. Both loss and storage modulus were decreased compared to the values obtained when film thickness was 2 μm (figure 8). It has been shown that the scale effect on the mechanical properties of the polymers might be the reason for the decrease in the values (Liu et al, 2009, Luo et al, 2003).

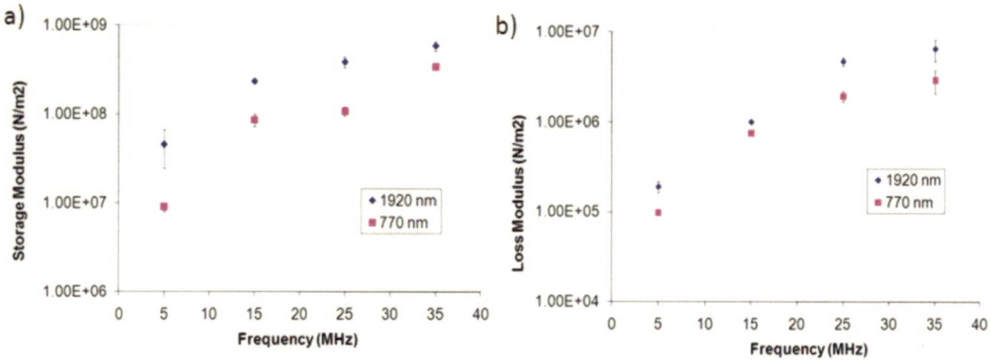

Fig. 8. a) Storage and b) loss modulus as a function of harmonic frequency for 770 nm and 1920 nm thick SU8 layer at 5, 15, 25 and 35 MHz (error bars are smaller than symbols when not visible)

c. Determination of mechanical and geometrical properties of an antibody layer

Third set of experiments were done by adsorbing an antibody layer on MTSM sensor surface under static conditions at 5, 15, 25 and 35 MHz. Antibodies play crucial importance in many applications such as biosensing (Hanbury et al. 1996) and drug delivery (Morrison et al., 1995). The sensor surface was saturated with antibody to form a uniform protein layer on the surface. Change in the frequency and magnitude responses at 15, 25 and 35 MHz are presented in figure 9. At the fundamental frequency (5 MHz), high fluctuations observed in sensor response are likely due to insufficient energy trapping as described by others (Li et al. 2004).

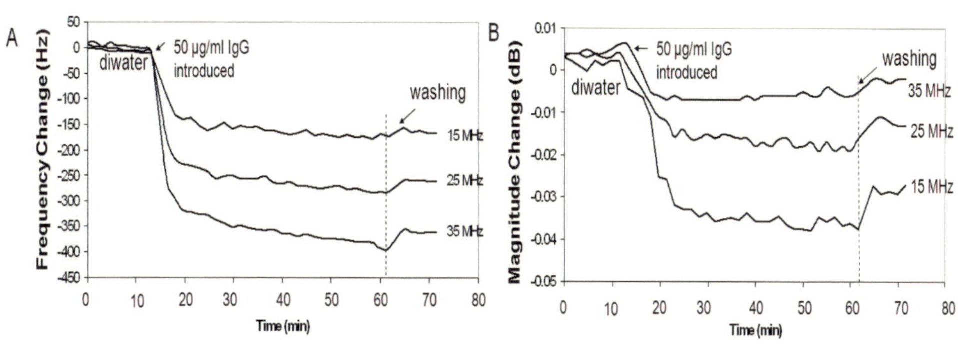

Fig. 9. Time response of A. resonant frequency and B. maximum magnitude responses of MTSM sensor to antibody binding at 15, 25 and 35 MHz

The properties of the medium were determined at t_1 = 10 and t_2 = 70 minutes. At t_1 = 10, the system is modeled as MTSM sensor loaded with semi-infinite Newtonian medium (DIwater) (fig 10A). The height of the column (2 mm) was much higher than the penetration depth of the acoustic wave at 5 MHz (~250 nm in DI water).
At t_2 = 10 min., the MTSM/GA determined properties of the layer at 15, 25 and 35 MHz are presented in table 5. The variations in the determined thickness values were very high (ranging from 300 nm to 5 μm due to the fact that column height was much larger than the penetration depth. Solution range for thickness values was set to be between 1 nm to 10 μm in genetic algorithm. Thus any thickness value larger than the penetration depth will satisfy the solution because the MTSM sensor is not sensitive to the changes beyond the penetration depth. However, the solutions were always higher than penetration depth as expected. Due to the high fluctuations in thickness values, it was not presented here. In contrast the solutions for ρ_1, η_1 and C_1 match with the literature values very well. (Literature values are ρ_1 = 1000 kg/m^3, η_1 = 0.001 kg/m.s and C_1 = 0 N/m^2 at room temperature (Greczylo and Deboswka 2005)).

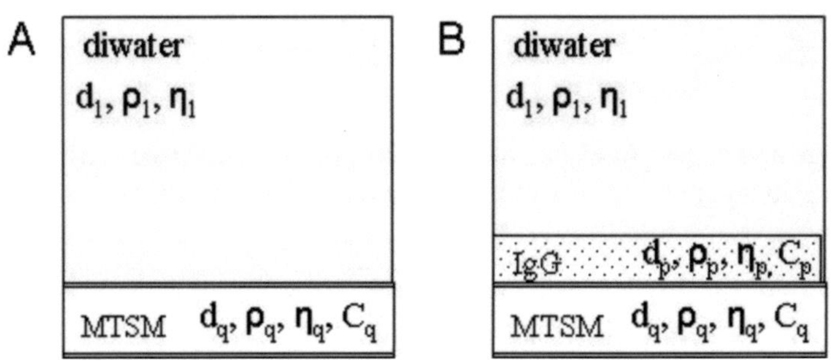

Fig. 10. Physical model for MTSM sensor system at A) t=10 and B) t=70

MTSM Frequency	Density (kg/m^3)	C^l (N/m^2)	η (kg/m.s)
15	1006±5	(2.00±1.00) x10^2	(1.05±.004) x10^{-3}
25	1003±2	(5.00±3.00) x10^2	(1.08±0.03) x10^{-3}
35	1004±4	(1.00±1.00) x10^2	(1.06±0.04) x10^{-3}

Table 5. Determined properties for semi-infinite Newtonian medium layer by MTSM/GA at 15, 25 and 35Hz

At t = 70 min., the physical model is presented in fig 10b. A viscoelastic layer (protein layer) with finite thickness and semi-infinite Newtonian medium were loaded on MTSM sensor. The properties for diwater layer were entered into the algorithm as known variables and the unknown properties (ρ_p, η_p, C_1 and d_p) of viscoelastic layer were determined using the MTSM/GA method. The results are presented in table 6. The thickness of the layer was determined to range from 10.3 to 11 nm for the harmonics. This number correlates well with the values presented by the other researches. Westphal et al (Westphal and Bornmann 2002) calculated the height of antibody layer as 9.2 nm. Furthermore Liao et al (Liao et al 2004) measured the average height of the antibody layer as 10.1±3.3 nm.

The density of the antibody layer was also determined by the MTSM/GA to be 1030±14 kg/m^3. This density value is close to the water density in which the antibodies were suspended. Hook et al. (2002) considered the density of antibody layer as 1050 kg/m^3 when the antibodies were not attached to gold surface. After the cross-linking, the density value was1300 kg/m^3, this is closer to the density value of dry protein. Voros (2004) also showed that the wet density of antibody layer is significantly different than the dry protein density value due to the solvent present in the adsorbed proteins. Therefore we believe that the determined value of the density is in a reasonable range.

MTSM Frequency	Thickness (nm)	Density (kg/m^3)	G^I (N/m^2)	G^{II} (N/m^2)
15	11±0.3	1050±10	(5.20±0.5) x10^4	(4.80±0.58) x10^5
25	10.4±0.6	1080±12	(5.00±0.13) x10^4	(9.50±1.40) x10^5
35	10.3±0.4	1040±14	(5.60±0.12) x10^4	(1.52±0.31) x10^6

Table 6. Determined properties for antibody layer by MTSM/GA at 15, 25 and 35 MHz

As seen from the table 6, the adsorbed antibody layer has low storage modulus (<1e5 N/m^2), and relatively higher loss modulus. While storage modulus was same for each harmonic, loss modulus changed with frequency. It has been experimentally shown that the adsorbed protein layers on TSM sensor, such as antibody, vesicles and cells do not behave like "rigid and thin" films (Voinova et al, 2002). Therefore the linear relationship between resonant frequency shift and mass deposition is not observed. Saluja et al. (2005) indicated low concentrations (less than 60 mg/ml) of antibody suspension behave like Newtonian medium. But it should not be expected that the properties of adsorbed layer will not be the same as the properties of antibody suspension. The effect of the binding between protein layer and gold layer should be considered. No literature value was found for direct comparison. Therefore we believe that MTSM/GA technique will lead to development of a quantitative tool for study of biological interfacial processes.

5. Conclusions

It was shown that MTSM sensor combined with genetic algorithm can be used to extract mechanical and geometrical properties of biological layers. The developed technique was first experimentally tested with SU8-2002 polymer layers with known properties having two different thicknesses. It was shown that the developed technique was successfully determined the mechanical and geometrical layers of thin polymer layers. MTSM/GA technique was then applied to extract the properties of antibody layer coated on MTSM sensor. The obtained data support our hypothesis about use of MTSM/GA technique can be a powerful tool for quantitative characterization of interfacial biological interfacial processes.

6. Acknowledgments

We are thankful to Dr. Moses Noh for providing supplies and micro-fabrication facilities for polymer coating.

7. Appendix I

A. The depth of penetration of a shear wave (δ)

The depth of penetration of a shear wave (δ) in a Newtonian medium is given by the equation shown below:

$$\delta_n = \frac{1}{\left(-\omega_n\left(\frac{\rho_m^2}{C_m^2+(\omega_n\eta_m)^2}\right)^{1/4}\right)Sin\left(-\frac{1}{2}\left(\arctan\left(\omega_n\frac{\eta_m}{C_m}\right)\right)\right)} \tag{2}$$

ρ_m = density of medium (kg/m³),
η_m = viscosity of medium (kg/m.s),
C_m = stiffness of medium (N/m²),
ω = angular frequency (rad/s) and
n = harmonic number

B. Basic terminologies of a genetic algorithm

Individual: A solution to the problem is called an individual.
Population: The total number of solutions is called population.
Chromosome: Each individual has a number of chromosomes that represent each parameter (i.e. variables to be determined) of the problem.
Genes: Each chromosome contains a fixed number of genes, the number of genes per chromosome determine the resolution of the total solution. The number of genes per chromosome is mostly determined by the broadness of the range in which each chromosome lies.
Fitness: Every individual has to be weighed according to its fitness. The individual fitness value determines its survival and breeding probability. A higher fitness individual has higher probability of survival.

C. Mason's transmission line model

As seen in fig. 11, the biological process consist of a piezoelectric layer (MTSM sensor) and a non-piezoelectric biological layer. In this model, each layer of load can be represented as a T-network of impedances.

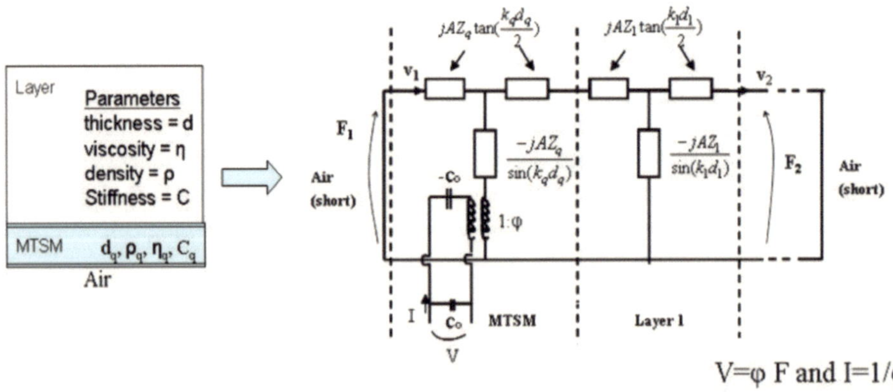

V=φ F and I=1/φ v

Fig. 11. Mason model representation of non-piezoelectric layers loaded on piezoelectric plate

8. Appendix II (Symbols)

F_1 = input force (N)
F_2 = output force (N)
v_1 = input particle velocity (m/s)
v_2 = output particle velocity (m/s)
A = area of active electrode of MTSM (m^2)
k = propagation constant (m^{-1})
d = thickness (m)
Z = acoustic impedance (acoustic ohm)
I = current (C)
C_0 = static capacitance of MTSM (F)
φ: transformer ratio

9. References

Alig et al, 1996. Ultrasonic shear wave reflection method for measurements of the viscoelastic properties of polymer films. *Rev. Sci. Instrumen.* 68. 1536-1542

Ayad M. M., Nagy L. Torad. 2009. Alcohol vapours sensor based on thin polyaniline salt film and quartz crystal microbalance. *Talanta.* 78. 1280-1285

Behling C, Ralf Lucklum and Peter Hauptmann. 1997. Possibilities and limitations in quantitative determination of polymer shear parameters by TSM resonators. *Sensors and Actuators A: Physical* V. 61, pp. 260-266

Cho N. J., J. Nelson D'Amour, Johan Stalgren, Wolfgang Knoll, Kay Kanazawa, Curtis W. Frank. 2007. Quartz resonator signatures under Newtonian liquid loading for initial instrument check. *Journal of colloid and Interface Science.* 315. 248-254

Cote G. L., R. M. Lec and M. Pishko. 2003. Emerging biomedical sensing technologies and their applications. *IEEE Sensors Journal.* 3. 251-266.

Desa J., Zhang Q., Ergezen E., and Lec. R., 2010 Microparticle manipulation on the surface of a piezoceramic actuator. Measurement Science and Technology. 21 (10) 105803

Dylkov M.S., Sanzharovskii A.T., Zubov P.I. 1966. The effect of thickness on the strength of polymer films. *Mekhanika Polimerov.* 2. 940-942

Edvardsson M., Michael Rodahl, Bengt Kasemo, and Fredrik Ho1o1k. 2005. A Dual-Frequency QCM-D Setup Operating at Elevated Oscillation Amplitudes. *Anal. Chem.* 77. 4918-4926.

Ergezen E, Appel M, Shah P, Kresh JY, Lec RM, Wootton DM, Real-Time Monitoring of Adhesion and Aggregation of Platelets using Thickness Shear Mode (TSM) Sensor, Biosensors and Bioelectronics, V. 23, #4, pp. 575-82, 2007.

Ergezen E. 2010 Multi-Resonant Thickness Shear Mode (MTSM) Measurement Technique for Quantitative Characterization of Biological Interfacial Processes Drexel University, PhD Thesis

Fredriksson, C., S. Kihlman, M. Rodahl, and B. Kasemo. 1998. The piezoelectric quartz Crystal mass and dissipation sensor: a means of studying cell adhesion. *Langmuir,* 14, 248-251.

Galipeau D.W., Vetelino J.V., Lec R. M. and Freger C.1991. The Study of Polyimide Film properties and Adhesion Using a Surface Acoustic Wave Sensors, ANTEC '91,

Conference Proceedings, Society of Plastic Engineers and Plastic Engineering, Montreal, pp. 1679-1984.

Gautam Garaia and B.B. Chaudhurib.2007. Adistributed hierarchical genetic algorithm for efficient optimization and pattern matching. *Pattern Recognition*. 40. 212-228

Greczylo T. and Deboswka E. 2005. Finding Viscosity of liquids from Brownian motion at students' laboratory. *European Journal of Physics*. 26. 827-833

Hanbury C. M., Miller G. W. and Harris B. R. 1996. Antibody characteristics for a continuous response fiber optic immunosensor for theophylline. *Biosensors and Bioelectronics*. 11. 1129-1138.

Hong S., Ertan Ergezen, Ryszard Lec, Kenneth A. Barbee. 2006. Real-time analysis of cell–surface adhesive interactions using thickness shear mode resonator. *Biomaterials*, 27, 5813-5820

Hook F., Larsson C., Fant C., 2002. Biofunctional Surfaces Studied by Quartz Crystal Microbalance with Dissipation Monitoring. *Encyclopedia of Surface and Colloid Science*. 774-790.

Jiang L., Hossenlopp J., Cernosek R., Josse F. 2003. Characterization of Epoxy Resin SU-8 Film Using Thickness-Shear Mode (TSM) Resonator. *Proceedings of IEEE International Frequency Control Symposium*. 996-982

Johannsmann D., Ilya Reviakine, Elena Rojas, and Marta Gallego. 2008. Effect of Sample Heterogeneity on the Interpretation of QCM(-D) Data: Comparison of Combined Quartz Crystal Microbalance/Atomic Force Microscopy Measurements with Finite Element Method Modeling. *Anal. Chem*. 80. 8891-8899.

K. Kanazawa and Nam-Joon Cho. 2009. Quartz Crystal Microbalance as a Sensor to Characterize Macromolecular Assembly Dynamics. *Journal of Sensors*. doi:10.1155/2009/824947

Katz A. and Ward M. D. 1996. Probing solvent dynamics in concentrated polymer films with a high frequency shear mode quartz resonator. *J. Applied Physics*. 80. 4153

D. Kroger, A. Katerkamp, R. Renneberg and K. Cammann. 1998. Surface investigations onthe development of a direct optical immunosensor. *Biosens. Bioelectron*. 13, 1141–1147.

M. Kunze, Kenneth R. Shull, and Diethelm Johannsmann. 2006. Quartz Crystal Microbalance Studies of the Contact between Soft, Viscoelastic Solids. *Langmuir*. 22. 169-173

Kwoun S., 2006. A Multi-Resonant Thickness Shear Mode (MTSM) Sensor for Monitoring The Formation of Biological Thin Films. PhD Thesis, Drexel University, pp. 125.

Li J, Thielemann C, Reuning U, Johannsmann D. 2004. Monitoring of integrin-mediated adhesion of human ovarian cancer cells to model protein surfaces by quartz crystal resonators: evauation in the impedance analysis mode. *Biosens Bioelectron*. 20. 1333–7.

Liao W., Wei F., Qian X. M., Zhao S. X. 2004. Characterization of protein immobilization on alkyl monolayer modified silicaon (111) surface. *Sensors and Actuators B*. 101. 361-367

Lin Z., Yip M. C., Joseph S. I., Ward D. M. 1993. Operation of an ultrasensitive 30 MHz quartz crystal microbalance in liquids. *Anal. Chem*. 65. 1546-1551.

Liu M., Sun J., Bock C., Chen Q. 2009. Thickness dependent mechanical properties of Polydimethylsiloxane. *J. Micromech. Microeng*. 19. 1-4

Lucklum R. and Hauptman P. 1997. Determination of polymer shear modulus with quartz crystal resonators. *Faraday Discussions*. 107. 123-140.

Lucklum R., Peter Hauptmann. 2000. The quartz crystal microbalance: mass sensitivity, viscoelasticity and acoustic amplification. Sensors and Actuators B 70 _2000. 30–36

Luo C., Schneider T., White R., Currie J., Paranjape M. 2003 A simple deflection testing method to determine Poission's ratio for MEMs applications. *J. Micromech. Microeng*. 13. 129-133

Mads Bruun Hovgaard, Mingdong Dong, Daniel Erik Otzen, and Flemming Besenbacher. 2007. Quartz Crystal Microbalance Studies of Multilayer Glucagon Fibrillation at the Solid-Liquid Interface. *Biophysical Journal*. 93. 2162-2169

Martin S.J and G.C. Frye. 1991. Polymer film characterization using quartz resonators. *Ultrasonic. Symp*. 393-398.

Morrison L.S. and Shin S. U. 1995. Genetically engineered antibodies and their applications to drug delivery. *Advance Drug Delivery Reviews*. 15.147-175.

Nam-Joon Cho, Kay K. Kanazawa, Jeffrey S. Glenn, and Curtis W. Frank. 2007. Employing Two Different Quartz Crystal Microbalance Models To Study Changes in Viscoelastic Behavior upon Transformation of Lipid Vesicles to a Bilayer on a Gold Surface. *Anal. Chem*. 79. 7027-7035.

Ndiritu J.G. and Daniell T. M. 2001. An improved genetic algorithm for rainfall-runoff model calibration and function optimization. *Mathematical and Computer Modeling*. 33. 695-706.

Newton M. I., Evans, C.R Simons, Hughes DC. 2007. Semen quality detection using time of flight and acoustic wave sensors. *Applied Physics Letters*. 90 (15) 154103

F. N. Nunalee, Kenneth R. Shull, Bruce P. Lee, Phillip B. Messersmith. 2006. Quartz Crystal Microbalance Studies of Polymer Gels and Solutions in Liquid Environments. *Anal. Chem*. 78. 1158-1166.

Roach, P; McHale, G; Evans, CR, Shirtcliffe NJ, Newton MI. 2007. Decoupling of the liquid response of a superhydrophobic quartz crystal microbalance. *Langmuir*, 23, 9823-9830

Rosenbaum J.F. 1998. Bulk acoustic Wave Theory and Devices. Boston, MA: Artech House Publishers

Saluja A., Kalonia S. D. 2005. Application of Ultrasonic Shear Rheometer to Characterize Rheological Properties of High Protein Concentration Solutions at Microliter Volume. *Journal of Pharmoceutical Sciences*. 94. 1161-1168

Sapper A., Wegener J., Janshoff A., 2006. Cell Motility Probed by Noise Analysis of Thickness Shear Mode Resonators. *Anal. Chem*. 78, 5184-5191.

Sivanandam S. N. and Deepa S. N. 2008. Introduction to Genetic Algorithms. Springer Berlin Heidelberg. New York. 34-35

Song Wang and Atam P. Dhawan. 2008. Shape-based multi-spectral optical image reconstruction through genetic algorithm based optimization. *Computerized Medical Imaging and Graphics*. 32. 429–441

Su H and Thompson M. 1996. Rheological and interfacial properties of nucleic acid films studied by thickness - shear mode sensor and network analysis. *Can. J. Chem*., 74, 344-358

Sun K., R. M. Lec, Cairncross R. A., Shah P., Brinker C. J. 2006. Characterization of Superhydrophobic Materials Using Multiresonance Acoustic Shear Wave Sensors. *IEEE Transactions on Ultrasonics Ferroelectrics and Frequency Control.* 53. 1400- 1403

Szabad, Z. Sangolola, B. and McAvoy, B. 2000. Genetic optimisation of manipulation forces for co-operating robots. *IEEE International Conference on Systems, Man, and Cybernetics*. 5. 3336-3341

Urbakh M. and Leonid Daikhin. 2007. Surface morphology and the quartz crystal microbalance response in liquids. *Colloids and Surfaces A. Physicochemical and Engineering Aspects*. 134. 75-84

Voinova M., Jonson M., Kasemo B., 2002. Missing Mass effect in biosensor's QCM Application. *Biosensors and Bioelectronics*. 17, 835-841

Voros J. 2004. The Density and Refractive Index of Adsorbing Protein Layers. *Biophysical Journal*. 87. 553-561

Yang D., Huang C., Lin Y., Tsaid D., Kao L., Chi C. Lin C., 2003 Tracking of secretory vesicles of PC12 cells by total internal reflection fluorescence microscopy. *Journal of Microscopy*, 209, 223-227

Young-Doo Kwon, Soon-Bum Kwon, Seung-Bo Jin and Jae-Yong Kim. 2003. Convergence enhanced genetic algorithm with successive zooming method for solving continuous optimization problems. *Computers and Structures*. 81. 1715-1725

Wegener, J, et al., 2000. Analysis of the composite response of shear wave resonators to the attachment of mammalian cells. *Biophys. J.* 78, 2821–2833.

Werner C. 2008. Interfacial Phenomena of Biomaterials. *Polymer Surfaces and Interfaces*. Dresden, Germany. Springer-Verlag Berlin Heidelberg. pp. 299

Westphal S. and Bornmann A. 2002. Bimolecular detection by surface plasmon enhanced Ellipsometry. *Sensors and Actuators B*. 84. 278-282

Zhang Q, Desa J., Lec R., Yag G., and Pourrezaei K. 2005, Combination of TSM and AFM for Investigating an Interfacial Interaction of Particles with Surfaces. *Joint IEEE International Frequency control Symposium (FCS) and Precise Tie and Time Interval (PTTI) Systems and Applications Meeting*. 4490454

Zhuang H., Pin Lu, Siak Piang Lim, and Heow Pueh Lee. 2008. Effects of Interface Slip and Viscoelasticity on the Dynamic Response of Droplet Quartz Crystal Microbalances. *Anal. Chem*. 80. 7347-7353

12

Analysis of Biological Acoustic Waves by Means of the Phase–Sensitivity Technique

Wojciech Michalski[1], Wojciech Dziewiszek[2] and Marek Bochnia[2]
[1]*Technical University of Wrocław,*
[2]*Medical University of Wrocław*
Poland

1. Introduction

The analysis of hearing mechanisms and research on the influence of various internal (pathologies, ageing) and external (trauma, vibration, noise) factors on sound perception are usually done using acoustic waves induced in the external ear canal. Stimuli which have been used for this purpose are: clicks, tone bursts, half-sine-waves, single tones or pairs of tones. The Corti organ's responses to the external stimuli have either an electric or acoustic character. In the former case, these are cochlear microphonics (CMs) picked up from the surface or from the inside of the cochlea, which are usually used as an indicator of damage to the organ of Corti in animals. In the latter case, these are acoustic waves that appear in the external ear canal as a result of stimulation. The acoustic waves have an important clinical value. Taking into account the presence of nonlinear distortions in the cochlea, the waves that appear after stimulation with a pair of tones are called distortion product otoacoustic emissions (DPOAE).

In studies on CMs, the origin of stimulating waves is often a single earphone (controlled by a generator of defined, often periodical, electrical signals) placed in the external auditory canal. In studies on DPOAE, a probe with two miniature earphones and one microphone is placed in the external auditory canal. The earphones are controlled by two generators of frequencies f_1 and f_2 and the microphone converts the returned DPOAE wave with a combination frequency, e.g. $f_3 = 2f_1 - f_2$, into an electrical signal.

The acoustic wave which induces CM signals is usually a periodic wave, while the waves inducing DPOAE signals consist of two pure tones. Thanks to the easy access to the output(s) of the generator(s) the phase-sensitive detection (PSD) technique can be used to measure both CM and DPOAE signals. Very weak (even below single microvolts) CM and DPOAE signals originating from the unimpaired cochlea can be measured in this way. Thanks to this technique signals obscured by other disturbing sources (even thousand times larger) can be measured accurately. This is possible because the phase-sensitive detector singles out the input signal with a specific reference frequency while signals with frequencies different from the reference are rejected. The fundamentals of this technique and its measuring potential are described in section 2.

In section 3, the authors' own experiments aimed at determining the effect various factors on the electrical function of the Corti organ are described. The factors include: vibration

(3.3), ototoxic medicines (3.4) and laser beams used in ear microsurgery (3.6). The role of the signal phase in the measurements is given special attention.

In section 4, experiments involving acoustic waves being nonlinear products of the Corti organ are presented. The individual subsections describe the way in which the PSD technique is applied (4.1), compare the latter with the previously used methods (4.2) and discuss the authors' own experiments in which the phase-sensitive technique is employed to measure DPOAE signals (4.3) and to measure simultaneously DPOAE signals and CMDP (cochlear microphonic distortion product) signals (4.4). It is shown that the phase of DPOAE signals plays an essential role in otoacoustic emission studies.

All the experiments described in sections 3 and 4 were carried out on coloured guinea pigs, each weighing 500-650 g, being under general ketamine/xylazine (15 mg/kg and 10 mg/kg body weight, respectively) anaesthesia. A Homoth measuring probe was placed in the external auditory meatus of the animals. The probe contained two mini earphones and a standard microphone. Prior to the measurements the probe had been graduated in a Brüel&Kraej artificial ear 4144, using a measuring amplifier 2607 made by the same company. Permission to carry out the experiments had been given by the Bioethical Committee in Wrocław.

Section 5 presents the final conclusions and discusses the future of the phase sensitive detection technique in investigations into the function of the cochlea exposed to various hazards.

2. Phase-sensitive detection technique

2.1 Fundamentals

The measurement apparatus based on the phase-sensitive detection technique is called a *lock-in amplifier* or a *lock-in nanovoltmeter*. Lock-in measurements require a frequency reference which should be strictly connected with a fixed frequency of the function generator used in the experiment. The reference signal can be either a square wave or a sinusoid. A block diagram of a typical lock-in amplifier is shown in fig. 1.

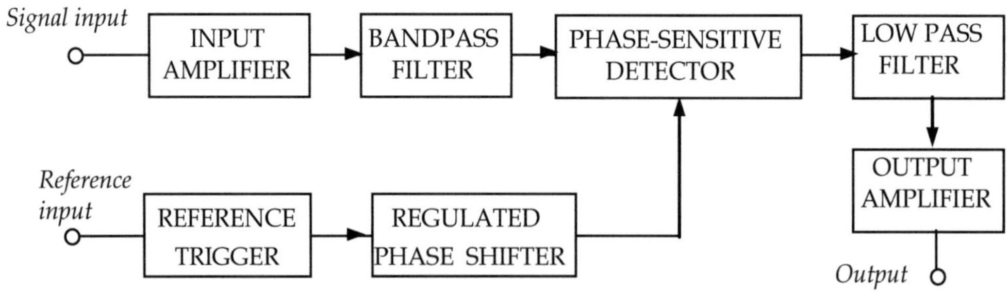

Fig. 1. Block diagram of lock-in amplifier with single phase-sensitive detector

Let us assume that the input signal can be described as:

$$V_{sig} = A_{sig} \cos(\omega_1 t + \alpha_0), \tag{1}$$

and the reference signal as:

$$V_{ref} = A_{ref}\cos(\omega_2 t + \beta_0) \tag{2}$$

The two signals have different amplitudes, frequencies and initial phases. At the phase-sensitive detector inputs there are signals with unchanged frequencies, but with different amplitudes and phases: $V_{sig} = A_{sig1}\cos(\omega_1 t + \alpha_{01})$ and $V_{ref} = A_{ref1}\cos(\omega_2 t + \beta_{01})$. The output of the PSD is simply the product of the two sine waves

$$V_{PSD} = A_{sig1}\cos(\omega_1 t + \alpha_{01})\cdot A_{ref1}\cos(\omega_2 t + \beta_{01})$$

$$= 0{,}5A_{sig1}\,A_{ref1}\left[\cos\left[(\omega_1 - \omega_2)t + (\alpha_{01} - \beta_{01})\right] + \cos\left[(\omega_1 + \omega_2)t + (\alpha_{01} + \beta_{01})\right]\right] \tag{3}$$

At the output of the PSD there are two signals: a slow-changing signal with differential frequency $(\omega_1 - \omega_2)$ and a signal with overall frequency $(\omega_1 + \omega_2)$. If the PSD output signals are passed through a low pass filter, the fast AC signal will be removed. When the frequencies of the two signals are approximately equal ($\omega_1 \approx \omega_2$), the filtered PSD output is a slowly changing DC signal proportional to the signal amplitude and cos(α_{01}-β_{01}). When $\omega_1 = \omega_2$, the filtered signal is exactly a DC signal. By adjusting the phase of the reference signal one can make (α_{01}-β_{01}) equal to zero, in which case only B_{sig} can be measured. This is true if both initial phases α_0 and β_0 do not change over time, otherwise cos(α_{01}-β_{01}) will change over time and V_{out} of the lock-in amplifier will not be a DC signal.
The phase dependency of the output voltage of the lock-in amplifier with one PSD unit can be eliminated by adding a second PSD multiplying the same measured signal by a reference signal shifted by 90^0. A block diagram of the lock-in amplifier with a double PSD is shown in fig. 2.

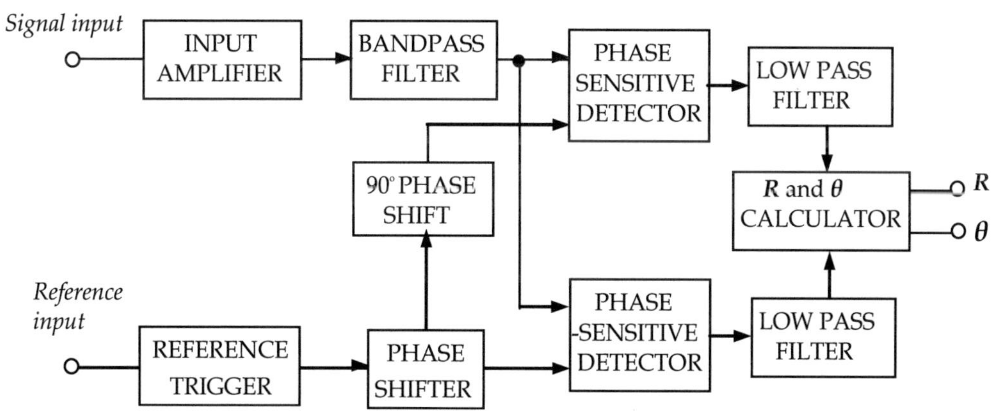

Fig. 2. Block diagram of lock-in amplifier with double PSD

The output of the second PSD, filtered by the low pass filter, is proportional to the signal amplitude and $\sin(\alpha_{01} - \beta_{01})$. Now there are two outputs: $X = A_{sig1}\cos(\alpha_{01} - \beta_{01})$ and $Y = A_{sig1}\sin(\alpha_{01} - \beta_{01})$. When

$$R = \sqrt{\left(A_{sig1}\cos(\alpha_{01} - \beta_{01})\right)^2 + \left(A_{sig1}\sin(\alpha_{01} - \beta_{01})\right)^2} = A_{sig1} \tag{4}$$

the phase dependency is removed. Phase difference (α_{01}-β_{01}) can be measured according to

$$\alpha_{01} - \beta_{01} = \tan^{-1}(Y/X) \tag{5}$$

The first lock-in amplifiers were based on analogue technology. The measured signal and the reference were analogue voltage signals and they were multiplied in an analogue PSD. The results of multiplication were filtered through a multistage RC filter. In such lock-ins the reference signal phase at the PSD input had to be manually adjusted to the phase of the measured signal so that $\cos(\alpha_{01} - \beta_{01}) = 1$. It was technically difficult to perform measurements by means of such lock-ins and it was practically impossible to register the amplitude and phase changes of the measured signals. Digital technology made it possible to build lock-in amplifiers in which both signal and reference inputs were multiplied and filtered digitally. Dual phase-sensitive detection eliminated the need for manual phase adjustments and enabled the simultaneous measurement of signal amplitude and phase. Such simultaneous measurements can be performed in *real time*, practically without any delay to the inducing signal. It also became possible to register short (below 0.1s) and slow changes in amplitude and phase over time.

2.2 Application of double-phase detection

The PSD technique offers greater measuring possibilities owing to the fact that:

1. the signal fed to the examined object may have various periodical waveforms,
2. the examined object can be linear or nonlinear,
3. the reference signal frequency can be equal to the frequency of the signal being delivered to the examined object, but it also can be an integral multiplicity (or submultiplicity) of this frequency.
4. two coherent signals can be introduced to the examined (usually nonlinear) object; the reference signal can be used at a frequency that is a linear combination of the frequencies of the inducing signals.

The basic experimental setup is shown in fig.3.

The generator used in the setup has two synchronous outputs. One of them (sync. output) supplies a TTL signal. Depending on the frequency of the reference signal one can measure the first harmonic, higher harmonics and subharmonics. In switch position 1 (fig.3), the reference signal is taken directly from the generator's synchronous output whereby the first harmonic can be measured. In switch position 2, the synchronous signal is multiplied by integral number n whereby the n-harmonic can be measured. In order to measure the n-subharmonic the generator's synchronic output must be divided by integral number n (the switch in position 3).

In the simplest case, the waveform of the signal directed to the examined object is sinusoidal. When the examined object is linear, using the PSD technique one can very precisely (with an accuracy of 1 nanovolt) measure the electrical response of the object. If the signal is a simple square wave or another periodical wave with frequency f, the examined linear object does not change the signal spectrum and the filtered PSD output is a DC signal proportional to the root mean square (rms) of the first component of the signal.

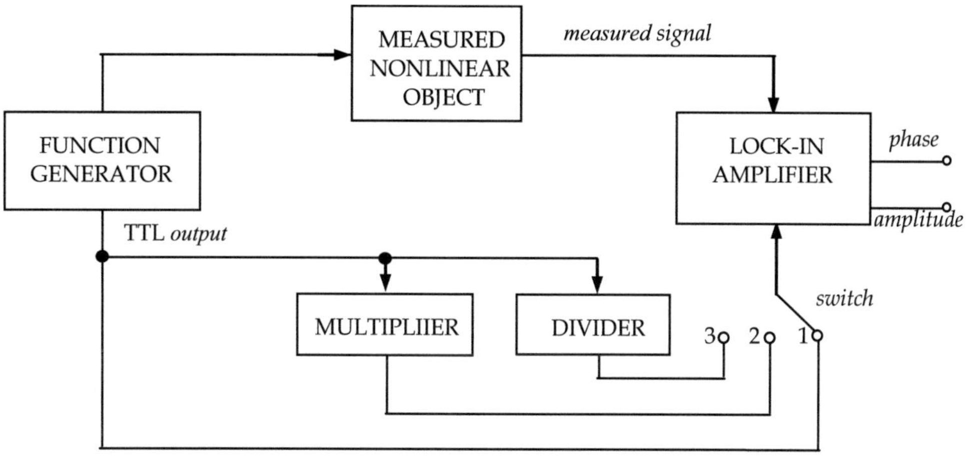

Fig. 3. Basic experimental set-up with lock-in amplifier for measuring first harmonic (1), higher harmonic (2) and subharmonic (3)

The situation becomes more complicated when the examined object is nonlinear. The signal spectrum at the PSD input differs from the one at the generator output (the nonlinear object changes the input signal spectrum). This is true for all the signal waveforms, including the sinusoidal one. Then the lock-in amplifier measures the rms of both the first harmonic and the *n*-harmonic (*n*-subharmonic). When harmonic or subharmonic distortion is measured, the function generator supplies a pure sinusoidal signal without any harmonics. The basic experimental setup shown in fig. 3 was used to carry out experiments described in section 3.

2.3 Nonlinear object testing with two synchronous signals

The double phase-sensitive detection technique and modern digital technologies offered new possibilities of examining nonlinear objects. An example of the measuring systems which have been developed is shown below (fig. 4). The main component of the setup is a digital sinus generator of three signals with synchronous frequencies. Two of the signals are delivered to the examined object while the third one serves as a reference signal. The frequency of the third signal is a linear combination of the frequencies of the other two signals.

Let us assume that the output-input function for the examined object can be described by the formula:

$$V_{out} = C \cdot V_{in}^3, \tag{6}$$

where C is a constant value. The input signal is the sum of signals with different amplitudes, phases and frequencies and so:

$$V_{out} = C\left[A_1 \cos(\omega_1 t + \alpha_{01}) + A_2 \cos(\omega_2 t + \alpha_{02})\right]^3 \tag{7}$$

After trigonometric conversions it is possible to receive a signal frequency spectrum at the object's output. The frequencies, amplitudes and phases of the particular spectral components (assuming that the examined object does not change its phase relations, i.e. it is characterized by pure resistances) are shown in table 1.

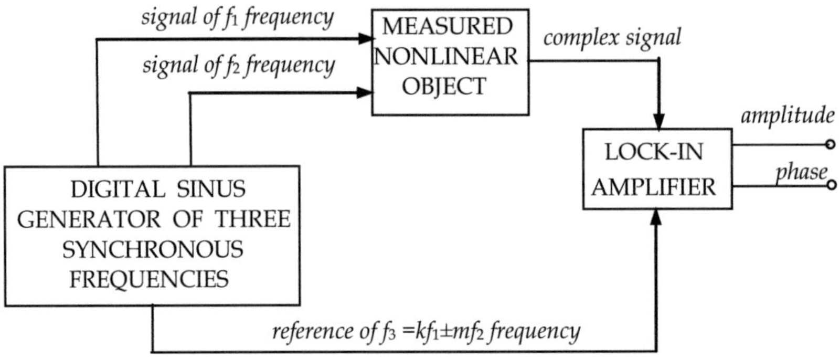

Fig. 4. Block diagram of experimental setup for more complicated studies of nonlinear objects

Frequencies of spectral components	Amplitudes of spectral components	Phase of spectral components
ω_1	$1{,}5A_1C\left(0{,}5A_1^2+A_2^2\right)$	a_{01}
ω_2	$1{,}5A_2C\left(0{,}5A_2^2+A_1^2\right)$	a_{02}
$3\,\omega_1$	$0{,}75C\,A_1^3$	$3\,a_{01}$
$3\,\omega_2$	$0{,}75C\,A_2^3$	$3\,a_{02}$
$2\,\omega_1+\omega_2$	$0{,}75C\,A_1^2\,A_2$	$2a_{01}+a_{02}$
$\omega_1+2\,\omega_2$	$0{,}75C\,A_1\,A_2^2$	$a_{01}+2\,a_{02}$
$2\,\omega_1-\omega_2$	$0{,}75C\,A_1^2\,A_2$	$2a_{01}-a_{02}$
$\omega_1-2\,\omega_2$	$0{,}75\,C\,A_1\,A_2^2$	$a_{01}-2\,a_{02}$

Table 1. Exemplary spectrum at output of object with 3rd-order nonlinearity, tested by pair of pure tones

The amplitude of each of the spectral components can be measured using this technique if a proper reference signal frequency is selected. The interpretation of phase shifts between the particular spectral components is much more complicated and requires taking into account the phase shifts introduced by the examined object. Moreover, the phase shifts introduced by the examined object may be a function of frequency and so they may be different for each spectral component. The technique was used to examine nonlinear distortions during the stimulation of the cochlea by a pair of pure tones. The results of this research are presented in section 4.

3. Using PSD technique to study cochlear potentials

3.1 Measuring techniques

Cochlear potentials are biopotentials of the inner ear. They are described as electrical signals arising in response to the acoustic stimulation (usually by a click or a tone) of the organ of

Corti. For the first time they were registered by Wever and Bray in 1930 (Wever &Bray, 1930). The discovery of the signals made it much easier to examine the function of the inner ear and made it possible to assess the impact of various external and internal factors on this function. It is widely believed that cochlear microphonics (CMs) are generated mainly by outer hair cells (OHCs). Therefore it seems reasonable to use CMs as an indication of the OHC function. On the basis of measurements performed over a long period (e.g. a few weeks or months) one can assess if given hearing damage is temporary or permanent. The CM signal originating from different places in the human ear (or the animal ear) can be recorded. In humans CMs are usually picked up from the round window during surgical procedures performed on patients with various hearing pathologies. There are much fewer reports describing the reception of CM signals from the promontory or the ear canal near the eardrum. The past and present studies of the mechano-electrical cochlear function (based on the reception of CMs) are conducted mainly on animals, using: *in vivo* preparations of anaesthetized animals with positive Preyer's reflex, *in vitro* preparations of the cochlea or *in vitro* preparations of the hair cells. As regards the research into the impact of various external and internal factors on the hearing organs, the *in vivo* studies seem to be most clinically valuable.

In the 1930s and 1940s CMs were measured at the cochlea's round window (Wever & Bray, 1930). In most animals the round window is relatively easily accessible and so measuring electrodes were usually placed on it or in its direct proximity. In the first years after the discovery of inner ear potentials, CM signals were measured by a single active probe. Several years later the first mapping of CMs on the cochlear surface was described (Thurlow, 1943). It was probably the first attempt ever to place the probe so close to the source of cochlear microphonics. CM potentials are continued to be measured at the cochlea's round window today (Brown, 2009). This measuring technique was not abandoned after the introduction of very sensitive (but invasive) procedures (Tasaki et al., 1952). Tasaki monitored CMs using a pair of active intracochlear electrodes in the basal turn (one electrode in scala tympani, the other in scala vestibuli). The electrodes were connected to a balanced differential amplifier. The reference electrode was placed on the neck muscles. This enables the measurement of the potentials very close to the organ of Corti and eliminates the auditory nerve potentials. The largest drawback is the mixing of perilymph and endolymph when the probe is introduced.

A new recording technique has been described by Carricondo (Carricondo at al., 2001). In this technique, CM potentials are recorded by subcutaneous electrodes in animals or by surface electrodes in humans. Two active electrodes are placed on the mandibular muscles while the reference electrode is located on the head's vertex. All the three electrodes are connected to a differential amplifier. The signal coming from the amplifier's output is filtered and subsequently averaged through in-phase synchronization with the sound stimuli.

3.2 History of CM studies in Wroclaw

In 1960 at the Wroclaw University of Technology an oscillograph was built and used as part of an experimental setup for registering cochlear microphonics (CMs). In the following years Jankowski and Giełdanowski started a series of experiments on animals - first on cats, later on guinea pigs (Jankowski et al., 1962). A Biopotentials Research Workshop was founded, where biopotentials were measured after damage to the inner ear or the skull, in acoustic trauma, in hypothermia or hypoxia, after the administration of certain drugs, and so on. Figure 5 shows a schematic of the experimental setup.

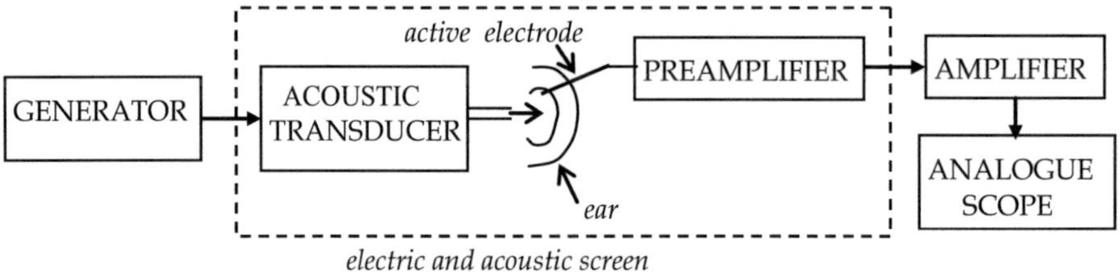

Fig. 5. Schematic of experimental setup used by Jankowski & Giełdanowski

The experimental animals (under urethane anaesthesia, which does not diminish CM voltages) underwent ear surgery: the bones were exposed and drilled until the round window was revealed. Platinum electrodes (platinum wires 0.1 mm in diameter, coated with PMMA) were used for CM measurements. The bare end (not coated with PMMA) of the platinum wire was brought into contact with the round window membrane (without damaging it). An injection needle was used as the other electrode. It was inserted into the muscles around the surgical wound. An example of an CM oscillogram from Ziemski's work (Ziemski, 1970) is shown in fig. 6.

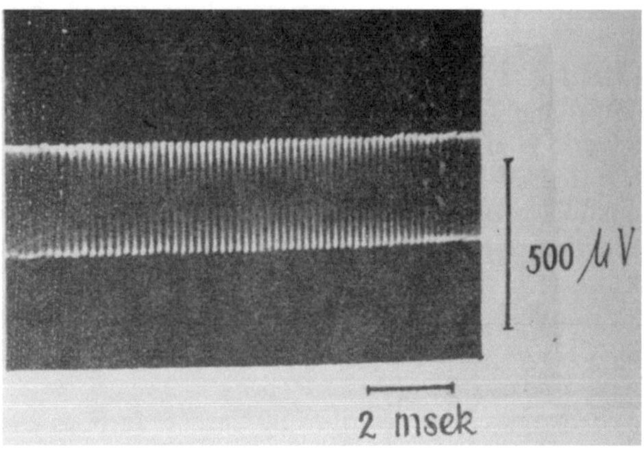

Fig. 6. Example of CM oscillograph record (stimulus tone parameters: f =4096 Hz, acoustic pressure level - 60dB)

In the middle of the 1990s a new surgical approach, making it possible to expose the whole cochlea in guinea pigs, was proposed. Skin was cut from the occipital part of the skull, around the angle of the mandible up to the place of about 0.5 cm above the angle of the animal's snout. The temporal muscles were removed, the mandible partially resected and the stylo-hyoid muscle cut. Bones were drilled, leaving the tympanic membrane intact. In the 1990s the present authors used the phase-sensitive detection technique to measure CMs. A patent was applied for in 1996 and the technique was patented in December 2000 (Patent PL no. 180060). A report on the use of a similar method was published by Kobayashi (Kobayashi, 1997). A schematic of the first setup used to measure CMs by the PSD technique is shown in fig. 7.

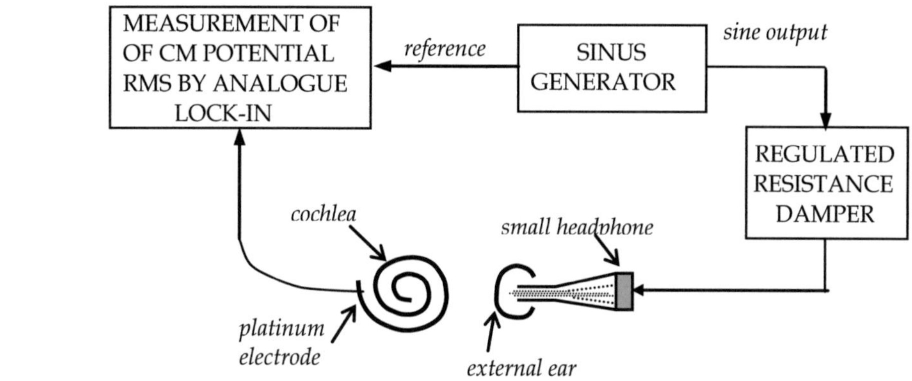

Fig. 7. Schematic of first setup used to measure CMs by phase-sensitive detection technique

Thanks to the new surgical approach combined with the phase-sensitive detection technique it became possible to create a map of CM potential amplitude and phase distributions on the cochlear surface. For this purpose, the active electrode would be fixed in six different points on the cochlea's surface (on the apex, one point at the third turn, two points at the second turn and two points at the basal turn). When the electrode was fixed in any of the six points, the frequencies of the stimulating acoustic wave would be successively selected from the measurable range. Three different acoustic wave pressures (60 dB, 80 dB and 95 dB) would be set for each of the measuring frequencies. This means that 18 different amplitude-phase (A-P) values of the CM signal were measured in each of the six points on the cochlea's surface. In this way 18 different A-P distribution patterns would be obtained for the guinea pig cochlea. Six of them (three for f = 260 Hz and three for f = 8 kHz) are shown in fig. 8. Phase in each measuring point is related to phase on apex at 60 dB.

It was found that CMs had a different phase and amplitude in the different parts of the cochlea, which was due to the fact that each of the six measuring points was located at a different distance from the CM sources inside the organ of Corti.

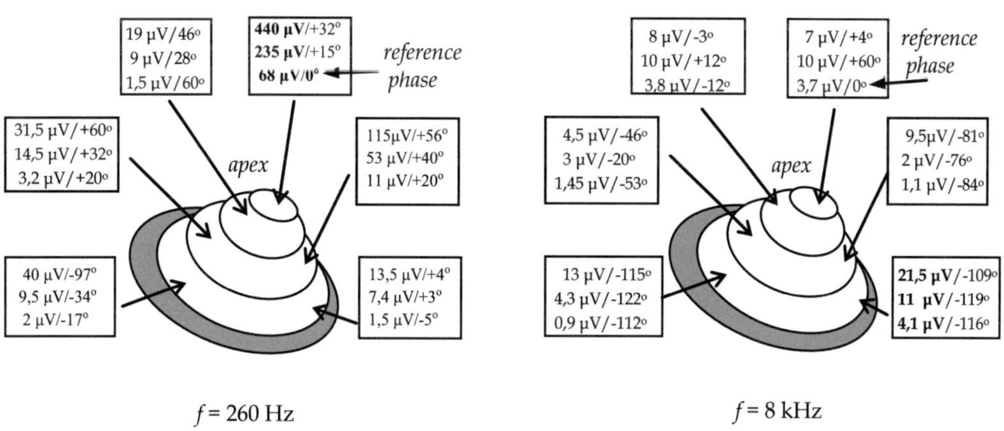

Fig. 8. CM potential amplitude and phase in six different points on cochlea surface, measured at 260 Hz and 8 kHz. At each of six points there are three amplitude/phase values: upper for 90 dB, central for 80 dB and lower for 60 dB

In order to unify CM measurements, two distinctive and universal measurement points were established: the cochlea's apex for the frequencies of 260, 500, 1000 and 2000 Hz and the cochlea's base for 4000 and 8000 Hz. Phase in each measuring point is related to phase on apex at 60 dB.

3.3 Influence of whole-body vibration on inner ear

Vibration is one of the most widespread injurious factors in the environment of civilized man (Palmer et al., 2000a, 2000b). The energy absorbed can have a pathological effect on all the tissues and organs of the body, although the consequences of exposure to vibration do not present a uniform clinical picture (Jones, 1996; Seidel & Heide, 1986). Because all machines and vibration devices also produce noise, usually the combined effect of the two factors is examined (Castelo Branco, 1999). There is a prevalent view that mechanical vibrations exert only a weak, additionally traumatic influence on the hearing organ (Seidel, 1993). Several experimental investigations into the harmfulness of vibration were carried out on animals (Hamernik et al., 1980, 1981). Changes in the hearing organ most often would be found in the hair cells (Rogowski, 1987). This made us undertake our own research in the 1990s. In order to determine the impact of long-term general vibration on the inner ear it was necessary to: 1) design and built noiseless vibration apparatus, 2) subject several groups of animals to general vibration (defined by controlled parameters over different periods of time) and 3) evaluate selected parts of the organ of hearing, using norms based on values derived from a control group.

In order to ensure proper experimental conditions, i.e. sinusoidal (10 Hz) vertical (5 mm) shaking, a device consisting of an electric impulse generator, a power amplifier and an impulse exciter was built (fig. 9). Experiments were carried out on young, coloured guinea pigs of both sexes weighing 240-360g. Fifty six animals with the normal Preyer reflex and without otoscopically detectable changes were used. The control group (group m0) consisted of 20 of the animals and served to establish functional and morphological norms. In order to avoid changes due to aging being interpreted as the effects of vibration, the control group was examined after a seven-month stay (6+1 months = duration of the longest experiment + a rest) in an animal house. The study group consisted of 36 guinea pigs divided into two subgroups of 18 animals each. Each subgroup was subjected to vibration over different periods, i.e. 30 (group m1) and 180 (m6) days. These were in fact respectively 22 days (5 days/week, 6 hours/day = 132 hours) and 132 days (792 hours). After the experiment and a one-month (30 day) rest, the animals which were in good general condition and without otoscopically detectable changes were qualified for functional and morphological investigations.

Cochlear microphonics were measured under urethane anaesthesia, using the PSD technique and the setup schematically shown in fig. 3 (the switch in position 1). CMs were picked up from the apex of the cochlea for the frequencies of 250 Hz, 500 Hz, 1 kHz and 2 kHz and from the region of the round window for 4 kHz and 8 kHz, using a platinum needle electrode. For the two study groups and the control group, a total of 6048 data values were taken for the bilaterally examined pulse wave frequencies (260 Hz-8 kHz) and intensities (55 dB-95 dB).

The results of the CM measurements were subjected to statistical analysis. The aim was to find out whether the experiment had any influence on CMs and, if so, what that influence was. The questions asked were: 1) are there statistically significant differences between the

CM voltages obtained from the control groups and the study groups, and 2) are there statistically significant differences in the CM voltages obtained within the study groups? The CM values obtained from the healthy animals showed considerable individual differences, and their distribution showed neither normalcy nor log-normalcy. Therefore all the experimental samples were examined using non-parametric tests. The K-S Lilliefors test showed: 1) for control group m0 compared with study groups m1 and m6, a significant decrease in CMs for the frequencies of 260 Hz, 1 kHz and 2 kHz, and 2) for m1 compared with m6, a decrease in CM for the frequencies of 260 Hz and 2 kHz. The Kruskall-Wallis test confirmed the results of the K-S Lilliefors test as regards the location and nature of the changes.

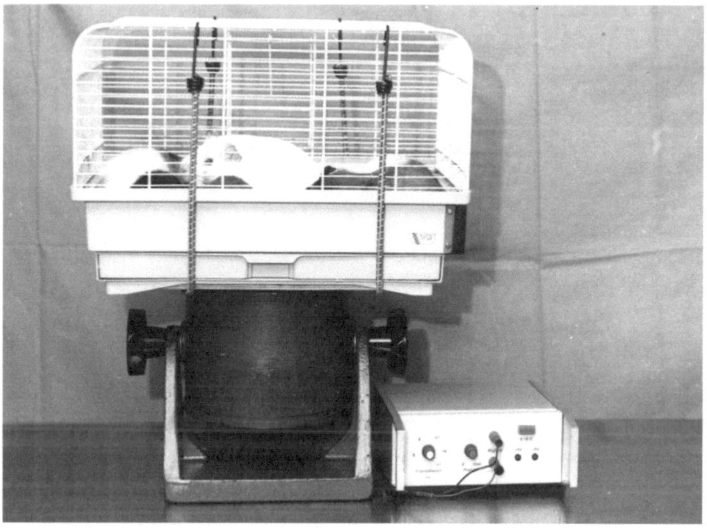

Fig. 9. Cage with animals exposed to vibrations

The results of the investigations indicated possible greater damage to the hair cells in the forth and third turnings of the cochlea. Further morphological examinations were needed to verify this observation. After the bilateral CM measurements the animals were decapitated and samples were prepared for SEM examinations of the sensorial epithelium. The samples were examined and photographed using a scanning DSM 950 microscope. The influence of general vibration on the organ of Corti was assessed on the basis of the condition of the hair cells, taking into consideration their disorganization, deformation, mutual adhesion and any reduction in the number of cilia.

SEM examinations were carried out on 20 cochleae from the control group animals and on all the animals in the two study groups. In the healthy animals, the sensorial epithelium was found to be normal in every case, but in each of the study groups the above mentioned damage was observed. It usually occurred in the OHC region of the apex, and its extent gradually increased in the direction of the cochlea's base (up to the second turning). OHC3 was found to be most susceptible to vibratory trauma. Cell damage decreased from the circumference to the modiolus, and the OHCs showed considerably greater resistance to vibration (fig.10). Undoubtedly, the observed damage to the sensorial epithelium resulted from mechanical vibration, and its severity clearly increased with the duration of the experiment. Consequently, the mechanism of deterioration in hearing in all the frequency ranges (especially at low and average frequencies) in persons subjected to whole-body vibration could be discovered by analyzing the observed changes.

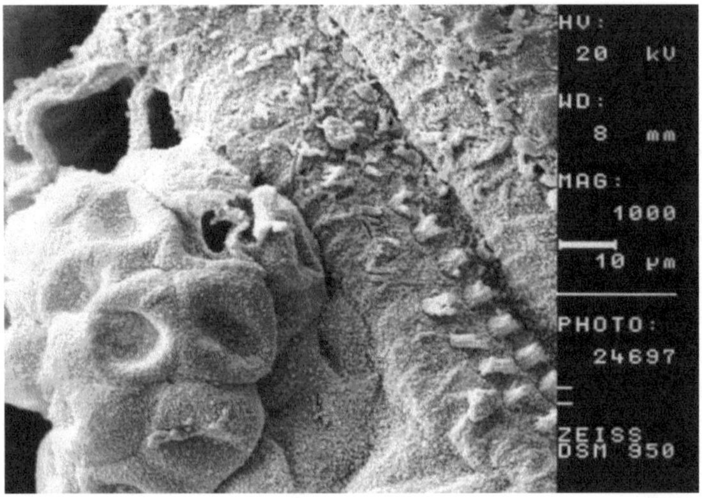

Fig. 10. Group M6, 4th cochlear turning: numerous lesions of hair cells and damage to Hensen's cells

3.4 Studies of gramicidin ototoxicity

Polypeptide antibiotics are used in a variety of clinical situations. Their molecules contain a specific chain of aminoacids and a non-aminoacidic part (e.g. fatty acids in polymyxins or glycopeptide in vancomycin). They are generally effective against Gram-positive bacteria, except for polymyxins which are effective against Gram-negative bacteria. They act by disrupting the selective permeability of bacterial cellular membranes. Despite their long history, polymyxins have had a limited clinical use due to the large number of side effects. Currently they are used primarily for topical treatment (Wadsten at all, 1985).

Since no descriptions of the effects of the systemic administration of gramicidin on the inner ear could be found in the literature, the authors decided to examine CMs and to compare the ototoxic effects after the systemic and topical administration of gramicidin. Also the inner ear of animals which received i.m. injections of gramicidin were examined using a DSM 950 scanning electron microscope (Bredberg at al., 1970; Davis, 1983) .

The research was conducted on 70 young, coloured guinea pigs. All the animals showed the positive Preyer reflex and no pathologies under otoscopic examinations. The experimental animals (G) were divided into 5 subgroups, depending on the drug administration mode and the administered dose. Each experimental subgroup (G1-G5) consisted of 8 randomly chosen animals. Subgroups G1-G3 received respectively 2, 5 and 10 mg of gramicidin/kg i.m., once per day, for 14 consecutive days. The animals from subgroups G4 and G5 were administered a 0.25% and 10% solution of gramicidin suspended on a haemostatic sponge placed on the round window.

The control group (K) consisted of 30 animals randomly divided into 2 subgroups (K1 and K2). The animals in control subgroup K1 were injected with normal saline solution once per day for 14 consecutive days. The animals in subgroup K2 were administered normal saline solution placed on the round window. One day after the last injection (the 15th day of the study) electrophysiological measurements were carried out on the animals in subgroups G1-G3 and K1. Then their cochleae were removed for SEM examinations. In the case of the animals belonging to subgroups G4, G5 and K2, CM measurements were performed after removing the haemostatic sponge from both ears and allowing the round windows with their surroundings to dry (Gale & Ashmore, 1977).

Cochlear microphonics (CMs) were investigated under urethane anaesthesia, using the PSD technique and the setup schematically shown in fig. 3 (the switch in position 1). CMs were picked up from the apex of the cochlea for the frequencies of 260 Hz, 500 Hz, 1 kHz and 2 kHz and from the region of the round window for 4 kHz and 8 kHz by means of a platinum needle electrode. As regards study subgroups G1-G5 and control subgroups K1 and K2, a total of 7560 data values were taken for the examined frequencies (260 Hz-8 kHz) and intensities (55 dB-95 dB). The results of the CM measurements were subjected to statistical analysis (the t-Student test).

Gramicidin administered systemically in a dose of 2 mg/kg led to a significant (38%) decline in CM voltage in K1 subgroup animals for the frequencies of 260 Hz and 2 kHz. For the other frequencies the drop in CMs amounted to about 15%, except for the 4 kHz at which a slight improvement was observed for sound levels between 55 and 70 dB. A significant drop in CMs was observed in subgroup G2 at 2 kHz and sound levels above 70 dB. At 95 dB the decline in CMs was 30% larger than in the G1 animals. The changes in the G2 animals relative to G1 were even more significant at 500 Hz, 1 kHz and 8 kHz. The animals receiving 10 mg/kg of gramicidin showed lower CMs than the ones registered in all the examined frequency ranges for control subgroup K1. The largest drop was registered at 2 kHz (31% lower than in the K1 control subgroup). The smallest changes were observed at 8 kHz. In subgroups G1-G3, the largest differences in CMs were observed at 4 kHz for all the sound levels.

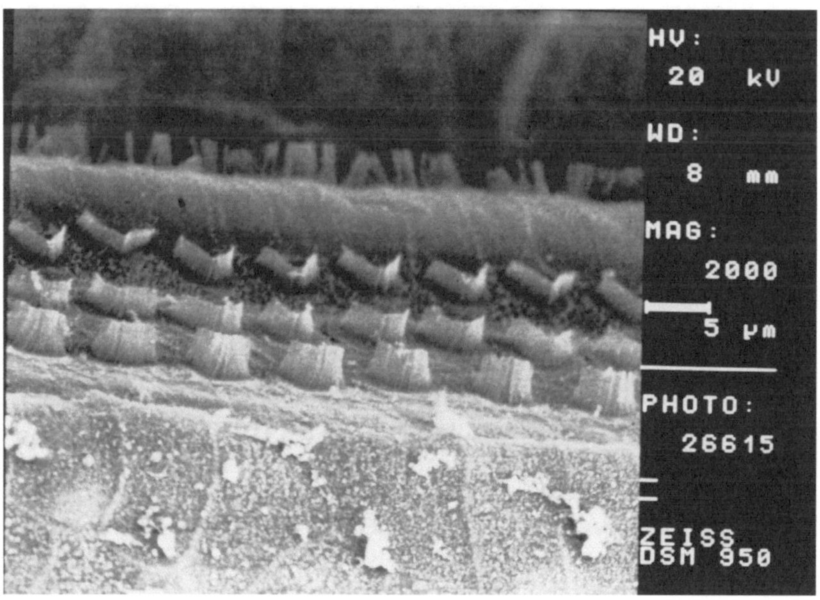

Fig. 11. Group K1, 2nd cochlear turn: unchanged sensory epithelium

In the animals receiving topical 0.25% gramicidin solution (G4), a significant drop in CMs (in comparison with control K2) was observed at 1 kHz and 2 kHz. In group G5 (where the animals were administered 10% gramicid in solution on the round window) a drop in CMs was observed also at 4 kHz and 8 kHz. At low sound levels the largest falls in CMs were observed in subgroup G4.

In the G1 and G2 animals no damage to the sensory epithelium was found under SEM. The destruction of cochlear hair cells occurred in the G3 animals. The changes were most visible in OHC3 cells in the cochlea's third turning.

To sum up, the systemic administration of gramicidin leads to greater disruptions of the bioelectric functions of the inner ear than local, topical administration (Linder at al., 1995).

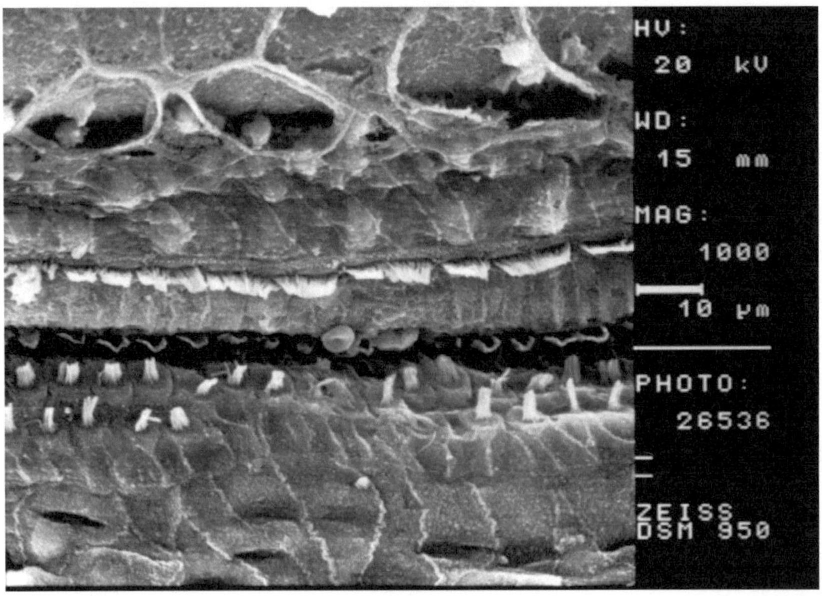

Fig. 12. Group G3, 3rd cochlear turn: numerous lesions in OHC3 cells and structural changes in cilia

3.5 CM amplitude and phase changes caused by changes in intensity of stimulating acoustic wave

Another important improvement in CM measurement came with the introduction of a lock-in amplifier with double phase-sensitive detection. In December 2003 a device for the phase-sensitive measurement of inner cochlea microphonic potentials was registered at the Patent Office. It was patented in November 2010. The device can measure harmonic, subharmonic and linear distortion products of the cochlea after dual-tone stimulation. Figure 13 shows a schematic of the measuring device.

The amplitude and phase of CMs in a given point on the surface of the cochlea depend on the intensity (*L*) and frequency (*f*) of the sound. When the frequency is fixed, the two CM potential parameters (amplitude and phase) depend on only parameter *L*. Typical changes in amplitude and phase over time registered at two different acoustic wave frequencies (260 and 8000 Hz) for the same guinea pig are shown in fig. 14. For this data, graphs of CM potential rms and phase depending on the level of sound intensity are shown in Fig. 15.

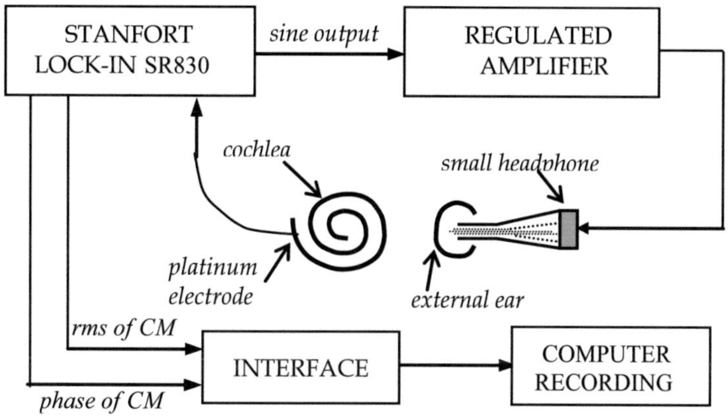

Fig. 13. Experimental setup for measuring CM potentials

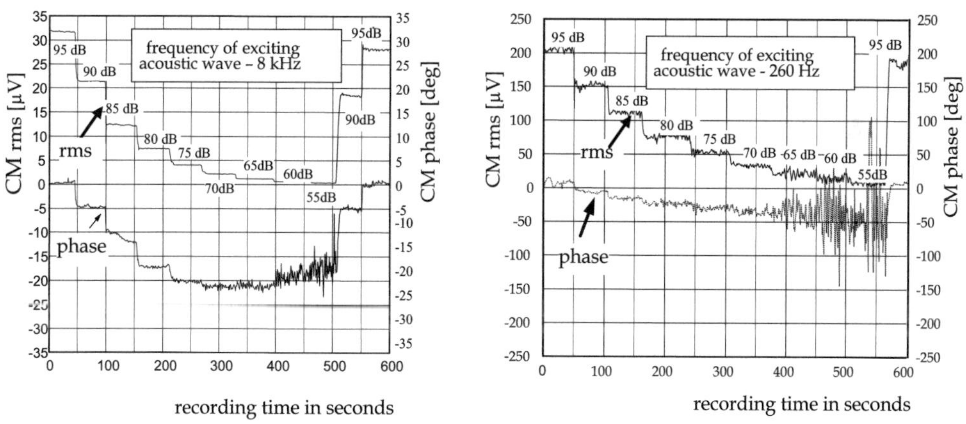

Fig. 14. Exemplary changes in CM rms and phase depending on sound intensity (sound levels were changed by 5 dB every 50 seconds)

Cochlear microphonic potentials are believed to be generated by the outer hair cells (OHCs). The latter are situated in three rows on the basilar membrane. All the OHCs have tiny strands (numbering about a hundred) called stereocillia. The apex of each single stereocillium lies in the tectorial membrane. In the resting state the stereocillia of each single cell form a conical bundle. During the acoustic excitation of the cochlea the stereocillia may dance about wildly. This alternating motion causes the channels in the stereocillia to open and close, providing a route for the influx of K^+ ions. The upper part of the OHCs acts as a resistor whose resistance changes according to the mechanical movements of the stereocillia. Changes in this resistance cause changes in extra-cellular currents. The measured CM potential is the result of the flow of extra-cellular currents through the input resistance of the lock-in amplifier.

The place theory suggests that a tone of a defined frequency excites mainly the OHCs located on the basilar membrane in a place specific for the given frequency (CF). The OHC electrical activity picked up from a given place on the cochlea surface is the vector sum of the extra-cellular currents generated by the particular OHC cells belonging to the given CF area (probably oval in shape). As the excitation wave intensity increases, extra-cellular currents are generated by an increasing number of OHC cells within the same CF area, which results in an increase in CM amplitudes. The phase changes registered then probably correspond to the shifts of the centre of the extra-cellular currents within the CF area.

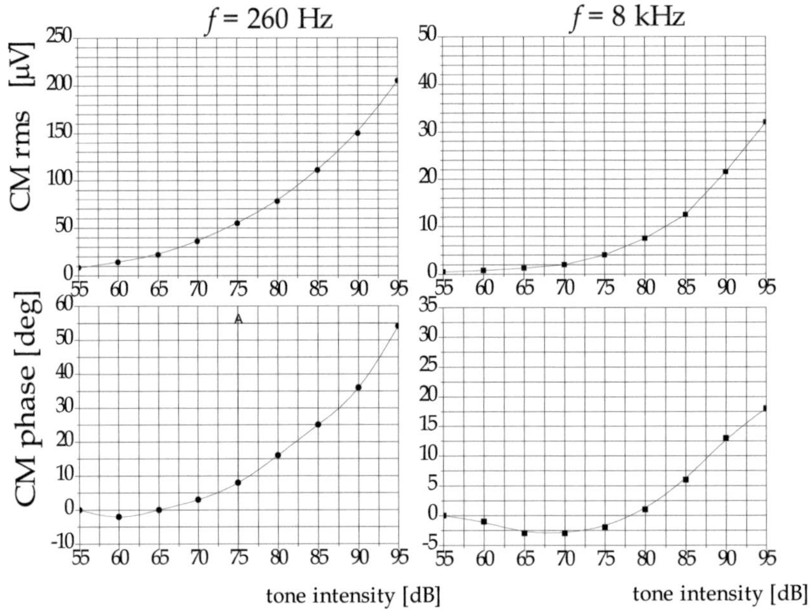

Fig. 15. Output-input characteristic obtained from traces shown in Fig. 14

3.6 Changes in amplitude and phase of CM potentials as result of laser irradiation

A focused laser beam can be a precise surgical scalpel. Perkins was the first to describe the use of a laser (an argon laser to be precise) in the surgical treatment of otosclerosis (Perkins, 1980). Since that time several kinds of laser (Ar, KTP, CO_2, Er) have been used in ear microsurgery. Vollrath and Schreiner were the first to use the rms of cochlear microphonics to estimate the effect of the argon laser beam on the electrical response of the cochlea in guinea pigs (Vollrath & Schreiner, 1982). The PSD technique enables the recording of the simultaneous changes in amplitude and phase of the CM potential during laser irradiation. The information about cochlear activity acquired in this way is more detailed.

Studies of the effect of Ar laser irradiation on the electrical activity of the cochlea have been described by us in several papers. We used the double PSD technique to record CM potentials prior to, during and after argon laser irradiation of the cochlea in guinea pigs. The goal of the studies was to determine safe laser parameters for argon laser stapedotomy, taking into account changes in not only the rms of CM potentials but also in their phase. In our experiments we used a CW argon laser with adjusted output power (0.1 - 3.0 W). An electronically controlled mechanical chopper was used to obtain laser light pulses differing in their parameters (the duration of a single laser pulse, the time interval between the successive pulses, the number of pulses in a series). Via a 200 μm optical lightguide the laser pulses would be delivered to the cochlear bone (near the round window) of an anaesthetized guinea pig with the surgically opened bulla. Exemplary traces selected from many different recordings are shown in fig. 16.

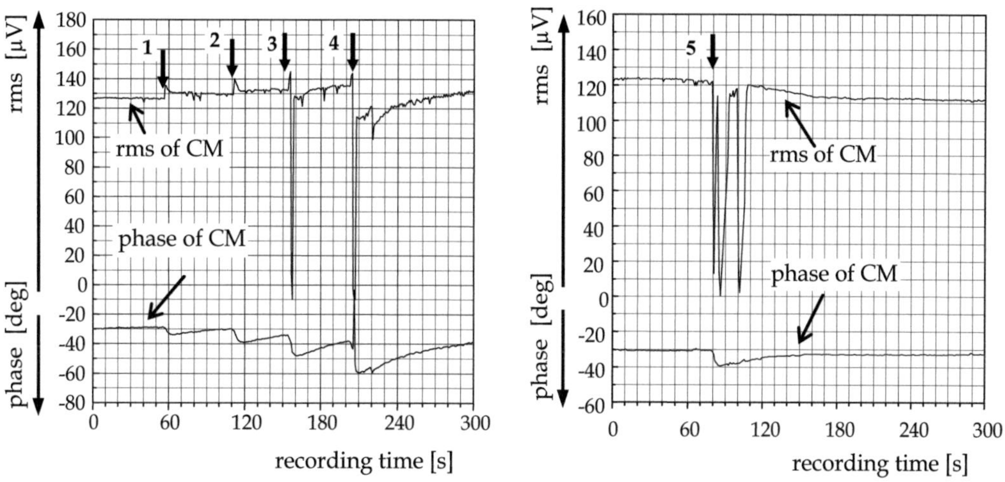

Fig. 16. Changes in rms and phase of CM potentials evoked by 80 dB acoustic wave of 1 kHz frequency during Ar laser pulse irradiation of 0.27 W (left) and 0.48 W (right) peak power. Irradiation parameters: 1 - single pulse of 0.5 s duration, 2 - single pulse of 0.5 s duration, 3 - single pulse of 1 s duration, 4 – two pulses of 1s duration with 1s interval between them, 5 – single pulse of 0.5s duration

It was found that in each registration the phase and amplitude of CM potentials changed during a laser pulse. The characteristic of the phase changes is always the same and diminishes relative to the initial (prior-to-irradiation) phase (in fig. 16 the initial phase was assumed to be equal to -30^0). The character of changes in CM rms depends on the peak power of the pulses used. Two characteristic peak power levels: P_1 and P_2 can be distinguished. When the peak power of the pulses is lower than P_1, laser irradiation results in a small increase in CM rms. This may be due to the slight increase in the temperature of the cochlea and to a biostimulating effect. After the first peak, but still below the second one (P_2), a sharp drop (even down to zero) in CM rms occurs. The drop is temporary and the cochlea quickly recovers its initial activity. Beyond P_2, changes in the electrophysiological activity of the cochlea are irreversible. As for today, the observed changes in the phase of CM potentials are hard to explain. It remains unknown why low-level laser radiation activates other groups of OHC cells in the CF area.

4. Double PSD technique in studies of DPOAE

4.1 Evoked otoacoustic emission

Evoked otoacoustic emissions (EOAE) are acoustic waves present in the external auditory canal after the cochlea is stimulated with an acoustic excitation wave. Depending on the excitation, different kinds of emission can be distinguished. If the stimulating signal is constant, then the emission is called *simultaneous evoked otoacoustic emission* (SEOAE). When pulse stimulating (clicks) sounds are used and the emission is registered between the clicks, the emission is called *transiently evoked otoacoustic emission* (TEOAE). If dual-tone stimulation (by two sinusoidal waves with respectively frequencies f_1 and f_2 and levels L_1 and L_2) is used, then the emission is called *distortion product otoacoustic emission* (DPOAE).

Otoacoustic emission was predicted by Gold as early as in 1948 (Gold, 1948). Thirty years later Kemp published a paper in which he described experiments proving the existence of this phenomenon (Kemp, 1978). He used clicks of 0.2 ms duration at a repetition rate of 16/s. In-between the successive pulses he recorded (with an electret microphone) acoustic wave pressure fluctuations at the outlet of the external acoustic canal. By applying an averaging procedure to the two-minute recordings he was able reduce the noise level to 0 dB SPL and reveal the backward signal which originated from the cochlea stimulated by the click. A few hundreds of works on this subject have been published since the first paper by Kemp. New experimental data are reported but their interpretations are not always explicit and mutually consistent. Despite the fact that the DPOAE mechanism is not yet fully understood, DPOAE signal estimation is a method of testing the human peripheral auditory function. The method is widely used in newborn hearing screening tests.

The presence of components which are absent in the stimulating acoustic wave is distinctive of DPOAE. The components result from the mechanical activity of the organ of Corti and are transmitted in the reverse direction through the middle ear and the tympanic membrane. Among the few possible products of cochlear nonlinearity, the acoustic wave $f_3 = 2f_1 - f_2$ is most widely examined because of its highest acoustic pressure level.

All the DPOAE acoustic waves are studied after their transduction into electric signals by a microphone. The microphone must be of high sensitivity and with a linear dynamic reserve (about 80 dB). The same requirements apply to the input preamplifier and the lock-in voltmeter amplifier since the measured DPOAE electrical signals cannot result from measuring system nonlinearity. The microphone placed in the external auditory canal transduces acoustic waves into electrical signals: both primary tones of 60-70 dB and reverse DPOAEs of 0 - 20 dB. Also floor noise occurs in the external ear canal. The apparatus used for measuring DPOAE must eliminate all undesirable signals with frequencies different than the frequency of the signal to be measured.

Otoemissions are examined after they have been converted in very accurate electric microphones. The biggest problem faced when examining DPOAEs is their extremely low level in comparison with the excitation waves. The difference may reach 30-60 dB. The phase-sensitive detection of DPOAE is therefore very useful. A basic experimental setup for measuring DPOAE signals is shown in fig. 17.

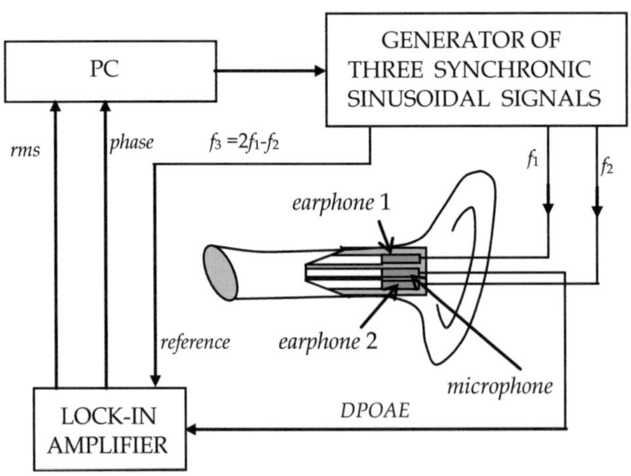

Fig. 17. Basic experimental setup for measuring DPOAE signals, using double PSD technique

The main unit of the experimental setup is a generator of three synchronous sinusoidal signals. The three frequencies are synchronized by a 18 MHz clock, whereby weak DPOAE signals can be measured using the PSD technique. The DPOAE response is measured by means of a probe which contains two miniature earphones and a low noise microphone. From the generator, pure tones with frequencies f_1 and f_2 are fed to the earphones. The two primary tones are digitally synthesized. The amplitudes, phases and frequencies of the tones are regulated by a PC with dedicated software. The software also enables the acquisition of the amplitude and phase of the DPOAE signals during measurements.

4.2 Previous techniques of measuring DPOAE

As mentioned earlier, DPOAE signals are of very low level, even if they are evoked in an unimpaired ear. Several signal processing techniques for the measurement of DPOAE signals under a large amount of noise and primaries of 60-70 dB have been developed. Initially, the Fast Fourier Transform (FFT) was used as the main signal processing tool for improving the signal-to-noise ratio in order to better estimate the level of DPOAE signals. In this method, the signals are first divided into data blocks and then averaged over time. For better reduction of the overall background noise long measurement time is required, which increases the amount of recorded data to be averaged. The FFT method requires about 10 seconds of block data. During long DPOAE recording, transient artefacts (e.g. talking, head movements) may occur, which when averaged together with the measuring signal may degrade the accuracy of the signal. Besides, the averaging method is incapable of measuring rapid changes of DPOAE signals.

In the first decade of the 21st century several novel methods of measuring DPOAE signals were developed (Ziarani & Konrad, 2004; Li et al., 2003). In comparison with the conventional methods, the new methods offer a shorter measurement time, which is of significance for clinical examinations. In addition, these methods are more immune to artefact and background noise. Thanks to the new methods it is possible to continuously record DPOAE signals. Besides offering the above advantages, the double PSD technique enables the simultaneous measurement of amplitude and phase of DPOAE signals. The two DPOAE parameters can be measured in a very short time, even below 10 ms.

In screening protocols typically a few pairs of primary tones with fixed acoustic levels are used and the responses are analyzed one after another (sequentially). In order to reduce the examination time the multiple-tone pairs method can be employed. In this method, DPOAE signals are evoked simultaneously by three or four pairs of two-tones. This method reduces measurement time but has a limited use .

4.3 DPOAE measurement using double PSD technique

Before the PSD technique was introduced to measure DPOAE signals it had been assumed that the amplitude and frequency of DPOAE signals depended on four acoustic parameters of the stimulating signals (primaries), i.e. the amplitude and frequency of each of the two signals. The DPOAE phenomenon itself is investigated according to the procedure described below. First the f_2/f_1 ratio (usually 1.22) and the stimulating signal intensity levels (e.g. L_2/L_1 = 60dB/65dB) are fixed. For the frequency of one of the stimulating waves (usually f_2) several discrete values are set while the frequency of the other wave is changed in small steps, e.g. 1/3 octave-bands centred around the fixed f_2 (Wagner at al., 2008). When examining the effect of different internal (e.g. age, gender) and external (e.g. industrial noise, medicines) factors, DPOAE is measured in the same stimulation conditions before and after the stimulus acts. Most experimental works in this field describe measurements of solely the amplitude of DPOAE signals. Some works also dealt with the phase of DPOAE signals, but it was measured in an indirect way, using signal processing methods. The phase-sensitive technique enables the simultaneous measurement of the amplitude and phase of DPOAE signals, with no need to use complex signal processing methods. The measurement takes place in real time.

Figure 18 shows an exemplary record of the simultaneous changes in the amplitude and phase of DPOAE signals caused by changes in the acoustic parameters of the stimulating waves. The recording was made in real time using the measuring setup shown in fig. 17. In the whole course of recording the combination frequency (f_3) remained constant at 3749 Hz while the other parameters were changed every 20 seconds in a specified sequence. The whole 980 second long recording time had been divided into seven 140 long time intervals in which parameter $k = f_2/f_1$ assumed the consecutive values: 1.10, 1.15, 1.20, 1.25, 1.39, 1.35, 1.40. The following seven combinations of stimulating wave levels: 1 - (55 dB, 55 dB), 2 - (55 dB, 60 dB), 3 - (60 dB, 55 dB), 4 - (60 dB, 60 dB), 5 - (65 dB, 60 dB), 6 - (60 dB, 65 dB) and 7- (65 dB, 65 dB) were fixed for each value of parameter k. In each of the combinations, the dB SPL of primary f_1 is in the first place. During the 980 second long recording the parameters of the primaries were changed 49 times in total.

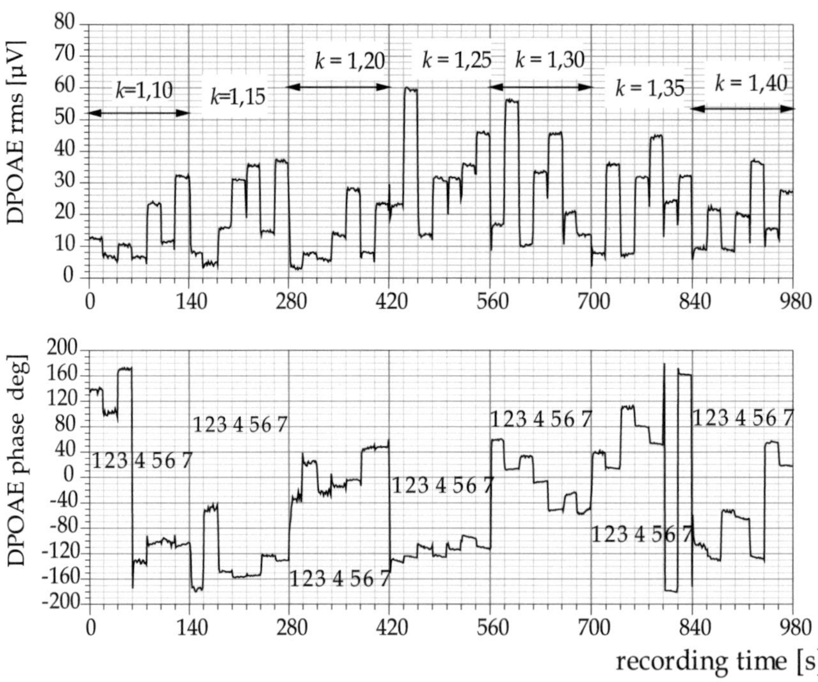

Fig. 18. Simultaneous changes in rms and phase of DPOAE signals, caused by fixed sequence of changes in parameters of primaries. Numbers 1 – 7 denote following combinations of primary frequencies L_{1dB} / L_{2dB} levels: 1 - (55 dB, 55 dB), 2 - (55 dB, 60 dB), 3 - (60 dB, 55 dB), 4 - (60 dB, 60 dB), 5 - (65 dB, 60 dB), 6 - (60 dB, 65 dB) and 7 - (65 dB, 65 dB). The same combinations of levels were used for each value of parameter $k=f_2/f_1$

The measurements showed that each change in the value of one of the parameters of the primaries results in a change of both the amplitude and phase of the DPOAE signal. Moreover, the character of the changes depends on the on the ontogenetic traits.

Thanks to the phase-sensitive technique one can determine the effect of the initial phase of each of the primaries on the amplitude and phase of DPOAE signals. For this purpose a generator of three synchronous sinusoidal signals was incorporated into the setup shown in fig. 17. The generator offers the possibility of fixing not only the amplitude and frequency of each of the primaries, but also the initial phase of each of the signals.

Five parameters of the primaries, i.e. the amplitude and frequency of each of the signals and the initial phase of one of the signals were fixed. The sixth parameter, i.e. the initial phase of the second primary was changed in a range of 0 – 360 degrees. The phase was changed in steps of 22.5 degrees. Exemplary measurements are shown in figs 19 and 21. Each of the figures comprises four panels. On the left side of each of the figures there are two panels showing experimentally determined changes in the amplitude and phase of DOPAE signals, caused by changes in the initial phase of primary f_1 (fig.19) or primary f_2 (fig(21). The data were obtained for combination frequency f_3 = 3749 Hz, parameter $k = f_2/f_1 = 1.25$, intensity levels L_1/L_2 = 65/55dB and the zero initial phase of primary f_1 (fig.19) or f_2 (fig.21).

The graphs in the panels on the right side of each of the figures were plotted on the basis of formulas (8) – (10), but the values of some of the constants in the formulas were matched to obtain agreement with the experimental traces.

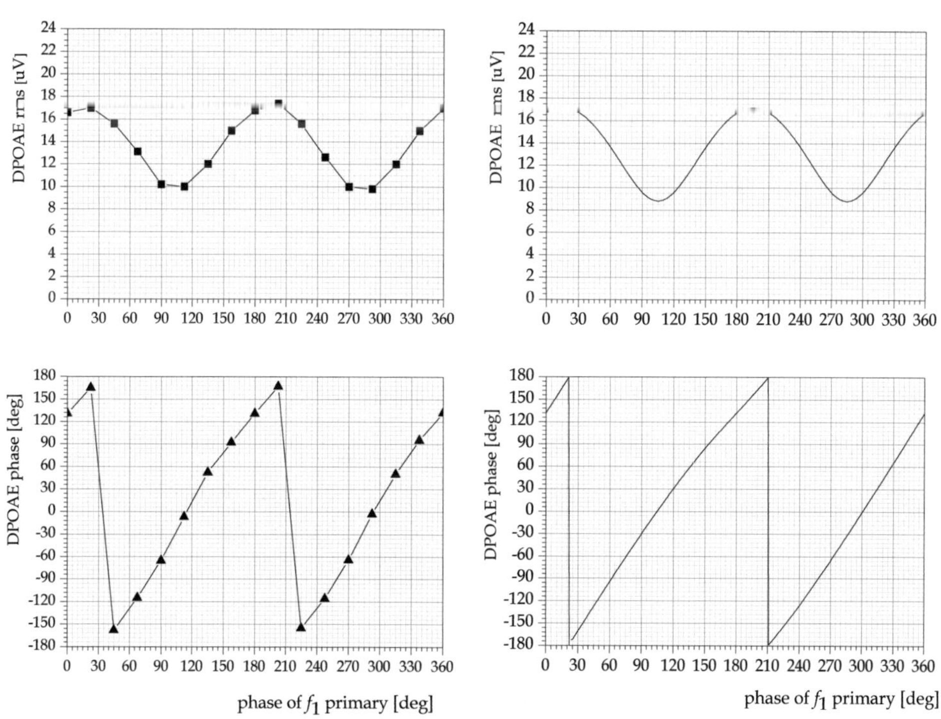

Fig. 19. Simultaneous changes in amplitude (upper panels) and phase (lower panels) of DPOAE signals, caused by changes in initial phase of primary f_1, obtained from experiment (left) and theoretically (right) (details in text)

Currently, it is generally believed that the DPOAE signal induced in the external acoustic canal by a double-tone is composed of two backward travelling waves (e.g. Knight &Kemp, 2000). The primary wave arises in the place where the two regions (CF_1 and CF_2) characteristic of frequency f_1 and f_2 overlap (but much more closer to CF_2). The wave propagates in the basilar membrane towards both the cochlea's base and its apex . The wave directed towards the apex bounces off in the region characteristic of frequency f_3 (CF_3) and propagates towards the base. Thus two waves with the same frequency f_3, but shifted in phase relative to each other, propagate towards the cochlea's base. Depending on the difference between the two waves, destructive or constructive amplitude interference occurs.

There are two different theories in the literature, concerning how the waves propagate backward from their generation places (He at all, 2007). According to one theory, the two waves propagate as compression waves to the cochlear base via the cochlear fluids. According to another theory, the two waves are transverse waves slowly propagating along the basilar membrane. Currently the prevailing view is that two backward waves, being transverse waves in the basilar membrane, arise in the cochlea excited by two tones. Taking into consideration this view and the previously determined sites where the backward waves arise, the schematic shown below (fig. 20) was drawn.

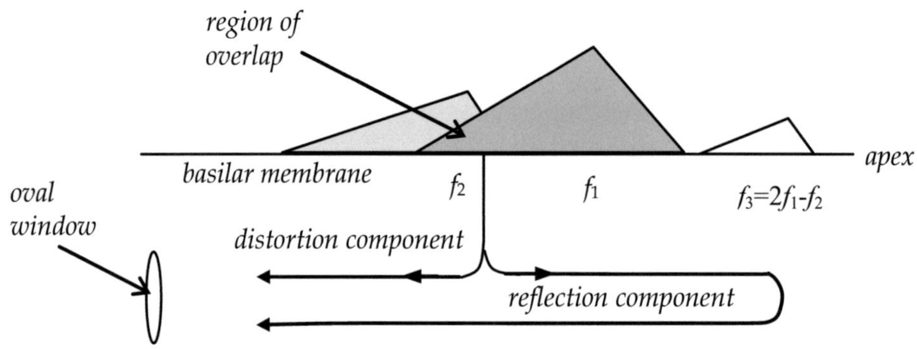

Fig. 20. Schematic diagram of source of two backward travelling waves whose interference produces DPOAE wave in auditory canal

The resultant wave near the oval window can be written as

$$A_w(t) = A_m \cos(\omega_3 t + \alpha_{10} + \kappa + 2\beta_1 - \beta_2), \tag{8}$$

where

$$A_m = A_{1m}\sqrt{1 + K^2 + 2K\cos\alpha_r}, \tag{9}$$

$$\kappa = \text{a}\tan\left(\frac{\sin\alpha_r}{K^{-1} + \cos\alpha_r}\right) \tag{10}$$

A_{1m} – the amplitude of the primary wave,
$K = A_{2m}/A_{1m}$,
α_{10} - the initial phase of the primary wave,

β_1, β_2 - the phases induced by the initial phases of the primaries,

α_r - a phase difference between the primary and secondary wave, due to the path length distance.

It follows from formula (9) that the amplitude of the resultant wave does not depend on the the initial phases of the primaries, and the phase of the resultant wave:

$$\Omega = \alpha_{10} + \kappa + 2\beta_1 - \beta_2 \tag{11}$$

(directly measured by the lock-in amplifier) is a linear function of the stimulating waves phase. However, experimental results do not corroborate the above dependence. The amplitude of the DPOAE signal turns out to be a function of the initial phases of the stimulating signals, and the measured phase is only approximately a linear function of the initial phases (panels on the left side of figs 19 and 21). If it is assumed that angle α_r changes in the same way as the initial phase of the primary, full agreement between the experimental traces and the ones determined from formulas (8) and (10) is obtained. This is shown in the panels on the right side of figs 19 and 21. The graphs on the right side of fig. 19 were plotted on the basis of formulas (8) and (10), assuming $\alpha_r = 2\beta_1$, while the graphs on the right side of fig. 21 were plotted assuming $\alpha_r = -\beta_2$.

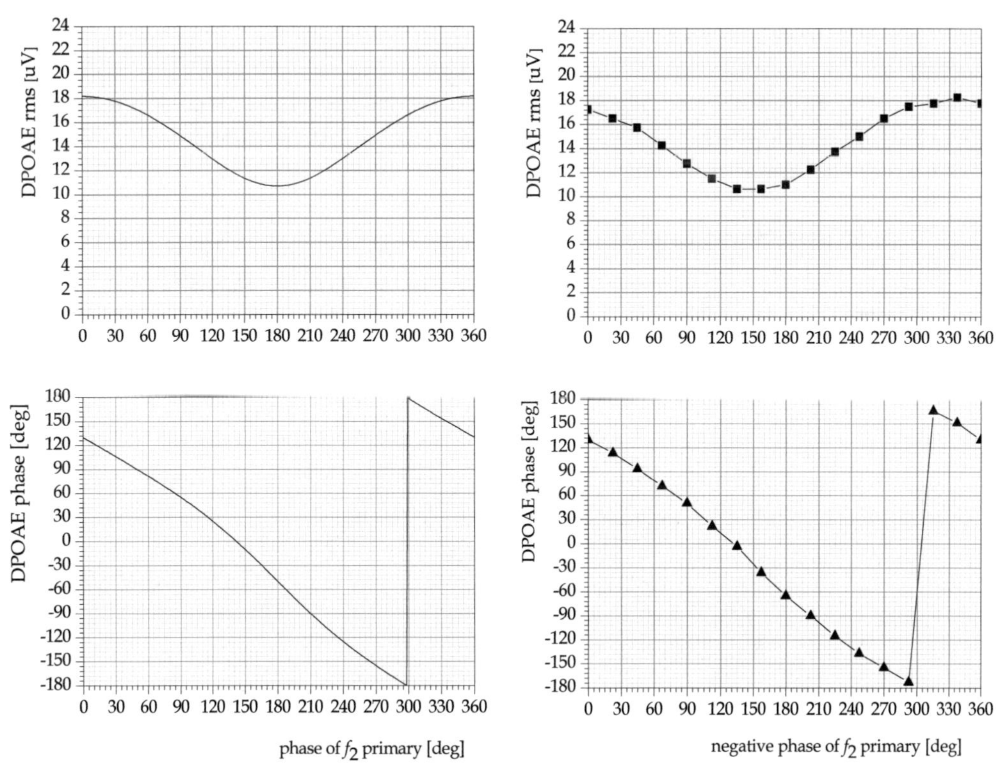

Fig. 21. Simultaneous changes in amplitude (upper panels) and phase (lower panels) of DPOAE signals, caused by changes in initial phase of primary f_2, obtained in experiment (left) and theoretically (right) (details in text)

It follows from formula (9) that the ratio of the maximum amplitude of the DPOAE wave (A_{wmax}) to the minimum value (A_{wmin}) amounts to

$$\frac{A_{w\max}}{A_{w\min}} = \frac{1+K}{1-K} \quad \Rightarrow \quad K = \frac{A_{2m}}{A_{1m}} = \frac{A_{w\max} - A_{w\min}}{A_{w\max} + A_{w\min}} \tag{12}$$

For the measurement conditions for which the traces shown in figs 19 and 21 were determined, it was calculated from formula (12) that K = 0.34 and K = 0.27 when respectively initial signal phase f_1 and f_2 is changed. This means that about 11.5% and 7.3% of the primary wave energy is reflected from region CF_3 in respectively the former and latter case.

The preliminary measurements shows that there is a certain mechanism in the Corti organ, which is responsible for the fact that a change in the phase of one of the stimulating signals (i.e. phase modulation) causes the amplitude modulation of the DPOAE signal. Further research is needed to explore this mechanism, but already at this stage one can say that the PSD technique proposed by the authors will play a major role in the exploration of this mechanism.

4.4 Simultaneous measurements of DPOAE and CMDP, using double PSD technique

Much of the experimental research reported in the world literature indicates that the main source of CM signals and DPOAE waves are OHCs. Thanks to the use of phase-sensitive detection in the measurement of each of the signals one can observe the simultaneous changes in the amplitude and phase of the two signals, resulting from changes in the parameters of the primaries. The measuring setup used for this purpose is shown in fig. 22. The setup incorporates two patents developed by the authors.

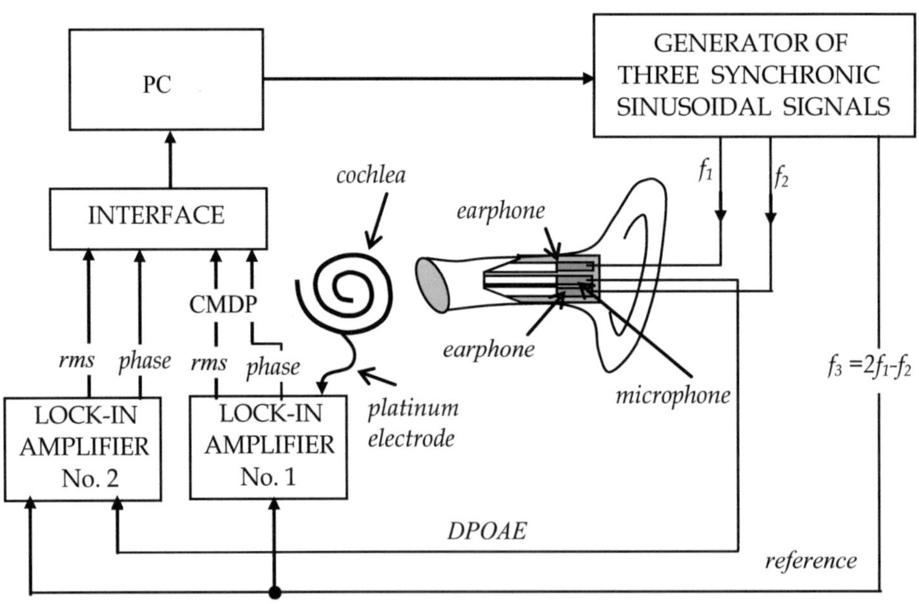

Fig. 22. Experimental setup for simultaneous measurement of amplitude and phase of DPOAE and CMDP signals

Two lock-in amplifiers, one for measuring the rms and phase of the DPOAE signal (amplifier No. 2) and the other for measuring the rms and phase of the CMDP signal (amplifier No. 1), have been incorporated into the setup. The CMDP signal is the distortion product in cochlear microphonics. The same signal (with combination frequency f_3) from the generator is fed to the reference inputs of each of the amplifiers. The reference input of lock-in No.2 can also be successively fed signals with stimulation frequencies f_1 and f_2 and combination frequency f_3 and the rms and phase of three CM signals with different frequencies can be measured. Measurements made in this way may provide a fuller picture of the cochlea functions.

The above setup was used to measure changes in the rms and phase of both DPOAE and CMDP signals, caused by changes in the excitation parameters. Four anaesthetized guinea pigs with the positive Preyer reflex were subjected to the experiments. Recordings were made at the following combination frequencies: 1312, 1875, 2671, 3749 and 5342 Hz. The parameters of the primaries were changed as in sect. 4.3, i.e. one of the parameters of the primaries was changed every 20 seconds in a specified way. Exemplary traces recorded for frequency f_3 = 1875 Hz are shown in fig. 23.

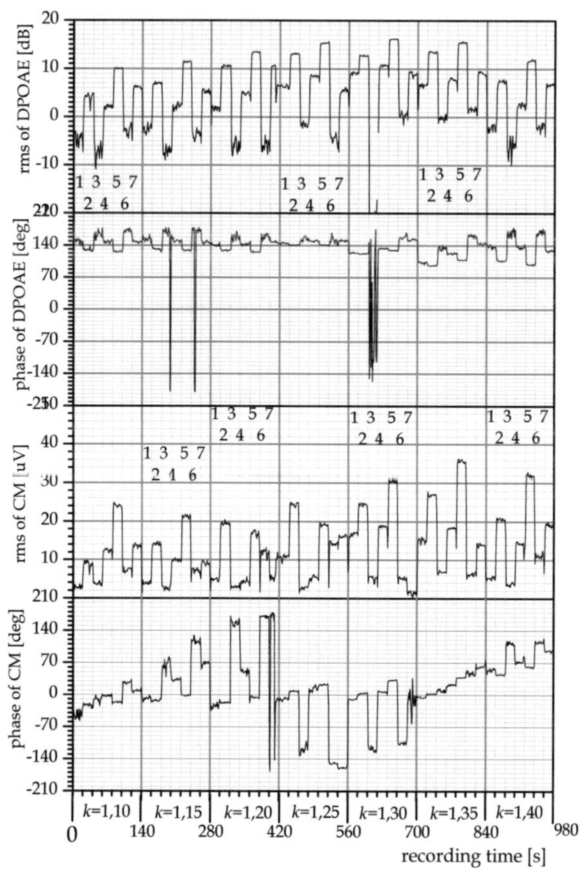

Fig. 23. Simultaneous changes in rms and phase of both DPOAE and CMDP, induced by changes in parameters of primaries

Nearly 100% correlation between the DPOAE rms and the CMDP rms was found, i.e. when after a change in one of the parameters of the primaries the DPOAE rms increased, then CMDP rms would also increase. Unfortunately, there was no such correlation between the phases of the two signals. This situations is well illustrated by the records of the changes, shown in fig. 23.

5. Conclusion

Practically all the ways of measuring biological acoustic waves in the cochlea, in which the phase-sensitive detection technique can be applied, have been described. Exemplary experimental results coming from many different measurement cycles carried out by the authors on guinea pigs in the last nearly 20 years were presented to demonstrate the measuring possibilities offered by the PSD technique. The latter's main advantage is that very weak (even below the ambient noise level) electrical signals can be measured in a very short time (in the order of milliseconds). A minor limitation of this technique is that it is applicable to objects to whose input periodical signals are fed from the outside.

In many investigations into the electrophysiological function of the cochlea it is essential not only to simultaneously measure the amplitude response, but also the phase response to the stimulation. This is undoubtedly another advantage of the PSD technique.

Much more difficult than the measurement of the phase is the interpretation of its changes. As for now, it is not always possible to interpret the observed changes, which particularly applies to DPOAE. This phenomenon has been known for over 30 years, but it still has not been fully explored. The great worldwide interest in this subject is reflected in the large number of publications devoted to it. The interest stems from the fact that for many years DPOAE measurements have been part of hearing screening tests during which DP-grams are recorded. This especially applies to newborns and people with mental disabilities, in which cases it is impossible to record audiograms. Besides gaining an insight into the nature of the DPOAE phenomenon, it is essential to determine the correlation between the DP-gram and the audiogram. In the authors' opinion, the phase-sensitive detection technique represents a new tool for investigating electrophysiological phenomena in the cochlea and it will contribute to the better understanding of the phenomena taking place in this organ.

6. References

Bredberg, G., Lindeman, H. H., Ades H. W., West R. & Engstrom H. (1970). Scanning electron microscopy of the organ of Corti. *Science*, Vol. 170, No.960, pp. 861-863, ISSN Print 0036-8075

Brown, D.J., Hartsock, J. J., Gill, R.M., Fitzgerald, H.E. & Salt, A.N. (2009). Estimating the operating point of the cochlear transducer using low-frequency biased distortion products. *J. Acoust. Soc. Am.*, Vol. 126, No. 4, pp.2129 – 2145, ISSN Print 0001-4966

Carricondo F., Sanjuan-Juaristi J., Gil-Loyzaga P. & Poch-Brotto Joaquin. (2001). Cochlear microphonic potentials: a new recording technique. *The Annals of otology, Rhinology & Laryngology*, Vol.110, No. 6, pp. 565 – 573, ISSN Print 0003-4894

Castelo Branco (1999). The clinical stages of vibroacoustic disease. *Aviation, Space, and Environmental Medicine* , Vol. 70, Suppl 3, A32-39, ISSN 0095-6562

Davis, H. (1983). An active process in cochlear mechanics. *Hearing Research*, Vol. 9, No,1, pp. 79-90, ISSN Print 0378-5955

Gale, J. A. & Ashmore J. F. (1997). An intrinsic frequency limit to the cochlear amplifier. *Nature*, Vol. 389, No.6646, pp.63-66, ISSN 0028-0836

Gold, T. (1948). The physical basis of the action of the cochlea. *Proceedings the Royal of Society*, Vol.135, No.881, pp. 492-498, ISSN 1471-2954

Hamernik, R.P., Henderson, D., Coling D. & Slepecky N. (1980). The interaction of whole body vibration and impulse noise. *J. Acoust. Soc. Am.*, Vol. 67, No.3, pp. 928-934, Hamernik, R.P., Henderson, D., Coling, D. & Salvi R. (1981). Influence of vibration on asymptotic threshold shift produced by impulse noise. *Audiology*, Vol.20, No.3. pp.259-269, ISSN Print 0020-6091

Hamernik, R.P., Henderson, D., Coling, D. & Salvi R. (1981). Influence of vibration on asymptotic threshold shift produced by impulse noise. *Audiology*, Vol. 20, No.3. pp.259-269, ISSN 1499-2027

He, W., Nuttall, A.L. & Ren, T. (2007). Two-tone distortion at different longitudinal locations on the basilar membrane. *Hear. Res.*, Vol. 228, No.1-2, pp.112-122, ISSN Print 0378-5955

Jankowski, W. Giełdanowski, J. & Birecki W. (1962). Effect of some vasoconstrictor drugs on the microphonic potential of the cichlea. *The Polish OtoLaryngology*, Vol.16, pp. 321 – 329, PL ISSN 0030-6657

Jones, C.M.(1996). ABC of work related disorders. Occupational hearing loss and vibration induced disorders. Occupational. *Br Med J* , Vol. 313, No. 7051, pp. 223-226, ISSN 0959-8138

Kemp, D.T. (1978). Stimulated acoustic emissions from within the human auditory system. *J. Acoust. Soc. Am.*, Vol. 64, No.5, pp. 1386-1391, ISSN Print 0001-4966

Knight, R.D. & Kemp D.T. (2000). Indications of different distortion product otoacoustic emission mechanisms from a detailed f_1, f_2 area study, *J. Acoust. Soc. Am.*, Vol. 107, No.1, pp. 457-473, ISSN Print 0001-4966

Kobayashi, T., Rong Y., Chiba T., Marcus C.D., Ohyama K. & Takasaki T. (1997). Ototoxic effect of erythromycin on cochlear potentials in the guinea pig. *Ann. Otol. Rhinol. Laryngol*, Vol.106, No.7, pp. 599-603, ISSN Print 0003-4894

Li, X., Wodlinder, H. & Sokolov Y. (February 2003). A new method for measuring DPOAEs and ASSRs, Available from http://www.hearing review.com/issues/articles/2003-02 05.asp

Linder, T.E., Zwicky, S. & Brändle P. (1995). Ototoxicity of ear drops: a clinical perspective. *Am. J. Otol.*, Vol. 16, No. 5, pp. 653-657, ISSN Print 0192-9763

Palmer, K.T., Griffin M.J., Bendall H., Pannett B. & Coggon D.(2000a). Prevalence and pattern of occupational exposure to hand transmitted vibration in Great Britain: findings from national survey. *Occup. Environ. Med.* , Vol. 57, No.4, pp. 218-228, ISSN Print 1351-0711

Palmer, K.T., Griffin M.J., Bendall H., Pannett B. & Coggon D. (2000b). Prevalence and pattern of occupational exposure to whole body vibration in Great Britain: findings from national survey. *Occup Environ Med*, Vol. 57, pp. 229-236, ISSN Print 1351-0711

Perkins, R. (1980). Laser stapedotomy for otosclerosis. *Laryngoscope*, Vol.90, No.2, pp.228-241, ISSN 1531-4995

Rogowski, M. & Chodynicki, S. (1987). Einfluβ von Vibrationen und Gentamycin auf das Gehörorgan des Meerschweinchens. *HNO-Prax* , Vol. 12, pp.219-223

Seidel, H., Heide R. (1986). Long-term effects of whole-body vibration: a critical survey of the literature. *Int Arch Occup Environ Health*, Vol.58, No.1, pp. 1-26, ISSN Print 0340-0131

Tasaki, I., Davis, H. & Legouix J.P. (1952). The space-time pattern of the cochlear microphonics (guinea pig) as recorded by differential electrodes. *J. Acoust. Soc. Am.*, Vol. 24, No.5, pp. 502 - 519, ISSN Print 0001-4966

Thurlow, W.R. (1943). Studies in auditory theory:II The distortion of distortion in the inner ear. *Journal of Experimental Physology: General*, Vol.32, No.4, pp. 344 - 350, ISSN 0096- 3445

Wadsten C.J., Bertilsson C.A., Sieradzki H. & Edström S. (1985). A randomized clinical trial of two topical preparations (framycitin/ gramicidin and oxytetracycline /hydrocortisone with polymyxin B) in the treatment of external otitis. *Arch Otorhinolaryngol* Vol. 242, No. 2, pp. 135-139. ISSN Print 0302-9530

Wever, E.G. & Bray, C.W. (1930). Auditory nerve impulses, *Science*, Vol.71, No.1834, p.215, ISSN Print 0036 8075

Vollrath, M., Schreiner, Chr.(1982). Influence of argon laser stapedotomy on cochlear potentials I: Alteration of cochlear microphonics (CM). *Acta Otolaryngol*, suppl **385**, pp.1-31, ISSN Print 365-5237

Ziarani, A.K. & Konrad A. (2004). A novel metod of estimation of DPOAE signals. *IEEE Transactions on biomedical engineering*, Vol.51, No.5, pp.864-868, ISSN 0018 9294

Ziemski, Z. (1970). Ototoxity of selected organic solvents of industrial plastics in experimental animals. *Papers of Medical University in Wroclaw*, Vol.15, No.1, pp.59-128

13

Photoacoustic Technique Applied to Skin Research: Characterization of Tissue, Topically Applied Products and Transdermal Drug Delivery

Jociely P. Mota, Jorge L.C. Carvalho, Sérgio S. Carvalho and Paulo R. Barja
UNIVAP
Brazil

1. Introduction

The photoacoustic (PA) effect basically consists in the production of acoustic waves due to the absorption of modulated (or pulsed) radiation by a sample. Graham Bell discovered the PA effect in 1880, when he noticed that the incidence of modulated light on a diaphragm connected to a tube produced sound. Thereafter, Bell studied the PA effect in liquids and gases, showing that the intensity of the acoustic signal observed depended on the absorption of light by the material.

In the nineteenth century, it was known that the heating of a gas in a closed chamber produced pressure and volume changes in this gas. However, there were many different theories to explain the PA effect. Rayleigh said that the effect was due to the movement of the solid diaphragm. Bell believed that the incidence of light on a porous sample expanded its particles, producing a cycle of air expulsion and reabsorption in the sample pores. Both were contested by Preece, who pointed the expansion/contraction of the gas layer inside the photoacoustic cell as cause of the phenomenon. Mercadier explained the effect conceiving what we call today *thermal diffusion mechanism*: the periodic heating of the sample is transferred to the surrounding gas layer, generating pressure oscillations.

The lack of a suitable detector for the PA signal made the interest in this area decline until the invention of the microphone. Even then, research in this field was restricted to applications in gas analysis up to 1973, when Rosencwaig started to use the PA technique in spectroscopic studies of solids and, together with Gersho, developed a mathematical model for the generation of the PA signal in solid samples - the Rosencwaig-Gersho (RG) Model (Rosencwaig & Gersho, 1976).

In condensed matter samples, one of the most important mechanisms for PA signal generation is the thermal diffusion, classically described by the RG model. According to this model, the (modulated) radiation absorbed by condensed matter samples is converted into heat, causing temperature modulation in the surrounding atmosphere. This eventually produces the mechanical effect of periodic expansion and contraction originating sound waves that can be detected by a microphone.

Since the publishing of the RG model and, soon after that, of the generalized theory for the PA effect by McDonald and Wetsel (1978), the PA technique has already proved its relevance in a large number of very different fields, from the polymerization of dental resins (Balderas-Lopez et al., 1999) to photosynthesis studies (Malkin & Puchenkov, 1997; Herbert et al., 2000).

1.1 Objectives

The purpose of this chapter is to present applications of the PA technique in skin research, both in the characterization of skin itself and in transdermal drug delivery studies. The basic experimental setup for such studies will be briefly presented, aiming to help those who may be interested in developing similar studies. Emphasis will be done to *in vivo* measurements, because of its importance in this field. Our objective is to show the usefulness of the PA technique in the biomedical field, particularly in skin research; finally, perspectives for future work in this field will be presented.

2. Photoacoustic measurements

2.1 Basic experimental setup

Figure 1 presents one scheme for a basic photoacoustic experimental setup.

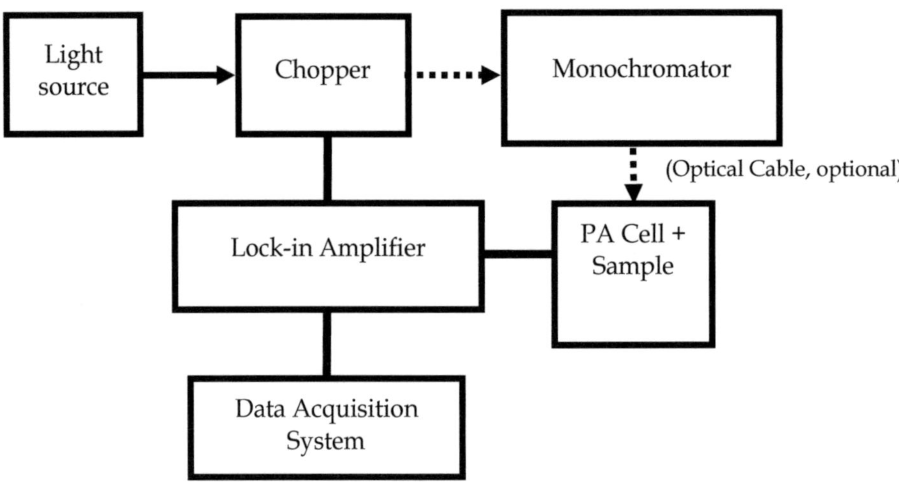

Fig. 1. Example of a basic photoacoustic experimental setup (scheme)

The experimental scheme in Figure 1 shows a (typically mechanical) chopper positioned in front of the light source, in order to modulate the radiation that comes into a monochromator (utilized in PA spectrocopy measurements). Light absorbed by the sample generates acoustic waves inside the PA cell; the PA signal is captured by a microphone (inside the PA cell) that sends it to the lock-in amplifier (also connected to the chopper, to receive information on the frequency modulation). The lock-in amplifier is connected to a microcomputer for data acquisition. In vivo, skin measurements are performed with an open-ended PA cell, in which it is the sample itself that closes the chamber.

2.2 Measurements as a function of time

The PA signal depends on the optical and thermal properties of the sample, which may vary with time due to different factors. When a sample undergoes changes in its composition or structure (as it occurs during the polimerization of a dental resin, for instance), the propagation of heat inside the sample is also modified, thereby altering the PA signal.

We must also mention the possibility of performing photosynthesis studies using PA measurements as a function of time (W.J. Silva et al., 1995). When PA measurements are performed in photosynthesizing samples as plant leaves, the PA signal presents, in addition to the photothermal component, a photobaric component, resulting from the gas exchanges associated to the photosynthesis process (Acosta-Avalos et al., 1996). This allows the study of the so-called photosynthetic induction, that is, the increase of the net photosynthetic rate that occurs when a plant is shifted from darkness to light (Sui et al., 2011).

As stated by Bodzenta et al. (2002) in their work on PA detection of drug diffusion into a membrane, PA measurements give the possibility for investigations in relatively long time periods. This makes the PA technique suitable for the monitoring of dehydration processes (Lopez et al., 2005) and of changes occurring in time in biological tissues such as skin. It is possible to study, for example, the kinetics of transdermal drug delivery through the analysis of PA measurements as a function of time. One example will be presented at the section 4 of the present chapter.

2.3 Studies on the modulation frequency: depth profile

In thermally thick samples (as skin tissue), only the light absorbed within the first thermal diffusion length (μ_T) of the sample/tissue contributes to the PA signal (Rosencwaig, 1980). As the thermal diffusion length depends on the modulation frequency (f) of the incident light by the relation

$$\mu_T = \sqrt{\frac{\alpha}{\pi f}} \qquad (1)$$

where α is the thermal diffusivity of the sample, it is possible to perform depth-profile studies, with the evaluation of the penetration depth of a product (or even a microorganism) in tissue. The possibility of performing depth-profile studies is particularly interesting in the characterization of multilayer systems (as skin itself).

The frequency dependence analysis of the PA signal can also be employed in the determination of the thermal properties (thermal diffusivity, thermal effusivity) of a sample or material (Balderas-Lopez & Mandelis, 2001), including biological tissues as porcine skin (Gao et al., 2005; Qiu et al., 2008).

2.4 Measurements as a function of the wavelength: Photoacoustic spectroscopy

Photoacoustic spectroscopy (PAS) is already incorporated to the roll of useful photothermal techniques since the 1980s (Rosencwaig, 1980; Vargas & Miranda, 1988). Besides the possibility of rendering depth-profile analysis in multi-layered samples, PAS presents at least two additional advantages over other spectroscopy techniques: i) as transmitted and reflected light do not interfere in PAS measurements, it is a "more direct" technique, representing a direct measurement of the light absorption by the sample; ii) it allows the study of optically opaque and highly scattering samples (which could not be analyzed by conventional optical spectroscopy).

In PAS measurements, the emission spectra of the light source is typically obtained through measurements using black carbon powder (or other black material) as the sample, with all the remaining measurements being normalized with respect to the lamp spectrum.

PAS can also be employed in skin research. In 2004, Benamar and co-workers presented a PAS study on the effect of dihydroxyacetone, frequently employed for artificial tan. Measurements were carried out in the presence and absence of dimethylisosorbide (a solvent for dihydroxyacetone), on excised human skin. By monitoring the PAS signal intensity with time in the UV (300-400nm) range, these authors demonstrated that dihydroxyacetone in combination with dimethylisosorbide enhances the process of tanning (Benamar et al., 2004).

Recently, Melo et al. (2011) applied PAS to evaluate the penetration rate of *Helicteres gardneriana* extract, topically applied for anti-inflammatory purposes. Experiments were conducted *ex vivo* in mice. Croton oil was applied into both mouse's right and left auricles to induce inflammatory response, and the left auricle was treated with the extract. The strong anti-inflammatory effect observed for the *Helicteres gardneriana* extract was associated with the deep percutaneous penetration observed for the formulation, according to PA data (Melo et al., 2011).

2.5 Photoacoustic imaging and tomography

Photoacoustic imaging is based on the production of acoustic waves following irradiation by a short pulse of light whose absorption generates local heating and transient thermoelastic expansion (Balogun et al., 2009). According to Beard (2009), haemoglobin "represents the most important source of endogenous contrast" in PA imaging. This makes the technique particularly indicated to studying tissue abnormalities as tumors and other diseases related to changes in the structure and oxygenation status of the vasculature (Beard, 2009).

Recently, Hu and Wang (2010) presented "PA tomography" as a method combining high spatial resolution and optical absorption contrast, important in microvascular imaging and characterization. Reviewing the "major embodiments of PA tomography" (microscopy, computed tomography and endoscopy), they have analyzed the methods employed in different studies, including hemodynamic monitoring, determination of hemoglobin concentration, evaluation of oxygen saturation, studies of blood flow and tumor-vascular interaction.

Besides being applied to soft tissues, PA imaging can also be employed to hard tissues. Li and Dewhurst (2010) have applied a PA imaging system with a near-infrared (NIR) pulsed laser to obtain images from both soft tissue and post-mortem dental samples. They have also performed simulations (based on the thermoelastic effect) to predict initial temperature and pressure fields within a tooth sample, observing that values are maintained below the corresponding safety limits. In this way, the results presented by Li and Dewhurst show that the PA technique can be sucessfully applied to image both soft and hard tissues.

3. Photoacoustic measurements and the characterization of skin

Biological materials are sometimes difficult to study employing conventional techniques that require previous preparation of the samples, because these materials can have its properties significantly altered by preparation processes as solubilization, for example. The PA technique does not require previous preparation; it can be described as a non-invasive technique that allows even *in vivo* measurements.

In general, biological tissues can be characterized as highly scattering samples; however, this is not a problem for PA mesurements, in which the signal is based in the direct absorption of radiation. As pointed by Cahen and co-workers (1980), "the relative insensitivity to scattered light of the PA signal makes such measurement an attractive way to measure biological samples *in vivo*". These features explain the potential of the PA technique in the study of opaque materials and complex biological systems such as skin. PA measurements can be employed to determine the absorption characteristics of the skin itself or topically applied products, as well as kinetic changes related to transdermal drug delivery.
Skin diseases can also be studied through PA measurements. In 2010, Swearingen et al. developed a PA methodology to determine the nature of skin lesions (pigmented and vascular) *in vivo*, which is important because misdiagnosis may even lead to cancerous lesions not receiving proper medical care. These authors irradiated skin with two laser wavelengths (422 and 530nm), with the relative response at these two wavelengths (422nm/530nm) indicating whether the lesion is pigmented or vascular, due to the distinct absorption spectrum of melanin and hemoglobin (Swearingen et al., 2010).

3.1 Skin type classification

Skin type classification is important not only for medical or clinical purposes, but also for pharmaceutical and cosmetic industries, following the idea that an objective, precise characterization of skin could be useful in the design of new topically applied products and in defining more specific skin treatments according to each skin type.
However, in dermatology, there is still no universal agreement about the best method for classifying skin, as even the widely accepted method proposed by Fitzpatrick (1988) - defining the so-called "skin phototypes" - is based in clinical, subjective analysis.
More recently, Baumann (2006a, 2006b) proposed a new skin type classification, according to which 16 different skin types are defined from the combination of four parameters, as skin can be characterized as: i) pigmented or nonpigmented; ii) dry or oily; iii) sensitive or resistant; and iv) wrinkled or tight. Baumann´s skin typing is based on an extensive research, performed with 1400 volunteers. However, it relies essentially on the response of volunteers to a questionnaire; therefore, it does not fulfill "per se" the need of an objective classification, which would require experimental evaluation.
PA measurements have a potentially important role to play in an experimental approach to skin type classification. In 2000, Schmidt and co-workers conducted non-contacting, *in vivo* PAS measurements in skin (performed in 50 volunteers), in the VIS-NIR range, seeking an objective determination of pigmentation, blood microcirculation and water content of human skin (Schmidt et al., 2000). According to these authors, strong spectral variations observed within the same skin type are probably based on the natural variability of human skin and in the subjective clinical evaluation of the skin type; nevertheless, PAS results obtained show good correlation between PA data and (clinically evaluated) skin type, indicating that skin type determination could indeed be performed through the analysis of PA measurements.

3.2 Skin pigmentation analysis employing photoacoustic measurements

In 2004, Viator and co-workers proposed a method for the determination of the epidermal melanin content employing a PA probe using a Nd:YAG (neodymium, yttrium, aluminum, garnet) laser at 532nm (Viator et al., 2004). Ten human subjects with skin phototypes I–VI were tested using the PA probe and visible reflectance spectroscopy (VRS); melanin content was evaluated through each of these methods, and a good linear fit (r^2=0.85) was obtained for the plot of PA x VRS.

Pigmentation skin level can also be evaluated through simple, direct PA measurements employing non-laser light sources. Actually, PA measurements have been performed at the Laboratory of Photoacoustic Technique Applied to Biological Systems (FASBio), at UNIVAP (Brazil). The objective of such *in vivo* measurements was to classify different skin types according to the amplitude of the PA signal, which can be associated to the corresponding pigmentation level of the skin.

In the following subsections, we present this straightforward PA approach to skin characterization according to the level of pigmentation, employing PA measurements in volunteers. Experimental results are compared both to Fitzpatrick and Baumann clinical skin type evaluations.

3.2.1 Materials and methods

The PA setup employed in such *in vivo* skin measurements consisted of a 250W tungsten halogen lamp as light source (with wavelength range 400nm$<\lambda<$700nm and light intensity of about 20W/m^2), a mechanical chopper (SRS, model SR540), a lock-in amplifier (SRS, model SR530) and a microcomputer for data acquisition.

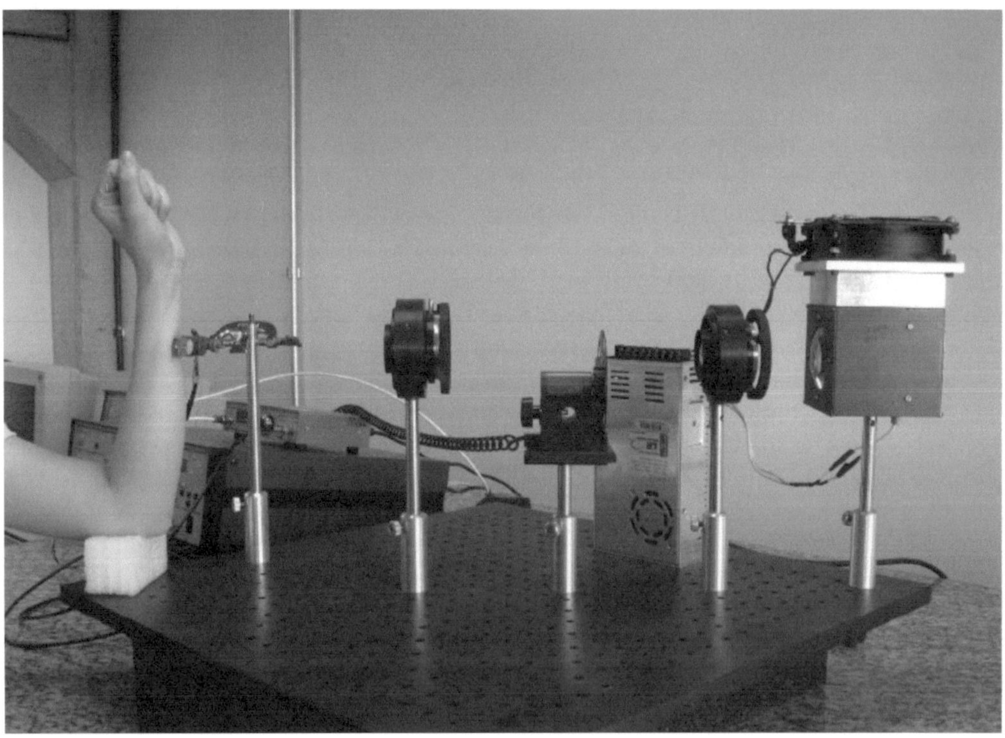

Fig. 2. Experimental setup with volunteer positioned for *in vivo* skin measurement

The double faced PA cell employed, with an electret microphone, was developed at UNIVAP; sensitivity was 15 mV/Pa for the frequency employed in skin measurements,

17Hz. The electret microphone structure was described by Marquezini et al. (1990). The PA cell has a cylindrical body and two opposite, parallel faces (one is closed by a thin glass layer and the other, by the sample itself). For the modulation frequency employed (17Hz), the thickness of the skin layer under study is about 30μm.

PA measurements were recorded as a function of time (200 readings for each measurement, in 0.5s intervals, up to a total of 100s per measurement). During measurements, one face of the PA cell was closed with a thin transparent window, while the forearm of the volunteer was gently pressed against the opposite face.

Figure 2 shows the PA experimental setup employed for in vivo skin measurements at the FASBio/UNIVAP, with a volunteer positioned for measurement.

Measurements were performed in 57 female volunteers, between 20 and 30 years-old. Initially, each volunteer answered a questionnaire according to their daily routine associated to skin care; volunteers were also clinically evaluated and, as a result, they were classified according to skin phototype, following Fitzpatrick classification (Fitzpatrick, 1988).

Before measurements, the skin area to be evaluated was cleaned with cotton embedded in alcohol 70%. The PA signal was then recorded for the inner and outer faces of both forearms. Volunteers were then classified according to the respective PA signal amplitude, and this classification was compared to the phototype classification.

3.2.2 Results and discussion

Initially, a comparison between the PA signal amplitude of the inner and outer faces was performed, showing a highly significant statistical difference (paired t-test, $p<0,005$), with higher PA amplitude being observed for the outer face of the forearm. This result can be attributed to the higher pigmentation level of the skin region continuously exposed to solar radiation, demonstrating that skin constitution and aspect are clearly influenced by the level of sun exposure.

After clinical evaluation of the volunteers for skin phototype (following Fitzpatrick classification), PA results were grouped according to the phototype of each volunteer. Results are presented in Table 1.

Skin phototype (Fitzpatrick)	PA signal amplitude (mV)
II	1.26 ± 0.05^{a}
III	1.59 ± 0.09^{ab}
IV	1.70 ± 0.10^{ab}
V	1.80 ± 0.10^{b}

Table 1. PA signal amplitude (mV) for the inner face of the forearm, for each skin phototype (average ± standard error). Different indexes (a, b) indicate significant statistical difference (comparison among groups: $p=0.009$, ANOVA)

Comparison among phototype groups was performed and significant statistical difference was verified (as we can see in Table 1), showing that the PA signal level (amplitude) for the inner forearm tends to scale with skin phototype, as defined by Fitzpatrick.

Afterwards, the PA signal amplitude for each volunteer (average values for the inner face of the forearm) allowed the division of the volunteers in two groups, "pigmented" (P) and "non-pigmented" (NP), following the Baumann proposal. As the average PA signal amplitude obtained for all measurements (inner face of the forearm) was 1.5mV, this was the cutoff value adopted for separating the volunteers into "P" (for PA signal amplitude above 1.5 mV) and "NP" (under 1.5 mV). Table 2 shows the division of each (clinically evaluated) phototype group into the (experimentally evaluated) P and NP groups. In this way, the PA technique allowed the comparison between two different skin classification forms.

Skin phototype	NP group (%)	P group (%)
II	93	7
III	46	54
IV	31	69
V	20	80

Table 2. Distribution of the volunteers of each phototype in the NP and P groups, according to the PA signal level

Table 2 shows that phototype II is highly related to the NP group, while phototypes IV and V concentrate in the P group. Phototype III appears in both groups, showing the variability of elements inside this classification.

The simple methodology presented here and the corresponding results obtained open perspectives for an objective, experimental classification of skin types, based upon PA measurements. Additional work in this field is currently being performed at FASBio/UNIVAP (Brazil).

4. Transdermal drug delivery

4.1 Topical application of drugs: advantages and requirements for evaluation

Topical application of drugs is known as an interesting alternative route to oral and intravenous administration, both aiming to systemic effects and local action, offering advantages such as ease of administration and lack of first-pass effect (Aqil et al., 2007). However, in studies of transdermal drug delivery (penetration of substances into skin), one must employ non-invasive techniques, in order to avoid second order effects that would at least bring difficulties to the interpretation of the results. *In vivo* measurements are particularly desirable in such studies, because the response of excised skin can be affected by dehydration, and the response of artificial skin differs significantly from that of *in vivo* skin at least in some cases - as when we talk about in-depth processes, in which even blood circulation may play a role.

As mentioned earlier in the present chapter, the PA technique can be applied without previous preparation of the samples and even for *in vivo* measurements; as such, transdermal drug delivery studies have been performed employing the PA technique in order to obtain the penetration rate of a wide range of different products topically applied to skin (Bernengo et al., 1998; Hahn et al., 2001; Savateeva et al., 2001; Pedrochi et al., 2005; Truite et al., 2007).

4.2 Substrates for transdermal drug delivery studies

Besides measurements in human skin, penetration rates of topically applied products are frequently evaluated through measurements performed in animal skin tissue. In this case, rabbit and pig skin are, by far, the most employed alternatives, because of the similarity to human skin.

As pointed by Simon & Malbach (2000), physiological and anatomical similarities between man and pig make this animal a good model for man in biomedical research. The correlation of quantitative data between pig skin and human skin can be frequently classified as very good (Benech-Kieffer et al., 2000); therefore, pharmacological (and even toxicological) skin research is often based on the knowledge of pig skin absorption and percutaneous permeation.

Recently, Nicoli et al. (2008) employed qualitative and quantitative analysis of stratum corneum lipids and permeation experiments to analyze the utilization of rabbit ear skin in transdermal permeation studies, using pig ear skin as a reference. Their results showed that the stratum corneum of both rabbit ear skin and pig ear skin present similar thickness. Probably due to its higher lipophilicity, rabbit ear skin was less permeable to hydrophilic compounds; however, the permeability to progesterone was comparable between isolated pig epidermis and rabbit ear skin. Nicoli and co-workers conclude that the rabbit ear skin can be sucessfully employed in skin permeatin studies.

4.3 Photoacoustic evaluation of topically applied products

Different pharmaceutical formulations for a topically applied drug may present very different transdermal delivery ratios, depending on the product composition (excipients usually play a major role in the penetration kinetics of topically applied products). These penetration rates may be evaluated through the analysis of the time-dependence of the PA signal after topical application of a given product in skin. This methodology can also be applied to the evaluation of sunscreens, that may be characterized in terms of their (photo)stability after topical application (in this case, the lower the rate, the better the product).

Gutierrez-Juarez et al. (2002) employed PA measurements in the analysis of substances topically applied to the human skin. To fulfill this purpose, these authors utilized a double-chamber PA cell; the absorption determination was obtained through the measurement of the thermal effusivity of the binary system substance–skin. The model employed by Gutierrez-Juarez and co-workers (that assumes that the effective thermal effusivity of the binary system corresponds to that of a two-phase system) was experimentally applied to study different topically applied substances, in different parts of the body. The corresponding relative concentrations of substances as a function of time were determined by fitting a sigmoidal function (for ketoconazol and sunscreen) or an exponential function (for nitrofurazona, vaseline and vaporub) to the experimental data.

Pedrochi and co-workers (2005) employed PAS measurements to evaluate the penetration rate of different sunscreens into human skin *in vivo*. Their results showed that the diminution rate of the sunscreen amount in the skin surface depends on the form of the product: sunscreens in cream form tend to present faster reduction after application in skin. This leads to the conclusion that sunscreens in gel form are more adequate (presenting longer protection against UV radiation).

Another transdermal drug delivery study is the work of Truite et al. (2007), which employed PAS measurements in the *ex vivo* determination of the penetration rate of different phytotherapic formulations (with and without salicylic acid) for treatment of vitiligo. Measurements were performed as a function of time in rabbits. PA depth monitoring showed that both formulations propagated through the skin up to the melanocytes region, leading the authors to suggest that the delivery of the active agents may occur even without the use of queratinolitic substances (that are not really recommended, since they are known to induce side effects in animals).

The PA technique can be employed to study the penetration kinetics of topically applied products not only as a function of product composition, but also according to the application method. Phonophoresis is the utilization of ultrasound (US) waves to enhance the delivery of topically applied substances (Byl, 1995). In physiotherapy practice, phonophoresis is one of the various strategies developed to overcome the skin's resistance in transdermal drug delivery, enhancing skin permeability (Duangit et al., 2011).

In the last years, comparative studies between massage and phonophoresis (in its different modes) as application methods for different anti-inflammatories have been in the front line of research at the FASBio/UNIVAP; in the experiments, transdermal drug delivery has been evaluated through PA measurements as a function of time after topical application of different drugs in the forearm region, using manual massage or phonophoresis.

Results indicate that different products present distinct absorption times (depending on the vehicle employed, for example); the application method also affects the typical time constant of drug penetration into skin, though not for all tested formulations.

In the following subsections, we present one experiment performed at FASBio/UNIVAP in which the penetration kinetics of the pharmaco *Cordia verbenacea* DC (Acheflan) in the human skin was evaluated through PA measurements as a function of time for each of the application methods: massage and phonophoresis.

4.3.1 Materials and methods

The pharmaco *Cordia verbenacea* DC (Acheflan) is a topic usage anti-inflammatory medication widely employed in medicine, having alpha-humulen and trans-caryophyllen as active agents. Our experiment aimed to evaluate the penetration kinetics of Acheflan in the human skin (massage *versus* phonophoresis) through *in vivo* PA measurements.

The survey was conducted in 10 volunteers (four men and six women) aged between 18 and 30 years. The following inclusion criteria were adopted: (i) absence of ulcers or any change in dermatology distal forearm and wrist; (ii) not being allergic to any component of the formula topically applied; (iii) absence of metal implants in the wrist or forearm; (iv) absence of stomach pain complaints; and (v) not being pregnant.

The protocol for cleaning prior to drug application consisted of cleaning the skin area to be evaluated (region near the distal forearm and right ulnar artery) with cotton soaked in 70% alcohol. The area of topical application was then demarcated and *Cordia verbenacea* DC (essential oil 5.0mg/g) was applied by rubbing the head of the ultrasound therapy equipment configured for continuous mode and intensity of 1.2W/cm^2, but switched off, for five minutes. This procedure was repeated on the opposite forearm with the ultrasound therapy equipment turned on, also for five minutes.

PA measurements employed the same experimental system described in section 3.2.1. During measurements, the volunteers were positioned adjacent to the assembly with an aluminum foil (60μm thick) sealing the PA cell and the distal forearm positioned in direct contact with it, as proposed by Bernengo et al. (1998). For an aluminum foil with this thickness, the cutoff frequency is approximately 7kHz; in the present study, the modulation frequency employed was 17Hz, so that the foil can be considered as thermally thin (Rosencwaig, 1980).

Each measurement series consisted of 40 readings (one each two seconds, for a total time of 80 seconds), repeated 10 times for each application form, with rest intervals of 100 seconds between successive series (total time of 30 minutes for each volunteer and application method). The software "SISCOMF" (developed at UNIVAP) was employed for data acquisition.

In order to analyze the typical time constant for the penetration of the applied drug, the experimental curves were fitted by the Boltzmann equation:

$$PA(t) = \frac{A_1 - A_2}{1 + e^{(t-t_0)/dt}} + A_2 \tag{2}$$

The Boltzmann curve is a S-shaped curve in which A_1 is the initial signal amplitude, A_2 the final signal amplitude, t_0 is the half-absorption time and dt, the time interval such as 67% of the penetration process occurs between t_0-dt e t_0+dt.

Analysis of the PA data was performed with the aid of the software Microcal Origin® 7.5 (employed for the generation of the fitting curves); statistical analysis was performed with GraphPad Instat® 3.0.

4.3.2 Results and discussion

Figure 3 shows an example of PA data measurements as a function of time fitted by a Boltzmann curve.

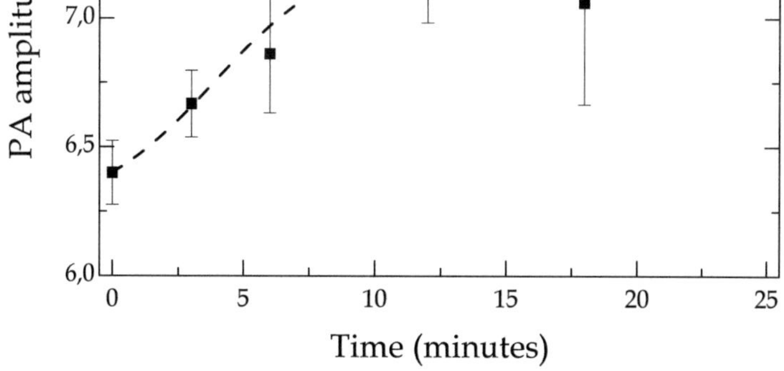

Fig. 3. Example of a Boltzmann curve (dashed line) fitting PA data obtained for one of the volunteers after phonophoresis application of the pharmaco *Cordia verbenacea* DC (Acheflan)

The experimental results obtained (average values for each parameter of the fitting curve) are summarized in Table 3; interpretation of such results must consider that the initial amplitude (A_1) of the PA signal corresponds to the system formed by the applied drug+skin, while the final amplitude (A_2) corresponds to the skin only, with the product having penetrated beyond the layer responsible for the generation of the PA signal (about 30μm, in the present case).

Application Method	A_1 (u.a)	A_2 (u.a)	t_0	Dt	t_s (t_0 + dt)
Massage	0,66 ± 0,04	0,70 ± 0,02	7 ± 2	4 ± 2	11 ± 2
Phonophoresis	0,65 ± 0,03	0,72 ± 0,04	5 ± 1	1,3 ± 0,3	7 ± 1

Table 3. Average values (± standard error) for A_1 and A_2 (in arbitrary units), t_0 and dt (in minutes), for each of the application methods (N=10)

In order to understand if penetration was effective, first of all, it is imperative to evaluate if the difference between A_1 and A_2 is statistically significant. Therefore, initially a paired t-test was carried on to verify if there was significant difference between A_1 and A_2 for each application method employed (indicating significant penetration of the applied product). This was verified for both application methods; however, this difference is more evident for phonophoresis application, in which the difference between the initial and the final signal presents $p=0{,}011$ ($p=0{,}066$ was found for massage application).
Statistical tests were also employed in the comparison between the two application methods (massage and phonophoresis); no statistical significance was found for t_0 and dt. Considering $t_S = (t_0 + dt)$ as the total effective penetration time for the epidermal layer under study, the results obtained (average ± error, N=10) are 11(±2) minutes (for massage application) and 7(±1) minutes (for phonophoresis application). The paired t-test for this parameter shows $p=0{,}073$.
Experiments performed at FASBio/UNIVAP show that the form of application can influence the kinetics of transdermal drug delivery, depending on the applied product. In the experiment presented here, significant penetration has been reported for both forms of administration (massage and phonophoresis); PA measurements showed that effective penetration is at least more evident after phonophoresis application, when compared to massage application.

5. Conclusion and perspectives

The possibility of performing *in vivo* studies brings great value to the use of the PA technique in experimental skin research. The potential and relevance of PA measurements in this field have been shown by a large range of experiments being performed by different research groups around the world - actually, the examples here presented must be seen as a sample of what has been done.
PA measurements *in vivo* are able to detect alterations in skin pigmentation. Even in skin regions normally protected from sun exposure, the PA signal level tends to follow clinical classification; actually, the use of the PA technique goes one step ahead, allowing comparative and quantitative research through simple, direct measurements.
Topical drug application has been employed in the treatment of many pathological processes; its efficiency is associated to the efficiency of transdermal drug delivery. PA

measurements have been sucessfully employed in transdermal drug delivery studies, allowing a quantitative analysis of the kinetics and effectivity of drug delivery. Different PA experiments point to the fact that gel formulations tend to be more adequate for topical use. Depending on the topically applied product, the form of application can also be determinant in the kinetics of transdermal drug delivery.
Measurements already performed indicate various perspectives for future research, such as: i) the determination of the skin oiliness level; ii) analysis of skin lesions; iii) studies on the photostability of sunscreens and even determination of the sun protection factor (SPF) of sunscreens (through PAS measurements); and iv) further studies on formulation and application form of a wide range of topically applied products.

6. Acknowledgment

The authors acknowledge Fapesp and CNPq for financial support of biomedical research being developed at FASBio/UNIVAP, São José dos Campos (SP), involving the characterization of human skin and transdermal drug delivery of topically applied products.

7. References

Acosta-Avalos, D.; Alvarado-Gil, J.J.; Vargas, H.; Frías-Hernández, J.; Olalde-Portugal, V.; Miranda, L.C.M. (1996) Photoacoustic monitoring of the influence of arbuscular mycorrhizal infection on the photosynthesis of corn (Zea mays L.). *Plant Science*, v.119, n.1-2, pp.183-190; doi:10.1016/0168-9452(96)04454-8

Aqil, M.; Abdul, A.; Yasmin, S.; Asgar, A. (2007) Status of terpenes as skin penetration enhancers. *Drug Discovery Today (Oxford)*, v.12, n.23/24, pp.1061-1067.

Balderas-Lopez, J.A.; Moreno-Márquez, M.M.; Martínez, J.L.; Sánchez-Sinencio, F. (1999) Thermal characterization of some dental resins using the photoacoustic phase lag discontinuites. *Superfícies y Vacio*, v.8, pp.42-45.

Balderas-Lopez, J.A.; Mandelis, A. (2001) Thermal diffusivity measurements in the photoacoustic open-cell configuration using simple signal normalization techniques. *Joumal of Applied Physics*, v.90, n.5. pp.2273-2279.

Balogun, O.; Regez, B.; Zhang, H.F.; Krishnaswamy, S. (2009) Real time, full-field imaging of photoacoustic generated signals for biomedical applications, *ICPPP15 - Book of Abstracts, 15th International Conference on Photoacoustic and Photothermal Phenomena*, pp.77, Leuven, Belgium, July 19-23, 2009.

Baumann, L. (2006a) New Skin Typing System. *Skin & Aging* , v.14, n.2, pp.60-64.

Baumann, L. (2006b) *The Skin Type Solution*. Bantam Books, ISBN-10: 0-553-80422-7, New York, NY.

Beard, P. (2009) High resolution spectroscopic photoacoustic imaging for characteristic tissue structure and function, *ICPPP15 - Book of Abstracts, 15th International Conference on Photoacoustic and Photothermal Phenomena*, pp.79, Leuven, Belgium, July 19-23, 2009.

Benamar, N.; Laplante, A.F.; Lahjomri, F.; Leblanc, R.M. (2004) Modulated photoacoustic spectroscopy study of an artificial tanning on human skin induced by dihydroxyacetone. *Physiological Measurement*, v.25, n.5, pp.1199-1210.

Benech-Kieffer, F.; Wegrich, P.; Schwarzenbach, R.; Klecak, G.; Weber, T.; Leclaire, J.; Schaefer, H. (2000) Percutaneous Absorption of Sunscreens in vitro: Interspecies

Comparison, Skin Models and Reproducibility Aspects. *Skin Pharmacology and Physiology*, v.13, pp.324-335; doi: 10.1159/000029940

Bernengo, J.C.; Gasquez, C; Falson-Rieg, F. (1998) Photoacoustics as a tool for cutaneous permeation studies. *High Temperature-High Pressure*, v.30, n.5, pp.619-624.

Bodzenta, J.; Bukowski, R.J.; Christ, A.; Pogoda, T. (2002) Photoacoustic detection of drug diffusion into a membrane: theory and numerical analysis. *International Journal of Heat and Mass Transfer*, v.45, n.22, pp.4515-4523; doi:10.1016/S0017-9310(02)00155-2

Byl, N.N. (1995) The Use of Ultrasound as an Enhancer for Transcutaneous Drug Delivery: Phonophoresis. *Physical Therapy*, v.75, n.6, pp.539-553.

Cahen, D.; Bults, G.; Garty, H.; Malkin, S. (1980) Photoacoustic in life sciences. *J. Biochemical Biophysical Methods*, v.3 (5), 1980, pp.293-310.

Duangjit, S.; Opanasopit, P.; Rojanarata, T.; Ngawhirunpat, T. (2011) Characterization and *In Vitro* Skin Permeation of Meloxicam-Loaded Liposomes versus Transfersomes. *Journal of Drug Delivery*, v.2011, ID 418316; doi:10.1155/2011/418316

Fitzpatrick, T.B. (1988) The validity and practicality of sun-reactive skin types I through VI. *Archives of Dermatology*, v.124, pp.869–871.

Gao, C.M.; Zhang, S.Y.; Zhang, Z.; Shui, X.J. (2005) Thermal diffusivity of porcine tissues characterized by photoacoustic piezoeletric technique. *Journal of Physics IV (France)*, v.125, pp.777-779; doi:10.1051/jp4:2005125179

Gutierrez-Juarez, G.; Vargas-Luna, M.; Cordova, T.; Varela, J.B.; Bernal-Alvarado, J.J.; Sosa, M. (2002) *In vivo* measurement of human skin absorption of topically applied substances by a photoacoustic technique. *Physiological Measurement*, v.23, pp.521-532.

Hahn, B.D.; Neubert, R.H.H.; Wartewig, S.; Lasch, J. (2001) Penetration of compounds through human stratum corneum as studied by Fourier transform infrared photoacoustic spectroscopy. *Journal of Controlled Release*, v.70, n.3, pp.393-398.

Herbert, S.K.; Han, T.; Vogelmann, T.C. (2000) New applications of photoacoustics to the study of photosynthesis. *Photosynthesis Research*, v.66, n.1-2, pp.13-31; doi: 10.1023/A:1010788504886

Hu, S.; Wang, L.V. (2010) Photoacoustic imaging and characterization of the microvasculature, *Journal of Biomedical Optics*, v.15, pp.011-101; doi:10.1117/1.3281673

Li, T.; Dewhurst, R.J. (2010) Photoacoustic imaging in both soft and hard biological tissue. *Journal of Physics: Conference Series*, v.214, 012028; doi: 10.1088/1742-6596/214/1/012028

Lopez, T.; Picquart, M.; Aguilar, D.H.; Quintana, P.; Alvarado-Gil, J.J.; Pacheco, J. (2005) Photoacoustic monitoring of dehydration in sol-gel titania emulsions. *Journal of Physics IV (France)*, v.125, pp.583-585; doi:10.1051/jp4:2005125134

McDonald, F.A.; Wetsel G.C. (1978) Generalized theory of the photoacoustic effect, *Journal of Applied Physics*, v.49, pp.2313.

Malkin, S.; Puchenkov, O.V. (1997) The photoacoustic effect in photosynthesis. In *Progress in Photothermal and Photoacoustic Science and Technology: Life and Earth Sciences* (Mandelis, A., and Hess, P., editors), SPIE, ISBN 0-8194-2450-1, Washington, USA.

Marquezini, M.V.; Cella, N.; Mansanares, A.M.; Vargas, H.; Miranda, L.C.M. (1990) Open photoacoustic cell spectroscopy. *Measurement Science & Technology*, v.2, pp.396-401.

Melo, J.O.; Pedrochi, F.; Baesso, M.; Hernandes, L.; Truiti, M.; Baroni, S.; Bersani-Amado, C. (2011) Evidence of Deep Percutaneous Penetration Associated with Anti-Inflammatory Activity of Topically Applied Helicteres gardneriana Extract: A Photoacoustic Spectroscopy Study. *Pharmaceutical Research*, v.28, n.2, pp.331-336; doi: 10.1007/s11095-010-0279-3

Nicoli, S.; Padula, C.; Aversa, V.; Vietti, B.; Wertz, P.W.; Millet, A.; Falson, F.; Govoni, P.; Santi, P. (2008) Characterization of Rabbit Ear Skin as a Skin Model for in vitro Transdermal Permeation Experiments: Histology, Lipid Composition and Permeability. *Skin Pharmacology and Physiology*, v.21, pp.218-226; doi: 10.1159/000135638

Pedrochi, F.; Sehn, E.; Medina, A.N.; Bento, A.C.; Baesso, M.L.; Storck, A.; Gesztesi, J.L. (2005) Photoacoustic Spectroscopy to Evaluate the Penetration Rate of Three Different Sunscreens into Human Skin *in vivo*. *Journal of Physics IV (France)*, v.125, pp.757-759; doi:10.1051/jp4:2005125174

Qiu, P.F.; Zhang, S.Y.; Shui, X.J. (2008) Photoacoustic study of thermal properties of biological tissues detected by PVDF film transducer. *European Physics Journal - Special Topics*, v.153, pp.487-490.

Rezende, D.V.; Nunes, O.A.C.; Oliveira, A.C. (2009) Photoacoustic Study of Fungal Disease of Açai (*Euterpe oleracea*) Seeds. *International Journal of Thermophysics*, v.30, n.5, pp.1616-1625; doi: 10.1007/s10765-009-0655-6

Rosencwaig, A.; Gersho, A. (1976) Theory of the photoacoustic effect with solids. *Journal of Applied Physics*, v.47, pp.64-69.

Rosencwaig, A. (1980) *Photoacoustics and Photoacoustic Spectroscopy*. John Wiley & Sons, ISBN 0-471-04495-4, New York, USA.

Savateeva, E.V.; Karabutov, A.A.; Oraevsky, A.A. (2001) *Proceedings of SPIE* 4256, pp.61-69.

Schmidt, W.D.; Fassler, D.; Zimmermann, G.; Liebold, K.; Wollina, U. (2000) Non-contacting diffuse VIS-NIR spectroscopy of human skin for evaluation of skin type and time-dependent microcirculation. *Progress in biomedical optics and imaging*, v.1, n.31, pp.91-102.

Silva, W.J.; Prioli, L.M.; Magalhães, A.C.N.; Pereira, A.C.; Vargas, H.; Mansanares, A.M.; Cella, N.; Miranda, L.C.M.; Alvarado-Gil, J.J. (1995) Photosynthetic O_2 evolution in maize inbreds and their hybrids can be differentiated by open photoacoustic cell technique. *Plant Science*, v.104, n.2, pp.177-181; doi:10.1016/0168-9452(94)04026-D

Simon, G.A.; Maibach, H.I. (2000) The Pig as an Experimental Animal Model of Percutaneous Permeation in Man: Qualitative and Quantitative Observations - An Overview. *Skin Pharmacology and Applied Skin Physiology*, v.13, pp.229-234; doi: 10.1159/000029928

Sui, X.; Sun, J.; Wang, S.; Li, W.; Hu, L.; Meng, F.; Fan, Y.; Zhang, A. (2011) Photosynthetic induction in leaves of two cucumber genotypes differing in sensitivity to low-light stress. *African Journal of Biotechnology*, v.10, n.12, pp. 2238-2247.

Swearingen JA, Holan SH, Feldman MM, Viator JA. (2010) Photoacoustic discrimination of vascular and pigmented lesions using classical and Bayesian methods. *Journal of Biomedical Optics*, v.15, n.1, 016019; doi: 10.1117/1.3316297

Truite, C.V.R.; Philippsen, G.S.; Ueda-Nakamura, T.; Natali, M.R.M.; Dias Filho; B.P.; Bento, A.C.; Baesso, M.L.; Nakamura, C.V. (2007) *Photochemistry and Photobiology*, v.83, pp.1529-1536; doi: 10.1111/j.1751-1097.2007.00197.x

Vargas, H.; Miranda, L.C.M. (1988) Photoacoustic and related photothermal techniques. *Physics Reports*, v.161, n.2, pp.43-101.

Viator, J.A.; Komadin, J.; Svaasand, L.O.; Aguilar, G.; Choi, B.; Nelson, J.S. (2004) A Comparative Study of Photoacoustic and Reflectance Methods for Determination of Epidermal Melanin Content. *Journal of Investigative Dermatology*, v.122, pp.1432-1439.

14

Acoustic–Gravity Waves in the Ionosphere During Solar Eclipse Events

Petra Koucká Knížová and Zbyšek Mošna
Institute of Atmospheric Physic, Czech Academy of Sciences
Czech Republic

1. Introduction

Terrestrial atmosphere shows a high variability over a broad range of periodicities, which mostly consists of wave-like perturbations characterized by various spatial and temporal scales. The interest for short time variability in ionospheric attributes is related to the role that ionosphere plays in the Earth's environment and space weather. Acoustic-gravity waves (AGWs), waves in the period range from sub-seconds to several hours, are sources of most of the short-time ionospheric variability and play an important role in the dynamics and energetics of atmosphere and ionosphere systems. Many different mechanisms are likely to contribute to the acoustic-gravity wave generation: for instance, excitation at high latitudes induced by geomagnetic and consequent auroral activity, meteorological phenomena, excitation in situ by the solar terminator passages and by the occurrence of solar eclipses.

During solar eclipse, the lunar shadow creates a cool spot in the atmosphere that sweeps at supersonic speed across the Earth's atmosphere. The atmosphere strongly responds to the decrease in ionization flux and heating. The very sharp border between sunlit and eclipsed region, characterized by strong gradients in temperature and ionization flux, moves throughout the atmosphere and drives it into a non-equilibrium state. Acoustic-gravity waves contribute to the return to equilibrium. At thermospheric heights, the reduction in temperature causes a decrease in pressure over the totality footprint to which the neutral winds respond. Thermal cooling and downward transport of gases lead to neutral composition changes in the thermosphere that have significant influence on the resulting electron density distribution. Although the mechanisms are not well understood, several studies show direct evidence that solar eclipses induce wave-like oscillations in the acoustic-gravity wave domain.

Many different mechanisms are likely to contribute to wave generation and enhancement at ionospheric heights. Hence, it is difficult to clearly separate or differentiate each contributing agent and to decide which part of wave field belongs to the in situ generated and which part comes from distant regions. First experimental evidence of the existence of gravity waves in the ionosphere during solar eclipse was reported by Walker et al. (1991), where waves with periods of 30–33 min were observed on ionosonde sounding virtual heights.

1.1 Ionospheric sounding

As the solar radiation penetrates Earth's atmosphere it forms pairs of charged particles. Under a normal day-time conditions the ionization solar flux increases immediately after sunrise, reaches maximum around local noon and decreases again till sunset. Under such conditions concentration of charged particles significantly grows in the atmosphere and forms atmospheric plasma called ionosphere. Due to the composition of the neutral atmosphere together with the changing efficiency of the incoming solar radiation, ionosphere is stratified into the layers denoted D, E, F1 and F2. After sunset, electrons and ions recombine rapidly in the D, E and F1 layer. Due to slower recombination processes of atomic ions that dominate at heights approximately above 150km altitude, F2 layer remains present all the night. Special stratification Es, sporadic E layer, occurs sometimes at heights of E layer (Davies, 1990).

Ionosphere significantly affects propagation of the electromagnetic waves. According to a frequency of the wave with respect to a concentration of the ionospheric plasma, wave propagates through the medium or it is reflected. Electromagnetic waves with frequency lower than plasma frequency of the particular plasma parcel are reflected, which allows to estimate plasma frequency. Higher frequency waves propagate through plasma. An instrument called ionosonde (or digisonde) transmits electromagnetic wave of a defined frequency and detects it after reflection from the ionosphere. Typical ionosonde sounding range is 1 MHz - 20 MHz. For each sounding wave ionosonde records time of flight τ on the path transmitter - reflection point - receiver. Time of flight is simply converted into a virtual height $h_{virtual} = \frac{\tau.c}{2}$ that corresponds to wave propagation in the vacuum (c stands here for speed of light). Virtual height is equal or higher than the corresponding real height. The output of the measurement is height-frequency characteristics called an ionogram. Real height electron concentration profiles can further be inverted from ionograms using for instance programs POLAN (Titheridge, 1985) or NHPC (Huang and Reinish, 1996). Ionosphere represents inhomogeneous and anisotropic medium which leads to a wave splitting into an ordinary and extraordinary wave modes. Hence, two reflection traces are recorded by the ionosonde (as seen on ionograms in Figure 1). However, the extraordinary mode is not further used for electron concentration profile inversion.

Figure 1 shows typical day-time and night-time ionograms recorded by a digisonde in the observatory Pruhonice. Together with the ionograms there are plots of the real height electron concentration derived by NHPC routine. On the day-time electron concentration profile, three ionospheric layers E, F1 and F2 are present while on the night-time profile there is only F2 layer detected by the ionosonde.

Sequences of ionograms are widely used for analyses of variability of atmospheric plasma ranging from detection of rapid changes with periods of minutes to the study of long-term trends.

2. Basic theory of AGWs in the Earth's atmosphere

Most of the wave-like oscillations in the atmosphere can be described/parametrized using basic acoustic-gravity wave theory in the atmosphere. Details can be found, for instance, in works of Davies (1990), Bodo et al. (2001), Hargreaves (1982), Yeh & Liu (1974) among others. Here, we show brief derivation of the dispersion relation that any wave motion of the AGW type must satisfy. In a plane-stratified, isothermal atmosphere under gravity that is constant with height, two frequency domains exist in the atmosphere where atmospheric waves can propagate, acoustic and gravity wave. Atmosphere represents compressible gas that once compressed and then released would expand and oscillate about its equilibrium state. Its oscillation frequency is known as an acoustic cut-off frequency

$$\omega_a = \frac{c}{2H} \tag{1}$$

where c is speed of sound

$$c = \sqrt{\gamma g H} \tag{2}$$

γ is the ratio of specific heats at constant pressure and constant volume, g is the gravitational acceleration, and H is the scale height. For diatomic gas $\gamma \sim 1.4$.

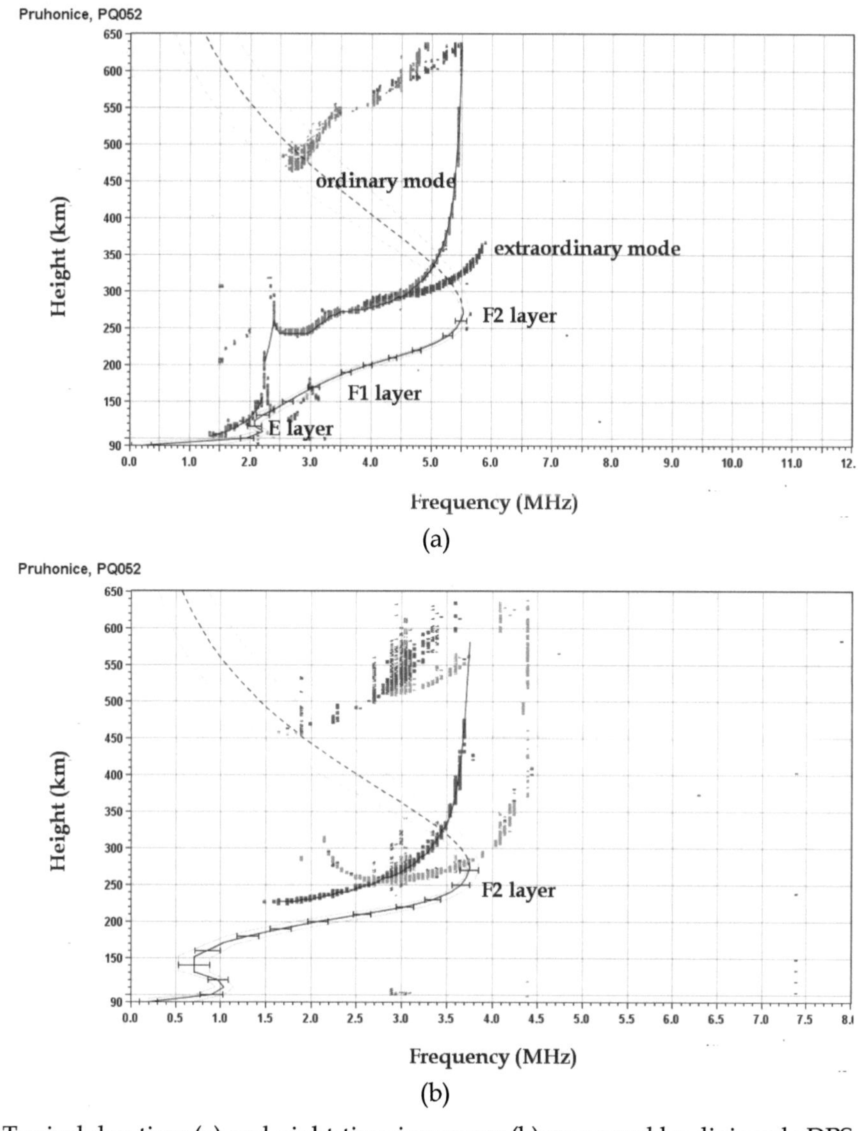

Fig. 1. Typical day-time (a) and night-time ionogram (b) measured by digisonde DPS 4 in the Observatory Pruhonice. On both plots, there is real hight electron concentration (solid line with error bars) provided as obtained by the NHPC routine.

Single element of fluid, parcel of the atmosphere, at height z with density ρ which is displaced in the vertical by Δz to a place where its density changes to ρ+Δρ, remains in pressure equilibrium with its surroundings. Displacement takes place adiabatically. This is valid when the motion is so slow that sound waves with speed

$$c = \sqrt{\frac{dp}{d\rho}} \tag{3}$$

where p stands for pressure can traverse the system faster than the time-scale of interest and the motion is so fast that the entropy is preserved. The parcel is no longer in equilibrium and starts to oscillate about its equilibrium height with buoyancy frequency.
The buoyancy force which acts on the parcel is balanced by inertial force (Newton's second law):

$$\rho \frac{d^2}{dt^2}(\Delta z) = -g\Delta\rho \tag{4}$$

where Δρ is the difference between internal and external densities.
Internal and external Δρ are derived as:

$$(\Delta\rho)_{internal} = \Delta p / c^2 = -\frac{g\rho}{c^2}\Delta z \tag{5}$$

which is due to compressibility of the fluid within the membrane and

$$(\Delta\rho)_{external} = -\frac{d\rho}{dz}\Delta z \tag{6}$$

is the change of background density at new position due to inhomogeneous nature of the atmosphere. Taking both the contributions of Δρ we get

$$\frac{d^2}{dt^2}(\Delta z) = \left(g\frac{d}{dz}(\ln\rho) + g^2/c^2 \right)\Delta z \tag{7}$$

which can be recast into

$$\frac{d^2}{dt^2}(\Delta z) + \omega_B^2 \Delta z = 0 \tag{8}$$

where

$$\omega_B^2 = -g\left(\frac{d}{dz}(\ln\rho) + g/c^2 \right) \tag{9}$$

If $\omega_B^2 > 0$, the solution is oscillatory and the fluid parcel will oscillate with characteristic buoyancy frequency ω_B called Brunt-Vaisala frequency.
More convenient form used for atmosphere is following:

$$\omega_B^2 = (\gamma - 1)g^2/c^2 + g/c^2 dc^2/dz \tag{10}$$

This approximation is valid in the atmosphere-ionosphere system of our interest.

In isothermal atmosphere ω_B reduces to

$$\omega_B^2 = (\gamma - 1) g^2 / c^2 \tag{11}$$

In the terrestrial atmosphere the buoyancy period depends on the height. The height variance of the acoustic cut-off and buoyancy frequencies in the isothermal atmosphere is shown in Figure 2.

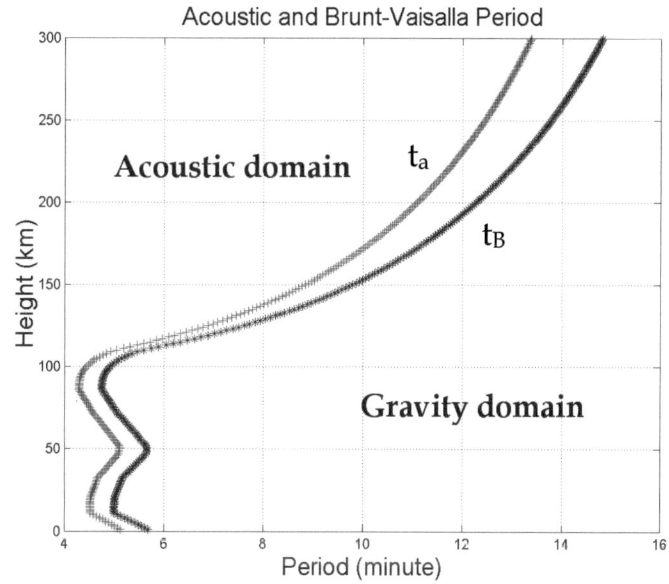

Fig. 2. Height dependence of acoustic cut-off period t_a and Brunt-Vaisala period t_B that represent limits dividing periods into acoustic and gravity wave domains. Period domain between acoustic cut-off and Brunt-Vaisala represents region where no AGW propagates.

Wave motion in the atmosphere can be described using mass conservation (continuity equation), and equation of motion:

$$\frac{\partial \rho}{\partial t} + \rho \nabla . \vec{u} + \left(\vec{u} . \nabla \right) \rho = 0 \tag{12}$$

$$\rho \left(\frac{\partial \vec{u}}{\partial t} + \left(\vec{u} . \nabla \right) \vec{u} \right) = -\nabla p + \rho g \tag{13}$$

where pressure gradients and gravity are the only forces causing the acceleration.
Oscillation takes place adiabatically

$$\rho \left(\frac{\partial p}{\partial t} + \vec{u} . \nabla p \right) = \gamma p \left(\frac{\partial \rho}{\partial t} + \vec{u} . \nabla \right) \tag{14}$$

where ρ, p, γ and $\vec{u}$ are parameters of the atmosphere - density, pressure, ratio of specific heats, and velocity.

Applying the perturbation approach we are searching for wave-like solutions for the perturbation quantities. Further simplification comes from the assumption that the background state is of constant temperature T in which p_0/ρ_0 must be a constant.

$$p_0 / \rho_0 = c^2 / \gamma \tag{15}$$

Then the system (12), (13) and (14) reduces to:

$$\frac{\partial \rho'}{\partial t} + \rho_0 \vec{\nabla}.\vec{u}' - \rho_0 u_z / H = 0 \tag{16}$$

$$\rho_0 \frac{\partial \vec{u}}{\partial t} + \nabla p' - \rho' g = 0 \tag{17}$$

$$\rho_0 \left(\frac{\partial p'}{\partial t} - p_0 u_z^{'} / H \right) = \gamma p_0 \left(\frac{\partial \rho'}{\partial t} - \rho_0 u_z^{'} / H \right) \tag{18}$$

where index 0 denotes stationary (non fluctuating) component and the apostrophe denotes perturbation. These are the basic governing equations for the gravity waves. For a non-trivial solution the following prescription of the dispersion relation must be satisfied:

$$\omega^4 - \omega^2 \omega_a^2 - k_x^2 c^2 \left(\omega^2 - \omega_g^2 \right) - c^2 \omega^2 k_z^2 = 0 \tag{19}$$

From disperse relation, it is evident that between buoyancy frequency and acoustic cut-off frequencies one cannot have both k_x and k_z real. Figure 2 shows two period domains with border limits of acoustic cut-off period and buoyancy period.

An attenuation or growth in the wave amplitude must occur in either the vertical or the horizontal directions. We suppose that there is no variation in amplitude in horizontal directions so that k_x is purely real and k_z has an imaginary component. At frequencies exceeding acoustic cut-off ω_a, expression (19) becomes simple and the waves may be termed as ACOUSTIC WAVES. At frequencies smaller than Brunt-Vaisala frequency where gravity plays an important role, the waves are called GRAVITY or INTERNAL GRAVITY WAVES. Brunt-Vaisala frequency and acoustic cut-off frequency divide the frequency spectrum into two domains in which ω_g forms the high frequency limit for one class $\omega<\omega_g$ normally called internal gravity waves and ω_a is the low frequency limit for another class $\omega>\omega_a$ called the acoustic waves. A gap in the frequency spectrum exists between ω_g and ω_a where no internal waves can propagate.

Important approximations can be obtained under the assumption $|k_z| >> 1/2H$ and $\omega << \omega_g$ then:

$$k_z^2 = \left(\omega_g^2 / \omega^2 \right) k_x^2 \tag{20}$$

These approximations apply to much of the observed gravity waves. From (20) we see that the angle of ascent of the phase α is:

$$tg\alpha = k_z / k_x = \omega_g / \omega = \tau / \tau_g \tag{21}$$

The motions of the air parcels are, in general, ellipses in the plane of propagation and have components transverse to the direction of wave propagation. The ratio of the horizontal displacement ξ to its vertical displacement ζ is:

$$\frac{\xi}{\zeta}=\frac{\frac{ck_x}{\omega}}{\left(\frac{ck_x}{\omega}\right)^2-1}\left(\frac{ck_z}{\omega}-i\sqrt{\left(\frac{\omega_a}{\omega}\right)^2-\left(\frac{\omega_g}{\omega}\right)^2}\right) \tag{22}$$

On the frequencies just above the acoustic cutoff the air motion is essentially vertical. With acoustic waves on high frequencies the motion is radial as in sound waves. The motion is circular with horizontal propagation at a frequency $\omega_a\sqrt{2/\gamma}$. Gravity wave propagation is limited to angles between

$$\phi_{min}=\sin^{-1}\left(\frac{\omega}{\omega_g}\right),\quad \phi_{max}=\pi-\sin^{-1}\left(\frac{\omega}{\omega_g}\right) \tag{23}$$

The sense of rotation of the air for gravity waves is opposite than for acoustic waves. As Φ approaches its asymptotic values the air motion becomes linear and transverse to the direction of propagation. Air parcel rotation is clockwise in case of acoustic waves while anticlockwise in case of gravity waves. Energy vector lies in the same quadrant as direction of propagation of acoustic waves. Energy flows up when phase travels down and vice versa in case of gravity waves propagation. This is important property since it accounts for the observed downward phase propagation when the source is below the level at which a disturbance is observed.

The horizontal u_x and vertical u_z components of the packet velocity are obtained from disperse relation:

$$u_x=\frac{c^2k_x\left(\omega^2-\omega_g^2\right)}{\omega\left(2\omega^2-\omega_a^2-c^2k^2\right)} \qquad u_z=\frac{c^2k_z\omega^2}{\omega\left(2\omega^2-\omega_a^2-c^2k^2\right)} \tag{24}$$

Due to coupling between neutral and charged components the initial wave-like oscillation in the neutral atmosphere induces wave-like perturbation in the ionosphere. Perturbation in the ion production is the most effective when solar ionizing rays are nearly in alignment with the initial wave front. Perturbations in the neutral atmosphere may cause perturbations in chemical processes. Presence of AGW influences the ionisation rate through changes in the local neutral density and temperature, and through changes in the ionisation radiation absorption (Hooke, 1970).

3. AGW in the ionospheric plasma

Acoustic-gravity waves are always present in the Earth's atmosphere. AGWs arise from many natural sources like convection, topography, wind shear, moving solar terminator, earthquakes, tsunami, etc. Increase in wave-like activity is associated also with human activity including coordinated experiments or unwilling accidents. AGWs influence on the upper atmosphere is not yet understood enough. They produce a great amount of variability and contribute to the background conditions in a specific parcel of the atmosphere. Gravity waves propagating from lower laying atmosphere have been long regarded as a very important source of the energy and momentum transfer in the upper atmosphere (Hines, 1960). The breaking of the upward propagating waves affects wind system, generates turbulence and heats the atmospheric gas.

Waves that reach upper atmosphere produce travelling atmospheric disturbances (TAD) or travelling ionospheric disturbances (TID) and even form the ionospheric inhomogenities which grow and finally break into the plasma instabilities observed by radar techniques that might cause scintillation of the communication signals propagating through the ionosphere. From the observation it is evident that the thermosphere is continuously swept by the acoustic-gravity waves. Statistically, the waves show a moderate preference for southward travel, with this preference being reduced or shifted to southeastward travel during disturbed times (Oliver et al., 1997). Experimental studies show that AGW activity in the ionosphere slightly increases during dawn and dusk periods of the day (Galushko et al., 1998; Somsikov & Ganguly, 1995; Sauli et al., 2005 among others). Influence of infrasonic waves generated by ground experimental sources on the ionosphere was reported for instance by Rapoport et al. (2004).

Solar eclipse represents well defined source of the AGW in the atmosphere and ionosphere systems. During solar eclipse event, solar ionization flux decreases producing well-defined cool spot in the atmosphere that moves through the Earth's atmosphere. Moving source in the atmosphere can emit both acoustic and gravity waves. Supersonic motion of the source forms wave field with bow wave. Both acoustic and gravity waves can be radiated in association with supersonic motion in the atmosphere. When the source is moving within atmosphere with subsonic velocity only gravity waves can be emitted (Kato et al., 1977).

4. Solar eclipse event – signatures in the ionospheric plasma

It has been proposed by Chimonas and Hines (1970) that solar eclipses can act as sources for AGWs. The lunar shadow creates a cool spot in the atmosphere that sweeps at supersonic speed across the Earth. The sharp border between sunlit and eclipsed regions, characterized by strong gradients in temperature and ionization flux, moves throughout atmosphere and drives it into a non-equilibrium state. Earth atmosphere shows variable sensitivity to the changes of ionization flux.

4.1 Experiments

Solar eclipse event represents phenomenon that can be precisely predicted, hence many observational campaigns are organised around the world. Effects of the solar eclipses on the ionospheric plasma are studied by mean of GPS techniques, radars, vertical ionospheric soundings etc. Study limitations lay mainly in the fact that there are no identical solar eclipse events. Moreover, solar eclipse induced effects are easily to be mixed with effects caused by geomagnetic field variations, diurnal changes of the ionosphere, seasonal variability of the atmosphere/ionosphere etc. In the upper atmosphere, AGWs can be observed either directly as neutral gas fluctuations or indirectly as induced ionospheric

plasma variations. Despite intensive research many questions in the problem of the generation and propagation remain to be understood.
Studies by Fritts and Luo (1993) suggest that perturbations generated by the eclipse induced ozone heating interruption may propagate upwards into the thermosphere-ionosphere system where they have an important influence. Temperature fluctuations and electron density changes propagate as a wave, away from the totality path, cf. Muller-Wodarg et al. (1998). By means of vertical ionospheric sounding, Liu et al. (1998) detected waves excited during solar eclipse event at F1 layer heights and their generation and/or enhancement attributed to changes of temperatures and variations of the height of the transition level for the loss coefficient and the height of the peak of electron production. Studies reported by Farges et al. (2001) suggest a longitudinal diversity of the disturbances with respect to pre-noon and postnoon phases. Xinmiao et al. (2010) reported synchronous oscillations in the Es and F layer during the recovery phase of the solar eclipse. Ivanov et al. (1998) found that during solar eclipse with maximum obscuration of about 70% the F-region electron density decreased by 6-8% compared to a control day and detected travelling ionospheric disturbances. Additionally, they detected strong variations in the difference group delays with a period about 40 minutes associated with the start and end of the eclipse. Oscillations in the ionosphere, similar to gravity waves, were observed following some solar eclipse events (Chimonas and Hines, 1970; Cheng et al., 1992; Liu et al., 1998; Sauli et al., 2006). Investigation of the latitudinal dependence of NmF2 (the maximum electron density of the F2 layer) indicated that the strongest response was at middle latitudes (Le et al., 2009). The response of the sporadic-E (Es) layer also differed in each solar eclipse event. A remarkable decrease in Es layer ionization was observed during the eclipse of 20 July 1963 (Davis et al., 1964). Enhancement of Es layer ionization has also been reported and it has been suggested that it is related to internal gravity waves generated in the atmosphere during the solar eclipse (Datta, 1972).

4.2 Processes induced by solar eclipse

During the solar eclipse, on the time scale shorter than day-night change, the ionosphere reconfigures itself into a state similar to that of night situation. Photochemical ionization falls heavily almost to a night-time level. With the decreasing solar flux, atmospheric temperature falls in the moon shadow creating a cool spot with well defined border. Then the increasing solar flux starts ionization processes and warms the atmosphere again to daytime level.
Such changes in the ionization cause variation in the reflection heights, decrease/increase in electron concentration at all ionospheric heights, decrease/increase in the total electron content, rising/falling of the layer height. Such effects are characteristic for the processes during sunrise/sunset in the ionosphere. However, supersonic movement of the eclipsed region represents a key difference from the regular solar terminator motion at sunrise and sunset times. These changes in the neutral atmosphere and ionosphere induced by solar eclipse force the evolution of the ionospheric plasma toward a new equilibrium state. The return to equilibrium is likely accompanied by the eclipse induced wave motions excited in the atmosphere. Any moving discontinuity of gas parameters such as temperature, pressure etc. will generate transit-like waves. In the upper ionosphere, waves can be generated by a strong horizontal electron pressure gradient. Possible mechanisms contributing to the wave generation in the region of solar terminator are in detail discussed by Somsikov & Ganguly (1995).

Solar eclipse induces changes in all atmospheric regions extend from the upper atmosphere down to ground level. Despite the low magnitude of the eclipse induced effects at ground level, Jones et al. (1992) reported wave-like oscillation related to eclipse on the microbarometer pressure records. The cooling effect of the Moon's shadow may induce the powerful meridional airflow in the atmosphere, which accelerates the ionized clouds in the Es layer and forms the wind shear to raise the observed Doppler frequency shift and foEs values, respectively (Chen et al., 2010).

5. Solar eclipse observed by vertical ionospheric sounding in midlatitudes

Vertical sounding measurements provide local information on the electron density distribution of the bottomside ionosphere. Electron concentration in the plasma and its corresponding plasma frequency are related via following equation:

$$f_p^2 = \frac{Ne^2}{4\pi\varepsilon_0 m} \qquad (25)$$

where f_P denotes plasma frequency and N, e, ε_0 and m stand for the electron concentration, the charge of electron, permittivity of free space, and the mass of the electron, respectively.

This section summarizes experimental results from the midlatitude ionospheric observatory Pruhonice (50N, 15E). At the observatory, the vertical sounding measurements were performed with ionosonde IPS 42 KEL Aerospace till the end of year 2003. Then this older equipment was replaced by digisonde DPS 4. Special campaigns of rapid sequence soundings were organized in order to study in detail ionospheric behavior during partial solar eclipses of 11 August 1999, 4 January 2011 and annular solar eclipse 3 October 2005. All three analyzed events were characterized by low geomagnetic activity; hence they represent a good occasion to observe mostly solar eclipse induced effects in the ionosphere. However, inconclusive results of the solar eclipse observations rise from the fact that different solar eclipses produce different plasma motions. Indeed, the travel cone geometry and its angular effects on the magnetized plasmas are different for each eclipse.

Solar eclipse of 11 August 1999 (as a total seen in place as close as 200 km from the measurement point) represents so far the event of the highest solar disc coverage observed in the Observatory Pruhonice. Figure 3 depicts sequence of raw ionograms measured during this event by IPS 42 KEL Aerospace equipment. The ionograms were recorded with the cadence of 1 minute. On the ionograms there is clearly seen that the eclipse event affects whole electron density profile. Critical frequencies in the E and F layer decrease before maximum disc occultation and then increase again. The electron density decrease in the E layer is much stronger than in the F layer due to different dominant type of the recombination. Electron density fall and increase occur simultaneously with occultation and de-occultation of the solar disc in the E and F1 layer while the F2 layer electron density reacts with slight delay. There are special structures of the spread F type developed on the profile after beginning of the solar disc occultation (clearly seen on ionograms at 9.14 UT and 9.16 UT). Shape of the F layer is affected as well. Unfortunately, effects in the F1 region cannot be discussed here in details because F1 layer is blanketed by strong sporadic E layer during part of the solar eclipse.

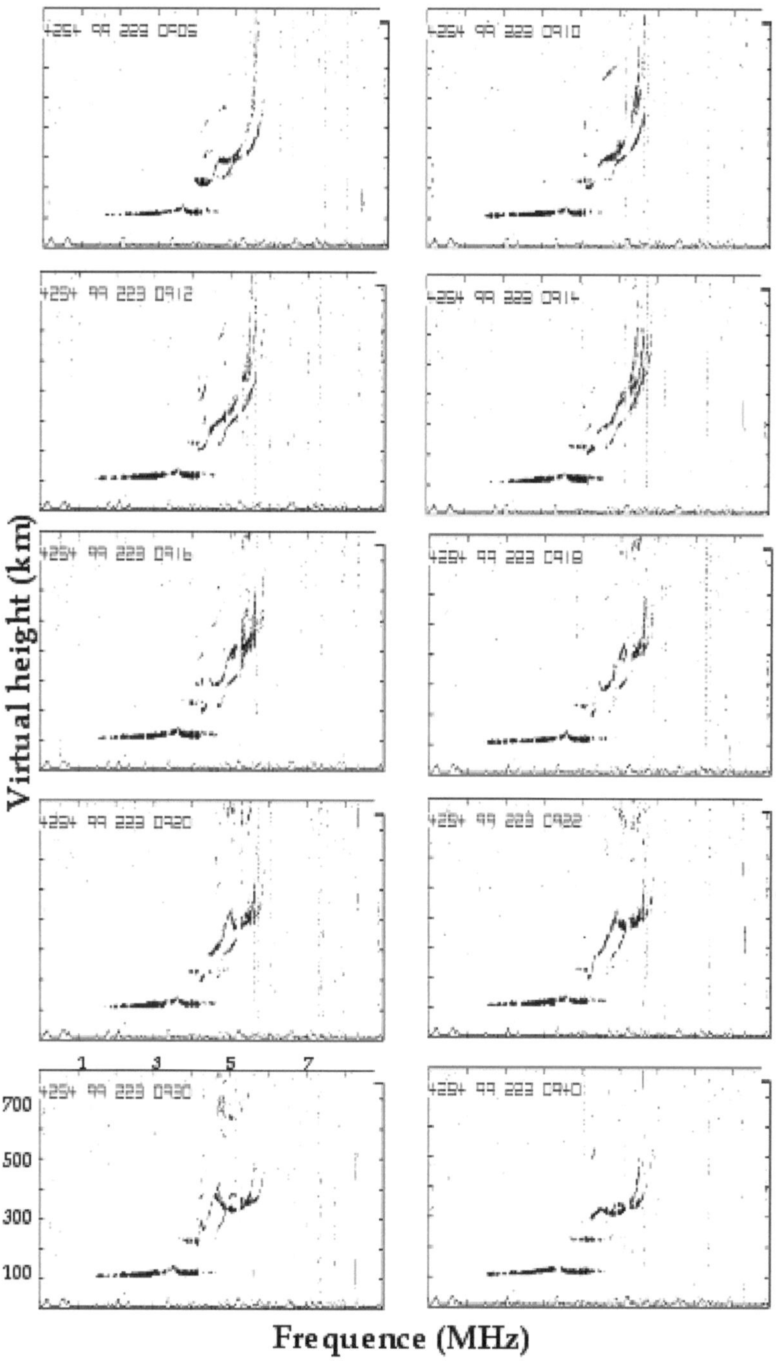

Fig. 3. Sequence of raw ionograms measured by the ionosonde KEL Aerospace IPS 42 at the observatory Pruhonice. During the special campaign ionograms were recorded with one-minute resolution in order to study rapid ionospheric changes during the solar eclipse.

Detail analysis of electron concentration by mean of spectral analysis reveals that within oscillation of electron concentration there occur several clear wave-like oscillations. It has been shown by Sauli et al. (2007) that wavelet spectral analysis is very convenient approach for such wave detection. The advantage of the wavelet based analysis is identification of the structure occurrence time which helps to associate particular wave-like structure to the agent. Figure 4 shows estimated wave parameters for selected structure that is coherent through all studied heights. Parametrization of the wave-like structure is based on AGW approximation described in Section 2. From Figure 4 it is evident that wave originates at height of about 200 km and propagates upward and downward from the source region.

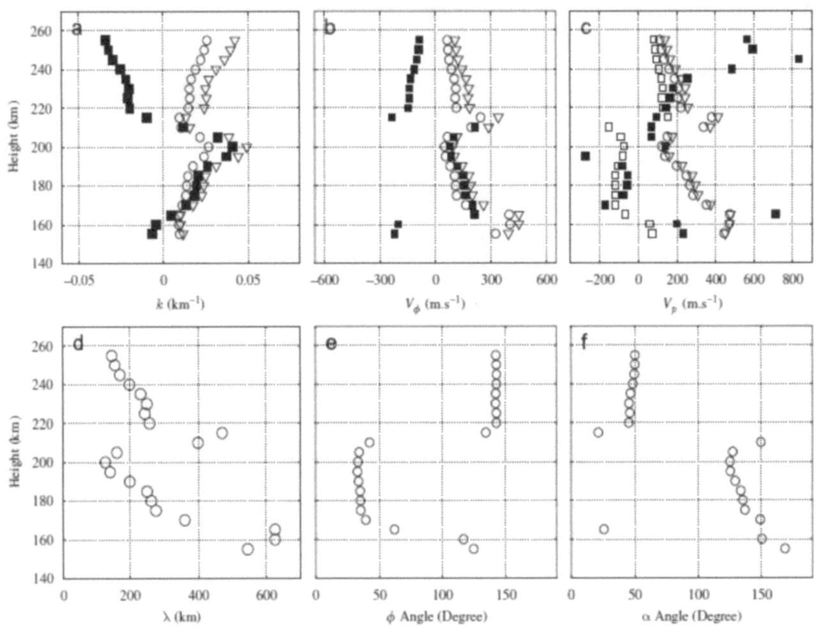

Fig. 4. Parameters of acoustic-gravity wave structure detected within ionospheric plasma during solar eclipse event 11 August 1999 (Sauli et al., 2007). Panels: wave vector (a), phase velocity (b), packet velocity (c), wave number (d), energy (e) and phase (f) angles. For the vectors of first row, the '□' correspond to the measured (full squares) and computed (empty symbols) z-components, the '○' correspond to the horizontal components while the '▽' are related to the modulus.

Another representation of the rapid changes in the ionospheric plasma is shown on the profilogram (Figure 5) measured during solar eclipse 3 October 2005 by DPS 4. Decrease in the plasma frequency at all heights is well developed. Within plasma frequency oscillation, several wave coherent structures were found that can be attributed to the eclipse event. These structures occur in the plasma at the maximum of the eclipse and after the event. In all cases we detected a component of upward energy progression. Due to the occurrence time and low geomagnetic activity the detected wave-like oscillations in the ionospheric plasma are likely signatures of bow shock and possibly waves excited by cooling of ozone in the lower laying atmosphere. Estimated velocities for one particular structure are shown in Figure 6.

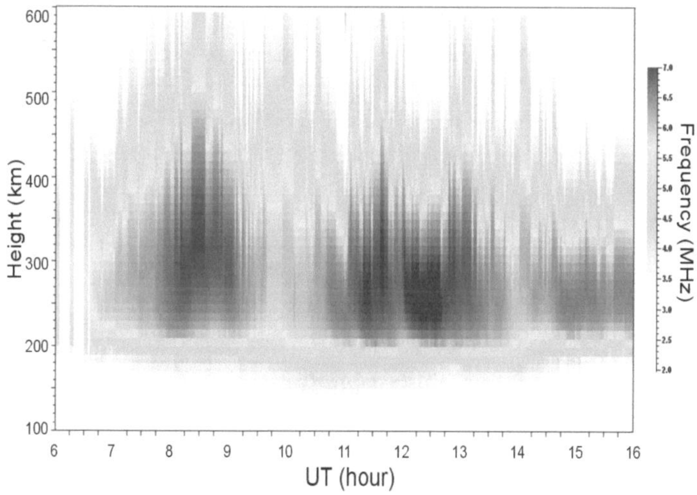

Fig. 5. Profilogram (height-time-plasma frequency development) during solar eclipse 3 October 2005 as measured by DPS 4. Ionograms were measured every 2 minutes. All ionograms were manually scaled and inverted into true-height profiles using True Height Profile Inversion Tool NHPC.

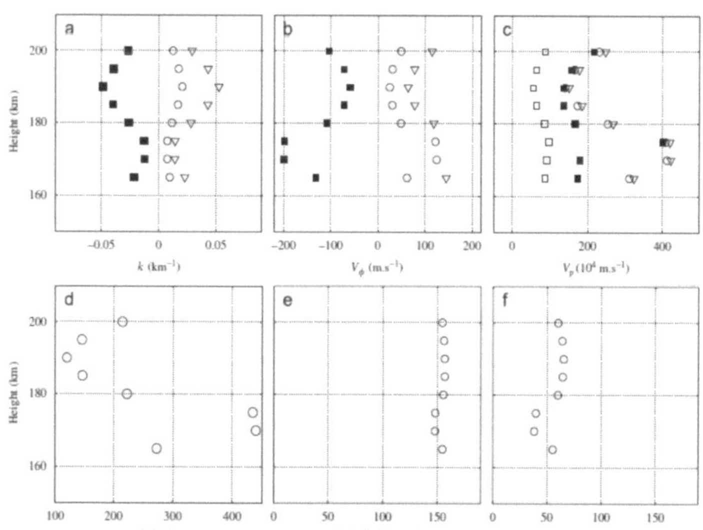

Fig. 6. Parameters of acoustic-gravity wave structure detected within ionospheric plasma during solar eclipse event 3 October 2005 (Sauli et al. 2007). Panels: wave vector (a), phase velocity (b), packet velocity (c), wave number (d), energy (e) and phase (f) angles. For the vectors of first row, the '□' correspond to the measured (black) and computed (empty) z-components, the '○' correspond to the horizontal components while the '∇' are related to the modulus.

Result of the annular eclipse is significantly different from the case of the total eclipse event of 1999 where the dominant AGW activity took place at the beginning of eclipse. The atmospheric cooling and decrease in radiation flux during an annular solar eclipse is not as strong as during a total eclipse and the ionospheric response occurs with time delay.

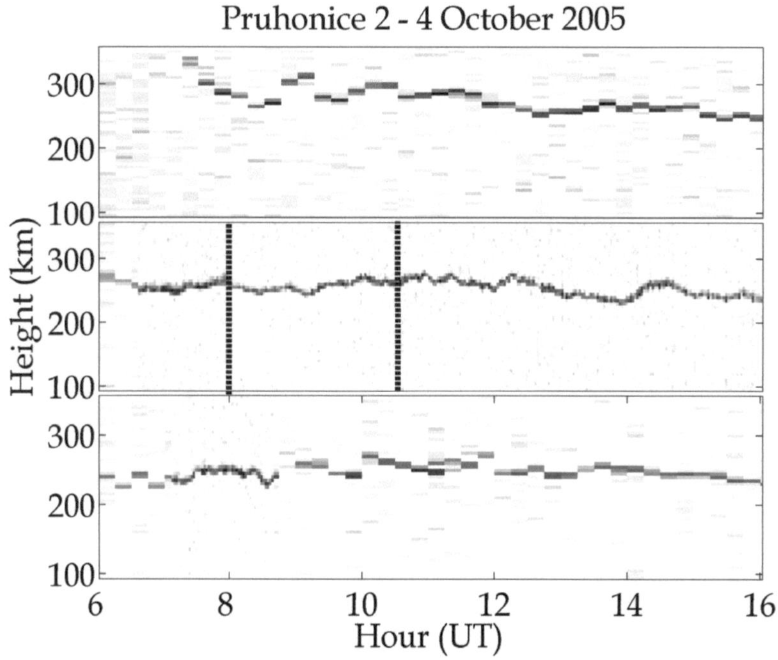

Fig. 7. Virtual reflection heights of plasma frequency in the range 4.2 - 4.3 MHz derived from raw ionograms. From up to bottom: day before, day of eclipse, day after eclipse. Vertical lines in middle panel depict beginning and end of the eclipse. Time resolution is different for day of solar eclipse (2 min) and days before/after (15 min).

In Figure 7 and Figure 8, there are plots of virtual reflection height variations at single frequency during three consecutive days, day of solar eclipse event and one day before and after the event. Variation of the reflection height during eclipse event of 3 October 2005 does not differ much from the corresponding variation during reference time span day before and day after. Wave-like oscillations excited by solar eclipse are of comparable magnitude as those induced by other sources preceding and consecutive day. On the contrary, clear difference in reflection height oscillation during reference days and solar eclipse event is perfectly seen in Figure 8. Records of virtual heights at fixed frequency from January 4, 2011 present strong ionospheric response which is exhibited as periodic changes in reflection height. Sharp changes in the reflection height develop immediately after the beginning of the solar disc occultation and last till the end of eclipse event. Higher wave-like activity remains remarkable whole day. In this partial solar eclipse event, wave-like oscillations can be very probably attributed to the solar eclipse.

Strong decrease in electron concentration in practically whole electron profile as well as the wave-like changes were observed during and after August 11, 1999 and January 4, 2011. Wave-like activity develops immediately after the start of the solar disc obscuration during partial solar eclipse. During annular solar eclipse, significant acoustic-gravity wave type bursts develop around and after maximum phase of the eclipse.

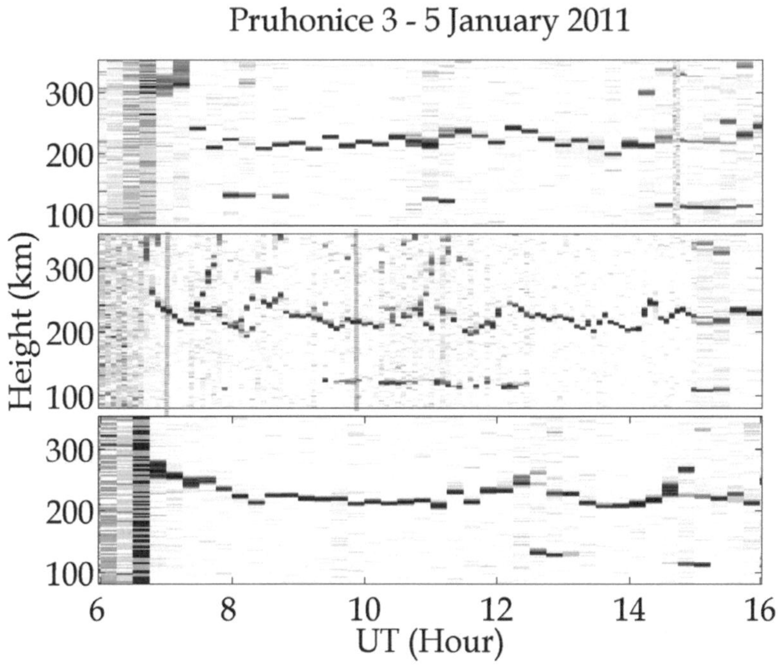

Fig. 8. Virtual reflection heights of plasma frequency range 3.4 - 3.5 MHz derived from raw ionograms. From up to bottom: day before, day of eclipse, day after eclipse. Vertical lines in middle panel depict beginning and end of the eclipse. Time resolution is different for day of solar eclipse (5 min) and days before/after (15 min).

6. Conclusion

Acoustic-Gravity waves play important role in the dynamic of the upper atmosphere. Vertical ionospheric sounding represents powerful tool that allows us to monitor acoustic-gravity wave activity in the ionosphere. Ionospheric observation of such a strong event as solar eclipse gives us an opportunity to better understand processes of creation and dissipation of the AGW in the area of the ionosphere. Although the acoustic-gravity waves are always present in the area of our interest, sharp temporally well-defined changes of solar flux during the solar eclipse give us a possibility to define sources of AGW.

It is rather uneasy to unambiguously assess causality between the solar eclipse events and the detected wave structures in the ionospheric plasma. Difficulties result from the fact that there are no two exactly identical solar eclipse events and from limitations of sounding techniques. Despite the fact that various AGW sources have been identified, many others remain to be found. Amongst irregular AGW bursts, regular increase in AGW activity were found to occur around sunrise and sunset hours, excited by Solar Terminator movement. Most of other sources (meteorological systems, geomagnetic and solar disturbances, etc.) and corresponding wave-like oscillations contribute to the irregular patterns of AGW activity observed in the ionospheric plasma.

As the solar eclipses, analyzed in the Section 5, occur sufficiently long time after the sunrise hours, one can assume that none of the reported waves are induced by solar terminator. During the analyzed sounding campaigns, no wave coming from auroral zone was expected, due to the quiet geomagnetic and solar activity. Additionally, meteorological analysis shows that meteorological systems very probably did not influence the ionosphere during studied events by means of AGW. The acoustic-gravity wave activity increases after a notably larger delay for the annular solar eclipse compared to the total solar eclipses: waves are found during the maximum phase of the eclipse only for the former while they occur during the initial phase for the latter. This discrepancy in gravity waves generation/occurrence can likely be explained by differences in the terrestrial atmosphere cooling: the border between sunlit and eclipsed region is much sharper in the case of total eclipse. Analyzing wave propagations, we observe predominantly upward propagating structures. The wave structure, that propagate upward and downward from the source region located around 200 km height, was created during an exceptional case related to the Solar eclipse of 11 August 1999.

7. References

Altadill, D., Sole, J.G. & Apostolov, E.M. (2001). Vertical structure of a gravity wave like oscillation in the ionosphere generated by the solar eclipse of August 11, 1999. *Journal of Geophysical Research*, 106 (A10), 21419–21428, ISSN 0148-0227.

Bodo, G., Kalkofen, W., Massaglia, S. & Rossi, P. (2001). Acoustic waves in a stratified atmosphere. III. Temperature inhomogenities. *Astronomy & Astrophysics*, 370, pp. 1088-1091, ISSN 0004-6361.

Chen, G., Zhao, Z., Zhou, C., Yang, G. & Zhang, Y. (2010). Solar eclipse effects of 22 July 2009 on Sporadic-E. *Annales Geophysicae*, 28, 353–357, ISSN 0992-7689.

Cheng, K., Huang, Y.N. & Chen, S.W. (1992). Ionospheric effects of the solar eclipse of September 23, 1987, around the equatorial anomaly crest region. *Journal of Geophysical Research*, 97, A1, 103–111. ISSN 0148-0227.

Chimonas, G. & Hines, C.O. (1970). Atmospheric gravity waves induced by a solar eclipse. *Journal of Geophysical Research*, 75, 4, pp. 875, ISSN 0148-0227.

Datta, R.N. (1972). Solar-eclipse effect on sporadic-E ionization. *Journal of Geophysical Research*, 77, 1, 260–262.

Davies, K. (1990). *Ionospheric radio.* Peter Peregrinus Ltd., ISBN 0 86341 186 X, London, United Kingdom.

Davis, J.R., Headrick, W.C. & Ahearn, J.L. (1964). A HF backscatter study of solar eclipse effects upon the ionosphere. *Journal of Geophysical Research*, 69 (1), 190–193. ISSN 0148-0227.

Farges, T., Jodogne, J.C., Bamford, R., Le Roux, Y., Gauthier, F., Vila, P.M., Altadill, D., Sole, J.G. & Miro, G. (2001). Disturbances of the western European ionosphere during the total solar eclipse of 11 August 1999 measured by a wide ionosonde and radar network. *Journal of Atmospheric and Solar-Terrestrial Physics*, 63, 9, pp. 915–924, ISSN 1364-6826.

Fritts, D.C. & Luo, Z. (1993). Gravity wave forcing in the middle atmosphere due to reduced ozone heating during a solar eclipse. *Journal of Geophysical Research*, 98, pp. 3011–3021, ISSN 0148-0227.

Galushko, V.G., Paznukhov, V.V., Yampolski, Y.M. & Foster, J.C. (1998). Incoherent scatter radar observations of EGW/TID events generated by the moving solar terminator. *Annales Geophysicae*, 16, pp. 821-827, ISSN 0992-7689.

Hargreaves, J.K. (1982). The upper atmosphere and solar-terrestrial relations. An introduction to the aerospace environment. Van Nostrand Reinhold, ISBN 0 521 32748 2, Cambridge, United Kingdom.

Hines, C.O. (1960). Internal atmospheric gravity waves at ionospheric heights. Canadian Journal of Physics, 38, pp. 1441–1481, ISSN 1208-6045.

Hooke, W.H. (1968). Ionospheric irregularities produced by internal atmospheric gravity waves. *Journal of Atmospheric and Solar-Terrestrial Physics*, 30, pp. 795-829, ISSN 1364-6826.

Huang X. & Reinish B.W (1996). Vertical electron density profiles from the digisonde network. *Advances in Space Research.*, 18, pp. 121□129, ISSN 0273-1177.

Ivanov, V.A., Ryabova, N.V., Shumaev, V.V., Uryadov, V.P., Nosov, V.E., Brinko, I.G. & Mozerov, N.S. (1998). Effects of the solar eclipse of 22 July 1990 at mid-latitude path of HF propagation. *of Atmospheric and Solar-Terrestrial Physics*, 60, 10, pp. 1013-1016, ISSN 1364-6826.

Jones, B.W., Miseldine, G.J. & Lambourne, R.J.A. (1992). A possible atmospheric-pressure wave from the total solar eclipse of 22 July 1990. *Journal of Atmospheric and Terrestrial Physics*, 54, 2, pp.113-115, ISSN 1364-6826.

Kato, S., Kawakami, T. & St. John, D. (1977). Theory of gravity wave emission from moving sources in the upper atmosphere. *Journal of Atmospheric and Terrestrial Physics*, 39, pp. 581–588, ISSN 1364-6826.

Le, H., Liu, L., Yue, X., Wan, W. & Ning, B. (2009). Latitudinal dependence of the ionospheric response to solar eclipses. *Journal of Geophysical Research*, 114, A07308, ISSN 0148-0227.

Liu, J.Y., Hsiao, C.C., Tsai, L.C., Liu, C.H., Kuo, F.S., Lue, H.Y. & Huang, C.M. (1998). Vertical phase and group velocities of internal gravity waves derived from ionograms during the solar eclipe of 24 October 1995. *Journal of Atmospheric and Solar-Terrestrial Physics*, 60, pp. 1679–1686, ISSN 1364-6826.

Muller-Wodarg, I.C.F., Aylward, A.D. & Lockwood, M. (1998). Effects of a mid-latitude solar eclipse on the thermosphere and ionosphere - a modeling study. *Geophysical Reseach Letters*, 25, 20, pp. 3787–3790, ISSN 0094-8276.

Oliver, W.L., Otsuka, Y., Sato, M., Takami, T. & Fukao, S. (1997). A climatology of F region gravity wave propagation over the middle and upper atmosphere radar. *Journal of Geophysical Research*, 102, A7, pp.14,499-14,512, ISSN 0148-0227.

Rapoport, V.O., Bespalov, P.A., Mityakov, N.A., Parrot, M. & Ryzhov, N.A. (2004). Feasibility study of ionospheric perturbations triggered by monochromatic infrasonic waves emitted with a ground-based experiment. *Journal of Atmospheric and Solar-Terrestrial Physics*, 66, pp. 1011-1017, ISSN 1364-6826.

Reinisch, B.W., Huang, X., Galkin, I.A., Paznukhov, V. & Kozlov, A. (2005). Recent advances in real-time analysis of ionograms and ionospheric drift measurements with digisondes, Journal of Atmospheric and Solar-Terrestrial Physics 67, pp. 1054-1062, ISSN 1364-6826.

Sauli, P., Abry, P., Boska, P. & Duchayne, L. (2006). Wavelet characterisation of ionospheric acoustic and gravity waves occuring during solar eclipse of August 11, 1999. *Journal of Atmospheric and Solar – Terrestrial Physics,* 68, pp. 586-598, ISSN 1364-6826.

Sauli, P., Roux, S.G., Abry, P. & Boska, J. (2007). Acoustic-gravity waves during solar eclipses: Detection and characterization using wavelet transforms. *Journal of Atmospheric and Solar-Terrestrial Physics,* 69, pp. 2465–2484, ISSN 1364-6826.

Sauli, P., Abry, P., Altadill, D. & Boska, P. (2005). Detection of the wave-like structures in the F-region electron density: two station measurement. *Studia Geophysica & Geodetica,* 50, pp. 131-146, ISSN 0039-3169.

Somsikov, V.M. & Ganguly, B. (1995). On the formation of atmospheric inhomogenities in the solar terminator region. *Journal of Atmospheric and Solar – Terrestrial Physics,* 57, 12, pp. 1513-1523, ISSN 1364-6826.

Titheridge J.E., 1985. Ionogram Analysis with the Generalised Program POLAN. *UAG Report-93,* 1985 (http://www.ips.gov.au/IPSHosted/INAG/uag_93/uag_93.html).

Walker, G.O., Li, T.Y.Y., Wong, Y.W., Kikuchi, T. & Huang, Y.N. (1991). Ionospheric and Geomagnetic effects of the solar eclipse of 18 march 1988 in East-Asia. *Journal of Atmospheric and Solar – Terrestrial Physics,* 53, 1–2, pp. 25–37, ISSN 1364-6826.

Xinmiao, Z., Zhengyu, Z., Yuannong, Z. & Chen, Z. (2010). Observations of the ionosphere in the equatorial anomaly region using WISS during the total solar eclipse of 22 July 2009. *Journal of Atmospheric and Solar – Terrestrial Physics,* 72, pp. 869–875, 1364-6826.

Yeh, K.C. & Liu, C.H. (1974). Acoustic-gravity waves in the upper atmosphere. *Reviews of Geophysics and Space Physics,* 12, 2, pp. 193-216, ISSN 8755-1209.

Part 3

Acoustic Waves as Manipulative Tools

15

Use of Acoustic Waves for Pulsating Water Jet Generation

Josef Foldyna
Institute of Geonics of the ASCR, v. v. i., Ostrava
Czech Republic

1. Introduction

The technology of a high-speed water jet cutting and disintegration of various materials attained considerable growth during the last decades. Continuous high-speed water jets are currently used in many industrial applications such as cutting of various materials, cleaning and removal of surface layers. However, despite the impressive advances made recently in the field of water jetting, substantial attention of number of research teams throughout the world is still paid to the improvement of the performance of the technology, its adaptation to environmental requirements and making it more beneficial from the economic point of view.
An obvious method of the water jetting performance improvement is to generate jets at ultra-high pressures. The feasibility of cutting metals with pure water jets at pressures close to 690 MPa was investigated already in early nineties of the last century (Raghavan & Ting, 1991). Such a high pressure, however, induces extreme overtension of high-pressure parts of the cutting system which has adverse effect on their lifetime.
An alternate approach, as shown in this chapter, is to eliminate the need for such high pressures by pulsing the jet. It is well known that the collision of a high-velocity liquid mass with a solid generates short high-pressure transients which can cause serious damage to the surface and interior of the target material. The liquid impact on a solid surface consists of two main stages (see Fig. 1). During the first stage, the liquid behaves in a compressible manner generating the so-called "water-hammer" pressures. These high pressures are responsible for most of the damage resulting from liquid impact on the solid surface. The situation shortly after the initial impact of the liquid on the solid surface is illustrated in Fig. 2. After the release of the impact pressure, the second stage of the liquid impact begins. Once incompressible stream line flow is established, the pressure on the central axis falls to the much lower Bernoulli stagnation pressure that lasts for relatively long time.
The force distribution on liquid jet impact on the solid surface can be summarized as follows: initially a small central area of the first contact is compressed under a uniform pressure. The magnitude of the impact pressure p_i on the central axis is given by

$$p_i = \frac{v\rho_1 c_1 \rho_2 c_2}{\rho_1 c_1 + \rho_2 c_2} \qquad (1)$$

where v is the impact velocity and ρ_1, ρ_2 and c_1, c_2 are the densities and the shock velocities in the liquid and the solid, respectively (de Haller, 1933).

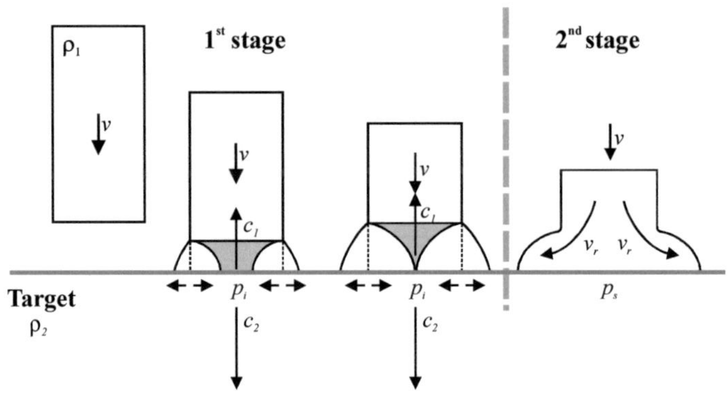

Fig. 1. Two stages of liquid impact on a solid target

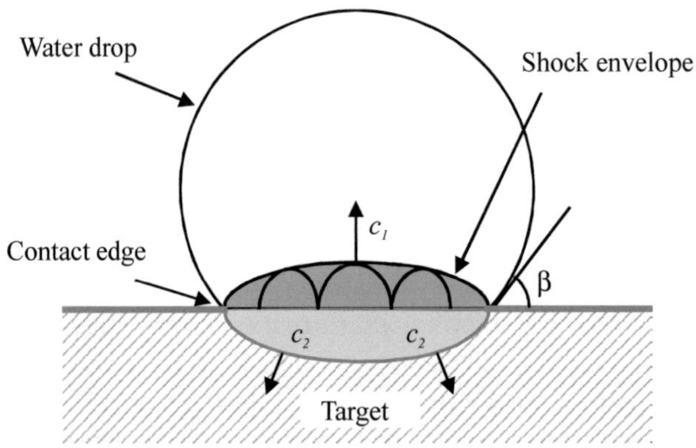

Fig. 2. Initial stage of impact between a water drop and a solid target with the contact edge moving faster than the shock velocity in the liquid. The liquid behind the shock envelope is compressed and the target beneath this area subjected to high pressure

The magnitude of the impact pressure is independent of the geometry of the drop (Thomas & Brunton, 1970), but the duration of the pressure is affected by the size and shape of the drop. For a sphere or cylinder the corresponding radius or half-width of the contact area R exposed to this pressure is given by

$$R = \frac{rv}{c_1} \tag{2}$$

where r is the radius of curvature of the drop or cylinder (liquid mass) in the region of contact (Bowden & Field, 1964).

The initial area of contact grows as the impact continues; there is very little reduction in pressure on the surface until appreciable outward flow begins. The outward flow of the liquid becomes possible when the limit of the compressible deformation of the liquid is exceeded. The limit is given by

$$\frac{v}{c_1} = \sin\beta \tag{3}$$

where β is the liquid/solid interface angle – see Fig. 2 (Hancox & Brunton, 1966).

At this stage there is a rapid fall in pressure along the periphery of contact. As the outward flow continues, the water-hammer compression at the centre of impact is relieved until the maximum pressure acting on the surface is the central stagnation pressure for the incompressible flow. The stagnation pressure is given by

$$p_s = \frac{1}{2}\rho_1 v^2 \tag{4}$$

When the liquid begins to flow away from the point of impact, there is evidence that the velocity of this tangential flow may be as much as five times the impact velocity (Thomas & Brunton, 1970). The velocity increase is thought to be connected with the shape of the head of the jet. It has been observed that an increase in velocity along the surface occurs only in cases where the jet head is inclined at an angle to the surface. Since spherical drops (and/or spherical heads of a train of pulses of pulsating jet) always provide a sloping interface to a plane solid surface it might be expected that high radial velocities will occur on impact. Therefore, there are additional shear forces associated with the high speed flow across the surface acting on the surface in addition to the normal forces. The shear forces acting on a roughened surface are large enough to cause local shear fractures, even in high strength materials (Hancox & Brunton, 1966).

Exploitation of above described effects associated with water droplet impingement on solids in a high-speed water jet cutting technology should lead to considerable improvement of its performance, better adaptation to more and more demanding environmental requirements, and consequently to more beneficial use of the technology also from the economical point of view. Generating sufficiently high pressure pulsations in pressure water upstream the nozzle exit enables to create a pulsating water jet that emerges from the nozzle as a continuous jet and it forms into a train of pulses at certain standoff distance from the nozzle exit. Such a pulsating jet produces all of the above mentioned effects associated with water droplet impingement on solids. In addition, the action of pulsating jet induces also fatigue stress in the target material due to the cyclic loading of the target surface. This further improves the efficiency of the pulsating liquid jet in comparison with the continuous one. Thus, destructive effects of the continuous high-speed water jet can be enhanced by the introduction of high-frequency pulsations in the jet, i.e. by generation of pulsating water jets.

Recently, a special method of the generation of the high-speed pulsating water jet was developed and tested extensively under laboratory conditions. The method is based on the generation of acoustic waves by the action of the acoustic transducer on the pressure liquid and their transmission via pressure system to the nozzle. The high-pressure system with integrated acoustic generator of pressure pulsations consists of cylindrical acoustic chamber connected to the liquid waveguide. The liquid waveguide is fitted with pressure liquid supply and equipped with the nozzle at the end. The acoustic actuator consisting of piezoelectric transducer and cylindrical waveguide is placed in the acoustic chamber (see Fig. 3). Pressure pulsations generated by acoustic actuator in acoustic chamber filled with pressure liquid are amplified by mechanical amplifier of pulsations and transferred by liquid waveguide to the nozzle. Liquid compressibility and tuning of the acoustic system are utilized for effective transfer of pulsating energy from the generator to the nozzle and/or nozzle system where pressure pulsations transform into velocity pulsations. The acoustic generator can be used for generation of both single and multiple pulsating water jets (e.g.

rotating) using commercially available cutting heads and jetting tools. Laboratory tests of the device based on the above mentioned method of the pulsating liquid jet generation proved that the performance of pulsating water jets in cutting of various materials is at least two times higher compared to that obtained using continuous ones under the same working conditions.

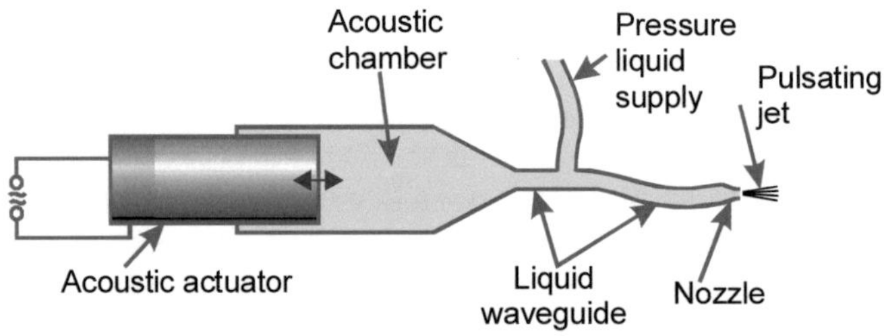

Fig. 3. Schematic drawing of the high-pressure system with integrated acoustic generator of pressure pulsations

However, further improvement of the apparatus for acoustic generation of pulsating liquid jet requires thorough study oriented at determination of fundamentals of the process of excitation and propagation of acoustic waves (and/or high-frequency pressure pulsations) in liquid via high-pressure system and their influence on forming and properties of pulsating liquid jet.

Problems related to the generation and propagation of pressure pulsations with frequency in the order of tens of kHz in liquid under pressure of tens of MPa and subsequent discharge of the liquid influenced by the pulsations through the orifice in the air (producing pulsating liquid jet with axial velocity in the order of hundreds meters per second) were not investigated in detail so far. Only partial information on this topic can be found in publications dealing with processes of a fuel injection for combustion in diesel engines (see e.g. Pianthong et al., 2003 or Tsai et al., 1999) and/or underwater acoustics (Wong & Zhu, 1995).

Therefore, the research on pulsating water jets was focused recently on the study of fundamentals of the process of excitation and propagation of acoustic waves (high-frequency pressure pulsations) in liquid via high-pressure system and their influence on forming and properties of pulsating liquid jet as well as on the visualization of the pulsating jets and testing of their effects on various materials. Results obtained in above mentioned areas so far are summarized in following sections.

2. Acoustic wave propagation in high-pressure system with integrated acoustic generator

The efficient transfer of the high-frequency pulsation energy in the high-pressure system to longer distances represents one of the basic assumptions for generation of highly effective pulsating water jets with required properties. To achieve that goal, the amplification of pressure pulsations propagating through the high-pressure system is necessary. The amplification can be accomplished by properly shaped liquid waveguide that is used for the pulsations transfer to the nozzle. In addition, maximum effects will be obtained if the entire high-pressure system from the acoustic generator to the nozzle is tuned in the resonance. To be able to study theoretically process of generating and propagation of pressure pulsations in the high-pressure system, both analytical and numerical models of the system with integrated acoustic generator were developed.

2.1 Analytical solution

The analytical solution of both pressure and flow oscillation waveforms in the conffuser-shaped tube with circular cross-section is based on linearized Navier-Stokes equations and wave equation for propagation of pressure wave. The wave equation incorporates both the standard kinematical viscosity and the kinematical second viscosity that is related to the liquid compressibility. Therefore, the irreversible stress tensor Π_{ij}, on the basis of which the wave equation is derived, can be written as follows:

$$\Pi_{ij} = 2\eta c_{ij} + \delta_{ij} \int_0^t \Theta(t-\tau) c_{kk}(\tau) d\tau \tag{5}$$

where the function Θ (dynamic second viscosity) is related to the voluminous memory, and c_{ij} represents the tensor of deformation velocity.

In the frequency domain (ω), equation (5) can be written in simplified form verified experimentally:

$$\Pi_{ij\omega} = 2\eta c_{ij\omega} + \delta_{ij} \frac{k}{\omega} c_{kk\omega} \tag{6}$$

whereby δ_{ij} represents Kronecker delta, and η dynamic viscosity. It is obvious from (6) that the dynamic second viscosity is frequency dependent. The kinematical second viscosity is then defined using following formula:

$$\xi = \frac{k}{\rho\omega} \tag{7}$$

where ρ represents liquid density.

2.1.1 Wave equation

If one considers linearized Navier-Stokes equations, the wave equation for pressure function can be written using the Laplace operator Δ in the following form:

$$\frac{\partial^2 p}{\partial t^2} - 2\gamma \frac{\partial}{\partial t}(\Delta p) - \int_0^t \rho^{-1}\Theta(t-\tau) \frac{\partial}{\partial \tau}(\Delta p)\, d\tau - v^2 \Delta p = 0 \tag{8}$$

where γ is kinematical viscosity, p pressure, t time and v speed of sound in water, respectively.

If Laplace transformation for zero initial conditions is applied in (8), following equation can be obtained:

$$s^2\sigma - s[2\gamma + \xi(s)]\Delta\sigma - v^2\Delta\sigma = 0 \tag{9}$$

where s represents parameter of the Laplace transformation according to time ($\xi(s)$ is the Laplace function of the second kinematical viscosity), and, at the same time, following is valid:

$$L\{p(t)\} = \sigma(s) \tag{10}$$

If following expression is denoted κ:

$$\kappa^2 = -s^2[v^2 + (2\gamma + \xi)]^{-1} \tag{11}$$

then it can be written

$$\kappa^2\sigma + \Delta\sigma = 0 \tag{12}$$

In the frequency domain it is valid that $\psi(i\omega) = k/\omega$.
The solution of (12) can be performed by the implementation of spherical coordinate system (r, φ, υ), see Fig. 4. Now, the wave equation can be written in the following form:

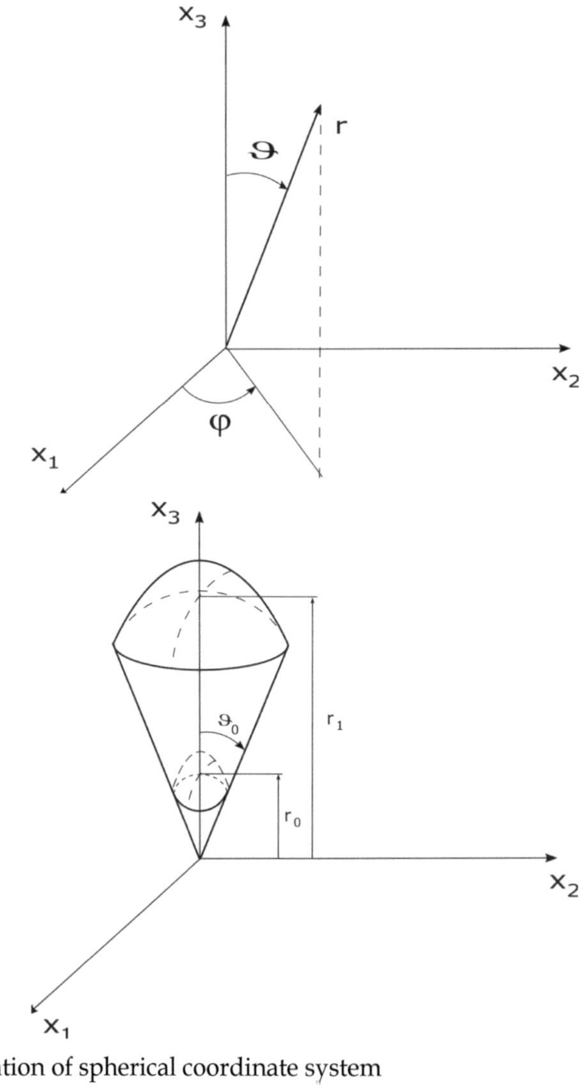

Fig. 4. Implementation of spherical coordinate system

$$\frac{\partial^2\sigma}{\partial r^2}+\frac{2}{r}\frac{\partial\sigma}{\partial r}+\frac{1}{r^2}\frac{\partial^2\sigma}{\partial\upsilon^2}+\frac{1}{r^2}\text{cotg}\frac{\partial\sigma}{\partial\upsilon}+\frac{1}{r^2}\frac{1}{\sin^2\upsilon}\frac{\partial^2\sigma}{\partial\varphi^2}+\kappa^2\sigma=0 \tag{13}$$

Let's assume the solution of (13) as a product of functions:

$$\sigma = Z(r)W(\cos\upsilon)\Phi(\varphi) \tag{14}$$

Then, individual particular integrals can be expressed as follows:

$$\Phi_p = A_n\cos n\varphi + B_n\sin n\varphi \tag{15}$$

$$W_p(\cos\upsilon) = M_{nm}P_m^n(\cos\upsilon) + N_{nm}Q_m^n(\cos\upsilon) \tag{16}$$

where *P, Q* are special Legendre polynomials:

$$Z_p = \frac{1}{\sqrt{r}}[F_mJ_D(\kappa r)+G_mY_D(\kappa r)] \tag{17}$$

$$D = \frac{\sqrt{1+4m(m+1)}}{2} \tag{18}$$

2.1.2 Transfer matrix

The objective is to determine transfer matrix ***P*** that can be used in solving pressure and flow pulsations in hydraulic systems in conffuser-type tubes. For this purpose, it is convenient to introduce the mean velocity of the liquid c_r in a direction of r using following formula:

$$\tilde{c_r} = \frac{1}{2\upsilon_0}\int_0^{\upsilon_0} c_r(r,\varphi,\upsilon)d\upsilon \tag{19}$$

The solution can be simplified by the assumption that the flow is rotary symmetrical. It can be derived under the above mentioned assumption that functions P_m^n, Q_m^n will be streamlined to the following form: P_m^0, Q_m^0. Further, considering that the pressure function p varies only a little with respect to the angle υ, the following relation for the mean velocity c_r can be written based on Navier-Stokes equations:

$$\frac{\partial\tilde{c_r}}{\partial t} = -\frac{2\upsilon_0}{\rho}\frac{\partial p}{\partial r} - \frac{2\upsilon_0\xi}{\rho^2\nu^2}\frac{\partial^2 p}{\partial r\partial t} \tag{20}$$

The continuity equation and component c_r in Navier-Stokes equations expressed in the spherical coordinate system were used in the above mentioned derivation. Withal, effects of dynamic viscosity were neglected. If we will keep considering zero initial conditions, it can be written after the Laplace transformation (20):

$$sw_r = \alpha\frac{\partial\sigma}{\partial r};\ L\{\tilde{c_r}(t)\} = w_r(s) \tag{21}$$

$$\alpha = -\frac{2v_0}{\rho}\left(1 + \frac{s\xi}{v^2}\right) \tag{22}$$

If all assumptions of the solution are considered, following can be written for Laplace images of both the pressure function and the velocity $\widetilde{c_r}$ and with respect to (21):

$$\sigma = \frac{1}{\sqrt{r}}[FJ_{0.5}(\kappa r) + GY_{0.5}(\kappa r)] \tag{23}$$

$$m = 0;\ D = \frac{1}{2} \tag{24}$$

$$w_r = \frac{\alpha}{s\sqrt{r}}\left[F\frac{\partial J_{0.5}(\kappa r)}{\partial \gamma} + G\frac{\partial Y_{0.5}}{\partial r}\right] \tag{25}$$

If we introduce for $r = r_0$ the state vector

$$\mathbf{u}^T = [w_r(r_0, s), \sigma(r_0, s)] \tag{26}$$

and for r the state vector

$$\mathbf{u}^T = [w_r(r, s), \sigma(r, s)] \tag{27}$$

the dependence in locations r and r_0 can be expressed by means of the transfer matrix:

$$\mathbf{u}(r, s) = \mathbf{P}\mathbf{u}(r_0, s) \tag{28}$$

Then, the matrix **P** will be derived from (26) and (27) by the elimination of integration constants F, G. If we designate:

$$\delta = \frac{\alpha}{r}\left[\frac{\partial J_{0.5}}{\partial r}(\kappa r_0)Y_0(\kappa r_0) - \frac{\partial Y_{0.5}}{\partial r}(\kappa r_0)J_{0.5}(\kappa r_0)\right] \tag{29}$$

following relation can be written for matrix **P**:

$$\mathbf{P} = \frac{1}{\delta\sqrt{r_0 r}}\begin{Vmatrix} \alpha\frac{\partial J_{0.5}}{\partial r}(\kappa r) & \alpha\frac{\partial Y_{0.5}}{\partial r}(\kappa r) \\ J_{0.5}(\kappa r) & Y_{0.5}(\kappa r) \end{Vmatrix}\begin{Vmatrix} Y_0(\kappa r_0) & -\alpha\frac{\partial Y_{0.5}(\kappa r_0)}{\partial r} \\ -J_{0.5}(\kappa r_0) & \alpha\frac{\partial J_{0.5}}{\partial r}(\kappa r_0) \end{Vmatrix} \tag{30}$$

In the frequency domain, $s = i\omega$ is substituted.

Both pressure and flow pulsations of hydraulic systems with conffuser-shaped tubes can be solved on the basis of the transfer matrix (30). Individual elements of the transfer matrix are dependent on values of the speed of sound and the second viscosity. Values of both these quantities depend on the static pressure and the value of second viscosity depends also on the frequency. The values can be determined experimentally using the transfer matrix.

2.1.3 Application of the transfer matrix

The transfer matrix derived in the previous section can be used in solving transmission of pressure and flow pulsations in complex hydraulic systems. Such a system can consist of cylindrical and confusser-shaped sections; the system can also be bifurcated.

The advantages of use of the transfer matrixes for determination of both pressure and flow oscillation waveforms in the hydraulic system are illustrated on the model of the conffuser-shaped tube with the circular cross-section. The tube consists of series of coaxial cylinders with various diameters connected with cone frustums and filled with water at a pressure of 30 MPa. The acoustic generator of pressure pulsations located at the end of the largest diameter cylinder vibrates at the frequency of 20 kHz. The cylindrical nozzle is situated on the other end of the tube. The length of the largest diameter cylinder L_1 (and thus also total length of the tube L) can be changed (see Fig. 5).

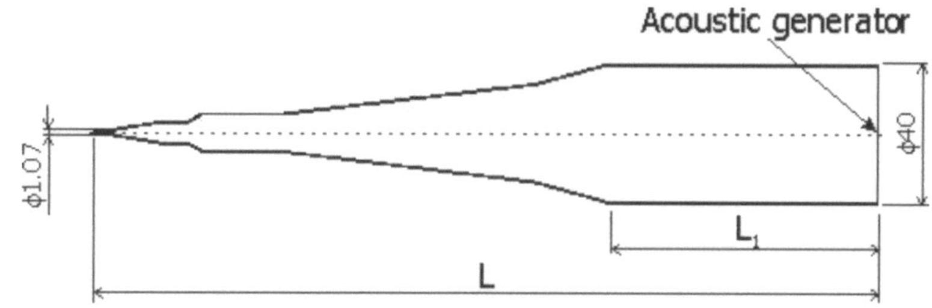

Fig. 5. Schematic drawing of the conffuser-shaped tube with circular cross-section

Effects of acoustic generator on the pressurized fluid in the tube are simulated by mean velocity:

$$\widetilde{c_r} = c_{r0}e^{i\omega t};\, w_{r0} = \frac{1}{s - i\omega} \tag{31}$$

The cylindrical nozzle at the end of the conffuser is represented by the linear hydraulic resistance, simulated by following equations:

$$p = \lambda c_r;\ \sigma = \lambda w_r \tag{32}$$

where λ represents the discharge coefficient.
Based on the transfer matrixes, a simulation model with the above described geometrical configuration was elaborated. Firstly, the natural frequency of the hydraulic system in question was determined using the simulation model. Next, the frequency was applied as oscillation frequency of the acoustic generator and propagation of a generated forced pressure waveform in the hydraulic system was investigated. Fig. 6 illustrates calculated forced pressure waveforms in the simulated geometry related to the phase angle.
It can be seen from the Fig. 6 that oscillations of the acoustic actuator generate a standing wave in the hydraulic system. The standing wave converts to the travelling wave at the area close to the nozzle exit. It is obvious from the presented solution that a properly designed conffuser-type tube can amplify the amplitude of pressure pulsations at the exit of the pulsating nozzle.
The presented analytical solution of the problem of the acoustic (pressure) wave propagation in the high-pressure system can be used for the determination of the optimal geometrical configuration of the high-pressure system to operate in the resonance mode. The solution can also be used in design of a transmission line for efficient transfer of high-frequency pulsation energy to longer distances in high-pressure systems for generation of pulsating high-speed water jets.

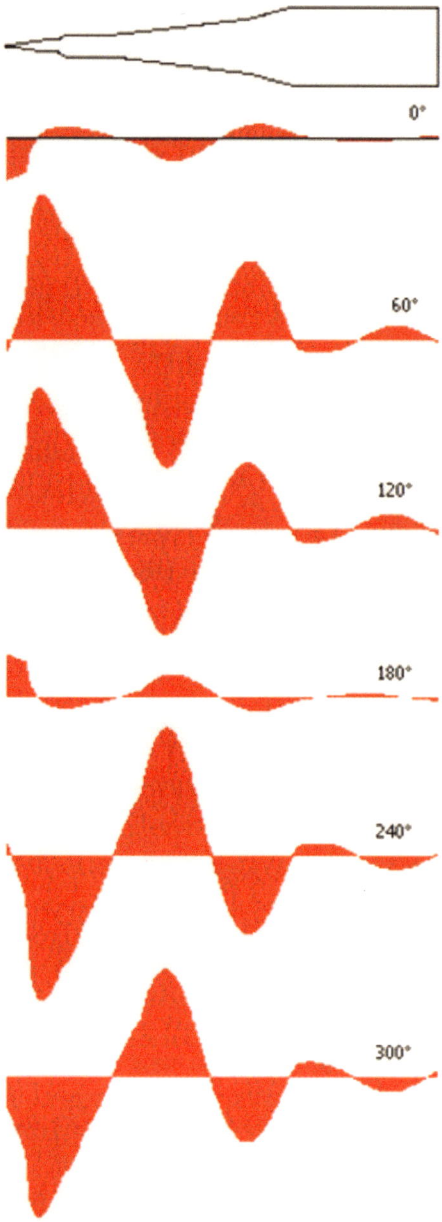

Fig. 6. Amplitudes of forced pressure waveforms in the simulated geometry related to the phase angle calculated from the analytical model

2.2 Numerical model

Computational Fluid Dynamics (CFD) models of selected geometrical configurations of the high-pressure system with integrated acoustic generator were created using CFD code ANSYS CFD to simulate numerically the influence of operating and configuration parameters of the acoustic generator and transmitting line on the generation and propagation of acoustic waves (pressure pulsations) in high-pressure system and properties of pulsating jet. The high-pressure system consisted of cylindrical acoustic chamber, liquid waveguide provided with high-pressure water supply and the nozzle. To simplify the model, acoustic actuator was substituted by vibrating wall of the acoustic chamber. The fluid flow in the model was solved as 3-D turbulent compressible unsteady flow of water. Water compressibility was taken into account in the numerical model using so called user defined function (UDF). The UDF covers calculations of both water density and speed of sound in water related to pressure:

$$\rho = \frac{\rho_{ref}}{1 - \frac{\Delta p}{K}} = \frac{\rho_{ref}}{1 - \frac{p - p_{op}}{K}} \tag{33}$$

$$a = 2.10^{-6} p + 1432 \tag{34}$$

where ρ is water density [kg.m^{-3}], ρ_{ref} is reference water density under normal conditions (998,2 kg.m^{-3}), p and p_{op} are real and operating pressures [Pa], K represents bulk modulus of water (2.2 . 10^9 Pa) and a is speed of sound in water [m.s^{-1}] determined experimentally.
The numerical simulation of a high-pressure system equipped with an acoustic generator was verified by the measurement of pressure pulsations in the high-pressure system upstream to the nozzle exit using dynamic pressure sensors. The pressure waveform in the numerical model was recorded at the same location as the pressure sensor was installed during the laboratory measurement. It was found out that numerical model provides information on the pressure waveform in high-pressure system that is in relatively very good agreement with experimental measurement. Comparison of results of numerical simulation and measurement also proved that the numerical model is able to simulate influence of geometry changes on the amplitude of dynamic pressure accurately and thus also to simulate pressure wave propagation and transmission in the high-pressure system.

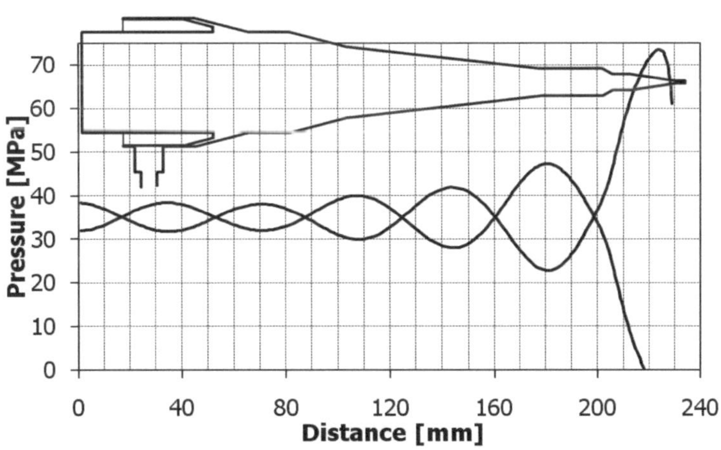

Fig. 7. Standing wave amplitudes along longitudinal axis of the high-pressure system

After the verification of plausibility of results of numerical simulation by the laboratory measurement, the model was used in studying of the process of propagation and transmission of acoustic waves in the high-pressure system from the acoustic actuator to the nozzle. An example of the behaviour of amplitudes of standing wave along the longitudinal axis of high-pressure system can be seen in Fig. 7. Figure 8 illustrates forced pressure waveforms in the simulated geometry related to the phase angle.

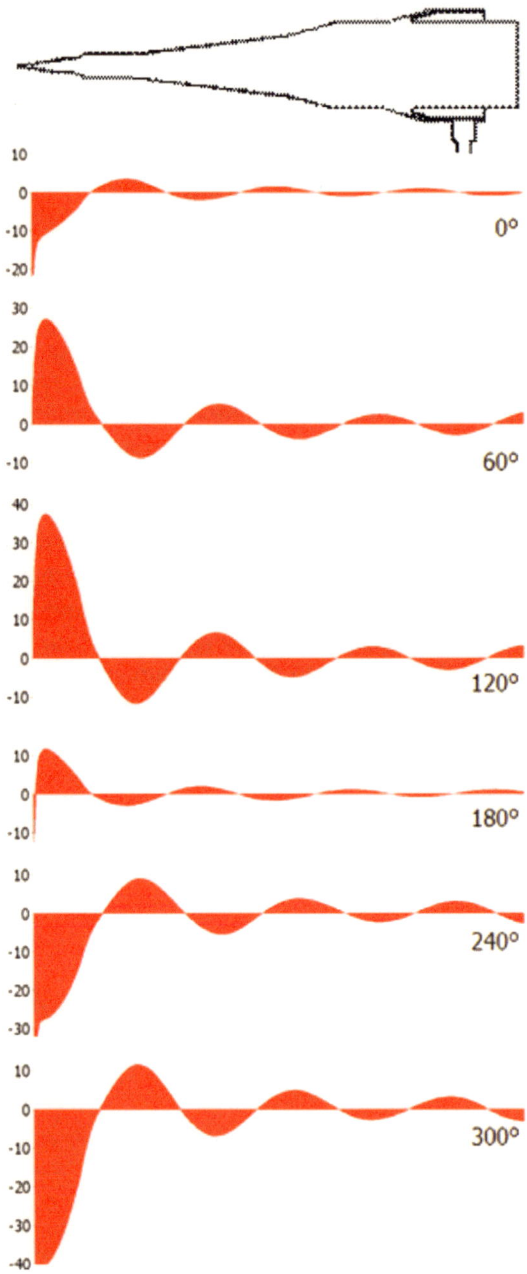

Fig. 8. Calculated forced pressure waveforms in the simulated geometry related to the phase angle recorded along the longitudinal axis of high-pressure system - from numerical model. (Scale indicates amplitude of dynamic pressure in MPa.)

Results obtained from the numerical simulation correspond to results obtained from the analytical one even if the numerical model used is not physically accurate in the close vicinity of the nozzle outlet where cavitation occurs. Cavitation model was not implemented in the numerical model with respect to the computational speed. Results of numerical modelling clearly indicate that the geometrical configuration of high-pressure system influences significantly propagation and transmission of pressure pulsations from the acoustic actuator to the nozzle. The amplitude of pressure waves increases towards the nozzle outlet due to the proper shaping of the liquid waveguide - its frustums act as mechanical amplifiers of the acoustic waves. At the same time, the amplitude of pressure pulsations close to the nozzle outlet (where it has crucial influence on the pulsating jet generation) changes significantly with respect to the geometrical configuration of the high-pressure system.

3. Visualization of pulsating jet

The use of visualization plays an important role in the study of behaviour of pulsating water jet. It enables not only the examination of characteristics of the jet such as mean velocity and break-up length of the pulsating jet but also to study the morphology and processes of formation of the pulsating jet and development of pulses in the jet. Furthermore, the visualization can be used to validate results obtained from numerical simulation of the process of generation of pulsating jets using CFD methods.

An original method of visualization of pulsating water jets based on the application of stroboscopic effect was elaborated for the above mentioned purposes. The method enables to obtain visual information not only on instantaneous structure of the pulsating jet but also on the mean structure of the jet. In addition, the stroboscopic effect allows observing process of formation of pulsating water jet by the naked eye. Special stroboscope for the pulsating jet visualization was developed where the frequency of stroboscope flashing is controlled by the frequency of pressure pulsations in the high-pressure system measured upstream from the nozzle exit. An example of the mean structure of pulsating jet with pulsating frequency of 20 kHz can be seen in Fig. 9. Exposure time of the photograph was 1/1000 s and the frequency of stroboscope flashing was about 20 kHz, therefore the figure represents superposition of 20 images of pulsating jet "frozen" by the stroboscope flashing.

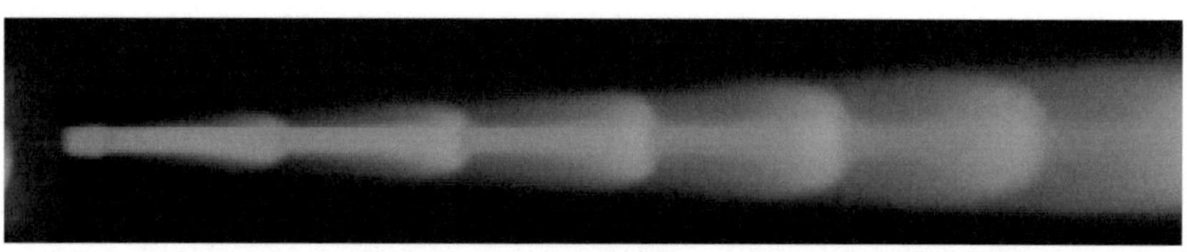

Fig. 9. The mean structure of the pulsating water jet generated at 30 MPa (illumination by the stroboscope)

An instantaneous structure of the pulsating water jet with the frequency of 20 kHz was studied using the high-speed camera LaVision VC-HighSpeedStar 5 equipped with image amplifier LaVision HighSpeed IRO. The record rate of the high-speed camera was 35 000 frames per second and the gate was set to 1μs. In addition, visualization of instantaneous structure of the pulsating water jet was also performed using Particle Image Velocimetry (PIV) system consisting of LaVision Imager Intense camera and New Wave Research laser,

Model Solo 120 with pulse duration 3-5 ns; the optical system was used to produce 1 mm thick sheet of light. An example of the pulsating jet visualized by high-speed camera can be seen in Fig. 10, the same jet visualized by PIV system is presented in Fig. 11.

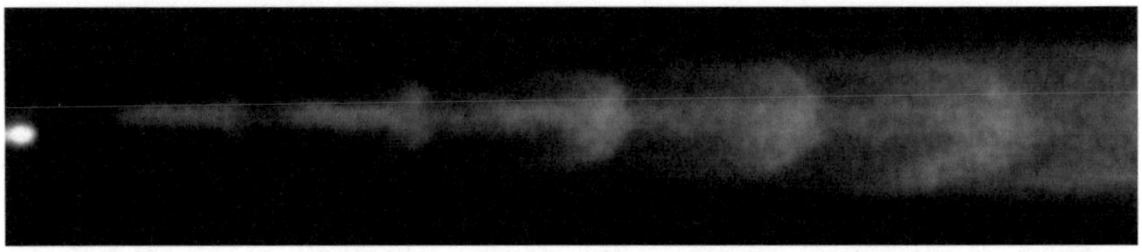

Fig. 10. The instantaneous structure of the pulsating water jet generated at 30 MPa (high-speed camera)

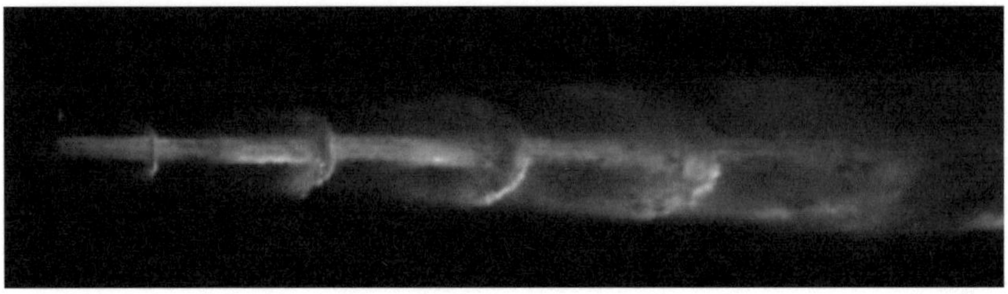

Fig. 11. The instantaneous structure of the pulsating water jet generated at 30 MPa (PIV system)

The fan (flat) pulsating water jet visualization was performed using the pulsed laser New Wave Research and digital camera Nikon D70s. Figure 12 shows the morphology of fan pulsating water jet generated at a pressure of 20 MPa.

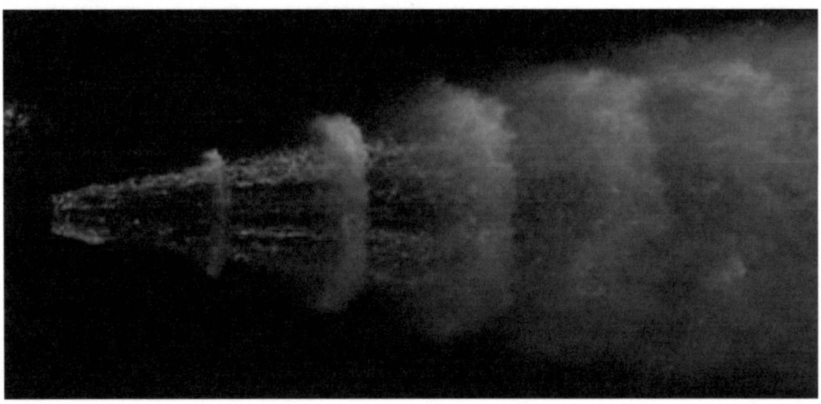

Fig. 12. The instantaneous structure of the fan pulsating water jet generated at 20 MPa (illumination by pulsed laser, camera Nikon D70s)

4. Effects of pulsating water jets on materials

Effects of pulsating water jets with the frequency of 20 kHz were tested on various types of materials, such as metals, rocks and concrete. Tested materials were exposed to the action of diverse types of jets: single round and fan pulsating jets as well as rotating pulsating jets. The effects of pulsating jets were evaluated in terms of cutting depth, rate of mass-loss or volume removal rate respectively and compared with the effects of continuous water jets under the same operating conditions.

Obtained results show clearly the supremacy of pulsating water jets over continuous ones in terms of their effects on material. Figures 13 to 15 illustrate the effects of various types of both pulsating and continuous jets on metal, rock and concrete samples. Differences in the surface structures created by pulsating and continuous water jets on individual materials are clearly visible in the above mentioned figures. An example of erosion effects of pulsating fan water jets (generated at various pressures) on aluminium samples at variable stand-off distance is presented in Figure 16.

a) b)

c) d)

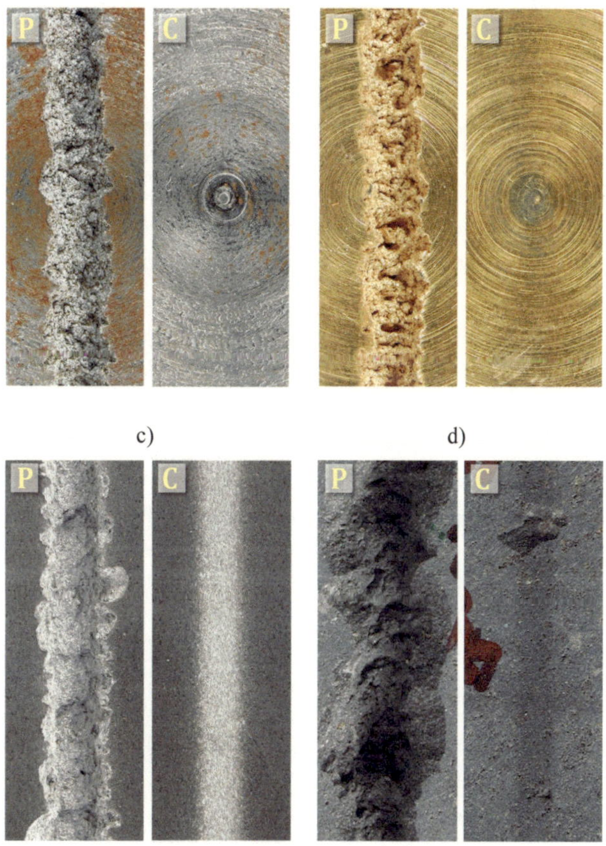

Fig. 13. Comparison of effects of pulsating (P) and continuous (C) water jets on samples of: a) mild steel (pressure 40 MPa, nozzle dia. 1.98 mm, traverse speed 0.03 m.min^{-1}, standoff distance 140 mm), b) brass (pressure 40 MPa, nozzle dia. 1.98 mm, traverse speed 0.03 m.min^{-1}, standoff distance 140 mm), c) duralumin (pressure 50 MPa, nozzle diameter 1.45 mm, traversing speed 0.05 m.min^{-1}, standoff distance 60 mm) and d) basalt (pressure 50 MPa, nozzle diameter 1.45 mm, traversing speed 1.0 m.min^{-1}, standoff distance 40 mm (P) and 20 mm (C))

Fig. 14. Comparison of effects of rotating pulsating (A) and rotating continuous (B) water jets on concrete sample (V_r - removed volume, pressure 30 MPa, nozzle diameter 2x1.47 mm, traversing speed 0.5 m.min^{-1}, standoff distance 40 mm (A) and 20 mm (B))

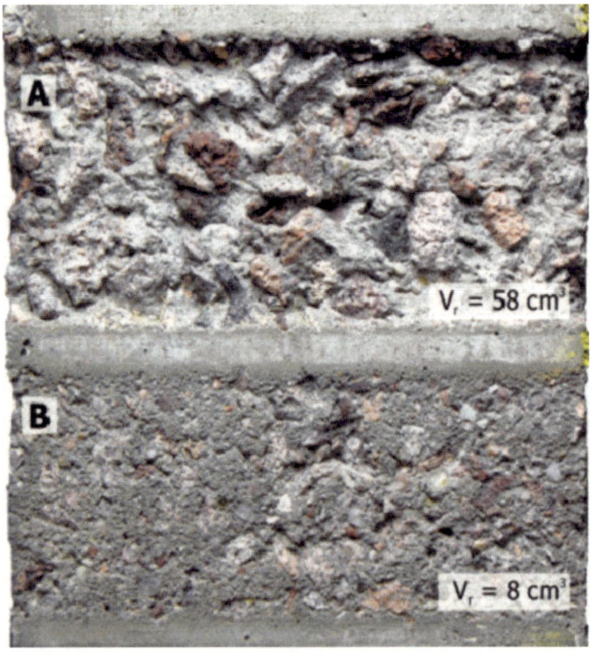

Fig. 15. Comparison of effects of fan pulsating (A) and fan continuous (B) water jets on concrete samples (V_r - removed volume, pressure 30 MPa, equivalent nozzle diameter 2.05 mm, traversing speed 0.2 m.min^{-1}, standoff distance 40 mm)

Results of the measurement of surface roughness characteristics on surfaces created by fan pulsating jets indicate that the characteristics are strongly influenced by both the standoff distance and the operating pressure. An example of the influence of a standoff distance on arithmetic mean roughness (SRa) and average maximum height roughness (SRz) can be seen in Figures 17 and 18, respectively. It should be pointed out that surface roughness (both SRa and SRz) produced by the pulsating fan water jet out of the range of "optimum" standoff distances (where the pulsating jet acts as a continuous jet) correspond to those produced by continuous jets reported by Kunaporn et al. (2009). On the other hand, the fan pulsating water jet produces surfaces with much higher values of the surface roughness (up to 20 times higher) within the "optimum" range of standoff distances (where the pulses are well-developed in the jet) compared to continuous jets.

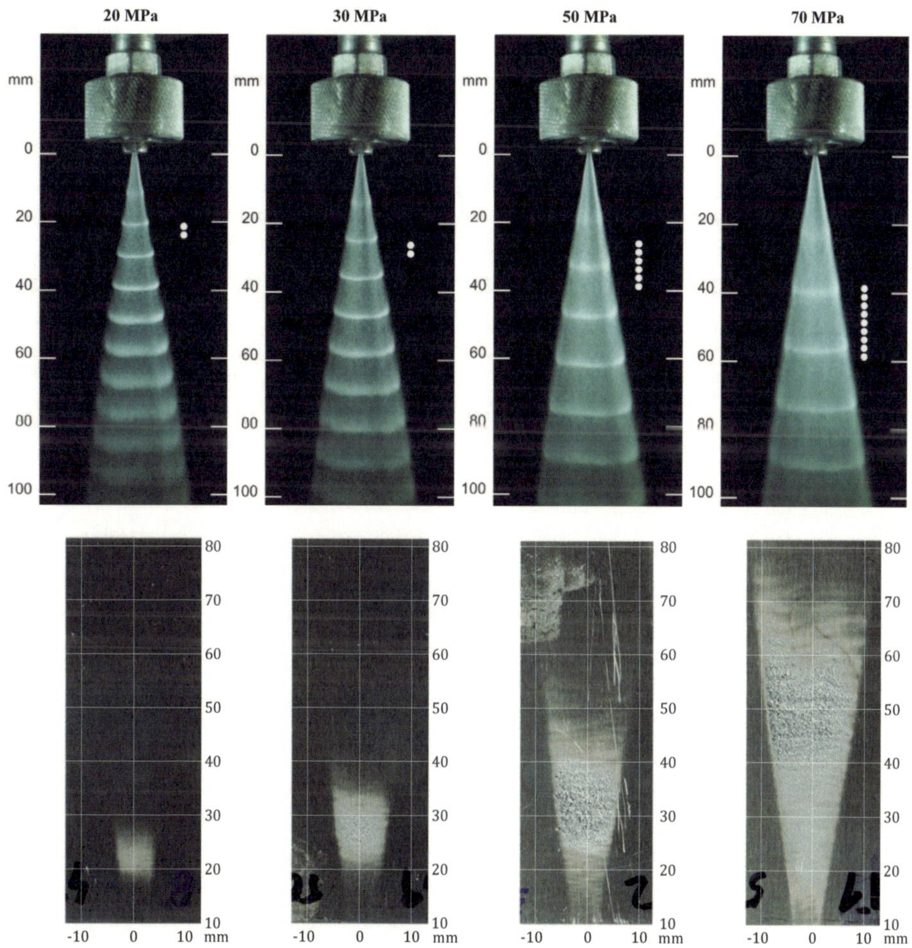

Fig. 16. Top: Pulsating fan water jets generated by nozzle with equivalent of diameter 1.10 mm and spraying angle of 10° at various operating pressures. Scale on the left side of photographs represents standoff distance in millimeters. Dots indicate the range of standoff distances where maximum erosion effects of pulsating fan water jet occur. Bottom: Erosion effects of the above pulsating fan water jets on duralumin samples. Scale on the right side of photographs indicates standoff distance in millimeters; scale on the bottom indicates width in millimeters

Results of the research of effects of pulsating jets on various materials obtained so far indicate that the pulsating jets can be used advantageously for the removal of surface layers of materials and/or "rough" cutting. However, further research will be necessary to be able to use the pulsating water jets in applications of precise cutting.

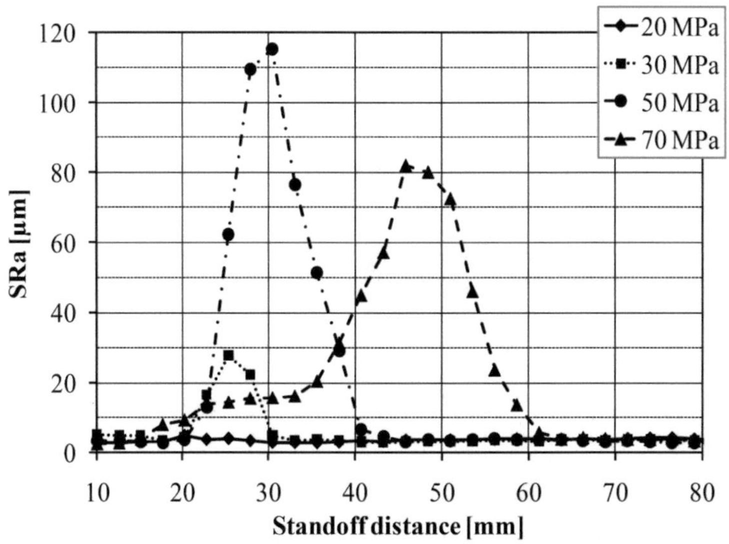

Fig. 17. Influence of a standoff distance on arithmetic mean roughness (SRa) of the surface created by the action of pulsating fan water jet generated by the fan jet nozzle with equivalent of diameter 1.10 mm and spraying angle of 10°

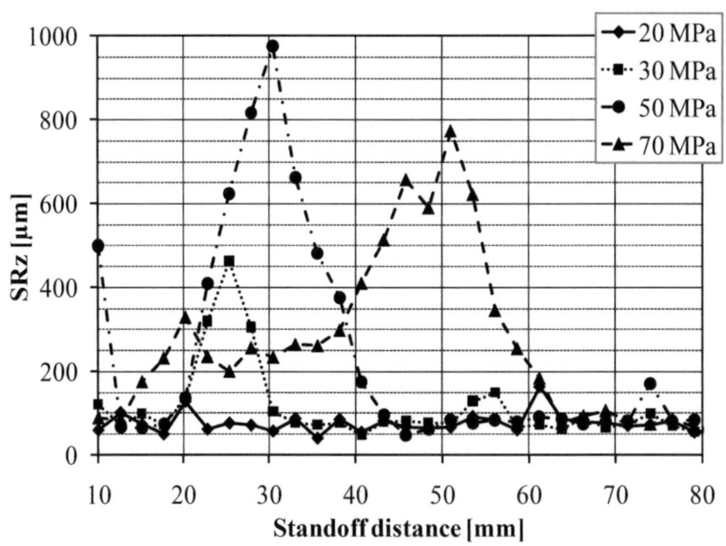

Fig. 18. Influence of a standoff distance on average maximum height roughness (SRz) of the surface created by the action of pulsating fan water jet generated by the fan jet nozzle with equivalent of diameter 1.10 mm and spraying angle of 10°

5. Conclusions

Presented results of the analytical solution and numerical simulation of the transmission of acoustic waves in high-pressure system represent the first step in gaining knowledge regarding processes of generation and propagation of high-frequency pressure pulsations in the liquid under high pressures and their influence on forming and morphology of pulsating liquid jets.

Results obtained from the visualization of pulsating water jets are used in studying of the characteristics of the jets and to verify results obtained from numerical simulation of the process of generating and forming of pulsating water jets. Laboratory and pilot tests of effects of pulsating water jets on various materials showed clearly the potential of pulsating jets to improve the performance of water jetting technology significantly.

It can be concluded that the research presented in the paper contributed to better knowledge of processes occurring in areas of generation and propagation of high-frequency pressure pulsations in the liquid under high pressure, their influence on forming and morphology of pulsating water jets and effects of the jets on materials. However, it is still necessary to further study problems of the efficient transfer of the high-frequency pulsation energy to longer distances in the high-pressure system. This will enable creation of the highly effective pulsating liquid jet with required properties.

6. Acknowledgements

The chapter has been done in connection with project Institute of clean technologies for mining and utilization of raw materials for energy use, reg. no. CZ.1.05/2.1.00/03.0082 supported by Research and Development for Innovations Operational Programme financed by Structural Founds of Europe Union and from the means of state budget of the Czech Republic. Presented work was also supported by the Academy of Sciences of the Czech Republic, project No. AV0Z30860518. Author is thankful for the support.

7. References

Bowden, F. P., & Field, J. E. (1964). The brittle fracture of solids by liquid impact, by solid impact, and by shock. *Proceedings of the Royal Society of London. Series A, Mathematical and Physical Sciences*, Vol. 282, No. 1390, pp. 331-352

de Haller, P. (1933). Untersuchungen über die durch Kavitation hervorgerufenen Korrosionen. *Schweizerische Bauzeitung*, Vol. 101, No. 21& 22, pp. 243-246,260-264

Hancox, N. L., & Brunton, J. H. (1966). The erosion of solids by the repeated impact of liquid drops. *Philosophical Transactions of the Royal Society of London. Series A, Mathematical and Physical Sciences*, Vol. 260, No. 1110, pp. 121-139

Kunaporn, S., Chillman A., Ramulu, M., & Hashish, M. (2009). Effect of waterjet formation on surface preparation and profiling of aluminum alloy. Wear, Vol. 265, No. 1-2, pp. 176-185

Pianthong, K., Zakrzewski, S., Behnia, M., & Milton, B. E. (2003). Characteristics of impact driven supersonic liquid jets. *Experimental thermal and fluid science*, Vol. 27, No. 5, pp. 589-598

Raghavan, C. & Ting, E. (1991). Hyper pressure waterjet cutting of thin sheet metal. *Proceedings of 6th American Water Jet Conference*, pp. 493-504, ISBN 1-880342-00-6, Houston, Texas, August, 1991

Thomas, G. P., & Brunton, J. H. (1970). Drop impingement erosion of metals. *Proceedings of the Royal Society of London. Series A, Mathematical and Physical Sciences*, Vol. 314, No. 1519, pp. 549-565

Tsai, S. C., Luu, P., Tam, P., Roski, G., & Tsai, C. S. (1999). Flow visualization of Taylor-mode breakup of a viscous liquid jet. *Physics of fluids*, Vol. 11, No. 6, pp. 1331-1341

Wong, G. S., & Zhu, S. (1995). Speed of sound in seawater as a function of salinity, temperature and pressure. *Journal of the Acoustical Society of America*, Vol. 97, No. 3, pp. 1732-1736

16

Molecular Desorption by Laser–Driven Acoustic Waves: Analytical Applications and Physical Mechanisms

Alexander Zinovev, Igor Veryovkin and Michael Pellin
Argonne National Laboratory
USA

1. Introduction

Analytical mass-spectrometry (MS) is a powerful, widely-used tool for materials analysis, helping to make progress in materials and environmental sciences, chemistry, biology, astrophysics, etc (Dass 2007). Often the sample to be studied (analyte) is a solid requiring: a) volatilization/desorption of the analyte atoms/molecules and b) their consequent conversion to the charged particles (ionization) prior to mass analysis. The last two decades have seen revolutionary advances in these techniques (Dass 2007) and the use of direct laser irradiation to achieve volatilization is one of the wide-spread methods (Lubman 1990) These pulsed laser-based techniques for the desorption/emission of the atoms, molecules and ions from the surface of solids has benefitted from fundamental study of the process beginning with the invention of the lasers (Honig and Woolston 1963) . A short laser pulse hitting a solid absorbing surface delivers high energy in a small volume inducing a variety of state changes. One consequence is the evaporation/desorption of surface atoms and molecules could be used for further analysis by MS technique. However, the increasing use of MS methods in analytical chemistry of organic and biomolecules revealed that this direct desorption process had significant drawbacks for the analysis of molecular solids. Most importantly, the high energy density produced during irradiation results in not only surface heating but also in excitation of internal vibrational and electronic states of desorbed molecules leading to their partial or even complete fragmentation (Lubman 1990). This difficulty was overcome for many samples by the development of Matrix Assisted Laser Desorption and Ionization (MALDI), which by imbedding the analyte in a specialized UV absorbing molecular solid (the "matrix") allows UV lasers to both desorb and ionize large organic and biomolecules without significant fragmentation (Cole 2010). Because MALDI combines both of the needed initial processes (desorption and ionization) it very quickly following the pioneering publication (Karas, Bachmann et al. 1985) became a key analytical tool. MALDI is now one of principle research tools in proteomics (Cole 2010) and its discovery was recognized with the Nobel Prize in chemistry in 2002.

Despite the success of the MALDI technique current active areas of research include quantification and analysis in the low mass region. Application of MALDI to analyte quantification while possible requires careful attention to matrix/analyte sample preparation, a detailed understanding of the crystallization process with regard to the analyte, and careful many spot analyses (to find the sample signal average which often varies by orders of magnitude as a function of laser position). (Duncan, Roder et al. 2011)

This desire to find a discriminative, sensitive and more easily quantifiable alternative to MALDI has lead us to re-examine another molecular desorption method that doesn't require the use of a matrix. The first observation of this method of molecular desorption (later called Laser Induced Acoustic Desorption, LIAD) belongs (to the best of our knowledge) to B. Lindner (Lindner and Seydel 1985). This observation has been followed by several studies including (Golovlev, Allman et al. 1997) where the abbreviation LIAD was introduced and by (Perez, Ramirez-Arizmendi et al. 2000) who applied the technique to several classes of MS problems.

LIAD has rather simple experimental layout: an analyte is deposited onto the front surface of thin metal foil (the substrate), which is irradiated from the back (i.e. the side opposite to both the analyte and the mass spectrometer) by a pulsed laser beam with power density insufficient to pierce the foil. Such irradiation results in the volatilization of the analyte. The volatilization is largely in the form of neutral molecules that can be utilized for further MS analysis using an appropriate post-ionization method. The method seems to be relatively insensitive to the sample preparation method. Commonly the sample preparation requires only evaporating a drop containing a few nano-moles of the analyte. Remarkably little desorption induced fragmentation is seen when a suitable "soft" ionization method such as VUV photoionization can be found. It is also useful to note that while the molecular signal depends on drive laser intensity the fragmentation observed varies only weakly.

The advantages of LIAD have been demonstrated by many researchers; however, the mechanism is not well understood. The first desorption mechanism was proposed (Golovlev, Allman et al. 1997). It was supposed that because the metal is opaque and completely blocks the direct interaction between the drive laser light and adsorbed, front-side molecules, the only possible way of energy transfer is the mechanical. In this model, the interaction of laser pulse with metal foil backside resulted in formation of acoustic waves that move through the foil inducing a front surface oscillation motion. The molecules that are sitting on this surface desorb due to a simple "shake-off" mechanism similar to those that we use to remove dust particles from our clothes by shaking it.

A difficulty with this model arises when considering the relatively strong surface binding energy experienced even by physisorbed molecules. In order to be efficiently desorbed from the surface, surface molecules need to achieve initial kinetic energies exceeding their surface binding energies (typically in the range of 0.05 - 0.5 eV for physically adsorbed molecules (Adamson and Gast 1997)). This corresponds to velocities of a few hundred m/s for molecules with masses of a few hundred atomic mass units. Unfortunately, acoustic vibrations have mass transfer velocities much lower than the speed of sound, and in elastic deformation mode, this velocity does not exceed a few m/s (Landau and Lifshits 1987). While laser-driven acoustic wave generation in metals is very well studied problem (Hutchins 1985), the physics of their generation in metal foils is crucial to understanding the need for development of new desorption mechanisms. In the next paragraphs we will give a brief theoretical overview and will present our experimental results on laser-driven acoustic waves in thin metal foils.

2. Laser-driven acoustic waves in thin metal foils

The interaction of pulse laser beam with metal surface is very complex phenomenon but our specific interest is in formation of the acoustic waves in irradiated material. To generate an acoustic wave a time dependent stress needs to be applied to the solid. A laser pulse is an excellent tool to generate this kind of the stress. There are two principal mechanisms of laser-induced stress formation in the solids: a) thermal stress, resulted from the non-uniform heating of the irradiated surface by the laser beam; b) mechanical stress due to mechanical impulse transferred from the leaving plasma plume formed on the surface during laser ablation. The parameters of the acoustic waves generated for these two processes are slightly different and will be discussed in details later. Here we will use acoustic wave theory but one should note that its' use is applicable only when the magnitude of the applied stress is small in comparison with the Young's modulus of the material. Large applied stresses can cause the development of the shock waves a phenomenon with different characteristics than acoustic waves (Menikoff 2007). Shock waves have been also hypothesized to be the driver of the LIAD phenomenon, and, as such cannot be entirely excluded from consideration, especially in some extreme cases. Nevertheless, during the last decade, laser-driven acoustic waves emerged in the literature as the "prime suspect" in the LIAD case.

2.1 Acoustic waves in metal foils

The general governing equation for the generation of elastic waves in solids can be derived by combining the equation of motion and the Hooke's law. In general case it is a differential tensor equation, which interconnects the stress tensor, applied to the body, the displacement of the body's elemental volumes (strain), and their elastic properties. In order to analyze the data in detail, the appropriate stress tensor needs to be determined, and the corresponding set of partial differential equations for strain and stress must be resolved (Pollard 1977). One can simplify this analysis by taking into account specifics of the experiments. For thin foils with h/R_0<<1 (where R_0 is the radius of the target foil, and h is its thickness), a round thin plate approximation can be applied to describe and analyze this problem (Smith 2000). The rise of the strain due to laser heating and the consequent development of the plasma plume can be considered as an external driving force. Generally speaking, the thin plate equation is a differential equation of the forth order,

$$\left(\frac{\partial^2}{\partial r^2}+\frac{1}{r}\cdot\frac{\partial}{\partial r}+\frac{1}{r^2}\cdot\frac{\partial^2}{\partial \theta^2}\right)^2 \xi + v_L^{-2}\cdot\left(2\cdot\alpha\frac{\partial \xi}{\partial t}-\frac{\partial^2 \xi}{\partial t^2}\right)=F(r,t) \quad (1)$$

where $\xi=\xi(r,t)$ is the surface displacement in the z direction (perpendicular to the sample surface), $F(r,t)$ is the external driving force caused by the laser irradiation, v_L is a parameter depending on the material density, ρ, thickness, h, and flexural rigidity, D. Furthermore,

$$v_L=\sqrt{D/\rho\cdot h}\ \text{, and , }\ D=(N\cdot h^3/12\cdot(1-\varepsilon^2)) \quad (2)$$

where ε is the Poisson ratio, N is Young's modulus and α is the oscillation decay constant. A vibrating thin plate is one of the most frequently analyzed mathematical problems, and its detailed analysis can be found elsewhere (McLachlan 1951; Smith 2000). In the discussion given below, we will remain in the framework of the analysis given by (Smith 2000).

Assuming separate solutions for the radial and tangential terms, Eq. (1) can be converted into a system of three differential equations of the second order. Being primarily interested in the foil displacement in normal (z) direction to the surface at the epicenter (r=0) and assuming that the external driving force $F(t)$ is a pulse function lasting a specific time τ, we can separate the variables in the Eq. (2) for $t>>\tau$. Under the assumption of harmonic motion for all modes, the surface displacement can be expressed as $\xi(r,t)=\xi_1(r)\cdot e^{i\omega_{m,n}t-\alpha\cdot t}$, where $\omega_{m,n}$ is the vibration frequency. The general governing equation can then be written in the following form:

$$\left(\frac{\partial^2}{\partial r^2}+\frac{1}{r}\cdot\frac{\partial}{\partial r}-\frac{n^2}{r^2}\pm k_{n,m}^2\right)\xi_1=0 \tag{3}$$

where n is an integer number, $k_{n,m}=\omega_{n,m}/v_L$ and v_L is the wave velocity as in Eq.(2).
Under the assumption that the oscillations are harmonic and decay exponentially, the following solution for the surface displacement can be obtained (McLachlan 1951):

$$\xi_n(r,t)=\xi_0\cdot\left(J_n(k_{n,m}r)+\chi\cdot I_n(k_{n,m}r)\right)\cdot\exp(i\omega_n t)\cdot\exp(-\alpha\cdot t) \tag{4}$$

where J_n and I_n are Bessel functions and χ is a constant. The term describing the angular dependence of the oscillations is omitted in Eq.(4). The fact that foils used in LIAD experiments are typically glued or welded on their perimeter corresponds in our analysis to the situation when edges of the round plate are fixed (i.e. non-vibrating), and is described by the following boundary conditions:

$$\xi(R_0,t)=0,\frac{d\xi(R_0,t)}{dt}=0\,, \tag{5}$$

that lead to an equation, whose solutions $j_{n,m}$ have tabulated values (Smith 2000). The corresponding vibration frequencies, $\omega_{n,m}$, can be then expressed as

$$\omega_{n,m}=\sqrt{j_{n,m}^4\cdot\frac{v_L^2}{R_0^4}-\alpha^2}\approx\frac{j_{n,m}^2\cdot h}{R_0^2}\cdot\sqrt{\frac{N}{12\rho\cdot(1-\varepsilon^2)}} \tag{6}$$

Because values of α are small compared to the first term under the square root sign in Eq.(6), it is a reasonable assumption that the frequency is proportional to the square root of the ratio of Young's modulus, N, to the density of the foil material, ρ. As described by Eq. (6), in the steady-state regime (driving force $F(t)$=0 for $t>>\tau$) the frequency spectra and decay times of the oscillation will remain the same while the laser intensity is varied, and only the amplitude should change.
For short times, ($t<\tau$), the approach to solving Eq.(1) is to assume that the external force is a delta-function in space $F(r,t)=F(t)\delta(r)$ (point source), allowing the solution of Eq.1 to be expressed in the form

$$\xi(r,t)=g(r)\cdot\xi(t) \tag{7}$$

where $g(r)$ and $\xi(t)$ represent the spatial and the time dependencies of the final solution, respectively. This approximation should help to develop a clearer understanding of physical problems related to the laser generation of acoustic waves in thin foils.

Using Eq. (7), Eq. (1) can be split into two independent equations, with the equation for $\xi(t)$ having the form

$$\frac{d^2\xi}{dt^2} - 2\cdot\alpha\cdot\frac{d\xi}{dt} + \omega_n^2\cdot\xi = F(t), \tag{8}$$

and the equation for $g(r)$ being a non-uniform Bessel equation with the $\boldsymbol{\delta(r)}$- function in the right side of it.

The exact solution of Eq. (8) can be expressed as the convolution of the Green function of the problem (8) and real time shape of $F(t)$. The governing equation for the Green function will be Eq.(8) with δ-function in the right side. It can be easily derived by applying the Laplace transform to the Eq.(8), subsequently solving the obtained linear equation in the s-space and returning back to the time space with using the inverse Laplace transform. As a result of these procedures, one can finally obtain the following equation

$$\xi_n(t) = \int_0^t \frac{e^{-\alpha\cdot(t-\tau)}}{\sqrt{\omega_n^2 - \alpha^2}} \cdot \sin\left[\sqrt{\omega_n^2 - \alpha^2}\cdot(t-\tau)\right]\cdot F(\tau)d\tau, \tag{9}$$

and the complete solution of Eq.(8) can then be obtained as the sum of the components of Eq.(9) over n.

Thus, the generation of acoustic waves in thin foils can be described by Eq.(9) which strongly depends on driving force $F(t)$ after whose cease the vibration evolves into decaying harmonic oscillations (Eq.(4)) with frequencies defined by Eq.(6). It is apparent that maximal surface velocities can be achieved only at the initial stage of acoustic wave generation when $t<\tau$. The application of Eq.(9) to analysis of laser-driven acoustic vibrations is complicated by the lack of the exact knowledge of the time profile of the driving force $F(t)$. Depending on regime of the surface irradiation, this force may be of different origins and, accordingly, have strongly varying magnitudes and time profiles. The appropriate mechanisms will be discussed in the following section.

2.2 Generation of the acoustic waves by laser pulses

2.2.1 The action of laser pulse on the metal surface: heating and plasma generation

Metals subjected to pulsed laser irradiation absorb energy within a very thin surface layer (the skin-depth for most metals is less than 10^{-7} m) so that the temperature of the irradiated surface can rise extremely fast. For moderate laser intensities (below the plasma formation threshold) and a Gaussian-shaped laser beam, the maximum surface temperature can be estimated using a well-known expression (Prokhorov, Konov et al. 1990)

$$T_{\max}(0) = 2.15\frac{AI_{\max}\tau^{1/2}}{(\pi c\rho \mathrm{K})^{1/2}}, \tag{10}$$

where $T_{max}(0)$ is the maximum surface temperature at z=0 (the z direction is orthogonal to the target surface), A is the laser radiation absorption coefficient, I_{max} is the peak laser power, τ is the laser pulse duration, and c, ρ, and K are the specific heat, density and thermal conductivity of the corresponding metal, respectively.

After cessation of the laser pulse, the adsorbed energy continues to diffuse into the bulk metal and along the metal surface, resulting in a temperature increase underneath the irradiated spot and outward from the spot along the surface. Temperature evolution at any moment, $t>\tau$, and for any position, $z>0$, proceeds according to the following equation (Prokhorov, Konov et al. 1990):

$$T(z,t)=\frac{2AI_{\max}\gamma^{1/2}}{\mathrm{K}}\cdot[t^{1/2}ierfc(\frac{z}{2(\gamma t)^{1/2}})-(t-\tau)^{1/2}\cdot ierfc(\frac{z}{2[\gamma(t-\tau)]^{1/2}})] \tag{11}$$

where γ is the thermal diffusivity of the metal, which can be expressed as $\gamma=\mathrm{K}/c\rho$.
The function $ierfc(x)$ is given by

$$ierfc(x)=\pi^{-1/2}\{\exp(-x^2)-x(1-erf(x))\} \tag{12}$$

where $erf(x)=\frac{2}{\sqrt{\pi}}\int_0^x \exp(-\xi^2)d\xi$

Eqs. (10) and (11) are the solutions of the one-dimensional heat diffusion equation and are valid only if the laser beam size, r_0, is significantly greater than both the foil thickness h and the thermal diffusion length l_{th} calculated as $l_{th}=(\gamma\cdot t)^{1/2}$.
The strong rise of the surface temperature given by Eq.(10) results in the surface melting and evaporation, as well as in plasma plume formation (ablation regime) (Miller and Haglund 1998). Despite the fact that laser plasma generation and evolution have been the focus of numerous studies, no general mechanisms exist that describe the plasma recoil pressure on the surface for a broad range of laser intensities (Phipps, Turner et al. 1988), due to the complexity of the phenomenon. For GW/cm^2 peak laser powers, hot and dense plasma is formed in the vicinity of the surface, which can screen the surface and prevent laser radiation from reaching it. In this case, the ablative pressure very weakly depends on the target material parameters (Phipps, Turner et al. 1988) and has a sub-linear dependence on laser intensity. In a semi-regulating, one-dimensional plasma model (which can be applied to our case as a simplified, first-order approximation), this equation is written, as follows (Gospodyn, Sardarli et al. 2002) :

$$P_{a,\max}\approx 7.26\cdot 10^8\cdot I^{3/4}\cdot(\lambda\cdot\sqrt{\tau})^{-1/4}, \tag{13}$$

where I is expressed in GW/cm^2, λ in microns, τ in nanoseconds and $P_{a,max}$ in Pa. While Eq. (13) was derived for an aluminum target in vacuum and for a supercritical plasma density, it exhibits only a weak dependence on the atomic mass, A, of the material irradiated ($A^{-1/8}$) (Gospodyn, Sardarli et al. 2002) and may be applicable to specific experiments only as an upper limit estimate. For lower laser intensities (<1 GW/cm^2), the plasma plume transmittance strongly varies with laser intensity (Song and Xu 1997), depending upon the plasma density and temperature. In this case the evaporated surface material is ionized only partially and the total mass of the evaporated atomic cloud are exponentially increasing with the surface temperature and, hence with the laser intensity (Murray and Wagner 1999).

2.2.2 Thermal and plasma driven acoustic waves in metal foils

The temperature rise, as heat is transported into the solid causes linear thermal expansion resulting in the development of thermoelastic acoustic waves in the irradiated metal. Eqs. (10) and (11) can be used to determine the driving force which produces the waves. In accordance with the general theory of thermal stresses in thin plates (Boley and Weiner 1960), a non-uniform heating of the surface is equivalent to a *negative loading pressure* and may be expressed as

$$P_T = -\frac{N \cdot \eta}{1-\varepsilon} \cdot \nabla_r^2 \left[\int_{-h/2}^{h/2} T(r,z) \cdot z dz \right] , \tag{14}$$

where the temperature distribution over z is described by the Eq. (11) and η is the linear thermal expansion coefficient.

If the loading force is negative (i.e., directed backwards, towards the heating laser beam, Eq. 14), it is not surprising that an initial depression observed in the foil surface is opposite to the heating laser beam. Similar results have been reported in the literature (Scruby 1987) for thicker metal samples where the thin plate approximation was not applicable. Maximum amplitudes and shapes of the observed depression vary for different metals and are defined by both the temperature profile (Eq. (11)) and by elastic properties of the material.

In the case of plasma formation, the situation becomes more complex. The amplitude of the driving force can be estimated using an expression similar to Eq.(13), whereas the time profile of generated stress pulse is the subject of experimental study (Krehl, Schwirzke et al. 1975). Laser plasmas formation and their interaction with the surface is very complex and multi-variable problem, which can only be solved in the framework of some model assumption (Mora 1982). This is why direct experimental studies of laser-driven surface vibrations should be an essential part of any acoustic wave related desorption phenomena.

2.2.3 Experimental observations of laser-generated acoustic waves in thin foils

Experimental studies of acoustic waves in solids due to pulsed laser irradiation have started with the advent of such lasers (White 1963). A great collection of experimental results and theoretical analyses of acoustic wave generation in solids driven by laser pulses has been accumulated since (Hutchins 1985), and these studies continue at present (Xu, Feng et al. 2008). Regrettably, there is a very limited data set, which could be used to interpret of LIAD experiments. To prove (or disapprove) the "shake-off" hypothesis of molecular desorption, direct measurements of thin foil surface velocities in back-side irradiation geometry are required. The scarcity of such data motivated us to setup a series of our own experiments aiming at measurements of thin foils vibrations under typical LIAD conditions. Experimental approaches to this problem are well known and described in the literature (Scruby and Wadley 1978; Royer and Dieulesaint 2000). Nevertheless, we will briefly describe below our system, in order to create a better stage for presentation and discussion of the original results.

2.2.3.1 Experimental technique: optical and electrical methods

One of the most popular and widely used methods to studies of acoustic waves is based on non-contact optical measurements. Figure 1a shows the experimental setup for measurements of surface displacement using interferometry-based approach. A He-Ne laser (1) (Melles-Griot, 543 nm, 0.5 mW) was used as the light source for a Michelson interferometer. It consisted of a beam splitter (5), an etalon and steering mirrors (3, 4), a focusing lens (6), a target (7), an imaging lens (9), an aperture (13), a focusing lens (14) and a photomultiplier (15). The target was back-irradiated by a pulsed laser (12) through a fused silica lens (8). Laser beam parameters were measured by intersecting the laser flux with two partially reflecting (8%) quartz plates (10, 11) directing reflected beams onto a fast photodiode (16) and an energy meter (17), respectively. A wedge-type optical attenuator (18) was used to balance the Michelson interferometer shoulders and to increase the contrast of the resulting interference pattern. Focusing lenses (6) and (8) were mounted on three-axis translation stages. In this arrangement, both the acoustic-wave generating and diagnostic laser beams could be independently focused and translated to different points on the target surface. A lens (9) formed the magnified interference pattern in the plane of an aperture (13) whose diameter was selected to be equal to the width of the dark band of the lowest interference order. A second lens (14) was used to collect the light, which passed through the aperture (13), and to direct it to the photomultiplier (PMT, 15).

The anode of the PMT was terminated with a 50 Ohm load to allow optical signals with rise times as short as 5 ns to be measured. This capability was confirmed by demonstrating that the shape of a 12 ns, NdYAG laser pulse was identical when measured by using this detection system and by a high-speed avalanche photodiode. The measurements bandwidth was limited by the PicoScope 3206 oscilloscope (200 MHz bandwidth and 200 Ms/s sampling rate), which was used for signal acquisition. The digitized PMT signal was transmitted to a PC via USB port and stored for further processing. The oscilloscope was triggered by a pulse from the fast photodiode (17). The measured lag between the trigger signal and the PMT signal was less than 40 ns. The maximum signal amplitude, corresponding to the peak-to-valley ratio of the interference pattern, was 80 mV. The minimum detectable signal was 5 mV at signal-to-noise ratio of about 3, which corresponds to a surface displacement of approximately 25 nm. However, because of the strong electrical noise generated by the Q-switch of the laser, the smallest surface displacement detectable in this series of experiments was about 40 nm.

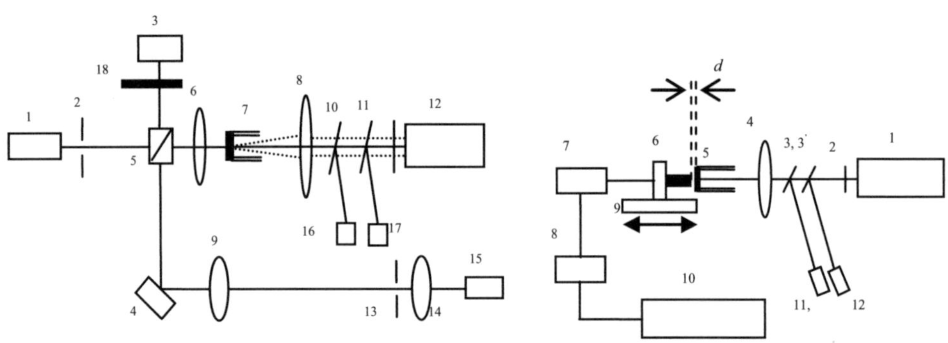

Fig. 1(a,b). Schematic drawing of the experimental setup for laser-driven acoustic wave studies. (a) Interferometry method; (b) Capacitance method

The same target irradiation scheme as described above was also used for the capacitance transducer measurements (Fig.1b). In contrast to the previous approach, a metal pin was placed in front of the target. This pin served as the second plate of a capacitor whose first plate was the target. These two capacitor electrodes were separated by a small gap *d*, typically about 100 μm. Both the target and the pin were fixed in optical mounts that allowed alignment in the plane of the sample surface. In addition, the mount of the pin was placed on a translation stage (9), driven by a picomotor, which could move the target (with the precision of 1 μm) in the direction orthogonal to the target surface. The pin had a diameter of 3 mm and its end was polished flat. In general, the design of this detector is

similar to that described in Ref.19. The pin was connected to the input of a miniature charge amplifier (7) powered by a constant (20 mA) direct current (DC) supply (8) and connected to the oscilloscope (10). The bandwidth of the charge amplifier was about 2 MHz, which provided a signal rise time of less than 400 ns. To increase the overall sensitivity of the detector, a positive bias potential of 100 V was applied to the target.
The sensitivity of the transducer to surface displacement can easily be expressed in terms of a planar capacitor

$$\Delta q = \frac{-\varepsilon VS}{d^2}\Delta d \tag{15}$$

where Δq is the change in the electric charge, V is the applied voltage, S is the surface area of the pin tip, d is the width of the gap between the electrodes and Δd is the change of the width. The sensitivity of the charge preamplifier was 10 mV/pC, which corresponded to a minimal detectable signal of about 5 mV (thus yielding reasonably good signal-to-noise ratio). With an applied bias potential of -100 V, the estimated detection limit of the transducer-based sensor was about 5 nm.

2.2.3.2 Experimental results: displacement and surface velocity of thin foils

Foils with various thicknesses (from 12.5 µm up to 100 µm) made from different materials were used in our experiments. Materials were selected to span the range from soft metals (Au, Al and Ni) to refractory metals (W, Mo and Ta) and semiconductors (Si). For each experiment, the front surface of the sample was mechanically polished to roughness of less than 0.250 µm (RMS). After polishing, the foils were glued with silver epoxy to the rim of a hollow quartz cylinder (8 mm outside diameter, 8 mm height, and 0.5 mm wall thickness). The epoxy was cured for 2 hours in an oven at temperature of 100º C. Due to the differences between thermal expansion coefficients of the foil materials and quartz, the foil stretched over the top of the quartz cylinder once the assembly cooled to room temperature. The tension was not very strong (according to our estimates, the total radial force did not exceed 1 N), and therefore, the foils in our experiments may be considered as supported at the edges. The silver epoxy also provided a conductive path between the sample and the instrument by placing a silver epoxy track along the quartz cylinder side.
Lasers generating both ultraviolet (UV) and infrared (IR) light were used for target irradiation. The UV light was generated by an ArF excimer laser (EX10-300, GAM, Inc.) having a wavelength 193 nm and a pulse duration of 15 ns. The output pulse energy could be varied from 0.4 to 4 mJ by using neutral density optical filters. The laser radiation was focused onto the backside of the target (opposite to the surface displacement sensors) using a fused silica lens with a focusing distance of f=300 mm. The irradiated spot on the target had rectangular dimensions of 100×500 µm, which corresponded to a UV laser power density in the range of 50–500 MW/cm². For IR irradiation of the target, a Q-switched Nd:YAG laser (Continuum) was used. This IR light had a wavelength of 1064 nm, a 12 ns pulse duration and an output energy in the range of 1–15 mJ/pulse. The IR laser beam had a Gaussian profile and produced a spot on the target surface with a nominal diameter of 500 µm, corresponding to peak power density of 40–600 MW/cm².
Waveforms representing foil oscillations were measured over the time range from 5 µs to 5 ms. For times much greater than the laser pulse duration ($t \gg \tau$), a decaying quasi-harmonic oscillations were observed for all materials. The measured time dependence of the displacement of the foils irradiated by laser pulses with different intensities exhibited

qualitatively similar behavior, although amplitudes, frequencies and decay times varied. These results suggest that each foil behave as a mechanical system able to oscillate in a free-running mode after the external force is removed. Fast Fourier Transform (FFT) analysis applied to the measured data has shown that the frequency spectra consisted of discrete lines (modes) appearing in the range of 10–100 kHz.

In contrast to the steady-state regime ($t>>\tau$) when different foils oscillated very similarly, the initial moment ($t<\tau$) of the evolving oscillation was distinctly different for each foil. Fig. 2 presents the time dependence of the surface displacement for different metals at low irradiation intensities (50 MW/cm^2) for "early" times in the range up to 50 μs. For all measurements at low intensities, the initial displacement is found to be negative, indicating that the surface is first depressed (i.e. towards to the driving laser beam and, correspondingly, away from the detector). Increasing the laser intensity leads to the initial surface vibration waveform displacement changing from negative to positive. This is due to the recoil pulse which occurs when material is ablated from the irradiated surface due to plasma formation. Fig. 3 shows the time dependences of the displacement of Ta foil surface for different laser irradiation intensities. The plasma formation threshold for this Ta foil was ~220 MW/cm^2, as determined by the observation of the plasma plume glow in a separate experiment. The FFT analysis of the experimental data at higher laser intensities revealed that the frequency spectrum of the foil oscillations in the non-steady-state regime contains much higher frequency components than found for the steady-state case. After a few tens of microseconds, the high frequency components disappear as the foil oscillation become harmonic as described by Eq.(4).

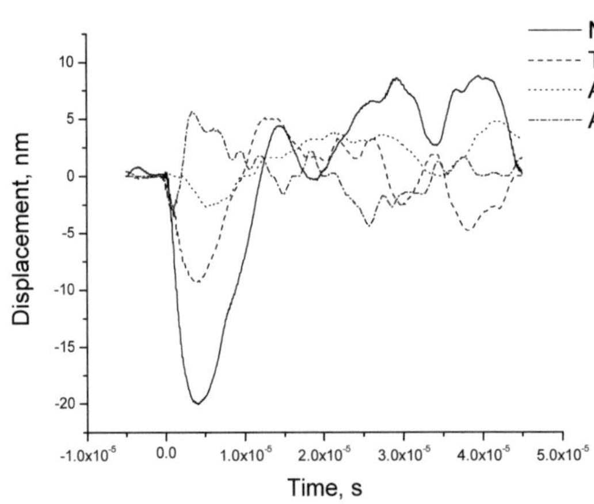

Fig. 2. Time dependent displacement for different metal foils at low laser intensities (50 MW/cm^2

The obtained results are in good agreement with time dependencies of surface displacement measured under slightly different experimental conditions (Hutchins 1985). This fact clearly demonstrates that laser-driven acoustic waves in thin metal foils have no experimental peculiarities distinguishing them from well-known acoustic wave mechanisms. This result allowed us to calculate surface velocities using the measured surface displacements (Fig. 4). As one can see from Fig.4, these velocities are indeed in the range of meters per second.

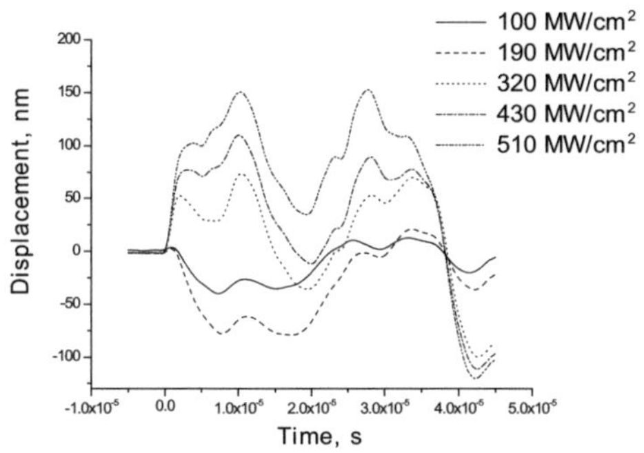

Fig. 3. Time dependent displacement of Ta foil (12.5 μm thick) at different laser intensities

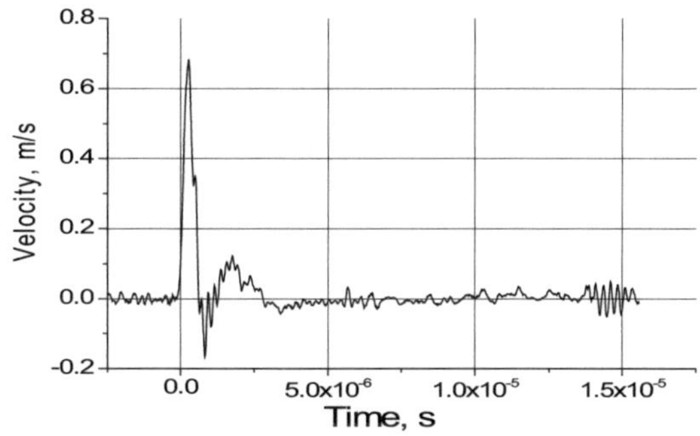

Fig. 4. Time dependence of Ta foil (12.5 mm thick) surface velocity at 400 MW/cm^2 driving laser intensity

This picture clearly confirms the statement above that the mass transfer velocity (or surface displacement velocity, in terms of our experiment) is much slower than the speed of sound in metals. In its turn, this result supports our hypothesis that the vibrational motion of the foil surface cannot serve as direct cause of molecular desorption, and that *the real physical mechanism of LIAD is not as simple as mechanical shake-off.*

3. Desorption of the molecules from back-irradiated thin metal foils

3.1 Laser desorption in modern MS: methods and applications

As shown in the previous sections, laser induced desorption phenomena play important role in modern MS. The primary role of the laser beam there is to deliver high energy density into some (typically small) volume of the analyte. Due to the local overheating this volume is volatilized forming hot and dense vapor plume, which might be partially ionized. This ionization phenomenon can be considered as a great advantage of laser desorption because there is no need for an additional ionization step, so that the desorbed ions may be directly analyzed by a mass-spectrometer. At the same time, this can be a significant drawback, because, due to collisions in the plume, organic molecules may fragment to the point that their mass analysis becomes meaningless (Miller and Haglund 1998). While using UV lasers in MS analyses of organic materials often produced encouraging results, it is well recognized in the literature that "general mechanism that is applicable to all organic solids at all UV wavelengths does not exist"(Srinivasan and Braren 1989). The introduction of MALDI gave strong indication that many problems, associated with laser desorption MS might have been solved. However despite popularity of MALDI in MS analyses of proteins, lipids and many other organics (Schiller, Suss et al. 2007), this method cannot be considered as universal because it requires to identify efficient matrix substances for different organic species, and often to develop specialized sample preparation protocols. And, regrettably, MALDI MS cannot be used to directly characterize mixtures of unknown molecules. This is why the search for more versatile and universal methods of molecular desorption/ionization is still on in the analytical mass spectrometry community. From this perspective, the ability of LIAD to volatilize different kinds of organic molecules without noticeable (or, very often, without any) fragmentation has attracted strong interest among researchers.

3.2 Laser-induced acoustic desorption

As described above, the acronym LIAD was suggested in the work conducted by the Chen's team (Golovlev, Allman et al. 1997) where the hypothesis about the acoustic wave nature of the desorption process was expressed. However, LIAD remained just an interesting observation until Kentamaa's team of researchers from Purdue University (Perez, Ramirez-Arizmendi et al. 2000) took on it and demonstrated successful applications of LIAD for the MS analysis of different organic species like cytosine, guanine, thimidine and some others. This work was followed by the series of studies where the applicability of LIAD to the MS analysis of a wide range of organic samples has been demonstrated. The LIAD volatilization method was successfully coupled with Fourier transform ion cyclotron resonance mass spectrometer and alanylglycine (Reid, Tichy et al. 2001), saturated hydrocarbons (Campbell, Crawford et al. 2004), polyethylene (Campbell, Fiddler et al. 2005) and even petroleum distillates (Crawford, Campbell et al. 2005) were analyzed. The great advantages of this technique were the ability to efficiently volatilize various organics and the simplicity of its incorporation into different classes of MS instruments, such as Linear Quadrupole Ion Trap MS (Habicht, Amundson et al. 2010) and Time-of-Flight MS (Zinovev, Veryovkin et al. 2007). Various approaches for the ionization of the desorbed molecules were successfully employed, among them: electron impact and chemical ionization (Crawford, Campbell et al. 2005), single-photon ionization (SPI) (Zinovev, Veryovkin et al. 2007), as well as Elecro- Spray Ionization (ESI) (Cheng, Huang et al. 2009). Moreover, the ability of the LIAD process to non-destructively eject from solid surfaces not only single molecules but also larger intact biological particles, such as viruses (Peng, Yang et al. 2006) and 1 μm size tungsten particles (Menezes, Takayama et al. 2005) have been demonstrated.

In our opinion, the wider spread of LIAD among analytical MS applications is now limited by the lack of an adequate theoretical concept able to explain the existing observations and to predict optimal experimental conditions for future measurements. The mechanical "shake-off" model was proposed only as a qualitative explanation of observed desorption process, and as such was never used to obtain any quantitative agreement between the observable LIAD parameters and the generated acoustic waves. Moreover, to date, there was no work published in the literature, which would be devoted to systematic studies of physical parameters of the molecules desorbed by LIAD. Since we have attempted such as study, we feel it would be beneficial for the research community if we describe here briefly our own experimental methods and experimental results on LIAD.

3.3 Energy and velocity distributions of desorbed molecules

Because the dominant fraction of the desorbed flux in LIAD are neutral molecules, it is very important to select an appropriate ionization method for the molecules as well as the type of their mass analysis technique. Single-photon ionization (SPI) is well suited for characterization of this phenomenon (Pellin, Calaway et al. 2001) because of its ability to efficiently ionize the desorbing flux with minimal fragmentation. SPI occurs following absorption of a single photon whose energy exceeds the ionization potential (IP) of the molecule of interest, creating a cation. For many molecules, particularly those with aromatic rings to stabilize the cation, fragmentation from photoionization is minimized, and thus the state of the initial LIAD flux can, in principle, be revealed(Lipson and Shi 2002). Currently, the shortest wavelength of commercially available energetic lasers suitable for SPI is 157 nm (F_2 laser), which corresponds to the photon energy of 7.9 eV. This energy limits the range of species that can be ionized by SPI to atoms and molecules with IPs less than 7.9 eV. Therefore, organic dyes were chosen as analytes for this LIAD study due to their low IPs, ability to form stable cations, and high photon absorption coefficients. We selected dyes that have IPs in the range from 5 eV to 7 eV and can be easily ionized by the F_2-laser radiation.

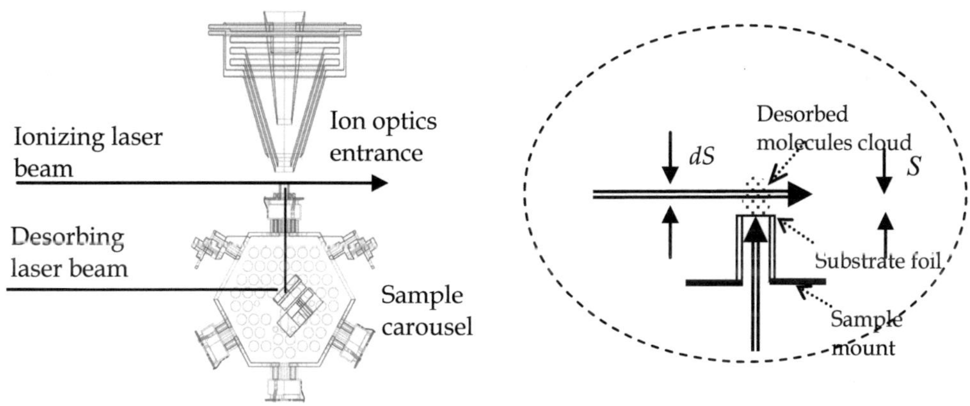

Fig. 5. Schematic drawing of target assembly and laser irradiation pathways. On the right the enlarged view of desorption/ionization scheme is displayed

A time-of-flight mass spectrometer (TOF MS) with a combined LIAD/SPI ion source was employed in our studies of the LIAD phenomenon. The experiments were conducted under ultra high vacuum conditions, with the residual gas pressure in the sample chamber less than 3×10^{-7} Pa. The schematic drawing of the target assembly in our instrument is shown in Fig. 5. The sample was mounted on one of six sample holders that were supported by a hexagonal carousel. This carousel was driven by an ultrahigh-vacuum compatible motion stage with closed-loop precision of better than 50 nm. The sample holders were secured on the carousel via three 30 mm long alumina ceramic insulators and connected (using vacuum feedthroughs) to a high-voltage pulser unit, which provided voltages necessary for the operating of TOF MS instrument. Each new sample was inserted into the UHV chamber through a vacuum loadlock. Using the motion stage, the sample was then positioned in the focal plane of the TOF MS source optics. The ion optics and operational principles of our instrument are described in more detail elsewhere (Veryovkin, Calaway et al. 2004). A dielectric mirror with 98% reflection at 248 nm was mounted in the center of the carousel, in order to deliver the laser beam to the back side of the sample. Note that we will use the convention that the front of the sample is the side facing the ion source and TOF, while the back side is the opposite. For desorption, an excimer KrF laser with wavelength 248 nm (EX10/300 GAM Laser Inc.) was used. The output energy of the laser pulse could be varied between 0.5 - 5 mJ by adjusting the laser discharge voltage and by an additional attenuation with a set of neutral optical filters. The driving laser beam was focused on the target back surface into a spot of rectangular shape ~200×800 μm^2 by the fused silica lens with focusing distance of 500 mm. The laser pulse duration was 7 ns, producing a peak power density on the irradiated surface ranging from 50 to 500 MW/cm^2. These laser intensities are close to those used in most of LIAD experiments (Perez, Ramirez-Arizmendi et al. 2000; Campbell, Crawford et al. 2004; Campbell, Fiddler et al. 2005; Crawford, Campbell et al. 2005) taking into account that the reflection coefficient in the UV is normally less than it is at visible wavelengths. Post-ionization of the desorbed molecules was performed with an F_2 laser, with output energy of 2 mJ/pulse, and pulse duration of 10 ns. The F_2 laser beam was focused just above the front target surface, with a waist of 400×2000 μm^2, using of a combination of MgF_2 spherical and cylindrical lenses. The F_2 laser radiation power density in the focal plane was ~10 MW/cm^2, which assured the saturation of the photoionization process for the investigated molecules (as verified by a laser power study).

For comparison with LIAD, direct laser desorption (LD) mass spectra were measured for the same samples, also using the F_2 laser for post-ionization. To this end, an N_2 laser (337 nm wavelength, 100 µJ/pulse energy and 7 ns pulse duration) was focused onto the target front surface using an in-vacuum Schwarzschild optical microscope (Veryovkin, Calaway et al. 2004). The beam spot size on the surface was about 50 µm in diameter.

The delay between the driving KrF (or N_2) laser pulses and the ionizing F_2 laser could be precisely controlled and varied from 0 to 1000 µs. The desorbed molecules that move away from the surface could therefore be ionized at a precisely defined moment in time and volume in space above the target after the desorption event, with the photoions then analyzed by the TOF MS. This approach allowed us to measure mass spectra for the (postionized) desorbed neutral molecules and determine their velocity distribution. Each mass spectrum was the sum of 128 individual acquired spectra. To prevent the rise of the average foil temperature due to adsorption of laser power, the repetition rate of the laser pulses was maintained at 8 Hz.

Foils from different materials with different thicknesses were used in the experiments. The foil preparation procedure was the same as described in paragraph 2.2.3.1. Before applying the analyte to the top surface of the foil, each substrate was cleaned in methanol-acetone solution (1:1) in an ultrasonic bath (10 minutes).

Organic dyes rhodamine B, fluorescein, methylanthracene (MA), coumarin-522 (N-Methyl-4-trifluoromethylpiperidino3,2-gcoumarin), and BBQ (4,4''-Bisbutyloctyloxy-p-quaterphenyl) were used as received (Eastman Kodak). The dyes were dissolved in methanol (for MA and BBQ, mixed xylenes were also used as solvent), and then the resulting solution (about 10^{-3} M) was used for sample preparation. One µl of the analyte solution was pipetted onto the foil surface, and then the quartz cylinder-foil assembly was spun at 4500 rpm for 30 seconds to coat the analyte uniformly over the surface. During spin-coating, a significant part of the solution (90% or more) was taken off the surface and, surface concentrations of the analyte could be estimated to be less than 0.5 nM/cm^2. After the sample preparation, the foil was introduced into the instrument via the loadlock for analysis.

In good agreement with the previously published results on LIAD, we detected strong and stable desorption signals from foils with thickness of 12.5 μm (Fig. 6). Thicker foils (25 μm) produced relatively weak signals for the range of acoustic wave driven laser intensities used in our experiments. For comprehensive experiments, a Ta foil with 12.5 μm thickness was chosen as optimal, not only because high desorption signals were detected from it, but also for its good mechanical strength, high melting point, and durability under powerful laser irradiation. The TOF mass spectra of different organic dye molecules desorbed from the front of the back-irradiated Ta foil surface and ionized by the 157 nm laser radiation are shown in Fig.6. This figure shows three major features, (1) all analytes display large parent molecular ion signals, (2) all spectra display a small number of peaks, with a few or none in the mass range below 100 Da., (3) the number of fragment ion peaks is specific to each molecular analyte. In order to characterize the desorption process in terms of the corresponding molecular fragmentation, a parameter ς, can be defined as a ratio of the sum of intensities of the fragment peaks A_f to the parent molecular peak intensity A_p , $\varsigma = \sum A_f / A_p$. As will be shown below, this parameter depends on laser intensities that drive the acoustic waves. For SPI , the parameter ς characterizes not only the peculiarities of the desorption process, but it also generally depends on the photoionization cross-section of the parent molecule and specifics of its photofragmentation such as possible decay channels and their activation energies.

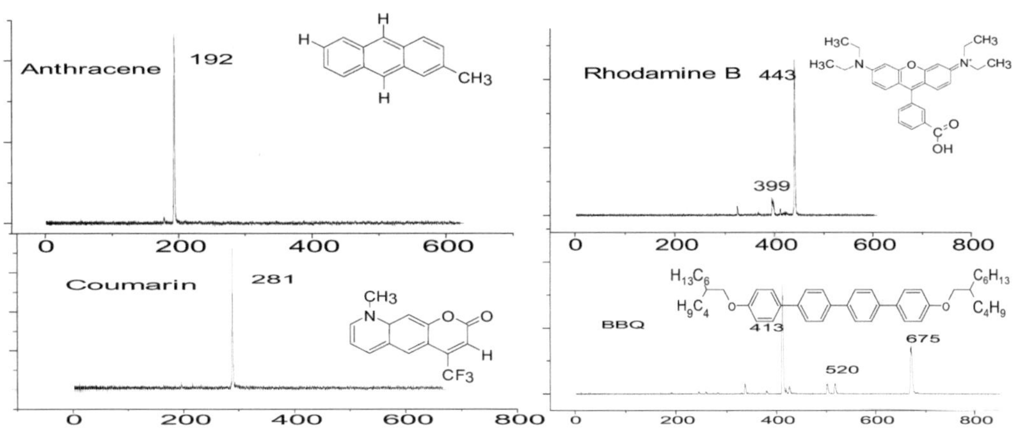

Fig. 6. Mass spectra of organic dyes in LIAD experiments

For rhodamine B, the parameter ς grew linearly with KrF laser intensity (Fig. 7), and a similar behavior was also detected for fluorescein, although values of ς were different (ς =0.8 for fluorescein at 300 MW/cm2 of KrF laser intensity). On the contrary, the same experiments conducted for BBQ, surprisingly revealed that the parameter ς decreased with increasing desorption laser power. This discrepancy in ς parameter dependency may indicate that the relationship between these two observations is not a trivial one. This raises the question of whether the desorption and fragmentation phenomena are driven by the same fundamental process. We note that fragmentation is not intrinsic to LIAD. In our experiments, we observed some indications of fragmentation for only three analytes out of five. The analysis of existing data from the literature shows that in some cases fragmentation was observed even with soft ionization of the desorbed molecules whereas in other cases fragmentation was very small or completely absent. A detailed study of fragmentation in LIAD has not been done yet and the presented here results are the first attempt to quantify this characteristic of LIAD.

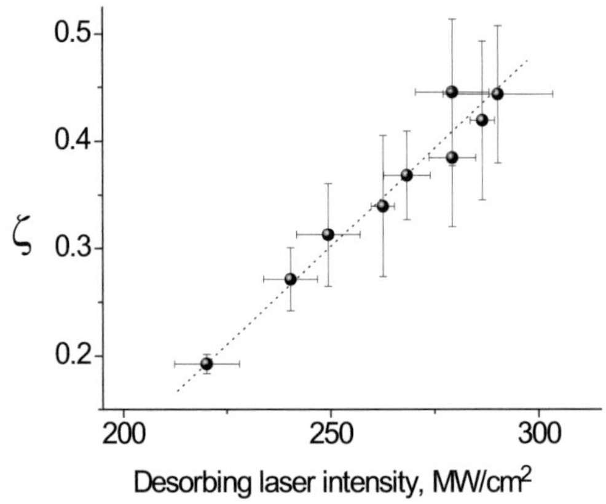

Fig. 7. Power dependence of fragmentation parameter

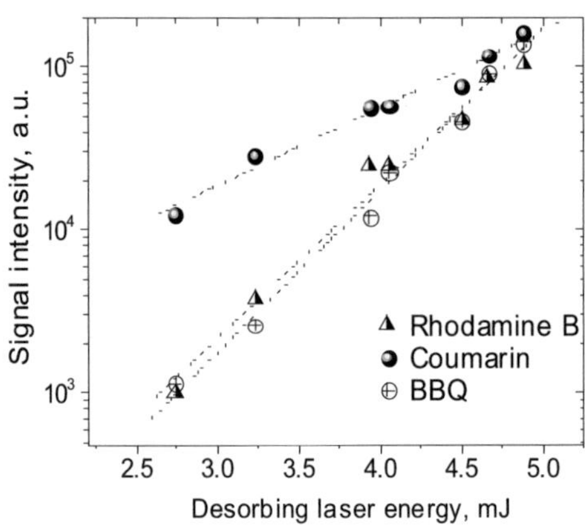

Fig. 8. Power dependence of total desorption yield for rhodamine B molecules

An important characteristic of any desorption phenomenon is its yield. This is why in order to understand the basic processes driving the phenomenon, one has to identify external parameters that have the strongest effect on the desorption yield and then measure a dependence of the yield for each parameter. In the case of LIAD, the dependence of key peak intensities in the mass spectra and the ς parameter on the driving laser intensity appears to be the most important for understanding this phenomenon. The overall desorption yield for all studied analytes strongly increased with desorption laser intensity (within our experimental range), displaying for most peaks approximately exponential dependency. Figure 8 demonstrates this dependence clearly. Plotted on a semi-logarithmic scale, these dependences appear linear, with different slopes for each analyte.
For the characterization of physical nature of the desorption processes, the knowledge of velocity distributions of desorbed neutral molecules is extremely important (Levis 1994). As

mentioned above, time-of-flight mass spectra of such molecules can be measured using the laser post ionization (LPI) technique (Spengler, Bahr et al. 1988). This experimental arrangement makes it possible to determine the distribution of the desorbed neutral molecules over their translational velocities (in the direction normal to the substrate surface) by varying the time delay between desorbing and post-ionizing laser pulses. Raw experimental data in this case are the dependencies of the observed signal on the laser delay time, called below as *signal-vs-delay* dependencies. In order to be able to directly compare LIAD and LD results, the energies of the desorbing N_2 (LD) and KrF (LIAD) lasers were adjusted such that the output molecular ion signals reached about the same intensities for the same gains of the TOF MS detector.

In good agreement with the previous experiments (Perez, Ramirez-Arizmendi et al. 2000), dramatic differences between LIAD and LD in widths of the *signal-vs-delay* dependencies and in positions of their maximums have been observed in these measurements. The most tempting explanation for this experimental finding was that the mean velocities of desorbed molecules in LIAD and LD processes were very different. Unfortunately, no actual energy or velocity distributions of desorbed molecules in LIAD process have been measured experimentally and reported in the literature to date. For the first time, this gap in knowledge can now be filled by processing our experimental data from the *signal-vs-delay* dependencies and converting them into kinetic energy distributions corresponding to our experimental conditions.

Our experiment geometry (Fig.5) is rather common, and the detailed discussion of the method as well as the appropriate conversion equations can be found elsewhere (Young, Whitten et al. 1989; Balzer, Gerlach et al. 1997). Typical values of the distance *S* and the thickness of ionization volume *dS* were set to 3 and 0.4 mm, respectively. While, as mentioned above, the value of *S* could be varied in our experiments between 1 mm and 5 mm, in order to achieve optimal compromise between the energy/velocity resolution and the signal-to-noise ratio, most of our data were obtained at S=3 mm. At this distance, the relative velocity resolution was $dv/v=dS/S$=0.4/3=0.13 and, correspondingly, the energy resolution was dE/E=0.18. A typical velocity distribution obtained for rhodamine B is demonstrated on Fig.10. It is apparent that the desorbed molecules in LIAD are very slow. The average velocities of rhodamine B and BBQ molecules were found to be 59±12 m/s and 47+9 m/s, respectively. For methylanthracene and coumarin522 molecules, the measured average velocities were 76±20 m/s and 70±18 m/s. It is important to recognize that these LIAD-desorbed molecules are much slower than what could be expected assuming the thermal mechanism of LIAD. In order to fit these rhodamine B data with Maxwellian distributions (dashed curve on Fig.10), physically unrealistic low temperatures of about 100 K, were required. On the other hand, the measured velocities are much higher than that of the laser-induced acoustic motion of the foil surface in normal direction, which, measured in the same experimental arrangement, did not exceed 1 m/s. This observation caused serious doubts on the validity of the "shake-off" mechanism considered by many as the most likely cause of LIAD. The measurements of velocity distribution of species desorbed in LD geometry were used for direct comparison with LIAD, and as the proof of validity of our experimental procedure. A smaller insert plot on Fig.10 demonstrates the velocity distribution of desorbed rhodamine B main fragment (399 amu) obtained in the LD irradiation scheme. The mechanisms of the velocity distribution formation in the LD process are well-known and discussed in many reviews (Levis 1994). According to a commonly used procedure described in (Natzle, Padowitz et al. 1988), this distribution can be fit by a two-temperature bi-modal velocity distribution: dashed lines in the insert plot represent spectral components with different temperatures, and the solid line corresponds to their sum. Being in a good agreement with general LD regularities, these results also confirm the validity of the experimental procedure used for both LD and LIAD.

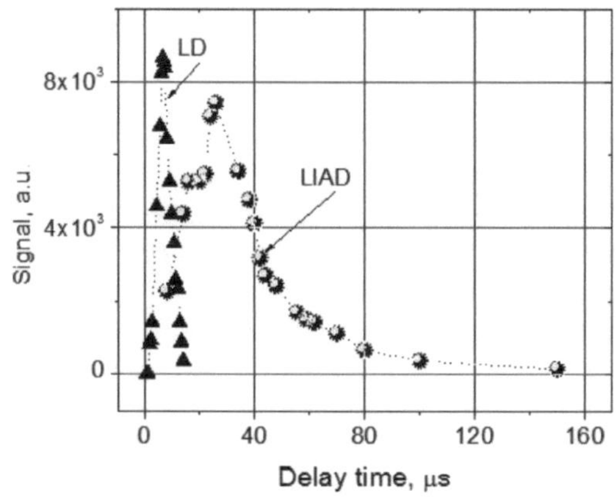

Fig. 9. The comparison of *signal-vs-delay* for LIAD and LD desorbed rhodamine B molecules

Velocity distributions were measured also at different fluences of the driving KrF laser. On Fig. 10, two distributions corresponding to laser fluences of 2.3 J/cm^2 and 3.4 J/cm^2 are plotted. Within the limits of accuracy of our measurements, no change of average velocity has been detected, which suggests that both thermal and "shake-off" models are not applicable for the explanation of LIAD process.

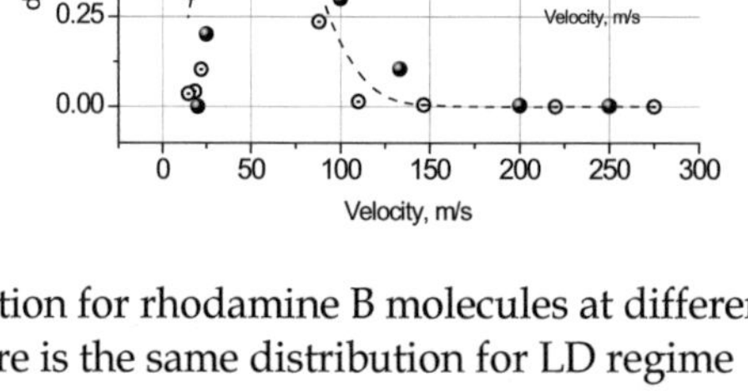

Fig. 10. Velocity distribution for rhodamine B molecules at different LIAD desorption laser intensities. On insert there is the same distribution for LD regime

In order to calculate the mean energies of desorbed molecules, we used the described above approach and took into account that the Jacobian of this variable transformation was given by S^2/t^3 (Balzer, Gerlach et al. 1997). The mean energies of molecules in LIAD experiments (as well as their mean velocities) showed no apparent trend with the increase of desorbing laser fluences keeping the average value at about 9 meV for rhodamine B, 9.5 meV for BBQ, 6.5meV for coumarine 522 and 7.5meV for methylanthracene. Typical energy distributions measured for these molecules are shown on Fig.11. Dashed line represents Maxwell distribution for rhodamine B corresponding to the 80 K temperature. This temperature was chosen to obtain the best fit at the distributions maximums. At the same time, high energy tails of the experimental energy distributions strongly deviate from the exponential law, which is apparent with the double logarithmic scale in Fig.11, and reveal behavior close to the power dependence.

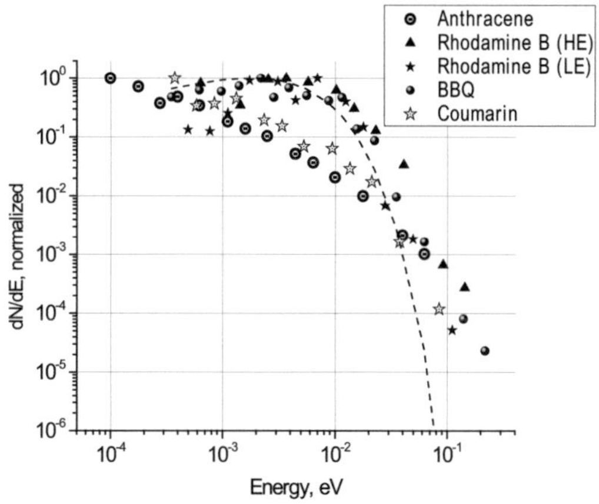

Fig. 11. Energy distribution of different organic dyes molecules. Dashed line represents equilibrium Boltzmann distribution at T=100 K

4. Is desorption process in LIAD really driven by acoustic waves?

4.1 "Shake-off "mechanism versus thermal origin of molecular desorption

The "shake-off" mechanism of molecular desorption in LIAD for a long time was considered as the only sensible explanation of the observed phenomenon. However, as was shown above, it contradicts both with general physical considerations and with the experimentally measured parameters of acoustic vibrations of the surface as well as with the observed energy and velocity spectra of desorbed molecules.

An important consequence of the backside irradiation is the heating of the front side. Considering this effect, we should keep in mind that this heating process in LIAD is distinctly different from the case of direct front-side laser irradiation (LD). In the latter case, due to the small value of skin depth in metals, the rate of the temperature rise is extremely high. In case of backside irradiation of the foils, the front surface temperature is governed by the heat conduction through the metal foil, which makes the heating rate much slower than that for the front side irradiation geometry. For one-dimensional heat conduction problem, the specific time of the temperature rise is defined by the heat propagation time $\tau = l^2 / \alpha$, where a is the thermal diffusivity of the metal and l is the foil thickness. For the foil thicknesses typical for LIAD (~10 μm) and the thermal diffusivity of the most metals $a \approx 0.2$ - 0.7 $cm^2 \cdot s^{-1}$, τ has the values in the range of a few μs. Our numerical calculations using the heat conduction equation showed that for Ta foil with the thickness of 12.5 μm and the driving laser fluence 3.5 J/cm^2 (corresponding under conditions of our experiments to the peak laser intensity of about 500 MW/cm^2), the front surface temperature rise is 375 K, which reaches its maximum 1.75 μs after the laser desorbing pulse ceases. If experiments start with room temperatures, the peak foil surface temperatures then can reach 668÷673 K. Because the melting point of rhodamine B is just 438 K, the thermal origin of the LIAD process can, in fact, come into the focus of our consideration. However, other experimental results obtained in this work, make this mechanism very unlikely. These results are:

1. rather small fragmentation of molecules in LIAD, compared to that for LD;
2. distinct differences in velocity and energy distributions between LIAD and LD;
3. slow velocities and low kinetic energies of desorbed molecules compared to those required by the thermal mechanism;
4. the apparent independence of mean energies of the desorbed molecules in LIAD on the driving laser fluence (also observed by by Kenttamaa et al. in their recent work (Shea, Petzold et al. 2007).

Moreover, the mechanical "shake-off" mechanism is also in contradiction with the observation (4), because the amplitudes and velocities of laser generated acoustic waves should increase with the driving laser fluencies. Thus we can conclude that both mechanisms of the direct energy transfer (acoustic waves and heat conduction) cannot serve as the primary explanation of the LIAD phenomenon, and apparently more complicated processes are involved here.

4.2 Stress and strain of the foil surface due to the laser irradiation

It is a well-known fact that a film deposition on a substrate surface results in many cases in the residual mechanical stress (due to the lattice parameters mismatch between the film and the substrate) and thus in some excessive potential energy stored in the film. This stress can be produced by two ways: one is the growth stress and the other is the induced (or extrinsic) stress (Freund and Suresh 2003). An external impact, such as acoustic or thermal wave generated by the laser irradiation of the substrate, should initiate the reconstruction of the film, which can result in releasing this excess energy or, possibly, generating an additional extrinsic stress. In both cases, it can cause formation of cracks in the film so that intermolecular bonds at the edges of the cracks can break. As a result of this crack formation process, excited electronic states can form at the crack edges and induce desorption of the molecules.

Despite a great variety of mechanisms of stress formation and evolution do exist, a quantitative description of this process is possible only in limited cases even for simple adsorbent-adsorbate systems. Accurate modeling of the film behavior requires precise knowledge of the thermal and mechanical properties of the film material and their interaction with substrate surface. For most organic materials, such data are unavailable, and thus only limited estimates may be done, based on the little amount of data published in the literature (Bondi 1968). Moreover, the deposited films of organic materials tend to consist of a mixture of micro-crystals with different crystallographic orientation and, therefore, there are many grain boundaries, defects, dislocations in such films that could significantly change their local mechanical properties.

A simplified numerical estimate of the stress energy, related to the thermal mismatch of the substrate foil and the analyte film on its top could be done, as follows. An equi-biaxial stress resulting from acoustic vibration of the back-irradiated metal foil may be estimated using the following expression (Boley and Weiner 1960; Freund and Suresh 2003)

$$\sigma_m = \frac{E \cdot h_f}{R \cdot (1-\nu)} \tag{16}$$

where E is the Young's modulus, v is the Poisson ratio, h_f is the thickness of organic film and R is the radius of curvature of the foil surface. From the other side, the stress, resulting from thermal mismatch has the following value (Boley and Weiner 1960)

$$\sigma_t = (\alpha_s - \alpha_f) \cdot \Delta T \cdot E / (1 - \nu), \tag{17}$$

where a_s and a_f are thermal expansion coefficients of the substrate and the film respectively, ΔT is the temperature increase. There is not much information in the literature on thermo-mechanical parameters for molecular crystals but based on existing data for anthracene (Bondi 1968) we can estimate the order of the generated stress values. Under the assumption that $a_f = 2.8 \cdot 10^{-4}$ K^{-1} , $a_s = 6.3 \cdot 10^{-6}$ K^{-1}, E= 13 GPa, $v \approx 0.25$, and ΔT=100 K, we can obtain σ_t=485 MPa. At the same time, the stress associated with acoustic vibrations is much lower: $\sigma_m \approx 1.7$ kPa (taking into account that maximal value of R is approximately 1 m, and the film thickness never exceeded 10^{-7} m,). This is a negligible value, in comparison with σ_t , which means that the thermal mismatch stress is the principal reason for cracks formation. The estimate of internal energy, stored in thermally strained organic films can be done with using of following expression (Boley and Weiner 1960)

$$G = \frac{E \cdot (1+\nu)^2}{2 \cdot (1-\nu^2)} \cdot (\alpha_s - \alpha_f)^2 \cdot \Delta T^2 \cdot 2 \cdot \pi \cdot r_0 \cdot h_f \tag{18}$$

The average energy per analyte molecule can easily be calculated

$$g_a = \frac{E \cdot (1+\nu)^2}{2 \cdot (1-\nu^2) \cdot \rho} \cdot \frac{(\alpha_s - \alpha_f)^2 \cdot \Delta T^2 \cdot M}{N_A} \tag{19}$$

Here M is the molar mass, ρ is the specific gravity and N_A is the Avogadro number. It is interesting to note that g_a does not depend on the analyte island size but strongly depends on the thermal and mechanical parameters. Again with the use of existing data for anthracene we can estimate the value of g_a, and for ΔT=100 K we will get g_a=0.025 eV . This is not enough to break intermolecular bonds but when thermally induced stress exceeds a critical value, the film can start to fracture and the stored energy is released in a small volume in the vicinity of the stress cracks. Due to the strong spatial nonuniformity of thermo-mechanical properties of molecular crystals (Bondi 1968) the physical mechanisms involved in this process are very complicated in nature; therefore we can give only a qualitative picture of this phenomenon. Presumably, the cracks are formed along grain boundaries, defects and interfaces. The increase of the desorption laser intensity that causes a rise in ΔT and, in accordance with Eq. 18, an increase in energy G, results in the formation of additional cracks. Some part of this excess energy can be then converted into the increased free surface energy, other part - into electronic excitations, but because these processes are obviously non-adiabatic, the crack formation most likely will be accompanied by breaking intermolecular bonds and forming new desorption sites.

4.3 Proposed mechanisms of the molecules desorption

The desorption process itself appears to be the most obscure part of the LIAD phenomenon. The formation of the electronically excited states on the surface due their mechanical fracture is considered to be the main physical nature of triboemission (Nakayama, Suzuki et al. 1992), also known as "Kramer effect" (Oster, Yaskolko et al. 1999; Oster, Yaskolko et al.

2001). The essence of this effect is the emission of charged particles and photons initiated by surface distortion (in particular by mechanical deformation, scratching, bending, etc.) and do not connected with thermal excitation. The luminescence of thin metal discs, irradiated from back side by the laser pulses (Abramova, Shcherbakov et al. 1999; Abramova, Rusakov et al. 2000) can also serve as evidence of formation of excited electronic states by the laser-driven stress in thin foils.

The mechanism of molecular desorption due to such laser-generated stress has been proposed earlier by Vertes (Vertes and Levine 1990; Vertes 1991) for MALDI, and was based on the thermal stress generation in the layer of organic film deposited on solid substrates. One should notice, however, a clear differences in the physical conditions between these sample volatilization methods. For MALDI, the absorption of the laser pulse energy occurs in an optically and thermally dense film, which experiences thermal stress due to its non-uniform and fast heating. In the case of LIAD, the laser radiation is absorbed by the back side of the metal foil substrate, opposite from where the sample was deposited. The amount of energy transmitted through the metal foil to the analyte layer on the front site should be so strongly attenuated, compared to the direct (front side) irradiation, that it cannot directly be sufficient for desorbing molecules with velocities observed in our experiments. On the other hand, a typical average thickness of the analyte film in LIAD can be estimated to be on the order of several molecular layers. Because of this, the specific density of energy stored in each analyte island due to intrinsic stress can be high, and during laser irradiation of the back side of the foil, the laser irradiation can simply trigger the release of this energy and to induce molecular desorption event.

The other experimental fact that could help in the interpretation of LIAD phenomenon is the similarity of the energy and velocity spectra of desorbed neutral molecules in case of LIAD and Electron Stimulated Desorption (ESD) (Young, Whitten et al. 1989). The primary mechanism of ESD is supposed to be the formation of the repulsive states in the surface due to electron excitation.

Thus we could hypothesize that in case of LIAD, due to complete opaqueness of metal foil for laser radiation, the only channel for energy transfer is the formation of acoustic and thermal field by laser pulse impact. As shown above, the result can be the mechanical distortion of the analyte film followed by film cracking and delamination and the consequent formation of excited and repulsive states electronic for analyte molecules.

This will increase the number of desorption sites and finally the total number of desorbed molecules as observed in the nonlinear laser intensity dependence. But because the formation of the any individual cracks is defined only by the intermolecular bonding forces in the vicinity of the crack, the translational kinetic energy of desorbed molecules should still remain independent of driving laser intensity, also matching our observations and other recent LIAD publications.

5. Conclusion

Traditionally, the name LIAD combine all desorption phenomena taking place when an opaque target is irradiated from the back side, ignoring differences in experimental conditions. In our opinion, it is not correct. As it was demonstrated in this work, under some experimental conditions (most commonly used in many present studies), the physical origin of the observed desorption phenomenon is not (and could not be) connected with the acoustic waves generated in the foil and most likely is defined by the film stress and cracking due to thermal and mechanical mismatch of the analyte and substrate. Therefore, ironically, the acronym LIAD in this case does not correctly reflect the physical nature of the process.

From the other hand, we cannot exclude that some strong change of the experimental conditions can also change the relationship between various physically possible mechanisms of molecular desorption, similarly to the case of direct laser desorption, when the thermal mechanism dominates under wide range of conditions and makes other possible mechanisms undetectable. One can expect, for example, that strong increase of laser power density (10 GW/cm^2 and above) will cause the corresponding increase of the foil temperature and formation of hot and dense plasma plume near its surface facing the laser. In some such cases and strongly depending on the foil material properties, the material motion could evolve from elastic into plastic regime of deformation. The acoustic wave relation could not then apply, and the velocity of the surface linear motion may strongly increase. "Large and heavy" objects weakly bound to the substrate surface may be then kinematically removed from it ("shaken-off"). This could indeed serve as an explanation of observations from recent experiments where the desorption of intact viruses and biological cells were reported (Peng, Yang et al. 2006). However, this regime cannot be connected with generation of the acoustic waves but most likely corresponds to a physically different mode of shock-wave generation (Menikoff 2007). One cannot exclude that under some experimental LIAD conditions only these shock-wave induced phenomena can be responsible for molecular desorption, particularly in the experiments where emission of ions was detected. In one of such pioneering experiments (Golovlev, Allman et al. 1997), where the laser generated pressure pulse was apparently much stronger than that in our work, because of confined ablation conditions (Fabbro, Fournier et al. 1990), the emission of both electrons and ions was observed. This may have been connected with significant surface disruption at the microscale generated by shock waves.

From another standpoint, the backside irradiation of very thin films (about two or three hundred nanometers), which also could be called LIAD, has demonstrated domination of the thermal mechanism in the desorption process (Ehring, Costa et al. 1996). It is clear that the thermal equilibrium between front and back sides in such thin films can establish within the time interval of tens of nanoseconds, and the absolute temperature difference between front and back sides is negligibly small. Thus, at some laser power densities, the front side temperature could reach the melting temperature of the metal that would be enough to cause an efficient thermal desorption of the most organic molecules.

To conclude, the variety of desorption and emission phenomena observed on the front side of thin metal foils whose back sides are subjected to pulsed laser irradiation, combined under the general name of LIAD, could, in fact, have a number of different physical origins depending on specific experimental conditions. According to our observations conducted at moderate laser power densities (0.1 - 1 GW/cm^2) and foils thicknesses of about 5 – 20 μm, that appear to be the most commonly used conditions in LIAD experiments, the predominant desorption mechanism is connected with the reorganization of the deposited analyte film and the consequent breaking of molecular bonds on the edges of these cracks.

6. Acknowledgment

This work is supported by the U.S. Department of Energy, BES-Materials Sciences, under Contract DE-AC02-06CH11357, by UChicago Argonne, LLC.

7. References

Abramova, K. B., A. I. Rusakov, et al. (2000). Photon emission from metals under fast nondestructive loading. *J. Appl. Phys.* 87(6): 3132-3136.

Abramova, K. B., I. P. Shcherbakov, et al. (1999). Emission processes accompanying deformation and fracture of metals. *Phys.Solid State* 41(5): 761-762.

Adamson, A. W. and A. P. Gast (1997). Physical Chemistry of Surface. New York, John Wiley &Sons, Inc.

Balzer, F., R. Gerlach, et al. (1997). Photodesorption of Na atoms from rough Na surface. *J. Chem. Phys.* 106(19): 7995-8012.

Boley, B. A. and J. H. Weiner (1960). Theory of thermal stresses. New York,, Wiley.

Bondi, A. (1968). Physical properties of molecular crystals, liquids and gases. New York, London, Sydney, Hohn Wiley & Sons.

Campbell, J. L., K. E. Crawford, et al. (2004). Analysis of saturated hydrocarbons by using chemical ionization combined with laser-induced acoustic desorption/Fourier transform ion cyclotron resonance mass spectrometry. *Analytical Chemistry* 76(4): 959-963.

Campbell, J. L., M. N. Fiddler, et al. (2005). Analysis of polyethylene by using cyclopentadienyl cobalt chemical ionization combined with laser-induced acoustic desorption/Fourier transform ion cyclotron resonance mass spectrometry. *Analytical Chemistry* 77(13): 4020-4026.

Cheng, S. C., M. Z. Huang, et al. (2009). Thin-Layer Chromatography/Laser-Induced Acoustic Desorption/Electrospray Ionization Mass Spectrometry. *Analytical Chemistry* 81(22): 9274-9281.

Cole, R. B. (2010). Electrospray and MALDI mass spectrometry: fundamentals, instrumentation, practicalities, and biological applications. Hoboken, N.J., Wiley.

Crawford, K. E., J. L. Campbell, et al. (2005). Laser-induced acoustic desorption/Fourier transform ion cyclotron resonance mass spectrometry for petroleum distillate analysis. *Anal. Chem.* 77(24): 7916-7923.

Crawford, K. E., J. L. Campbell, et al. (2005). Laser-induced acoustic desorption/Fourier transform ion cyclotron resonance mass spectrometry for petroleum distillate analysis. *Analytical Chemistry* 77(24): 7916-7923.

Dass, C. (2007). Fundamentals of contemporary mass spectrometry. Hoboken, N.J., Wiley-Interscience.

Duncan, M. W., H. Roder, et al. (2011). Quantitative matrix-assisted laser desorption/ionization mass spectrometry. *Briefings in Functional Genomics* 7(5): 355-370.

Ehring, H., C. Costa, et al. (1996). Photochemical versus thermal mechanisms in matrix-assisted laser desorption/ionization probed by back side desorption. *Rapid Communications in Mass Spectrometry* 10(7): 821-824.

Fabbro, R., J. Fournier, et al. (1990). Physical Study of Laser-Produced Plasma in Confined Geometry. *Journal of Applied Physics* 68(2): 775-784.

Freund, L. B. and S. Suresh (2003). Thin Film Materials. Stress, Defect Formation and Surface Evolution. Cambridge, Cambridge University Press.

Golovlev, V. V., S. L. Allman, et al. (1997). Laser-induced acoustic desorption of electrons and ions. *Applied Physics Letters* 71(6): 852-854.

Gospodyn, J. P., A. Sardarli, et al. (2002). Ablative generation of surface acoustic waves in aluminum using ultraviolet laser pulses. *Journal of Applied Physics* 92(1): 564-571.

Habicht, S. C., L. M. Amundson, et al. (2010). Laser-Induced Acoustic Desorption Coupled with a Linear Quadrupole Ion Trap Mass Spectrometer. *Analytical Chemistry* 82(2): 608-614.

Honig, R. E. and J. R. Woolston (1963). Laser-induced emission of electrons, ions and neutral atoms from solid surfaces. *Applied Physics Letters* 2(7): 138-139.

Hutchins, D. A. (1985). Mechanisms of Pulsed photoacoustic generation. *Can. J. Phys.* 64: 1247-1263.

Karas, M., D. Bachmann, et al. (1985). Influence of the Wavelength in High-Irradiance Ultraviolet-Laser Desorption Mass-Spectrometry of Organic-Molecules. *Analytical Chemistry* 57(14): 2935-2939.

Krehl, P., F. Schwirzke, et al. (1975). Correlation of Stress-Wave Profiles and Dynamics of Plasma Produced by Laser Irradiation of Plane Solid Targets. *Journal of Applied Physics* 46(10): 4400-4406.

Landau, L. D. and E. M. Lifshits (1987). Fluid mechanics. New York, Pergamon.

Levis, R. J. (1994)."Laser-Desorption and Ejection of Biomolecules from the Condensed-Phase into the Gas-Phase. *Annu. Rev. Phys. Chem.* 45: 483-518.

Lindner, B. and U. Seydel (1985). Laser Desorption Mass-Spectrometry of Nonvolatiles under Shock-Wave Conditions. *Analytical Chemistry* 57(4): 895-899.

Lipson, R. H. and Y. J. Shi (2002). Ultraviolet Spectroscopy and UV lasers. New York-Basel, Marcel Deccer, Inc.,.

Lubman, D. M. (1990). Lasers and mass spectrometry. New York, Oxford University Press.

McLachlan, N. W. (1951). Theory of vibrations. New York, Dover Publications.

Menezes, V., K. Takayama, et al. (2005). Laser-ablation-assisted microparticle acceleration for drug delivery. *Applied Physics Letters* 87(16): -.

Menikoff, R. (2007). Empirical Equations of State for Solids. *Shock Wave Science and Technology Reference Library*. Y. Horie. Berlin, Springer. 2: v. <1-3>.

Miller, J. C. and R. F. Haglund (1998). Laser ablation and desorption. San Diego, Academic Press.

Mora, P. (1982). Theoretical-Model of Absorption of Laser-Light by a Plasma. *Physics of Fluids* 25(6): 1051-1056.

Murray, T. W. and J. W. Wagner (1999). Laser generation of acoustic waves in the ablative regime. *Journal of Applied Physics* 85(4): 2031-2040.

Nakayama, K., N. Suzuki, et al. (1992). Triboemission of Charged-Particles and Photons from Solid-Surfaces during Frictional Damage. *J. Phys. D: Appl. Phys.* 25(2): 303-308.

Natzle, W. C., D. Padowitz, et al. (1988). Ultraviolet-Laser Photodesorption of No from Condensed Films - Translational and Internal Energy-Distributions. *Journal of Chemical Physics* 88(12): 7975-7994.

Oster, L., V. Yaskolko, et al. (1999). Classification of exoelectron emission mechanisms. *Physica Status Solidi a-Applied Research* 174(2): 431-439.

Oster, L., V. Yaskolko, et al. (2001). The experimental criteria for distinguishing different types of exoelectron emission mechanisms. *Physica Status Solidi a-Applied Research* 187(2): 481-485.

Pellin, M. J., W. F. Calaway, et al. (2001). Laser Post Ionization for Quantative Elemental Analysis. *ToF-SIMS: Surface Analysis by Mass Spectrometry*. J. Vickerman and D. Briggs, Surface Spectra Ltd. and IM Publications 375.

Peng, W. P., Y. C. Yang, et al. (2006). Laser-induced acoustic desorption mass spectrometry of single bioparticles. *Angew. Chem., Int. Ed.* 45(9): 1423-1426.

Perez, J., L. E. Ramirez-Arizmendi, et al. (2000). Laser-induced acoustic desorption/chemical ionization in Fourier-transform ion cyclotron resonance mass spectrometry. *Int. J. Mass Spectrom.* 198(3): 173-188.

Phipps, C. R., T. P. Turner, et al. (1988). Impulse coupling to targets in vacuum by KrF, HF and CO_2 single-pulse lasers. *Journal of Applied Physics* 64(3): 1083-1096.

Pollard, H. F. (1977). Sound waves in solids. London, Pion.

Prokhorov, A. M., V. I. Konov, et al. (1990). Laser Heating of Metals. Bristol, Philadelphia, New York, Adam Higler.

Reid, G. E., S. E. Tichy, et al. (2001). N-terminal derivatization and fragmentation of neutral peptides via ion-molecule reactions with acylium ions: Toward gas-phase Edman degradation? *Journal of the American Chemical Society* 123(6): 1184-1192.

Royer, D. and E. Dieulesaint (2000). Elastic waves in solids. Berlin ; New York, Springer.

Schiller, J., R. Suss, et al. (2007). Maldi-Tof Ms in Lipidomics. *Frontiers in Bioscience* 12: 2568-2579.

Scruby, C. B. (1987). An introduction to acoustic emission. *Journal of Physics E (Scientific Instruments)* 20(8): 946-953.

Scruby, C. B. and H. N. G. Wadley (1978). Calibrated Capacitance Transducer for Detection of Acoustic-Emission. *Journal of Physics D-Applied Physics* 11(11): 1487-1494.

Shea, R. C., C. J. Petzold, et al. (2007). Experimental investigations of the internal energy of molecules evaporated via laser-induced acoustic desorption into a Fourier transform ion cyclotron resonance mass spectrometer. *Analytical Chemistry* 79(5): 1825-1832.

Smith, S. T. (2000). Flexures: elements of elastic mechanisms. Amsterdam, Gordon & Breach.

Song, K. H. and X. Xu (1997). Mechanisms of absorption in pulsed excimer laser-induced plasma. *Applied Physics a-Materials Science & Processing* 65(4-5): 477-485.

Spengler, B., U. Bahr, et al. (1988). Postionization of Laser-Desorbed Organic and Inorganic-Compounds in a Time of Flight Mass-Spectrometer. *Anal. Instrum.* 17(1-2): 173-193.

Srinivasan, R. and B. Braren (1989). Ultraviolet-Laser Ablation of Organic Polymers. *Chemical Reviews* 89(6): 1303-1316.

Vertes, A. (1991). Laser Desorption of Large Molecules: Mechanisms and Models. *Methods and Mechanisms for Producing Ions from Large Moleculres*. K. G. Standing and W. Ens. New York, Plenum Press.

Vertes, A. and R. D. Levine (1990). Sublimation Versus Fragmentation in Matrix-Assisted Laser Desorption. *Chem. Phys. Lett.* 171(4): 284-290.

Veryovkin, I. V., W. F. Calaway, et al. (2004). A new time of flight instrument for quantative surface analysis. *Nucl. Instrum. Methods Phys. Res., Sect. B* 219-220: 473.

White, R. M. (1963). Generation of elastic waves by transient surface heating *Journal of Applied Physics* 34(12): 3559-&.

Xu, B. Q., J. Feng, et al. (2008). Laser-generated thermoelastic acoustic sources and Lamb waves in anisotropic plates. *Applied Physics a-Materials Science & Processing* 91(1): 173-179.

Young, C. E., J. E. Whitten, et al. (1989). Electron-Stimulated Desorption of Neutrals from 6063 Aluminum - Velocity Distributions Detected by 193 Nm Non-Resonant Laser Ionization. *Surface and Interface Analysis* 14(10): 647-655.

Zinovev, A. V., I. V. Veryovkin, et al. (2007). Laser-driven acoustic desorption of organic molecules from back-irradiated solid foils. *Analytical Chemistry* 79(21): 8232-8241.

17

Excitation of Periodical Shock Waves in Solid–State Optical Media (Yb:YAG, Glass) at SBS of Focused Low–Coherent Pump Radiation: Structure Changes, Features of Lasing

N.E. Bykovsky and Yu.V. Senatsky
Lebedev Physical Institute, Russian Academy of Sciences, Moscow Russia

1. Introduction

During several last decades much attention was paid to the processes that occur in solid-state optical media under the interaction with high-power focused laser radiation. A great number of studies were devoted to the phenomena of optical breakdown, structure changes, stimulated scatterings, generation of hypersonic waves in transparent dielectrics under the action of nanosecond (ns) and picosecond (ps) laser pulses (Manenkov & Prokhorov, 1986; Nelson et al., 1982; Ready, 1971; Robinson et al., 1984; Stuart et al., 1995). Recent interest in these studies was stimulated by the appearance of lasers with femtosecond (fs) pulses (Gordienko et al., 2010; Merlin, 1997; Sakakura et al., 2007).
An experimental study of a small region with high pressure and temperature gradients formed in a medium at focusing high-power laser radiation had been performed, as a rule, outside the laser cavity. In our experiment (Basiev et al., 2004), a region with such properties happened to be formed directly in the 2-mirror laser cavity, when Yb:YAG samples were pumped by the focused wide-band (0,89-0,95 μm) radiation from a pulsed LiF: F_2^+ color center laser (ccl). Thus, in contrast to many studies on ytterbium lasers, conditions for generation in Yb-doped samples in this experiment had been distinguished by the very high intensity (over 1 GW/cm^2) of the pump, which moreover had a low coherence. Experiments on pumping of Yb-doped and non-doped samples of different optical media (YAG, glass, LiF et al.) by powerful low-coherent radiation from LiF: F_2^+ ccl were continued in subsequent papers (Bykovsky, 2005, 2006; Bykovsky & Senatsky, 2008a,b, 2010). At intensities $I \geq 1$ GW/cm^2 interaction of ccl pump radiation with the medium in the focal region was essentially nonlinear. The interaction of ccl pulses with samples was accompanied by excitation of stimulated Brillouin scattering (SBS) and stimulated Raman scattering (SRS) of pump radiation. The scattering generated hypersonic waves of high amplitude, which were converted into a periodic sequence of shock waves with sharp pressure jumps on their fronts propagating along the direction of pump. Pressure jumps were so large that they caused a phase transition in an optical medium, which was observed near the sample surface in the form of small domains with spatial modulation of the refractive index caused by the interference of hypersonic waves.

Under ccl pumping due to heat release and generation of intensive hypersonic waves a region with strong temperature, pressure and refractive index gradients and at the same time with a high-level of inversion was formed in the focus of the pump laser in Yb-doped materials. Despite the strong optical inhomogeneity of the medium, Yb lasing in 10-15 ns pulses was observed in Yb:YAG (with 20% concentration of Yb^{3+} ions) and Yb:glass (with 10% Yb^{3+} concentration) during the action of the 20-30 ns pump pulse and after it (due to inversion remaining in the medium). During the SBS of pump radiation the hypersonic wave spatial structure served as a resonator for Yb lasing. Shock waves (with phonon energies up to 1000 cm^{-1}) affected the generation dynamics. The Yb lasing was distinguished by some specific characteristics such as a surprisingly wide spectrum (up to 50 nm) and a high directivity of the emission. In addition to the wide-band generation on the shock-waves grating there was also observed Yb lasing on resonator modes. After the end of the pump pulse another sequence of shock waves diverging outward the focal region affected the build-up of generation between 2 mirrors in the cavity. The line spectra of Yb generation in the resonator contained twisted spectral lines with structures of small-scale spots.

Description of these unusual phenomena observed under the interaction of short intense pump laser pulse of low coherence with optical media and their explanation are presented in this chapter. The optical scheme and the parameters of LiF:F_2^+ ccl are considered. The features of SBS and SRS and the appearance of the periodical shock waves in the optical medium at low-coherent pumping are discussed. The interpretation to observed specific optical damage is given. The mechanism of generation of broadband, high-directional short laser pulses in the spatial structure of thin layers with inversion produced in the region of propagation of intense hypersonic waves in the medium is discussed. Conditions for generation in a 2 mirror resonator containing active medium with a strong refractive index gradient are considered. The interpretation of the observed twisted lines with small-scale structures in generation spectra as well as temporal, spatial-angular characteristics of Yb lasing in the resonator is given.

2. LiF: F_2^+ color center pump laser

Optical pumping of Yb-doped materials can be performed only into the single Yb^{3+} ion absorption band at $^2F_{7/2}$ -$^2F_{5/2}$ transition near 0.9 μm (Figs.1, 2). In this connection, the use of flash lamps as sources of broadband radiation for pumping an Yb-doped medium is ineffective. At the present moment, the most effective and widely used sources of Yb-doped materials pumping are semiconductor laser diodes operating within the spectral range near 0.94 μm. Along with semiconductor diodes solid-state laser pump sources have been used to investigate Yb-doped active media. Cr:LiSAF, Ti:Sa, Nd:YAG (0.94 μm transition) lasers have been used to pump Yb:S-FAP and Yb:YAG (Bykovsky et al., 2000; Kanabe et al., 2000; Marshall et al., 1997, as cited in Bykovsky & Senatsky, 2008b).

In our work a LiF: F_2^+ color center laser (ccl) was used to pump an Yb:YAG crystal (Basiev et al., 2004). The lithium fluorine color center (LiF: F_2^+, LiF: F_2^-) lasers are the sources of radiation in the near IR (0.8-1.3 μm) , and they effectively convert the neodymium and ruby laser radiation into this spectral range (Basiev et al., 1982). Possible room temperature operation, high conversion efficiency (up to 30%), and a large generation tuning range (more than 1000 cm^{-1}) make such lasers very attractive for certain practical applications. The pumping of Yb-doped samples by the ccl had been, undoubtedly, inferior to laser diodes pumping in efficiency. However, at focusing ccl radiation in the Yb-doped medium the pump power densities typical of semiconductor diodes could be easily exceeded. This had been, of course, of interest for research. Therefore, just the ccl had been used later in experiments on study lasing in Yb-doped samples as well as nonlinear interaction of pump radiation with optical materials. The possible generation region of LiF: F_2^+ ccl extends from 0.83 to 1.1 μm (Basiev et al., 1982) and completely covers the absorption band of Yb:YAG. In particular experimental conditions the actual emission band of ccl depends on the selective properties of the resonator, and usually covers only part of the noted wavelength range.

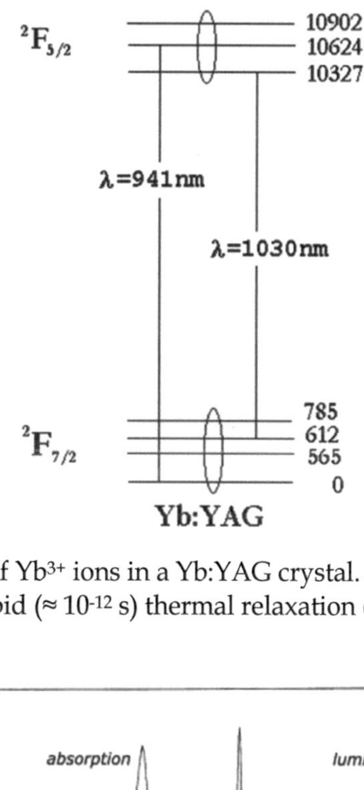

Fig. 1. Energy level diagram of Yb^{3+} ions in a Yb:YAG crystal. Ovals combine the Stark components of levels with rapid (≈ 10^{-12} s) thermal relaxation (Krupke, 2000)

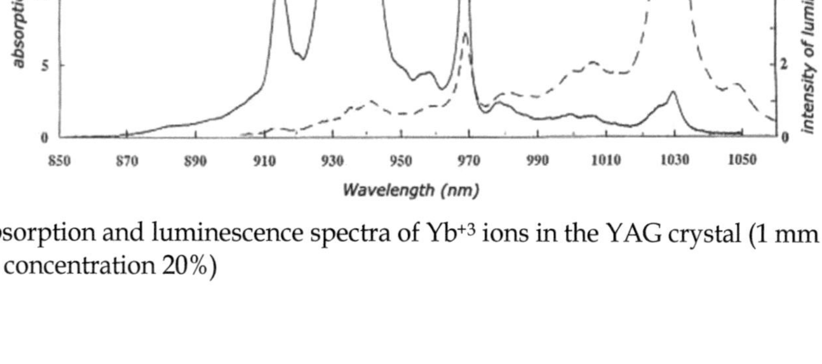

Fig. 2. Absorption and luminescence spectra of Yb^{+3} ions in the YAG crystal (1 mm plate, Yb^{+3} ions concentration 20%)

Figure 3 illustrates the scheme of the ccl with elements of radiation transport and diagnostics. A LiF: F_2^+ crystal (40x20x6 mm) was placed in the resonator formed by a plane mirror M_1 (≈ 100% reflection at 0.9 μm) and a glass plate M_2. The length of the ccl resonator made 30 cm. The LiF: F_2^+ crystal was pumped through glass plate by a ruby laser operating in the single shot regime with pulse duration ≈ 30 ns and the energy up to 1 J. A multimode radiation at the wavelength of 0.694 μm was focused into the LiF: F_2^+ crystal by a lens L_1 with the focal length F_1=500 mm. The ccl multimode radiation was, in its turn, focused on the studied samples by a lens L_2 with the focal length F_2=120 mm. The Yb-doped samples were placed into a compact two-mirror resonator.

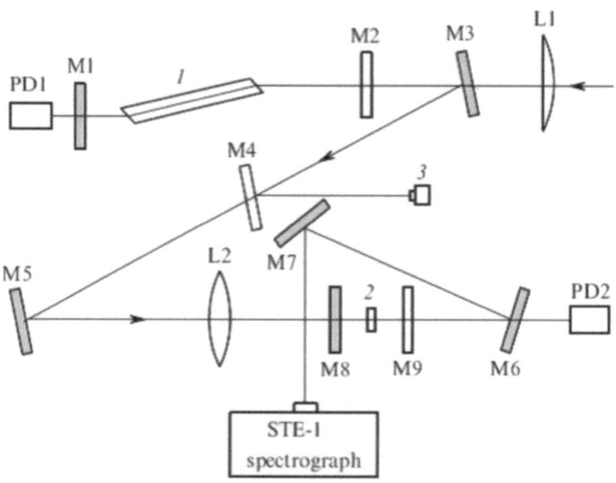

Fig. 3. Scheme of the experimental setup: (1) active element of a ccl; (2) Yb-dopped plate in the resonator with mirrors M8 and M9; (3) calorimeter; (M1, M2) ccl resonator mirrors; (M3 - M7) steering mirrors for ccl and Yb laser radiation; (L1 – L2) lenses; (PD1, PD2) photodiodes

The ruby laser and ccl pulses were registered with the help of photodiodes and a two-channel oscilloscope, and the laser energy was measured by calorimeters. The ccl radiation spectra were analyzed by the STE-1 spectrograph operating in the near IR region. The ccl emitted pulses of 20-30 ns duration and the energy of 100-150 mJ. For the most cases, the ccl pulse shape repeated the shape of the ruby laser pulse, Fig. 4a. Since the ccl resonator round-trip time made ≈ 2 ns, then during the ruby laser pumping the radiation made not more than 15 round trips in the ccl cavity. Due to high amplification in the LiF: F_2^+ medium the ccl pulse was formed in several round trips inside the resonator. Though the LiF: F_2^+ crystal was cut at a Brewster angle the ccl radiation was weakly polarized.

Figure 4b presents the densitograms of LiF: F_2^+ laser spectrum. The LiF: F_2^+ laser emission was observed within the range 0.89-0.95 μm. The lines of an argon spectral lamp were used as the wavelengths markers. Large spectrum width of ccl and short time of radiation development in the resonator speak about low coherency of the ccl emission. The ccl multimode radiation divergence was ≈ $2x10^{-3}$ rad. This allowed focusing the ccl pump at the sample into a ≈250μm spot. Moving samples along the axis of the focused pump beam one could change the size of the focal region within the limits of 250÷1000 μm, and the power density in the medium within the range of 0.5÷5 GW/cm^2. Such a range of power density variation allowed one to carry out both experiments on Yb lasing (see Section 4-6) and experiments on nonlinear interaction of ccl radiation with optical materials (see Section 3). In some of experiments, the ccl energy density was close to the damage threshold of Yb:YAG, glass and other studied materials. The material being damaged, the further experiments used a fresh part of a sample.

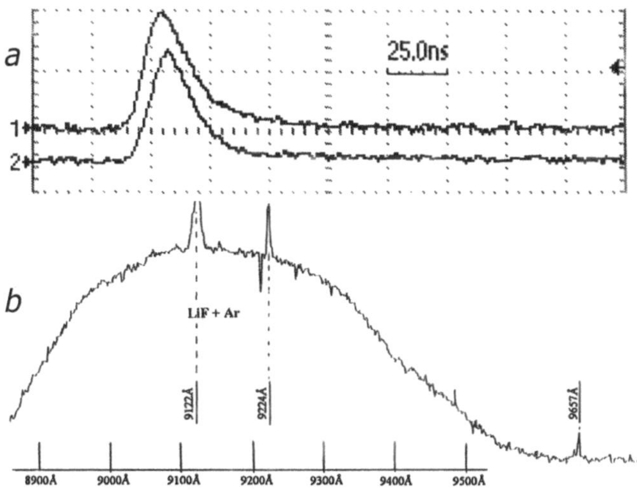

Fig. 4. Oscillograms of the ruby laser (1) and ccl (2) pulses - (a); densitograms of the LiF: F_2^+ laser spectrum with the argon spectral lamp reference lines - (b)

3. Interaction of ccl radiation with optical media

The experiments on nonlinear interaction of ccl radiation with optical media have been mainly performed using non-doped samples out of the cavity, Fig. 5. The ccl radiation was focused at samples by L_2, F=120 mm, and, so, within the sample thickness of 1-3 mm the pump power density changed insignificantly. The lens was tilted at 10^0 to the direction of ccl beam so that reflections from lens surfaces would not come back to the ccl resonator. As the samples there were used plates and slabs with polished surfaces made of the following materials: crystalline quartz (10x10x20mm), Yb:YAG crystal (1-2 mm thick plates), YAG crystal (4,5x30x30 mm), calcite (2 mm plates), LiF crystal (5 mm plate), 2 mm plexiglas plates, glass cube (20x20x30 mm). The ccl radiation was directed onto samples at a normal or at some angle (including the Brewster angle) to the sample's surface. The ccl pulse energy coming to samples varied within the range from 50 to 120 mJ.

In all materials a strong scattering of radiation was registered under the action of a ccl pulse. A diagnostic complex consisting of photodiodes, oscilloscopes, calorimeters and a spectrograph was arranged on the stand (Fig.5) in order to study the scattering of the low coherent wideband (0,89-0,95 μm) radiation of LiF: F_2^+ ccl. Due to very large spectrum width (tens of nm) and complicated space-angular structure of the scattered radiation components, one comes across difficulties in obtaining the spectral data on the scattered radiation, and it was not done. Strong scattering from the ccl focusing region in a wide angular range (tens of degrees) in forward and opposite directions (relative to the ccl beam) was observed for all samples at the ccl power density ≥ 1 GW/cm². The scattering intensity grew with the growth of the ccl intensity. The strongest scattering was observed at an angle of 180^0, i.e. back to the ccl aperture. The scattering was also observed at 90^0 to the ccl beam. Using photodiodes FD with ≈1 ns time resolution, the shape and duration of the ccl pulse and scattered radiation were registered by two-channel oscilloscopes of 1ns and 4ns resolution. To collect the scattered radiation to photodiodes the spherical mirrors were used, Fig. 5. The registered time and space-angular characteristics of scattering evidenced the rise of stimulated scattering (SBS and SRS) of wideband low-coherent ccl radiation in the samples.

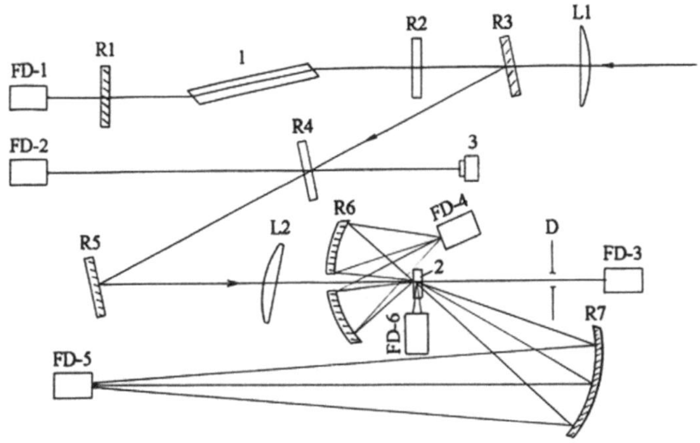

Fig. 5. Scheme of the experiment on the study of nonlinear phenomena in the crystal and glass samples: R1, R2 - LiF: F_2^+ ccl mirrors; R3 - R5 ccl radiation transport mirrors; R6 - R7 - mirrors to collect scattered radiation at diagnostic system; L1, L2 - lenses; FD - photodiodes; 1 - LiF: F_2^+ active element; 2 - sample; 3 - calorimeter; D - diaphragm

Figure 6 shows oscillograms of the ccl and scattered radiation pulses. A series of oscillograms in Fig.6 (left) illustrates the SBS of ccl pump radiation in YAG. Strong back scattering typical of SBS was observed in the YAG sample, placed at different positions to the pump. The duration of backscattered pulses was always smaller than the pump pulse duration (20÷30 ns) and constituted 8÷15 ns. At the oscillogram of the ccl pulse which passed the sample and the diaphragm D, one observes a pump «cut-off» typical for SBS, Fig. 6 (left, I, b). This pulse «cut-off» was connected with the reflection of the ccl beam incident onto the sample under the occurrence of a periodic grating formed by hypersonic waves.

Figure 6 (right) illustrates oscillograms of scattering in the YAG and crystalline quartz samples in forward and backward directions, as well as at 90^0 to the ccl beam. Mirror R6 (Fig.5) focused the backward scattering at angles from 6^0 to 43^0, and mirror R7 - forward scattering at 5^0 to 25^0 in respect to the pump beam. The photodiode FD 6 recorded the 90^0 scattering. The data obtained on the scattering of wide-band low-coherent ccl radiation allow one to consider the pulses observed as different forms of the stimulated scattering. The scattering backward to the ccl aperture proves to be the SBS of pump, as noted above. The forward scattering at large angles may be attributed to the anti-Stokes component of SRS, and the backward scattering at large angle is due to the Stokes component of SRS. Scattering at 90^0, apparently, is due to the SBS and SRS of pump in the transverse direction.

Profiles of the scattered pulses observed in the oscilloscope, Fig.6 (right) illustrate the dynamics of SBS and SRS in samples and the influence of different components of stimulated scattering on each other. It should be noted that the pump intensities at which SBS and SRS were observed in our experiments (0.5 GW/cm^2), had been almost an order lower than those typical of scattering at a coherent pumping.

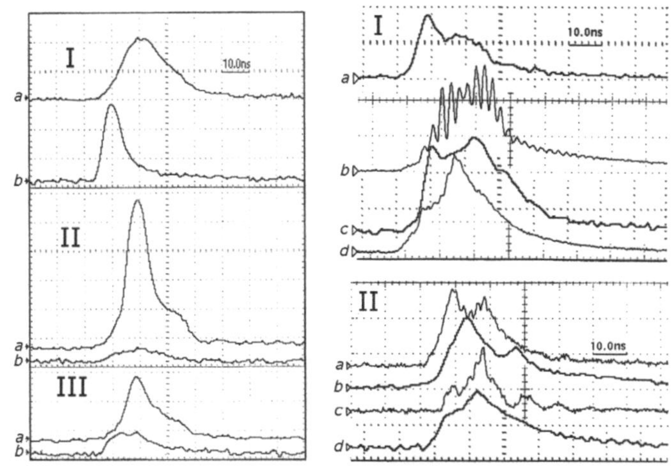

Fig. 6. (Left) SBS of ccl radiation in YAG slab (4,5x30x30mm). Oscillogams of SBS pulses (a) and ccl pulses behind the diaphragm D (b) for different orientations of the slab: I - pumping along the slab long side; II - pump beam along the short side; III - sample at Brewster angle to the pump. (Right) Oscillograms of scattering for YAG (I) and crystalline quartz (II): a–SBS pulse; b–scattering at 90^0 to the pump; c–anti-Stokes SRS signal, d–Stokes SRS signal

After the action of ccl pulses of 4-5 GW/cm^2 intensity onto samples multiple marks of such an action in the form of volume and surface damages, and changes of the material structure were observed. Microphotographs of structural changes were collected, data were systematized. The material of transparent dielectrics was found to be damaged in different ways. Cracks in the medium, local regions of structure changes, tracks of self-focusing, and some other types of optical damage were observed near the surface and in the depth of a sample in the direction of the pump beam. Figures 7-9 demonstrate microphotographs of structural changes in YAG and LiF samples. One can observe plane formations of several micron thicknesses with a sharp boundary, Figs.7, 9. Most of such objects were concentrated near the sample surface at depths < 200 μm. Fig.8 demonstrates cracking at the LiF crystal surface. All these figures are illustrative for the energy release in the subsurface layers of samples. The optical damage in transparent dielectrics caused by high-power radiation was observed and studied for a few decades. The structure changes observed in the ccl experiments were compared with literature on the interaction of highly coherent radiation with transparent media. It was found that the low-coherent ccl radiation causes structural changes of some new kinds, which are not described in literature, for example, flat regions with structural changes bordered by a sharp boundary. The effects of nonlinear interaction of ccl pump radiation with transparent optical media (stimulated scattering, structural changes) were interpreted under the assumption that the SBS of low-coherent ccl radiation causes a succession of hypersonic shock waves. The formation of hypersonic shock waves at elastic nonlinearity at SBS was considered theoretically (Polyakova, 1966, 1968). But those results have not been so far supported by SBS experiments using the coherent pumping.

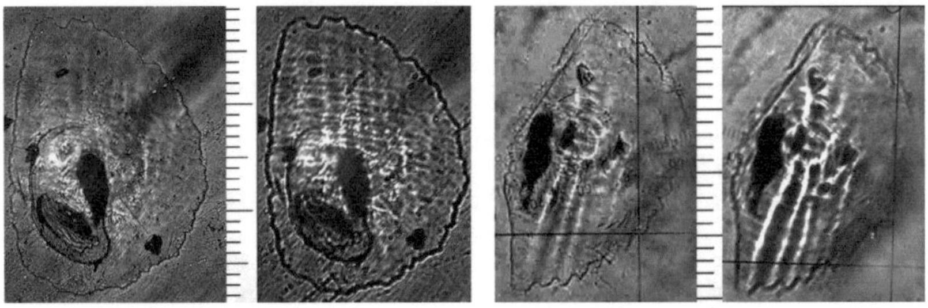

a b

Fig. 7. Structure changes in YAG plate in 2 different ccl shots (a, b): left - sample surface, right- 7-10 μm under the surface. Wavy lines on the photos are supposed to be caused by the hypersonic waves interference inside the sample, connected with the aberration (coma) of the focusing lens, L_2. Scale ~ 11,8 μm/mark

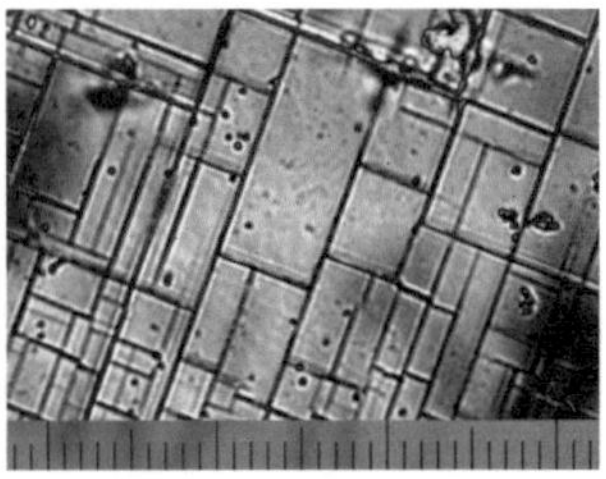

Fig. 8. Cracking at the LiF crystal surface. Scale ~ 11,8 μm/mark

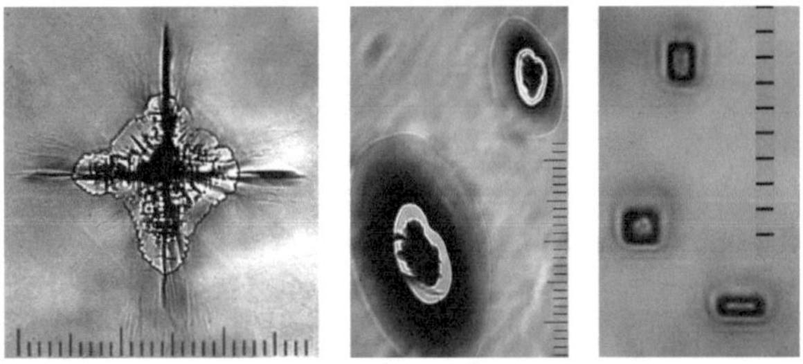

a b c

Fig. 9. Cross-like cracking having an additional plane formation with a sharp boundary inside the LiF crystal (a); ellipse-like cracking in the YAG crystal (b); a group of rectangular damages in the LiF crystal (c). Scale ~11,8 μm/mark

Structural changes in samples subjected to ccl pulses allow one to suppose the effect of hypersonic shock waves. Actually, a microphotograph of Fig.7a shows a region of the medium located at the depth of 7-10 μm under a YAG plate surface, where one can clearly see a quasi-periodic structure of stripes "imprinted" into an oval region with a sharp boundary. Figure 7b demonstrates another region, where a similar structure with step 10-15 μm is shown. These structures happened to be visible thanks to the refractive index gradient. However, such structures could not be formed in a medium as a result of the interference of low-coherent ccl radiation. At the same time, these structures could be the result of interference of intensive hypersonic waves propagating at a small angle to each other in the region of SBS of pump radiation. There are some other experimental data, which support a hypothesis of origin of intensive hypersonic waves in the medium. For example, a sharp boundary outlining structural changes in an YAG crystal, Fig.7 is the evidence of matter displacement deep down caused by the motion of dislocations in a surface region probably due to the phase transition in the low-lying layers in the region of hypersonic waves propagation. Note, that no apparent disturbances of the sample surface were observed. It is seen that some elements at the surface inside the oval region are near the same as they are outside: the pre-radiation scratches across the region boundary being continued, Fig.7.

To interpret the phenomena observed (stimulated scattering, structural changes) we consider a physical picture of the ccl radiation interaction with the medium, and a possible mechanism of the hypersonic waves buildup. The pump radiation spectral width of ≈ 600 Å permits one to suppose that a ccl pulse represents a sequence of ultra-short pulses (USP) of fs durations distributed chaotically within the 20÷30 ns envelope of the pump pulse. Energy of a single USP is estimated as 0.1÷1 μJ. Power density of separate ultra-short spikes could exceed 10^{10} W/cm^2. The interaction of ps and fs pulses with optical media was intensively studied, see for example (Nelson et al.,1982). The interaction of a single fs pulse (10^{13}W/cm^2 intensity) with transparent dielectrics (quartz, LiF, CaF_2, and others) produced generation of coherent optical phonons in the THz region leading to Raman scattering of probe pulses with ps delay (Gordienko et al., 2010; Merlin, 1997). The hypersonic waves (GHz region) were excited in liquids and solid-state media due to SBS on a space grating at the interference of two identical laser beams (10^{10} W/cm^2 intensity, ps durations) propagating at an angle to each other (Nelson et al., 1982; Robinson et al., 1984). In our conditions samples were irradiated by a succession of the chaotically distributed USP with a spectral range, which may be characterized by the central wavelength λ_p and the wave vector $k_p = 2\pi/\lambda_p$.

The appearance of shock waves at low-coherent pumping and their absence at the coherent pumping may be due to different conditions of SBS of pump radiation for those two cases. As is known, the SBS of coherent pump is developed on density fluctuations, i.e. hypersonic waves of the thermal origin (Ritus, 1982; Starunov & Fabelinsky, 1969). Aside from thermal fluctuations, there exist the stationary inhomogeneities in any solid-state medium, such as micro-inclusions, lattice inhomogeneities, dislocations. The stationary inhomogeneities may be the source of excitation of hypersonic waves and SBS. In fact, the interference of the incident and reflected pump radiation near inhomogeneity gives rise to a standing light wave with a spatial period $\lambda_p/2n_p$ where n_p is the refractive index of the medium at λ_p. The standing wave amplitude is determined by the coefficient of the pump radiation reflection from an inhomogeneity. The standing wave causes modulation of the refractive index, which gives rise to a spatial structure with the period $\Lambda=\lambda_p/2n_p$, which is resonant to the incident pump radiation and can be considered as the "seed" of the hypersonic wave. This structure starts moving due to the absorption of the momentum of light at scattering of pump photons. However, the stationary inhomogeneities do not exert influence on the development of hypersonic waves at the quasi-continuous coherent pumping, because the constant generation of the perturbations near the inhomogeneity is a drawback to the development of hypersonic waves and SBS.

At the pulsed low-coherence pumping, not only fluctuations but stationary inhomogeneities may be the source of SBS and hypersonic waves as well. Let an interaction of a single USP (a spike of pump) with the radiation reflected by an inhomogeneity produces a spatial structure with the period $\Lambda=\lambda_p/2n_p$ in a medium at the distance corresponding to the spike coherence length (for YAG, $n_p \approx 1.82$, $\Lambda \approx 0.25$ μm). For USP with durations ranging 0.01-0.1 ps the coherence length makes 5-50 μm. If the spike duration t_{sp} is less than half period of the hypersonic wave $t_{sp} < T/2$ ($T = \lambda_p/2n_p v_s \approx 50$ ps, $v_s \approx 5\text{x}10^5$ cm/s -sound velocity in YAG) then after the spike end the periodical perturbations should propagate in the direction of pump and backward in the form of two travelling hypersonic waves, whose lifetime in the solid-state medium is $\geq 10^{-9}$ s for the GHz frequencies (Ritus, 1982; Starunov & Fabelinsky, 1969). In this time interval, the hypersonic waves may fall under influence of many other USP pulses, which follow the first spike. The pump radiation affects the hypersonic waves, and as a result it is partially scattered with changing frequency due to the Doppler shift both to the Stokes and anti-Stokes sides. In the anti-Stokes case an atom moving towards pump is decelerated by absorbing momentum from a counter propagating pump photon. That stops eventually the counter-propagating hypersonic wave. At Stokes scattering, however, momentum from a co-propagating pump photon is transferred to an atom in the direction of its motion, supporting the hypersonic wave propagation. Therefore, only one from the two hypersonic waves, i.e. the wave moving along the pump direction is sustained. Herewith the scattered light wave at the Stokes frequency ω_s propagates towards the pump and stimulates emission (at frequency ω_s) of the excited atoms, which absorbed a pump photon (at frequency ω_p). The difference of the photons energy, $\Delta\omega=\omega_p-\omega_s$, determines the amplification of the hypersonic wave, co-propagating the pump. At the same time this wave is extended towards the pump source. In other words, the scattered light wave deflates energy from the pump to the co-propagating hypersonic wave and extends it toward the pump source by half of the coherence length of the USP. The following next pump spikes running over the hypersonic wave should extend it up to the sample input surface.

Because the spatial structure of the co-propagating hypersonic wave, which is resonant with pump, reflects effectively the incident radiation, the maximum photon energy and momentum are transferred to atoms in a "leading" edge of this structure facing the pump source. The energy is accumulated there as a result of a multiple run of USP pulses on a hypersonic wave, which stops near the surface unable to propagate further. Thus, the hypersonic wave amplitude rises. This provides jumps of pressure at the wave fronts, which make the sound velocity higher at the maxima and the leading fronts steeper. The hypersonic wave is transformed into a periodic shock wave due to this growing "anharmonicity". Thus concentration of acoustic energy near the input surface of the sample occurs.

A proposed model of shock waves formation at SBS of low-coherent ccl radiation makes it possible to interpret the experimental data on stimulated scattering and structure changes in the medium. A directed action of high-energy phonons causes structural changes near the sample surface, as shown on microphotographs of Figs.7-9. Release of acoustic energy and the rise of tension in a sub-surface layer may be evidenced by cracking of a LiF crystal surface, Fig.8. Other evidences may be the flat regions of structural changes, Figs.7, 9. The analysis of experimental data on lasing in the Yb-doped samples (see section 4-6) made it possible to estimate pressure at the shock wave fronts (tens of GPa). The pressure above 10 GPa is higher than the pressure of phase transitions in some materials. One may assume that structural changes of YAG crystal with a sharp boundary, Fig.7, 9 are rather due to phase transitions in a medium.

It should be noted that propagation of a hypersonic wave produced at some inhomogeneity deep in the medium may be stopped inside the sample by a "barrier" in the form of a dislocation, a crack, and the like. The wave would be partially stopped by the barrier. Its acoustic energy is released in the sample near the side of the barrier. In the case of acoustic energy exceeding the crystal lattice deformation threshold or the phase transition threshold there occurs local splitting of medium, shift of the barrier's wall. In this way one can explain the observed structural changes in crystals occurring along cleavage planes, Fig.9. Figure 9a demonstrates a typical cross-like cracking of a LiF crystal having an additional plane formation with a sharp boundary. Figure 9b shows an ellipse-like splitting in the YAG crystal. Sharp boundaries in a region of structural changes may be interpreted as phase transition boundaries in the medium. Figure 9c illustrates a group of rectangular damages, which seems to be due to the concentration of acoustic energy on defects (dislocation walls) inside the sample. The formation of tensions around damages, which seen in the form of blurred fringes, Fig.9c, is also the evidence of a considerable release of energy in that region.

The comparison of mechanisms for hypersonic wave formation at SBS of the coherent and incoherent pump may be supplemented by the following considerations. The efficiency of the pump energy transfer into the acoustic energy proves to be higher in the case of the low-coherence pumping. At similar pump energy the light–acoustic wave interaction proceeds more effectively when pumping by USP with higher intensities and at small distances (USP coherence lengths) as compared with the case of a high-coherence radiation. Rather large amplitude of light wave scattered on stationary inhomogeneities (as compared to fluctuation scattering) and high intensity of the low-coherence spikes (compared to average level of the pump intensity) should make the SBS threshold lower and stimulate formation of the intensive hypersonic waves. This explains the appearance of intensive hypersonic waves at low-coherence pumping. At the coherent pumping, the hypersonic wave seeded by a small-dimensional fluctuation of pressure, needs large distances for its development, and cannot "grow" to a shock wave. Note, that at the low-coherence pumping, the hypersonic wave may be caused even by a large-scale inhomogeneity, like a plane back surface of a sample. In that case, the produced hypersonic wave has a wide wave front. Even in a thin sample (1-2mm) this may cause structural changes of medium (phase transitions) comparable with dimensions of the pump spot.

Sharpening of the hypersonic wave fronts provides the appearance of high-frequency harmonics up to the optical phonons excitation. Moreover, as steepness of the shock waves increases, the momentum transferred to waves from successive USP pumping increases too, thus exciting oscillations of atoms (coherent phonons) with frequencies inversely proportional to the USP duration. The USP of the pump being scattered on coherent phonons (like probe pulses in works Gordienko et al., 2010; Merlin, 1997) give rise to the stimulated Stokes or anti-Stokes scattering. As optical phonons are concentrated on shock-wave fronts, hence the stimulated scattering is linked to fronts, Fig.10. Due to a large shift of the Stokes wavelength (λ_s), the SRS may be amplified only at the angle α to the pump: $\cos\alpha = n_s\lambda_p/n_p\lambda_s$, Fig.10. The anti-Stokes SRS is developed in the opposite direction. The respective directions and angles of scattering are illustrated in Fig.10. It should be noted that high-intensity optical phonons excited at shockwave fronts stir up the molecules at amplitudes and frequencies close to the possible limit in the crystal lattice, and hence make easier structural changing of the medium, i.e. its phase transition to a more dense state.

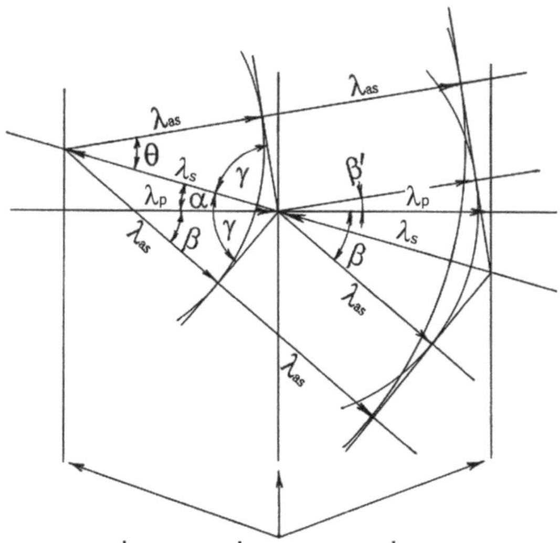

Fig. 10. Illustration to Stokes (λ_s) and anti-Stokes (λ_{as}) scattering at hypersonic shock wave fronts: a – angle between Stokes and pump waves; β, β' - angles between anti-Stokes and pump: $n_s\lambda_p/n_p\lambda_s = \cos a$; $n_s\lambda_{as}/n_{as}\lambda_s = \sin\gamma = \cos\theta$; $\theta - a = \beta'$ and $\beta = 90° - (\gamma - a)$, $\beta' = 90° - (\gamma + a)$.

4. Experiments on pumping Yb-doped media by LiF: F_2^+ ccl

Experiments on pumping Yb-doped media were performed using a setup of Fig. 3. Samples: plane - parallel 2-mm-thick Yb :YAG (20% of Yb) or 3-mm Yb-doped phosphate glass (10% of Yb) plates were mounted in a resonator of length L ≈ 20 mm formed by plane mirrors M8 and M9. Samples were pumped by pulses of focused ccl radiation through mirror M8 with the reflectance ≈100% at ≈1 μm transmitting 80%-90% of pump radiation. The pump beam propagated close to the normal to the resonator mirrors. The reflectance of mirror M9 at ≈ 1 μm was 30% (for Yb: YAG) or 70% (for Yb: glass). The energy, spectrum, shape of Yb-laser pulses were registered by calorimeters, photodiodes, and STE-1 spectrograph. Experiments were performed at room temperature in the single-shot regime. In most experiments samples in the resonator were exposed to the ccl radiation of energy 60 - 70 mJ. The focused ccl beam produced the radiation intensity distribution I (r) (r-radius of the excited region) in the subsurface layers of samples with the maximum at the pump beam axis. The minimal size $2r$ of the focal spot on the sample surface was ≈ 250 μm. The ccl intensity was varied within 0.5 - 5 GW cm^{-2} by moving the lens. Lasing in a Yb :YAG crystal was observed at pump intensities exceeding 0.5 GW cm^{-2}, Yb laser energy did not exceed 0.5 mJ. Figures 11 and 12 show Yb-lasing oscillograms recorded for one shot. The first of a series of laser pulses appeared during the pump pulse, while the next pulses appeared after the end of the pump and were delayed by tens and hundreds of ns, up to 1.2 microsecond (μs). Spectra were recorded by STE-1 on an IR film or on the image converter. The optical scheme of the STE-1 imaged (1:1) the vertical slit of ≈100μm width on the film. At uniform illumination of the slit by the monochromatic radiation of a spectral lamp the film registered separate narrow vertical lines, Fig.4b. In case of Yb-lasing we observed much more complex spectra, Figs 11-14. The lasing spectra were recorded within an angle of ≈10^{-2} rad, which was determined by the slit height and a distance (≈1m) from the sample. Laser radiation going along the resonator axis was directed to the lower part of the slit. No focusing of radiation on the slit was performed. Figs 11, 12 demonstrate lines in the region of Yb transitions in YAG at 1.03 and 1.05 µm. At low pump intensities line lasing appeared first at ≈1.05 µm. As the pump intensity was increased, line lasing was observed simultaneously at 1.03 and 1.05 µm (Fig. 11), and at the maximum pump it was observed at 1.03 µm, Fig.12. Fig.11 demonstrates a noticeable inclination of the spectral line at 1.03 µm from the vertical direction. This means that the lasing wavelength in this line was varied along the slit height. Structure of small-scale (50-200µm) spots was observed in 1.05 µm line, Fig.11. Along with narrowband lasing, broadband unstructured lasing was observed virtually in each pulse. Spectral bands up to 20 nm width were extended to the blue (up to 1.0 µm) and red (up to 1.06 µm) parts of the spectrum far beyond the regions near 1.03 and 1.05 µm in which Yb :YAG lasing is usually observed. As pump was increased, the broadband spectra were shifted to the blue (Fig. 12).

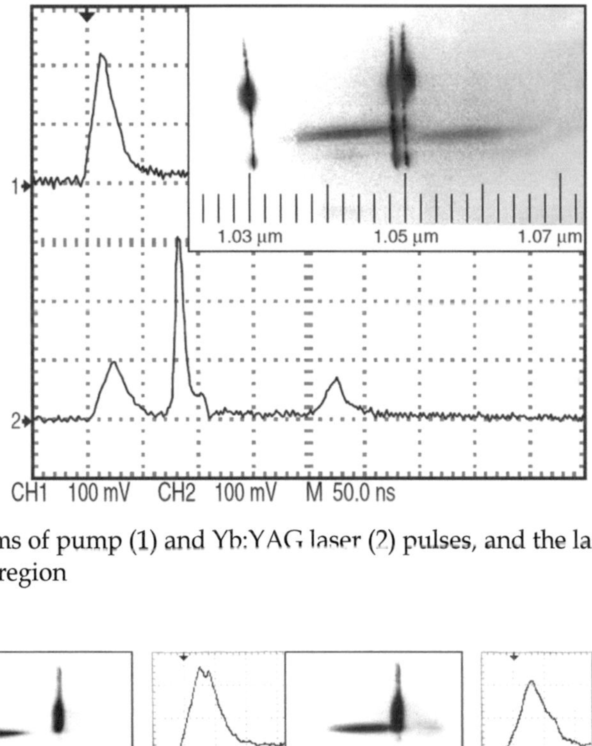

Fig. 11. Oscillograms of pump (1) and Yb:YAG laser (2) pulses, and the lasing spectrum in the 1.03 - 1.06-µm region

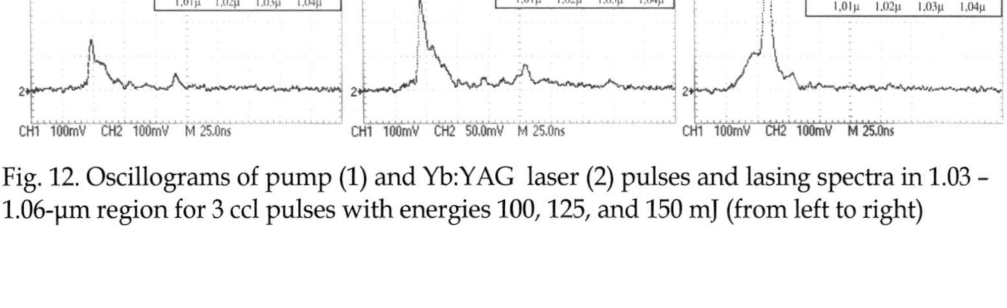

Fig. 12. Oscillograms of pump (1) and Yb:YAG laser (2) pulses and lasing spectra in 1.03 - 1.06-µm region for 3 ccl pulses with energies 100, 125, and 150 mJ (from left to right)

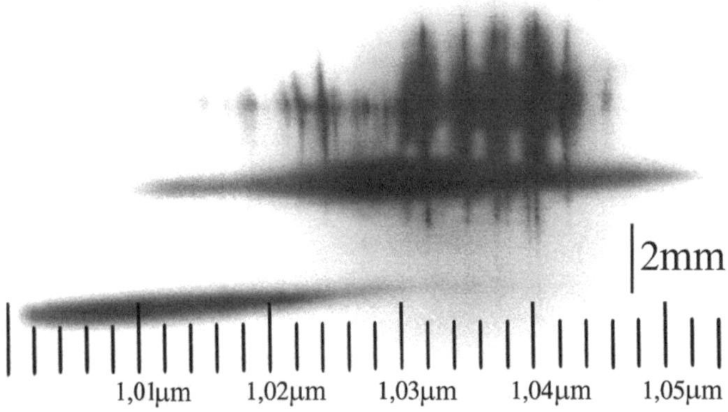

Fig. 13. Lasing spectra of Yb: glass in the 1.00 – 1.05-μm region

Lasing in the Yb: glass (Figs 13,14) appeared at pump intensities 4-5 GW cm^{-2} and was accompanied, as a rule, by the local damage of samples. The lasing energy did not exceed 0.1 mJ. The oscillograms of the glass lasing (similar to the Yb :YAG lasing) exhibit several ns pulses delayed with respect to the pump, Fig.14. The spectrograms demonstrate line (in 1.02-1.05 μm region) and broadband (1.00-1.06 μm) lasing spectra. Structures of small-scale spots are clearly visible in Yb:glass line spectra, Fig.14. The total spectrum of two broad bands of the Yb:glass emission exceeded 50 nm, Fig. 13. The width of stripes of broadband lasing in vertical direction was about 1mm, Fig.13. The line and broadband lasing spectra both in glass and Yb:YAG were recorded from the slit parts located at different heights, Figs 11-14. Note, that spectra of Yb lasing were recorded when the slit was in the far-field diffraction zone with respect to the position of the radiation source. So, the observed narrow stripes and small spots of generation, Figs 11-14, indicated a high directivity of radiation. The angular divergence of radiation, estimated from the size of spots on the slit, did not exceed usually 10^{-3}-10^{-4} rad. At the same time, the angular divergence caused by diffraction from the lasing region with r ≈100 μm should be ≈10^{-2} rad. Thus, the angular directivity of Yb-lasing could exceed the diffraction limit by one- two orders of magnitude, i.e. the generation should occur in highly directed beams of small transverse size.

At pump intensities from 0.5 to 5 GWcm^{-2} used in our experiments, we observed intense pulses of scattered ccl radiation in Yb-doped crystal and glass samples similar to observations of scattering (SBS and SRS) in non-doped samples described in the part 3. Consider the features of the inversion distribution in the Yb-doped medium upon SBS of pump radiation. The dynamic grating of hypersonic waves produced in samples near the surface played the role of an additional external mirror for the ccl and efficiently reflected incident radiation, preventing its propagation inside a crystal or glass sample. The penetration depth of hypersonic waves into the medium is $l \approx 1/\eta$, (η = 10^2- 10^3 cm^{-1} is the hypersonic waves attenuation coefficient) (Ritus, 1982; Starunov & Fabelinsky, 1969). Thus, the excitation region of SBS (and SRS) in the medium was a surface layer of thickness l < 100 μm. The heat release caused by the dissipation of energy of hypersonic waves and optical phonons excited in the medium upon SRS of ccl radiation occurred in this layer. Inversion was also mainly produced in this thin layer. The maximum energy density stored in Yb :YAG containing 20% of Yb (N_0 = 2.9 x10^{21} cm^{-3}) upon excitation of all Yb^{3+} ions to the $^2F_{5/2}$ metastable level was estimated as ≈500 J cm^{-3}. Even for such an extremely large energy in the volume of an Yb :YAG crystal cylinder of diameter 250 μm and length 100 μm, the energy stored in inversion does not exceed, according to estimates, 2.5 mJ, which corresponds to the low level of Yb-lasing energy in experiments. A part of the pump energy was absorbed by Yb ions beyond the region of SBS excitation, and the other part escaped from the sample. So, the population of Yb ions on the metastable level decreased over the sample length along the pump beam. Only a small fraction (estimated as 10%) of the ccl output energy incident on samples was spent to excite ytterbium. A grater part (up to 70%) of the ccl energy was transformed to stimulated scattering due to nonlinear interaction with the medium and was spent to form intense hypersonic waves and optical phonons accompanied by heat release in the medium. The thermal energy required for heating and melting of the above-mentioned Yb :YAG crystal cylinder was estimated as 30 mJ. At the same time, the thermal energy in the medium during the formation of inversion and generation and due to stimulated scattering did not exceed, according to estimates, 10 mJ, i.e. it was insufficient for melting of the material. This suggests that structural changes in the surface layer of samples (Fig.7-9) could be produced by intense hypersonic waves appearing during the SBS of pump radiation. Thus, the Yb-doped medium during the SBS of pump proved to be divided into a layer of thickness ≈100 μm with the very high inversion and the main part of the sample to which pump radiation hardly penetrated.

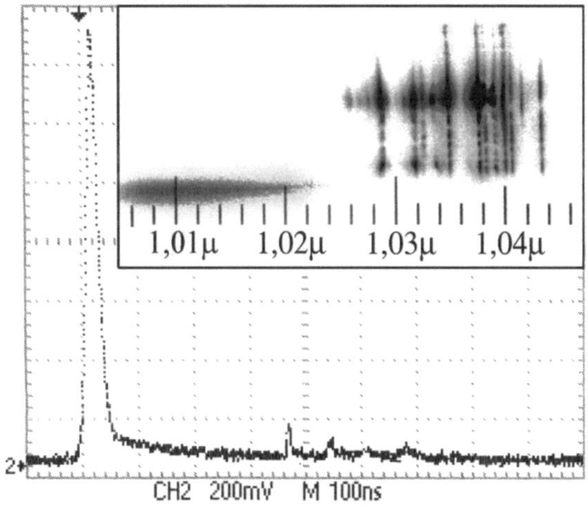

Fig. 14. Oscillograms of Yb: glass laser pulses and spectra in the 1.00 - 1.04-μm region

Figures 11-14 demonstrate the integrated (per pulse) temporal pictures and broadband and narrowband lasing spectra of ytterbium. These two types of emission correspond to different Yb-laser pulses appearing at different stages of the development of lasing. We can distinguish two such stages: the first one is the development of Yb-lasing during irradiation of a sample by the pump pulse, and the second one is Yb-lasing after the end of the pump pulse. At the first stage of duration t_1 = 20–30 ns, during the action of a ccl pulse onto the sample the generation could occur without a resonator, just due to the feedback at the dynamic hypersonic wave grating. The broadband unstructured spectra correspond to this stage. At the second stage of duration t_2 up to 1.2 μs there was generation in the Yb laser resonator which was characterized by line spectra, Figs. 11-14.

5. Features of Yb- lasing: broadband spectra, data interpretation

We attribute broadband lasing in the crystal and glass to Yb-laser pulses appearing at the first stage. Consider the conditions of inversion formation and lasing at the first stage by using a model assuming the appearance of shock hypersonic waves in the medium during the SBS of pump radiation. Excitation of Yb ions to the $^2F_{5/2}$ metastable level and formation of hypersonic wave grating (Fig. 15) occurred simultaneously in a thin layer of the medium. The grating period Λ is related to the pump wavelength λ_p and the refractive index of the medium n_p by the expression $\Lambda = \lambda_p/(2n_p)$ (Ritus, 1982; Starunov & Fabelinsky, 1969). By assuming that $\lambda_p \approx 0.92$ µm, we obtain for YAG $\Lambda \approx 0.25$ µm. The period of the hypersonic wave is $T=\lambda_p/(2n_p v_s)$, where v_s is the sound speed. For the glass, we have $T \approx 70$ ps, and for the YAG crystal, T ≈ 50 ps. Intensive hypersonic waves strongly affected the spatial distribution of inversion. Indeed, conditions were produced for excitation of high-frequency vibrations of the medium (optical phonons) and SRS on the fronts of shock waves. The scattering parameters show that a broad phonon spectrum was excited upon SRS, including phonons with energies ≈1000 cm^{-1} near the high-frequency boundary of the phonon spectrum in the medium (Gorelik, 2007). The energy of such phonons is sufficient for the population of the Stark components of the $^2F_{7/2}$ level lying above the ground level of Yb (Fig.1). The redistribution of the population of Stark levels occurs during several ps. Due to the phonon population on the fronts of shock waves, the inversion for transitions from the $^2F_{5/2}$ level to the Stark components of the $^2F_{7/2}$ level was decreased, preventing the development of lasing. At the same time, due to the rapid decrease of pressure and dynamic cooling of the medium, which occurred behind the fronts of shock waves (Fig. 15), the population of the Stark components of the $^2F_{7/2}$ level was rapidly depleted, resulting in the inversion jump. As a result, a structure of thin (<0.3 µm) periodical layers with the high inversion was established in the medium for a short time (t_{inv} < T). During the propagation of a hypersonic wave, the inversion regions were displaced. After the time T, the same inversion distribution over layers again reestablished. Thus, upon the SBS and SRS of pump radiation, amplification could appear at transitions between the $^2F_{5/2}$ level and Stark components of the lower $^2F_{7/2}$ level, including transitions in the short-wavelength region (< 1.03 µm), which are not observed usually in Yb- lasers.

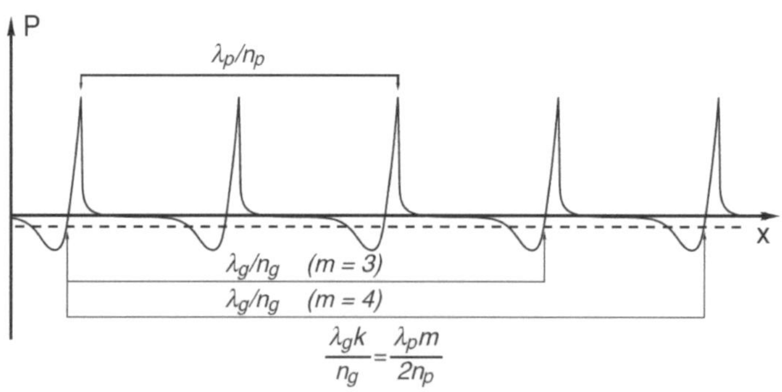

Fig. 15. Illustration to conditions for inversion formation and stimulated emission in the region of shock hypersonic waves in the active medium. The dependence of the pressure profile P in the hypersonic wave on the coordinate x along the pump direction is shown. The dashed straight line indicates the pressure at which inversion is produced. The possible relations between the pump, λ_p and laser, λ_g wavelengths satisfying condition (1) are given

The formation of inversion in the region of propagation of hypersonic waves produces the conditions for generation of short radiation pulses in the medium. Indeed, a distributed feedback (DFB) laser (Kogelnik & Shank, 1971) appeared in fact in our experiments in the hypersonic wave grating region. Such a laser operates without an external resonator and can emit ps pulses (Bor & Muller, 1986; Katarkevich et al, 1996; Kogelnik & Shank, 1971). Unlike dye DFB lasers based on a photo-induced sinusoidal grating in the medium, the high-contrast DFB structure was formed under our conditions due to modulation of the medium parameters (refractive index, inversion) during the propagation of a sequence of intense hypersonic waves. The possibility of creating a DFB laser on a hypersonic wave grating was discussed already in pioneering paper (Kogelnik & Shank, 1971). The length of the DFB structure in the region of SBS is ≈ 100 μm; in this case, the travel time t′ of a photon in the structure is much shorter than the round-trip transit time τ in the Yb- laser resonator ($\tau = 2L/c \approx 100$ ps). The possibility of lasing using the hypersonic wave grating without the resonator was confirmed by experiments. Lasing during the ccl pump pulse was observed both in samples mounted inside the resonator and in samples without the resonator, in particular, in samples mounted at an angle to the pump beam axis. The generation of ps pulses in dye DFB lasers on a stationary grating pumped by ns pulses occurs due to self-Q-switching mechanism: `blowing away' of the gain grating by the structure emission itself (Bor & Muller, 1986; Katarkevich et al, 1996). Unlike this, upon the SBS of ccl pump radiation, a DFB structure moving in the medium appears with period Λ which can depend on the pump intensity (pressure in the medium). Under these conditions, the Q-switching in the DFB structure can be determined, apart from the mechanism considered above, by the rapid movement of the grating in the medium and by variations in its parameters as well. Pulses of ps duration, $t < t_{inv}$ should be generated in our experiments successively in these periodic inversion layers during their movement. Besides, the change in Λ should lead to scanning of the lasing spectral band over the gain profile in the medium. The temporal picture of Yb- lasing at the first stage (during pump) was not investigated in detail due to the lack of recording equipment with a proper time resolution. The shape and duration of Yb-laser pulses were determined by the resolution of the oscilloscope (1- 4 ns).

By considering the conditions for appearing of stimulated emission in the moving layer structure, we can interpret features of broadband lasing spectra of the glass and Yb :YAG crystal (Figs 11-14). The Yb emission appears at wavelengths $\lambda_g > \lambda_p$. This emission is amplified along the normal to the layer structure when the path difference between the layers involved in lasing is equal to the quantity multiple of λ_g. This synchronism condition between hypersonic waves and lasing can be written in the form

$$\frac{\lambda_g k}{n_g} = \frac{\lambda_p m}{2n_p} \tag{1}$$

Here, n_g is the refractive index of the medium for λ_g; and k and m are integers. It is important that the refractive index depends on pressure P in the medium: $n_g, n_p \sim P$. Pressure depends on the intensity of hypersonic waves and the heat release in the medium upon pumping. The maxima of the spatial structure of the laser field are located in regions with a decreased density, behind the fronts of shock waves, Fig. 15. Here n_g can take values lower than the refractive index n_0 at the normal pressure. At the same time, the refractive index n_p on the fronts of shock waves can be larger than at the normal pressure. Under these conditions for $n_g < n_p$, relation (1) is fulfilled for $k = 1$ for small values of m ($m > 2$, $m = 3, 4...$, see, Fig. 15). It follows from the relation $\Lambda = \lambda_p/(2n_p)$ that, as the intensity of hypersonic waves and heat release increase (with increasing P and n_p) at the fixed λ_p, the period Λ of the grating decreases. This in turn leads to the blue shift of λ_g. Correspondingly, as P is decreased, the wavelength λ_g should shift to the red. Experiments confirm these conclusions. The shift of broadband spectra with changing the pump intensity is illustrated in Fig. 12. One can see that the spectrum of Yb:YAG shifts to the blue by 10 nm with increasing the ccl pulse energy from 100 to 150 mJ. Note that the shift of the emission region of DFB lasers with changing of the pressure was observed in dye lasers (Bor & Muller, 1986).

The inhomogeneous distribution of the pump with the maximum at the beam axis gave rise to the transverse gradient of pressure $P(r)$ in the region of propagation of hypersonic waves. Because of this gradient, for regions with different pump intensities (at the beam axis and its periphery) relation (1) is fulfilled for different values of m. These values can correspond to the regions of the broadband spectrum shifted in the wavelength scale and over the slit height, Figs 11-14. In the axial region with the maximum pressure ($r \approx 0$, the lower part of spectrograms), lasing was developed in the short-wavelength part of the spectrum. The spectra in regions with the lower pump ($r > 0$) are shifted to the red (Figs 11-14).

The regions of the broadband spectrum should be located on the wavelength scale in accordance with possible transitions in Yb: glass and Yb:YAG. According to the energy level diagram presented in Fig. 1, the emission of Yb:YAG in the spectral region under study (Figs 11, 12) can be determined by transitions between the two lower components of the metastable $^2F_{5/2}$ level with energies 10327 and 10624 cm^{-1} and three components of the $^2F_{7/2}$ level with energies 565, 612, and 785 cm^{-1}. In each spectral region, lasing in the DFB structure was built up in the active medium within the part of the gain line corresponding to the Yb transition which was resonant for the particular period of hypersonic waves. Thus, a large width of lasing spectra of Yb-doped materials (Figs 11 - 14) could be caused not only by the generation of short pulses in the DFB structure upon emission of thin layers of the active medium but, as explained above, by the shifts of λ_g with changing the pump intensity as well. This is a picture of lasing in Yb-doped media in the region of the propagation of longitudinal hypersonic waves at SBS of pump radiation. Thus, the data on Yb broadband lasing confirm the concept of intense (shock) hypersonic wave formation at powerful low-coherent pumping of the optical media.

The interpretation of data on the "sub-diffraction" angular divergence of Yb:YAG and Yb:glass generation had required the introduction of a new concept for the photon, which means its existence not in the form of a "traveling" wave, but with fixed positions of maxima, minima, and nodes along the photon propagation direction (Bykovsky, 2006; Bykovsky & Senatsky, 2008a,b, 2010). It was assumed that the photon is a combination of the two pinpoint dipoles, which move jointly with light velocity and rotate with frequency ω in the opposite directions. The phase of the photon and its field are connected with the phase of this rotation (Bykovsky, 2006). Such an approach helps to describe the effects observed. The field produced by the photon motion along the coordinate x is described by the function: $E = E_0 \sin(kx - \varphi_0)$, where $k = 2\pi/\lambda$, and λ is the distance, at which the photon changes its phase by 2π; E_0 -the field amplitude, φ_0 - the initial phase. The new concept does not contradict to the properties of the well-known electromagnetic "traveling" wave: the traveling wave may be "constructed" from separate photons following each other along the same direction and having a continuous relative phase shift from 0 to 2π. If we assume that the excited atoms in a medium irradiate photons along the same direction and with the initial phases $+\pi/2$ or $-\pi/2$ only, and that emitted photons are correlated by the relation $\Delta x = c\Delta t$, where Δx is the distance between the neighbor photons and Δt - the interval between the moments of their emission, then such a combination of photons should form a traveling wave. This assumption follows from an equation describing a field of a neighbor photon irradiated from a point distanced by Δx: $E = E_0 \sin[k(x \pm \Delta x) \pm \varphi_0]$. Taking into account that $k\Delta x = kc\Delta t = \omega\Delta t$ we obtain: $E = E_0 \sin(kx \pm \omega\Delta t \pm \varphi_0)$ - the traveling wave relation. The traveling wave amplitude increases in proportion to the number of equal-phase photons. To provide the traveling wave formation it is necessary that the size of a homogeneous region with inversion, from which a large number of photons is spontaneously irradiated into 4π, exceeded the wavelength λ. Organization of photons in the form of the traveling wave makes it possible to deplete inversion with maximum efficiency, as it is taken off continuously from all points on the path of the wave propagating in the active medium. If the inversion is non-homogeneous and structured with the period $\lambda/2$, such a medium will emit photon in trains, whose maxima should spatially coincide with the inversion maxima. Photons having phases differing by $2\pi n$ form a photon train described by the function: $E_n = E_0 \sin(kx \pm 2\pi n \pm \varphi_0)$. That should not be a traveling wave. `The instant photograph' of the filed distribution in a photon train should represent the contrast periodic structure of antinodes and nodes and can noticeably differ from `the instant photograph' of a traveling sinusoidal wave. However, the summary field produced by many photon trains, which are correlated by the relation $\Delta x = c\Delta t$, becomes "traveling". This "traveling" is like the Christmas-tree garland, whose running lights are due to the time shift of neighboring lamps burning moments.

Because the field produced by the photon is described by a harmonic function, the field superposition of the photon trains propagating at different angles and with different phases gives rise to the wave effects: interference and diffraction. The diffraction is connected with spatial confinement of the light wave front and caused by the interference of photons with different phases contained in the aperture of a beam. The intensity peaks are formed in the directions defined by the beam wave front profile, dimension and shape of a confined region, and the light wavelength. By the angular width of these peaks one may judge about the light beam divergence. The diffraction divergence of the traveling wave arises because the traveling wave restricted by an aperture contains within this aperture at the length of one spatial period the photons with different phases ranging from 0 to 2π. In general the phase dispersion of photons distributed in the interval Δx along the beam direction can be estimated as $\Delta\varphi = 2\pi\Delta x/\lambda$. When the beam is limited by an aperture of diameter d, the direction to the first minimum of the far-field intensity distribution (half the angle α of the diffraction divergence) can be written in the form $\alpha \sim \lambda\Delta\varphi/(2\pi d)$. In the case of a traveling plane monochromatic wave, the phase dispersion $\Delta\varphi$ in the beam can achieve $\pm\pi$. This gives the known estimation $\alpha \sim \lambda/d$ of the divergence angle for a beam limited by an aperture of diameter d. If the phase dispersion of photons is decreased, the diffraction divergence angle α decreases proportionally to $\Delta\varphi$. The group of such phased photons will form a beam with the sharp angular radiation pattern, which was observed in our experiments. Formation of beams with sub diffraction divergence occurs, apparently, in experiments on super-resolution microscopy, when the size of the fluorescent or light scattering object along the direction of beam does not exceed the light wavelength (Galbraith, C.& Galbraith, J., 2011; Tychinskii, V., 2008).

Unlike a usual laser with the extended active medium in the resonator, in our experiments radiation was formed in a structure consisting of thin inversion layers. The regions with reduced pressure behind the front of shock waves, shown in Fig. 15, represent as if the `instant photograph' of the spatial arrangement of such a structure. The photons with the wavelength resonant with the structure, $\lambda_g/n_g = m\Lambda$, will occupy different spatial positions in the `photograph' with respect to the layers of thickness $l_{inv} \ll \lambda_g/n_g$. And only a small part of the total number of photons in the `resonator' of our DFB laser will coincide in position and phase with the inversion maxima in the medium and, therefore, will be amplified efficiently during the lifetime $t_{inv} < T$ of the layers of the given spatial configuration, forming a beam on the basis of photon trains. During the movement of the inversion layers structure (following the movement of the hypersonic wave), new groups of phased photons will be formed, and their frequency will shift with changing the structure parameters. In such a way the model considered above explains the appearance of the sharp angular radiation pattern of the broadband lasing of ytterbium in the crystal and glass. In the case of a `thick' sinusoidal grating, photons occupying different positions in space (which is equivalent to the large phase dispersion within the photon train) are amplified. This leads to the increase in the radiation divergence. Thus, when a periodic spatial grating of thin inversion layers is produced in the active medium, the high-directional stimulated radiation of an ensemble of excited atoms can be observed in the optical range.

6. Features of ytterbium lasing: line spectra, data interpretation

Line spectra observed in the crystal and glass relate to the second stage of Yb-lasing. The second stage is the Yb-lasing in the resonator after the end of pump due to inversion remained in the medium. Of special interest the details of Yb spectra are. Twisted, inclined and structured lines in spectra were observed in the region 1.03–1.05 μm, Figs.11-14,16. Figures 11, 13, 14 showed an inclination of some lines to the short-wavelength side of the spectrum in respect to the vertical position of the slit. Nearly all line spectra showed quasi-periodic structures of small-scale (50–200 μm) spots (Figs. 11, 14, 16). Twisted lines consisting of separate spots are shown at Fig.16. This picture indicates a wave-like change in the generation wavelength in separate lasing spots along the height of the slit.

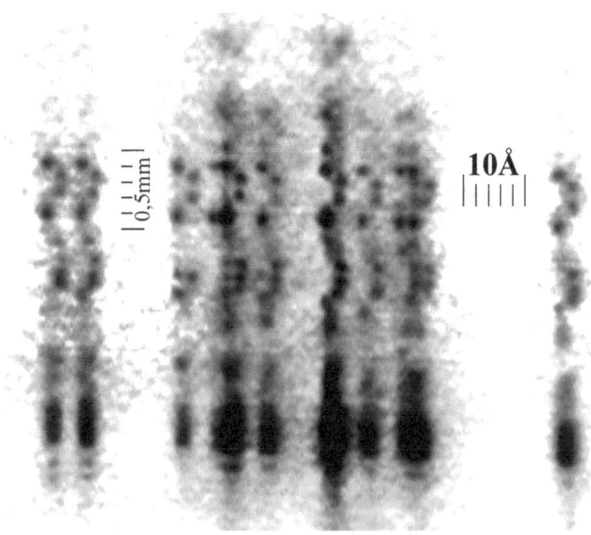

Fig. 16. Twisted lines with multiple spots of Yb:glass generation at several transitions near 1.04 μm; time delay from the pump pulse ≈ 50 ns

Consider the conditions for the development of lasing in a resonator at the end of pump, under disappearance of longitudinal hypersonic waves and relaxation of the excitation region in the active medium. Lasing developed in the resonator in the presence of the region of strong optical inhomogeneity near the active element surface- a thin layer with the inhomogeneous inversion distribution and gradient of the refractive index, n produced by pressure, temperature, and density gradients. To the rest part of the sample the pump penetrated weakly. The refractive index profile was formed during the action of the pump on a medium and it also changed after the end of pump during the relaxation of the excited region. This profile affected the development of Yb-lasing. An experimental study of a non-stationary refractive index profile (so-called "transient lens") arising under short laser pulse interaction with a medium is a rather complicated task. Here we discuss in a qualitative form processes that affected formation of the index profile, $n(r)$ assuming that pumping was symmetrical over the azimuth angle. Consider again a part of the medium in the form of a cylinder of 250 μm diameter and 100 μm length, which fits the size and configuration of the focal region at tight ccl focusing. The heat load occurred in such a cylinder during inversion formation, Yb-lasing, and mainly, due to pump radiation scattering. The heat load was estimated to be <10 mJ, which allows for the average thermal energy density in the cylinder, $< 2 \times 10^3$ J/cm^3. Estimations for the average pressure, P_{av} and temperature, T_{av} just after the end of pump give $P_{av} \approx 1$ GPa and $T_{av} \approx 10^3$ K. In actuality, due to the bell-shaped profile of the pump beam, $P(r)$ and $T(r)$ profiles with maxima exceeding averaged values arose in the medium. The maximum pressure near the axis of the excitation region at the moment near the end of pump pulse may be estimated by using the value P_{pt}, at which a phase transition occurs. Structure modifications similar to phase transitions had been observed in Yb:YAG and Yb:glass samples near the axis, where hypersonic waves of maximum intensity propagated (section 3). We use data for glass $P_{pt} \approx 10$ GPa, $\delta n/\delta P = -0.13 \times 10^{-5}$ bar^{-1}, and $\delta n/\delta T = -6 \times 10^{-6}$ K^{-1} (Alcock & Emmony, 2002; Koechner, 2006; Mak et al., 1990). At $P(0) \approx P_{pt}$ the drop of $n(r)$ along r from axis to periphery of the cylinder due to the drop in pressure in the medium is $\Delta n(P_{pt}) = \delta n/\delta P \times P_{pt} = 0.13$. Changes of Δn due to temperature gradients are significantly less. For example, at $\Delta T = 10^3$ K, $\Delta n(T) = \delta n/\delta T \times T = -6 \times 10^{-3}$. For YAG $\delta n/\delta T \approx 8 \times 10^{-6}$ K^{-1} (Koechner, 2006; Mak et al., 1990) and $\Delta n(T) \approx 8 \times 10^{-3}$ .Thus, the main contribution to the $n(r)$ profile was made just by pressure. The $n(r)$ profile defines a sign, optical power and aberrations of a "transient" lens in the medium. Optical power of the lens after the end of pump can be estimated at the assumption that the index had been changed by $\Delta n \approx 0.1$ at $\Delta r \approx 150$ μm. That gradient corresponds to appearance in the medium of a positive lens with diameter $2r \approx 300$ μm and $f \approx 1$cm. A simplified scheme for the case of lasing in the resonator with a "transient" lens is given in Fig. 17.

The evolution of profiles $P(r)$ and $n(r)$ after pumping was determined by unloading of the high-pressure region in the focal volume. A problem of unloading of a small cylindrical or spherical region in a solid-state optical medium, where the energy had been stored after ns laser pulse irradiation, had been considered in several studies (Bullough & Gilman, 1966; Conners & Thompson, 1966; Sharma & Rieckhoff, 1970). A typical process of relaxation is the propagation of an elastic dilatational wave across the medium. The time of traveling, t_{tr} of an elastic wave with the sound velocity across the excitation region can be evaluated by $t_{tr} = r/v_s$. For the glass, at $r_p = 250$ µm and $v_s = 4.5 \times 10^5$ cm/s, $t_{tr} \approx 50$ ns. The propagation of the dilatational wave along radial direction outward the center of the pumped region had led to the pressure profile deformation. With the drop of pressure at the axis, the gradient dn/dr

decreased accordingly. So, in the course of relaxation there occurred smoothing of profiles $P(r)$ and $n(r)$. It should be noted that profile $n(r)$ could include a small-scale modulation as well (imitated at Fig.17). The sources for such modulation might be pressure and density perturbations that had been arising in a medium at propagation of hypersonic waves during and after pumping. Note, that typical time of thermal relaxation of the medium in a cylindrical region with radius r_p is $\tau = r_p{}^2/4\mu \gg t_{tr}$. Here r_p is the radius of the pumped region and μ is the coefficient of the temperature diffusivity. For Yb:YAG at r_p = 250 μm, μ = 0.046 cm^2s^{-1} (Koechner, 2006), $\tau \approx$ 3 ms. Hence, the time of thermal relaxation of medium is several orders higher than dynamic unloading. That is, roughly, a picture of formation and evolution of the refractive index profile in the medium in the region of ccl focusing defined by a non-uniform heat load and dynamics of the pressure profile during pumping and under the medium relaxation. So, a sequence of acoustic waves diverging in radial direction outward the focal region should affect the build-up of Yb-lasing in the resonator.

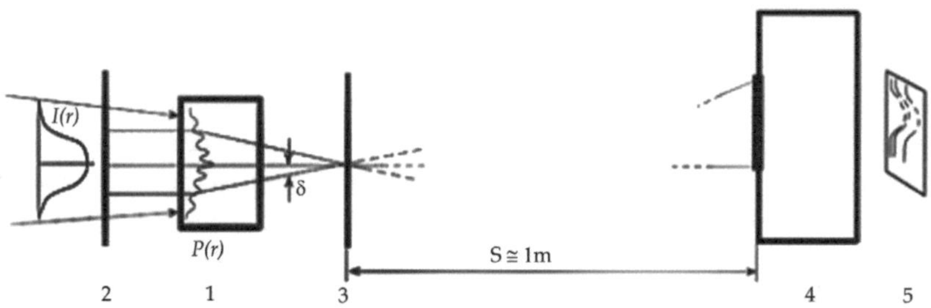

Fig. 17. Scheme of the experiment on excitation of Yb-doped medium in the resonator by focused ccl radiation: 1 - active element; 2, 3 - mirrors; 4 - spectrograph STE-1 located at distance S≈1m from the resonator; 5 - a picture imitating spectra registered; δ –angle between a generation trajectory and resonator axis. The dimensions of the resonator with active element are exaggerated relative to the scale of the scheme. Profiles of pump beam intensity, *I(r)* and pressure in the medium, *P(r)* are out of scale as well

The interpretation of data on the Yb-lasing in the resonator has required also the introduction of a new concept for the photon electromagnetic field distribution in space, considered in section 5. This new approach allows the existence of light beams of an aperture $d \approx \lambda$ and with a "sub-diffraction" angle of divergence. Within the framework of this new concept the picture of Yb lasing in the resonator looks as follows. The location of line spectra in the upper part of the slit evidences that Yb radiation emergent from the resonator was deflected from the axis, i.e., off-axis oscillations were built up in the resonator. For the arrangement of mirrors and active element given in our experiment off-axis lasing could be developed over trajectories with reflections at angles δ_1, δ_2 … at the mirror (3), Fig. 17. Such trajectories located at distances $r_{pt} < r < r_p$ from the axis, should contain curvilinear segments in the area of the optical inhomogeneity. We suppose that beams with aperture of several λ (small generation channels) constituted lasing along these off-axis trajectories. At the high value of dn/dr and the very high level of inversion, in each of the channels the selection of emitted photons (in accordance with their phases) took place in such a way that only a group of photons being approximately in the same phase φ_0 overcame the generation threshold. The photons different by phase from φ_0 were scattered due to diffraction and left the channel aperture. In the given channel with a coordinate r a group of photons of the same phase made several round trips in the resonator during the time t', and a beam of the high directivity with the generation wavelength λ_g corresponding to one of the longitudinal modes was formed. So, generation in different channels over the pumped area developed at their own frequencies, and this radiation, reaching the slit at different points gave us an integrated picture of lasing beams and the spectrum in the form of lines composed of small spots, which dimensions characterized beam divergence in each of the channels.

For Yb-doped plates placed in the middle of the resonator with L = 20 mm and for trajectories located in plates at r < 500 μm, angles δ should lie inside the spectrograph registration angle ≈10^{-2} rad. Beams of generation left resonator at angles δ and reached the slit. The greater was the distance r between the generation trajectory inside the sample and the resonator axis, the greater was the shift of the corresponding generation beam spot upward in vertical direction on the slit, Fig. 17. So, under the propagation of beams of high-directivity from resonator at angles δ, one, actually, observed at the slit a projection (a magnified "image") of a spatial distribution of generation channels in the active medium over the radial coordinate. Since data on the Yb emission intensity distribution over the azimuth angle were absent, one can speak only about partial mapping on the slit of the spatial distribution of generation channels in the lasing area. The magnification coefficient k of such an image may be estimated from comparison of dimensions of the possible generation area ($r \leq$ 500 μm) and the height of line spectra at the slit (≈1 cm), $k \geq 20$. With account of k factor, to observed structures of small-scale spots in line spectra should correspond sources of Yb emission with dimensions of several wavelengths only.

Structures of multiple spots in Yb:YAG and Yb:glass lines (Figs. 11, 14, 16) may find explanation when one takes into consideration the specific spatial configuration of the field of thermo-elastic stresses in the area of ns laser pulse focusing into the medium. The quasi-periodic, alternating in sign, oscillating character of the amplitude of the tangential component of stress in dependence on the radial coordinate inside a small spherical region of a solid-state optical medium at ns laser pulse focusing was ascertained (Conners & Thompson, 1966; Sharma & Rieckhoff, 1970). In spherical ring zones round the centre of the focal region the tangential tensile stresses in the medium are consequently replaced by compressing stresses, and then, again, by the tensile ones, etc. The spatial period of such oscillations calculated for the case of ≈70 mJ laser pulse focusing into the glass constituted 30 μm (Sharma & Rieckhoff, 1970). In conditions of our experiment one can expect the occurrence of oscillating (in space and time) profiles of thermo-elastic stresses in Yb-doped media within the area of ccl radiation focusing as well. Alternating in sign stresses should result in a small-scale modulation at the profile $n(r)$, Fig.17. The oscillation of the $n(r)$ profile should stimulate the Yb generation in ring zones, which may fall into separate generation channels. The structures of bends consisting of generation spots, Figs.16 may be considered as a kind of an "image" of the distribution over the radial coordinate of tangential stress peaks in the focal region projected with a magnification on the slit. It is possible to say that such images were taken by the high-speed photography method. The "illumination" for this high-speed photography came from the Yb laser pulse itself. The "exposure" time of a single frame corresponded to the duration of the generation pulse. Time delay between the pump pulse and the "shooting" moments makes ≈ 50ns (Figs. 16) and over 300 ns (Fig.14). Splitting of a single bend into several generation spots with wavelength shifts (Fig.16) is in agreement with the model of generation channels. The shift of a spot to the long-wavelength side corresponds to increase of pressure (stresses) in the medium, and the shift to the short-wavelength—to decrease of pressure. In the course of medium relaxation, stress amplitudes and, correspondingly, spot wavelength shifts should have been reduced. At ≈50ns delay (twisted lines, Fig.16) spots have noticeable wavelength shifts which correspond to high stress amplitudes. Structures at moments over 300 ns after pump (Fig.14) correspond to a smoothed picture of $n(r)$ profile. Note, that structures recorded with such big delays indicate the continued acoustic "ringing" in the medium. Estimated from Fig. 11, 14, 16 (at k = 20) period of stress spatial oscillations in YAG and glass varies from 15 to 40 μm and is in a qualitative agreement with the calculations (Sharma & Rieckhoff, 1970). Acoustic vibrations frequencies corresponding to these values constitute 10^8–10^9 Hz. Attenuation of phonons at these frequencies at a room temperature for YAG is smaller than 0.1 db/μs and for glass—about 10 cm^{-1} (Dutoit, 1974; Zhu et al., 1991). These data confirm that "ringing" of the unloading medium in the focal region in YAG and glass may continue over several μs.

The Yb:YAG line spectra near 1,03 μm were registered usually with 10÷50 ns delays after the end of ccl pulse and even together with the trailing edge of the ccl pulse. Observations of line spectra near 1,03μm, emitted soon after the pump pulse, like Fig.11, reveal noticeable inclination of spectral lines. This bending means that the wavelength of lasing at the same longitudinal mode (the number of nodes is preserved) changes from the centre to periphery of the excited region in the active medium. This corresponds to the development of lasing in some sites of the medium with a pressure (refractive index) gradient from the beam centre to its periphery. It can be easily shown that λ_g shifts to the blue if the refractive index gradient decreases from the beam axis to its periphery, Fig. 11. The pressure drop ΔP in the medium after the end of pump can be estimated from this line spectral shift. The frequency ω_q of the longitudinal mode of the resonator with the number of wavelengths λ_q over the resonator length $2L$ equal to q is described by the expression

$$\omega_q = \frac{\pi c q}{(L-\Delta l)n_0 + \Delta l \cdot n(r)} \tag{2}$$

where Δl is the longitudinal size of the optical inhomogeneity of radius r; n_0 is the averaged refractive index outside the nonlinearity region; and $n(r)$ is the refractive index in the nonlinear region. It follows from (2) that the change in the mode frequency ω_q during the displacement along the radius from r_1 to r_2 is

$$\Delta\omega_q = \frac{\pi c q \Delta l[n(r_1) - n(r_2)]}{[(L-\Delta l)n_0 + \Delta l \cdot n(r_1)][(L-\Delta l)n_0 + \Delta l \cdot n(r_2)]} \tag{3}$$

By substituting the expression for $q = \dfrac{2[(L-\Delta l)n_0 + \Delta l \cdot n(r)]}{\lambda_q(r)}$ into (3) and assuming that $\Delta l n(r) < L n_0$, we obtain the dependence of the change in the refractive index, Δn on r

$$\Delta n = \frac{\Delta\lambda_q(r) L n_0}{\lambda_q \Delta l} \tag{4}$$

Here, $\Delta\lambda_q(r)$ is the wavelength shift along the radius. For $\Delta\lambda_q(r) \approx 1.4 \times 10^{-7}$ cm, $\lambda_q \approx 10^{-4}$ cm, L ≈ 2 cm, $n_0 \approx 1$, and $\Delta l \approx 10^{-2}$ cm, the change in the index is $\Delta n \approx 0.28$. By assuming that the change in the index is produced only by the change in pressure along the radius and using the value of *dn/dP* for glass (Alcock & Emmony, 2002), we obtain $\Delta P \approx 30$ GPa. So, this estimates the pressure, which arises at the axis of the focal region after the end of pump. Similar estimations are possible for the pressure jumps, which arise during the propagation of dilatational waves outward the center of the focal region. Knowing a value of the wavelength shift in a bend at Fig.16, $\Delta\lambda_q(r)$, one can estimate changes of the index, Δn and pressure, ΔP in the medium, using the expression (4). As seen from Fig.16, for glass at $\Delta\lambda_q(r) \approx 0.3\text{x}10^{-7}$ cm, $\lambda_q \approx 10^{-4}$ cm, $L \approx 2$ cm, $n_0 \approx 1$, $\Delta l \approx 10^{-2}$ cm, $\Delta n \approx 0.05$, and $\Delta P \approx 5$ GPa. The estimated ΔP considerably exceeds the glass fracture strength < 0.1 GPa (Sharma & Rieckhoff, 1970), which was measured usually for applied static load. The role of tensile stresses in laser damage of transparent dielectrics was discussed in many publications (Koldunov et al., 2002; Sharma & Rieckhoff, 1970; Strekalov, 2000). It was considered that due to oscillating character of the stress amplitude in the focal area there must be observed laser damage of the medium in the form of periodically spaced spherical rings (Sharma & Rieckhoff, 1970; Strekalov, 2000). It is known that material strength sharply grows under pulsed load as compared to the static load. Under the high-speed deformation (in the ns range), strength of material becomes comparable to the theoretical limit-tens GPa (Kanel et al., 2007). So a fast periodic change of the stress sign in the medium should decelerate the development of material destruction in the form of rings. This is the reason why the laser damage in the form of multiple rings usually was not observed in many experiments when ns pulses of laser radiation were focused into the volume of transparent dielectrics. The damage in the form of rings was not observed in our experiments as well. One of the few observations of the multiple ring damage (Martinelli, 1966) relates to the case when glass samples were exposed to focused free-running laser radiation. Anyway, according to calculations (Sharma & Rieckhoff, 1970) oscillating profile of thermo-elastic stresses should occur in the region of ns laser pulse focusing. The presented material provides experimental data which confirm in the quality form the calculated (Sharma & Rieckhoff, 1970) picture of thermo-elastic stresses distribution and elastic wave propagation across the medium.

7. Conclusion

The study of the interaction of powerful ns pulses of low-coherence radiation of the LiF: F_2^+ color center laser (ccl) with optical materials (Yb:YAG, glass, et al.) was carried out. Efficient SBS of low-coherence pump, accompanied by SRS and formation of hypersonic waves reaching the intensity of shock waves were found. A physical model of excitation of SBS and hypersonic waves at scattering of ultrashort pulses of low-coherence pump at stationary inhomogeneities in optical materials is presented. It is shown that ns laser pulse, whose duration is much higher than its inverse spectral width, causes SBS much more efficient than a pulse of high coherence with the same duration and energy. Unlike SBS of a coherent radiation caused by a pressure fluctuation, scattering of low-coherence pump may be caused by any stationary inhomogeneities in a medium: cracks, dislocations, micro-inclusions, or just by a plane back surface of a sample. An effective energy contribution of light pulses into hypersonic waves on a small coherence length near the input surface of a sample leads to their transformation into a periodic succession of high-pressure shock waves, which results in structure changes of a crystal lattice (phase transition) in that region. The appearance of structural changes in optical materials that are specific to the interaction of powerful pulses of low coherence radiation with matter was found. The mechanisms of structural changes based on the action of intense hypersonic waves were considered.

Nanosecond pulses of Yb lasing in the region 1.00-1.06 μm with the spectral width up to 20 nm in Yb :YAG and 50 nm in the Yb: glass samples were observed. The divergence of the broadband laser radiation (10^{-3} -10^{-4} rad) was one or two orders of magnitude smaller than the diffraction limit respectively to the source of Yb radiation in a sample. The mechanism of generation of broadband laser pulses of short duration and high directivity in the spatial structure of thin layers with inversion produced in the region of the propagation of intense hypersonic waves in the medium is discussed. The interpretation of experimental data is based on a new concept of the spatial distribution of the electromagnetic field of a photon not in the form of a "traveling" wave but with the field structures located in fixed positions along the photon propagation direction. The new approach allows the existence of light beams of an aperture $d \approx \lambda$ and with a "sub-diffraction" angle of divergence. Such beams must consist of groups of phase-synchronized photons with a small phase difference distanced by an interval $\approx \lambda$. This synchronized group is no longer a "traveling" wave, its angle of divergence is defined by the phase difference of photons in a group.
Spectral lines in 1.03–1.05 μm region structured by 50–200 μm spots as well as lines with inclinations were found at Yb lasing in a resonator. Structures of multiple spots in spectra reflect the specific spatial configuration of the field of thermo-elastic stresses in the unloading region of ccl pulse focusing after the end of the pump. Inclinations of spectral lines reflect the pressure gradient from the center to the periphery of the region of ccl focusing. Basing on inclinations of spectral lines the pressure in the region of shock hypersonic wave propagation was estimated. Estimations show that for some of the studied media the pressure values may exceed the phase transition threshold.

8. Acknowledgment

Authors thank O. Yaremchuk for the help in preparing this article for publication.

9. References

Alcock, R. & Emmony, D. (2002). Sensitivity of reflection transducers *J. Appl. Phys.*, Vol. 92, No.3 (August 2002), pp. 1630-1643, ISSN 0021-8979.

Basiev, T. et al. (1982). Solid-state tunable lasers based on color centers in ionic crystals *Bulletin of the Academy of Sciences of the USSR, ser. phys.*, Vol. 46, No.8, pp.145-154.

Basiev, T.T, Bykovsky, N.E, Konyushkin, V.A & Senatsky, Yu.V. (2004). Use of a LiF colour centre laser for pumping an Yb:YAG active medium. *Quantum. Electron*. Vol.34, No.12, pp. 1138-1142, ISSN 0368-7147.

Bor, Z. & Muller, A. (1986). Picosecond distributed feedback dye lasers *IEEE J. Quantum Electron.*, 22 (8), 1524-1533. ISSN: 0018-9197.

Bullough, R. & Gilman, J.J. (1966). Elastic Explosions in Solids Caused by Radiation. *J. Appl. Phys*. Vol.37, pp.2283-2288 , ISSN 0021-8979.

Bykovsky, N. & Senatsky, Yu. (2008a). Broadband collimated generation in YAG:Yb crystal and ytterbium glass under LiF:F2+color center laser pumping. *Laser Phys. Lett.* Vol.5, Iss. 9. pp. 664–670, ISSN 1612-2011.

Bykovsky, N.& Senatsky, Yu. (2008b). Spectra, temporal structure, and angular directivity of laser radiation of a Yb:YAG crystal and ytterbium glass pumped by low-coherence radiation from a F2+:LiF colour centre laser. *Quantum Electron*. Vol.38, No.9, pp. 813-822, ISSN 0368-7147.

Bykovsky, N.E. & Senatsky, Yu.V. (2010). Twisted lines and small-scale structures in generation spectra of Yb-doped media excited by focused radiation of LiF:F 2+ color center laser. *Laser Physics* Vol.20, No.2, pp. 478-486, ISSN 1054-660X.

Bykovsky, N.E. (2005). *Preprint FIAN* No. 16. (http://ellphi.lebedev.ru/12/pdf16/pdf).

Bykovsky, N.E. (2006). *Preprint FIAN* No. 36. (http://ellphi.lebedev.ru/17/pdf36/pdf).

Conners, G. & Thompson, R. (1966). A Continuum Mechanical Model for Laser-Induced Fracture in Transparent Media. *J. Appl. Phys.* Vol.37, pp.3434-3441, ISSN 0021-8979.

Dutoit, M. (1974). Microwave phonon attenuation in yttrium aluminum garnet and gadolinium gallium garnet. *J. Appl. Phys.* Vol.45, pp.2836-2841, ISSN 0021-8979.

Galbraith, C. & Galbraith, J. (2011). Super-resolution microscopy at a glance *Journal of Cell Science* Vol.124, №10, pp.1607-1611 ***ISSN***: 0021-9533

Gordienko,V., Mikheev P. & Potemkin F. (2010). Generation of coherent terahertz phonons by sharp focusing of a femtosecond laser beam in the bulk of crystalline insulators in a regime of plasma formation.*JETP Letters,* Vol.92, No.8, pp.502-506, ISSN 0021-3640.

Gorelik, V. (2007). Optics of globular photonic crystals. *QuantumElectron,* Vol.37(5), pp. 409-432, ISSN 0368-7147.

Kanel, G., Fortov, V. & Razorenov, S. (2007). Shock waves in condensed-state physics. *Phys. Usp.* Vol.50, pp.771-791, ISSN: 0038-5670.

Katarkevich, V., Kurstak V., Rubinov, A., & Efendiev T. (1996). Kinetics of the operation of a distributed-feedback dye laser with nanosecond excitation. *Quantum Electron.*,Vol. 26, pp.1061- 1064, ISSN 0368-7147.

Koechner, W. (2006). *Solid State Laser Engineering*, 6th ed., Springer, Berlin.

Kogelnik, H., & Shank, C. (1971). Stimulated emission in a periodic structure. *Appl. Phys. Lett.*, Vol.18 (4), pp.152-155, ISSN 0003-6951.

Koldunov, M., Manenkov, A. & Pokotilo, I. (2002). Mechanical damage in transparent solids caused by laser pulses of different durations. *Quantum Electron.*, Vol.32. No.4, pp.335–340, ISSN 0368-7147.

Krupke W.F. (2000). Ytterbium solid-state lasers. The first decade *IEEE J. Sel. Top. Quantum Electron.*, Vol.6 (6), pp.1287 - 1296, ISSN1077-260X.

Mak, A & Soms, L. (1990). *Nd:Glass Lasers,* Nauka, Moscow.

Manenkov, A. & Prokhorov, A. (1986). Laser-induced damage in solids. *Sov. Phys. Usp.*, Vol.148, pp.104–122. ISSN: 0038-5670.

Martinelli, J. (1996). Laser-Induced Damage Thresholds for Various Glasses. *J. Appl. Phys.*, Vol.37, pp.1939-1941, ISSN 0021-8979.

Merlin R. (1997). Generating coherent THz phonons with light pulses. *Solid State Commun.*, Vol.102, pp.207-220, ISSN 0038-1098.

Nelson K.A., Miller R.J.D. & Fayer M.D., (1982). Optical generation of tunable ultrasonic waves. *J. Appl. Phys.*, Vol.53, pp.1144-1149, ISSN 0021-8979.

Polyakova A.L. (1968). Elastic Nonlinearity in Stimulated Mandel'shtam-Brillouin Scattering. *JETP Letters,* Vol. 7, Iss. 2, pp. 57-59, ISSN: 0021-3640.

Polyakova, A. (1966). Nonlinear Effects in a Hypersonic Wave. *JETP Letters,* Vol.4, Iss. 4, p.90-92, ISSN: 0021-3640.

Ready, J. F. (1971). *Effects of High Power Laser Radiation,* Academic, New York, London.

Ritus, A. (1982). Study of Mandel'shtam-Brillouin light scattering in crystals and glasses respecting to problems of quantum electronics and fiber optics. *Trudy FIAN* 137, pp. 3-80.

Robinson M.M., Yan Y.-X., Gamble E.B., Williams Jr., L.R., Meth J.S, & Nelson K.A. (1984). Picosecond impulsive stimulated Brillouin scattering: Optical excitation of coherent transverse acoustic waves and application to time-domain investigations of structural phase transitions. *Chem.Phys. Lett.*, Vol.112, pp.491-496, ISSN 0009-2614.

Sakakura, M., Terazima, M., Shimotsuma, Y., Miura, K. & Hirao, K, (2007) Observation of pressure wave generated by focusing a femtosecond laser pulse inside a glass *Opt. Exp.* 15, 5674-5686, ISSN 1094-4087.

Sharma B.S. & Rieckhoff, K.E. (1970). Laser-induced dielectric breakdown and mechanical damage in silicate glasses. *Can. J. Phys.* Vol.48(10), 1178–1191, ISSN 0008-4204.

Starunov V.S. & Fabelinskii I.L. (1970) Stimulated Mandel'shtam-Brillouin scattering and stimulated entropy (temperature) scattering of light. *Sov. Phys. Usp.*, Vol.13, pp.428–428. ISSN: 0038-5670.

Strekalov, V. (2000).Mechanical damage of transparent dielectrics by focused laser radiation *J. Tech. Phys. Lett.* Vol.26, No. 24, pp.19-22. ISSN: 0320 - 0116 .

Stuart, B, Feit, M, Rubenchik, A., Shore, B. & Perry, M. (1995). Laser-Induced Damage in Dielectrics with Nanosecond to Subpicosecond Pulses. *Phys. Rev. Lett.* Vol.74, pp.2248–2251, ISSN 0031-9007.

Suguira, H., Ikeda, R., Kondo, K. & T. Yamadaya, T. (1997). Densified silica glass after shock compression. *J. Appl. Phys.*, Vol.81, pp.1651-1656, ISSN 0021-8979.

Tychinskii, V. (2008). Super-resolution and singularities in phase images *Phys.Usp.* Vol.178, №11, pp. 1205-1214. ***ISSN***: 0038-5670.

Zhu, T., Maris, H., & Tauc, J. (1991). Attenuation of longitudinal-acoustic phonons in amorphous SiO_2 at frequencies up to 440 GHz. *Phys. Rev. B* Vol.44, pp.4281–4289, ISSN 1098-0121.Dictionary -

18

An Optimal Distribution of Actuators in Active Beam Vibration – Some Aspects, Theoretical Considerations

Adam Brański
Rzeszow University of Technology
Poland

1. Introduction

The reduction of the effects of mechanical vibration fall into the of vibration isolation, design for vibration or vibration control (de Silva, 2000). The vibration control is subdivided into two group: passive control and active one. The core of the vibration control is to detect the level of vibration in a system and to counteract the effects of the vibration, so it needs two devices.
Hence, the passive devices do not require external power for their operation. Hence, passive control is relatively simple, reliable and economical. But it has limitations namely, the control force depends entirely on the natural dynamics and it may not be adjust on line. Furthermore, in a passive device, there is no supply of power from an external source. It leads to the incomplete control, particularly in complex and high-order systems.
The shortcomings of passive control can be overcome using an active one. In this case, the system response is directly sensed on line and on that basis, the specific control actions are applied to any locations of the system. But the active control needs external power, namely to apply control forces to vibrating system through actuators and to measure vibration response using sensors.
Two different types of actuators can be applied (Shimon et al., 2005). The first, inertial actuators, make up a piezoelectric material to vibrate large masses. Their vibrations are used to counteract the vibrations of the structure (Jiang et al., 2000). The advantages and disadvantages are enumerated in above reference.
The second type of actuators is a layer of smart or intelligent materials. The sensors also belong to these materials; together they are well–known as piezoelectric elements (Tylikowski & Przybyłowicz, 2004). It was shown that these elements can offer excellent potential for an active vibration reduction of the structure vibrating with low frequencies (Croker, 2007; Fuller at al, 1997; Hansen & Snyder, 1997; Kozień, 2006; Przybyłowicz, 2002; Wiciak, 2008). As a general, piezoelectric elements are glued to the host structure. It makes the advantage, namely their incorporating into the structure is that the actuating mechanism becomes part of the structure. Both sensors and actuators are relatively light, compared to the structure, and can be made in arbitrary shape. The disadvantage is that they once bonded and they cannot be used again. In recent years the measure of the vibration with the sensors are replaced by touch less measures. For this reason, hereafter in research the sensors are omitted and only second type actuators will be considered. Nowadays actuators

are used to very original structures for example to the satellite boom (Moshrefi-Torbati et al., 2006) or to sun plate (Qiu et al., 2007).
To make the reduction more effective, many problems should be solved.

- dynamic effects (mass loading and stiffness) of the actuators on the structure vibration (Charette et al., 1998; Gosiewski & Koszewnik, 2007; Hernandes et al., 2000; Q. Wang & C. Wang, 2001)
- dynamic effects of the glue (between actuators and structure) on the structure vibration (Pietrzakowski, 2004; Sheu et al., 2008).
- actuators' geometric-technical features (Frecker, 2003; Hong et al., 2007; Wang, 2007),
- orientation of the actuators on the structure (Bruant et al., 2010; Ip & Tse, 2001; Qiu et al., 2007),
- appropriate actuators distribution on the structure (Bruant et al., 2010),
- others, but they play a minor part.

Reviewing the literature, it appears that the actuators distribution play a major part. Now, a question arises about an optimal distribution of actuators. In the recent year, a great number of papers has been published on this subject. It is obvious that there are a lot of optimization techniques; an excellent survey is given in (Bruant et al., 2010). Two main approaches are distinguished to this problem.
First of them is the coupling of the optimization of actuators/sensors locations and controller parameters. In this case the following criterions are taken into account for the optimization:

- quadratic cost function of the measure error and the control energy (Bruant et al., 2001),
- maximization of dissipation energy during the control (Yang, 2005),
- spatial H_2 norm of the closed-loop transfer matrix from the disturbance to the distributed controlled output (Liu et al., 2006),
- simultaneous simple H_∞ controller (Guney & Eskinat, 2007).

As can be seen, the optimization criterions are dependent on the choice of controllers. Therefore, the optimal location obtained using one controller may not be a suitable choice for another one.
At the latter approach, the optimal location is obtained independently of the controller definition. In this case, the following criterions are used:

- maximization controllability/observability criterion using the gramian matrices (Bruant & Proslier, 2005; Jha & Inman, 2003),
- modal controllability index based on singular value analysis of the control vector (Dhuri, & Seshu, 2006),
- maximization of the control forces transmitted by the actuators to the structure (Q. Wang & C. Wang, 2001),
- using the H_2 norm (Halim & Reza Moheimani, 2003; Qiu et al., 2007).

In the quoted references, it was not provided the actuators distribution in explicite; only the general rules (criterions) were formulated. However, this problem was partially solved; it was proved in (Brański & Szela, 2007; Brański & Szela, 2008; Szela, 2009; Brański & Szela 2010; Brański & Lipiński, 2011) that the most effective actuators distribution was on the structure sub-domains with the largest curvatures; such distribution was called quasi-optimal one. As the research object, a right-angled triangle plate with clamped-free-free boundary conditions was taken into account. The quasi-optimal distribution was deduced based on the heuristic reasons and the conclusions were confirmed only numerically. Furthermore, the problem was solved merely for the separate modes.

Basing on the quasi-optimal distribution of the actuators, the protection beam vibration is achieved (Brański et al., 2010; Brański & Lipiński, 2011). In this case always the separate modes were considered. The problem was solved based on heuristic reasons and was confirmed analytically. In the latest own research, the results presented in (Brański et al., 2010) were substantiated analytically (Brański & Lipiński, 2011).
In this chapter, the above attitude to the optimal actuators distribution is continued and extended. First at all, the optimal problem is formulated. For this purpose, the optimization criterion is defined. It is assumed that a measure of the vibration reduction is a reduction coefficient (Szela, 2009; Brański & Szela 2010) and here it becomes the objective function. This attitude is quite similar to the maximization of the control forces transmitted by the actuators to the structure (Q. Wang & C. Wang, 2001).
Dynamics effects of the glue and actuators are also considered. Furthermore, the solution of active vibrations reduction is derived for general solution, not only for separate modes. Since analytical solution was attained with separation of variables method, first of all the modes of the problem are derived. Next, the orthogonality condition of the modes is derived too.
The simple supported beam is chosen as the research object. The study of beams is very important in a variety of practical cases, noteworthy, the vibration analysis of structures like bridges, tall buildings, and so on. Loosing a bit on generality, it is considerably easier to realize the aim of the paper. It is assumed that the beam is excited with evenly spread and harmonic force. The material inner damping coefficients of all elements of the research system are taken into account. It seems that all main factors having the influence on the beam vibration were considered.
To solve the problem analytically, a few simplifications are made. Namely, the energy provided to the system is in the form of voltage applied to the surface of the actuators. Assuming that the charge is homogenously distributed, as a result of piezoelectric effect, the actuators interact with the beam with moments for couple of forces homogenously distributed along the actuators' edges. Next, these moments are replaced with the couple of forces and finally, they are counteracted the vibrations.
All problems were considered only theoretically; no calculations are run. It seems that presented considerations will be the base to many numerical simulations and experiments. To the author's knowledge, the theoretical description of the optimal actuators distribution on even simple structure like the beam, up to now have not been brought up.

2. Active beam vibration reduction with additional elements

In this problem, the additional elements make the concentrated masses and actuators and all constitute the mechanical set beam-actuators-masses. Adding actuators (and the glue at the same time) is the technical necessity but they introduce to the mechanical set the additional dynamics effects namely, local stiffness and concentrated masses. As far as concentrated masses are concerned, adding them is substantiated as follows. The proposed optimal distribution of the actuators needs asymmetrical beam vibrations and these ones may be ensured by at least one concentrated mass.

2.1 Uniform beam vibration with damping

There are four theories (models) for the transversely vibrating uniform beam (Han et al., 1999): Euler-Bernoulli, Rayleigh, shear and Timoshenko. The first of them, called the classical beam theory, is applied here. It is simple and provides reasonable results for formulated problem.

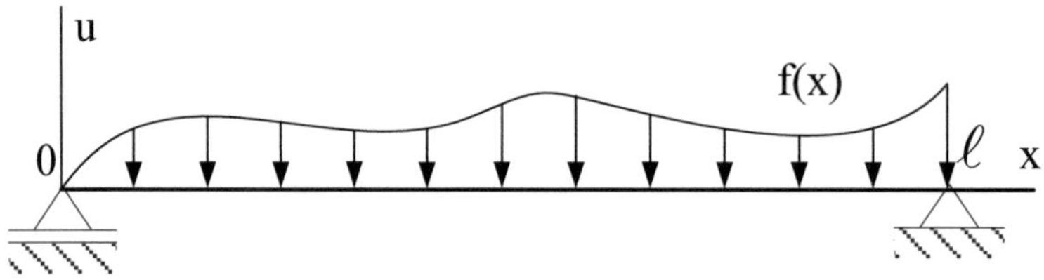

Fig. 1. The geometry of the simple supported beam

Let be the beam as depicted in Fig. 1. The Bernoulli-Euler equation governs transverse vibration (or bending or lateral vibration) of the beam has a following standard form (Kaliski, 1986; Pietrzakowski, 2004),

$$EJD^4u + \mu EJD^4(D_t u) + \rho S D_t^2 u = -f \tag{1}$$

where $u = u(x,t)$ - beam deflection at the point x and the time t, $f = f(x,t)$ - load force, $D^4(.) = \partial^4(.)/\partial x^4$, $D_t(.) = \partial(.)/\partial t$; hereafter the rest symbols are jointly explained.
To solve Eq. (1) explicitly, four boundary conditions, at the ends of the beam, are needed. In general, boundary conditions represent displacement, slope, moment and shear respectively. Here, it is assumed that the beam is simple supported, then both displacement and the bending moment equal zero

$$u(0,t) = 0, \qquad D^2u(0,t) = 0 \tag{2}$$

$$u(\ell,t) = 0, \qquad D^2u(\ell,t) = 0 \tag{3}$$

To solve over determined problem, one needs to know initial conditions. But here, the harmonic steady state plays a major part, so that the initial conditions are omitted.

2.2 Beam vibration with concentrated masses

To solve the intended problem, Eq. (1) must be rounded out. First of all, to obtain asymmetric modes and consistently asymmetric general vibration, a few concentrated masses are added to the beam (Low & Naguleswaran, 1998; Majkut, 2010; Naguleswaran, 1999). They are marked by $\{m_r\}$, and their distribution is described with set of coordinates $\{x_r\}$, see Fig. 2, hence

$$\sum_r m_r \delta(x - x_r) = m_1 \delta(x - x_1) + m_2 \delta(x - x_2) + \ldots + m_r \delta(x - x_r) + \ldots \tag{4}$$

where $r = 1, 2, \ldots, n_r$, $\delta(.)$ – Dirac's delta function.

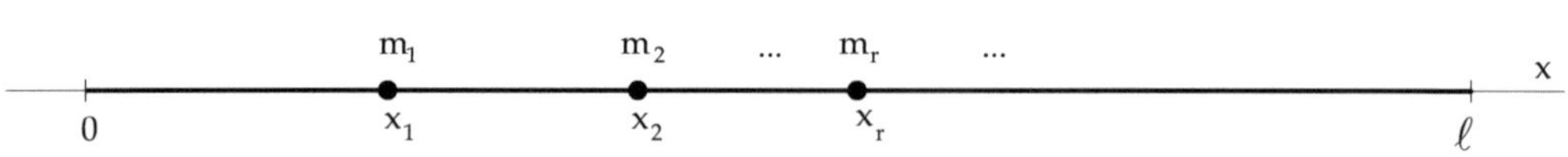

Fig. 2. Distribution of the concentrated masses

Furthermore, the dynamic effects of the actuators and glue on the beam vibration are introduced. The location and length of separate actuators, and the glue layers simultaneously, are denoted commonly with coordinates $\{x_s\}$ and $\{\ell_s\}$ respectively and they are arranged as depicted in Fig. 3.

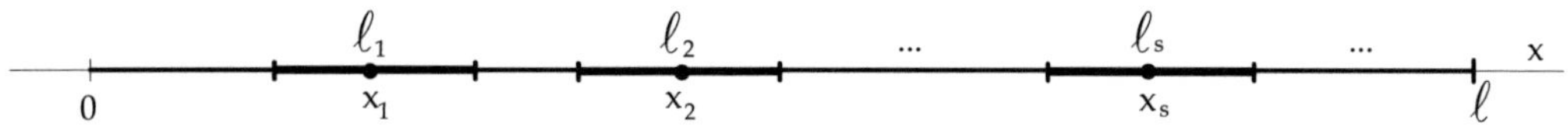

Fig. 3. Distribution of actuators and glue layers

For simplicity, let $P=\{E,J,h,\rho,S,\mu\}$ means the physical and geometrical parameters of the beam, actuators and glue, i.e. {Young's modulus, surface moment of inertia, thickness, mass density, surface of the rectangular cross-section, inner damping factor} respectively. Furthermore all parameters are supplemented with following index $\vartheta=\{b,a,g\}=\{[b]eam,[a]ctuator,[g]lue\}$, for example $S_\vartheta = bh_\vartheta$ means the surface of the rectangular cross-section, b – beam / glue layer width. Moments of inertia are calculated relatively of y-axis, see Fig. 4, where the neutral axis displacement d is neglected , hence $J_b=(bh_b^3)/12$, $J_g=(bh_g^3)/12+S_g(h_b/2+h_g/2)^2$, $J_a=(bh_a^3)/12+S_a(h_b/2+h_g+h_a/2)^2$.

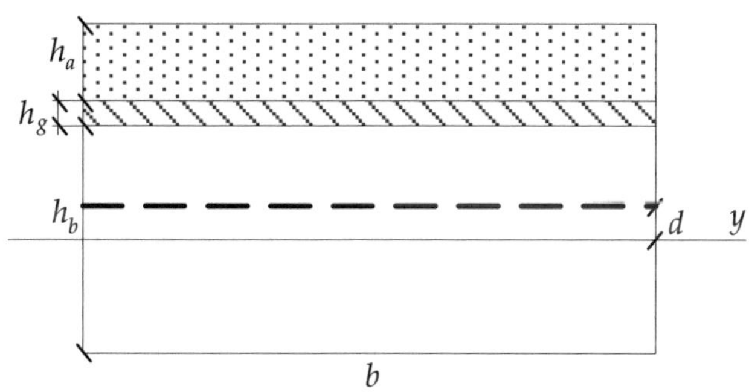

Fig. 4. Cross-sections of the set beam–actuator–glue

The parameters of the set beam-actuators-glue may be written as

$$P = P_b + \sum_s P_s H(x_{1s}-x_{2s}) = P_b + \sum_s P_s \langle H\rangle^0 \tag{5}$$

where $s=1,2,\ldots,n_s$, $P_s=P_a+P_g$, $\langle H\rangle^0 = H(x_{1s}-x_{2s}) = H(x-x_{1s})-H(x-x_{2s})$, $H(x-x_{1s})$ – Heaviside step function in point x_{1s} and so on, $\{x_{1s},x_{2s}\}=\{x_s-\ell_s/2, x_s+\ell_s/2\}$.
For n_s aktuators (n_s glue layers) and n_r concentrated masses, Eq. (1) takes the form

$$\left(E_bJ_b+\sum_s E_sJ_s\langle H\rangle^0\right)D^4u+\left(\mu_b E_b J_b+\sum_s \mu_s E_sJ_s\langle H\rangle^0\right)D^4(D_t u)+\left(\rho_b S_b+\sum_s \rho_s S_s\langle H\rangle^0\right)D_t^2 u+$$
$$+\sum_r m_r\delta(x-x_r)D_t^2u=-f \tag{6}$$

The Eq. (6) may be written down quite similar like Eq. (1), namely

$$EJD^4u+\mu EJD^4(D_t u)+(\rho S+\alpha_r)D_t^2u=-f \tag{7}$$

where hereafter

$$EJ = E_bJ_b + \sum_s E_sJ_s\langle H\rangle^0, \qquad \mu EJ = \mu_b E_b J_b + \sum_s \mu_s E_s J_s \langle H\rangle^0,$$

$$\rho S = \rho_b S_b + \sum_s \rho_s S_s \langle H\rangle^0, \qquad \alpha_r = \sum_r m_r \delta(x - x_r) \tag{8}$$

On the ground of the EJ, ρS and α_r form, Eq. (7) can not be understood in a classical manner. To solve it, some methods may be applied. One of them is presented in (Ercoli & Laura, 1987; Kasprzyk & Wiciak, 2007; Majkut, 2010); another attitude may be found in (C.N. Bapat & C. Bapat, 1987) and it is applied here.

Fig. 5. Geometry of the set beam–one actuator–one mass

At the latter attitude, the beam is divided into some uniform elements. The division may not be coincidental. To clearly explain this problem, for simplicity consider a set beam-one actuator (and glue)-one concentrated mass, Fig. 5. The division is imposed out of the change of physical properties namely, properties of the actuators (and glue) and concentrated masses. So, the beam is divided into $j = 1,2,...,n_j = 4$ elements. All elements may be considered separately and the solution to Eq. (7) can be expressed as

$$u(x,t) = \sum_j u_j(x,t) \tag{9}$$

where $u_j(x,t)$ is the solution on j -element and it is fulfilled the following equation

$$E_jJ_j D^4u_j + \mu_j E_jJ_j D^4\left(D_t u_j\right) + \left(\rho_j S_j + \alpha_r\right) D_t^2 u_j = -f_j \tag{10}$$

To find $u_j(x,t)$ with the separation of variables method, the eigenvalues and eigenfunctions for each element are needed.

2.3 Eigenvalues and eigenfunctions problem

In this problem it is assumed that $f_j(x,t) = 0$ and $\mu_j = 0$, hence based on Eq. (10) one obtains

$$E_jJ_j D^4u_j + \rho_jS_j D_t^2 u_j = 0 \tag{11}$$

where E_jJ_j and ρ_jS_j may be different on the separate elements, but here, as depicted in Fig. 5, is

$$E_1J_1 = E_3J_3 = E_4J_4 = E_bJ_b, \qquad E_2J_2 = E_bJ_b + E_aJ_a + E_gJ_g \tag{12}$$

$$\rho_1S_1 = \rho_3S_3 = \rho_4S_4 = \rho_bS_b, \qquad \rho_2S_2 = \rho_bS_b + \rho_aS_a + \rho_gS_g \tag{13}$$

The boundary conditions for the j –element consist of boundary conditions of the problem and coupling conditions between neighboring elements. The concentrated mass m_r is

considered in coupling conditions between third and forth elements and therefore it is omitted in Eq. (11).
Let the solution be represented by a product of spatial and temporal functions

$$u_j(x,t) = X_j(x)\,T(t) \tag{14}$$

Substituting (14) into (11) gives

$$E_jJ_j\,D^4X_jT + \rho_jS_j\,X_j\,D_t^2T = 0 \tag{15}$$

or

$$\frac{E_jJ_j}{\rho_jS_j}\frac{D^4X_j}{X_j} = -\frac{D_t^2T}{T} = \omega^2 \tag{16}$$

hence

$$D^4X_j - \lambda_j^4X_j = 0 \tag{17}$$

$$D_t^2T + \omega^2\,T = 0 \tag{18}$$

where the dispersion relationship is given by

$$\lambda_j^4 = \omega^2\,\frac{\rho_jS_j}{E_jJ_j} = \frac{\omega^2}{\gamma_j} \tag{19}$$

The Eq. (17) is very important and the solution to it is

$$X_j(x) = A_jK_1(\lambda_jx) + B_jK_2(\lambda_jx) + C_jK_3(\lambda_jx) + D_jK_4(\lambda_jx) \tag{20}$$

where Krylov functions are defined as, (Kaliski, 1986),

$$K_1(z) = \left(\mathrm{ch}(z)+\cos(z)\right)/2, \qquad K_2(z) = \left(\mathrm{sh}(z)-\sin(z)\right)/2,$$
$$K_3(z) = \left(\mathrm{ch}(z)-\cos(z)\right)/2, \qquad K_4(z) = \left(\mathrm{sh}(z)+\sin(z)\right)/2 \tag{21}$$

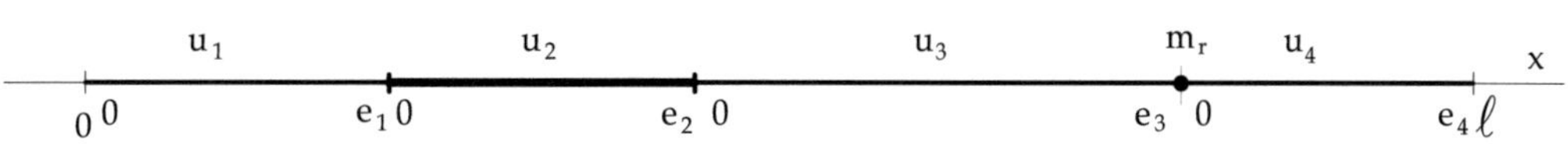

Fig. 6. Geometry of the set beam-one actuator-one mass in local coordinates

The boundary conditions in local coordinates, $x \in [0, e_j]$, to the separate j-element have the form, Fig. 6,

- boundary conditions at the left end of the 1st–element

$$X_1(0) = 0, \qquad D^2X_1(0) = 0 \tag{22}$$

- coupling conditions between 1st and 2nd–elements

$$X_1(\lambda_1 e_1) = X_2(\lambda_2 0), \qquad DX_1(\lambda_1 e_1) = DX_2(\lambda_2 0),$$
$$E_1 J_1 D^2 X_1(\lambda_1 e_1) = E_2 J_2 D^2 X_2(\lambda_2 0), \qquad E_1 J_1 D^3 X_1(\lambda_1 e_1) = E_2 J_2 D^3 X_2(\lambda_2 0) \tag{23}$$

- coupling conditions between 2nd and 3rd–elements

$$X_2(\lambda_2 e_2) = X_3(\lambda_3 0), \qquad DX_2(\lambda_2 e_2) = DX_3(\lambda_3 0),$$
$$E_2 J_2 D^2 X_2(\lambda_2 e_2) = E_3 J_3 D^2 X_3(\lambda_3 0), \qquad E_2 J_2 D^3 X_2(\lambda_2 e_2) = E_3 J_3 D^3 X_3(\lambda_3 0) \tag{24}$$

- coupling conditions between 3rd and 4th–elements

$$X_3(\lambda_3 e_3) = X_4(\lambda_4 0), \qquad DX_3(\lambda_3 e_3) = DX_4(\lambda_4 0), \qquad E_3 J_3 D^2 X_3(\lambda_3 e_3) = E_4 J_4 D^2 X_4(\lambda_4 0)$$

and

$$E_3 J_3 D^3 X_3(\lambda_3 e_3) + m_r \omega^2 X_3(\lambda_3 e_3) = E_4 J_4 D^3 X_4(\lambda_4 0)$$

or

$$E_3 J_3 D^3 X_3(\lambda_3 e_3) = m_r \omega^2 X_4(\lambda_3 0) + E_4 J_4 D^3 X_4(\lambda_4 0) \tag{25}$$

- boundary conditions at the right end of the 4th–element

$$X_4(\lambda_4 e_4) = 0, \quad D^2 X_4(\lambda_4 e_4) = 0 \tag{26}$$

Since $\lambda_1 \neq \lambda_2 \neq \lambda_3 \neq \lambda_4$ then, to calculate them, the Eq. (19) must be used. It is convenient to express $\{\lambda_2, \lambda_3, \lambda_4\}$ as a function λ_1, hence

$$\lambda_1^4 \gamma_1 = \lambda_2^4 \gamma_2 = \lambda_3^4 \gamma_3 = \lambda_4^4 \gamma_4 = \omega^2 \tag{27}$$

or

$$\lambda_2^4 = \lambda_1^4 (\gamma_1/\gamma_2), \; \lambda_3^4 = \lambda_1^4 (\gamma_1/\gamma_3), \; \lambda_4^4 = \lambda_1^4 (\gamma_1/\gamma_4) \tag{28}$$

Substituting Eq. (20) into boundary conditions (22) it appears that $A_1 = 0$, $C_1 = 0$. In the same way, the rest of conditions given by Eqs. (23) – (26) lead to the set of algebraic equations and it may be written in the matrix form

$$\mathbf{A}\,\mathbf{x} = 0 \tag{29}$$

The matrix $\mathbf{A}$ is too large, to presented it in explicit form. Hence, its elements fall into blocks so that the matrix $\mathbf{A}$ can be written as

$$\mathbf{A} = \begin{bmatrix} \mathbf{A}_1' & \mathbf{A}_2'' & \mathbf{0} & \mathbf{0} \\ \mathbf{0} & \mathbf{B}_1'' & \mathbf{B}_2''' & \mathbf{0} \\ \mathbf{0} & \mathbf{0} & \mathbf{C}_1''' & \mathbf{C}_2'''' \\ \mathbf{0} & \mathbf{0} & \mathbf{0} & \mathbf{D}_1'''' \end{bmatrix} \tag{30}$$

In the current boundary problem, the separate blocks take the form

$$\begin{bmatrix} \mathbf{A}_1' & \mathbf{A}_2'' \\ \mathbf{0} & \mathbf{B}_1'' \end{bmatrix} = \left[\begin{array}{cc|cccc} K_2' & K_4' & -1 & 0 & 0 & 0 \\ \lambda_1 K_3' & \lambda_1 K_1' & 0 & 0 & 0 & -\lambda_2 \\ \lambda_1^2 E_1 J_1 K_4' & \lambda_1^2 E_1 J_1 K_2' & 0 & 0 & -\lambda_2^2 E_2 J_2 & 0 \\ \lambda_1^3 E_1 J_1 K_1' & \lambda_1^3 E_1 J_1 K_3' & 0 & -\lambda_2^3 E_2 J_2 & 0 & 0 \\ & & K_1'' & K_2'' & K_3'' & K_4'' \\ & \mathbf{0} & \lambda_2 K_2'' & \lambda_2 K_3'' & \lambda_2 K_4'' & \lambda_2 K_1'' \\ & & \lambda_2^2 E_2 J_2 K_3'' & \lambda_2^2 E_2 J_2 K_4'' & \lambda_2^2 E_2 J_2 K_1'' & \lambda_2^2 E_2 J_2 K_2'' \\ & & \lambda_2^3 E_2 J_2 K_4'' & \lambda_2^3 E_2 J_2 K_1'' & \lambda_2^3 E_2 J_2 K_2'' & \lambda_2^3 E_2 J_2 K_3'' \end{array}\right] \tag{31}$$

$$\begin{bmatrix} \mathbf{B}_1'' & \mathbf{B}_2''' \\ \mathbf{0} & \mathbf{C}_1''' \end{bmatrix} = \left[\begin{array}{c|cccc} & -1 & 0 & 0 & 0 \\ \mathbf{B}_1'' & 0 & 0 & 0 & -\lambda_3 \\ & 0 & 0 & -\lambda_3^2 E_3 J_3 & 0 \\ & 0 & -\lambda_3^3 E_3 J_3 & 0 & 0 \\ & K_1''' & K_2''' & K_3''' & K_4''' \\ \mathbf{0} & \lambda_3 K_2''' & \lambda_3 K_3''' & \lambda_3 K_4''' & \lambda_3 K_1''' \\ & \lambda_3^2 E_3 J_3 K_3''' & \lambda_3^2 E_3 J_3 K_4''' & \lambda_3^2 E_3 J_3 K_1''' & \lambda_3^2 E_3 J_3 K_2''' \\ & \lambda_3^3 E_3 J_3 K_4''' & \lambda_3^3 E_3 J_3 K_1''' & \lambda_3^3 E_3 J_3 K_2''' & \lambda_3^3 E_3 J_3 K_3''' \end{array}\right] \tag{32}$$

$$\begin{bmatrix} \mathbf{C}_1''' & \mathbf{C}_2'''' \\ \mathbf{0} & \mathbf{D}_1'''' \end{bmatrix} = \left[\begin{array}{c|cccc} & -1 & 0 & 0 & 0 \\ \mathbf{C}_1''' & 0 & 0 & 0 & -\lambda_4 \\ & 0 & 0 & -\lambda_4^2 E_4 J_4 & 0 \\ & -m_r \omega_\nu^2 & -\lambda_4^3 E_4 J_4 & 0 & 0 \\ \mathbf{0} & K_1'''' & K_2'''' & K_3'''' & K_4'''' \\ & K_3'''' & K_4'''' & K_1'''' & K_2'''' \end{array}\right] \tag{33}$$

where the symbols in matrices are given by

$$\{K_\upsilon\} = \{K_1, K_2, K_3, K_4\},$$

$$K_\upsilon' = K_\upsilon(\lambda_1 e_1), \qquad K_\upsilon'' = K_\upsilon(\lambda_2 e_2), \qquad K_\upsilon''' = K_\upsilon(\lambda_3 e_3), \qquad K_\upsilon'''' = K_\upsilon(\lambda_4 e_4) \tag{34}$$

The unknowns are collected in column matrix

$$\mathbf{x} = [B_1, D_1, A_2, B_2, C_2, D_2, A_3, B_3, C_3, D_3, A_4, B_4, C_4, D_4]^T \tag{35}$$

To solve of the homogeneous matrix equation (29), one assumes that $\det \mathbf{A}(\lambda_1) = 0$ and it gives the set $\{\lambda_{1\nu}\}$, $\nu = 1,2,...,n$. Based on Eq. (28) one can calculate $\{\lambda_{2\nu}, \lambda_{3\nu}, \lambda_{4\nu}\}$ and finally, based on Eq. (19), the frequency $\{\omega_\nu\}$ of the system beam-actuator-mass.

Now, the unknowns put down in column matrix, Eq. (35), should be determined. Let the main matrix elements **A** be written as two suffix quantities $A_{\alpha\beta}$, where α and β label the rows and columns respectively. Let $M_{\alpha\beta}$ be the minor of the $A_{\alpha\beta}$ element. The general solution to Eq. (29) is

$$B_1 : D_1 : A_2 : ... = (-1)^{\alpha+1} M_{\alpha 1} : (-1)^{\alpha+2} M_{\alpha 2} : (-1)^{\alpha+3} M_{\alpha 3} : ... \quad (36)$$

Substituting $\{\lambda_{j\nu}\}$ and unknowns $\mathbf{x}$ to Eq. (20), the ν–eigenfunctions (ν–modes) assigned to the j–element are obtained. The solution to Eq. (10) is given by

$$X(x) = \sum_j X_j(x) = \sum_{j\nu} X_j(\lambda_{j\nu} x) = \sum_{j\nu} X_{j\nu}(x) = \sum_\nu X_\nu(x) \quad (37)$$

where $\sum_{j\nu}(...) = \sum_j \sum_\nu (...)$ and the separate modes are equal

$$X_{j\nu}(x) = A_j K_1(\lambda_{j\nu} x) + B_j K_2(\lambda_{j\nu} x) + C_j K_3(\lambda_{j\nu} x) + D_j K_4(\lambda_{j\nu} x) \quad (38)$$

2.4 Orthogonality of modes

Orthogonality condition of the uniform beam modes may be found in (Kaliski, 1986; de Silva, 2000). First of all, based on twice integration by parts, one has

$$\int_0^\ell X_\nu(x) D^4 X_\mu(x) dx = \left(X_\nu(x) D^3 X_\mu(x) - DX_\nu(x) D^2 X_\mu(x) \right)\Big|_0^\ell + \int_0^\ell D^2 X_\nu(x) D^2 X_\mu(x) dx \quad (39)$$

The separate modes $X_\mu(x)$, $X_\nu(x)$, fulfill the following modal equations

$$EJ D^4 X_\mu(x) = \omega_\mu^2 \rho S X_\mu(x) \quad (40)$$

$$EJ D^4 X_\nu(x) = \omega_\nu^2 \rho S X_\nu(x) \quad (41)$$

Multiplying above equations by $X_\nu(x)$ and $X_\mu(x)$ respectively, integrate both in range o integration $x \in [0, \ell]$, use Eq. (39), subtract the second result from the first one, one obtains (for simplicity an argument (x) is omitted)

$$\left(\omega_\nu^2 - \omega_\mu^2\right) \rho S \int_0^\ell X_\nu X_\mu dx = EJ\left[\left(X_\mu D^3 X_\nu - DX_\mu D^2 X_\nu \right) - \left(X_\nu D^3 X_\mu - DX_\nu D^2 X_\mu \right) \right]\Big|_0^\ell \quad (42)$$

For standard boundary conditions, the right-hand-side equals zero.

The procedure outlined above can be used to the problem presented in Fig. 6, but Eq. (39) must be applied to the separate j–element, namely

$$\int_0^{e_j} X_{j\nu}(x) D^4 X_{j\mu}(x) dx = \left(X_{j\nu}(x) D^3 X_{j\mu}(x) - DX_{j\nu}(x) D^2 X_{j\mu}(x) \right)\Big|_0^{e_j} + \\ + \int_0^{e_j} D^2 X_{j\nu}(x) D^2 X_{j\mu}(x) dx \quad (43)$$

Considering both boundary conditions of the problem and coupling conditions between neighboring elements, Eqs. (22)–(26), instead of Eq. (42) one has

$$\left(\omega_\nu^2-\omega_\mu^2\right)\Big(\rho_1 S_1\int_0^{e_1} X_{1\nu}X_{1\mu}\,dx + \rho_2 S_2\int_0^{e_2} X_{2\nu}X_{2\mu}\,dx + \rho_3 S_3\int_0^{e_3} X_{3\nu}X_{3\mu}\,dx + \rho_4 S_4\int_0^{e_4} X_{4\nu}X_{4\mu}\,dx + $$
$$+ m_r X_{4\nu}(0)X_{4\mu}(0)\Big) = E_4 J_4\cdot$$
$$\cdot\Big[\left(X_{4\mu}(e_4)D^3X_{4\nu}(e_4)-DX_{4\mu}(e_4)D^2X_{4\nu}(e_4)\right)-\left(X_{4\nu}(e_4)D^3X_{4\mu}(e_4)-DX_{4\nu}(e_4)D^2X_{4\mu}(e_4)\right)\Big] \quad (44)$$

Because of Eq. (26), the right-hand-side is zero, hence

$$\left(\omega_\nu^2-\omega_\mu^2\right)\Big(\rho_1 S_1\int_0^{e_1} X_{1\nu}X_{1\mu}\,dx + \rho_2 S_2\int_0^{e_2} X_{2\nu}X_{2\mu}\,dx + \rho_3 S_3\int_0^{e_3} X_{3\nu}X_{3\mu}\,dx + \rho_4 S_4\int_0^{e_4} X_{4\nu}X_{4\mu}\,dx + $$
$$+ m_r X_{4\nu}(0)X_{4\mu}(0)\Big) = 0 \quad (45)$$

Fig. 7. Geometry of set with n_j–elements

The orthogonality condition, Eq. (45), may be generalized in a simple way. Let the system beam-actuators-masses be divided into n_j–elements as depicted in Fig. 7. In this case one has

$$\left(\omega_\nu^2-\omega_\mu^2\right)\Big[\sum_j\Big(\rho_j S_j\int_j X_{j\nu}X_{j\mu}\,dx + m_j X_{j\nu}(0)X_{j\mu}(0)\Big) + m_{n_j+1}X_{n_j\nu}(e_{n_j})X_{n_j\mu}(e_{n_j})\Big]=0 \quad (46)$$

Since the term $\omega_\nu^2-\omega_\mu^2$ is canceled for $\mu=\nu$, the general orthogonality condition is given by

$$\sum_j\Big(\rho_j S_j\int_j X_{j\nu}X_{j\mu}\,dx + m_j X_{j\nu}(0)X_{j\mu}(0)\Big) + m_{n_j+1}X_{n_j\nu}(e_{n_j})X_{n_j\mu}(e_{n_j}) = \begin{cases}0, & \nu\neq\mu\\ \beta_\nu^2, & \nu=\mu\end{cases} \quad (47)$$

The Eq. (47) in particular case is used in deriving the solution to the forced vibration problem.

2.5 Forced vibrations with damping

A point departure for further consideration is Eq. (7); for j –element one has

$$E_j J_j D^4 u_j + \mu_j E_j J_j D^4\left(D_t u_j\right) + \rho_j S_j D_t^2 u_j = -f_j \quad (48)$$

The solution to Eq. (48) is forced vibrations with damping. Let be the load force in the form

$$f_j(x,t) = f_j(x)\exp(i\omega_f t) \quad (49)$$

where $i=(-1)^{1/2}$, ω_f – excited frequency.
Applying separation of variables method, the solution to Eq. (48) is assumed as

$$u_j(x,t) = X_{jf}(x)\exp(i\omega_f t) \quad (50)$$

Substituting Eqs. (49) and (50) to Eq. (48) one obtains

$$\gamma_j(1+i\mu_j\omega_f)D^4X_{jf}(x) - \omega_f^2 X_{jf}(x) = -\frac{1}{\rho_j S_j} f_j(x) \tag{51}$$

The solution of the above equation is given by

$$X_{jf}(x) = \sum_\nu C_{j\nu} X_{j\nu}(x) \tag{52}$$

where $C_{j\nu}$ – constants, $X_{j\nu}(x)$ – Eq. (38).
After some calculation, the constants $C_{j\nu}$ are expressed by

$$C_{j\nu} = \frac{1}{\rho_j S_j} \frac{1}{\alpha_{j\nu}^2} \frac{1}{\beta_\nu^2} I_{j\nu} = C_{j\nu}^* I_{j\nu} \tag{53}$$

where

$$C_{j\nu}^* = \frac{1}{\rho_j S_j} \frac{1}{\alpha_{j\nu}^2} \frac{1}{\beta_\nu^2}, \quad \frac{1}{\alpha_{j\nu}^2} = \frac{1}{(1+i\mu_j\omega_f)\omega_\nu^2 - \omega_f^2}, \quad I_{j\nu} = -\int_j f_j(x) X_{j\nu}(x) dx \tag{54}$$

In the end, the problem of the forced j –element beam vibration with damping, excited with the force $f_j(x)$ is solved; in the harmonic steady state it is given by

$$X_f(x) = \sum_j X_{jf}(x) = \sum_{j\nu} C_{j\nu} X_{j\nu}(x) \tag{55}$$

In current problem, two form of the forces have the practical meaning namely, the force with constant amplitude $f_{j0}(x) = f_0$ and the force acting at discrete point $f_{ja}(x_i)$. The former may be interpreted as the spread excitation forced, for example with plane acoustic wave, but the latter is the control force due to actuators, henceforth

$$f_j(x) = f_0 + f_{ja}(x_i) \tag{56}$$

2.6 Interaction between beam and actuators

It is assumed that the actuator is perfectly bonded to the beam surface. Exciting actuator, the interaction between actuator and the beam is appeared. The interaction process is explained in (Hansen & Snyder, 1997; Fuller at al, 1997) in detail and references cited therein. Assuming the spatially uniform actuator, it provides boundary induction solely in terms of the external line moment distributed along its edges (Burke & Hubbard, 1991; Sullivan et al., 1996). So, the bending moment in y –direction is given by the formula (Hansen & Snyder 1997), Fig. 8,

$$M_x = -M_0\left(<x - x_1>^{-2} - <x - x_2>^{-2}\right) \tag{57}$$

where $<.>^{-1} = \delta(.)$ and $<.>^{-2} = D\delta(.)$ – doublet function, M_0 – line moment amplitude

$$M_0 = C_a \frac{d_{31}}{h_a} V \tag{58}$$

where C_a - constant depending on geometry and mechanical properties of the actuator and plate, d_{31} – piezoelectric material strain constants, V - voltage in the direction of polarization.

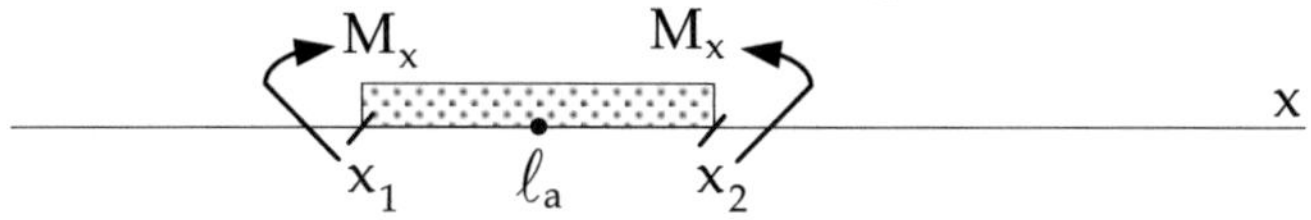

Fig. 8. External line moments of the actuator

The problem is to determine of the C_a, because it depends on the analysis method of the mutual interaction between beam-actuator (Hansen & Snyder 1997; Pietrzakowski, 2004). Let the static force coupling model is taken into account. If relatively thin actuator compared with beam thickness is assumed (so uniform normal stress distribution is accepted) and furthermore by ignoring the neutral axis displacement d, see Fig. 4, the constant C_a is come down to the form

$$C_a = \frac{E_b h_b E_a h_a (h_b + h_a)}{2(E_b h_b + E_a h_a)} \tag{59}$$

Since the beam vibration equation is the forces equation then to consider the action of actuator with the beam, moments M_x are replaced with two couples of forces, Fig. 9,

$$M_x = f_a \ell_a / 2 \tag{60}$$

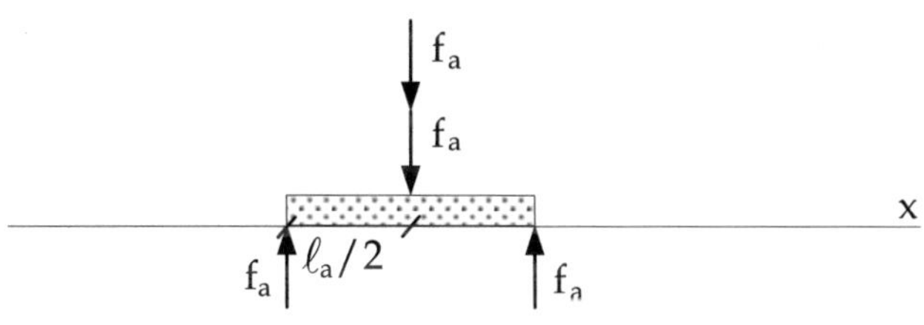

Fig. 9. External pair of forces of the actuator

Next, the separate forces are considered in Eq. (56).

2.7 Beam vibration reduction through actuators

For the problem presented in Fig. 5, the total load of the beam, described by Eq. (56), is given by

$$f_j(x) = -f_0 + \left(f_{js}\delta(x - x_{1s}) - 2f_{js}\delta(x_s) + f_{js}\delta(x + x_{2s})\right) \tag{61}$$

where the symbol f_{ja} is replace by f_{js} in order to express, in the future, the interaction sum of actuators and the glue on the beam.
An expression in brackets is the sum of interacting forces actuator-beam. Hence, the integral $I_{j\nu}$, Eq. (54), for $f_j(x)$ expressed by above equation is given by

$$I_{j\nu} = -\int_j f_j(x) X_{j\nu}(x) dx = -f_0 \int_j X_{j\nu}(x) dx + f_{js} \int_j (\delta(x - x_{1s}) - 2\delta(x_s) + \delta(x + x_{2s})) X_{j\nu} dx =$$
$$= -f_0 \int_j X_{j\nu}(x) dx + f_{js} \left[X_{j\nu}(x_{1s}) - 2X_{j\nu}(x_s) + X_{j\nu}(x_{2s}) \right] \tag{62}$$

The expression in square bracket constitutes the second-order central finite difference. Since the distance between nodes ℓ_s is constant, then the difference can be transformed into

$$\frac{1}{\ell_s^2} \left[X_{j\nu}(x_{1s}) - 2X_{j\nu}(x_s) + X_{j\nu}(x_{2s}) \right] = D^2 X_{j\nu}(x_s) \tag{63}$$

where

$$D^2 X_{j\nu}(x_s) = \pm \kappa_{j\nu}(x_s) \tag{64}$$

The $\kappa_{j\nu}(x_s)$ is the curvature of the mode $X_{j\nu}(x)$ at the point $x = x_s$ (Brański & Szela, 2007; Brański & Szela, 2008). The sign of the $\kappa_{j\nu}(x_s)$ is contractual namely, if the bending of the beam is directed upwards, the sign is positive and vice versa. Substituting Eq. (64) into Eq. (62), one obtains

$$I_{j\nu} = -f_0 \int_j X_{j\nu}(x) dx + f_{js} \ell_s^2 \kappa_{j\nu}(x_s) = I_{j\nu 0} + I_{j\nu s} \tag{65}$$

Next, substituting Eq. (65) into Eq. (52) through Eq. (53), the reduction vibration is obtained

$$X_{jf}(x) = \sum_\nu C_{j\nu}^* I_{j\nu} X_{j\nu}(x) = \sum_\nu C_{j\nu}^* \left(I_{j\nu 0} + I_{j\nu s} \right) X_{j\nu}(x) = \sum_\nu A_{jf\nu} X_{j\nu}(x) \tag{66}$$

Note, that the amplitude $A_{jf\nu}$ is the direct quantity which is liable to the reduction, in explicit form is

$$A_{jf\nu} = C_{j\nu}^* I_{j\nu} = C_{j\nu}^* \left(I_{j\nu 0} + I_{j\nu s} \right) \tag{67}$$

At the same time, together with the vibration reduction amplitude $A_{jf\nu}$, the curvature is subjected to the reduction and based on Eq. (66) is

$$\kappa_{jf} = \pm D^2 X_{jf} = \sum_\nu \kappa_{jf\nu} = \pm \sum_\nu A_{jf\nu} \kappa_{j\nu} \tag{68}$$

Furthermore, the reduction of the $A_{jf\nu}$ leads to the reduction of the shear force $Q_{jf}(x)$ and bending moment $M_{jf}(x)$ (Brański et al., 2010; Kaliski, 1986; Kozień, 2006),

$$Q_{jf}(x) = \pm E_j J_j D\kappa_{jf}(x) = \pm E_j J_j \sum_\nu D\kappa_{jf\nu}(x) = \pm E_j J_j \sum_\nu A_{jf\nu} D\kappa_{j\nu}(x) \tag{69}$$

$$M_{jf}(x) = \pm E_j J_j \kappa_{jf}(x) = \pm E_j J_j \sum_\nu \kappa_{jf\nu}(x) = \pm E_j J_j \sum_\nu A_{jf\nu} \kappa_{j\nu}(x) \tag{70}$$

As can be seen, the vibration reduction undergo on the following amplitudes: of the beam vibration $X_{jf}(x)$ – Eq. (66), of the shear force $Q_{jf}(x)$ – Eq. (69) and of the bending moment $M_{jf}(x)$ – Eq. (70). Hereafter, the notion e.g. "shear force reduction" is used instead of "the reduction of the amplitude of the shear force", and so on.

The Eqs (66), (69) and (70) and may be written commonly

$$\Psi_{jf}(x) = \sum\nolimits_{\nu} \Psi_{jf\nu}(x) = C_j \sum\nolimits_{\nu} A_{jf\nu}\, \Phi_{j\nu}(x) \tag{71}$$

where

$$\Psi_{jf} = \{X_{jf}, Q_{jf}, M_{jf}\}, \quad \Phi_{j\nu} = \{X_{j\nu}, D\kappa_{j\nu}, \kappa_{j\nu}\}, \quad C_j = \{1, \pm E_j J_j, \pm E_j J_j\} \tag{72}$$

Omitting for simplicity the index "f", for entire system beam-actuators one has

$$\Psi(x) = \sum\nolimits_{j} \Psi_j(x) = \sum\nolimits_{j\nu} \Psi_{j\nu}(x) \tag{73}$$

or

$$\Psi(x) = \sum\nolimits_{j\nu} C_j A_{j\nu}\, \Phi_{j\nu}(x) = C A \Phi(x) \tag{74}$$

where

$$A = C^{*} I = C^{*}\left(I_0 + \sum\nolimits_{s} f_s \ell_s^2 \kappa(x_s)\right) = C^{*}\left(I_0 + I_{\Sigma}\right) \tag{75}$$

$$\Phi(x) = \{X(x), D\kappa(x), \kappa(x)\} \tag{76}$$

It is appeared from Eq. (75) that the active vibration reduction depends on the following parameters:

- ω_f – excited frequency, it is contained in C_{ν}^{*},
- x_s – distribution of the actuators on the beam,
- $\kappa(x_s)$ – value of the beam curvature at the point of the actuators distribution,
- f_s – interacting forces between beam-actuators or more generally – mechanical properties of the actuators,
- ℓ_s – actuators lengths or more generally - geometrical properties of the actuators,
- n_s – number of actuators.

As mentioned above, the optimal actuators distribution described with $\{x_s\}$ has an important meaning and finding of the $\{x_s\}$ is the aim of the chapter.

3. Optimal actuators distribution problem

Before the optimization problem will be formulated, any coefficients of the vibration reduction should be defined.

3.1 Reduction and effectiveness coefficients

Let be the difference between any quantities of the beam vibration

$$\Delta\Psi(x) = \Psi(x) - \Psi_R(x) \tag{77}$$

where $\Psi(x)$, $\Psi_R(x)$ - quantities calculated without and with actuators respectively; $\Psi(x)$, $\Psi_R(x)$ are given together by Eq. (74), where

$$A = C^{*} I_0 \tag{78}$$

and

$$A_R = C^*(I_0 + I_\Sigma) \tag{79}$$

The difference $\Delta\Psi(x)$ is interpreted as the quantity of the vibration reduction and it is the first measure of this reduction namely, the quantity reduction coefficient.
The second measure of the vibration reduction is defined as

$$R_\psi(x) = \frac{\Delta\Psi(x)}{\Psi(x)} = \frac{\Psi(x) - \Psi_R(x)}{\Psi(x)} \tag{80}$$

It is called as the reduction coefficient and it may be expressed in per cent. Note, that if the reduction coefficient equals one, the vibration reduction is total, $\Psi_R(x) = 0$.
An effectiveness of the vibration reduction is defined as a quotient of some vibration reduction measure by an amount of the energy W provided to the system in order to excite actuators. Hence, thirst measure of the vibration reduction may be defined by so called the effectiveness coefficient

$$E_\Psi(x) = R_\Psi(x)/W \tag{81}$$

The energy W provided to the system is translated into couples of forces, Fig. 9. Therefore, the energy W may be replaced by forces $f_R = \sum_s 4f_s$, hence

$$E_\Psi(x) = R_\Psi(x)/f_R \tag{82}$$

The Eqs. (77) - (82) define the appropriate factors of the vibration reduction at the point x. In many cases, it is convenient to calculate mean values of these coefficients at whole beam domain or at the beam sub-domains. First of them is the mean quantity reduction coefficient and it is defined by the formula

$$\Delta\Psi_m = \frac{1}{n_i}\sum_i (\Psi(x_i) - \Psi_R(x_i)) \qquad i = 1,2,\ldots,n_i \tag{83}$$

Consequently, the mean reduction coefficient and the mean effectiveness coefficient are defined respectively

$$R_{\Psi m} = \Delta\Psi_m/\Psi_m, \qquad \Psi_m = \sum_i \Psi(x_i)/n_i \tag{84}$$

$$E_{\Psi m} = R_{\Psi m}/f_R \tag{85}$$

The coefficients defined above may constitute the base to formulate the optimization problem; hereafter the $R_\Psi(x)$ is chosen.

3.2 Formulation of the optimization problem

In this chapter, one formulates the following problem: find the optimal actuators distribution $\{x_s\}$ which maximize of the reduction coefficient $R_\Psi(x)$; hence $R_\Psi(x)$ is assumed as an objective function. In this case the maximal value of $R_\Psi(x)$ equals one and it

means p-reduction; such instance is considered in (Brański et al., 2010; Brański & Lipiński, 2011) and it seems that it is possible only in for separate mode.
Let the energy provided to the actuators be constant and hence, the f_s is always constant. Now, for clarity of the disquisition, rewrite the effectiveness coefficient in explicit form

$$R_\Psi(x) = \frac{\Psi(x) - \Psi_R(x)}{\Psi(x)} = \frac{CA\Phi(x) - CA_R\Phi(x)}{CA\Phi(x)} \tag{86}$$

Working out on above assumption, the $R_\Psi(x)$ will be maximal, if $\Psi_R(x)$ is minimal. Hence, the optimal condition $R_\Psi(x) = R_{\Psi;max}(x)$ leads to the next condition $\Psi_R(x) \equiv \Psi_{R;min}(x)$. Note, that the $\Psi_R(x)$ depends on the reduction amplitude A_R. So, the $\Psi_R(x) \equiv \Psi_{R;min}(x)$, if the amplitude A_R is minimal and instead of the above condition, it leads to

$$A_R = A_{R;min} \tag{87}$$

3.3 Heuristic analysis of the optimization problem

Note, that the amplitude A_R comprises the factor $C^* \neq 0$, but it is constant and this is the factor $I_0 + I_\Sigma = I_R$ which is changed. In practice, instead of the condition (87), the following condition of the reduction must be fulfilled

$$I_R = I_0 + I_\Sigma = I_0 + \sum_s f_s \ell_s^2 \kappa(x_s) = I_{min} \tag{88}$$

For future considerations the sign of I_R is very important. The vibrations are reduced with actuators, if the I_Σ is positive, but must be fulfilled the following condition: $I_R = I_0 + I_\Sigma \geq 0$; $I_R = 0$ assures the total reduction. If this condition is not fulfilled, the actuators excite vibrations and thereby they are not accomplished owns role. Note, that the sign of I_0 is always negative, see Eq. (65). Then, the sign of I_Σ must be positive and it depends on the signs both forces f_s and curvatures $\kappa(x_s)$.
From physical point of view, the sign of $\kappa(x)$ is changed and as established above; it is positive, if the bending of the beam is directed upwards. Then, for many actuators one has

$$I_\Sigma = f_1 \ell_1^2 (+)\kappa(x_1) + f_2 \ell_2^2 (-)\kappa(x_2) + \ldots = \sum_s f_s \ell_s^2 (-1)^{s+1} \kappa(x_s) > 0 \tag{89}$$

To obtain the positive sign of I_Σ, the signs of f_s should alternates; this problem clearly expressed in the following way

$$I_\Sigma = (+)f_1 \ell_1^2 (+)\kappa(x_1) + (-)f_2 \ell_2^2 (-)\kappa(x_2) + \ldots = \sum_s (-1)^{s+1} f_s \ell_s^2 (-1)^{s+1} \kappa(x_s) > 0 \tag{90}$$

First at all, it is possible if the signs of $\kappa(x_s)$ and f_s are the same, namely positive or negative, and they take their extremes. To fulfill this requirement, the actuators should be specially distributed on the beam. An idea of description of the sign of $\kappa(x_s)$ is advance determined. The value of $\kappa(x_s)$ is determined by means of the distribution of the actuators; they are bended at $\{x_s\}$. Hence, the distribution has a great significance; this problem was solved in (Brański & Szela, 2007; Brański & Szela, 2008; Brański et al., 2010; Brański & Szela 2010; Szela, 2009). Interpreting Eq. (88) through Eq. (90) it is appear that the actuators ought to be bonded on the beam sub-domains in which the curvatures reach their extremum and consequently the highest and lowest values respectively, see Fig. 10. This is so called quasi-optimal actuators distribution and it is described with $x_Q \equiv x_s$ points, their number is $n_Q \equiv n_s$.

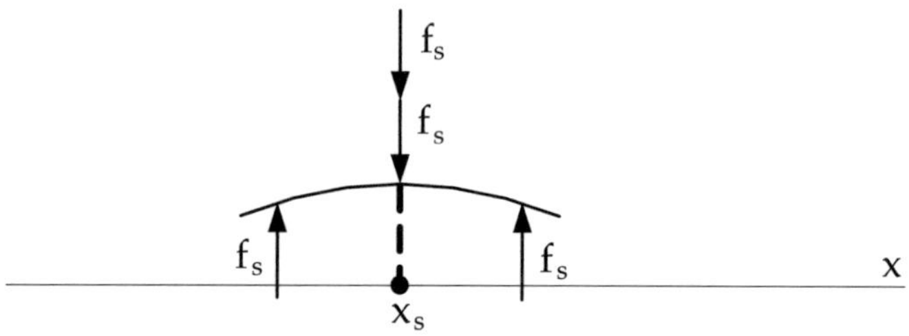

Fig. 10. Optimal distribution of the actuators

As far as signs and values of $\{f_s\}$ are concerned, it was assumed that the added energy exciting actuators is constant. So, the values $\{f_s\}$ of the separate actuators are known and always are constant, while the sign f_s springs from the physical interpretation of the interaction beam-actuator. As can be seen in Fig. 10, the forces $2f_s$ are placed at the point of local extreme, namely at the x_s, with the opposite direction to the bending of the beam $X(x)$. At the same time, the forces f_s on the actuator edges are in the direction of the beam bending and let assume that this sign of f_s is positive. Another way, the vectors f_s and the beam bending $X(x)$ are in the same direction. In such sign convection, both f_s and $\kappa(x_s)$ in Eq. (90) have the same signs and all terms are positive. Furthermore, the actuators distribution described with $x_Q \equiv x_s$ ensures the maximum of the reduction coefficient.
The heuristic analysis described above was substantiated numerically for the separate beam and triangular modes and the details may be found in own papers.

3.4 Analytical analysis of the optimization problem

The aim of this section is to work out of the analytical method, which will describe such distribution of the actuators in order to assure the maximum of the reduction coefficient. It is expected that the analytical method will confirm the quasi-optimal distribution which has been found above with heuristic method. Therefore the assumptions are the same like in heuristic method, namely n_s, f_s and ℓ_s are settled.
Let the distribution of actuators be marked with the set of unknown coordinates $\{x_s\}$ for the moment; that are exactly these coordinates $\{x_s\} \equiv \{x_Q\}$ of which are looked for. One starts from Eq. (88), hence

$$I_R = I_0 + I_\Sigma = I_0 + f_1\,\ell_1^2\,\kappa(x_1) + f_1\,\ell_2^2\,\kappa(x_2) + \ldots \tag{91}$$

Since the $\kappa(x)$ is the function which changes the sign, it is appropriate to search the points x_s which assure the extreme $I_R(x)$, not minimum only. The function $I_R(x)$ can have the extreme only at points x_s, at which $DI_R(x)$ is equal to zero or $f_1\,\ell_1^2\,D\kappa(x_1) + f_1\,\ell_2^2\,D\kappa(x_2) + \ldots = 0$ does not exist (Fichtenholtz, 1999). Because I_0 is constant than a necessary condition for existing extreme value is

$$DI_R(x_s) = 0 \quad (92)$$

where $DI_R(x_s) = DI_R(x = x_s)$ and hence

$$f_1 \ell_1^2 D\kappa(x_1) + f_1 \ell_2^2 D\kappa(x_2) + \ldots = 0 \quad (93)$$

Because of $f_s \ell_s^2 \neq 0$ then instead of Eq. (93) one has

$$D\kappa(x_s) = 0 \quad (94)$$

From the condition (94), a set of stationary points $\{x_s\}$ is obtained. The sufficient condition for existing extreme demands, in order to the function be determined on either side of the point x_s and $D\kappa(x)$ must change sign at this point (turning point); it is sufficient condition formulated in the first form. This condition is expressed in the other form

$$D^2\kappa(x_s) \neq 0 \quad (95)$$

If this condition is not fulfilling, then this point should be omitted.
One still needs to consider the biggest and the lowest values of $\kappa(x)$; they are in hypothetical points $\{x_{max}, x_{min}\}$. In order to find them, the values of $\kappa(x)$ at the stationary points $\{x_s\}$ are calculated and they are compared to the values calculated at the end points of the appropriate interval. In the future consideration, the n_s points among stationary $\{x_s\}$ and $\{x_{max}, x_{min}\}$ ones, at which $\kappa(x)$ takes in turn its absolute values, are taken into account. The problem of the signs of the $\kappa(x_s)$ and f_s is quite the same as in heuristic analysis.
Analytical analysis was applied for p-reduction and for the separate beam modes (Brański & Lipiński, 2011). As pointed out there, the analytical solution to the optimal actuators distribution problem confirms the results obtained with heuristic solution.

4. Conclusion

Deriving the shape of $\kappa(x)$, the influence both masses and stiffness of the actuators and glue on the shape of $X(x)$, and consistently on the shape of $\kappa(x)$, were omitted; if not, an adaptation method must be applied. But after determining shape of $\kappa(x)$, all these parameters were considered.
As can be note, the actuators optimal distribution is attained assuming that the added energy to excite actuators is constant. It is translated into constant f_s. Having the optimal distribution, the reduction coefficient may be improved by adding more energy or in order words, by increasing f_s. This way, presented optimal method corresponds to that one presented in (Q. Wang & C. Wang, 2001), namely "maximization of the control forces transmitted by the actuators to the structure".
Based on theoretical considerations, and numerical ones presented in own papers, the following conclusion may be formulated.

1. The optimization problem of the actuators distribution assuring the maximal active vibration reduction of the beam, measured with reduction coefficient, may be solved both heuristically and analytically. In analyzed problem, it turned out that both methods give the same results.
2. The following algorithm of analytical method may be worked out:

- to search of stationary points $\{x_s\}$ of the beam curvature,
- to search of $\{x_{max}, x_{min}\}$ points of the beam curvature,
- n_Q points among stationary $\{x_s\}$ and $\{x_{max}, x_{min}\}$ ones, at which $\kappa(x)$ takes in turn its maximum absolute values, are selected, they are denoted by $\{x_Q\}$,
- to bond the actuators at the $\{x_Q\}$ points,
- to determine the value of the reduction coefficient,
- to increase the value f_s, through the energy increase which excites actuators, until the reduction coefficient will attain its maximum.

It seems that proposed optimization method is very simple and may be useful in many technical problems of active vibration reduction. This work is a starting point for many computer simulations and experiments.

5. References

Bapat, C.N. & Bapat, C. (1987). Natural frequencies of a beam with non-classical boundary conditions and concentrated masses, *Journal of Sound and Vibrations*, Vol.112, No.1, pp. 177–182.

Brański, A. & Szela, S. (2007). On the quasi optimal distribution of PZTs in active reduction of the triangular plate vibration. *Archives of Control Sciences*, Vol.17, No.4, pp. 427–437.

Brański, A. & Szela, S. (2008). Improvement of effectiveness in active triangular plate vibration reduction. *Archives of Acoustics*, Vol.33, No.4, pp.521-530.

Brański, A.; Borkowski, M. & Szela, S. (2010). The idea of the selection of PZT-beam interaction forces in active vibration protection problem. *Acta Physica Polonica*, Vol.118, pp.17-22.

Brański, A. & Szela S. (2010). Quasi-optimal PZT distribution in active vibration reduction of the triangular plate with P-F-F boundary conditions. *Archives of Control Sciences*, Vol.20, No.2, pp.209–226.

Brański, A. & Lipiński, G. (2011). Analytical determination of the PZT's distribution in active beam vibration protection problem. (in press in *Acta Physica Polonica*).

Brański, A. & Szela, S. (2011). Evaluation of the active plate vibration reduction via the parameter of the acoustic field. (in press in *Acta Physica Polonica*).

Bruant, I.; Coffignal, G.; Lene, F. & Verge, M. (2001). A methodology for determination of piezoelectric actuator and sensor location on beam structure. *Journal of Sound and Vibrations*, Vol.243, No.5, pp.861–882.

Bruant, I. & Proslier, L. (2005). Optimal location of actuators and sensors in active vibration control, *Journal Inteligent Material System Structures*, Vol.16, pp.197–206.

Bruant, I.; Gallimard, L. & Nikoukar, S. (2010). Optimal piezoelectric actuator and sensor location for active vibration control, using genetic algorithm. *Journal of Sound and Vibrations*, Vol.329, pp.1615-1635.

Burke, S.E. & Hubbard, J.E. (1991). Distributed transducer vibration control of thin plate, *J.A.S.A.*, Vol.90, No.2, pp.937–944.

Charette, F.; Berry, A. & Guigou, C. (1998). Dynamic effect of piezoelectric actuators on the vibration response of a plate. *Journal Inteligent Material System Structures*, Vol.8, pp.513–524.

Croker, M.J. (2007). *Handbook of noise and vibration control*, John Wiley & Sons.

Dhuri, K.D. & Seshu, P. (2006). Piezo actuator placement and sizing for good control effectiveness and minimal change in original system dynamics. *Smart Material Structure*, Vol.15, pp.1661–1672.

Ercoli, L. & Laura, P.A.A. (1987). Analytical and experimental investigation on continuous beam carying elastically mounted masses. *Journal of Sound and Vibrations*, Vol.114, No.3, pp.519–533.

Fichtenholtz, G.M. (1999). Differential and integral calculus, PWN, Warsaw.

Frecker, M. (2003). Recent advances in optimization of smart structures and actuators. *Journal Intelligent Material System Structures*, Vol.14, pp.207-215.

Fuller, C.R.; Elliot, S.J. & Nielsen, P.A. (1997). *Active control of vibration*, Academic Press, London.

Gosiewski, Z. & Koszewnik, A. (2007). The influence of the piezoelements placement on the active vibration damping system, *Proceedings Active Noise and Vibration Control Method*, pp.69–79, Krakow,

Guney, M. & Eskinat, E. (2007). Optimal actuator and sensor placement in flexible structures using closed-loop criteria, *Journal of Sound and Vibrations*, Vol.312, pp.210–233.

Halim, D. & Reza Moheimani, S.O. (2003). An optimization approach to optimal placement of collocated piezoelectric actuators and sensors on a thin plate. *Mechatronics*, Vol.13, pp.27–47.

Han, S.M., Benaroya, H. & Wei, T. (1999). Dynamics of transversely vibrating beams using four engineering theories, *Journal of Sound and Vibrations*, Vol.225, No.5, pp.935–988.

Hansen, C.H. & Snyder, S.D. (1997). *Active control of noise and vibration*, E&FN SPON, London.

Hernandes, J.A.; Almeida, S.F.M. & Nabarrete, A. (2000). Stiffening effects on the free vibration behavior of composite plates with PZT actuators, *Composite Structures*, Vol.49, pp.55–63.

Hong, C.; Gardonio, P. & Elliott, S.J. (2007). Active control of resiliently mounted beams using triangular actuators, *Journal of Sound and Vibrations*, Vol.301, pp.297–318.

Ip, K.H. & Tse, P.C. (2001). Optimal configuration of a piezoelectric path for vibration control of isotropic rectangular plate, *Smart Material Structure*, Vol.10, pp.395–403.

Jha, A.K. & Inman, D.J. (2003). Optimal sizes and placement of piezoelectric actuators and sensors for an inflated torus. *Juornal Inteligent Material System Structures*, Vol.14, pp.563–576.

Jiang, T.Y.; Ng, T.Y. & Lam, K.Y. (2000). Optimization of a piezoelectric ceramic actuator. *Sensors and Actuators*, Vol.84, pp.81–94.

Kaliski, S. (1986). *Vibrations and Waves*, PWN, Warsaw.

Kasprzyk, S. & Wiciak, M. (2007). Differential equation of transverse vibrations of a beam with a local stroke change of stiffness, *Opuscula Mathematica*, Vol.27, No.2, 245–252.

Kozień, M. (2006). *Acoustic radiation of plates and shallow shells*, PK, Monograph 331, ISS 0860-097X, Krakow.

Liu, W.; Hou, Z. & Demetriou, M.A. (2006). A computational scheme for the optimal sensor/actuator placement of flexible structures using spatial H2 measures, *Mechanical Systems and Signal Processing*, Vol.20, pp.881–895.

Low, K.H. & Naguleswaran, S. (1998). On the eigenfrequencies for mass loaded beams under classical boundary conditions, *Journal of Sound and Vibrations*, Vol.215, No.2, pp.381–389.

Majkut, L. (2010). Eigenvalue based inverse model of beam for structural modification and diagnostics. Part I: Theoretical formulation, *Latin American Journal of Solids and Structures*, Vol.7, pp.423436.

Moshrefi-Torbati, M.; Keane, A.J.; Elliott, S.J.; Brennan, M.J.; Anthony, D.K. & Rogers, E. (2006). Active vibration control (AVC) of a satelite boom structure using optimally positioned stacked piezoelectric actuators, *Journal of Sound and Vibrations*, Vol.292, pp.203-220.

Naguleswaran, S. (1999). Lateral vibration of a uniform Euler-Bernoulli beam carrying a particle at an intermediate point, *Journal of Sound and Vibrations,* Vol.227, No.1, pp.205–214.

Pietrzakowski, M. (2004). *Active damping of transverse vibration using distributed piezoelectric elements*, Monograph 204, PW, ISSN 0137-2335, Warsaw.

Przybyłowicz, P.M. (2002). *Piezoelectric vibration control of rotating structures,* Monograph 197, PW, ISSN 0137-2335. Warsaw.

Qiu, Z.; Zhang, X.; Wu, H. & Zhang, H. (2007). Optimal placement and active vibration control for piezoelectric smart flexible cantilever plate, *Journal of Sound and Vibrations,* Vol.301, pp.521-543.

Sheu, W.J.; Huang, R.T. & Wang, C.C. (2008). Influence of bonding glues on the vibration of piezoelectric fans, *Sensors and Actuators A,* Vo.148, pp.115–121.

Shimon, P.; Richer, E. & Hurmuzlu, Y. (2005). Theoretical and experimental study of efficient control of vibration in a clamped square plate, *Journal of Sound and Vibrations,* Vol.282, pp.453-473.

de Silva, C.W. (2000). *Vibration, Fundamentals and practice,* CRC Press.

Sullivan, J.M.; Hubbard, J.E. & Burke, S.E. (1996). Modeling approach for two-dimensional distributed transducers of arbitrary spatial distribution, *J.A.S.A.,* Vol.99, No.5, pp.2965–2974.

Szela, S. (2009). *Distribution method of the actuators in an active vibration reduction of the triangular plate,* AGH, Krakow, Ph.D. thesis, in polish.

Tylikowski, A. & Przybyłowicz, P.M. (2004). *Non-classical piezoelectric materials in vibrations stability and dumping,* PW, Warsaw.

Wang, F. (2007). *Shape optimization for piezoceramics,* Paderborn, Ph.D. thesis.

Wang, Q. & Wang, C. (2001). A controllability index for optimal design of piezoelectric actuators in vibration control of beam structures, *Journal of Sound and Vibrations,* Vol.242, No.3, pp.507–518.

Wiciak, J. (2008). *Vibration and Structural Acoustic Control-Selected Aspects,* AGH, Monografh 175, ISSN 0867-6631, Krakow.

Yang, Y.; Jin, Z. & Soh, C.K. (2005). Integrated optimal design of vibration control system for smart beam using genetic algorithms, *Journal of Sound and Vibrations,* Vol.282, pp.1293–1307.

Part 4

Acoustic Wave Based Microdevices

19

Multilayered Structure as a Novel Material for Surface Acoustic Wave Devices: Physical Insight

Natalya Naumenko
Moscow Steel and Alloys Institute (Technological University)
Russia

1. Introduction

Since 70-ies, when the first delay lines and filters employing surface acoustic waves (SAW) were designed and fabricated, the use of SAW devices in special and commercial applications has expanded rapidly and the range of their working parameters was extended significantly (Hashimoto, 2000; Ruppel, 2001, 2002). In the last decade, their wide application in communication systems, cellular phones and base stations, wireless temperature and gas sensors has placed new requirements to SAW devices, such as very high operating frequencies (up to 10 GHz), low insertion loss, about 1 dB, high power durability, stable parameters at high temperatures etc.

The main element of a SAW device is a piezoelectric substrate with an interdigital transducer (IDT) used for generation and detection of SAW in the substrate. The number of single crystals utilized as substrates in SAW devices did not increase substantially since 70-ies because a new material must satisfy the list of strict requirements to be applied in commercial SAW devices: sufficiently strong piezoelectric effect, low or moderate variation of SAW velocity with temperature, low cost of as-grown large size crystals for fabrication of 4-inch wafers, long-term power durability, well developed and non-expensive fabrication process for SAW devices etc. Today only few single crystals are utilized as substrates in SAW devices: lithium niobate, $LiNbO_3$ (LN), lithium tantalate, $LiTaO_3$ (LT), quartz, SiO_2, lithium tetraborate, $Li_2B_4O_7$ (LBO), langasite, $La_3Ga_5SiO_{14}$ (LGS) and some crystals of LGS group (LGT, LGN etc.) with similar properties.

The SAW velocities in these single crystals do not exceed 4000 m/s, which limit the highest operating frequencies of SAW devices by 2.5-3 GHz because of limitations imposed by the line-resolution technology of IDT fabrication. The minimum achievable insertion loss and maximum bandwidth of SAW devices depend on the electromechanical coupling coefficient, which can be evaluated for SAW as $k^2 \approx 2\Delta V/V$, where ΔV is the difference between SAW velocities on free and electrically shorted surfaces. The largest values of k^2 can be obtained in some orientations of LN and LT. Ferroelectric properties of these materials are responsible for a strong piezoelectric effect. As a result, k^2 reaches 5.7% in LN and 1.2% in LT, for SAW. For leaky SAW (LSAW) propagating in rotated Y-cuts of both crystals, the coupling is higher and can exceed 20% for LN and 5% for LT. However, LSAW attenuates because of its leakage into the bulk waves when it propagates along the crystal surface. As a result, insertion loss of a SAW device increases. Attenuation coefficient depends on a crystal cut and IDT geometry. For example, in 36° to 48° rotated YX cuts of LT and in 41° to 76° YX rotated YX cuts of LN, high electromechanical coupling of LSAW can be combined with low attenuation coefficient via simultaneous optimization of orientation and electrode structure (Naumenko & Abbott, US patents, 2003, 2004). When these substrates are utilized in radiofrequency (RF) SAW filters with resonator-type structures, low insertion loss of 1dB or even less can be obtained. Today such low loss filters are widely used in mobile communication and navigation systems. The main drawback of these devices is high sensitivity of the characteristics to variations of temperature because the typical values of temperature coefficient of frequency (TCF) vary between -30 ppm/°C and -40 ppm/°C for LT and between -60 ppm/°C and -75 ppm/°C for LN.

Contrary to LN and LT, quartz is characterized by excellent temperature stability of SAW characteristics but low electromechanical coupling coefficient, k^2<0.15%. Hence, even in resonator-type SAW filters with very narrow bandwidths, about 0.05%, where the loss of radiated energy is minimized due to the energy storage in a resonator, the best insertion loss achieved in a SAW device with matching circuits is only 2.5-4 dB.

In some orientations of LBO, LGS and other crystals of LGS group, zero TCF is combined with a moderate electromechanical coupling coefficient. However, these crystals have limited applications in commercial SAW devices because low SAW velocities restrict high-frequency applications on LGS and LBO dissolves in water and acid solutions, which prohibits application of conventional wafer fabrication processes to this material and finally results in an increased cost of SAW devices.

Hence, none of available single crystalline materials provides a combination of large piezoelectric coupling, zero TCF and high propagation velocity. A strong need in such material exists today, especially for application in SAW duplexers and multi-standard cellular phones, where the temperature compensation is the key issue because of necessity to divide a limited frequency bandwidth into few channels with no overlapping allowed in a wide range of operating temperatures. As an alternative to conventional SAW substrates, layered or multilayered (stratified) materials were studied extensively since 80-ies but only in the last decade some of these structures found commercial applications in SAW devices, due to the recent successes of thin film deposition technologies and development of robust simulation tools for design of SAW devices on layered structures.

2. Multilayered structures as materials for SAW devices

As described above, the increasing requirements to the substrate materials, on one side, and rapid development of thin film deposition technologies, on the other side, gave rise to the novel class of materials for SAW devices - layered or multilayered structures. One or more films of different materials deposited on a regular substrate can improve its characteristics significantly. A proper combination of a substrate and a thin film helps to overcome the limitations of the conventional SAW substrates. In this section, a brief overview of the layered structures will be given. The examples presented here are currently investigated by different research groups as promising compound SAW substrate materials or found already some applications in SAW devices. The focus is made on the recent achievements in material research mostly motivated by challenges of the rapidly developing market of communication devices. Based on this overview, a generalized multilayered structure, which includes all described examples, will be derived and a method of investigating this structure will be presented and then applied to few different structures, for which more detailed discussion of SAW propagation characteristics will be given.

The first example of a layered structure is a dielectric isotropic silicon dioxide (SiO_2) film deposited on one of rotated YX-cuts of LN or LT characterized by a high electromechanical coupling. Today these structures attract attention of researchers and SAW designers as the most promising candidates for application in SAW duplexers required in most of popular mobile phone systems (Kovacs et al., 2004; Kadota, 2007; Nakai et al., 2008; Nakanishi et al., 2008]. These RF devices must separate the transmitted and received signals in a narrow frequency interval and in a wide range of operating temperatures, e.g. between -30°C and +85°C. Therefore, substrate materials combining high propagation velocity, high electromechanical coupling and low TCF are strongly required. Due to the opposite signs of TCF in SiO_2 and LT or LN, a layered SiO_2/LT or SiO_2/LN structure allows to obtain the desired combination of characteristics. When it is utilized in a resonator-type filter, the electrodes of IDT and metal gratings are commonly built at the interface between SiO_2 and LN or LT. Besides, a heavy metal, such as copper, is utilized as an electrode material (Nakai et al., 2008). A layered structure with such electrode configuration is schematically presented as *Type 1* in Fig.1. The location of electrodes at the interface helps to keep a high electromechanical coupling and combine it with a large reflection coefficient in SAW resonators. A large metallization thickness effectively reduces resistive losses and results in a high *Q* factor of a SAW resonator. Another advantage of using heavy electrodes is a reduced propagation loss, which is achieved due to the transformation of LSAW into SAW. SAW devices with low insertion loss of 1-2 dB and low TCF of -7ppm/°C, have been successfully realized on such substrates (Kadota, 2007).

A thickness of SiO_2 film in SiO_2/LT and SiO_2/LN structures can vary within a wide range, from a few percent of a SAW wavelength up to a few wavelengths, to provide the required combination of electromechanical coupling, TCF and propagation loss. Moreover, SiO_2 film helps to isolate the working surface of a SAW device from environmental influences and facilitates packaging of SAW chip.

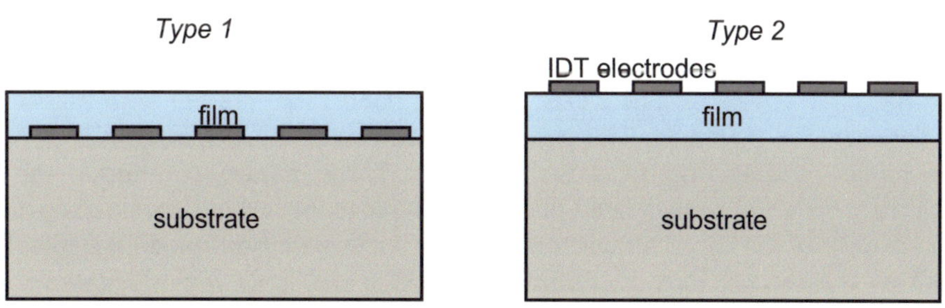

Fig. 1. Two typical structures with one thin film

Another electrode configuration in a structure with one thin film is schematically presented as *Type 2* in Fig. 1, with IDT located on the top surface. It is typical for a piezoelectric film on a non-piezoelectric substrate or on a substrate with low electromechanical coupling coefficient. As a piezoelectric film, zinc oxide ZnO is widely used (Kadota & Minakata, 1998; Nakahata et al., 2000; Emanetoglu et al., 2000; Brizoual et al., 2008). ZnO films are cheap and provide sufficiently high values of electromechanical coupling. The film deposition technique (e.g. magnetron sputtering) has been well developed for this material. Another piezoelectric film, which is extensively studied as a promising material for high-frequency SAW devices is aluminum nitride (AlN) (Benetti et al., 2005, 2008; Fujii et al., 2008; Omori et al., 2008). It is characterized by chemical stability, mechanical strength, high acoustic velocity and good dielectric quality. Some other piezoelectric films, like CdS or GaN, were investigated previously but did not receive as much attention as ZnO or AlN.

A piezoelectric film is usually combined with silica glass, silicon, sapphire or diamond substrate. Silica glass is cheap, the use of silicon as a wafer enables simple integration of IF and RF components in one chip, sapphire is characterized by high SAW velocities, up to 6000 m/s, and diamond provides the highest SAW velocities among all materials, up to 11000 m/s, and is being used for high frequency SAW devices in the GHz range. For example, SAW resonator with center frequency about 4.5 GHz was built on AlN/diamond structure characterized by SAW velocity about 10000 m/s and $k^2 \approx 1\%$ (Omori et al., 2008). A combination of SAW velocity about 5500 m/s and electromechanical coupling about 0.25% can be obtained in AlN/sapphire structure (Ballandras et al., 2004). To reduce TCF of a SAW device, ZnO film can be combined with quartz or LGS. For example, nearly zero TCF and k^2 about 1.8% was achieved for SAW in ZnO/quartz structure, via optimization of quartz orientation (Kadota et al., 2008).

One more structure, which can be referred to the *Type 1*, recently found application in SAW devices. It is a thin plate of a piezoelectric crystal, such as LN or LT with thickness 10-15 wavelengths, which is directly bonded to a dielectric or semiconductor wafer. The bonding technology (Eda et al., 2000) provides excellent contact between the two materials and allows fabrication of SAW devices with reproducible characteristics on a thin LN or LT plate bonded to a thick silicon or glass wafer. In these structures, high values of electromechanical coupling coefficients typical for LN and LT are combined with improved TCFs, due to low thermal expansion coefficients (TCE) determined by massive silicon, glass or sapphire wafer (Tsutsumi et al., 2004). An example of bonded wafer will be numerically investigated in section 4.

The quality of a contact between LN or LT plate and a silicon wafer can be improved if a thin SiO_2 film is deposited between these materials (Abbott et al., 2005). Such two-layered structure with IDT on the top surface is schematically shown as *Type 3* in Fig. 2. With silicon as a substrate, SiO_2 as the first film and LN or LT as the second film (plate), this structure can give the same advantages as LT/Si or LN/Si bonded wafers but with higher quality contact between the materials. The presence of additional SiO_2 film results in spurious acoustic modes propagating in a SAW device. These modes deteriorate the device performance and should be simulated properly to achieve the desired device characteristics.

Another example of the *Type 3* structure is a silicon wafer with isotropic SiO_2 as the first film and ZnO as the second film. Optimization of SiO_2 and ZnO film thicknesses enables obtaining of a structure with TCF=0 (Emanetoglu et al., 2000). A high frequency SAW device can be built if SiO_2 and ZnO films are deposited atop of a diamond or a sapphire substrate. With ZnO as the first film and isotropic SiO_2 as the second film, the preferential location of IDT electrodes is at the substrate-film interface (*Type 4*). Alternatively, IDT can be built on ZnO surface and then buried in SiO_2 overlay (*Type 5*). For example, Nakahata (Nakahata et al., 2000) reported on a SAW resonator using SiO_2/ZnO/diamond structure with two different electrode configurations (*Type 4* and *Type 5*). Zero TCF, high velocity about 10000 m/s and $k^2 \approx 1.2\%$ were obtained for shear horizontally (SH) polarized SAW mode. A resonator with center frequency about 2.5 GHz, temperature compensated characteristics and low insertion loss was fabricated on this structure.

Nakahata (Nakahata et al., 1995) reported one more example of the two-layered structure, which can be referred to *Type 4*. It is a ZnO film on a silicon wafer with thin isotropic diamond layer between them. The following SAW characteristics have been obtained: velocity $V \approx 8050$ m/s, $k^2 \approx 1.42$ %, TCF≈0.

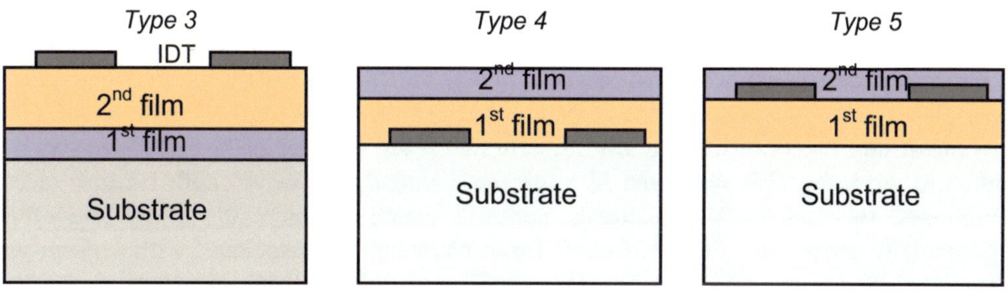

Fig. 2. Three typical structures with two thin films

The examples described above are not aimed at comprehensive survey of layered structures potentially applicable in SAW devices but demonstrate that a variety of layered structures can be referred to a few basic types. A unified approach to analysis of acoustic modes in different layered structures would be beneficial for optimization of SAW devices, because such approach allows comparing characteristics of the same SAW design built on different combinations of film and substrate materials.

The simulation of SAW characteristics is an important part of the SAW device design procedure. In a specified structure, such simulation must take into account orientation of each material if it is anisotropic, film thicknesses, a thickness and shape of IDT electrodes, electrode width to pitch ratio etc. Besides, the accurate analysis of all modes propagating in the investigated structure is required, including the main SAW or LSAW mode and all spurious modes generated by IDT in the specified frequency interval. A number of spurious modes grows with a number of layers and increasing of their thicknesses, which makes the simulation procedure more complicated. Moreover, with increasing film thickness SAW changes its nature and eventually transforms into a new type of acoustic wave. However, the characteristics of any acoustic mode change continuously with this transformation.

The variation of film thicknesses within wide range helps to obtain a variety of *novel* materials with different combinations of characteristics demanded for SAW devices of different applications. After a proper combination of materials is selected, the geometrical parameters of a multilayered structure must be optimized to satisfy the desired electrical specification, including frequency bandwidth, insertion loss, out-of-band rejection, shape factor of frequency response or Q factor of a SAW resonator, temperature deviation of frequency etc. It is a common practice to optimize film and electrode thicknesses and other geometrical parameters of IDTs simultaneously with orientations of anisotropic materials included in the layered structure, to achieve the best SAW device performance.

The challenges described above require a robust, fast and universal numerical technique, which could be applied to different types of multilayered structures, with film thicknesses varying within wide range and allowing transformation of SAW into boundary waves, plate modes or other types of acoustic waves. Such technique is described in the next section.

3. An advanced numerical technique for analysis of acoustic waves in multilayered structures

The most popular numerical technique used for simulation of SAW characteristic in multi-layered structures is *Transfer Matrix Method* (TMM) (Adler, 1990). It is based on the matrix formalism suggested by Stroh (Stroh, 1965) for solution of a SAW problem in anisotropic media. For each material of a multilayered structure, TTM assumes building of a *fundamental acoustic tensor* dependent on the material constants and the analyzed orientation. Then the characteristics of the partial acoustic modes are found as the eigen vectors and eigen values of the matrix associated with this tensor. Finally, the *transfer matrix* is calculated, which characterizes the change of acoustic fields within the analyzed layer. The method is fast and convenient and does not impose any limitations on the number of layers. However, it is known to work unstable when the film thickness exceeds 3-5 wavelength because of bad conditioned matrices built of elements, some of which exponentially decay and others exponentially grow with film thickness. These elements are associated with *incident* and *reflected* modes. As suggested by Tan (Tan, 2002) and Reinhardt (Reinhardt et al., 2003), a separate treatment of these two groups of partial modes helps to avoid the instability and extends the range of the analyzed thicknesses from zero to infinite value.

Another limitation of the previously reported numerical techniques developed for analysis of SAW in multilayered structures is their focusing on a certain type of acoustic waves, which is related to a fixed type of a multilayered structure, with analytically defined boundary conditions at the interfaces and external boundaries. For example, acoustic waves propagating in a substrate with a thin film of finite thickness and *stress-free* boundary conditions at the top surface are different from acoustic waves propagating in the same combination of materials when the film thickness tends to infinite value. In practice, the results obtained with different versions of software using fixed boundary conditions often diverge and do not allow seeing how the wave characteristics change continuously with variation of film thickness within wide range. For example, the software *FEMSDA* (Endoh et al., 1995; Hashimoto et al., 2007, 2008), which is very popular among SAW device designers, includes separate versions for analysis of SAW in a substrate with a thin film and for investigation of boundary waves. In the first version, a film thickness providing robust calculations does not exceed a half-wavelength.

To overcome the described limitation, an advanced numerical technique was developed (Naumenko, 2009, 2010). It can be applied to a variety of multilayered structures and types of acoustic waves. The universal character of the software is achieved due to characterization of the air as a dielectric medium with a very small density and elastic stiffness constants and treatment of this medium as an example of a dielectric. The same numerical methods are applied to this and other materials, which compose a layered structure. With such approach, it is not necessary to fix the stress-free boundary conditions at the top surface of a thin film or at the boundaries of a finite-thickness plate to find acoustic waves propagating in these structures. The stress-free boundary conditions are automatically simulated with high accuracy when the adjacent medium is specified as the air.

The developed technique refers to a multilayered structure schematically shown in Fig. 3, in which a metal film or IDT is located between N upper and M lower layers, where M and N can vary between one and ten or more if necessary.

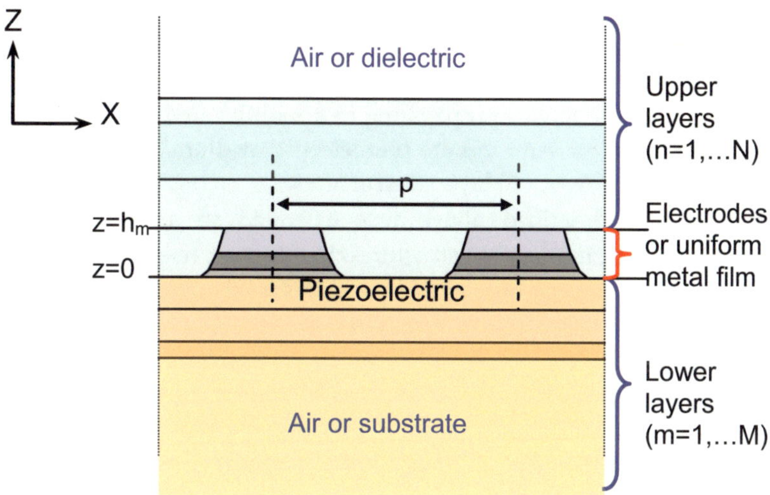

Fig. 3. Schematic drawing of analyzed multilayered structure

Analysis starts from the uppermost or lowermost half-infinite material, in which the wave structure is calculated. It can be a dielectric, a piezoelectric material or the air. In each adjacent finite-thickness layer, the transformation of the wave structure is deduced via separate treatment of incident and reflected partial modes. It means that the reflection and transmission matrix coefficients replace the transfer matrix to escape numerical noise at film thicknesses exceeding 3-5 wavelengths. For the structures with few dielectric (isotropic) films, the variation of the dielectric permittivity within each film characterized by the finite thickness h and dielectric permittivity ε_{film} is taken into account via the well known recursive equation (Ingebrigtsen, 1969):

$$\varepsilon(z+h)=\varepsilon_{film}\frac{\left[\varepsilon(z)+\varepsilon_{film}\cdot th(kh)\right]}{\left[\varepsilon_{film}+\varepsilon(z)\cdot th(kh)\right]}, \tag{1}$$

where k is the wave number. Analysis of the lower and upper multilayered half-spaces is considered completed when the wave structure has been determined at z=0 and z=h_m, where h_m is a metal film thickness, and the surface impedance matrices $\hat{Z}_{UP}(k)$ and $\hat{Z}_{LOW}(k)$ have been calculated at the upper and lower boundaries of the metal film. Each of these matrices characterizes the ratio between the vectors of displacements $\mathbf{u}$ and normal stresses $\mathbf{T}$ at the analyzed interface, $\hat{Z}=\mathbf{u}\mathbf{T}^{-1}$. A piezoelectric material is characterized by the generalized 4-dimensional displacement and stress vectors, with added electrostatic potential ϕ and normal electrical displacement D, respectively. The matrices $\hat{Z}_{UP}(k)$ and $\hat{Z}_{LOW}(k)$ comprise the information about the layers located above and below the metal film and enable simple formulation of electrical boundary conditions at z=0 and z=h_m. If the mass load of metal film is included in the analysis of the upper N layers, then the function of *effective dielectric permittivity* (EDP) $\varepsilon_s(k)$ can be calculated at z=0 (Ingebrigtsen, 1969; Milsom et al., 1977). This function relates the electric charge σ at the surface to the electrostatic potential φ,

$$\sigma = |k| \cdot \varepsilon_s(k) \cdot \varphi \tag{2}$$

and can be used for extraction of the velocities and electromechanical coupling coefficients of SAW and other acoustic modes propagating in a multilayered structure. EDP function was originally introduced for semi-infinite piezoelectric medium, but it is also an efficient tool for analysis of acoustic waves in layered structures.

The numerical method described above was extended to a periodic metal grating sandwiched between two multilayered structures. In this case, the spectral domain analysis (SDA) of the upper and lower multi-layered half-spaces is combined with the finite-element method (FEM) applied to simulation of the electrode region. To some extent, the developed *SDA-FEM-SDA* technique (Naumenko, 2010) can be considered as an advanced FEMSDA method. In this case, the function of Harmonic Admittance $Y(f,s)$ (Blotekjear et al, 1973; Zang et al, 1993) is calculated. Similar to EDP function, Harmonic Admittance relates the electric charge on the electrodes of the infinite periodic grating Q to the applied harmonically varying voltage V_e,

$$Q = (j\omega)^{-1} Y(f,s) \cdot V_e \tag{3}$$

and depends on frequency f and the normalized wave number, $s = p / \lambda$, where p is a pitch of the grating. This function can be used for simulation of a SAW resonator and calculation of its main parameters: resonant and anti-resonant frequencies, reflection coefficient etc. Also it comprises the information about SAW and other acoustic modes, which can be generated in the analyzed layered structure, and their characteristics can be extracted from $Y(f,s)$.

It should be mentioned that a numerical procedure of finding eigen modes of the fundamental acoustic tensor can be successfully applied to the air as an example of isotropic medium and the calculated SAW characteristic do not differ noticeably from that obtained with stress-free conditions set analytically at the film/air interface. The method and software SDA-FEM-SDA enable analysis of electrodes composed of few different metal layers and having a complicated profile, with different edge angles in metal layers. The gaps between electrodes may be empty or filled with a dielectric material. Due to these options, some important physical effects can be simulated, such as the effect of a sublayer (e.g. titanium) often used for better adherence of electrode metal to the substrate or the effect of nonrectangular electrode profile on a SAW device performance.

The developed numerical technique can be applied to different types of multilayered structures and different acoustic waves can be investigated, for example,

a. SAW and LSAW in a piezoelectric substrate;
b. SAW and LSAW in a substrate with one or few thin films (e.g. Love modes);
c. plate modes generated by IDT and propagating in a thin plate (e.g. Lamb waves);
d. boundary waves propagating along the interface between two half-infinite media;
e. acoustic waves propagating in a thin piezoelectric plate bonded to a thick wafer.

In addition, a continuous transformation between different types of acoustic waves can be observed. It gives a physical insight into the mechanisms of wave transformation with increasing film thickness. An example of wave transformation will be considered in Section 5.

4. Multilayered structures: Examples of analysis

In this section, few examples of application of the developed numerical technique to the structures of practical importance are presented.

4.1 SiO_2/42°YX LT with Al film at the interface

The first example is a dielectric film on a piezoelectric substrate, which can be referred to the *Type 2* structure shown in Fig.1. The calculated characteristics of LSAWs propagating in SiO_2/42°YX LT with uniform Al film atop of the structure are presented in Fig. 4.

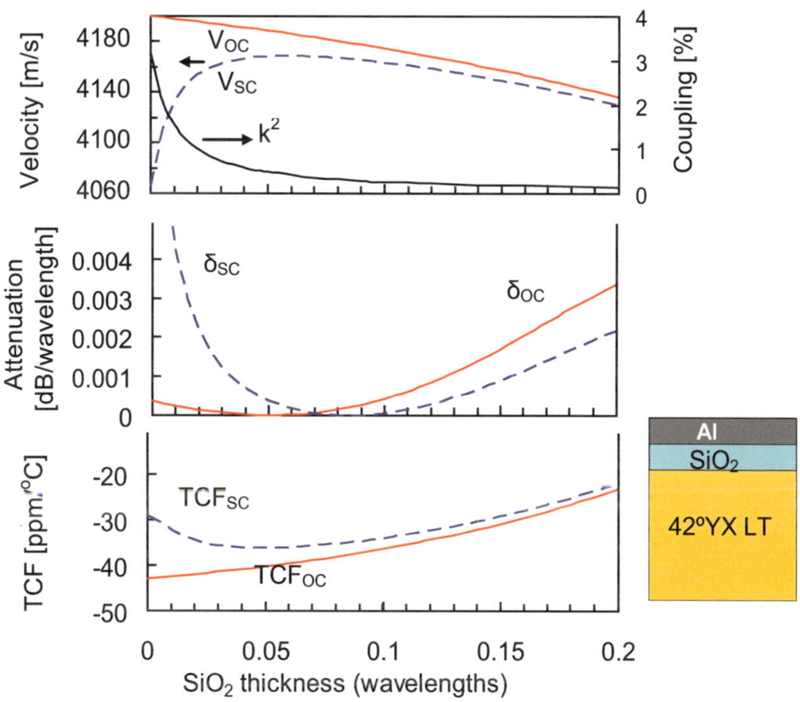

Fig. 4. Characteristics of leaky SAW propagating in SiO_2/42°YX LT with uniform Al film (h_{Al}=5%λ) atop of SiO_2 film, as functions of normalized SiO_2 film thickness. OC or SC electrical conditions are analyzed

SiO_2 is an isotropic dielectric film. LSAW velocities V, attenuation coefficients δ (in dB/λ, where λ is LSAW wavelength) and TCF are presented as functions of the normalized SiO_2 film thickness, h/λ. These characteristics have been calculated for the open-circuited (OC) and short-circuited (SC) electrical conditions in Al film. The finite thickness of a metal film (h_{Al}=5%λ) was taken into account. The difference between the OC and SC velocities determines the electromechanical coupling coefficient k^2, which decreases rapidly with increasing dielectric film thickness. The behavior of attenuation coefficients depends on the electrical condition. The functions $\delta_{OC}(h_{SiO2})$ and $\delta_{SC}(h_{SiO2})$ reach nearly zero values at h_{SiO2}=5%λ and h_{SiO2}=8%λ, respectively. Therefore, the variation of SiO_2 film thickness can be used for minimization of propagation losses in a SAW device. Due to the opposite signs of TCF in SiO_2 film and LT substrate, in the layered structure the absolute value of TCF reduces with increasing film thickness but does not reach zero in the investigated interval of film thicknesses. However, larger SiO_2 thicknesses are not considered because the electromechanical coupling coefficient becomes too low for practical applications. Further improvement of TCF is possible if IDT is located at the SiO_2/LT interface (*Type 1* structure). Such example will be considered in Section 5.

4.2 ZnO/sapphire: Existence of high-velocity SAW

The next example, ZnO film on a sapphire substrate, is a layered structure with a piezoelectric film on a non-piezoelectric substrate, which is potentially useful for high-frequency SAW device applications. Also this example demonstrates that a deposition of a thin film on a substrate can result in the existence of a high-velocity SAW, which cannot exist in a crystal without a thin film.

The typical SAW velocities in layered structures using a sapphire substrate are about 5500 m/s. Leaky SAWs, which have higher velocities, propagate with certain attenuation dependent on orientation of a substrate. Two types of LSAW can exist in crystals and layered structures: common or low-velocity LSAW, with velocities confined in the interval between that of the slow quasi-shear and fast quasi-shear *limiting* bulk acoustic waves (LBAWs), and high-velocity LSAW with velocities between that of the fast quasi-shear and quasi-longitudinal LBAWs. The *limiting* BAW is a bulk wave, which propagates in the sagittal plane (i.e. the plane, which is normal to the substrate surface and parallel to the propagation direction of SAW or LSAW) and is characterized by the group velocity parallel to the substrate surface. Usually a high velocity LSAW is not suitable for SAW device applications because of its fast attenuation. In some crystals with strong acoustic anisotropy, a high-velocity LSAW degenerates into the quasi-longitudinal LBAW in selected orientations, and low-attenuated LSAW can propagate around such orientations (Naumenko, 1996). For example, such waves exist in some orientations of quartz and LBO.

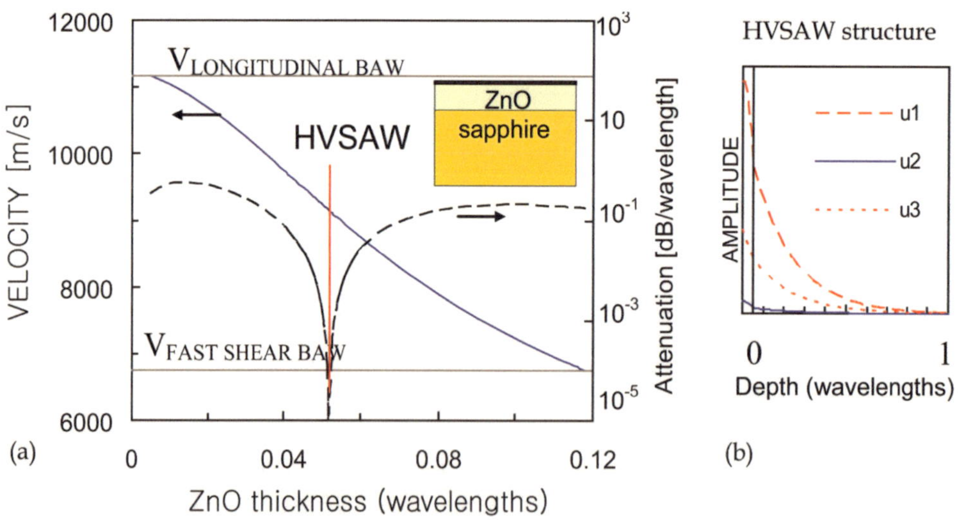

Fig. 5. (a) Velocity and attenuation coefficients of high-velocity leaky waves propagating in sapphire, Euler angles (0°, -20.3°, 0°), with ZnO film, as functions of normalized ZnO thickness, and (b) Displacements as function of depth, for HVSAW existing at h_{ZnO}=5.16%λ

Fig. 5, a shows the calculated velocities and attenuation coefficients of high-velocity LSAWs propagating in ZnO/sapphire structure, when the sapphire orientation is defined by the Euler angles (0°, -20.3°, 0°). The thickness of a metal film deposited atop of ZnO is not taken into account. The velocities of the fast quasi-shear and quasi-longitudinal LBAWs in sapphire are shown as $V_{FAST\ SHEAR\ BAW}$ and $V_{LONGITUDINAL\ BAW}$, respectively. The non-attenuated SAW solution, which occurs on a high-velocity LSAW branch at h_{ZnO}≈5.16%λ (Fig.5, a), is not a quasi-bulk wave described above. Fig.5,b shows the mechanical displacements, which follow the propagation of this wave. The analyzed solution is a sagittally polarized surface wave, which attenuates exponentially into the depth of sapphire, similar to Rayleigh SAW. Such *high velocity SAW* (HVSAW) can not exist without perturbation of a free crystal surface, e.g. by deposition of a thin film or a metal grating. The existence of this type of waves was revealed via numerical analysis of experimental data on SAW modes in ZnO/SiC structure (Didenko et al., 2000) and confirmed by other examples of layered structures, which support propagation of these waves, such as ZnO/diamond, Zno/sapphire etc. (Naumenko & Didenko, 1999).
The HVSAW found in ZnO/sapphire may be attractive for applications in high-frequency SAW devices because it combines a high propagation velocity exceeding 9000 m/s with electromechanical coupling about 0.3%. With deposition of a metal film or a periodic metal grating the wave with attractive properties does not disappear but a combination of cut angle and ZnO thickness should be optimized properly to provide low LSAW attenuation.

4.3 Al grating on 46°YX LT/Si bonded wafer

In this example, SAW modes are investigated in LT/Si structure with a periodic Al grating atop of LT, which can be obtained experimentally by bonding LT plate to a silicon wafer. In such structure, the TCF may be dramatically reduced compared to regular LT wafer.

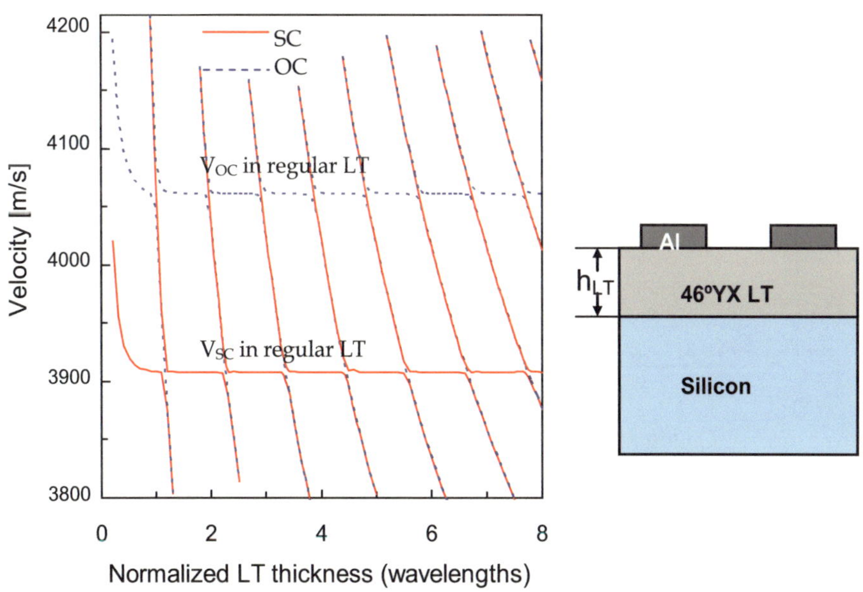

Fig. 6. Velocities of acoustic modes propagating in 46°YX LT plate bonded to silicon wafer, with OC and SC Al grating, as functions of normalized LT thickness, h_{Al}=9%λ

The results shown in Fig.6 were obtained with the software SDA-FEM-SDA because the effect of a periodic metal grating is different from that of a uniform metal film. When LT thickness is about 1-2 wavelengths, the velocity of the SAW mode propagating in the layered structure is nearly the same as in a regular LT substrate with electrode thickness h_{Al}=9%λ but the wave characteristics are perturbed by interactions with multiple plate modes, the number of which grows with increasing LT thickness. It should be noted that in a regular 46°YX LT substrate the acoustic wave propagating with nearly the same velocity has a leaky wave nature. The bonding of LT plate to a silicon wafer results in the transformation of LSAW into SAW. The leakage of the wave becomes impossible because in silicon the shear BAW propagates faster than the analyzed SAW mode.

4.4 Al grating on 46°YX LT/SiO_2/Si bonded wafer

The next example differs from the previous one by the additional SiO_2 film between LT and Si wafer. A silicon dioxide layer is required to be deposited on the LT wafer to enable a stronger bond to silicon (Abbott et al., 2005). The presence of additional SiO_2 film impacts the acoustic and electrical properties of a bonded wafer and SAW resonators built on its surface. The spectrum of acoustic modes propagating in LT/SiO_2/Si bonded wafer looks more complicated than in LT/Si structure and depends on SiO_2 and LT thicknesses. In Fig. 7, the velocities of acoustic modes are shown as functions of the normalized SiO_2 film thickness when LT thickness is fixed, h_{LT}=6λ.

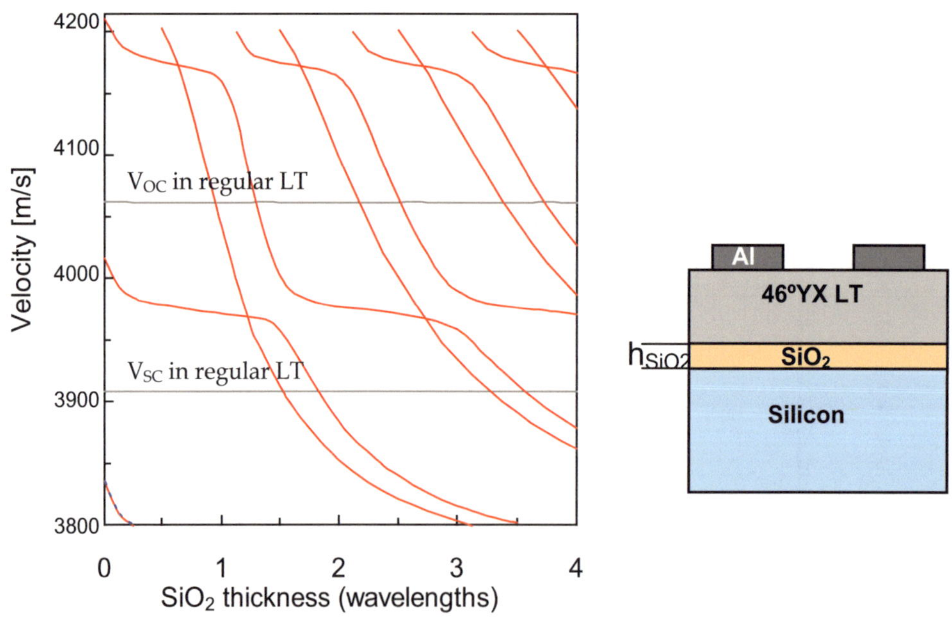

Fig. 7. Velocities of acoustic modes propagating in 46°YX LT/SiO_2/Si structure with Al grating, as functions of normalized thickness of SiO_2 film, when h_{LT}=6λ and h_{Al}=9%λ

4.5 Cu grating buried in SiO_2, on 46°YX LT/SiO_2/Si bonded wafer

This numerical example refers to the same bonded structure as described above but demonstrates the effect of additional SiO_2 overlay deposited over periodic metal grating electrodes. Such layer is aimed at further improvement of the temperature characteristics of SAW devices.

In this case, Cu electrodes are investigated because this metal provides higher reflection coefficients in SAW resonators and hence lower insertion loss in RF SAW filters than Al, when electrodes are buried in SiO_2 overlay. The velocities of acoustic modes are shown in Fig.8 as functions of the normalized intermediate SiO_2 film thickness, while the thicknesses of SiO_2 overlay, LT plate and Cu electrodes are fixed, h_{OVL}=0.25%λ, h_{LT}=4λ and h_{Cu}=2.5%λ, respectively. The dispersion of SAW velocities in 46°LT/SiO_2/Si (Fig. 8,a) and SiO_2/46°LT/SiO_2/Si (Fig. 8,b) structures demonstrates that in practice very accurate simulation is required to account for all spurious modes, because these modes may affect the admittance of a SAW resonator.

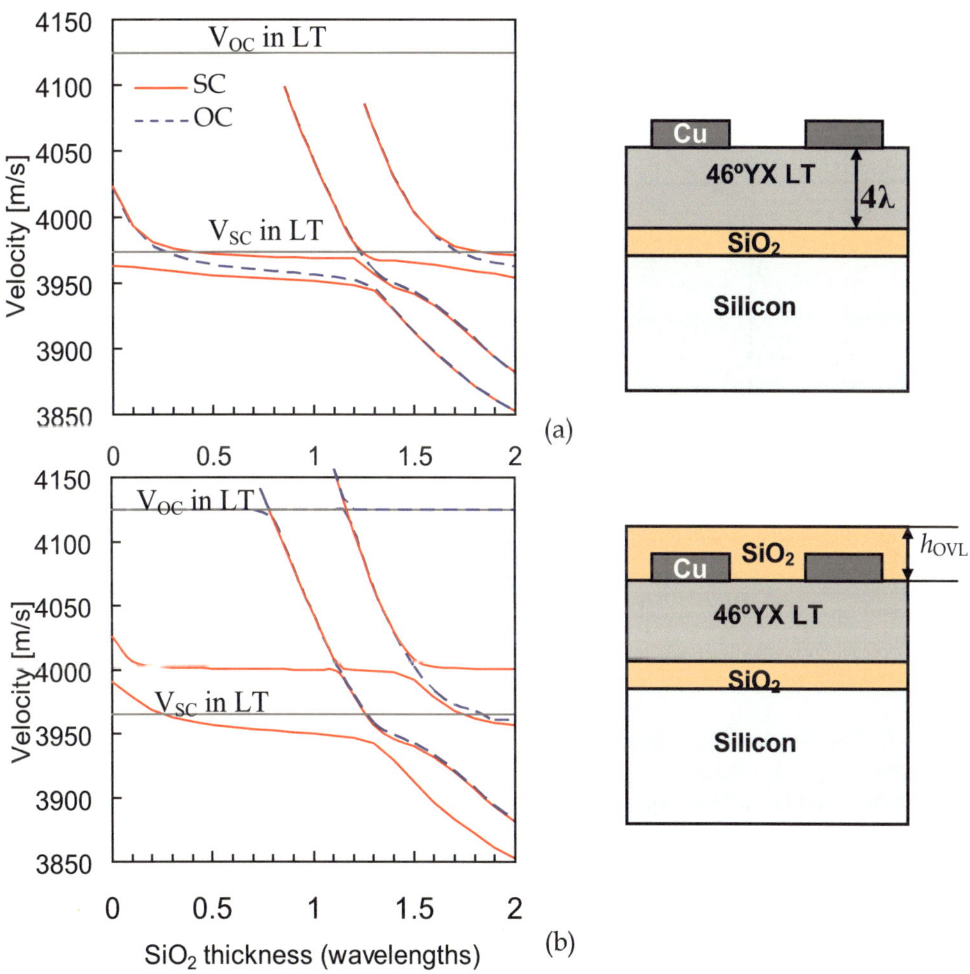

Fig. 8. Velocities of acoustic modes as functions of normalized thickness of intermediate SiO_2 film, when h_{LT}=4λ and h_{Cu}=2.5%λ, (a) in 46°LT/SiO_2/Si structure with Cu grating, and (b) in SiO_2/46°LT/SiO_2/Si structure with Cu grating atop of 46°LT plate and overlay thickness h_{OVL}=0.25%λ

Fig. 9 demonstrates an example of calculated admittance of a periodic Cu grating used with different multilayered structures described above. In addition to the main SAW mode, which exhibits resonance and anti-resonance, the multiple spurious modes propagate in the analyzed layered structures and can disturb a SAW resonator performance. The frequencies of the spurious responses are very sensitive to the thicknesses of the layers in a multilayered structure. However, the accurate simulation of these modes enables optimization of the layered structure to minimize the effect of the spurious modes on a resonator performance.

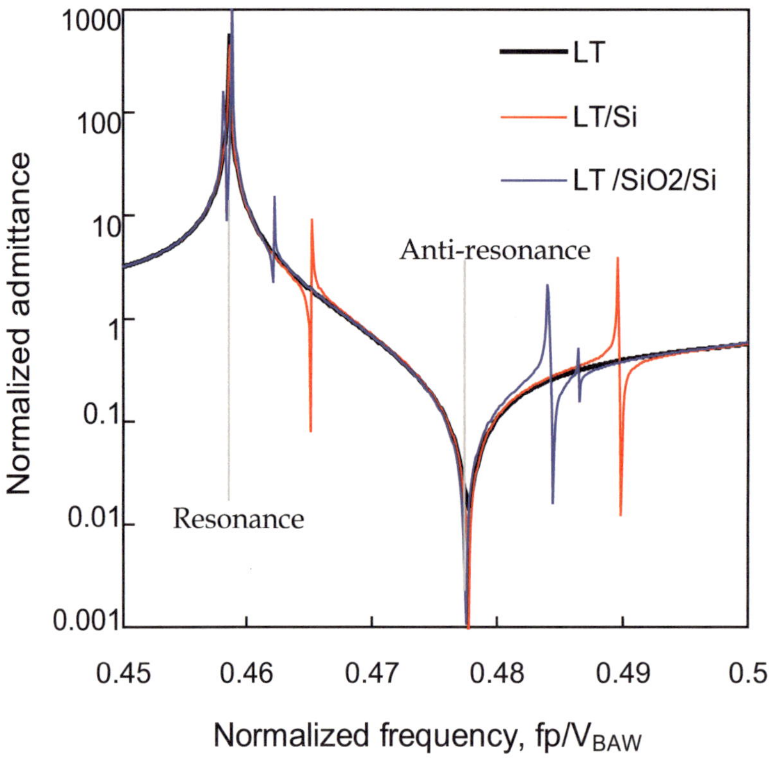

Fig. 9. Admittance of infinite periodic grating with Al electrodes (h_{Al}=10%λ), as function of normalized frequency, when the grating is built on regular 42°YX LT substrate, on LT plate (h_{LT}=5.5λ) bonded to Si substrate and atop of LT/SiO_2/Si structure, with h_{SiO2}=2λ. V_{BAW}=4214.636 m/s

The effect of spurious modes on a resonator performance can be minimized by the variation of rotation angle of LT plate. If one of rotated YX cuts of LN is used as a piezoelectric plate, the insertion loss of a resonator SAW device can be reduced in a wider bandwidth.

The examples presented in this section illustrate possible applications of the developed numerical technique and demonstrate that being a part of design tools for SAW device simulation it can be also an efficient tool for analysis of acoustic modes in multilayered structures.

5. Transformation of acoustic waves in anisotropic layered structures

In isotropic or highly symmetric materials, acoustic waves are characterized by mechanical displacements either belonging to the sagittal plane, $u=(u_1, 0, u_3)$, or normal to this plane, $u=(0, u_2, 0)$, i.e. such waves are Rayleigh-type or SH polarized modes. Similar solutions occur in some symmetric orientations of materials belonging to the lower symmetry classes. Such solutions have been extensively studied analytically. To the best author's knowledge, the most comprehensive overview of different types of acoustic waves existing in substrates with thin films, in thin plates and at the boundary between two half-infinite media was made by Viktorov (Viktorov, 1967, 1981). Some statements, which refer to acoustic waves propagating in isotropic structures, are listed below.

a. In isotropic substrate with isotropic thin film, SH-polarized Love waves can propagate if the shear BAW propagates faster in a substrate than in a film.
b. Along the boundary between two rigidly connected isotropic half-infinite media, the sagittally polarized Stonely waves can propagate. These waves are usually trapped near the interface between two media, with penetration depth about one wavelength.
c. SH-polarized boundary waves can exist if additional thin film with lower shear BAW velocity is added between two half-infinite media.
d. In isotropic thin plates, two types of waves can exist: sagittaly polarized Lamb waves (symmetric and anti-symmetric modes) and SH-polarized plate modes. With increasing plate thickness, higher-order modes appear and their number increases.
e. With plate thickness decreasing to zero, the first-order symmetric Lamb mode degenerates into the longitudinal BAW. With increasing plate thickness, this mode finally degenerates into two Rayleigh SAWs propagating along the boundaries of the plate.
f. Higher-order plate modes arise from the shear and longitudinal BAWs at certain cut-off thicknesses and have a structure of standing waves propagating between two interfaces.

The layered structures used in SAW devices must include at least one anisotropic material to provide a piezoelectric coupling of SAW with IDT. Anisotropy results in mixed polarizations of acoustic modes propagating in thin films, plates and along the interface between two media. The transformation of each mode with increasing film thickness is unique and requires separate investigation. Whereas analytical study of such waves is possible only in some symmetric orientations, the numerical technique presented in this chapter enables calculation of the wave characteristics and analysis of displacements associated with different acoustic modes. Its application to multilayered structures can reveal the mechanisms of wave transformation. The understanding of these mechanisms helps to select properly the thicknesses of metal and dielectric or piezoelectric layers to ensure the propagation of a required acoustic wave.

An example of such investigation is presented here. It refers to 42°YX LT with SiO_2 film. Similar structure was considered in Section 4.1, but in the present example a periodic grating is analyzed instead of a thin metal film and this grating is located at the interface between LT substrate and SiO_2 film. As a metal of the grating, copper is considered. Such structure is of great practical importance as potential material for RF SAW devices with improved temperature characteristics.

The calculated velocities of acoustic modes propagating in LT substrate with copper grating are shown in Fig.10,a as functions of the Cu electrode thickness. Fig.10,b shows the velocities of acoustic waves propagating in SiO_2/Cu grating/LT structure, as functions of the normalized SiO_2 thickness, with Cu thickness fixed, h_{Cu}=0.2λ. The analyzed Cu thicknesses look too high for application in SAW resonators but provide better insight into the wave transformation mechanisms, which is a purpose of this investigation.

When the metal thickness is small, two acoustic waves exist in LT, SAW1 and LSAW. With increasing Cu thickness, the velocities of both modes decrease rapidly and at h_{Cu}=0.075λ LSAW transforms into the second SAW mode, SAW2. With further increasing of electrode thickness, two SAW modes interact with each other. To avoid discontinuities in the characteristics of two SAW modes, these modes are distinguished by their velocities: $V_{SAW2}>V_{SAW1}$.

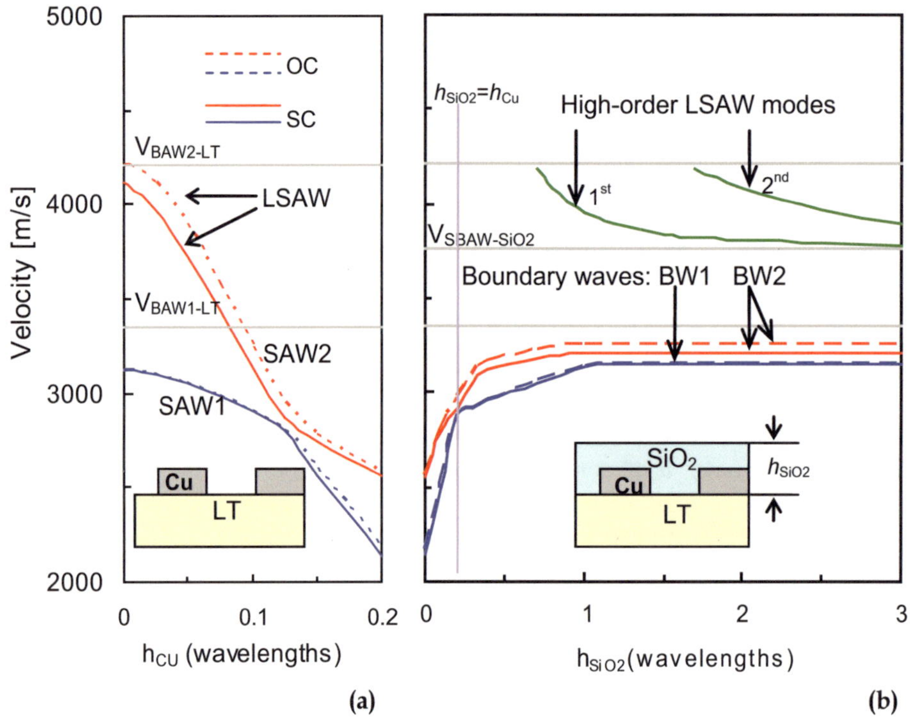

Fig. 10. SAW and leaky SAW velocities in 42°YX LT, (a) with Cu film, as functions of film thickness; (b) with Cu and SiO_2 films, as functions of normalized SiO_2 film thickness

If metal thickness is fixed (h_{Cu}=0.2λ) and the gaps between electrodes are filled with SiO_2, the velocities of two SAW modes grow rapidly (Fig.10,b). Another interaction between SAW1 and SAW2 occurs at $h_{SiO2}\approx h_{Cu}$, i.e. when the top surface of the whole structure becomes flat. With increasing SiO_2 film thickness two SAW modes finally transform into the boundary waves, BW1 and BW2. The boundary waves propagate with velocities lower than that of the shear BAW in SiO_2. The wave BW2 shows electromechanical coupling sufficient for application in resonator SAW devices, k^2=3.49%. For this mode, TCF grows from -31 ppm/°C in LT substrate up to 10 ppm/°C in a layered structure with SC grating and from -43 ppm/°C up to 6 ppm/°C with OC grating. At h_{SiO2}>0.7λ, higher-order plate modes arise from the fast shear limiting BAW in LT.

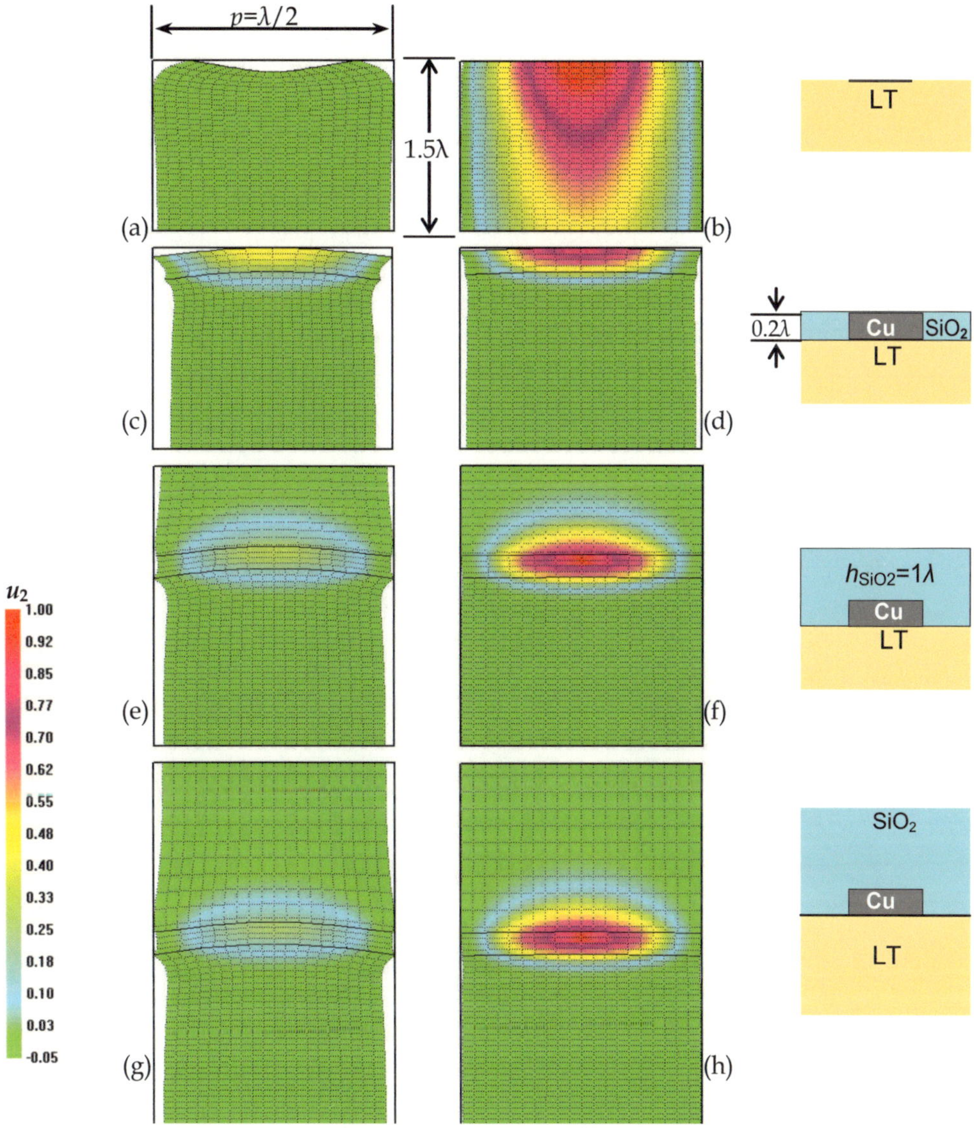

Fig. 11. Transformation of two SAW modes in 42°YX LT with Cu grating with increasing SiO_2 film thickness; (a), (c), (e), (g) SAW1, (b), (d), (f), (h) SAW2

Fig.11 demonstrates how the acoustic fields associated with the modes SAW1 and SAW2 change with increasing SiO_2 film thickness. OC electrical condition is considered, by way of example. The components of mechanical displacements in the sagittal plane are revealed as perturbations of the regular mesh and the values of SH components are presented as colored diagrams. When Cu thickness is small (Fig.11,a,b), SAW1 is nearly perfect Rayleigh wave and LSAW is a quasi-bulk SH-type wave slowly attenuating with depth. When $h_{Cu}>0.075\lambda$, LSAW transforms into SAW2, which is also SH-type wave. At Cu thicknesses about 0.12λ, both SAW modes are perturbed by interaction between them. In the interval of

Cu thicknesses between 0.12λ and 0.2λ, the modes SAW1 and SAW2, which have been determined as the lower-velocity and higher velocity modes, exchange their polarizations. After the second interaction, which occurs at $h_{SiO2}=h_{Cu}=0.2\lambda$, SAW1 and SAW2 turn back into Rayleigh-type and SH-type waves, respectively. However, at $h_{Cu}=0.2\lambda$ (Fig.11,c,d) both waves still have mixed polarizations. With increasing SiO_2 film thickness, SAW1 and SAW2 transform into the boundary waves BW1 and BW2, respectively (Fig. 11, g, h), with acoustic waves localized in Cu grating and around it. The boundary waves have mixed polarizations, which would be impossible in isotropic substrate with isotropic thin film, but due to specific features of the analyzed LT orientation, BW1 is nearly sagittally polarized wave and BW2 is nearly pure SH wave. BW2 penetrates deeper into SiO_2 film than into LT substrate. A numerical analysis reveals that SiO_2 thickness about 1.5λ is sufficient for transformation of SAW into the boundary wave.

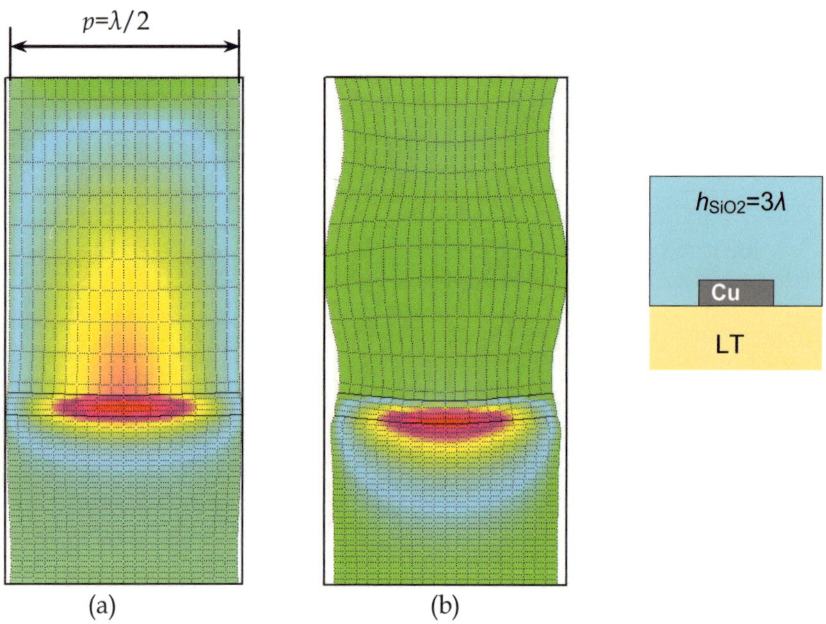

Fig. 12. Acoustic fields associated with two higher-order modes propagating in 42°YX LT with Cu grating and SiO_2 film when $h_{SiO2}=3\lambda$. (a) 1st mode; (b) 2nd mode

The acoustic fields associated with propagation of the two higher-order modes (Fig.10,b) have been also investigated. These modes have leaky wave nature. Fig.12 illustrates the structure of these modes at $h_{SiO2}=3\lambda$. The first mode, which exists when $h_{SiO2}>0.7\lambda$, has SH polarization deeply penetrating into SiO_2 (Fig.12,a). With increasing SiO_2 thickness, this mode degenerates into the SH BAW propagating in SiO_2. The second mode, which exists at $h_{SiO2}>1.6\lambda$, looks as a combination of SH-type SAW in LT substrate with Cu grating and sagittally polarized quasi-bulk wave propagating in SiO_2 (Fig.12,b), with the amplitude of SH polarization component much higher than the amplitudes of two other components. This example demonstrates the effect of anisotropy on the propagation of acoustic waves in multilayered structures and reveals another application of the numerical technique described in this chapter. Similar investigation can be performed for other structures of practical importance or serve to study the wave processes in multilayered structures.

6. Conclusion

In this chapter, some layered and multilayered structures, which look promising as substrates for modern SAW devices developed for applications in cellular phones, communication and navigation systems have been overviewed. A universal numerical technique, which enables fast and accurate analysis of these and other structures have been presented and, by way of example, applied to some multilayered structures of practical interest. The physical insight into the mechanisms of SAW transformation with increasing film thickness in a multilayered structure was provided via simulation of acoustic fields in one of the structures.

7. References

Abbott B.P., Naumenko N.F. & Caron J. (2005). Characterization of bonded wafer for RF filters with reduced TCF, *Proceedings of IEEE Ultrasonics Symposium*, pp.926-929, Rotterdam, the Netherlands, Sept. 2005.

Adler E. L. (1990). Matrix methods applied to acoustic waves in multi-layers, *IEEE Trans. Ultrason., Ferroelectr., Freq. Control*, Vol. 37, No 6, pp. 485-490.

Ballandras S., Reinhardt A., Laude L., Soufyane A., Camou S. & Ventura P. (2004). Simulations of surface acoustic wave devices built on stratified media using mixed finite element/boundary integral formulation, *J. Appl. Phys.*, Vol. 96, No 12, pp.7731-7741.

Benetti M., Cannatta D., Pietrantonio F. D. & Verona E. (2005). Growth of AlN piezoelectric film on diamond for high-frequency surface acoustic wave devices, *IEEE Trans. Ultrason., Ferroelectr., Freq. Control*, Vol. 52 , pp. 1806–1811.

Benetti M., Cannata D., Di Pietrantonio F., Verona E., Almaviva S., Prestopino G., Verona C. & Verona-Rinati G. (2008). Surface acoustic wave devices on AlN/single-crystal diamond for high frequency and high performances operation, *Proceedings of IEEE Ultrasonics Symposium*, pp. 1924-1927, Beijing, China, Nov. 2008.

Blotekjear K., Ingebrigsten K. & Skeie H. (1973). A method for analyzing waves in structures consisting of metal strips on dispersive media, *IEEE Trans. Electronic Devices*, vol. 20, no. 12, pp. 1133–1138.

Brizoual L. L., Sarry F. , Elmazria O. , Alnot P. , Ballandras S. & Pastureaud T. (2008). GHz frequency ZnO/Si SAW device, *IEEE Trans. Ultrason., Ferroelectr., Freq. Control*, Vol. 55, No 2, pp. 442–450.

Didenko I. S., Hickernell F. S. & Naumenko N. F. (2000). The theoretical and experimental characterization of the SAW propagation properties for zinc oxide films on silicon carbide, *IEEE Trans. Ultrason., Ferroelectr., Freq. Control*, Vol. 47, No. 1, pp.179-187.

Eda K., Onishi K., Sato H., Taguchi Y. & Tomita M. (2000). Direct bonding of piezoelectric materials and its applications, *Proceedings of IEEE Ultrasonics Symposium* , pp. 299-309, San Juan, Puerto Rico, Sep. 2000.

Emanetoglu N.W., Patounakis G., Muthukumar S. & Lu Y. (2000). Analysis of temperature compensated modes in $ZnO/SiO_2/Si$ multilayer structures, *Proceedings of IEEE Ultrasonics Symposium* , pp. 325-328, San Juan, Puerto Rico, Sep. 2000.

Endoh G., Hashimoto K. & Yamagouchi M. (1995). SAW propagation characteristics by Finite Element Method and Spectral Domain Analysis, *Jpn. J. Appl. Phys.*, Vol. 34, No 5B, pp. 2638-2641.

Fujii S., Kawano S., Umeda T., Fujioka M. & Yoda M. (2008). Development of a 6GHz resonator by using an AlN diamond structure, *Proceedings of IEEE Ultrasonics Symposium*, pp. 1916-1919, Beijing, China, Nov. 2008.

Hashimoto, K. Y. (2000). *Surface Acoustic Wave Devices in Telecommunications: Modeling and Simulation*, Springer, ISBN 354067232X, USA.

Hashimoto K., Omori T. & Yamaguchi M. (2007). Extended FEM/SDA Software for Characterizing Surface Acoustic Wave Propagation in Multi-layered Structures, *Proceedings of IEEE Ultrasonics Symposium*, pp. 711-714, Rome, Italy, Nov. 2008.

Hashimoto K., Wang Y., Omori T., Yamaguchi M., Kadota M., Kando H. & Shibahara T. (2008). Piezoelectric Boundary Waves: Their Underlying Physics and Applications, *Proceedings of IEEE Ultrasonics Symposium*, pp. 999-1005, New York, USA, Oct. 2007.

Ingebrigtsen K.A. (1969). Surface waves in piezoelectrics, *J. Appl. Phys.*, Vol. 40, No 7, pp.2681-2686.

Kadota M. & Minakata M. (1998). Piezoelectric Properties of ZnO Films on a Sapphire Substrate Deposited by an RF-Magnetron-Mode ECR Sputtering System, *Jpn. J. Appl. Phys.*, Vol.37, pp.2923-2926.

Kadota M. (2007). High Performance and Miniature Surface Acoustic Wave Devices with Excellent Temperature Stability Using High Density Metal Electrodes, *Proceedings of IEEE Ultrasonics Symposium*, pp.496-506, New York, USA, Oct. 2007.

Kadota M., Nakao T., Murata T. & Matsuda K. (2008). Surface Acoustic Wave Filter in High Frequency Range with Narrow Bandwidth and Excellent Temperature Property, *Proceedings of IEEE Ultrasonics Symposium*, pp. 1584-1587, Beijing, China, Nov. 2008.

Kovacs G., Ruile W., Jakob M., Rosler U., Maier E., Knauer U. & Zottl H. (2004). A SAW Duplexer with Superior Temperature Characteristics for US-PCS, *Proceedings of IEEE Ultrasonics Symposium*, pp. 974-977, Montreal, Canada, Aug. 2004.

Milsom R. F., Reilly N.C.H. & Redwood M. (1977). Analysis of generation and detection of surface and bulk acoustic waves by interdigital transducers, *IEEE Trans. Sonics and Ultrasonics*, Vol. 24, pp. 147-166.

Nakahata H., Higaki K., Fujii S. , Hachigo A., Kitabayashi H. , Tanabe K., Seki Y. & Shikata S. (1995). SAW Devices on Diamond, *Proceedings of IEEE Ultrasonics Symposium*, pp.361-370, Seattle, USA, Nov. 1995.

Nakahata H., Hachigo A., Itakura K., Fujii S. & Shikata S. (2000). SAW Resonators on SiO_2/ZnO/Diamond Structure in GHz Range, *Proceedings of IEEE Freq. Contr. Symposium*, pp. 315-320, Kansas City, USA, June 2000.

Nakai Y., Nakao T. , Nishiyama K. & Kadota M. (2008). Surface Acoustic Wave Duplexer composed of SiO_2 film with Convex and Concave on Cu-electrodes/LiNbO3 Structure, *Proceedings of IEEE Ultrasonics Symposium*, pp.1580-1583, Beijing, China, Nov. 2008.

Nakanishi H., Nakamura H. , Hamaoka Y., Kamiguchi H. & Iwasaki Y. (2008). Small-sized SAW Duplexers with Wide Duplex Gap on a SiO_2/Al/$LiNbO_3$ structure by using Novel Rayleigh-mode Spurious Suppression Technique, *Proceedings of IEEE Ultrasonics Symposium*, pp. 1588-1591, Beijing, China, Nov. 2008.

Naumenko N. F. (1996). Application of exceptional wave theory to materials used in surface acoustic wave devices", *J. Appl. Phys.*, Vol.79, pp. 8936-8943.

Naumenko N.F. & Didenko I.S. (1999). High-velocity surface acoustic waves in diamond and sapphire with zinc oxide film, *Appl. Phys. Lett.*, Vol. 75, No 19, pp. 3029-3031.

Naumenko N.F. & Abbott B.P. (2003). Surface Acoustic Wave Devices Using Optimized Cuts of a Piezoelectric substrate, *US Patent #6,556,104 B2.* Int.Cl. H03H 9/64. Apr.29, 2003

Naumenko N.F. & Abbott B.P. (2004). Surface Acoustic Wave Filter , *US Patent 6,833,774 B2.* Int.Cl. H03H 9/16. Dec. 21, 2004.

Naumenko N.F. (2009). Transformation of Surface Acoustic Waves into Boundary Waves in Piezoelectric/Metal/Dielectric Structures, *Proceedings of IEEE Ultrasonics Symposium*, pp.2635-2638, Rome, Italy, Sep. 2009.

Naumenko N.F. (2010). A Universal Technique for Analysis of Acoustic Waves in Periodic Grating Sandwiched Between Multi-Layered Structures and Its Application to Different Types of Waves, *Proceedings of IEEE Ultrasonics Symposium*, San Diego, USA, Oct. 2010.

Naumenko N.F. & Abbott B.P. (2010). Insight into the Wave Transformation Mechanisms via Numerical Simulation of Acoustic Fields in Piezoelectric/Metal Grating/ Dielectric Structures, *Proceedings of IEEE Ultrasonics Symposium*, San Diego, USA, Oct. 2010.

Omori T., Kobayashi A., Takagi Y., Hashimoto K. & Yamaguchi M. (2008). Fabrication of SHF range SAW devices on AlN/Diamond-substrate, *Proceedings of IEEE Ultrasonics Symposium*, pp. 196-200, Beijing, China, Nov. 2008.

Reinhardt A., Pastureaud T., Ballandras S. & Laude L. (2003). Scattering matrix method for modeling acoustic waves in piezoelectric, fluid, and metallic multilayers, *J. Appl. Phys.*, Vol. 94, No 10, pp.6923-6931.

Ruppel C.C.W. & Fjeldy T. A. Eds. (2001). *Advances in Surface Acoustic Wave Technology, Systems and Applications. Volume 2* . ISBN 981-02-4538-6. World Scientific Publishing Co. Pte Ltd, Singapore.

Ruppel, C. C. W., Reindl, L. & Weigel, R. (2002). SAW devices and their wireless communication applications, *IEEE Microwave Magazine, ISSN 1527-3342* 3(2): pp. 65–71.

Stroh A. N. (1962). Steady state problems in anisotropic elasticity, *J. Math. Phys.* Vol. 41., pp. 77-103.

Tan E. L. (2002). A robust formulation of SAW Green's functions for arbitrarily thick multilayeres at high frequencies, *IEEE Trans.Ultrason., Ferroelectr., Freq. Control,* Vol. 49, No 7, pp. 929-936.

Tsutsumi J. , Inoue S., Iwamoto Y., Miura M., Matsuda T., Satoh Y., Nishizawa T., Ueda M. & Ikata O. (2004). A Miniaturized 3x3-mm SAW Antenna Duplexer for the US-PCS band with Temperature-Compensated $LiTaO_3$/Sapphire Substrate, *Proceedings of IEEE Ultrasonics Symposium*, pp. 954-958, Montreal, Canada, Aug. 2004.

Viktorov, I.A. (1967). *Rayleigh and Lamb Waves: physical theory and applications*. Plenum Press, New York.

Viktorov I.A. (1981). *Surface sound waves in solids*. Moscow: Nauka (in Russian).

Zhang Y., Desbois J. & Boyer L. (1993). Characteristic parameters of surface acoustic waves in a periodic metal grating on a piezoelectric substrate, *IEEE Trans.Ultrason., Ferroelectr., Freq. Control*, vol. 40, pp. 183-192, May 1993.

20

SAW Parameters Analysis and Equivalent Circuit of SAW Device

Trang Hoang
Faculty of Electrical-Electronics Engineering, University of Technology, HoChiMinh City VietNam

1. Introduction

Surface Acoustic Wave (SAW) devices, using interdigital electrodes, play a key role in today's telecommunication systems and are widely used as electronic filters, resonators, delay lines, convolvers or wireless identification systems (ID tags).
During the last three decades, demands set by the expansion of the telecommunication industry and many applications in sensor have resulted in the introduction of a new generation of the SAW devices. Consequently, the design of high performance SAW devices requires precise and efficient models, simulation tools. Several methods have been proposed for modeling, analyzing SAW devices. These include the impulse model, the equivalent circuit models, the coupling-of-mode (COM) model, P-matrix model, angular spectrum of waves models [1] and the Scattering Matrix approach that was presented by Coldren and Rosenberg [2]. While the impulse model is only a first order model, the other models include second order effects, e.g. reflections, dispersion, and charge distribution effects. Purely numerical methods have also been and are being developed by many authors [3]-[35], [41].
In this chapter, the method for calculating the SAW parameters, including modeling and simulation, is given.
Section 2 gives the calculation of SAW properties and analyses of different SAW device structures.
Section 3 presents the equivalent circuit of SAW delay line based on Mason model.
The equivalent circuit of SAW delay line based on Couple-Of-Mode theory is presented in section 4.
Based on section 3 and 4, section 5 shows comparison between using the equivalent circuit of SAW delay line device based on Mason model and COM theory. This model is useful and fast model for designing the SAW device.

2. Calculation of SAW parameters

2.1 SAW parameters

The most important parameter for SAW device design is the center frequency, which is determined by the period of the IDT fingers and the acoustic velocity. The governing equation that determines the operation frequency is:

$$f_0 = v_{SAW} / \lambda \tag{1}$$

where
λ is the wavelength, determined by the periodicity of the IDT and v_{SAW} is the acoustic wave velocity . For the technology being used in this research:

$$\lambda = p = \text{finger width} \times 4 \tag{2}$$

with the finger width (as shown in Figure 1) is determined by the design rule of the technology which sets the minimum metal to metal distance.
v_{SAW} is surface acoustic wave velocity.

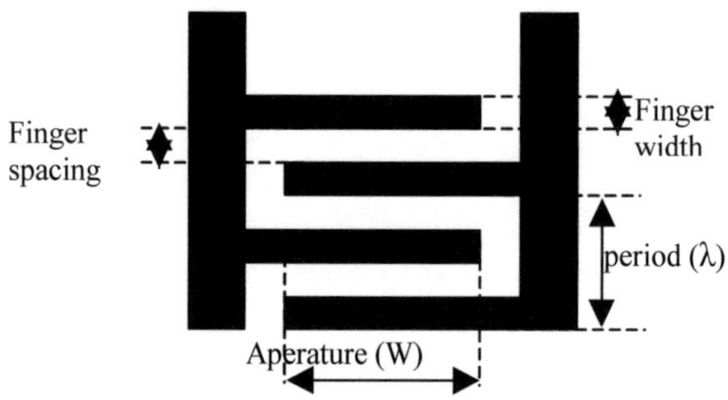

Fig. 1. IDT parameters

By using matrix method or Finite Element Method (FEM) in section 2.2, the velocity v of acoustic wave is derived in two cases:
- Wave velocity V_0 is velocity in case of free surface.
- Wave velocity Vs is velocity in case of short-circuit surface.

Therefore, the electromechanical coupling coefficient K is calculated approximately by Ingebrigtsen [54] as:

$$K^2 = 2\frac{V_o - V_s}{V_o} \tag{3}$$

By using the matrix method or FEM and approximation of coupling factor as in (3), the SAW parameters in different structures AlN/Si, AlN/SiO_2/Si and AlN/Mo/Si are calculated and analysed in three next sections.

2.2 Matrix method and Finite Element Method (FEM). The choice between them

Matrix method

The SAW propagation properties on one layer or multilayer structure are obtained by using matrix approach, proposed by J.J.Campbell and W.R.Jones [50], K.A.Ingebrigtsen [54], and then developed by Fahmy and Adler [31], [32], [33] and other authors [51], [52], [53]. The numerical solution method is based on characterizing each layer by means of a transfer matrix relating the mechanical and electrical field variables at the boundary planes. The boundary conditions for multilayer are based on the mechanical and electrical field variables those quantities that must be continuous at material interfaces. This matrix method is used to calculate the wave velocity and therefore, the electromechanical coupling factor. A general view and detail of this approach are given as follows and also presented in [50]-[53].

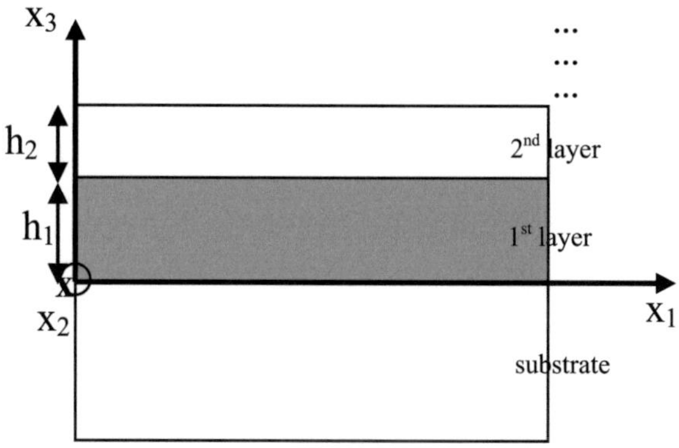

Fig. 2. Multilayer structure

Constitutive equations:

$$c_{ijkl}\frac{\partial^2 u_l}{\partial x_j \partial x_k} + e_{kij}\frac{\partial^2 \phi}{\partial x_j \partial x_k} = \rho \frac{\partial^2 u_i}{\partial t^2} \tag{4}$$

$$e_{jkl}\frac{\partial^2 u_l}{\partial x_j \partial x_k} - \varepsilon_{jk}\frac{\partial^2 \phi}{\partial x_j \partial x_k} = 0 \tag{5}$$

where

$c_{ijkl}, e_{ijk}, \varepsilon_{jk}, \rho$ are elastic tensor, piezoelectric tensor, dielectric tensor and mass density, respectively, of the considered material.

U is the particle displacement.

ϕ is the scalar electric potential.

The boundary conditions are shown in Table 1

Position	Mechanical conditions	Electrical conditions
$x_3=0$	$U_i^S = U_i^{1^{st}}$ $T_{3i}^S = T_{3i}^{1^{st}}$	Boundary is open $\phi^S = \phi^{1^{st}}$, $D^S = D^{1^{st}}$ Boundary is short $\phi^S = \phi^{1^{st}} = 0$
$x_3=h_1$	$U_i^{1^{st}} = U_i^{2^{nd}}$ $T_{3i}^{1^{st}} = T_{3i}^{2^{nd}}$	Boundary is open $\phi^{1^{st}} = \phi^{2^{nd}}$, $D^{1^{st}} = D^{2^{nd}}$ Boundary is short $\phi^{1^{st}} = \phi^{2^{nd}} = 0$
$x_3=h_1 + h_2$	$T_{3i}^{2^{nd}} = 0$	Boundary is open $D^{2^{nd}} = -\varepsilon.k.\phi^{2^{nd}}$ Boundary is short $\phi^{2^{nd}} = 0$

Table 1. Boundary conditions

where D: electronic displacement,

$$D_k = \frac{\partial \phi}{\partial x_k} \tag{6}$$

The general solution for U_l and ϕ (1) and (2) may be written as follows:

$$U_l = \sum_{m=1}^{n} C_m A_l^{(m)} \exp[ik(b^{(m)}x_3 + x_1 - vt)] \tag{7}$$

where l =1, 2, 3

$$\phi = \sum_{m=1}^{n} C_m A_4^{(m)} \exp[ik(b^{(m)}x_3 + x_1 - vt)] \tag{8}$$

The coefficients C_m are determined from boundary conditions.
By substituting (7) and (8) in every layer into the boundary conditions, we have general form

$$[H]\begin{bmatrix} C_1^S \\ \cdots \\ C_4^S \\ C_1^{1^{st}} \\ \cdots \\ C_8^{1^{st}} \\ C_1^{2^{nd}} \\ \cdots \\ C_8^{2^{nd}} \end{bmatrix} = 0 \tag{9}$$

Phase velocity is determined from the condition:

$$\text{Det(H)=0} \tag{10}$$

(use approximation to solve (10))
Figure 3 shows the wave velocity of structure AlN/SiO_2(1.3μm)/Si(4μm).

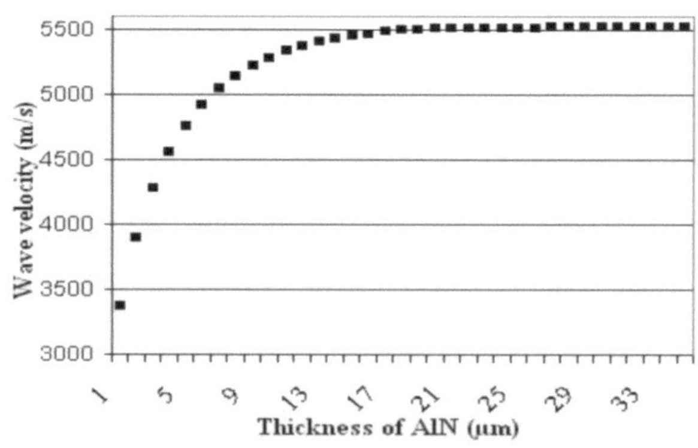

Fig. 3. Wave velocity in structure AlN/SiO_2(1.3μm)/Si(4μm) with different thicknesses of AlN

Finite Element Method (FEM)

In the design procedure of SAW devices, simple models like Equivalent Circuit Model coming from Smith Model and COM Model as presented above are used to achieve short calculation time and to get a general view of response of SAW devices. They are a good approach for designing SAW devices, for getting the frequency response, impedance parameters and transfer characteristics of SAW device. They could allow the designer to determine the major dimensions and parameters in number of fingers, finger width, and aperture. However, they are subjected to some simplifications and restrictions.

Field theory is the most appropriate theory for the design SAW devices as it involves the resolution of all the partial differential equations for a given excitation. The Finite Element Model (FEM) is the most appropriate numerical representation of field theory where the piezoelectric behaviour of the SAW devices can be discretized [45], [46]. Besides, nowadays, FEM tools also provide 3D view for SAW device, such as COMSOL® [47], Coventor® [48], ANSYS® [49].

The typical SAW devices can include a lot of electrodes (hundreds or even thousands of electrodes). In fact, we would like to include as many IDT finger pairs as possible in our FEM simulations. This would however significantly increase the scale of the device. Typically finite element models of SAW devices require a minimum of 20 mesh elements per wavelength to ensure proper convergence. A conventional two-port SAW devices consisting of interdigital transducers (IDT) may have - especially on substrate materials with low piezoelectric coupling constants - a length of thousands of wavelengths and an aperture of hundred wavelengths. Depending on the working frequency, the substrate which carries the electrode also has a depth of up to one hundred wavelengths. Taking into account that FEM requires a spatial discretization with at least twenty first order finite elements per wavelength and that an arbitrary piezoelectric material has at least four degrees of freedom, this leads to 8×10^8 unknowns in the three dimensional (3-D) case. Hence, the 3-D FEM representation of SAW device with hundreds of IDT fingers would require several million elements and nodes. The computational cost to simulate such a device is extremely high, or the amount of elements could not be handled by nowadays computer resources.

Fortunately, SAW devices consist of periodic section. M.Hofer et al proposed the Periodic Boundary Condition (PBC) in the FEM that allows the reduction of size of FE model tremendously [45], [46].

A good agreement between FEM and analytic method is obtained via the results in case of SAW with AlN thickness of 4µm, wavelength of 8µm presented in Table 2.

	Matrix method	FEM	Difference between Matrix method and FEM (%)
f_0 (MHz)	771.13	775.48	0.56
f_s (MHz)	770.26	774.57	0.56
v_0 (m/s)	6169.02	6203.87	0.56
v_s (m/s)	6162.07	6196.54	0.56
K (%)	4.74	4.86	2.4

Table 2. Comparison between matrix method and FEM

From this table, matrix method and FEM give the same results. However, FEM would takes a long time and require a trial and error to find the results. Consequently, to reduce time, the matrix method proposed to be used to extract the parameters of SAW devices; FEM is

used to get a 3D view and explain some results that can not be explained by equivalent circuit. This point will be presented in next sections.
The three next sections present and analyse SAW parameters in different structures AlN/Si, $AlN/SiO_2/Si$ and AlN/Mo/Si.

2.3 Wave velocity, coupling factor in AlN/Si structure

Figure 4 shows the dependence of Rayleigh wave velocity V_0 and Vs on the normalized thickness as respect to the wavelength, khAlN of AlN layer in SAW device AlN/Si substrate, where normalized thickness is defined by:

$$kh = \frac{2\pi h}{\lambda} \tag{11}$$

In this graph, when the normalized thickness of AlN, khAlN is larger than 3, the wave velocity reaches the velocity of the Rayleigh wave in AlN substrate $v_{(AlN\ substrate)}$=6169 (m/s). This could be explained that the wave travels principally in AlN layer when khAlN is larger than 3, because for low frequency the wave penetrates inside the other layer and this work is in the case where the wave are dispersive. It is better to be in the frequency range where the Rayleigh wave is obtained to have a constant velocity.

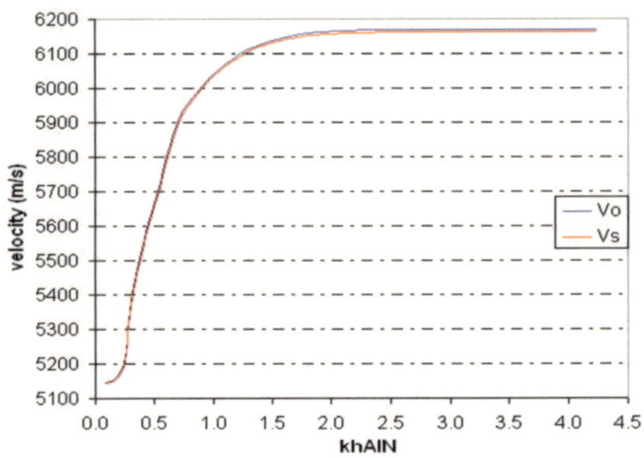

Fig. 4. Calculated values of wave velocity V_0 and Vs in SAW device AlN/Si substrate depend on the normalized thickness khAlN of AlN layer

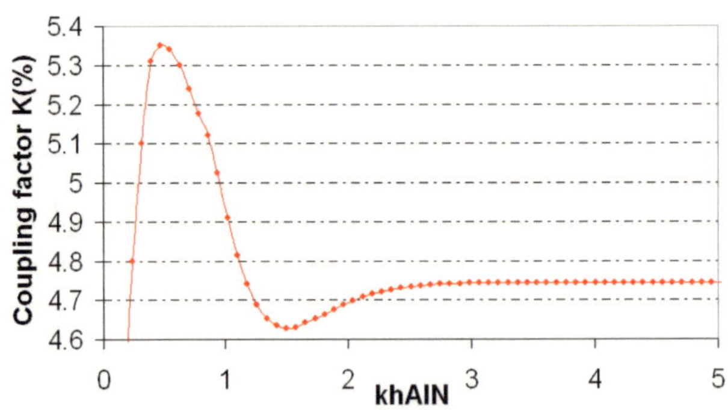

Fig. 5. Calculated values of coupling factor K(%) in SAW device AlN/Si substrate depends on the normalized thickness khAlN of AlN layer

The coupling factor K for this kind of device is shown in Figure 5. When normalized thickness of AlN layer is larger than 3, the coupling factor K still remain at 4.74% by that the wave travels principally in AlN layer.
In this configuration, K is at its maximum value of 5.34% when khAlN=0.55.

2.4 Wave velocity, coupling factor in AlN/SiO_2/Si structure

Wave velocity and coupling factor in structure AlN/SiO_2/Si are also presented in Figure 6 and Figure 7, respectively.

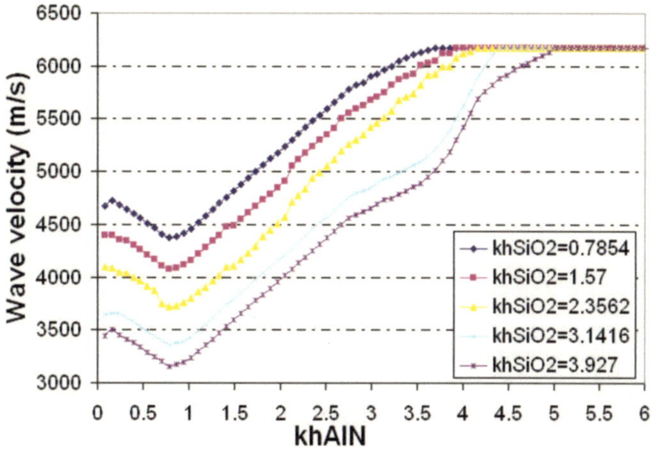

Fig. 6. Dependence of wave velocity in SAW device AlN/SiO_2/Si substrate on the normalized thickness khAlN of AlN layer and khSiO_2

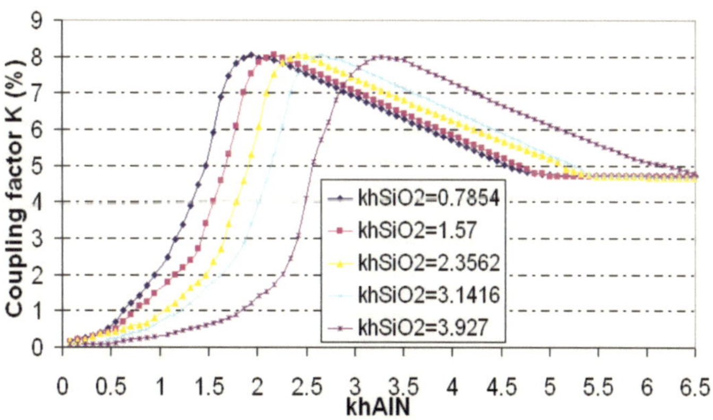

Fig. 7. Dependence of coupling factor K(%) in SAW device AlN/ SiO_2/Si substrate on the normalized thickness khAlN of AlN layer and khSiO_2

In this configuration, as results in Figure 6, when $khAlN < 6$, with the same thickness of AlN layer, an increase in thickness of SiO_2 would decrease the wave velocity. When $khAlN > 6$, the wave velocity reaches the velocity of the Rayleigh wave in AlN substrate v(AlN substrate)=6169 (m/s). A same conclusion is formulated also for coupling factor for this kind of structure, AlN/SiO_2/Si, in Figure 7; when $khAlN > 6$, K remains at the value of 4.7%.

To understand the above behavior, we use FEM method to display displacement profile along the depth of multilayer AlN/SiO_2/Si. These results obtained from FEM method in case of khSiO_2=0.7854, khAlN=5 and khAlN=0.2 are compared as in Figure 8.

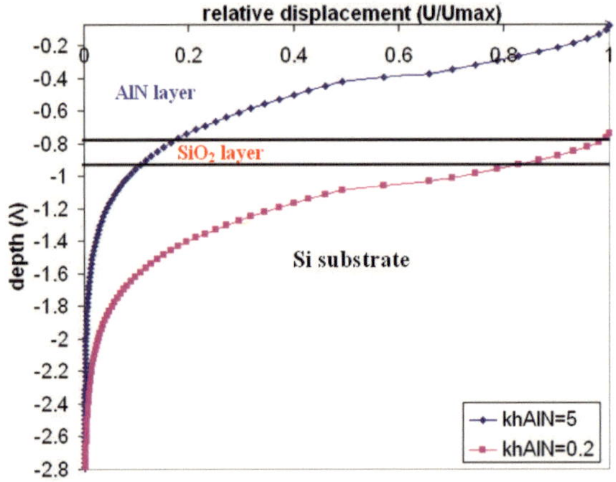

Fig. 8. Displacement profile along the depth of the multilayer AlN/SiO_2/Si, khSiO_2=0.7854

From Figure 8, we note that wave travels principally in AlN layer for a khAlN value of 5. By this reason, from a khAlN value of larger than 5, the coupling factor K remains at 4.7% and wave velocity remains at 6169m/s. For khAlN=0.2, wave travels principally in SiO_2 layer and Si substrate that are not piezoelectric layer. Consequently, the coupling factor K reaches the value of 0%.

In conclusion, the values of wave velocity and coupling factor depend on wave propagation medium, in which constant values of wave velocity and coupling factor indicate a large contribution of AlN layer, and coupling factor value of near 0% indicates a large contribution of SiO_2 layer and Si substrate.

2.5 Wave velocity, coupling factor in AlN/Mo/Si structure

For our devices, a thin Mo layer will be also deposited below the AlN layer to impose the crystal orientation of AlN. Besides this dependence, the Mo layer also has influences on

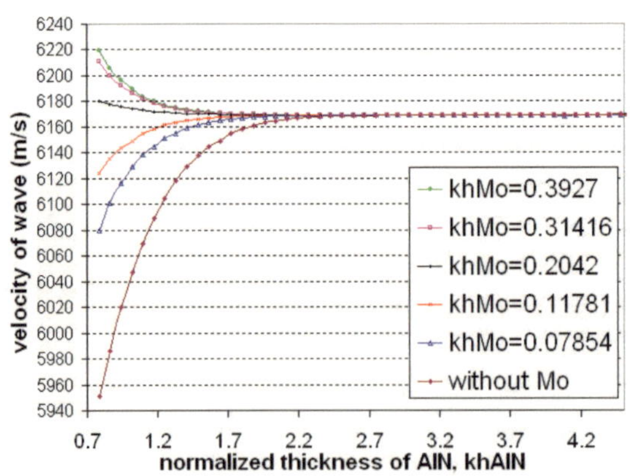

Fig. 9. Wave velocity AlN/Mo/Si substrate depends on the normalized thickness khAlN and khMo

wave velocity and coupling factor K. These influences are shown in Figure 9 and Figure 10, respectively.

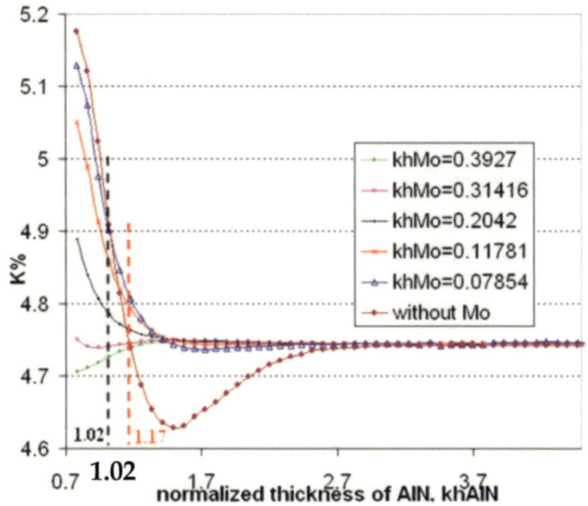

Fig. 10. Coupling factor K(%) in SAW device AlN/Mo/Si substrate depends on the normalized thickness khAlN and khMo

From Figure 9, the use of Mo layer would increase the wave velocity with any thickness of AlN layer and Mo layer. In case of coupling factor K as in Figure 10, the Mo layer, however, could decrease K when the khAlN is less than 1.02. When the normalized thickness of AlN layer khAlN is in the range from 1.17 to 2.7, the Mo layer would increase the coupling factor K. And when the khAlN is larger than 2.7, the Mo has no influence on wave velocity and coupling factor. The reason of this effect could be explained by the displacement profile in AlN/Mo/Si structure, as shown in Figure 11 for thickness AlN value of khAlN=2.7. We could note that when $khAlN \geq 2.7$, the first interesting point is that the wave travels principally in AlN layer and Si substrate, the second one is that the relative displacement U/Umax in Mo layer will be smaller than 0.5. These points would explain the reason why when $khAlN \geq 2.7$ the use of Mo has no influence on wave velocity and coupling factor.

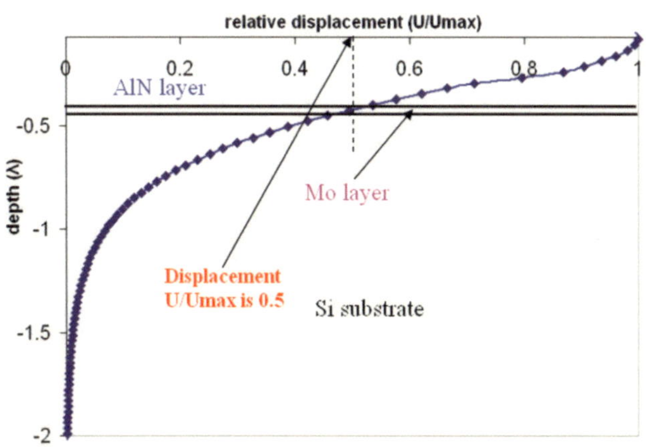

Fig. 11. Displacement profile along the depth of the multilayer AlN/Mo/Si, khAlN=2.7

3. Equivalent circuit for SAW delay line based on Mason model

3.1 Why Equivalent Circuit model is chosen?

Actual devices exist in a three-dimensional physical continuum. Their behaviour is governed by the laws of physics, chemistry, biology, and electronics. From a general point of view, the analysis of devices can be carried out by using some equations of laws of physics, chemistry … For example; the analysis of piezoelectric resonators or transducers and their application to ultrasonic system can be solved by using the wave equation [36],[37]. But through analysis, equivalent electrical circuit representations of devices can be extracted. So, they can be readily expressible with Equivalent Electric Circuit. Below is the presentation of advantages and disadvantages of equivalent circuit.

Advantages:

- There are an immensely powerful set of intellectual tools to understand electric circuits.
- The equivalent circuit approach has distinct advantages over the direct physical, chemical equations approach (such as direct wave equations approach).
- Many theories, problems of electric circuits have already been solved such as microwave network theory, integrated circuit etc.
- Electric circuit approach is intrinsically correct from an energy point of view [56].
- A further advantage of electric circuit model is that it permits efficient modelling of the interaction between the electric and non-electric components of a microsystem. Both the electrical and mechanical portions of a system are represented by the same means. With software like Simulink, the block diagram is easily constructed, easily to build a more complex system but when we would like to connect a mechanical element to electrical circuits, Simulink can not do that. The analogies between electrical and mechanical elements are presented clearly by Warren P.Mason [57], [58].

Disadvantages:

- Care must be taken to make sure whether the boundary conditions are compatible with those used in the original derivation of the equivalent circuit [58].

In many systems, both commercial and industrial, pressure measurement plays a key role. Since pressure is a normal stress (force per unit area), pressure measurement can be done by using piezoelectric material which can convert stress into voltage. Equivalent circuits such as Mason's model [36] provide a powerful tool for the analysis and simulation of piezoelectric transducer elements. Most of the analogous circuits which have appeared in the literature implement transducers as the circuit elements. This model simulates both the coupling between the mechanical and electrical systems and the coupling between the mechanical and acoustical systems [39]. The mechanical, electrical and acoustic parts of piezoelectric transducer can be varied and analysed about behaviour by implementing equivalent circuits on computer tools such as Ansoft®, Spice, ADS, etc. For IDT composing of N periodic sections, Smith et al [41] developed the equivalent circuit model based on Berlincourt et al [40] work about equivalent circuit for Length Expander Bar with parallel electric field and with perpendicular electric field and based on the equivalent circuit for electromechanical transducer presented by Mason [36]. "Smith model" henceforth will be used to indicate this model. From this model, some models for SAW device in literature have been implemented. However, these models would include only IDTs [42], [43]. In SAW pressure sensor, one of sensitive parts is propagation path. It should be included in the model. The hybrid model based on Smith model for SAW pressure sensor which includes the IDTs and propagation path have been constructed.

Another equivalent model is based on the Coupling-Of-Modes (COM) theory. An excellent recent review of COM theory used in SAW devices was written by K.Hashimoto [10]. Based on the COM equations, as the force and voltage analogy can be used, the relationships between the terminal quantities at the one electrical port and two acoustic ports for an IDT have been done. K.Nakamura [44] introduced a simple equivalent circuit for IDT based on COM approach that is developed in section 4.

In conclusion, the equivalent-circuit model is chosen because it can allow fast design. This allows the designer to determine the major dimensions and parameters in number of fingers, fingers width, aperture, delay line distance, frequency response, impedance parameters and transfer characteristics of SAW device.

3.2 Equivalent circuit for IDT including N periodic sections

Based on Berlincourt et al [39] about equivalent circuit for Length Expander Bar with parallel electric field and with perpendicular electric field and based on the equivalent circuit for electromechanical transducer presented by Mason [36], Smith and al [41] have developed the equivalent circuit for IDT composed of N periodic sections of the form shown in Figure 12.

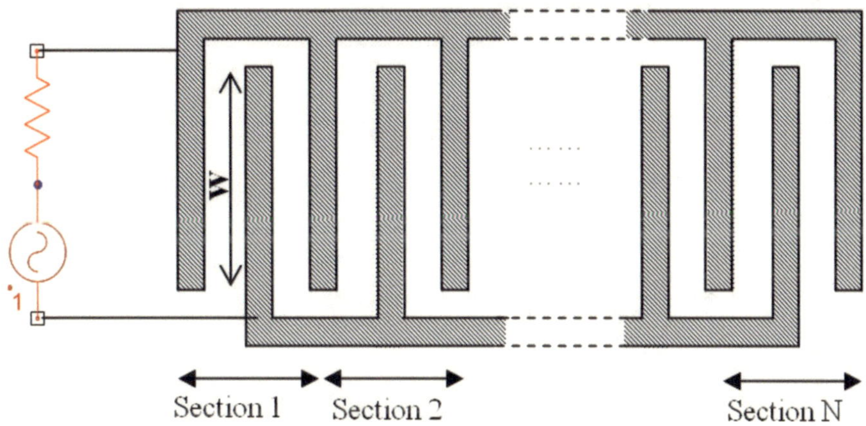

Fig. 12. Interdigital transducer diagram

One periodic section as shown in Figure 13 (a) can be presented by analogous one-dimensional configurations: "crossed-field" model as in Figure 13 (b), and "in-line" model as in Figure 13 (c). In "crossed-field" model, the applied electric field is normal to the acoustic propagation vector; while in "in-line field" model, the electric field is parallel to the propagation vector.

The important advantage of two one-dimensional models is that each periodic section can be represented by equivalent circuit of Mason, as shown in Figure 14 for "crossed-field" model and Figure 15 for "in-line field" model. The difference between these two equivalent circuits is that in "crossed-field" model, the negative capacitors are short-circuited.

Where:

$$\alpha = \frac{\theta}{4} = \frac{\pi}{2}\frac{\omega}{\omega_0} \tag{12}$$

With periodic section transit angle

$$\theta = 2\pi \frac{\omega}{\omega_0}$$

$$R_0 = \frac{2\pi}{\omega_0 C_s k^2} \tag{13}$$

R_0 is electrical equivalent of mechanical impedance Z_0 [59]
k: electromechanical coupling coefficient
C_0=Cs/2 with Cs : electrode capacitance per section
ω_0 is center angular frequency

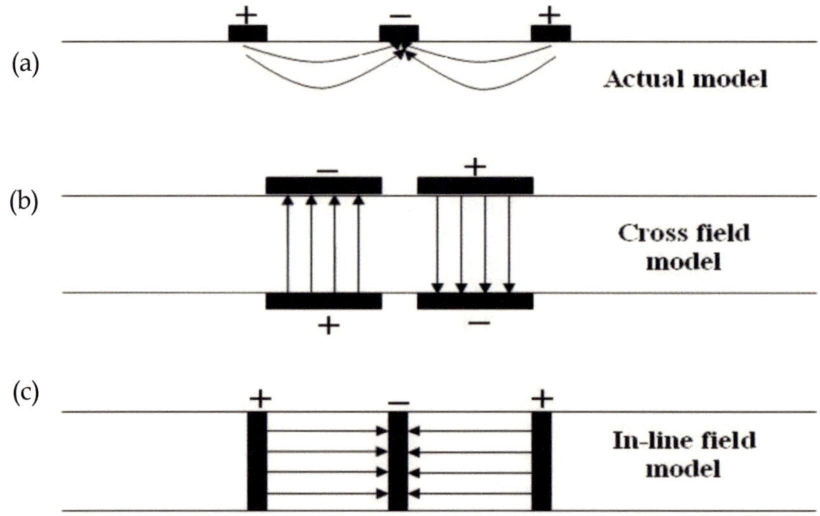

Fig. 13. Side view of the interdigital transducer and 2 analogous one-dimensional configurations (a) Actual model, (b) "crossed-field" model, (c) "in-line field" model

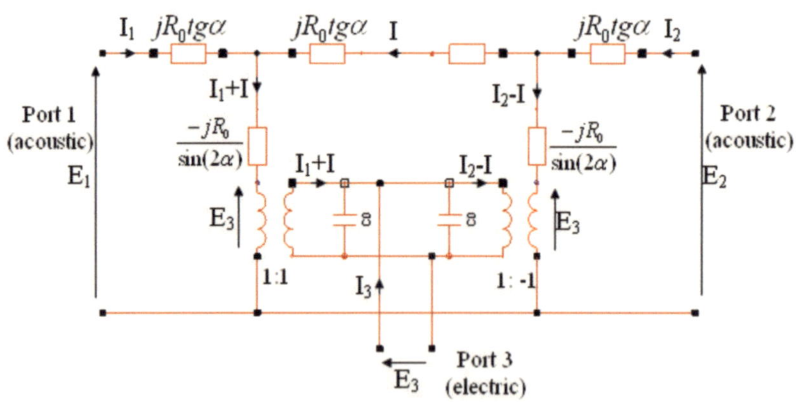

Fig. 14. Mason equivalent circuit for one periodic section in "crossed-field" model

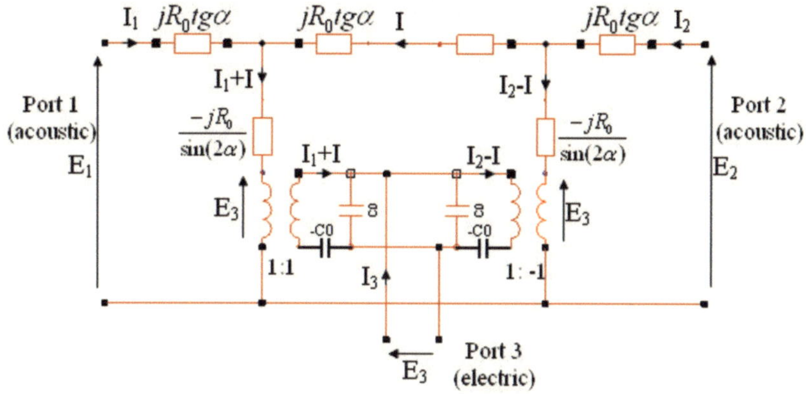

Fig. 15. Mason equivalent circuit for one periodic section in "in-line field" model

One periodic section can be represented by the 3-port network [y] matrix. The [y] matrix of one periodic section for 2 models as follows (see Appendix, section Appendix 1), with $G_0=R_0^{-1}$, R_0 is expressd by (13):

- for the "crossed-field" model:

$$
\begin{aligned}
y_{11} &= -jG_0 \cot g(4\alpha) \\
y_{12} &= \frac{jG_0}{\sin(4\alpha)} \\
y_{13} &= -jG_0 tg\alpha \\
y_{33} &= j(2\omega C_0 + 4G_0 tg\alpha)
\end{aligned}
\tag{14}
$$

- for the "in-line field" model:

$$
\begin{aligned}
y_{11} &= -jG_0 \cot g\alpha \left(\frac{G_0}{\omega C_0} - \cot g(2\alpha) \right) \left[2 - \frac{\left(\frac{G_0}{\omega C_0} - \frac{1}{\sin(2\alpha)} \right)^2}{\left(\frac{G_0}{\omega C_0} - \cot g(2\alpha) \right)^2} \right] \\
y_{12} &= jG_0 \frac{\cot g\alpha \left(\frac{G_0}{\omega C_0} - \frac{1}{\sin(2\alpha)} \right)^2}{2\left(\frac{2G_0}{\omega C_0} - \cot g\alpha \right)\left(\frac{G_0}{\omega C_0} - \cot g(2\alpha) \right)} \\
y_{13} &= -jG_0 \frac{tg\alpha}{1 - \frac{2G_0}{\omega C_0} tg\alpha} \\
y_{33} &= \frac{j2\omega C_0}{1 - \frac{2G_0}{\omega C_0} tg\alpha}
\end{aligned}
\tag{15}
$$

In IDT including N periodic sections, the N periodic sections are connected acoustically in cascade and electrically in parallel as represented in Figure 16.

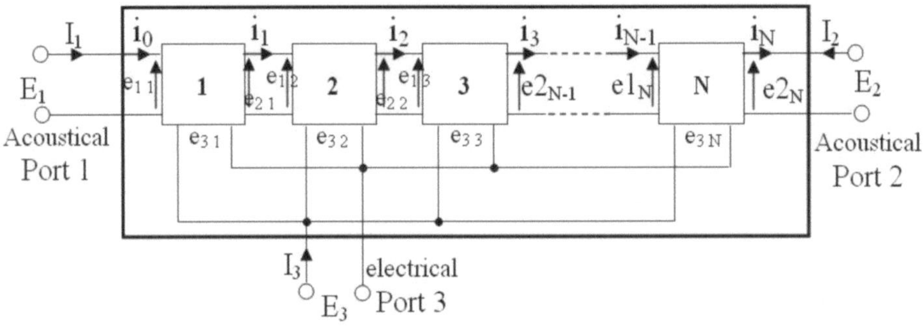

Fig. 16. IDT including the N periodic sections connected acoustically in cascade and electrically in parallel

Matrix [Y] representation of N-section IDT for two models, "crossed-field" model and "in-line" model are in (16) and (17), respectively (the calculation development is presented in Appendix, section Appendix 1):

- In "crossed-field" model:

$$
\begin{aligned}
Y_{11} &= -jG_0 \cot g(4N\alpha) \\
Y_{12} &= \frac{jG_0}{\sin(4N\alpha)} \\
Y_{13} &= -jG_0 tg\alpha \\
Y_{33} &= jN(2\omega C_0 + 4G_0 tg\alpha)
\end{aligned}
\tag{16}
$$

- In "in-line field" model:

$$
\begin{aligned}
Y_{11} &= -\frac{Q_{11}}{Q_{12}} \\
Y_{12} &= \frac{1}{Q_{12}} \\
Y_{13} &= -jG_0 \frac{tg\alpha}{1-\frac{2G_0}{\omega C_0} tg\alpha} \\
Y_{33} &= \frac{j2\omega N C_0}{1-\frac{2G_0}{\omega C_0} tg\alpha}
\end{aligned}
\tag{17}
$$

It was shown in the literature that the crossed field model yielded better agreement than the experiment when compared to the in-line model when K is small. In section 2, K is always smaller than 7.2%. Besides, in section stated above, the "crossed-field" model is simpler than "in-line field" model in term of equations of all element of [Y] matrix. Consequently, the "crossed-field" model is selected henceforth for the calculating, modeling the devices.

3.3 Equivalent circuit for propagation path

The delay line SAW device could be used for pressure sensor application. The sensitive part of this kind of device will be the propagation path. To model the pressure sensor using

SAW, it is necessary to construct the model for propagation path. Based on the equivalent circuit for electromechanical transducer presented by Mason [36], equivalent circuit of propagation path is presented as in Figure 17.

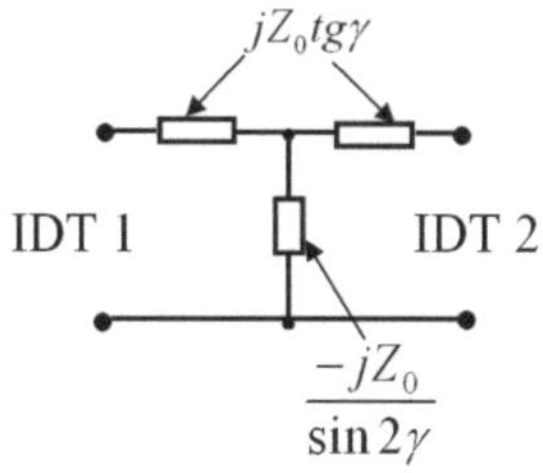

Fig. 17. Equivalent circuit of propagation path, based on Mason model

Where

$$\gamma = \frac{\omega l}{2v} \tag{18}$$

with v is SAW velocity, l is propagation length.

3.4 Equivalent circuit for SAW delay line

Due to the piezoelectric effect, an RF signal applied at input IDT stimulates a micro-acoustic wave propagating on its surface. These waves propagate in two directions, one to receiving IDT and another to the medium. The approximations as follows are assumed to construct the equivalent circuit for SAW delay line:

- Assume that the IDT radiates the wave into a medium of infinite extent. Experimentally, an infinite medium is approximated either by using absorber, such as wax, polyimide to provide acoustic termination, or by using a short RF pulse measurement. The condition of infinite medium means that no wave reflects back to input IDT. This is created for SAW device model by connecting the acoustic characteristic admittance Y_0 to one terminal of IDT.
- Assume that the wave propagating to receiving IDT has no attenuation during propagation way between two IDTs. So, the propagation path between two IDTs can be expressed as the no-loss transmission line.

Based on these two approximations, the [Y] matrix representation of IDT in section 3.2, and propagation path representation in section 3.3, the SAW delay line can be expressed as equivalent circuit as in Figure 18.

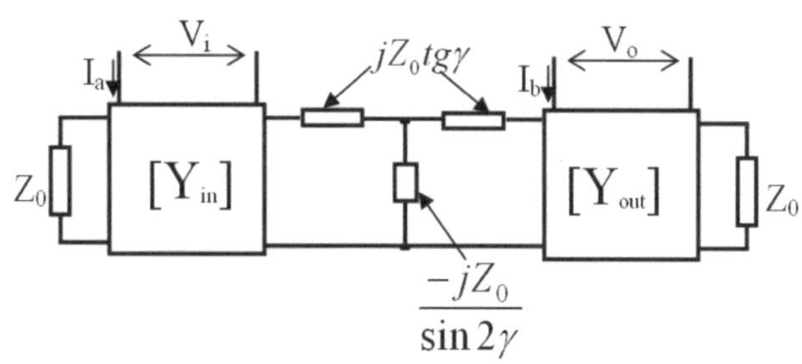

Fig. 18. Equivalent circuit of SAW delay line, based on Mason model

Section 3 gives the equivalent circuit of SAW delay line, including IDT input, IDT output and propagation path. All of calculation developments are presented in appendix, section 2. In this appendix, a new equivalent circuit of IDT including N periodic section plus one finger, which we call it "N+1/2", also are developed and presented. Another representation of SAW delay line is [ABCD] matrix representation which also proposed in appendix, section Appendix 4. [ABCD] matrix representation has one interesting property that in cascaded network, the [ABCD] matrix of total network can be obtained easily by multiplying the matrices of elemental networks.

4. Equivalent circuit for IDT based on the Coupling-Of-Mode theory

The Coupling-Of-Modes formalism is a branch of the highly developed theory of wave propagation in periodic structure, which has an history of more than 100 years. This theory covers a variety of wave phenomena, including the diffraction of EM waves on periodic gratings, their propagation in periodic waveguides and antennas, optical and ultrasonic waves in multi-layered structures, quantum theory of electron states in metal, semiconductors, and dielectrics.... Theoretical aspects of the wave in periodic media and applications were reviewed by C.Elachi [4], in which it included theories of waves in unbounded and bounded periodic medium, boundary periodicity, source radiation in periodic media, transients in periodic structures, active and passive periodic structures, waves and particles in crystals. An excellent recent review of COM theory used in SAW devices was written by K.Hashimoto [10].

A simple equivalent circuit for IDT based on COM approach was proposed by K.Nakamura [29]. This model would be useful to analyze and design SAW devices. Based on the COM equations, the relationships between the terminal quantities at the one electrical port and two acoustic ports for an IDT have been done.

4.1 COM equation for particle velocities

Consider an IDT including N periodic sections with periodic length of L as shown in Figure 19.

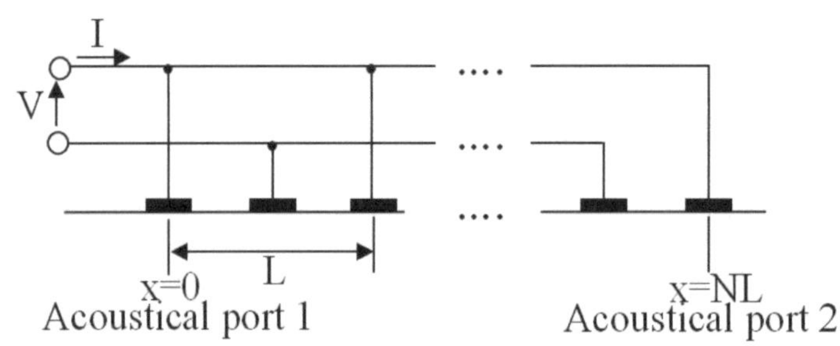

Fig. 19. IDT including N periodic sections

The particle velocities $v^+(x)$ and $v^-(x)$ of the wave propagating in the +x and –x directions in the periodic structure can be expressed as follows with the time dependence exp(jωt) term:

$$v^+(x) = A^+(x)e^{-jkx} \tag{19}$$

$$v^-(x) = A^-(x)e^{jkx} \tag{20}$$

Where k is the wave number

$$k = \omega / V_{SAW} \tag{21}$$

The amplitude $A^{+}(x)$ and $A^{-}(x)$ obey the following coupled-mode equations [60]:

$$\frac{dA^{+}(x)}{dx} = -jK_{11}A^{+}(x) - jK_{12}e^{j2\delta x}A^{-}(x) + j\zeta e^{j\delta x}V \tag{22}$$

$$\frac{dA^{-}(x)}{dx} = jK_{12}e^{-j2\delta x}A^{+}(x) + jK_{11}A^{-}(x) - j\zeta e^{-j\delta x}V \tag{23}$$

Where V is the voltage applied to the IDT,
ζ is the constant associated with the convention from electrical to SAW quantities,
K_{11} and K_{12} are coupling coefficients, sum of the coupling coefficient coming from the piezoelectric perturbation and that coming from the mechanical perturbation.

$$\delta = k - k_0 \text{ , with } k_0 = \frac{2\pi}{L} \tag{24}$$

The solution to (22) and (23) can be expressed as

$$v^{+}(x) = \left(h_1 e^{-j\beta_1 x} + ph_2 e^{-j\beta_2 x} + q\zeta V\right)e^{-jk_0 x} \tag{25}$$

$$v^{-}(x) = \left(ph_1 e^{-j\beta_1 x} + h_2 e^{-j\beta_2 x} + q\zeta V\right)e^{jk_0 x} \tag{26}$$

Where the subscripts 1 and 2 indicate the elementary waves with wavenumbers $k_0 + \beta_1$ and $k_0 + \beta_2$ in the +x direction, and the magnitudes h_1 and h_2, respectively.

$$\beta_1, \beta_2 = \pm\sqrt{(\delta + K_{11})^2 - K_{12}{}^2} \tag{27}$$

$$p = \frac{\beta_1 - \delta - K_{11}}{K_{12}} \tag{28}$$

$$q = \frac{1}{\delta + K_{11} + K_{12}} \tag{29}$$

4.2 Equivalent circuit for IDT based on COM theory

From the equations (25) and (26), the particle velocities at the both ends of the IDT can be expressed as:

$$v^{+}(0) = h_1 + ph_2 + q\zeta V \tag{30}$$

$$v^{-}(0) = ph_1 + h_2 + q\zeta V \tag{31}$$

$$v^{+}(NL) = \pm\left(e^{-j\beta_1 NL}h_1 + e^{j\beta_1 NL}ph_2 + q\zeta V\right) \tag{32}$$

$$v^{-}(NL) = \pm\left(e^{-j\beta_1 NL} ph_1 + e^{j\beta_1 NL} h_2 + q\zeta V\right) \tag{33}$$

The upper and lower signs in (32) and (33) correspond to the cases N=i and N=i+0.5, respectively, where i is an integer. Consequently, the total particle velocities at the two acoustical ports can be expressed as:

- Particle velocity at port 1 (x=0):

$$v_1 = v^{+}(0) + v^{-}(0) = (1+p)(h_1 + h_2) + 2q\zeta V \tag{34}$$

- Particle velocity at port 2 (x=NL):

$$v_2 = -\left[v^{+}(NL) + v^{-}(NL)\right] = \mp\left[(1+p)(e^{-j\beta_1 NL} h_1 + e^{j\beta_1 NL} h_2) + 2q\zeta V\right] \tag{35}$$

The two forces at two acoustic ports are considered to be proportional to the difference of v^+ and v^-. For the simplicity, these forces can be expressed as follows:

$$F_1 = v^{+}(0) - v^{-}(0) = (1-p)(h_1 - h_2) \tag{36}$$

$$F_2 = v^{+}(NL) - v^{-}(NL) = \pm\left[(1-p)(e^{-j\beta_1 NL} h_1 - e^{j\beta_1 NL} h_2)\right] \tag{37}$$

From these equations, h_1 and h_2 are the terms of F_1 and F_2 as follows:

$$h_1 = \frac{e^{j2\beta_1 NL}}{(1-p)(e^{j2\beta_1 NL} - 1)} F_1 \mp \frac{e^{j\beta_1 NL}}{(1-p)(e^{j2\beta_1 NL} - 1)} F_2 \tag{38}$$

$$h_2 = \frac{1}{(1-p)(e^{j2\beta_1 NL} - 1)} F_1 \mp \frac{e^{j\beta_1 NL}}{(1-p)(e^{j2\beta_1 NL} - 1)} F_2 \tag{39}$$

The current I at the electrical ports can be expressed as:

$$\begin{aligned} I &= \eta \int_0^{NL} \left[(1+p)(h_1 e^{-j\beta_1 x} + h_2 e^{-j\beta_2 x}) + 2q\zeta V\right] dx + j\omega N C_s V \\ &= j\eta\left\{(1+p)\left[\frac{h_1}{\beta_1}(e^{-j\beta_1 NL} - 1) + \frac{h_2}{\beta_2}(e^{j\beta_1 NL} - 1)\right]\right\} + 2q\zeta\eta NLV + j\omega N C_s V \end{aligned} \tag{40}$$

where η is the constant associated with the convention from SAW to electrical quantities, therefore associated with the coupling factor K.

C_s is the capacitance for one electrode pair.

By substituting equations (38) and (39) in (34), (35) and (40), the following equations can be obtained:

$$I = (j\omega N C_s + 2q\zeta\eta NL)V + \frac{\eta(1+p)}{j\beta(1-p)} F_1 \mp \frac{\eta(1+p)}{j\beta(1-p)} F_2 \tag{41}$$

$$v_1 = 2q\zeta V + \frac{1+p}{1-p}\frac{1}{j\tan 2\theta} F_1 \mp \frac{1+p}{1-p}\frac{1}{j\sin 2\theta} F_2 \tag{42}$$

$$v_2 = \mp 2q\zeta V \mp \frac{1+p}{1-p}\frac{1}{j\sin 2\theta}F_1 + \frac{1+p}{1-p}\frac{1}{j\tan 2\theta}F_2 \tag{43}$$

Where

$$\theta = \beta NL / 2 \tag{44}$$

$$\beta \equiv \beta_1 = -\beta_2 \tag{45}$$

From these equations, the matrix as follows can be obtained:

$$\begin{bmatrix} I \\ v_1 \\ v_2 \end{bmatrix} = \begin{bmatrix} (j\omega NC_s + 2q\zeta\eta NL) & \frac{\eta(1+p)}{j\beta(1-p)} & \mp\frac{\eta(1+p)}{j\beta(1-p)} \\ 2q\zeta & \frac{1+p}{1-p}\frac{1}{j\tan 2\theta} & \mp\frac{1+p}{1-p}\frac{1}{j\sin 2\theta} \\ \mp 2q\zeta & \mp\frac{1+p}{1-p}\frac{1}{j\sin 2\theta} & \frac{1+p}{1-p}\frac{1}{j\tan 2\theta} \end{bmatrix} \begin{bmatrix} V \\ F_1 \\ F_2 \end{bmatrix} \tag{46}$$

In the acoustic wave transducer using piezoelectric effect, the force and voltage analogy can be used. Therefore, the COM-based circuit of IDT as matrix in (46) can be considered as the reciprocal circuit. The reciprocity theorem states that if a voltage source E acting in one branch of a network causes a current I to flow in another branch of the network, then the same voltage source E acting in the second branch would cause an identical current I to flow in the first branch. By using this theorem in this case, replacing V and F_1 together, the same value I requirement leads the following equations:

$$2q\zeta = \frac{\eta(1+p)}{j\beta(1-p)} \tag{47}$$

$$\eta = 2j\zeta \tag{48}$$

From (46), (47), and (48), the matrix as in (46) becomes:

$$\begin{bmatrix} I \\ v_1 \\ v_2 \end{bmatrix} = \begin{bmatrix} j\omega C_T + \frac{\phi^2}{j2\theta Z_0} & \frac{\phi}{j2\theta Z_0} & \mp\frac{\phi}{j2\theta Z_0} \\ \frac{\phi}{j2\theta Z_0} & \frac{1}{jZ_0\tan 2\theta} & \mp\frac{1}{jZ_0\sin 2\theta} \\ \mp\frac{\phi}{j2\theta Z_0} & \mp\frac{1}{jZ_0\sin 2\theta} & \frac{1}{jZ_0\tan 2\theta} \end{bmatrix} \begin{bmatrix} V \\ F_1 \\ F_2 \end{bmatrix} \tag{49}$$

Where

$$Z_0 = \frac{1-p}{1+p} = \frac{1}{q\beta} \tag{50}$$

$$\phi = \eta NL = 2j\zeta NL \tag{51}$$

$$C_T = NC_s \tag{52}$$

Consequently, the simple equivalent circuit obtained for IDT with N electrode pairs is shown in Figure 20:

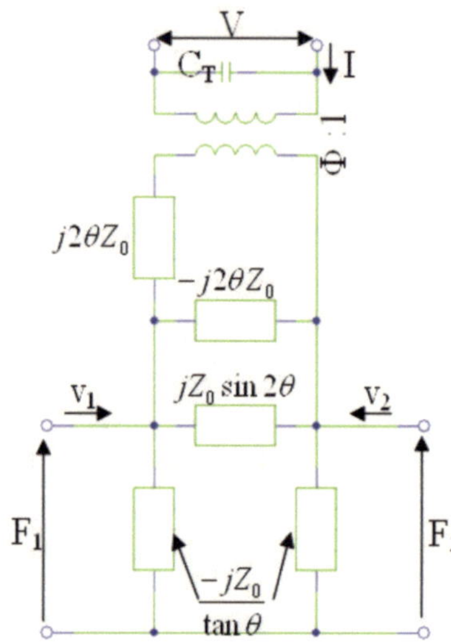

Fig. 20. Equivalent circuit IDT based on COM theory

4.3 Equivalent circuit for propagation path based on COM theory

In SAW devices, the propagation path should be taken into account. It is necessary to determine the equivalent circuit for a propagation path of distance l between 2 IDTs. This propagation path is a uniform section of length l, with a free surface or a uniformly metallized surface. In this case, $K_{11}=K_{12}=0$, and $\beta=\delta$.
Consequently, from equations (38) and (39), h_1 and h_2 can be expressed as:

$$h_1 = \frac{e^{j2\beta l}}{e^{j2\beta l}-1}F_1 - \frac{e^{j\beta l}}{e^{j2\beta l}-1}F_2 \tag{53}$$

$$h_2 = \frac{1}{e^{j2\beta l}-1}F_1 - \frac{e^{j\beta l}}{e^{j2\beta l}-1}F_2 \tag{54}$$

And, the particle velocities are expressed as:

$$v^+(x) = h_1 e^{-j(k_0+\beta_1)x} = h_1 e^{-jkx} \tag{55}$$

$$v^-(x) = h_2 e^{j(k_0+\beta_1)x} = h_2 e^{jkx} \tag{56}$$

If the v_1, v_2, F_1, and F_2 are defined as:

$$v_1 = v^+(0) + v^-(0) \tag{57}$$

$$v_2 = -[v^+(NL) + v^-(NL)] \tag{58}$$

$$F_1 = v^+(0) - v^-(0) \tag{59}$$

$$F_2 = v^+(NL) - v^-(NL) \tag{60}$$

Then, by expressing h_1 and h_2 in terms of F_1 and F_2 based on equations (53) and (54), the v_1 and v_2 become as follows:

$$v_1 = h_1 + h_2 = \frac{e^{j2\beta l}+1}{e^{j2\beta l}-1}F_1 - \frac{2e^{j\beta l}}{e^{j2\beta l}-1}F_2 \tag{61}$$

$$v_2 = h_1 e^{-jkl} + h_2 e^{jkl} = \frac{2e^{j\beta l}}{e^{j2\beta l}-1}F_1 - \frac{e^{j2\beta l}+1}{e^{j2\beta l}-1}F_2 \tag{62}$$

Using the relation between complex number and trigonometry, the v_1 and v_2 can be expressed as follows:

$$v_1 = \frac{1}{jZ_0^{'}\tan 2\theta^{'}}F_1 - \frac{1}{jZ_0^{'}\sin 2\theta^{'}}F_2 \tag{63}$$

$$v_2 = -\frac{1}{jZ_0^{'}\sin 2\theta^{'}}F_1 + \frac{1}{jZ_0^{'}\tan 2\theta^{'}}F_2 \tag{64}$$

Consequently, the equivalent circuit for propagation path can be represented by the π-circuit of Figure 21:

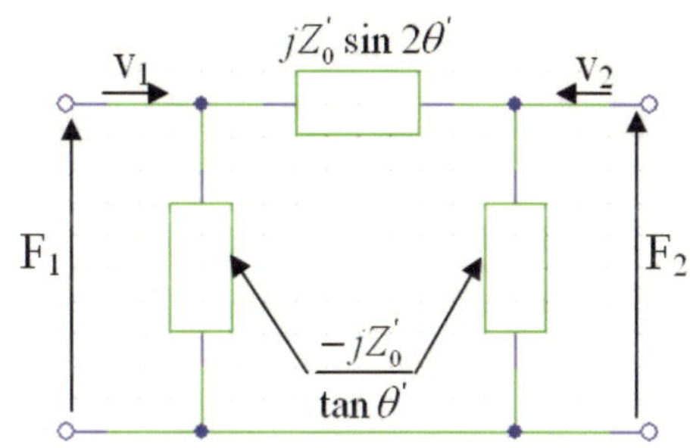

Fig. 21. Equivalent circuit of propagation path based on COM theory

Based on Mason model, the equivalent circuit of propagation path was presented in Figure 17, which has star form. In Figure 21, the circuit has triangle form. By using triangles and stars transformation theory published by A.E. Kennelly, equivalent circuit of propagation in these two figures is the same. Consequently, the approachs that are based on Mason model and COM theory can get the same equivalent circuit of propagation path.

4.4 Equivalent circuit for SAW delay line based on COM theory

Based on section 4.2 and 4.3, equivalent circuit of SAW delay line based on COM theory is presented in Figure 22.

In this model, some parameters must to be calculated or extracted. SAW velocity v, piezoelectric coupling factor K could be calculated from section 2. The periodic length L (or wavelength λ) is determined by design and fabrication.

The parameters K_{11} and K_{12} are coupling coefficients. They are sum of the coupling coefficient coming from the piezoelectric perturbation and that coming from the mechanical perturbation, and their equations for calculation are complicated [55]. Exact equations for

K_{11} and K_{12} were given by Y.Suzuki et al [55], but it seems so complex that their usefulnesses could be limited. However, from this work of Y.Suzuki et al [55], we propose the K_{11} and K_{12} could be expressed as follows:

$$K_{11}= O_{11}K^2k_0 \tag{65}$$

$$K_{12}= O_{12}K^2k_0 \tag{66}$$

Where k_0 is stated by (24) $k_0 = \frac{2\pi}{L}$ and K is piezoelectric coupling factor.

O_{11} is so-called self-coupling constant of finger, and O_{12} is so-called coupling constant between fingers. O_{12} could also presents the reflective wave between two fingers.

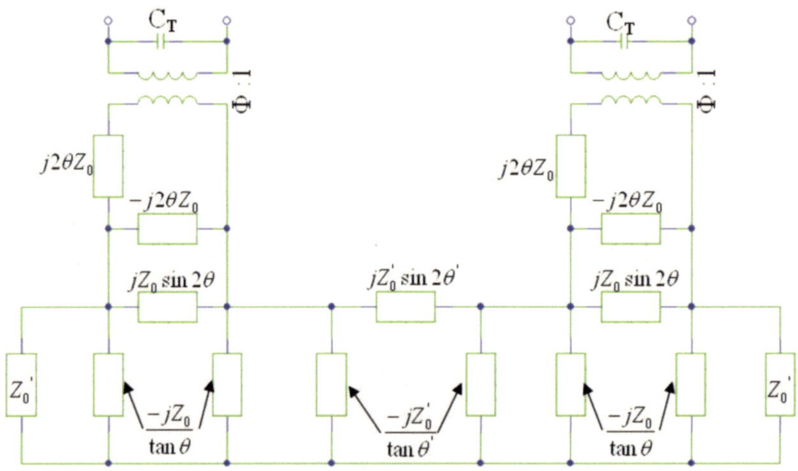

Fig. 22. Equivalent circuit of SAW delay line based on COM theory

(a)

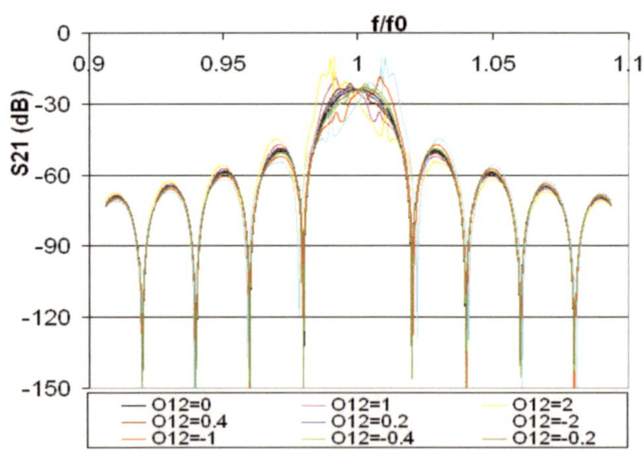

Fig. 23. Effect of O_{12} on S21(dB), N=50, v_{SAW}=5120m/s, λ=8μm, K=0.066453, O_{11}=0

Figure 23 shows the effects of O_{12} on S21(dB) of SAW device N=50, v_{SAW}=5120m/s, λ=8μm, K=0.066453 when O_{11}=0. S_{21} is the transmission coefficient in the scattering matrix representation [28].

Figure 24 shows the effects of O_{11} on S21(dB) of SAW device N=50, v_{SAW}=5120m/s, λ=8μm, K=0.066453 when O_{12}=0. So, O_{11} coefficient shifts the center frequency of SAW device, the positive value of O_{11} reduces the center frequency f_0 of device, the negative on will increase the f_0.

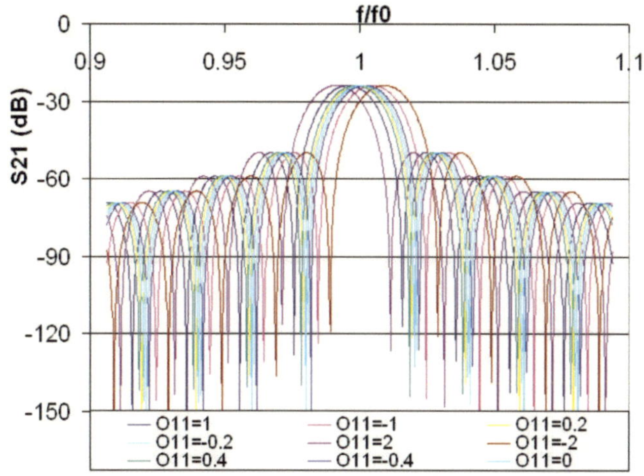

Fig. 24. Effect of O_{11} on S21(dB), N=50, v_{SAW}=5120m/s, λ=8μm, K=0.066453, O_{12}=0

The effect of K_{11} and K_{12} could be explained by their measurement method [61]. K_{11} could be derived from the measurement of frequency response, therefore the usefulness of its calculation could be limited. Meanwhile, K_{12} can be extracted from FEM. It is shown in literature that K_{12} depends on the thickness of finger with respect to the wavelength. In our work, the ratio thickness/wavelength (its maximum value is 300nm/8μm) is too small that its effect can be ignored. In conclusion, in our work, value of K_{11} and K_{12} are 0.

5. Comparison of equivalent circuit of SAW device based on Mason model and COM thoery

Figure 25 presents the comparison between hybrid model and COM model in that O_{11}= O_{12}=0, distance between 2 IDTs is 50λ. These models could be the same, except that a

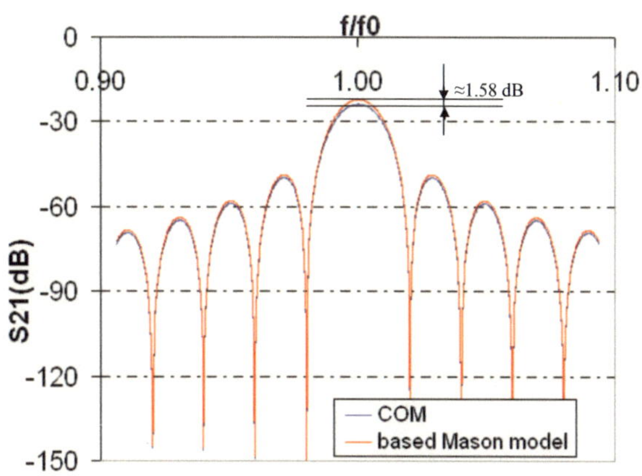

Fig. 25. Comparison between Hydrid model and COM model (O_{11}=O_{12}=0)

small difference in the peak value of S21 (dB) occurs. This difference could be explained by using "crossed-filed" model instead of actual model as in Figure 13.

6. Conclusion

The model used for SAW pressure sensor based on delay line are presented. For usefulness and reduction of time in design process, the equivalent circuit based on COM model, in which K_{11}, K_{12}=0 is proposed to be used.

Acoustic wave properties in different structures of AlN/SiO_2/Si, AlN/Si, and AlN/Mo/Si are analyzed. The wave velocity, coupling factor could depend on the wave propagation medium.

From analyses of these structures, the range in which there is a weak dependence of the wave velocity, coupling factor on the AlN layer thickness could be known. The SAW devices should be fabricated in this range to facilitate manufacturing.

For AlN/Si structure, this range is $khAlN \geq 3$.

For AlN/Mo/Si, if this kind of SAW device is fabricated in the range from $khAlN \geq 2.7$ to facilitate manufacturing, the use of Mo layer is useless. Consequently, to take full advantage of using Mo layer in term of wave velocity and coupling factor, it should be required to control the fabrication process carefully to obtain the required AlN thickness from khAlN=1.02 to khAlN=2.7.

For AlN/SiO_2/Si, this range is $khAlN \geq 5$ for khSiO_2=0.7854, for thicker SiO_2 layer, this range changes based on Figure 6 and Figure 7. Besides, using SiO_2 layer would reduce temperature dependence of frequency. To choose the thickness of SiO_2 layer, it would consider the effect of temperature dependence and analyses of wave velocity, coupling factor.

7. Appendix: Development of calculation for equivalent circuit of SAW device

7.1 Appendix 1. Equivalent circuit for normal IDT including N periodic sections

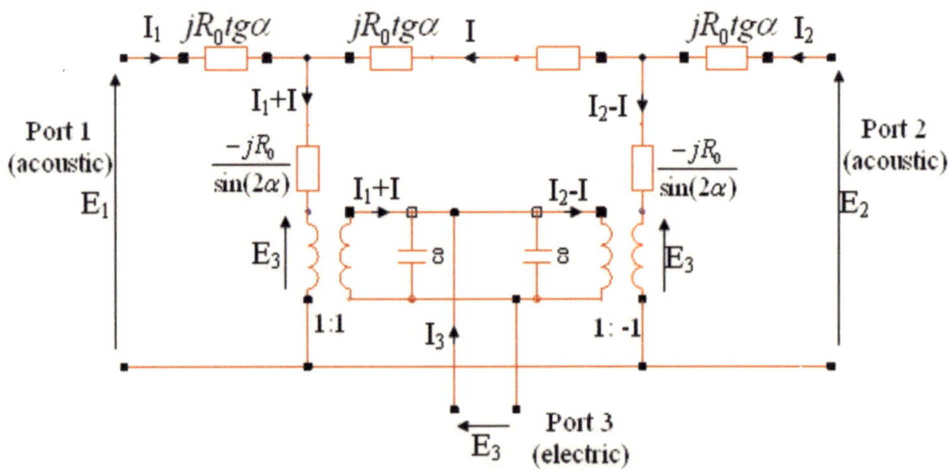

Fig. Appendix.1. Mason equivalent circuit for one periodic section in "crossed-field" model

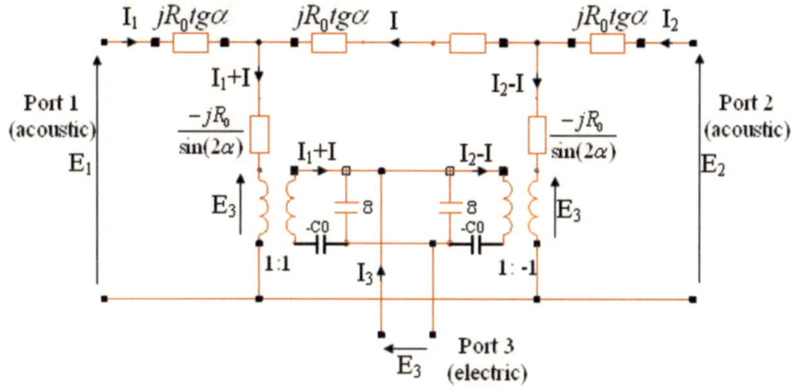

Fig. Appendix.2. Mason equivalent circuit for one periodic section in "in-line field" model

One periodic section can be expressed by the 3-port network as follows:

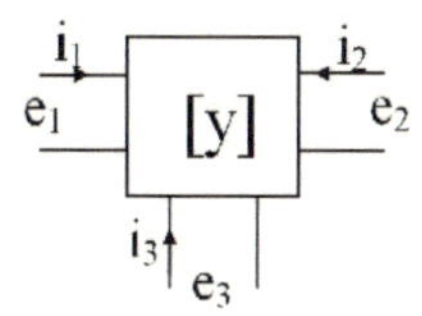

Fig. Appendix.3. One periodic section represented by 3-port network, admittance matrix [y]

$$\begin{bmatrix} i_1 \\ i_2 \\ i_3 \end{bmatrix} = \begin{bmatrix} y_{11} & y_{12} & y_{13} \\ y_{21} & y_{22} & y_{23} \\ y_{31} & y_{32} & y_{33} \end{bmatrix} \begin{bmatrix} e_1 \\ e_2 \\ e_3 \end{bmatrix} \qquad \text{(Appendix.1)}$$

By the symmetrical properties of one periodic section (the voltage applied at port 3 will result in stress of the same value at port 1 and 2), the [y] matrix in (Appendix.1) becomes (Appendix.2) for Figure Appendix.4 and becomes (Appendix.3) for Figure Appendix.5.

$$\begin{bmatrix} i_1 \\ i_2 \\ i_3 \end{bmatrix} = \begin{bmatrix} y_{11} & y_{12} & y_{13} \\ y_{12} & y_{11} & -y_{13} \\ y_{13} & -y_{13} & y_{33} \end{bmatrix} \begin{bmatrix} e_1 \\ e_2 \\ e_3 \end{bmatrix} \qquad \text{(Appendix.2)}$$

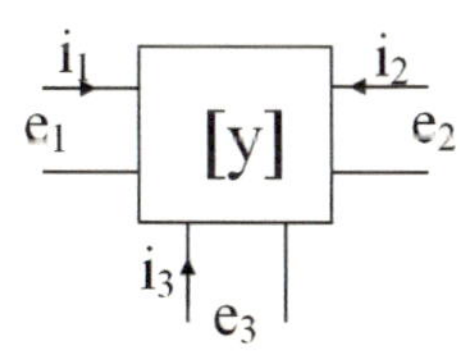

Fig. Appendix.4. 3-port network representation of one periodic section, with the change of sign between Y_{13} and Y_{23} to ensure that acoustic power flows symmetrically away from transducer

$$\begin{bmatrix} i_1 \\ i_2 \\ i_3 \end{bmatrix} = \begin{bmatrix} y_{11} & y_{12} & y_{13} \\ -y_{12} & -y_{11} & y_{13} \\ y_{13} & -y_{13} & y_{33} \end{bmatrix} \begin{bmatrix} e_1 \\ e_2 \\ e_3 \end{bmatrix} \qquad \text{(Appendix.3)}$$

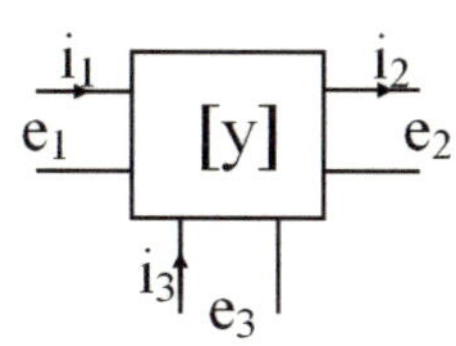

Fig. Appendix.5. 3-port network representation of one periodic section, with the no change of sign between Y_{13} and Y_{23}

Applying circuit theory, definitions of [y] matrix elements are presented:

$$y_{11} = \left.\frac{i_1}{e_1}\right|_{\substack{e_2=0\\e_3=0}} \quad ; \quad y_{12} = \left.\frac{i_1}{e_2}\right|_{\substack{e_1=0\\e_3=0}}$$
$$y_{13} = \left.\frac{i_1}{e_3}\right|_{\substack{e_1=0\\e_2=0}} \quad ; \quad y_{33} = \left.\frac{i_3}{e_3}\right|_{\substack{e_1=0\\e_2=0}} \qquad \text{(Appendix.4)}$$

And using trigonometric functions as follows:

$$tg\alpha - \frac{2}{\sin(2\alpha)} = -\cot g\alpha$$
$$tg\alpha - \frac{1}{\sin(2\alpha)} = \frac{1}{2}(tg\alpha - \cot g\alpha) \qquad \text{(Appendix.5)}$$
$$tg\alpha \frac{3\cos(2\alpha) + 1 - tg\alpha \sin(4\alpha) - tg\alpha \sin(2\alpha)}{tg\alpha \sin(4\alpha) + tg\alpha \sin(2\alpha) - \cos(2\alpha)} = -tg(4\alpha)$$

The [y] matrix can be obtained for 2 models as follows:
- for the "crossed-field" model:

$$y_{11} = -jG_0 \cot g(4\alpha)$$
$$y_{12} = \frac{jG_0}{\sin(4\alpha)} \qquad \text{(Appendix.6)}$$
$$y_{13} = -jG_0 tg\alpha$$
$$y_{33} = j(2\omega C_0 + 4G_0 tg\alpha)$$

- for the "in-line field" model:

$$y_{11} = -jG_0 \cot g\alpha \left(\frac{G_0}{\omega C_0} - \cot g(2\alpha) \right) \left[2 - \frac{\left(\frac{G_0}{\omega C_0} - \frac{1}{\sin(2\alpha)} \right)^2}{\left(\frac{G_0}{\omega C_0} - \cot g(2\alpha) \right)^2} \right]$$

$$y_{12} = jG_0 \frac{\cot g\alpha \left(\frac{G_0}{\omega C_0} - \frac{1}{\sin(2\alpha)} \right)^2}{2\left(\frac{2G_0}{\omega C_0} - \cot g\alpha \right)\left(\frac{G_0}{\omega C_0} - \cot g(2\alpha) \right)} \qquad \text{(Appendix.7)}$$

$$y_{13} = -jG_0 \frac{tg\alpha}{1 - \frac{2G_0}{\omega C_0} tg\alpha}$$

$$y_{33} = \frac{j2\omega C_0}{1 - \frac{2G_0}{\omega C_0} tg\alpha}$$

In IDT including N periodic sections, the N periodic sections are connected acoustically in cascade and electrically in parallel as Figure Appendix.6.

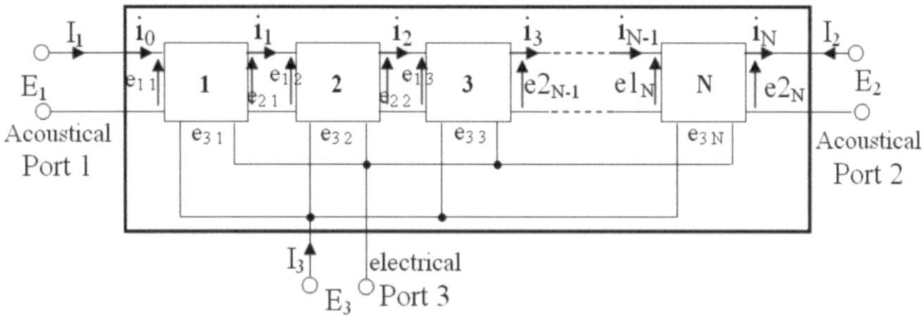

Fig. Appendix.6. IDT including the N periodic sections connected acoustically in cascade and electrically in parallel

Because the symmetric properties of the IDT including N section like these of one periodic section, and from (Appendix.2), (Appendix.3), Figure Appendix.4 and Figure Appendix.5, the [Y] matrices of N-section IDT are represented as follows:

$$[Y]=\begin{bmatrix} Y_{11} & Y_{12} & Y_{13} \\ -Y_{12} & -Y_{11} & Y_{13} \\ Y_{13} & -Y_{13} & Y_{33} \end{bmatrix}$$ corresponds to model

(a)

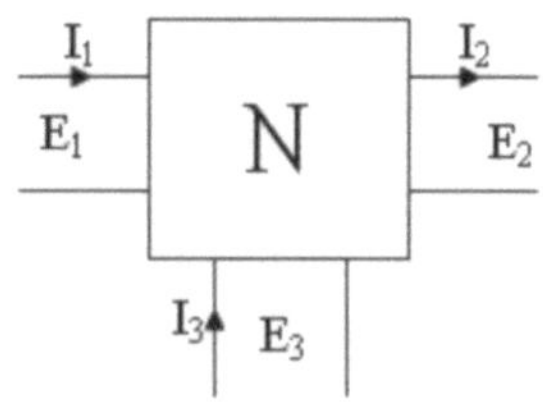

$$[Y]=\begin{bmatrix} Y_{11} & Y_{12} & Y_{13} \\ Y_{12} & Y_{11} & -Y_{13} \\ Y_{13} & -Y_{13} & Y_{33} \end{bmatrix}$$ corresponds to model

(b)

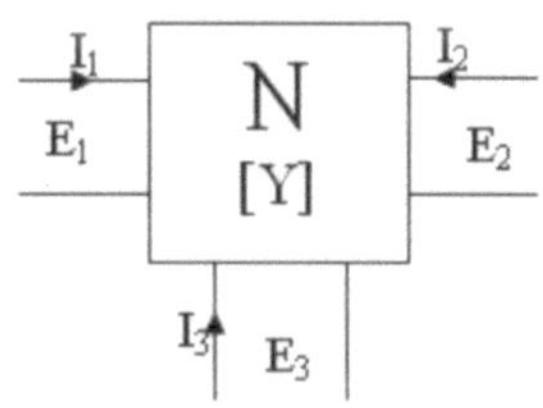

Fig. Appendix.7. The [Y] matrices and the model corresponsive models

Since the periodic sections are identical, the recursion relation as follows can be obtained:

$$e_{1\,m}=e_{2\,m-1} \qquad \text{(Appendix.8)}$$

$$e_{3\,N}=e_{3\,N-1}=e_{3\,N-2}=\ldots=e_{3\,2}=e_{3\,1}=E_3 \qquad \text{(Appendix.9)}$$

$$i_{1\,m}=i_{2\,m-1} \qquad \text{(Appendix.10)}$$

With m is integer number, m=1,2, …, N-1, N
The total transducer current is the sum of currents flowing into the N sections.

$$\begin{aligned} I_3 &= i_{3\,1}+i_{3\,2}+\ldots+i_{3\,N-1}+i_{3\,N} \\ &= \left(y_{13}e_{1\,1}-y_{13}e_{2\,1}+y_{33}e_{3\,1}\right)+\left(y_{13}e_{1\,2}-y_{13}e_{2\,2}+y_{33}e_{3\,2}\right)+\ldots \\ &\quad +\left(y_{13}e_{1\,N-1}-y_{13}e_{2\,N-1}+y_{33}e_{3\,N-1}\right)+\left(y_{13}e_{1\,N}-y_{13}e_{2\,N}+y_{33}e_{3\,N}\right) \end{aligned} \qquad \text{(Appendix.11)}$$

By applying (Appendix.8), (Appendix.9) and boundary conditions ($e_{11} = E_1$, $e_{2N}=E_2$), (Appendix.11) becomes:

$$I_3= y_{13}e_{1\,1}-y_{13}e_{2\,N}+ Ny_{33}E_3 = y_{13}E_1-y_{13}E_2+ Ny_{33} E_3 \quad \text{(Appendix.12)}$$

From Figure Appendix.7, the Y_{13} and Y_{33} can be expressed as:

$$Y_{13}=y_{13} \quad \text{(Appendix.13)}$$

$$Y_{33}= Ny_{33} \quad \text{(Appendix.14)}$$

Because the N periodic sections are connected acoustically in cascade and electrically in parallel, the model as in Figure Appendix.5 should be used to obtain the [Y] matrix of N-section IDT.
From (Appendix.3) for one section, the i_1 and i_2 can be expressed

$$i_1= y_{11}e_1+y_{12}e_2+ y_{13}\, e_3,\; i_2= -y_{12}e_1-y_{12}e_2+ y_{13}\, e_3 \quad \text{(Appendix.15)}$$

Equations (Appendix.15) can be represented in matrix form like [ABCD] form in electrical theory as follows:

$$\begin{bmatrix} e_2 \\ i_2 \end{bmatrix} = [K]\begin{bmatrix} e_1 \\ i_1 \end{bmatrix} + [L]e_3 \quad \text{(Appendix.16)}$$

Where

$$[K]=\begin{bmatrix} -\dfrac{y_{11}}{y_{12}} & \dfrac{1}{y_{12}} \\ \dfrac{y_{11}^2-y_{12}^2}{y_{12}} & -\dfrac{y_{11}}{y_{12}} \end{bmatrix} \quad \text{(Appendix.17)}$$

$$[L]=\begin{bmatrix} -\dfrac{y_{13}}{y_{12}} \\ \dfrac{y_{11}y_{13}+y_{12}y_{13}}{y_{12}} \end{bmatrix} \quad \text{(Appendix.18)}$$

By applying (Appendix.16) into N-section IDT as in Figure Appendix.6 and using (Appendix.9), the second recursion relation is obtained as follows:

$$\begin{bmatrix} e_m \\ i_m \end{bmatrix} = [K]\begin{bmatrix} e_{m-1} \\ i_{m-1} \end{bmatrix} + [L]E_3 \quad \text{(Appendix.19)}$$

Where m is integer number, m=1,2, …, N-1, N
Starting (Appendix.19)(Appendix.19) by using with m=N, and reducing m until m=1 gives the expression:

$$\begin{bmatrix} e_N \\ i_N \end{bmatrix} = [Q]\begin{bmatrix} e_0 \\ i_0 \end{bmatrix} + [X]E_3 \quad \text{(Appendix.20)}$$

Where

$$[Q]=[K]^N \tag{Appendix.21}$$

$$[X]=\begin{bmatrix} X_1 \\ X_2 \end{bmatrix}=\sum_{n=0}^{N-1}[K]^n[L] \tag{Appendix.22}$$

Solving (Appendix.20) and using the boundary conditions ($e_0 = E_1$, $i_0 = I_1$) gives:

$$I_1=-\frac{Q_{11}}{Q_{12}}E_1+\frac{1}{Q_{12}}E_2-\frac{X_1}{Q_{12}}E_3 \tag{Appendix.23}$$

Consequently,

$$Y_{11}=-\frac{Q_{11}}{Q_{12}} \tag{Appendix.24}$$

$$Y_{12}=\frac{1}{Q_{12}} \tag{Appendix.25}$$

$$Y_{13}=-\frac{X_1}{Q_{12}} \tag{Appendix.26}$$

The Y_{13} is known by (Appendix.13), so (Appendix.26) and matrix [X] don't need to be solved.
To solve (Appendix.24) and (Appendix.25), matrix [Q] should be solved.
In "crossed-field" model, matrix [Q] can be represented in a simple form as follows:

$$[K]=\begin{bmatrix} \cos(4\alpha) & -jR_0\sin(4\alpha) \\ -jG_0\sin(4\alpha) & \cos(4\alpha) \end{bmatrix} \tag{Appendix.27}$$

$$[K]^2=\begin{bmatrix} \cos(8\alpha) & -jR_0\sin(8\alpha) \\ -jG_0\sin(8\alpha) & \cos(8\alpha) \end{bmatrix} \tag{Appendix.28}$$

$$[K]^3=\begin{bmatrix} \cos(12\alpha) & -jR_0\sin(12\alpha) \\ -jG_0\sin(12\alpha) & \cos(12\alpha) \end{bmatrix} \tag{Appendix.29}$$

...... etc. Consequently, matrix [Q] will be given:

$$[Q]=[K]^N=\begin{bmatrix} \cos(N4\alpha) & -jR_0\sin(N4\alpha) \\ -jG_0\sin(N4\alpha) & \cos(N4\alpha) \end{bmatrix} \tag{Appendix.30}$$

From (Appendix.24) and (Appendix.35), Y_{11} and Y_{12} in "cross-field" model can be expressed:

$$Y_{11}=-jG_0\cot g(N4\alpha) \tag{Appendix.31}$$

$$Y_{12}=\frac{jG_0}{\sin(N4\alpha)} \tag{Appendix.32}$$

In conclusion, matrix [Y] representation of N-section IDT is:

- In "crossed-field" model, from (Appendix.6), (Appendix.13), (Appendix.14), (Appendix.31) and (Appendix.32):

$$\begin{aligned} Y_{11} &= -jG_0 \cot g(4N\alpha) \\ Y_{12} &= \frac{jG_0}{\sin(4N\alpha)} \\ Y_{13} &= -jG_0 tg\alpha \\ Y_{33} &= jN(2\omega C_0 + 4G_0 tg\alpha) \end{aligned} \qquad \text{(Appendix.33)}$$

- In "in-line field" model, from (Appendix.7), (Appendix.13), (Appendix.14), (Appendix.24) and (Appendix.25):

$$\begin{aligned} Y_{11} &= -\frac{Q_{11}}{Q_{12}} \\ Y_{12} &= \frac{1}{Q_{12}} \\ Y_{13} &= -jG_0 \frac{tg\alpha}{1 - \frac{2G_0}{\omega C_0} tg\alpha} \\ Y_{33} &= \frac{j2\omega N C_0}{1 - \frac{2G_0}{\omega C_0} tg\alpha} \end{aligned} \qquad \text{(Appendix.34)}$$

Where [Q] can be calculated from (Appendix.17) and (Appendix.21).

7.2 Appendix 2: Equivqlent circuit for "N+1/2" model IDT

In case IDT includes N periodic sections (like in section 3.2 plus one finger (in color red) as shown in Figure Appendix.8 that we call "N+1/2" model IDT.

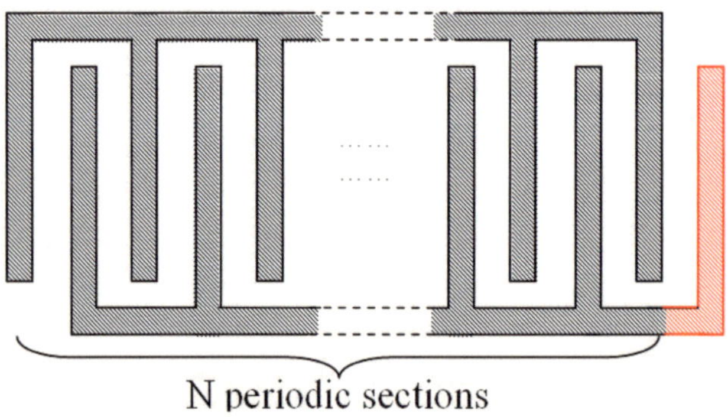

Fig. Appendix.8. "N+1/2" model IDT

The equivalent circuit for this model is shown in Figure Appendix.9 and the matrix [Yd] representation is shown as in Figure Appendix.10 (letter "d" stands for **d**ifferent from model [Y] in section 3.2.

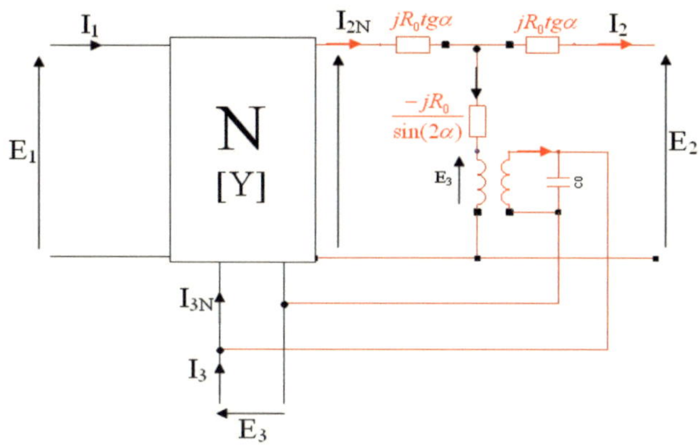

Fig. Appendix.9. Equivalent circuit of "N+1/2" model IDT

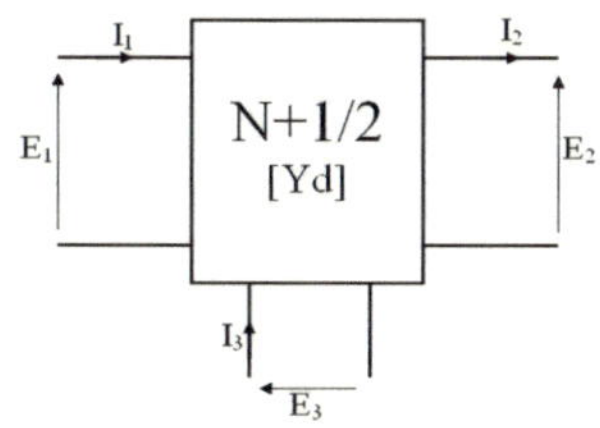

Fig. Appendix.10. [Yd] matrix representation of "N+1/2" model IDT

The form of matrix [Yd] is:

$$[Yd]=\begin{bmatrix} Yd_{11} & Yd_{12} & Yd_{13} \\ Yd_{21} & Yd_{22} & Yd_{23} \\ Yd_{31} & Yd_{32} & Yd_{33} \end{bmatrix} \tag{Appendix.35}$$

The elements of [Yd] matrix for "crossed-field" model are given as follows:

$$Yd_{11}=jG_0\left\{\frac{1}{\sin^2(4N\alpha)(\cot g(2\alpha)+\cot g(4N\alpha))}-\cot g(4N\alpha)\right\} \tag{Appendix.36}$$

$$Yd_{12}=\frac{jG_0}{\sin(4N\alpha)}\left\{\cos(2\alpha)-\frac{\sin(2\alpha)[\cot g(4N\alpha)\cos(2\alpha)-\sin(2\alpha)]}{\cos(2\alpha)+\cot g(4N\alpha)\sin(2\alpha)}\right\} \tag{Appendix.37}$$

$$Yd_{13}=jG_0\left\{\frac{(-tg\alpha+2\cot g(4N\alpha)\sin^2\alpha+\sin(2\alpha))\sin(2\alpha)}{\sin(4N\alpha)(\cos(2\alpha)+\cot g(4N\alpha)\sin(2\alpha))}+\frac{2\sin^2\alpha}{\sin(4N\alpha)}-tg\alpha\right\} \tag{Appendix.38}$$

$$Yd_{21}=-jG_0\frac{1}{\sin(4N\alpha)(\cos(2\alpha)+\cot g(4N\alpha)\sin(2\alpha))} \tag{Appendix.39}$$

$$Yd_{22}=jG_0\frac{\cot g(4N\alpha)\cos(2\alpha)-\sin(2\alpha)}{\cos(2\alpha)+\cot g(4N\alpha)\sin(2\alpha)} \tag{Appendix.40}$$

$$Yd_{23} = jG_0 \frac{-tg\alpha + 2\cot g(4N\alpha)\sin^2(2\alpha) + \sin(2\alpha)}{\cos(2\alpha) + \cot g(4N\alpha)\sin(2\alpha)} \quad \text{(Appendix.41)}$$

$$Yd_{31} = -jG_0 tg\alpha \quad \text{(Appendix.42)}$$

$$Yd_{32} = -jG_0 \sin(2\alpha) \quad \text{(Appendix.43)}$$

$$Yd_{33} = j\omega C_0(2N-1) + jG_0\{\sin(2\alpha) + (4N+1)tg\alpha\} \quad \text{(Appendix.44)}$$

7.3 Appendix 3: Scattering matrix [S] for IDT

The scattering matrix [S] of a three-port network characterized by its admittance matrix [Y] is given by [3]:

$$S = \Pi_3 - 2Y(\Pi_3 + Y)^{-1} \quad \text{(Appendix.45)}$$

Where Π_3 is the 3x3 identity matrix.
After expanding this equation, the scattering matrix elements for a general three-port network are given by the following expressions:

$$\begin{aligned} S_{11} = \frac{1}{M}\{&(1+Y_{33})(1-Y_{11}+Y_{22}-Y_{11}Y_{22}+Y_{12}Y_{21})+ \\ &+Y_{13}[Y_{31}(1+Y_{22})-Y_{21}Y_{32}]+Y_{23}[Y_{32}(Y_{11}-1)-Y_{12}Y_{31}]\} \end{aligned} \quad \text{(Appendix.46)}$$

$$S_{12} = -\frac{2}{M}[Y_{12}(1+Y_{33})-Y_{13}Y_{32}] \quad \text{(Appendix.47)}$$

$$S_{13} = -\frac{2}{M}[Y_{13}(1+Y_{22})-Y_{12}Y_{23}] \quad \text{(Appendix.48)}$$

$$S_{21} = -\frac{2}{M}[Y_{21}(1+Y_{33})-Y_{23}Y_{31}] \quad \text{(Appendix.49)}$$

$$\begin{aligned} S_{22} = \frac{1}{M}\{&(1+Y_{33})(1+Y_{11}-Y_{22}-Y_{11}Y_{22}+Y_{12}Y_{21})+ \\ &+Y_{13}[Y_{31}(Y_{22}-1)-Y_{21}Y_{32}]+Y_{23}[Y_{32}(Y_{11}+1)-Y_{12}Y_{31}]\} \end{aligned} \quad \text{(Appendix.50)}$$

$$S_{23} = -\frac{2}{M}[Y_{23}(1+Y_{11})-Y_{13}Y_{21}] \quad \text{(Appendix.51)}$$

$$S_{31} = -\frac{2}{M}[Y_{31}(1+Y_{22})-Y_{21}Y_{32}] \quad \text{(Appendix.52)}$$

$$S_{32} = -\frac{2}{M}[Y_{32}(1+Y_{11})-Y_{12}Y_{31}] \quad \text{(Appendix.53)}$$

$$\begin{aligned} S_{33} = \frac{1}{M}\{&(1-Y_{33})(1+Y_{11}+Y_{22}+Y_{11}Y_{22}-Y_{12}Y_{21})+ \\ &+Y_{13}[Y_{31}(Y_{22}+1)-Y_{21}Y_{32}]+Y_{23}[Y_{32}(Y_{11}+1)-Y_{12}Y_{31}]\} \end{aligned} \quad \text{(Appendix.54)}$$

where

$$\begin{aligned} M &= \det(\Pi_3 + Y) \\ &= (1+Y_{33})[(1+Y_{11})(1+Y_{22}) - Y_{12}Y_{21}] - Y_{23}[Y_{32}(1+Y_{11}) - Y_{12}Y_{31}] - \\ &\quad - Y_{13}[Y_{31}(1-Y_{22}) - Y_{21}Y_{32}] \end{aligned} \quad \text{(Appendix.55)}$$

For model IDT including N identical sections, these equations can be further simplified. In case of Figure Appendix.7 (b):

$$\begin{aligned} Y_{11} &= Y_{22} \\ Y_{21} &= Y_{12} \\ Y_{31} &= Y_{13} \\ Y_{23} &= Y_{32} = -Y_{13} \end{aligned} \quad \text{(Appendix.56)}$$

Therefore, S_{ij}'s take the following form

$$S_{11} = S_{22} = \frac{1}{M}\left\{(1+Y_{33})(1-Y_{11}^2+Y_{12}^2) + 2Y_{13}^2(Y_{11}+Y_{12})\right\} \quad \text{(Appendix.57)}$$

$$S_{12} = S_{21} = -\frac{2}{M}\left[Y_{12}(1+Y_{33}) + Y_{13}^2\right] \quad \text{(Appendix.58)}$$

$$S_{13} = S_{31} = -\frac{2}{M}Y_{13}(1+Y_{11}+Y_{12}) \quad \text{(Appendix.59)}$$

$$S_{23} = S_{32} = -S_{13} \quad \text{(Appendix.60)}$$

$$S_{33} = \frac{1}{M}\left\{(1-Y_{33})[(1+Y_{11})^2 - Y_{12}^{\ 2}] + 2Y_{13}^2(1+Y_{11}+Y_{12})\right\} \quad \text{(Appendix.61)}$$

Where

$$M = (1+Y_{33})[(1+Y_{11})^2 - Y_{12}^{\ 2}] - 2Y_{13}^2(1+Y_{12}) \quad \text{(Appendix.62)}$$

7.4 Appendix 4: Equivalent circuit for SAW device base on Mason model, [ABCD] Matrix representation

7.4.1 Appendix 4.1: [ABCD] Matrix representation of IDT

In SAW device, each input and output IDTs have one terminal connected to admittance G_0. Therefore, one IDT can be represented as two-port network. [ABCD] matrix (as in Figure Appendix.11) is used to represent each IDT, because [ABCD] matrix representation has one interesting property that in cascaded network, the [ABCD] matrix of total network can be obtained easily by multiplying the matrices of elemental networks.

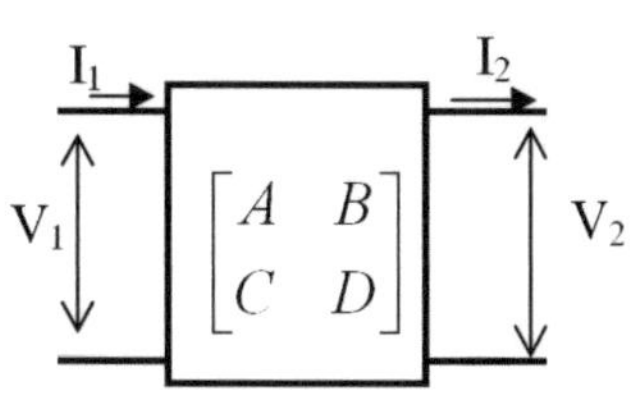

Fig. Appendix.11. [ABCD] representation of two-port network for one IDT

To find the [ABCD] matrix for one IDT in SAW device, the condition that no reflected wave at one terminal of IDT, and the current-voltage relations by [Y] matrix in section are used as follows:

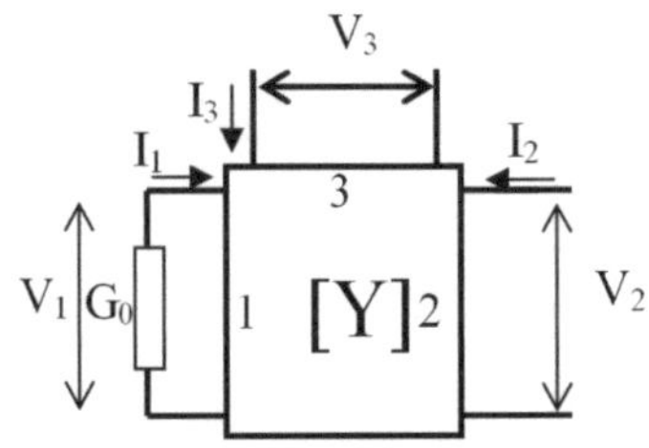

Fig. Appendix.12. Two-port network for one IDT

$$\begin{bmatrix} I_1 \\ I_2 \\ I_3 \end{bmatrix} = \begin{bmatrix} Y_{11} & Y_{12} & Y_{13} \\ Y_{12} & Y_{11} & -Y_{13} \\ Y_{13} & -Y_{13} & Y_{33} \end{bmatrix} \begin{bmatrix} V_1 \\ V_2 \\ V_3 \end{bmatrix} \quad \text{(Appendix.63)}$$

And $I_1 = -G_0 V_1$ (Appendix.64)

From these current-voltage relations, the V_3 and I_3 are given:

$$V_3 = \frac{Y_{11}^2 - Y_{12}^2 + Y_{11}G_0}{Y_{12}Y_{13} + Y_{11}Y_{13} + Y_{13}G_0} V_2 - \frac{G_0 + Y_{11}}{Y_{12}Y_{13} + Y_{11}Y_{13} + Y_{13}G_0} I_2 \quad \text{(Appendix.65)}$$

$$I_3 = \frac{-(Y_{13}Y_{12} + Y_{13}Y_{11} + Y_{13}G_0)^2 + (Y_{11}Y_{33} - Y_{13}^2 + Y_{33}G_0)(Y_{11}^2 - Y_{12}^2 + Y_{11}G_0)}{(G_0 + Y_{11})(Y_{12}Y_{13} + Y_{11}Y_{13} + Y_{13}G_0)} V_2 - \frac{Y_{11}Y_{33} - Y_{13}^2 + Y_{33}G_0}{Y_{12}Y_{13} + Y_{11}Y_{13} + Y_{13}G_0} I_2 \quad \text{(Appendix.66)}$$

From (Appendix.65) and (Appendix.66), equivalence between port 3 in Figure Appendix.12 equals to port 1 in Figure Appendix.11, and consideration of direction of current I_2 in Figure Appendix.11 and Figure Appendix.12, [ABCD] matrix representation for two-port network of IDT in obtained:

$$A = \frac{Y_{11}^2 - Y_{12}^2 + Y_{11}G_0}{Y_{12}Y_{13} + Y_{11}Y_{13} + Y_{13}G_0} \quad \text{(Appendix.67)}$$

$$B = \frac{G_0 + Y_{11}}{Y_{12}Y_{13} + Y_{11}Y_{13} + Y_{13}G_0} \quad \text{(Appendix.68)}$$

$$C = \frac{-(Y_{13}Y_{12} + Y_{13}Y_{11} + Y_{13}G_0)^2 + (Y_{11}Y_{33} - Y_{13}^2 + Y_{33}G_0)(Y_{11}^2 - Y_{12}^2 + Y_{11}G_0)}{(G_0 + Y_{11})(Y_{12}Y_{13} + Y_{11}Y_{13} + Y_{13}G_0)} \quad \text{(Appendix.69)}$$

$$D = \frac{Y_{11}Y_{33} - Y_{13}^2 + Y_{33}G_0}{Y_{12}Y_{13} + Y_{11}Y_{13} + Y_{13}G_0} \quad \text{(Appendix.70)}$$

In case of "crossed-field" model, the [ABCD] can be further simplified:

$$A = \frac{\sin(4N\alpha) - j\cos(4N\alpha)}{tg\alpha[1 - \cos(4N\alpha) - j\sin(4N\alpha)]} \qquad \text{(Appendix.71)}$$

$$B = \frac{A}{G_0} \qquad \text{(Appendix.72)}$$

$$D = \frac{\sin(4N\alpha)}{1 - \cos(4N\alpha) - j\sin(4N\alpha)}[N(2\omega C_0 Z_0 \cot\alpha + 4)(\cot(4N\alpha) + j) + tg\alpha] \qquad \text{(Appendix.73)}$$

$$C = -\frac{1}{B} + G_0 D \qquad \text{(Appendix.74)}$$

One interesting property of [ABCD] of "crossed-field" mode is:

$$AD\text{-}BC=1 \qquad \text{(Appendix.75)}$$

This means [ABCD] matrix is reciprocal.
In SAW device, the ouput IDT is inverse of input IDT. By the reciprocal property of [ABCD], the [ABCD] matrix of output IDT can be easily obtained:

$$A_{output} = D_{input} \qquad \text{(Appendix.76)}$$

$$B_{output} = B_{input} \qquad \text{(Appendix.77)}$$

$$C_{output} = C_{input} \qquad \text{(Appendix.78)}$$

$$D_{output} = A_{input} \qquad \text{(Appendix.79)}$$

in which N is replaced by M (number of periodic sections in output IDT)
Consequently, the [ABCD] matrix of output IDT is:

$$A_{out} = \frac{\sin(4M\alpha)}{1 - \cos(4M\alpha) - j\sin(4M\alpha)}[M(2\omega C_0 Z_0 \cot\alpha + 4)(\cot(4M\alpha) + j) + tg\alpha] \qquad \text{(Appendix.80)}$$

$$B_{out} = \frac{1}{G_0}\frac{\sin(4M\alpha) - j\cos(4M\alpha)}{tg\alpha[1 - \cos(4M\alpha) - j\sin(4M\alpha)]} \qquad \text{(Appendix.81)}$$

$$D_{out} = \frac{\sin(4M\alpha) - j\cos(4M\alpha)}{tg\alpha[1 - \cos(4M\alpha) - j\sin(4M\alpha)]} \qquad \text{(Appendix.82)}$$

$$C_{out} = -\frac{1}{B_{out}} + G_0 A_{out} \qquad \text{(Appendix.83)}$$

At the center frequency f_0, the [ABCD] matrix becomes infinite since $\alpha=0.5\pi(f/f_0)= 0.5\pi$. However, [ABCD] elements may be calculated by expanding for frequency very near frequency f_0.
By setting:

$$\alpha = \frac{\pi}{2}\frac{f - f_0}{f_0} + \frac{\pi}{2} = \frac{x}{2N} + \frac{\pi}{2} \qquad \text{(Appendix.84)}$$

Where $$x = N\pi\frac{f-f_0}{f_0} \qquad \text{(Appendix.85)}$$

By using the limit of some functions as follows:

$$\lim_{x\to 0}[\sin(4N\alpha)] = \lim_{x\to 0}[\sin(2x)] \approx 2x \qquad \text{(Appendix.86)}$$

$$\lim_{x\to 0}[\cos(4N\alpha)] = \lim_{x\to 0}[\cos(2x)] \approx 1 \qquad \text{(Appendix.87)}$$

$$\lim_{x\to 0}[tg\alpha] = \lim_{x\to 0}[-\cot(\frac{x}{2N})] \approx -\frac{2N}{x} \qquad \text{(Appendix.88)}$$

The [ABCD] matrix of input IDT is obtained:

$$A \approx \frac{2x-j}{j4N} \qquad \text{(Appendix.89)}$$

$$B \approx \frac{1}{G_0}\frac{2x-j}{j4N} \qquad \text{(Appendix.90)}$$

$$C \approx 2\pi fC_0x - 4NG_0 - j\left(\pi fC_0 + \frac{4NG_0}{2x-j}\right) \qquad \text{(Appendix.91)}$$

$$D \approx 2\pi fC_0Z_0x - 4N - j\pi fC_0Z_0 \qquad \text{(Appendix.92)}$$

7.4.2 Appendix 4.2: [ABCD] matrix representation of propagation path

Based on equivalent circuit star model of propagation path in section 3.3, [ABCD] matrix representation of propagation way can be obtained clearly:

$$A_{path} = D_{path} = \cos 2\theta \qquad \text{(Appendix.93)}$$

$$B_{path} = C_{path} = j\sin 2\theta \qquad \text{(Appendix.94)}$$

With $$\theta = \frac{\pi fl}{v} \qquad \text{(Appendix.95)}$$

Where l is the length of propagation path between two IDTs.
So, [ABCD] matrix representations of input IDT, propagation way and output IDT are obtained. They are cascaded as Figure Appendix.13:

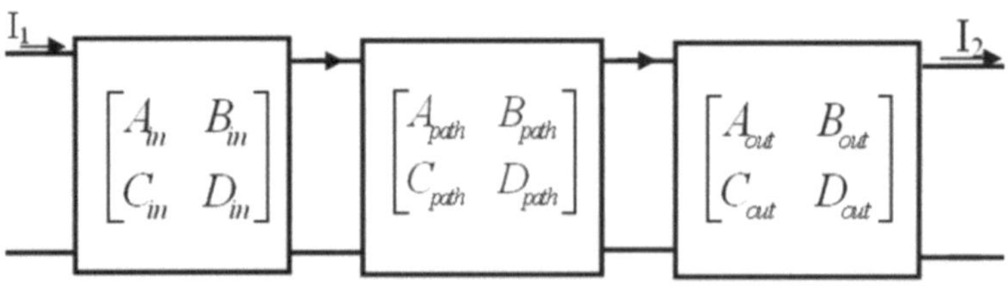

Fig. Appendix.13. Cascaded [ABCD] matrices of input IDT, propagation way and output IDT

And the [ABCD] equivalent matrix of SAW device is shown in Figure Appendix.14

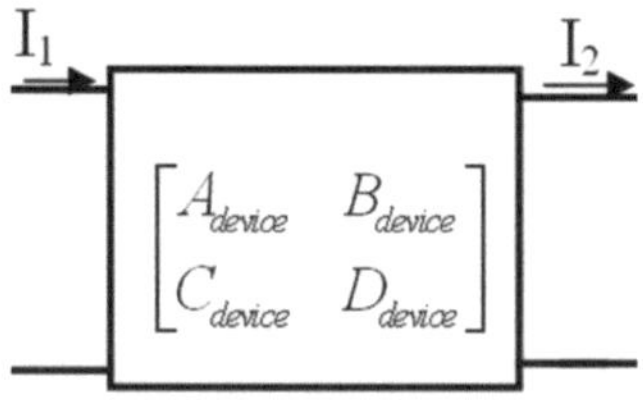

Fig. Appendix.14. [ABCD] matrix of SAW device

[ABCD] matrix of delay line SAW is

$$\begin{bmatrix} A_{device} & B_{device} \\ C_{device} & D_{device} \end{bmatrix} = \begin{bmatrix} A_{in} & B_{in} \\ C_{in} & D_{in} \end{bmatrix} \begin{bmatrix} A_{path} & B_{path} \\ C_{path} & D_{path} \end{bmatrix} \begin{bmatrix} A_{out} & B_{out} \\ C_{out} & D_{out} \end{bmatrix} \quad \text{(Appendix.96)}$$

$$A_{device} = A_{in}A_{path}A_{out} + B_{in}C_{path}A_{out} + A_{in}B_{path}C_{out} + B_{in}D_{path}C_{out} \quad \text{(Appendix.97)}$$

$$B_{device} = A_{in}A_{path}B_{out} + B_{in}C_{path}B_{out} + A_{in}B_{path}D_{out} + B_{in}D_{path}D_{out} \quad \text{(Appendix.98)}$$

$$C_{device} = C_{in}A_{path}A_{out} + D_{in}C_{path}A_{out} + C_{in}B_{path}C_{out} + D_{in}D_{path}C_{out} \quad \text{(Appendix.99)}$$

$$D_{device} = C_{in}A_{path}B_{out} + D_{in}C_{path}B_{out} + C_{in}B_{path}D_{out} + D_{in}D_{path}D_{out} \quad \text{(Appendix.100)}$$

Where $[ABCD]_{in}$ is calculated from (Appendix.71), (Appendix.72), (Appendix.73) and (Appendix.74).
$[ABCD]_{out}$ is calculated from (Appendix.80), (Appendix.81), (Appendix.82) and (Appendix.83).
$[ABCD]_{path}$ is calculated from (Appendix.93) and (Appendix.94).

8. References

[1] C.C.W.Ruppel, W.Ruile, G.Scholl, K.Ch.Wagner, and O.Manner, Review of models for low-loss filter design and applications, *IEEE Ultransonics Symposium*, pp.313-324, 1994.

[2] L.A.Coldren, and R.L.Rosenberg, Scattering matrix approach to SAW resonators, *IEEE Ultrasonics Symposium*, 1976, pp.266-271.

[3] R.W.Newcomb, Linear Multiport Synthesis, McGraw Hill, 1966.

[4] C.Elachi, Waves in Active and Passive Periodic Structures: A Review, *Proceedings of the IEEE*, vol.64, No.12, December 1976, pp.1666-1698

[5] M.Hikita, A.Isobe, A.Sumioka, N.Matsuura, and K.Okazaki, Rigorous Treatment of Leaky SAW's and New Equivalent Circuit Representation for Interdigital Transducers, *IEEE Transactions on Ultrasonics, Ferroelectrics, and Frequency Control*, Vol.43, No.3, May 1996.

[6] L.F.Brown, and D.L.Carlson, Ultrasound Transducer Models for Piezoelectric Polymer Films, *IEEE Transactions on Ultrasonics, Ferroelectrics, and Frequency Control*, Vol.36, No.3, May 1989.

[7] K.Hashimoto, and M.Yamaguchi, Precise simulation of surface transverse wave devices by discrete Green function theory, *IEEE Ultrasonics Symposium*, 1994, pp.253-258.

[8] K.Hashimoto, G.Endoh, and M.Yamaguchi, Coupling-of-modes modelling for fast and precise simulation of leaky surface acoustic wave devices, *IEEE Ultrasonics Symposium*, 1995, pp.251-256.

[9] K.Hashimoto, and M.Yamaguchi, General-purpose simulator for leaky surface acoustic wave devices based on Coupling-Of-Modes theory, *IEEE Ultrasonics Symposium*, 1996, pp.117-122.

[10] K.Hashimoto, Surface Acoustic Wave Devices in Telecommunications, *Modelling and Simulation*, Springer, 2000, ISBN: 9783540672326.

[11] P.M.Smith, and C.K.Campbell, A Theoretical and Experimental Study of Low-Loss SAW Filters with Interdigitated Interdigital Transducers, *IEEE Transactions on Ultrasonics, Ferroelectrics, and Frequency Control*, Vol.36, No.1, January 1989, pp.10-15.

[12] C.K.Campbell, Modelling the Transverse-Mode Response of a Two-Port SAW Resonator, *IEEE Transactions on Ultrasonics, Ferroelectrics, and Frequency Control*, Vol.38, No.3, May 1991, pp.237-242.

[13] C.K.Campbell, P.M.Smith, and P.J.Edmonson, Aspects of Modeling the Frequency Response of a Two-Port Waveguide-Coupled SAW Resonator-Filter, *IEEE Transactions on Ultrasonics, Ferroelectrics, and Frequency Control*, Vol.39, No.6, November 1992, pp.768-773.

[14] C.K.Campbell, Longitudinal-Mode Leaky SAW Resonator Filters on 64^0 Y-X Lithium Niobate, *IEEE Transactions on Ultrasonics, Ferroelectrics, and Frequency Control*, Vol.42, No.5, September 1995, pp.883-888.

[15] C.K.Campbell, and P.J.Edmonson, Conductance Measurements on a Leaky SAW Harmonic One-Port Resonator, *IEEE Transactions on Ultrasonics, Ferroelectrics, and Frequency Control*, Vol.47, No.1, January 2000, pp.111-116.

[16] C.K.Campbell, and P.J.Edmonson, Modeling a Longitudinally Coupled Leaky-SAW Resonator Filter with Dual-Mode Enhanced Upper-Sideband Suppression, *IEEE Transactions on Ultrasonics, Ferroelectrics, and Frequency Control*, Vol.48, No.5, September 2001, pp.1298-1301.

[17] J.Munshi, and S.Tuli, A Circuit Simulation Compatible Surface Acoustic Wave Interdigital Transducer Macro-Model, *IEEE Transactions on Ultrasonics, Ferroelectrics, and Frequency Control*, Vol.51, No.7, July 2004, pp.782-784.

[18] M.P.Cunha, and E.L.Adler, A Network Model For Arbitrarily Oriented IDT Structures, *IEEE Transactions on Ultrasonics, Ferroelectrics, and Frequency Control*, Vol.40, No.6, November 1993, pp.622-629.

[19] D.R.Mahapatra, A.Singhal, and S.Gopalakrishnan, Numerical Analysis of Lamb Wave Generation in Piezoelectric Composite IDT, *IEEE Transactions on Ultrasonics, Ferroelectrics, and Frequency Control*, Vol.52, No.10, October 2005, pp.1851-1860.

[20] A APPENDIX. Bhattacharyya, Suneet Tuli, and S.Majumdar, SPICE Simulation of Surface Acoustic Wave Interdigital Transducers, *IEEE Transactions on Ultrasonics, Ferroelectrics, and Frequency Control*, Vol.42, No.4, July 1995, pp.784-786.

[21] C.M.Panasik, and APPENDIX.J.Hunsinger, Scattering Matrix Analysis Of Surface Acoustic Wave Reflectors And Transducers, *IEEE Transactions On Sonics And Ultrasonics*, Vol.SU-28, No.2, March 1981, pp.79-91.

[22] W.Soluch, Admittance Matrix Of A Surface Acoustic Wave Interdigital Transducer, *IEEE Transactions on Ultrasonics, Ferroelectrics, and Frequency Control*, Vol.40, No.6, November 1993, pp.828-831.

[23] W.Soluch, Scattering Matrix Approach To One Port SAW Resonators, *IEEE Frequency Control Symposium*, 1999, pp.859-862.

[24] W.Soluch, Design of SAW Synchronous Resonators on ST Cut Quartz, *IEEE Transactions on Ultrasonics, Ferroelectrics, and Frequency Control*, Vol.46, No.5, September 1999, pp.1324-1326.
[25] W.Soluch, Scattering Matrix Approach To One-Port SAW Resonators, *IEEE Transactions on Ultrasonics, Ferroelectrics, and Frequency Control*, Vol.47, No.6, November 2000, pp.1615-1618.
[26] W.Soluch, Scattering Matrix Approach To STW Resonators, *IEEE Transactions on Ultrasonics, Ferroelectrics, and Frequency Control*, Vol.49, No.3, March 2002, pp.327-330.
[27] W.Soluch, Scattering Analysis Of Two-Port SAW Resonators, *IEEE Transactions on Ultrasonics, Ferroelectrics, and Frequency Control*, Vol.48, No.3, May 2001, pp.769-772.
[28] W.Soluch, Scattering Matrix Approach To STW Multimode Resonators, Electronics Letters, 6th January 2005, Vol.41, No.1.
[29] K.Nakamura, A Simple Equivalent Circuit For Interdigital Transducers Based On The Couple-Mode Approach, *IEEE Transactions on Ultrasonics, Ferroelectrics, and Frequency Control*, Vol.40, No.6, November 1993, pp.763-767.
[30] K. Nakamura, and K.Hirota, Equivalent circuit for Unidirectional SAW-IDT's based on the Coupling-Of-modes theory, *IEEE Trans on Ultrasonics, Ferroelectrics, and Frequency Control*, Vol.43, No.3, May 1996, pp.467-472.
[31] A.H.Fahmy, and E.L.Adler, Propagation of acoustic surface waves in multilayers: A matrix description. *Applied Physics Letter*, vol. 22, No.10, 1973, pp. 495-497.
[32] E.L.Adler, Matrix methods applied to acoustic waves in multilayers, *IEEE Transactions on Ultrasonics, Ferroelectrics, and Frequency Control*, Vol.37, No.6, November 1990, pp.485-490.
[33] E.L.Adler, SAW and Pseudo-SAW properties using matrix methods, *IEEE Transactions on Ultrasonics, Ferroelectrics, and Frequency Control*, Vol.41, No.5, September 1994, pp.699-705.
[34] G.F.Iriarte, F.Engelmark, I.V.Katardjiev, V.Plessky, V.Yantchev, SAW COM-parameter extraction in AlN/diamond layered structures, *IEEE Transactions On Ultrasonics, Ferroelectrics, And Frequency Control*, Vol. 50, No. 11, November 2003.
[35] M.Mayer, G.Kovacs, A.Bergmann, and K.Wagner, A Powerful Novel Method for the Simulation of Waveguiding in SAW Devices, *IEEE Ultrasonics Symposium*, 2003, pp.720-723.
[36] W.P.Mason, Electromechanical Transducer and Wave Filters, second edition, D.Van Nostrand Company Inc, 1948.
[37] W.P.Mason, Physical Acoustics, Vol 1A, Academic Press, New York 1964.
[38] S.D.Senturia, Microsystem Design, Kluwer Academic Publishers, 2001, ISBN 0-7923-7246-8.
[39] W.Marshall Leach, Controlled-Source Analogous Circuits and SPICE models for Piezoelectric transducers, *IEEE Transactions on Ultrasonics, Ferroelectrics, and Frequency Control*, Vol.41, No.1, January 1994.
[40] D.A.Berlincourt, D.R.Curran and H.Jaffe, Chapter 3, Piezoelectric and Piezomagnetic Materials and Their Function in Tranducers.
[41] W.R.Smith, H.M.Gerard, J.H.Collins, T.M.Reeder, and H.J.Shaw, Analysis of Interdigital Surface Wave Transducers by Use of an Equivalent Circuit Model, *IEEE Transaction on MicroWave Theory and Techniques*, No.11, November 1969, pp.856-864.
[42] C. K. Campbell, Surface acoustic wave devices, in *Mobile and Wireless Communications*, New York: Academic, 1998.

[43] O.Tigli, and M.E.Zaghloul, A Novel Saw Device in CMOS: Design, Modeling, and Fabrication, *IEEE Sensors journal*, vol. 7, No. 2, February 2007, pp.219-227.

[44] K.Nakamura, A Simple Equivalent Circuit For Interdigital Transducers Based On The Couple-Mode Approach, *IEEE Transactions on Ultrasonics, Ferroelectrics, and Frequency Control*, Vol.40, No.6, November 1993, pp.763-767.

[45] M.Hofer, N.Finger, G.Kovacs, J.Schoberl, S.Zaglmayr, U.Langer, and R.Lerch, Finite-Element Simulation of Wave Propagation in Periodic Piezoelectric SAW Structures, *IEEE Transactions on Ultrasonics, Ferroelectrics, and Frequency Control*, Vol.53, No.6, June 2006, pp.1192-1201.

[46] M. Hofer, N. Finger, G. Kovacs, J. Schoberl, U. Langer, and R. Lerch, Finite-element simulation of bulk and surface acoustic wave (SAW) interaction in SAW devices, *IEEE Ultrasonics Symposium*, 2002.

[47] Online: http://www.comsol.com/

[48] Online: http://www.coventor.com/

[49] Online: http://www.ansys.com/

[50] J.J. Campbell, W.R. Jones, A method for estimating optimal crystal cuts and propagation directions for excitation of piezoelectric surface waves, *IEEE Transaction on Sonics Ultrasonics.* SU-15 (4), 1968, pp.209–217.

[51] E.Akcakaya, E.L.Adler, and G.W.Farnell, Apodization of Multilayer Bulk-Wave Transducers, *IEEE Transactions on Ultrasonics Ferroelectrics and Frequency Control*, Vol.36, No.6, November 1989, pp 628-637.

[52] E.L.Adler, J.K.Slaboszewicz, G.W.FARNELL, and C.K.JEN, PC Software for SAW Propagation in Anisotropic Multilayers, *IEEE Transactions on Ultrasonics Ferroelectrics and Frequency Control*, Vol.37, No.2, May 1990, pp.215-223.

[53] E.L.Adler, SAW and Pseudo-SAW Properties Using Matrix Methods, *IEEE Transactions on Ultrasonics Ferroelectrics and Frequency Control*, Vol.41, No.6, pp.876-882, September 1994.

[54] K. A. Ingebrigtsen, Surface waves in piezoelectrics, *Journal of Applied Physics*, Vol.40, No.7, 1969, pp.2681-2686.

[55] Y.Suzuki, H.Shimizu, M.Takeuchi, K.Nakamura, and A.Yamada, Some studies on SAW resonators and multiple-mode filters, *IEEE Ultrasonics Symposium Proceedings*, 1976, pp.297-302.

[56] S.D.Senturia, Microsystem Design, Kluwer Academic Publishers, 2001, ISBN 0-7923-7246-8.

[57] W.P.Mason, Electromechanical Transducer and Wave Filters, second edition, D.Van Nostrand Company Inc, 1948.

[58] W.P.Mason, Physical Acoustics, Vol 1A, Academic Press, New York 1964.

[59] W.R.Smith, H.M.Gerard, J.H.Collins, T.M.Reeder, and H.J.Shaw, Analysis of Interdigital Surface Wave Transducers by Use of an Equivalent Circuit Model, *IEEE Transaction on MicroWave Theory and Techniques*, No.11, November 1969, pp.856-864.

[60] Y.Suzuki, H.Shimizu, M.Takeuchi, K.Nakamura, and A.Yamada, Some studies on SAW resonators and multiple-mode filters, *IEEE Ultrasonics Symposium Proceedings*, 1976, pp.297-302.

[61] K.Nakamura, A Simple Equivalent Circuit For Interdigital Transducers Based On The Couple-Mode Approach, *IEEE Transactions on Ultrasonics, Ferroelectrics, and Frequency Control*, Vol.40, No.6, November 1993, pp.763-767.

21

Sources of Third–Order Intermodulation Distortion in Bulk Acoustic Wave Devices: A Phenomenological Approach

Eduard Rocas and Carlos Collado
Universitat Politècnica de Catalunya (UPC), Barcelona
Spain

1. Introduction

Acoustic devices like Bulk Acoustic Wave (BAW) resonators and filters represent a key technology in modern microwave industry. More specifically, BAW technology offers promising performance due to its good power handling and high quality factors that make it suitable for a wide range of applications. Nevertheless, harmonics and 3IMD arising from intrinsic nonlinear material properties (Collado et al., 2009) and dynamic self-heating (Rocas et al., 2009) could represent a limitation for some applications.

Driven by the need for highly linear devices, there is a demand for further development of accurate models of BAW devices, capable of predicting the nonlinear behavior of the device and its impact on a circuit. Many authors have attempted to model the nonlinearities of BAW devices by using different approaches, mostly involving phenomenological lumped element models. Although these models can be useful because of their simplicity, they are mainly limited to narrow-band operation and they usually cannot be parameterized in terms of device-independent parameters (Constantinescu et al., 2008). Another approach consists of extending all the material properties on the constitutive equations to the nonlinear domain and introducing the nonlinear relations to the model implementation, which leads to several possible nonlinear sources increasing model complexity (Cho et al., 1993; Ueda et al., 2008). On the other hand, (Feld, 2009) presents a one-parameter nonlinear circuit model to account for the intrinsic nonlinearities. Such a model does not include the self-heating mechanism and can underestimate the 3IMD by more than 20 dB.

In this work, a model that uses several nonlinear parameters to predict harmonics and 3IMD distortion is presented. Its novelty lies in its ability to predict the nonlinear effects produced by self-heating in addition to those due to intrinsic nonlinearities in the material properties. The model can be considered an extension of the nonlinear KLM model (originally proposed by Krimholtz, Leedom and Matthaei) (Krimholtz et al., 1970) to include the thermal effects due to self-heating caused by viscous losses and electrode losses. For this purpose a thermal domain circuit model is implemented and coupled to the electro-acoustic model, which allows us to calculate the dynamic temperature variations that change the material properties. In comparison to (Rocas et al., 2009), this work describes the impact that electrode losses produce on the 3IMD, presents closed-form expressions derived from the circuit model and validates the model with extensive measurements that confirm the necessity to include dynamic self-heating to accurately predict the generation of spurious signals in BAW devices.

2. Nonlinear generation mechanisms

The origin of nonlinearities in BAW resonators has been controversial and there still exists no consensus (Nakamura et al., 2010). However, recent results point to several underlying causes which combine in different ways to give rise to a wide range of nonlinear effects (Rocas et al., 2009). We summarize the nonlinear effects of a stiffened elasticity, and then address the nonlinearity due to self-heating caused by viscous losses and electrode losses. We develop a circuit model to describe self-heating effects, and compare the measured results with simulations. Closed-form expressions for a simple one-layer BAW device model are then extracted to better understand the nonlinear generation mechanisms.

2.1 Nonlinear stiffened elasticity

Nonlinear elasticity has been proposed as the predominant contribution to the measured second harmonics and as a potential source of the observed 3IMD products (Collado et al., 2009) in two-tone experiments.

The approach described in (Collado et al., 2009) starts by considering a nonlinear stress-strain relation under electric field described by a nonlinear stiffened elasticity $c^D(T)$ in the form of the polynomial

$$c^D(T) = c_0^D + \Delta c_1^D T + \Delta c_2^D T^2 \tag{1}$$

where T is the stress. As detailed in (Collado et al., 2009), (1) translates into a nonlinear distributed capacitance $C_d(v)$ in the equivalent electric model of the acoustic transmission line (Auld, 1990), in which the voltage v is equivalent to force. In the equivalent electric model (1) transforms into:

$$C_d(v) = C_{d,0} + \Delta C_1 v + \Delta C_2 v^2. \tag{2}$$

Equation (2) leads to the nonlinear acoustic Telegrapher's equations which can be used to obtain the maximum voltage amplitude occurring along a resonating transmission line as shown in (Collado et al., 2009; Collado et al., 2005). When the device is driven by two tones at frequencies ω_1 and ω_2, standing waves with maximum force amplitudes $V_{\omega 1}$ and $V_{\omega 2}$ are trapped in the line. Then, as detailed in (Collado et al., 2009), the nonlinear capacitance (2) is responsible for generating 3IMD signals that result from adding the contributions due to Δc_1^D and Δc_2^D:

$$V_{\omega_{12}} = A_1 Q_L V_{\omega_1}^2 V_{\omega_2}^* \Delta C_1^{\ 2} \tag{3}$$

$$V_{\omega_{12}} = A_2 Q_L V_{\omega_1}^2 V_{\omega_2}^* \Delta C_2 \tag{4}$$

where $\omega_{12} = 2\omega_1 - \omega_2$, Q_L is the loaded quality factor and A_1 and A_2 are constants that depend on the geometry of the device and on its materials. Identical results would be obtained for the 3IMD at $2\omega_2 - \omega_1$ (which we will denote as ω_{21}).

2.2 Self-heating

Third-order intermodulation distortion due to dynamic self-heating is a well known process in microwave power amplifiers (Camarchia et al., 2007; Parker et al., 2004; Vuolevi et al., 2001) but has received less attention in passive devices (Rocas et al., 2010). What makes it different from the 3IMD caused by intrinsic nonlinearities is its dependence on the envelope frequency of the signal. For the particular case of a two-tone experiment, in which the envelope is a sinusoid, the thermal generation of 3IMD has a low-pass dependence on the envelope frequency due to the slow dynamics related with heating effects.

Recent results of two-tone 3IMD tests in BAW resonators as a function of the tones spacing reveal the important impact of self-heating effects in thin-Film Bulk Acoustic Resonators (FBAR) (Collado et al., 2009; Feld, 2009; Rocas et al., 2008) and Solidly Mounted Resonators (SMR) (Rocas et al., 2009). Heating produced by viscous damping in the acoustic domain and by ohmic loss in the electric domain produce local temperature oscillations which affect the temperature-dependent material properties.

If $\omega_1 = \omega_0 - \Delta\omega/2$ and $\omega_2 = \omega_0 + \Delta\omega/2$ are the input signals for a two-tone test, dissipation occurs as a result of electric and acoustic losses, and the quadratic dependence of the dissipated power on the signal amplitude

$$P_d \propto \left(V_1 \cos\left(\left(\omega_0 - \frac{\Delta\omega}{2} \right) t \right) + V_2 \cos\left(\left(\omega_0 + \frac{\Delta\omega}{2} \right) t \right) \right)^2 \tag{5}$$

gives rise to several frequency components of the dissipated power:

$$\begin{aligned} P_d \propto\ & \frac{1}{2}V_1^2 + \frac{1}{2}V_2^2 + \frac{1}{2}V_1^2 \cos(2\omega_0 t - \Delta\omega t) + \\ & \frac{1}{2}V_2^2 \cos(2\omega_0 t + \Delta\omega t) + V_1 V_2 \cos(2\omega_0 t) \\ & + V_1 V_2 \cos(\Delta\omega t). \end{aligned} \tag{6}$$

These frequency components produce temperature variations on the device at the same frequencies. These temperature variations $K(\omega)$ can be written in terms of the dissipated power and the thermal impedance as (Parker et al., 2004)

$$K(\omega) = Z_{th}(\omega) P_d(\omega). \tag{7}$$

It is important to point out that the temperature variation at the envelope frequency ($\Delta\omega = \omega_2 - \omega_1$) is the most relevant for the generation of spurious signals because of the low-pass filter character of the thermal impedance $Z_{th}(\omega)$. These slow temperature oscillations induce low frequency changes of the material properties, and consequently, generate undesired 3IMD.

In addition to being able to calculate the temperature oscillations, we also need to determine how these oscillations influence the device performance. For the specific case of BAW devices, there is consensus in assuming that the detuning of BAW devices with temperature is due to the variation of multiple material properties with temperature (Lakin et al., 2000; Ivira et al., 2008; Petit et al., 2007). We reflect this in our model by adding a temperature-dependent term to the stiffened elasticity in (1)

$$c^D(T,K) = c_0^D + \Delta c_1^D T + \Delta c_2^D T^2 + \Delta c_K^D K \tag{8}$$

where K represents the temperature, the equivalent capacitance is

$$C_d(v,K)=C_{d,0}+\Delta C_1 v+\Delta C_2 v^2+\Delta C_K K, \tag{9}$$

where each of the nonlinear terms ΔC_1, ΔC_2 and ΔC_K are related to their counterparts Δc_1^D , Δc_2^D, Δc_K^D respectively, as detailed in Appendix I.
The term ΔC_K generates 3IMD, whose maximum voltage $V_{\omega 12}$ can be found in a similar way as the contribution of ΔC_1 in (3) and ΔC_2 in (4) (see details in Appendix I):

$$V_{\omega_{12}}=A_T Q_L \Delta C_K P_{d\Delta\omega} Z^*_{th\Delta\omega} V_{\omega_1}, \tag{10}$$

where A_T is a constant that depends on the device geometry and material parameters, Q_L is the loaded quality factor, $Z_{th,\Delta\omega}$ is the thermal impedance (7) evaluated at $\Delta\omega$, and $P_{d,\Delta\omega}$ is the $\Delta\omega$ frequency component of the dissipated power in (6). Equation (10) describes the 3IMD signal due to self-heating effects, inside the acoustic transmission line, in terms of the dissipated power. As detailed in the following sub-sections, the dissipated power is due to both electric and acoustic loss, thus both effects contribute to the 3IMD in (10).

2.2.1 3IMD due to viscous losses

Viscosity is introduced in the model as a complex elasticity (Auld, 1990), which translates into a shunt resistance $R_{d,\eta}$ in series with the shunt capacitance C_d in a transmission line implementation. Appendix II details a model transformation to go from the original $R_{d,\eta}$ to an equivalent model in which the viscosity is implemented as a conductance G_d in parallel with the capacitance C_d. The equivalent model allows for an easier extraction of the closed-form expressions.
The instantaneous dissipated power due to viscous damping at each position z along the transmission line of length l (thickness of the piezoelectric layer) is

$$\frac{\partial P_{d,\Delta\omega}(z)}{\partial z}=G_d V_{\omega_1} V_{\omega_2} \cos^2\left(\frac{\pi z}{l}\right), \tag{11}$$

which can be integrated along l to obtain the total dissipated power

$$P_{d,\Delta\omega}=\frac{1}{2} l G_d V_{\omega_1} V^*_{\omega_2}. \tag{12}$$

Equation (12) can be combined with (10) to obtain the peak 3IMD voltage ($V_{\eta,\omega 12}$) due to the viscous damping

$$V_{\eta,\omega_{12}}=\frac{1}{2} A_T l G_d Q_L \Delta C_K Z^*_{th\Delta\omega} V^2_{\omega_1} V^*_{\omega_2} \tag{13}$$

2.2.2 3IMD due to loss in the electrodes

There is certain agreement in considering ohmic losses as a significant dissipation mechanism (Thalhammer et al., 2005) in addition to the viscous damping. As it will be discussed in section II.B.3, electrodes losses are introduced in the circuit model as parasitic series resistances at the input and at the output ports, and their values are determined by fitting the model to the measured scattering parameters in the linear regime. Their contribution to the 3IMD can be calculated by the use of (10) and the power dissipated in the parasitic resistances $P_{\rho\Delta\omega}$:

$$V_{\rho,\omega_{12}} = A_T Q_L \Delta C_K P_{\rho\Delta\omega} Z^{*}_{th\Delta\omega} V_{\omega_1} \tag{14}$$

Whereas the parasitic resistance and distributed conductance can be obtained from the measured scattering parameters, that is, they produce distinguishable measurable effect, examination of (13) and (14) looks like both self-heating mechanisms produce the same experimental observable so they may not be distinguishable. This is true if a two-tone experiment at a fixed frequency is performed, but the two effects have different frequency dependence that can be distinguished if the central frequency ω_0 of the 2 tones is swept while keeping the tones spacing $\Delta\omega$ constant. This happens because the frequency pattern of the dissipation due to ohmic losses is different than that produced by viscous losses, as shown in Fig 1. This information is extremely useful to validate the model with 3IMD measurements by looking at the frequency dependence of the 3IMD.

Note that (13) and (14) keep the same definition of thermal impedance $Z_{th,\Delta\omega}$. This is because the electrodes and the piezoelectric layer are thin and made of good thermal conductors, so that the thermal impedance between those layers is negligible, as will be verified with the temperature simulations shown in Section III.B.2.

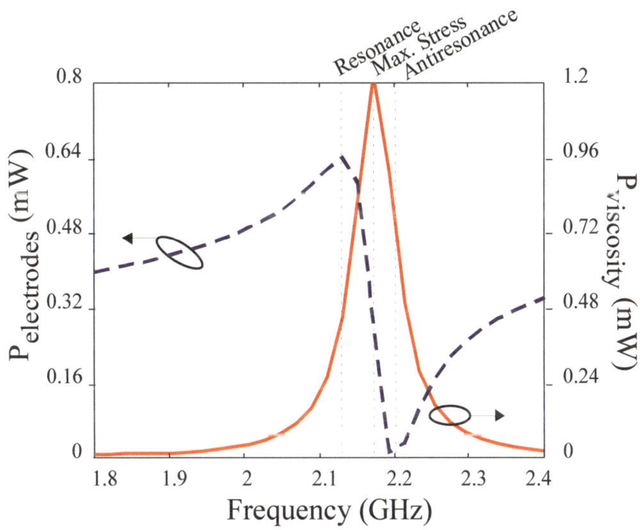

Fig. 1. Simulations of the dissipated power, for an input power of 20 dBm, due to acoustic viscous damping (solid line) and electrode electric losses (dashed line)

2.2.3 Circuit model with self-heating effects

A circuit model implementation to reproduce thermal effects should be capable of predicting dynamic temperature variations. To achieve this, we extend the nonlinear KLM model (Collado et al., 2009) to include the thermal domain (Rocas et al., 2009).

The procedure starts with the one dimensional heat equation along the z direction:

$$\frac{\partial^2 K}{\partial z^2} = \frac{\rho C_p}{k_{th}} \frac{\partial K}{\partial t} - \frac{P_d}{k_{th}}, \tag{15}$$

where the equivalent distributed parameters can be identified as the volumetric heat capacitance

$$C_{d,th} = \rho C_p \tag{16}$$

and the thermal resistance

$$R_{d,th} = \frac{1}{k_{th}} \tag{17}$$

with C_p and k_{th} being the material-specific heat capacity and thermal conductivity, respectively.

With the above-mentioned distributed parameters, a thermal distributed model can be constructed as a cascade of sections of series resistances and shunt capacitance, where each section corresponds to a specific thickness and area. Figure 2 shows a segment with R_{th}= $R_{d,th} \cdot \Delta z/A$ and C_{th}= $C_{d,th} \cdot A \cdot \Delta z$, where A is the area of the cross-section perpendicular to the z direction. In such a thermal equivalent circuit the equivalents of voltage and current are the temperature and heat respectively.

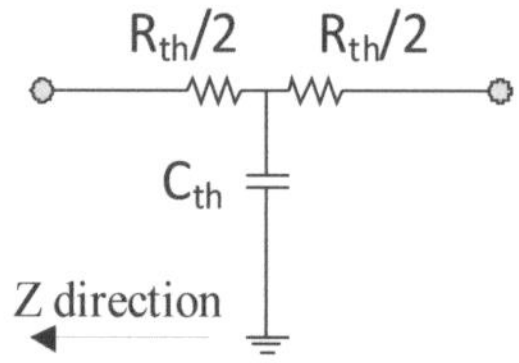

Fig. 2. Implementation of a Δz section of thermal equivalent circuit

The thermal model of a multilayer SMR can be implemented as a cascade of the previously described sections for each material, as shown in Fig. 3. The boundary conditions are the ambient temperature, modeled as a voltage source under the substrate, and the parallel combination of the radiation and convection resistances, terminated with a voltage source at ambient temperature on the upper side of the device (Larson et al., 2002).

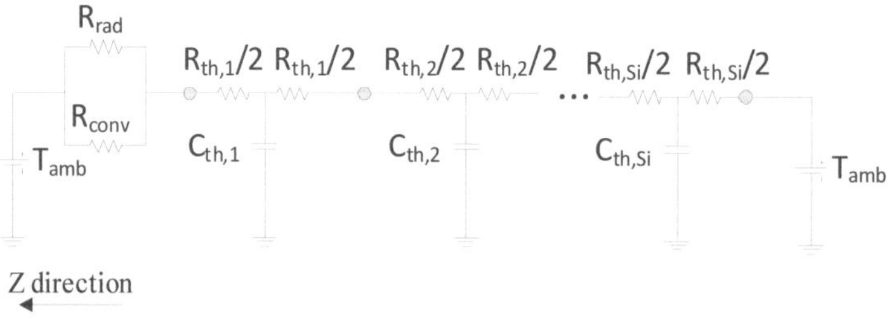

Fig. 3. Thermal model of the upper and lower materials' stacks with boundary conditions

As it can be seen from Fig. 3, the thermal impedance seen from any point along the line has a low-pass filter behavior, which means that for faster variations of the heat source, smaller temperature variations are produced.

The piezoelectric layer is implemented as a cascade of cells, in which the dissipated power due to viscous damping is directly coupled to its correspondent thermal cell. A current source is used because current is the analogue of heat in the thermal domain. The temperature rise is used to modify the distributed acoustic capacitance $C_{NL}(T,K)$, as shown in Fig. 4.

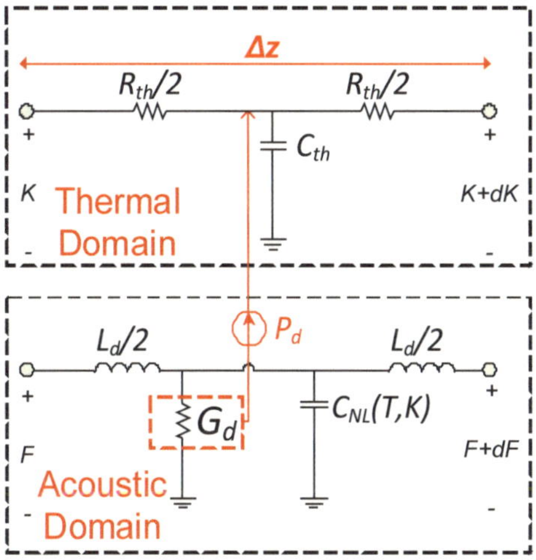

Fig. 4. Implementation of a section of the piezoelectric layer with the acoustic and thermal domains coupled by the generated heat at G_d and the temperature K. L_d is the acoustic distributed inductance $L_d = \rho \cdot A \cdot \Delta z$.

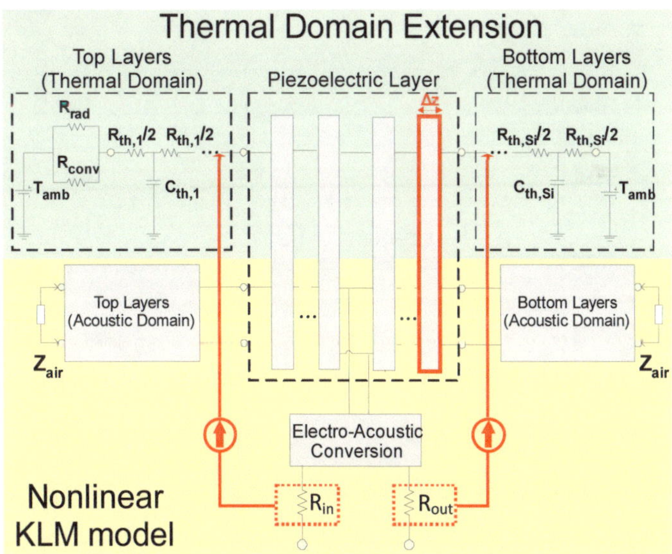

Fig. 5. Complete circuit model with thermo-acoustic model of the piezoelectric layer, top and bottom layers, and lossy electrodes. Electric losses, in the electrodes, and viscous losses, in the piezoelectric layer, produce dissipation that is coupled to the thermal domain to reproduce temperature rise. The temperature rise is used to change the material properties

On the other hand, the parasitic electrodes losses are implemented by use of a lumped resistor at the input and output of the modeled device as shown in Fig. 5. As done for the viscosity, the dissipation in each resistor is coupled to the thermal model as a heat source. In

fact, dissipation in the input and output resistors is coupled to the correspondent top and bottom thermal sections that model the electrodes. The complete model can be seen in Fig. 5, where a cell of the piezoelectric layer like that in Fig. 4, is highlighted in red.
In the figure above, the electric-acoustic conversion box includes those elements of the KLM model whose purpose is the electro-acoustic signal conversion (Krimholtz et al., 1970). Additionally, the material layers above and below the piezoelectric are shown as simplified blocks for clarity.

2.2.4 Comparison of formulation and nonlinear simulations

We use the circuit model of Fig. 5, with only a piezoelectric layer, to check the accuracy of the formulation described in the previous section. The circuit model has been simulated, reproducing a two-tone experiment, with Harmonic Balance techniques by use of a commercial CAD software. A simple model is implemented making use of 100 cells to reproduce a 1.25 μm thick and $2.33 \cdot 10^{-8}$ m^2 piezoelectric layer with a quality factor of 1800. The electrodes losses and viscous losses are coupled to a low-pass thermal impedance.

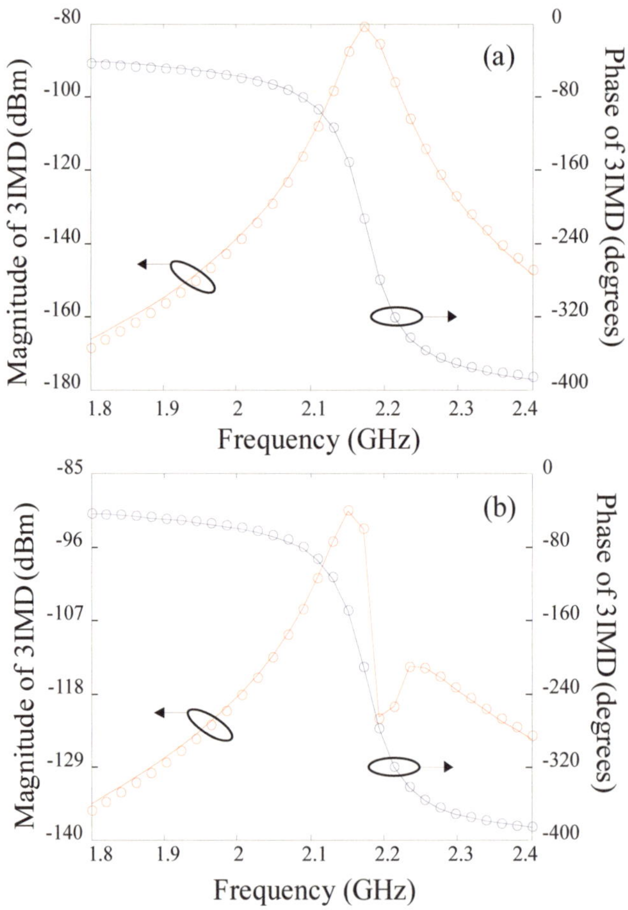

Fig. 6. Comparison of the magnitude and phase of $2\omega_1$-ω_2 calculated with equation (13) (circles) in Fig.6a (viscous losses, no electrode losses) and equation (14) (circles) in Fig.6b (electrode losses, no viscous losses), vs. simulation with the circuit model (solid lines)

In the first set of simulations we keep the tones spacing constant at $\Delta\omega/2\pi$ = 220 Hz and sweep the central frequency ω_0 in a 600 MHz range around the resonance frequency, which is 2.18 GHz. By doing this, we can distinguish the 3IMD produced by viscous self-heating from that produced by electric self-heating by analyzing the resulting frequency dependence. In the former case, we do not connect the dissipation in the electrodes to the thermal domain (Fig.6a), whereas in the latter case we do not connect the dissipation in the piezoelectric layer to the thermal domain (Fig.6b). The 3IMD frequency dependences are a direct consequence of the frequency dependences of the dissipated power. More specifically, a minimum at the anti-resonance frequency appears in Fig. 6.b because there is minimum current flowing through the electrodes at anti-resonance, which can be used in experimental measurements to identify different sources of self-heating effects.
In the second set of simulations we keep the central frequency constant at 2.18 GHz and we change the separation between tones from 100 Hz to 1 MHz. This allows us to reproduce the low-pass filter behavior of the thermal impedance. Figure 7 shows the results of the second set of simulations for a wide range of separation between tones when the self-heating effects are due to viscous losses, where it is clear the low-pass filter behavior of the temperature induced effects. A very similar plot was obtained for electrode losses, which is not shown for simplicity.

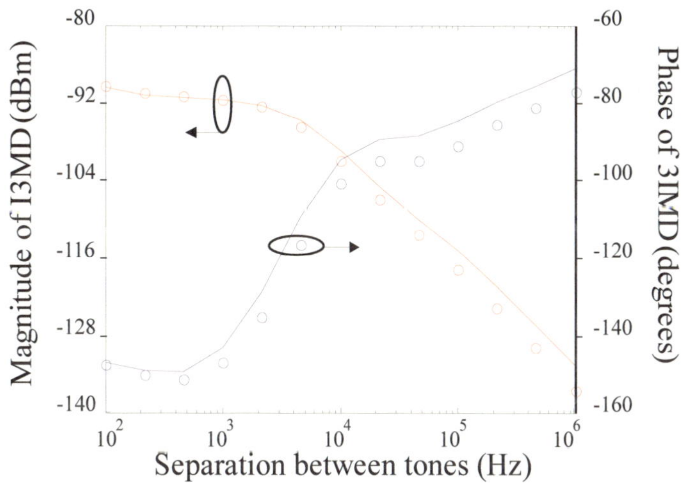

Fig. 7. Magnitude and phase of equation (13) (circles) and simulations with the circuit model (traces) for a wide range of separation between tones

Figures 6 and 7, in addition to giving useful qualitative information about the 3IMD generation due to the self-heating mechanism, show that the formulation of equations (13) and (14) is in very good agreement with the simulations, so that these expressions can be used for a better understanding of the temperature-induced 3IMD in BAW resonators.

3. Experimental results

Four state-of-the-art rectangular Solidly-Mounted Resonators (SMR) from a commercial manufacturer, with different areas summarized in Table 1, have been measured. The resonators have a 1.25 µm thick aluminum nitride layer and a W - SiO_2 Bragg mirror (alternating layers of W and SiO_2), and show quality factors around 1800.

From (8) it is clear that several sources, characterized by $\Delta c^D{}_1$, $\Delta c^D{}_2$ and $\Delta c^D{}_K$, can generate 3IMD. Therefore, we follow a step-by-step procedure that includes several experiments to determine which nonlinear source is designed responsible for each observable at each experiment by use of the circuit model:

- We first adjust the linear model to the measured S-parameters of the devices, so that the electric and viscous losses can be quantified. The procedure consists of a fine tuning of the material properties.
- Second harmonic measurements are performed along the frequency range of interest to extract the intrinsic nonlinear parameter $\Delta c^D{}_1$.
- The term $\Delta c^D{}_2$ also contributes to the 3IMD generation. We use the literature value in (Łepkowski et al., 2005) because this contribution cannot be independently extracted from measurements.
- Third-order intermodulation distortion measurements with sweeping the tone spacing are conducted to quantify the frequency dependence of the thermal impedance, and set the temperature coefficient of stiffened elasticity $\Delta c_K{}^D$ accordingly.

3.1 Linear modeling of the devices under test

Broadband one-port S-parameters measurements of the devices have been performed after an on-wafer OSL calibration. The measurements have been done at a power level of -10 dBm to ensure the linear regime and are used to fit the linear parameters of the circuit model. The only differences between the devices are the resonator area and electrode losses.

Electric losses due to the resistivity of the electrodes are modeled as lumped parasitic resistances, so their values are dependent on the resonator area and have a broadband effect on the linear device response. Table 1 summarizes the device areas and electric resistances.

On the other hand, acoustic losses due to viscosity only have an observable effect at those frequencies where there is substantial electro-acoustic coupling, that is around resonance and anti-resonance. By using the model transformation in Appendix II, the acoustic losses that fit all devices can be described with the same material viscosity value $\eta = 0.033$ N·s·m^{-2}, what verifies the validity of the linear model.

Resonator	Area (m^2)	Electric Resistance R_s (Ω)
A1	6.41e-8	0.28
A2	4.88e-8	0.37
A3	2.33e-8	0.42
A4	1.25e-8	1

Table 1. Tested devices

3.2 Nonlinear characterization

The nonlinear behavior of a BAW resonator arises from different contributions due to intrinsic nonlinear material properties and self-heating mechanisms.

3.2.1 Intrinsic nonlinearities

Intrinsic nonlinearities due to the stiffened elasticity, as shown in (1), predominate above other intrinsic nonlinear material contributions. As a consequence, the second harmonic has the same frequency dependence as the mechanical stress in the piezoelectric layer (Collado et al., 2010). In fact, as done in (Collado et al., 2010), the second harmonic is used to extract a unique unitless value $\Delta c^{D}{}_{1} = 10.5$ of the stiffened elasticity that fits the second harmonic for all the devices.

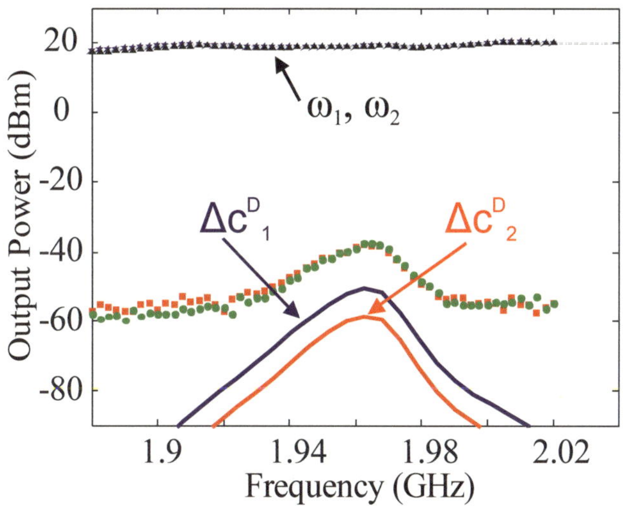

Fig. 8. 3IMD measurements (squares are $2\omega_1$-ω_2, circles are $2\omega_2$-ω_1) and circuit simulations of the $\Delta c^{D}{}_{1}$ and $\Delta c^{D}{}_{2}$ contributions to the 3IMD ($2\omega_1$-ω_2 and $2\omega_2$-ω_1 overlap). Both measurements and simulations are done for $\Delta f = 220$ Hz. The intrinsic nonlinearities are not sufficient to explain the measurements. Measurements (squares and triangles) and simulations (dashed lines) of ω_1 and ω_2 are also presented

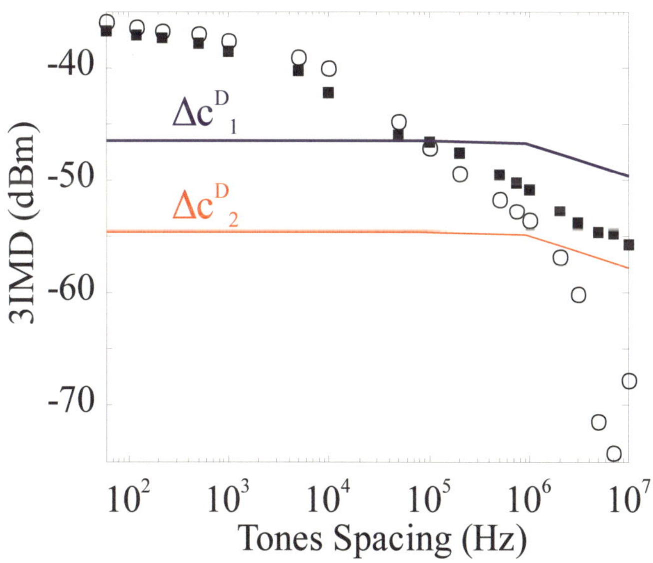

Fig. 9. 3IMD measurements (squares are $2\omega_1$-ω_2, circles are $2\omega_2$-ω_1) and circuit simulations of the $\Delta c^{D}{}_{1}$ and $\Delta c^{D}{}_{2}$ contributions to the 3IMD ($2\omega_1$-ω_2 and $2\omega_2$-ω_1 overlap), for several separations between tones. The intrinsic contributions cannot reproduce the envelope frequency-dependent 3IMD level

The parameter $\Delta c^D{}_1$ is responsible for second harmonic generation, which in turn mixes with the fundamental frequencies ω_1 and ω_2, and gives rise to 3IMD. On the other hand, $\Delta c^D{}_2$ directly generates a certain level of 3IMD distortion. We use the literature value of $\Delta c^D{}_2 = -1 \cdot 10^{-10}\ N^{-1}m^2$ (Łepkowski et al., 2005).

With the above-mentioned values of intrinsic nonlinearity, simulations of 3IMD are performed obtaining values which are below the measured levels, as shown in Fig. 8. This shows that other contributions exist. Figure 9 shows 3IMD measurements at different envelope frequencies, centered at the frequency where the 3IMD is maxima. The measurements reveal a strong dependence with the envelope frequency that cannot be accounted for intrinsic nonlinearities. The dependence of the 3IMD level on the envelope frequency suggests that the 3IMD is dominated by a thermal effect.

3.2.2 Self-heating

The thermal model, as presented in section II.B.3, is implemented by using the literature values of thermal conductivity and specific heat for each layer. The materials stack is composed of more than ten layers. Dissipation on the electrodes and viscous losses are coupled to the thermal domain by means of current sources. By using the model we can determine the temperature distribution along the materials stack, even at different envelope frequencies as shown in Fig. 10. Note that the temperature is almost the same in the piezoelectric layer and the electrodes, which validates the hypothesis in Section II.B.2, of negligible thermal impedance between the electrodes and the piezoelectric layer, to obtain the closed-form expressions.

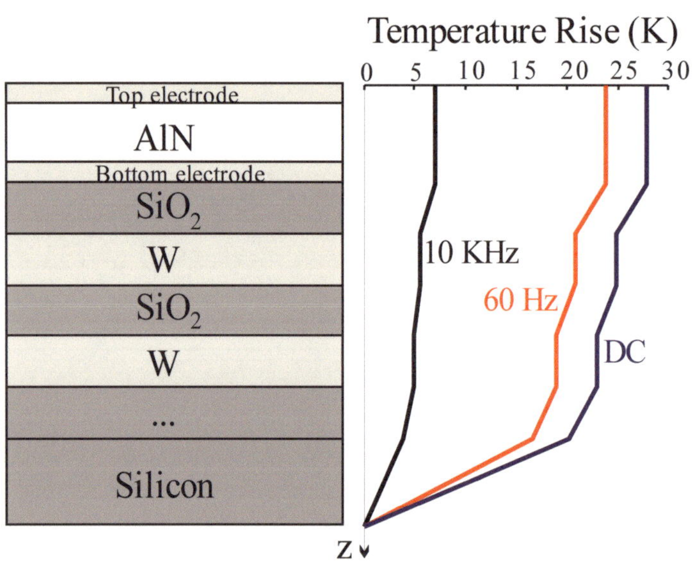

Fig. 10. Simulations of the distribution of the z temperature variation, inside the resonator, calculated with the circuit model. Simulations are done for resonator A1 at P_{in} = 30 dBm, for different separation between tones

A value of $\Delta c^D{}_K/c^D{}_0$= -15 ppm/K is found to fit the 3IMD best for all resonators. Figure 11 shows the intrinsic and self-heating 3IMD contributions independently as well as the sum of all contributions, which match the measurements for all the tone separations. The model reproduces very well the measured 3IMD values except the asymmetry between lower and higher 3IMD that appears for envelope frequencies around 6 MHz. This asymmetry is considered to be a consequence of a cancellation between different 3IMD contributions and is currently under investigation.

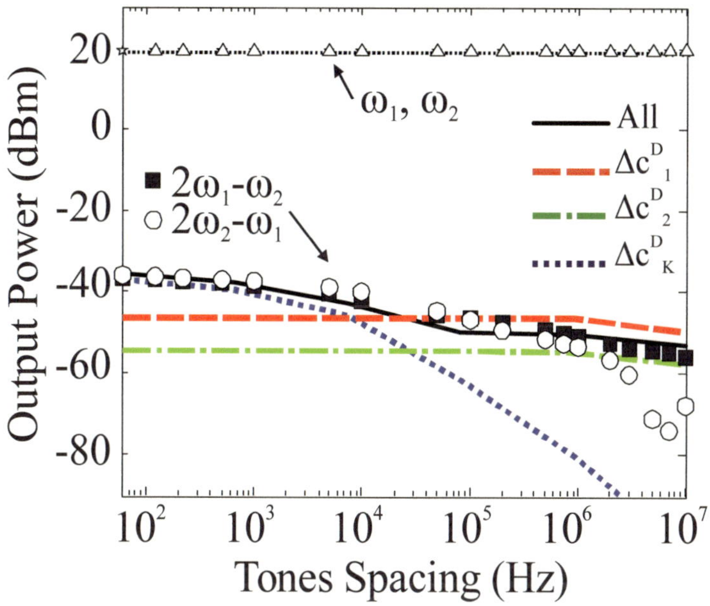

Fig. 11. Measurements (stars, triangles, squares and circles are ω_1, ω_2, $2\omega_1$-ω_2 and $2\omega_2$-ω_1 respectively) and simulations (thin dotted line is ω_1 and ω_2, solid line is $2\omega_1$-ω_2 and $2\omega_2$-ω_1 overlapped) of the fundamentals and the 3IMD for different tones spacing for A1. The dotted line, the dash-dot line and the dashed line are simulations of the 3IMD contribution due to Δc^D_K, Δc^D_2 and Δc^D_1 respectively. The solid line is the 3IMD simulation with all the nonlinear contributions

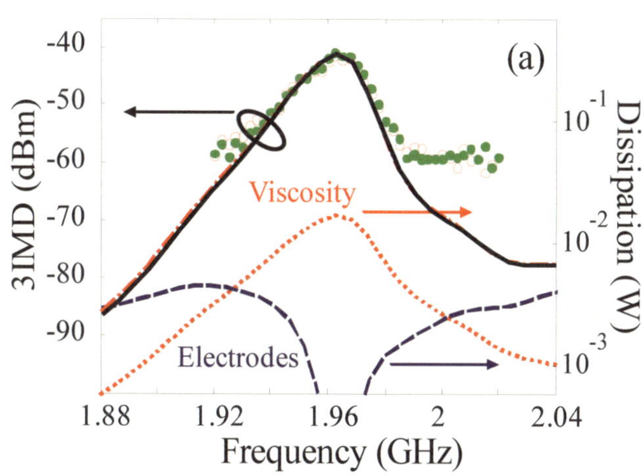

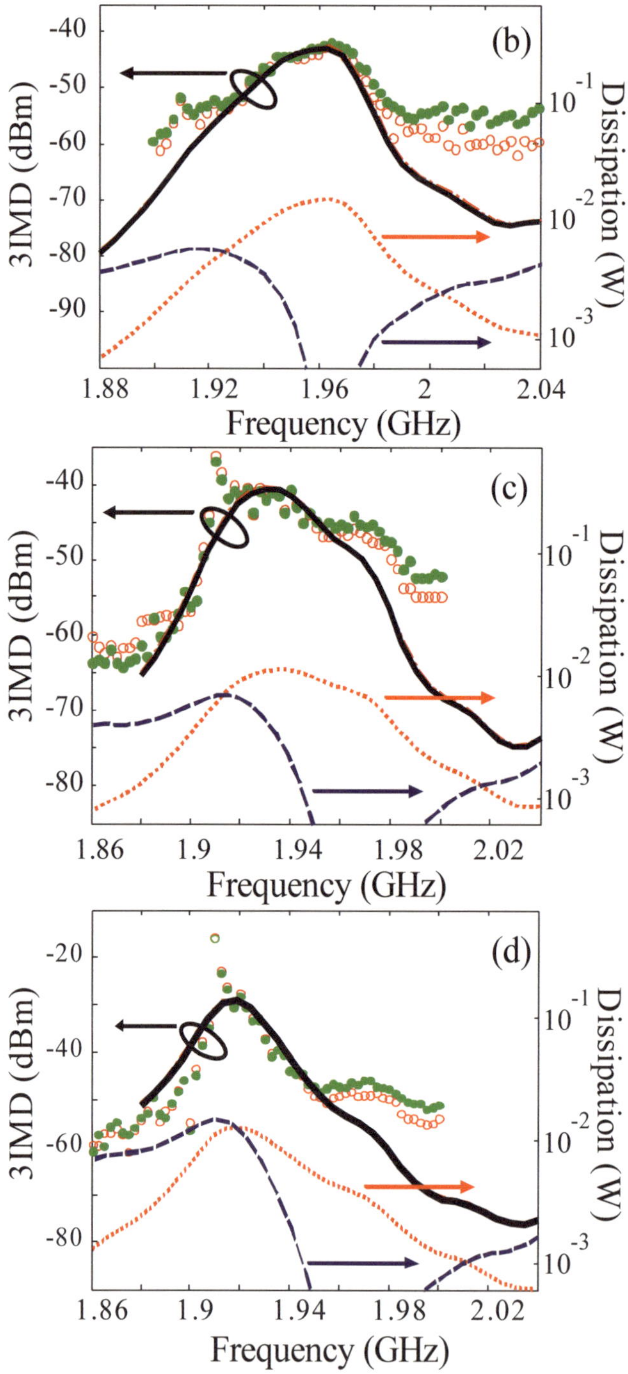

Fig. 12. Measurements (filled circles are $2\omega_1$-ω_2, empty circles are $2\omega_2$-ω_1) and simulations (solid line is $2\omega_1$-ω_2, dash-dot line is $2\omega_2$-ω_1) of the 3IMD for resonators A1, A2, A3 and A4 in Fig.12.a, Fig.12.b, Fig.12.c and Fig.12.d respectively. The figures also show the dissipation in the electrodes (dashed line) and in the piezoelectric layer due to viscous losses (dotted line)

Figure 12 shows measurements and simulations of the 3IMD about the frequency range of interest, for each resonator, by use of the circuit model. In this experiment, the two tones are swept around the resonating frequencies keeping the separation between tones constant at 100 Hz. The results show good agreement between simulations and measurements above the nonlinear system baseline level, which is around -60 dBm.

The dashed and dotted lines in Fig. 12 show the simulated dissipated power due to electrodes losses and viscosity respectively, which have different frequency dependences according to the maximum electric current and mechanical stress, respectively. 3IMD measurements for resonators A3 and A4 in Fig. 12 show a peak at the antiresonant frequency that is underestimated by the simulations. The frequency dependence in that range points to a possible electric-field contribution to the 3IMD. This contribution is below the system nonlinear baseline level for resonators A1 and A2, and the area scaling has not been successfully reproduced by use of an electric-field dependent permittivity or stiffened elasticity in the acoustic transmission line, so further research is needed.

4. Conclusion

The role of self-heating and material nonlinearities in the generation of 3IMD in bulk acoustic wave devices has been evaluated through measurements, models and equations. Self-heating is found to have a very significant contribution to 3IMD and thus thermal considerations are critical in the device design. The presented circuit model implementation offers the possibility to predict 3IMD in BAW resonators, given their materials stack and geometry. With such information one can use the resonator model to accurately predict 3IMD in filters. Further research will be performed to investigate the relation between the electric-field contribution to 3IMD and the cancellation shown in the measurements. The development of a 3D equivalent thermal model, to take into account complex heat dissipation through the substrate, will also be investigated.

5. Appendix I – 3IMD equations

At each elemental section, and following a similar process than that described in (Collado et al., 2009), the nonlinear capacitance acts as an infinitesimal nonlinear current generator at $2\omega_1 - \omega_2$ (and $2\omega_2 - \omega_1$), when ω_1 and ω_2 are at resonance:

$$\frac{\partial I_{nl,\omega_{12}}(z)}{\partial z} = \frac{1}{2} j\omega_{12}\Delta C_K K^*_{\Delta\omega} V_{\omega_1} \cos\left(\frac{\pi z}{l}\right) \tag{18}$$

where $K_{\Delta\omega} = Z_{th}(\Delta\omega)\, P_d(\Delta\omega)$.

Therefore the broadband energy balance all over the acoustic transmission line leads to

$$V_{\omega_{12}} = -j\frac{1}{2} Q_L \frac{\Delta C_K}{C_{d,0}} P_{d\Delta\omega} Z^*_{th\Delta\omega} V_{\omega_1} \left(\frac{1}{j\dfrac{C_{eq} Q_L}{2W_{0,v}} - 1} \right) \tag{19}$$

6. Appendix II - Model transformation

Losses are introduced as a complex elasticity by means of the viscous damping coefficient η:

$$c \rightarrow c + \eta \frac{\partial}{\partial t} \tag{20}$$

The inverse damping coefficient can also be understood as the conductance per unit length $G_d=\eta^{-1}$. With that, the acoustic telegrapher equations, making use of the analogy between the acoustic and electric domains, can be written as:

$$\frac{\partial V}{\partial z} = -L_d j\omega I \tag{21}$$

and

$$\frac{\partial I}{\partial z} = -\frac{1}{Ac^D + j\omega A\eta} j\omega V. \tag{22}$$

The shunt admittance of the acoustic transmission line implementation, given by (22) and in which $A \cdot c^D = C_d^{-1}$, is a shunt capacitance in series with a resistance. To transform this to be a capacitance in parallel with the loss term, we introduce eq. 8 in eq. 22 and expand the shunt admittance in as a Taylor series. The result is a conductance value in parallel with a nonlinear capacitance of the form:

$$C_d(v,K) = C_{d,0} + \Delta C_1 v + \Delta C_2 v^2 + \Delta C_K K \tag{23}$$

whose terms are related with the material linear and nonlinear properties as follows:

$$G = \omega^2 C_{d,0}{}^2 A\eta \tag{24}$$

$$\Delta C_1 = \frac{c_1^D}{(Ac_0^D)^2} \tag{25}$$

$$\Delta C_2 = -\frac{c_2^D}{A^3 (c_0^D)^2} \tag{26}$$

$$\Delta C_1 = -\frac{c_K^D}{A(c_0^D)^2} \tag{27}$$

7. Appendix III - Broadband loaded quality factor

The loaded quality factor can be defined as (Russer, 2006)

$$Q_L = \frac{Q_0}{1+\beta} \tag{28}$$

where β relates the dissipated power in the acoustic resonator P_{res}, that is the acoustic transmission line, and the externally dissipated power P_{ext} as follows:

$$\beta = \frac{P_{ext}}{P_{res}}. \tag{29}$$

By circuit analysis of the KLM circuit model, it can be found that β is

$$\beta = \frac{\mathrm{Re}\left(\dfrac{1}{\left(50 + jX_m + \dfrac{1}{j\omega C_0}\right)^*} \right)}{\mathrm{Re}\left(\dfrac{1}{\left(Z_{in} - jX_m - \dfrac{1}{j\omega C_0}\right)^*} \right)} \tag{30}$$

where Z_{in} is the input impedance of the device and X_m is the series reactive term of the KLM model (Krimholtz et al., 1970). Q_0 in (28) represents the unloaded quality factor, that is obtained from S-parameters using (Feld et al., 2008)

$$Q_0 = \omega\tau \frac{|S_{11}|}{1 - |S_{11}|^2} \tag{31}$$

8. References

Auld B. A., Acoustic Fields and Waves in Solids (Krieger, Malabar, Florida), Vol. I, 1990

Camarchia V., Cappelluti F., Pirola M., Guerrieri S. D., Ghione G. 2007. Self-Consistent Electrothermal Modeling of Class A, AB, and B Power GaN HEMTs Under Modulated RF Excitation. *IEEE Transactions on Microwave Theory and Techniques*, vol. 55, no. 9, Sept. 2007, pp. 1824-1831.

Cho Y., Wakita J. 1993. Nonlinear equivalent circuits of acoustic devices. *Ultrasonics Symposium, 1993. Proceedings, IEEE 1993*, pp. 867-872 vol. 2, 31 Oct-3 Nov 1993

Constantinescu F., Nitescu M., Gheorghe A. G. 2008. New Nonlinear Circuit Models for Power BAW Resonators. *ICCSC 2008. 4th IEEE International Conference on Circuits and Systems for Communications*, pp. 599-603, 26-28 May 2008

Collado C., Rocas E., Padilla A., Mateu J., O'Callaghan J. M., Orloff N. D., Booth J. C., Iborra E., Aigner R. 2010. First-order nonlinearities of bulk acoustic wave resonators. *IEEE Transactions on Microwave Theory and Techniques*, submitted.

Collado C., Rocas E., Mateu J., Padilla A., O'Callaghan J. M. 2009. Nonlinear Distributed Model for Bulk Acoustic Wave Resonators. *IEEE Transactions on Microwave Theory and Techniques*, vol. 57, no. 12, Dec. 2009,pp. 3019-3029

Collado C., Mateu J. and O'Callaghan J. M. 2005. Analysis and Simulation of the Effects of Distributed Nonlinearities in Microwave Superconducting Devices. *IEEE Trans. Appl. Supercond.* vol. 15, No. 1, March 2005, pp. 26-39.

Feld D. A. 2009. One-parameter nonlinear mason model for predicting 2nd & 3rd order nonlinearities in BAW devices. *2009 IEEE International Ultrasonics Symposium (IUS)*, pp. 1082-1087, 20-23 Sept. 2009

Feld D. A., Parker R., Ruby R., Bradley P., Shim D. 2008. After 60 years: A new formula for computing quality factor is warranted. *2008 IEEE International Ultrasonics Symposium*, pp. 431-436, 2-5 Nov. 2008

Ivira B., Benech P., Fillit R., Ndagijimana F., Ancey P., Parat G. 2008. Modeling for temperature compensation and temperature characterizations of BAW resonators at GHz frequencies. *IEEE Transactions on Ultrasonics, Ferroelectrics and Frequency Control*, vol. 55, no. 2, February 2008, pp. 421-430.

Krimholtz R., Leedom D.A., Matthaei G.L. 1970. New equivalent circuits for elementary piezoelectric transducers. *Electronics Letters* , vol. 6, no. 13, pp. 398-399, June 1970

Lakin K. M., McCarron K. T., McDonald J. F. 2000. Temperature compensated bulk acoustic thin film resonators. *2000 IEEE International Ultrasonics Symposium,* pp. 855-858 vol. 1, Oct 2000

Larson III J. D., Oshrnyansky Y. 2002. Measurement of effective kt2, q, Rp, Rs vs. Temperature for Mo/AlN FBAR resonators. *2002 IEEE International Ultrasonics Symposium*, pp. 939- 943 vol. 1, 8-11 Oct. 2002

Łepkowski S. P., Jurczak G. 2005. Nonlinear elasticity in III-N compounds: Ab initio calculations. *Physical Review B,* vol. 72, 245201, 2005, pp 1-12.

Nakamura H., Hashimoto K.-y., Ueda M. 2010. Nonlinear effects in SAW and BAW components. *2010 IEEE International Ultrasonics Symposium (IUS),* Short Course, 11-14 Oct. 2010

Parker A. E., Rathmell J. G. 2004. Self-heating process in microwave transistors. *Workshop on Applications in Radio Science,* Hobart, TAS, Australia, 18-20 Feb. 2004.

Petit D., Abele N., Volatier A., Lefevre A., Ancey P., Carpentier J.-F. 2007. Temperature Compensated Bulk Acoustic Wave Resonator and its Predictive 1D Acoustic Tool for RF Filtering. *2007 IEEE International Ultrasonics Symposium,* pp. 1243-1246, 28-31 Oct. 2007

Rocas E., Collado C., Orloff N. D., Mateu J., Padilla A., O'Callaghan J. M., Booth J. C. 2010. Passive intermodulation due to self-heating in printed transmission lines. *IEEE Transactions on Microwave Theory and Techniques,* accepted for publication.

Rocas E., Collado C., Booth J. C., Iborra E., Aigner R. 2009. Unified model for Bulk Acoustic Wave resonators' nonlinear effects. *2009 IEEE International Ultrasonics Symposium (IUS),* pp. 880-884, 20-23 Sept. 2009

Rocas E., Collado C., Mateu J., Campanella H., O'Callaghan J. M. 2008. Third order Intermodulation Distortion in Film Bulk Acoustic Resonators at Resonance and Antiresonance. *IEEE MTT-S International Microwave Symposium Digest,* pp.1259-1262, June 2008.

Russer P. 2006. Electromagnetics, microwave circuit, and antenna design for communications engineering (Artech House), 2006

Thalhammer R., Aigner R. 2005. Energy loss mechanisms in SMR-type BAW devices. *2005 IEEE MTT-S International Microwave Symposium Digest,* pp. 4, 12-17 June 2005

Ueda M., Iwaki M., Nishihara T., Satoh Y., Hashimoto K.-Y. 2008. A circuit model for nonlinear simulation of radio-frequency filters using bulk acoustic wave resonators. *IEEE Transactions on Ultrasonics, Ferroelectrics and Frequency Control,* vol. 55, no. 4, pp.849-856, April 2008

Vuolevi J. H. K., Rahkonen T., Manninen J. P. A. 2001. Measurement technique for characterizing memory effects in RF power amplifiers. *IEEE Transactions on Microwave Theory and Techniques,* vol. 49, no. 8, Aug. 2001, pp. 1383-1389.

22

Shear Mode Piezoelectric Thin Film Resonators

Takahiko Yanagitani
Nagoya Institute of Technology
Japan

1. Introduction

1.1 Shear mode bulk acoustic wave devices and sensors

Acoustic microsensor technique, well known as QCM (Quartz crystal microbalance) or TSM (Thickness shear mode) sensor, is an effective method to detect small mass loading on the sensor surface. This sensor can be operated even in liquid by using shear mode resonance. Therefore, shear mode piezoelectric film resonators are attractive for liquid microsensor technique such as biosensors and immunosensors.

Shear wave has some unique features compared with the longitudinal wave, for example, it has extremely low velocity in the liquid. Longitudinal wave velocity in the water is 1492.6 m/s, whereas, shear wave velocity in the water is 20-60 m/s at 20-200 MHz (Matsumoto et al., 2000). Therefore, shear mode vibrating solid maintains its vibration even in the liquid, because the difference of acoustic impedance which determines the refection coefficient of solid / liquid interface is very large in the case of shear wave.

The complex refection coefficient Γ of the interface is given as

$$\Gamma = \frac{Z_l - Z_s}{Z_l + Z_s} \tag{1}$$

where Z_s and Z_l are the complex acoustic impedance of solid and liquid.

Complex acoustic impedance can be written as

$$Z = R + jX = \left(\rho(c + j\omega\eta) \right)^{1/2} \tag{2}$$

R and X represent the real part and imaginary part of the acoustic impedance and ρ, c and η represent mass density, stiffness constant and viscosity in the medium, respectively.

Acoustic wave equation gives dispersion relation of

$$\left(\frac{\omega}{v} - j\alpha \right)^2 (c + j\omega\eta) = \rho\omega^2 \tag{3}$$

where v is velocity and α is attenuation factor (B. A. Auld, 1973).

According to (2) and (3), acoustic impedance gives

$$R = \frac{\rho v \omega^2}{\omega^2 + \alpha^2 v^2}, \quad X = \frac{\rho v^2 \omega \alpha}{\omega^2 + \alpha^2 v^2} \tag{4}$$

Longitudinal and shear wave velocities of water were reported as 1492.6 m/s (Kushibiki et al., 1995) and 35 m/s (Matsumoto et al., 2000), respectively, at 100 MHz. Attenuations of longitudinal and shear wave in the water were also measured to be $\alpha/f^2 = 2.26\times10^{-14}$ neper·s²/m (Kushibiki et al, 1995) and $\alpha/f^2 = 2.12\times10^{-9}$ neper·s²/m, (Matsumoto et al., 2000) respectively. By substituting these values into Eq. (4), the complex longitudinal wave and shear wave acoustic impedance of the water can be estimated to be $1489000+j800$ N·s/m³ and $14510+j17340$ N·s/m³ at 100 MHz, respectively.
From these values and Eq. (1), when quartz resonator is immersed in water, the reflection coefficient of acoustic energy $|\Gamma|^2$ in an X-cut quartz vibrating in thickness extensional mode (Z_s= 15.23×10^{-6} N·s/m³) is estimated to be only 68 % whereas that in an AT-cut quartz vibrating in thickness shear mode (Z_s= 8.795×10^{-6} N·s/m³) is 98 %. This is because an AT-cut quartz has been used as a QCM or TSM sensor operating in liquid. Sensitivity of the QCM mass sensor is determined by the ratio of the mass and the entire mass of the vibrating part in the sensor, at constant sensor active area (Sauerbrey, 1959). Therefore, it is important to decrease thickness of the vibrating part of sensor. Shear mode thin film is promising for high sensitivity mass sensor.

1.2 Piezoelectric thin film for shear mode excitation

Piezoelectric thin film, which excites shear wave, is expected to provide higher sensitivity and IC compatibility, but it is not straightforward. To excite shear wave by standard sandwiched electrode configuration, polarization axis in the film must be tilted or parallel to the film plane. Although perovskite ferroelectric films have large piezoelectricity, their polarization axis is generally normal to the film surface due to the nature of crystal growth, difficultly of in-plane polarization treatment and domain control. Trigonal piezoelectric material such as $LiNbO_3$, $LiTaO_3$ and quartz are difficult to crystallize (tend to form amorphous structure) or to obtain a strong preferred orientation in polycrystalline film.
6mm wurtzite AlN and ZnO film can be easily crystallized, but they tend to develop their polarization axis (c-axis) perpendicular to the substrate plane. This c-axis oriented film cannot excite shear wave in the case of standard sandwiched electrode structure.
Crystalline orientation control for both in-plane and out-of-plane direction is necessary to excite shear wave. One solution is to use an epitaxial growth technique. However, the combinations of the shear mode piezoelectric film and substrate are limited due to the lattice mismatch. a-plane ZnO or AlN/r-plane sapphire (Mitsuyu et al., 1980; Wittstruck et al., 2003), a-plane ZnO/42° Y-X $LiTaO_3$ (Nakamura et al., 2000) where c-axis in the film is parallel to the substrate plane have been reported.
Ion beam orientation control technique (Yanagitani & Kiuchi, 2007c), which enables in-plane and out-of-plane orientation without use of epitaxial growth, is introduced in the third section. This technique is a good candidate for obtaining c-axis parallel films which excites pure shear wave without any excitation of longitudinal wave.

2. Electromechanical coupling properties of wurtzite crystal

Elastic and piezoelectric properties of wurtzite crystals vary with direction due to the crystal anisotropy. Electromechanical coupling changes as a function of the angle between the c-axis and the applied electric field direction (Foster et al., 1968; Auld, 1973).

The analytical model of a thin film resonator is shown in Fig. 1. The electric field is applied in the x_3 direction. The c-axis is assumed to lie in the x_1-x_3 plane and be inclined at an angle β with respect to the x_3 direction.

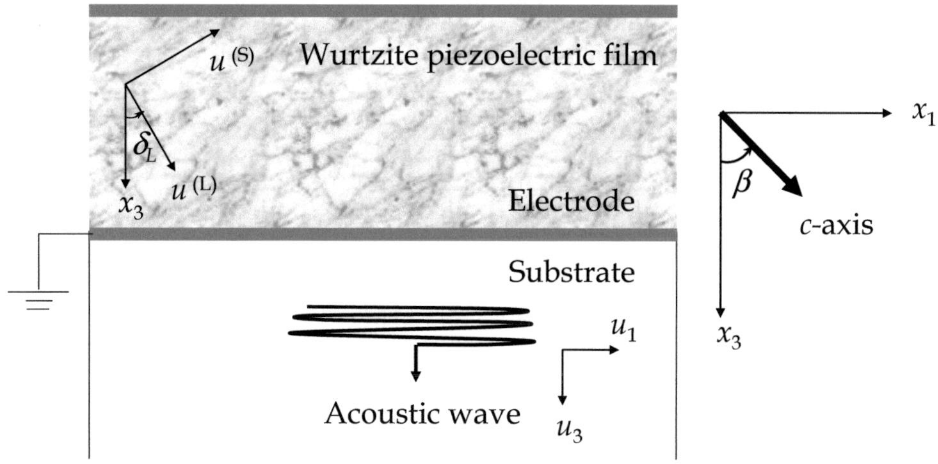

Fig. 1. Analytical model of a thin film resonator

The physical constants of the crystal in each direction are determined by the transformed coordinate of each constant tensor. Bond's method (Bond, 1943) for transforming the elastic and piezoelectric constant tensor is introduced below, which can be applied to the constant tensor with abbreviated subscript notation. For example, the transformation matrix $[a]$ of a clockwise rotation through an angle β about the x_2-axis is described by:

$$[a]=\begin{bmatrix}\cos\beta & 0 & -\sin\beta\\ 0 & 1 & 0\\ \sin\beta & 0 & \cos\beta\end{bmatrix} \quad (5)$$

The dielectric constant ε' transforms as

$$[\varepsilon']=[a]\ [\varepsilon]\ [a]^T. \quad (6)$$

The 6×6 transformation matrix of coefficients M is defined as

$$[M]=\begin{bmatrix}a_{xx}^2 & a_{xy}^2 & a_{xz}^2 & 2a_{xy}a_{xz} & 2a_{xz}a_{xx} & 2a_{xx}a_{xy}\\ a_{yx}^2 & a_{yy}^2 & a_{yz}^2 & 2a_{yy}a_{yz} & 2a_{yz}a_{yx} & 2a_{yx}a_{yy}\\ a_{zx}^2 & a_{zx}^2 & a_{zz}^2 & 2a_{zy}a_{zz} & 2a_{zz}a_{zx} & 2a_{zx}a_{zy}\\ a_{yx}a_{zx} & a_{yy}a_{zy} & a_{yz}a_{zz} & a_{yy}a_{zz}+a_{yz}a_{zy} & a_{yx}a_{zz}+a_{yz}a_{zx} & a_{yy}a_{zx}+a_{yx}a_{zy}\\ a_{zx}a_{xx} & a_{zy}a_{xy} & a_{zz}a_{xz} & a_{xy}a_{zz}+a_{xz}a_{zy} & a_{xz}a_{zx}+a_{xx}a_{zz} & a_{xx}a_{zy}+a_{xy}a_{zx}\\ a_{xx}a_{yx} & a_{xy}a_{yy} & a_{xz}a_{yz} & a_{xy}a_{yz}+a_{xz}a_{yy} & a_{xz}a_{yx}+a_{xx}a_{yz} & a_{xx}a_{yy}+a_{xy}a_{yx}\end{bmatrix} \quad (7)$$

Finally, using the above transformation matrix, transformed elastic constant and piezoelectric constant tensors c' and e' are obtained:

$$[c'] = [M]\,[c]\,[M]^T,\ [e'] = [M]\,[e]\,[M]^T \tag{8}$$

In the x_2 axis rotation of a hexagonal (6mm) crystal, the transformed stiffness and piezoelectric constant tensors c' and e' are given by

$$[c'] = \begin{bmatrix} c'_{11} & c'_{12} & c'_{13} & 0 & c'_{15} & 0 \\ c'_{12} & c'_{22} & c'_{23} & 0 & c'_{25} & 0 \\ c'_{13} & c'_{23} & c'_{33} & 0 & c'_{35} & 0 \\ 0 & 0 & 0 & c'_{44} & 0 & c'_{46} \\ c'_{15} & c'_{25} & c'_{35} & 0 & c'_{55} & 0 \\ 0 & 0 & 0 & c'_{46} & 0 & c'_{66} \end{bmatrix},\ [e'] = \begin{bmatrix} e'_{11} & e'_{12} & e'_{13} & 0 & e'_{15} & 0 \\ 0 & 0 & 0 & e'_{24} & 0 & e'_{26} \\ e'_{31} & e'_{32} & e'_{33} & 0 & e'_{35} & 0 \end{bmatrix} \tag{9}$$

In case, wave propagation toward x_3 direction is only focused, the term of $\partial/\partial x_1$ and $\partial/\partial x_2$ can be ignored. Thus, the wave motion equation for the x_3 direction is given by mechanical displacement component u_1, u_2 and u_3:

$$\frac{\partial T_{31}}{\partial x_3} = \rho \frac{\partial^2 u_1}{\partial t^2} \tag{10a}$$

$$\frac{\partial T_{33}}{\partial x_3} = \rho \frac{\partial^2 u_3}{\partial t^2} \tag{10b}$$

$$\frac{\partial T_{32}}{\partial x_3} = \rho \frac{\partial^2 u_2}{\partial t^2} \tag{10c}$$

where

$$T_{31} = c'^E_{55} \frac{\partial u_1}{\partial x_3} + c'^E_{35} \frac{\partial u_3}{\partial x_3} + e'_{35} \frac{\partial \varphi}{\partial x_3} \tag{11a}$$

$$T_{33} = c'^E_{35} \frac{\partial u_1}{\partial x_3} + c'^E_{33} \frac{\partial u_3}{\partial x_3} + e'_{33} \frac{\partial \varphi}{\partial x_3} \tag{11b}$$

$$T_{32} = c'^E_{44} \frac{\partial u_2}{\partial x_3} \tag{11c}$$

As div $\boldsymbol{D} = 0$, the electrostatic equation is given by

$$\frac{\partial D_3}{\partial x_3} = e'_{35} \frac{\partial^2 u_1}{\partial x_3^2} + e'_{33} \frac{\partial^2 u_3}{\partial x_3^2} - \varepsilon'^S_{33} \frac{\partial^2 \varepsilon}{\partial x_3^2} = 0 \tag{12}$$

In Eqs. (10)-(12), T_{31} and T_{33} are stress components, D_3 is the electric displacement, $c_{33}{}^E$, $c_{35}{}^E$ and $c_{55}{}^E$ are the stiffness constants with constant electric field, e_{33} and e_{35} are piezoelectric constants, $\varepsilon_{33}{}^S$ and $\varepsilon_{35}{}^S$ are dielectric constants with constant strain, and φ is the electric potential.

Equation (10c) describes a pure shear wave with a u_2 displacement component in the x_2 direction and propagates along the x_3 direction with a phase velocity of $\sqrt{c_{44}/\rho}$. Eqs. (10a) and (10b) represent a quasi-longitudinal wave and quasi-shear wave. These waves incorporate u_1, u_3, and φ, which are coupled with each other. It is well known that Eqs (10a), and (10b) have plane-wave solutions:

$$\begin{pmatrix} u_1 \\ u_3 \\ \varphi \end{pmatrix} = \begin{pmatrix} A \\ B \\ C \end{pmatrix} \exp\left\{ j\omega \left(t - \frac{x_3}{v} \right) \right\} \tag{13}$$

Substituting Eq. (13) into Eqs. (11) and (12), the simultaneous equations are obtained

$$\begin{pmatrix} c_{55}^D - \rho v^2 & c_{35}^D & 0 \\ \overline{c}_{35} & c_{33}^D - \rho v^2 & 0 \\ -e_{35} & -e_{33} & \varepsilon_{33}^S \end{pmatrix} \begin{pmatrix} A \\ B \\ C \end{pmatrix} = 0, \tag{14}$$

where

$$\begin{aligned} c_{33}^D &= c_{33}'^E + (e_{33}')^2 / \varepsilon_{33}'^S, \\ c_{35}^D &= c_{35}'^E + (e_{33}' e_{35}') / \varepsilon_{33}'^S, \\ c_{55}^D &= c_{55}'^E + (e_{35}')^2 / \varepsilon_{33}'^S. \end{aligned} \tag{15}$$

A, B and C are all nonzero when the coefficient matrix in Eq. (14) is zero. From this condition, we obtain the phase velocity $v^{(L,\,S)}$ of a quasi-longitudinal wave and quasi-shear wave:

$$v^{(L,S)} = \left[\frac{c_{33}^D + c_{55}^D}{2\rho} \pm \sqrt{\left(\frac{c_{33}^D - c_{55}^D}{2\rho} \right)^2 + \left(\frac{c_{35}^D}{\rho} \right)^2} \right]^{\frac{1}{2}} \tag{16}$$

Figure 2 shows the calculated results of phase velocity of a quasi-longitudinal wave and quasi-shear wave for a ZnO crystal as function of the angle β between the c-axis and x_3 direction. Physical constants in a ZnO single crystal reported by Smith were used in the calculation (Smith, 1969).

The general solutions for u_1, u_3 and φ are given by

$$\begin{pmatrix} u_1 \\ u_3 \\ \varphi \end{pmatrix} = \begin{pmatrix} A_1 \\ B_1 \\ C_1 \end{pmatrix} \exp\left\{ j\omega \left(t - \frac{x_3}{V^{(+)}} \right) \right\} + \begin{pmatrix} A_2 \\ B_2 \\ C_2 \end{pmatrix} \exp\left\{ j\omega \left(t - \frac{x_3}{V^{(-)}} \right) \right\} \tag{17}$$

and

$$\frac{B_1}{A_1} = -\frac{A_2}{B_2} \tag{18}$$

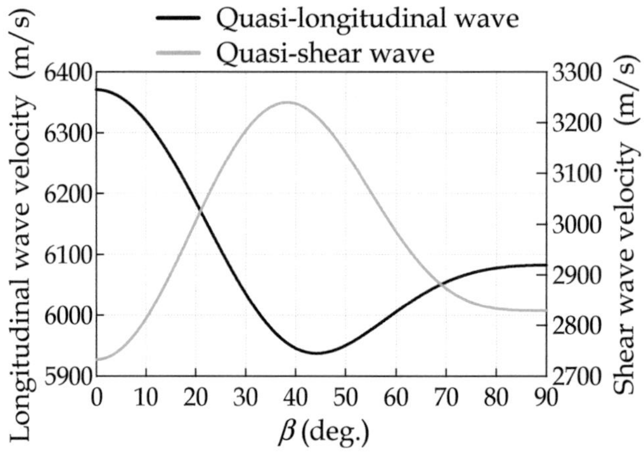

Fig. 2. Phase velocity of quasi-longitudinal wave and quasi-shear wave for a ZnO crystal as function of the angle β between the c-axis and x_3 direction

is derived from Eqs. (14) and (16). It can be seen that the displacement components of the quasi-longitudinal wave and quasi-shear wave are perpendicular to each other. From Eqs. (14) and (16), the angle δ_L between the quasi-longitudinal wave displacement u_3 and the x_3 direction and the angle δ_S between the quasi-shear wave displacement u_1 and the x_1 direction are given by

$$\delta_L = \tan^{-1}\left(\frac{A_1}{B_1}\right),\ \delta_S = \tan^{-1}\left(\frac{B_2}{A_2}\right) \tag{19}$$

The extensional and shear effective piezoelectric constants $e^{(L)}_{eff}$ and $e^{(S)}_{eff}$ are defined as

$$e^{(L)}_{eff} = e'_{35}\sin\delta_L + e'_{33}\cos\delta_L,\ e^{(S)}_{eff} = e'_{35}\cos\delta_S - e'_{33}\sin\delta_S \tag{20}$$

Thus, the quasi-longitudinal and quasi-shear-mode electromechanical coupling coefficients $k^{(L)}$ (transformed k_{33}) and $k^{(S)}$ (transformed k_{15}) are

$$\left(k^{(S)}\right)^2 = \left(e^{(S)}\right)^2 / e'_{33}\rho\left(V^{(-)}\right)^2,\ \left(k^{(L)}\right)^2 = \left(e^{(L)}\right)^2 / e'_{33}\rho\left(V^{(+)}\right)^2 \tag{21}.$$

Finally, Figs. 3 (a) and (b) show the calculated angle δ and the electromechanical coupling coefficients (k values) of the quasi-longitudinal and quasi-shear waves for the ZnO crystal as function of the angle β (Foster et al., 1968)

From these figures, we can see a relatively large shear-mode electromechanical coupling k_{15} = 0.39 at c-axis tilt angle of β = 28°. Several author reported FBAR (film bulk acoustic resonator)-type viscosity sensor and biosensor, consisting of c-axis tilted wurtzite films (Weber et al., 2006; Link et al., 2007; Wingqvist et al., 2007, 2009, 2010; Yanagitani, 2010, 2011a). However, the thickness extensional mode (longitudinal wave mode) also has the coupling of k_{33} = 0.155 and the displacement inclination angle of δ_S = 4.1° at angle of β = 28°. This indicates that the resonator excites both thickness extensional and shear mode (longitudinal and shear wave modes), and the shear displacement direction is not perpendicular to the propagation direction. Larger δ_S values may result in energy leakage

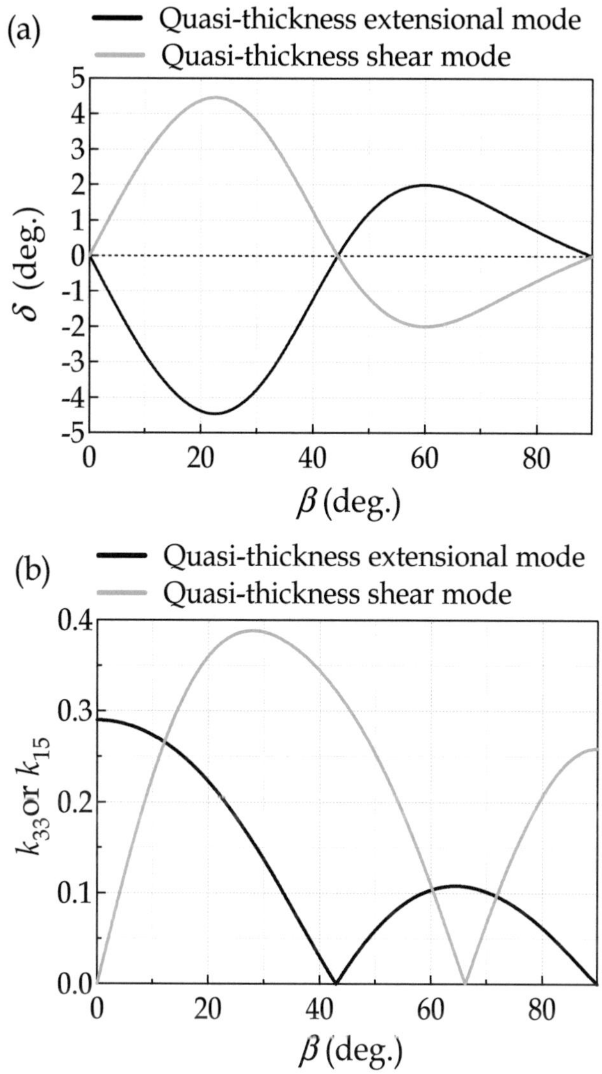

Fig. 3. (a) Angle δ between the wave displacement u and the x direction and (b) electromechanical coupling coefficient of the quasi-longitudinal and quasi-shear waves for the ZnO crystal as function of the angle β between the c-axis and x_3 direction

due to mode conversion in the reflection plane. This induces the decrease of Q value. Both of the no extensional mode coupling and small δ_S values of 0.38° can be obtained at $\beta=43°$, however, it is difficult to adjust such as large c-axis tilt angle in a large area deposition. One option is to use a pure-shear-mode ($\beta= 90°$) resonator to satisfy both the conditions of no extensional coupling and $\delta_S = 0°$. Pure shear mode excitation can be achieved by two electric field-orientation combination. One is to apply the cross-electric field to c-axis parallel film by sandwiched electrode (Yanagitani et al., 2007d), and the other is to apply the in-plane electric field to c-axis normal film by IDT electrode (Corso et al., 2007; Milyutin et al., 2008, 2010). Of course, the latter is the easiest way to obtain pure shear mode because deposition technique of c-axis normal film has been well established, but effective electrometrical coupling is weak (k_{eff}=0.04-0.06) (Corso et al., 2007; Milyutin et al., 2008). The former has large electrometrical coupling (k_{15}=0.24) (Yanagitani et al., 2007a), and recently the c-axis parallel oriented film can be easily obtained by using ion beam orientation control technique (presented in next section), even in a large area (Kawamoto et al., 2010).

3. Ion beam orientation control technique for shear mode piezoelectric films

3.1 Ion beam orientation control of wurtzite thin film by ion beam irradiation

Polycrystalline films tend to grow in their most densely packed plane parallel to the substrate plane. Bravais proposed the empirical rule that the growth rate of the crystal plane is proportional to the surface atomic density. Namely, the lattice plane with higher surface atomic density grows more rapidly. Curie argued that the growth rate perpendicular to a plane is proportional to the surface free energy (Curie, 1885).

Ion bombardment during film deposition can modify this preferred orientation of the films. This is usually explained by a change in anisotropy of the growing rate of the crystal plane in the grain, which is reflected by the difference in the degree of the ion channeling effect or ion-induced damage in the crystal plane (Bradley et al., 1986; Ensinger, 1995; Ressler et al., 1997; Dong & Srolovitz, 1999). For example, during ion beam irradiation, the commonly observed <111> preferred orientation in a face-centered cubic film changes to a <110> preferred orientation, which corresponds to the easiest channeling direction (Van Wyk & Smith, 1980; Dobrev, 1982). In-plane texture controls have also been achieved by optimizing the incident angle of the ion beam (Yu et al., 1985; Iijima et al., 1992; Harper et al., 1997; Kaufman et al., 1999; Dong et al., 2001; Park et al., 2005).

In wurtzite films, for example, the surface energy densities of the (0001), $(11\bar{2}0)$ and $(10\bar{1}0)$ planes of the ZnO crystal are estimated to be 9.9, 12.3, 20.9 eV/nm^2, respectively (Fujimura et al., 1993). The (0001) plane has the lowest surface density. Thus, the ZnO film tends to grow along the [0001] direction. When wurtzite crystal is irradiated with ion beam, the most densely packed (0001) plane should incur more damage than the $(10\bar{1}0)$ and $(11\bar{2}0)$ planes, which correspond to channeling directions toward the ion beam irradiation. We can therefore expect that the thermodynamically preferred (0001) oriented grain growth will be disturbed by ion damage so that the damage-tolerant $(10\bar{1}0)$ or $(11\bar{2}0)$ orientated grains (c-axis parallel oriented grain) will preferentially develop instead.

On this basis, in-plane and out-of-plane orientation control of AlN and ZnO films by means of ion beam-assisted deposition technique, such as evaporation (Yanagitani & Kiuchi, 2007c) and sputtering (Yanagitani & Kiuchi, 2007e, 2011b) was achieved. c-axis parallel oriented can be obtained even in a conventional magnetron sputtering technique using a low pressure discharge (<0.1 Pa) (Yanagitani et al., 2005) or RF substrate bias (Takayanagi, 2011), which leads ion bombardment on the substrate. Figure 4 shows the XRD patterns of the ZnO films deposited with various ion energy and amount of flux in ion beam assisted evaporation (Yanagitani & Kiuchi, 2007c). Table 1 shows the ion current densities in the case of "Large ion flux" and "Small ion flux" in Fig 4. The tendency of the $(10\bar{1}0)$ orientation is enhanced with increasing ion energy and amount of ion irradiation, demonstrating that the ion bombardment induced the (0001) orientation to change into a $(10\bar{1}0)$ orientation, which corresponds to the ion channeling direction.

Ion energy	A: Large ion flux	B: Small ion flux
0.05 keV	0-5 μA/cm^2	
0.25 keV	30-50 μA/cm^2	
0.5 keV	190 μA/cm^2	140 μA/cm^2
0.75 keV	220 μA/cm^2	130 μA/cm^2
1.0 keV	240 μA/cm^2	120 μA/cm^2

Table 1. Ion current densities in "Large ion flux" and "Small ion flux"

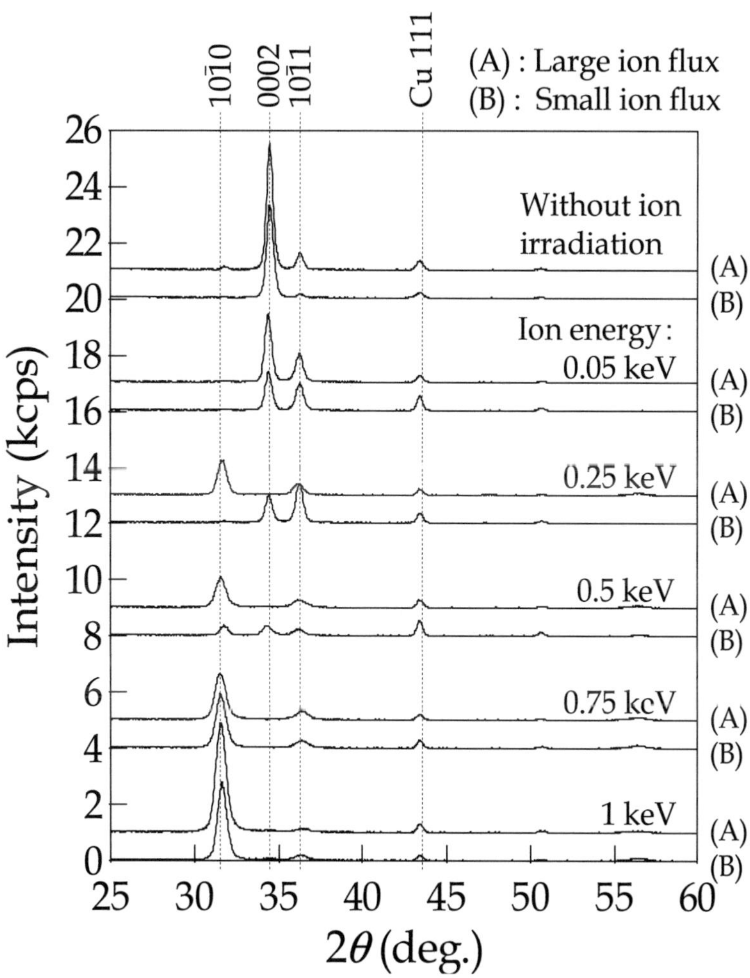

Fig. 4. 2θ-ω scan XRD patterns of the ZnO films deposited without ion irradiation, and with ion irradiation of 0-1 keV with "Large ion flux" and "Small ion flux" (Yanagitani & Kiuchi, 2007c)

Figure 5 shows the XRD patterns of the samples deposited under the conditions that various RF and DC bias are applied to the substrate. Although any dramatic change in usual (0001) preferred orientation is not occurred in the case of positive or negative DC bias, (0001) orientation changed to $(11\bar{2}0)$ and $(10\bar{1}0)$ orientation with the increase of RF bias power which induces the bombardment of positive ion on substrate. Interestingly, the order of the appearance of the (0001) to $(11\bar{2}0)$ and $(10\bar{1}0)$ corresponds to the order of increasing surface atomic density, which may be the order of damage tolerance against ion bombardment.

In order to excite shear wave in the c-axis parallel film, c-axis is required to orient not only in out-of-plane direction but also in in-plane direction. The ion beam orientation control technique allows us to control even in in-plane c-axis direction and polarization by the direction of beam incident direction (Yanagitani et al., 2007d).

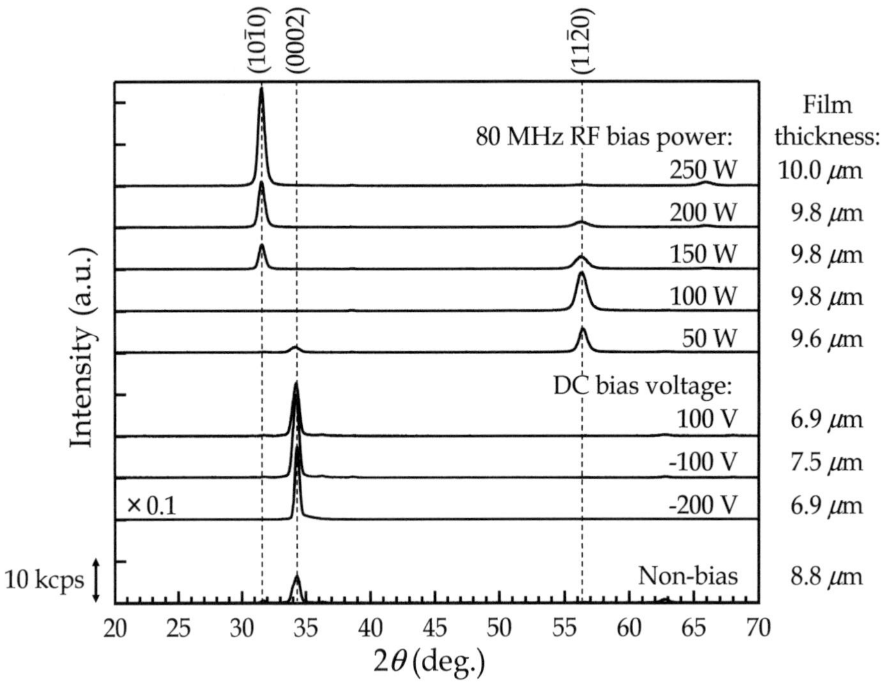

Fig. 5. 2θ-ω scan XRD patterns of the samples deposited without bias, with 80 MHz RF bias of 50 to 250 W, or with -200 to 100 DC bias. All samples were measured at the center of the bias electrode (Takayanagi et al., 2011)

4. Method for determining *k* values in piezoelectric thin films

4.1 *k* value determination using as-deposited structure (HBAR structure)

A method for determining piezoelectric property in thin films is described in this section. In general, electromechanical coupling coefficient (*k* value) in thin film can be easily determined by series and parallel resonant frequency of a FBAR consisting of top electrode layer/piezoelectric layer/bottom electrode layer or SMR (Solidly mounted resonator) consisting of top electrode layer/piezoelectric layer/bottom electrode layer/Bragg reflector. In case thickness of electrode film is negligible small compared with that of piezoelectric film. *k* of the piezoelectric film can be written as follows (Meeker, 1996):

$$k^2 = \frac{\pi}{2}\frac{f_s}{f_p}\tan\left(\frac{\pi}{2}\frac{f_p - f_s}{f_p}\right) \tag{22}$$

where f_p and f_s are the parallel resonant frequency and series resonant frequency, respectively.
However, it takes considerable time and effort to fabricate FBAR structure which have self-standing piezoelectric layer. It is convenient if k value can be determined from as deposited structure, namely so-called an HBAR (high-overtone bulk acoustic resonator) or composite resonator structure consisting of top electrode layer/piezoelectric layer/bottom electrode layer/thick substrate. Methods for determining the k value of the films from HBAR structure are more complex than that for the self-supported single piezoelectric film structure (FBAR structure). Several groups have investigated methods for the determination of k_t value from the HBAR structure (Hickernell, 1996; Naik, et al., 1998; Zhang et al., 2003). One of the easiest ways of k determination is to use a conversion loss characteristic of the HBAR structure. When the thickness of electrode layers is negligible small compared with that of piezoelectric layer, capacitive impedance of resonator is equal to the electrical source impedance, and k value of the piezoelectric layer is smaller than 0.3, conversion loss CL is approximately represented by k value at parallel resonant frequency (Foster et al., 1968):

$$CL \approx 10\log_{10}\frac{\pi}{8k^2}\cdot\frac{Z_s}{Z_p} \tag{23}$$

where, Z_s and Z_p is acoustic impedance of the substrate and piezoelectric layer, respectively.
However, various inhomogeneities sometimes exist in the film resonator, such as non-negligible thick and heavy electrode layers, thickness taper, or the piezoelectrically inactive layer composed of randomly oriented gains growing in the initial stages of the deposition. In this case, the k values of the film can be determined so as to match the experimentally measured conversion losses (CL) of the resonators with theoretical minimum CL by taking k value as adjustable parameter. The theoretical CL in this case can be calculated by Mason's equivalent circuit model including electrode layer, film thickness taper and piezoelectrically inactive layer. This method allows various inhomogeneous effect of film to be taken into account (Yanagitani et al., 2007b, 2007c).

4.2 Experimental method to estimate conversion loss of HBAR structure

The experimental CL of HBAR can be determined from reflection coefficients (S_{11}) of the resonators, which can be obtained using a network analyzer with a microwave probe. The inverse Fourier transform of S_{11} frequency response of the resonator gives the impulse response of the resonator in the time domain. In the HBAR structure, the impulse response is expected to include echo pulse trains reflected from the bottom surface of the substrate, and the insertion loss of resonator can be obtained from the Fourier transform of the first echo in this impulse response. This experimental insertion loss $IL_{experiment}$ includes doubled CL in the piezoelectric film and round-trip diffraction loss DL and round-trip propagation loss PL in the silica glass substrate. Therefore, CL can be expressed as

$$CL = \frac{1}{2}\left(IL_{experiment} - DL - PL\right), \tag{24}$$

where diffraction loss DL can be calculated according to the method reported by Ogi *et al.* (Ogi et al., 1995). This method is based on integration of the velocity potential field in the divided small transducer elements, which allows calculation of the DL with electrode areas of various shapes. The round-trip propagation loss PL is given as

$$PL = 2d_s \frac{\alpha_s}{f^2}, \tag{25}$$

where d_s is the thickness of the substrate, α_s represents the shear wave attenuation in the substrate, for example, $\alpha_s / f^2 = 19.9\times10^{-16}$ (dB·s^2/m) in silica glass substrate (Fraser, 1967).

4.3 Conversion loss simulation in HBAR by Mason's equivalent circuit model

Electromechanical coupling coefficient k can be estimated by comparing an experimental CL with a theoretical CL of the HBAR. One-dimensional Mason's equivalent circuit model is convenient tool for simulating theoretical CL of the resonator. Generally, in case non-piezoelectric elastic solid vibrates in thickness mode, its can be described as T-type equivalent circuit (Fig. 6 (a)) where F_1 and F_2 are force and v_1 and v_2 are particle velocity acting on each surface of elastic solid. Piezoelectric elastic solid can be represented as the Mason's three ports equivalent circuit which includes additional electric terminal concerning electric voltage V and current I (Fig. 6 (b)) (Mason, 1964). Here, γ is propagation constant, Z is acoustic impedance and d_p is thickness of elastic solid. To take account of attenuation of vibration, mechanical quality factor Q_m is defined as $Q_m = c_r / c_i$ where c_r and c_i are real part and imaginary part of elastic constant, respectively. Using mechanical quality factor Q_m, propagation constant γ and acoustic impedance Z are given as:

$$\gamma = j\omega \sqrt{\frac{\rho}{c_r\left\{\,1 + j\,(1/Q_m)\,\right\}}}, \quad Z = S\,\sqrt{\rho c_r\left\{\,1 + j\,(1/Q_m)\,\right\}} \tag{26}$$

where ρ is density of the elastic solid and S is electrode area of the resonator.
Static capacitance C_0 and ratio of transformer ϕ_0 in the circuit are given as:

$$C_0 = \varepsilon_{11}^S \frac{S}{d_p}, \quad \phi_0 = \left[\frac{C_0 v_p Z_p}{d_p}\left(\frac{k_{15}^2}{1 - k_{15}^2}\right)\right]^{\frac{1}{2}}, \tag{27}$$

where d is the thickness of the layers, ε_{11}^S is permittivity, and v is the velocity of the shear wave. Subscript p, $e1$, $e2$ and s respectively represent piezoelectric layer, top electrode layer, bottom electrode layer and substrate. k value affects the equivalent circuit through the ratio of transformer ϕ_0.

Equivalent circuit for the over-moded resonator structure is given in Fig. 7 by cascade arranging non-piezoelectric and piezoelectric part as described in Figs. 6 (a) and (b). Substrate thickness is assumed infinite to ignore reflection waves from bottom surface of the substrate in this case. When the surface of the top electrode is stress-free, the acoustic input port is shorted. As top electrode part circuit can be simplified, three-port circuit in Fig. 7 is transformed to the two-ports circuit shown in Fig. 8 (Rosenbaum, 1988).

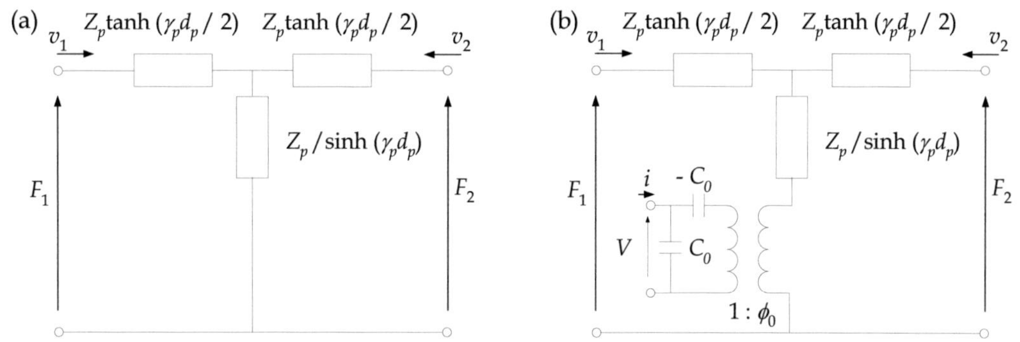

Fig. 6. Equivalent circuit model of (a) non-piezoelectric (b) piezoelectric elastic solid

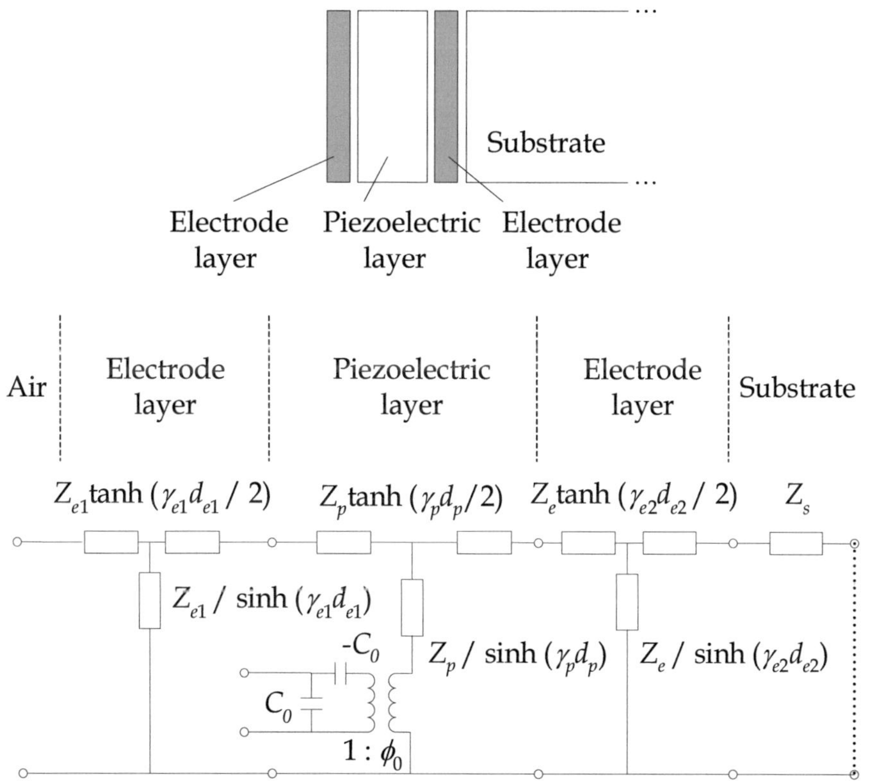

Fig. 7. Equivalent circuit model of the over-moded resonator structure

It is convenient to derive whole impedance of the circuit by using *ABCD*-parameters (Paco et al., 2008) As shown in Eqs. (28)-(32), *ABCD*-parameters of whole circuit is derived multiplying each circuit element.

$$F_{Transformer} = \begin{bmatrix} 1/\phi_0 & 0 \\ 0 & \phi_0 \end{bmatrix}, \; F_{Electric\ port} = \begin{bmatrix} 1 & 0 \\ j\omega C_0 & 1 \end{bmatrix} \cdot \begin{bmatrix} 1 & -1/j\omega C_0 \\ 0 & 1 \end{bmatrix}, \; F_{Substrate} = \begin{bmatrix} 1 & Z_s \\ 0 & 1 \end{bmatrix} \quad (28)$$

$$F_{Piezo+Electrode\ layer} = \begin{bmatrix} 1 & Z_p / \sinh(\gamma_p d_p) \\ 0 & 1 \end{bmatrix} \cdot \begin{bmatrix} 1 & 0 \\ 1 / \left\{ Z_{e1} \tanh(\gamma_{e1} d_{e1} / 2) + Z_p \tanh(\gamma_p d_p / 2) \right\} & 1 \end{bmatrix} \cdot \begin{bmatrix} 1 & Z_p \tanh(\gamma_p d_p / 2) \\ 0 & 1 \end{bmatrix} \quad (29)$$

$$F_{Counter-electrode} = \begin{bmatrix} 1 & Z_{e2} \tanh(\gamma_{e2} d_{e2} / 2) \\ 0 & 1 \end{bmatrix} \cdot \begin{bmatrix} 1 & 0 \\ \sinh(\gamma_{e2} d_{e2}) / Z_{e2} & 1 \end{bmatrix} \cdot \begin{bmatrix} 1 & Z_{e2} \tanh(\gamma_{e2} d_{e2} / 2) \\ 0 & 1 \end{bmatrix} \quad (30)$$

$$F_{Substrate} = \begin{bmatrix} 1 & Z_s \\ 0 & 1 \end{bmatrix} \quad (31)$$

$$F_{Over-moded\ resonator} = F_{Electric\ port} \cdot F_{Transformer} \cdot F_{Piezo+Elecrode\ layer} \cdot F_{Counterelectrode} \cdot F_{Substrate} \quad (32)$$

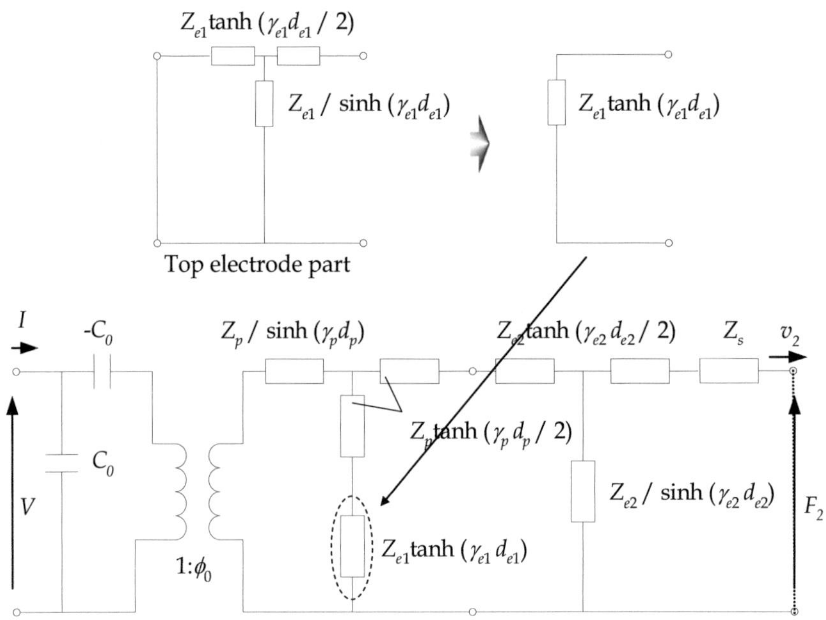

Fig. 8. Simplification of equivalent circuit model for over-moded resonator structure

Insertion loss *IL* is expressed as the ratio of the signal power delivered from a source into load resistance to the power delivered from a source into the inserted network. *IL* of the resonators can be calculated with the following equation using conductance of the electrical source G_S (0.02 S), input conductance G_f, and susceptance B_f of the circuit model, which can be derived from *ABCD*-parameter to Y-parameter conversion of eq. (32):

$$IL = 20\log_{10} \frac{\Re e\left\{ \left(G_f + jB_f\right) \dfrac{G_S^2}{\left(G_S + G_f\right)^2 + B_f^2} \right\}}{G_S/4}. \quad (33)$$

Hence the CL is

$$CL = \frac{IL}{2} = 10\log_{10}\frac{4G_S G_f}{\left(G_S + G_f\right)^2 + B_f^2}. \tag{34}$$

4.4 *k* value determination from conversion loss curves

Figure 9 (a) shows the pure shear mode theoretical and experimental CL curves of the c-axis parallel film HBAR as an example. By comparing experimental curve with theoretical curves

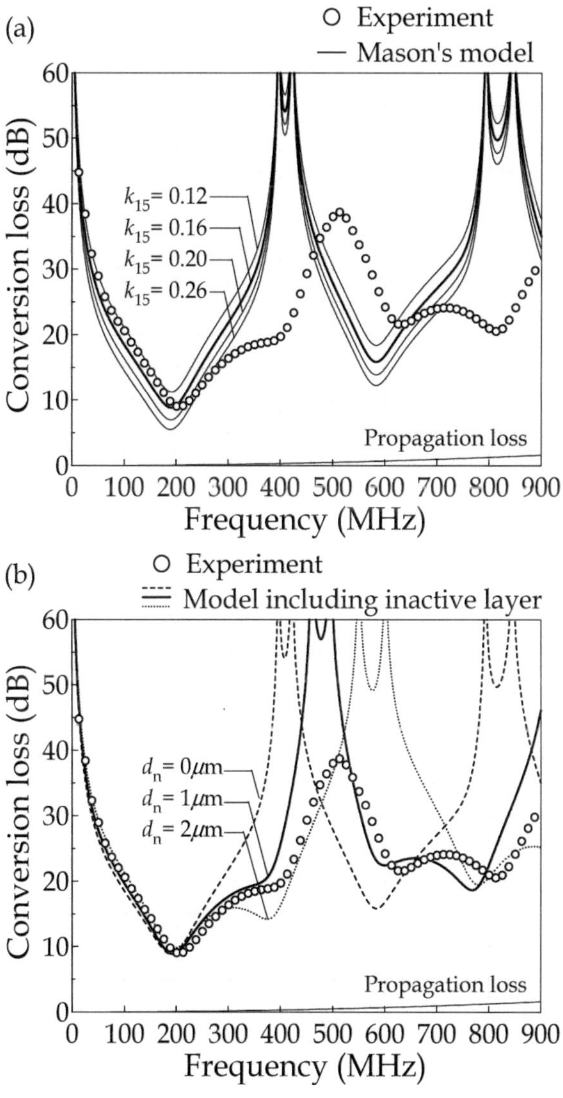

Fig. 9. Frequency response of the experimental shear mode CL (open circles). (a) The simulated shear mode CL curves (solid line) in various k_{15} values and (b) the curve simulated by the model including various thickness of piezoelectrically inactive layer (Yanagitani & Kiuchi, 2007c)

at minimum *CL* point (near the parallel resonant frequency), we can determine the k_{15} value of the film. As shown in Fig. 9 (b), effective thickness of the piezoelectrically inactive layer d_n in the initial stages of the deposition also can be estimated from comparison of the curves. Figure 10 shows the correlation between k_{15} value and crystalline orientation of the film. FWHM values of ψ-scan and ϕ-scan curve of the XRD (X-ray diffraction) pole figure show the degree of crystalline orientation in out-of plane and in-plane, respectively. Thicker films tend to have large k_{15} values even though they have same degree of crystalline orientation as thinner one. This kind of correlations and inhomogeneities characterization in wafer can be easily obtained from as-deposited film structure, by using present *k* value determination method.

4.5 Conclusion

In this chapter, shear mode piezoelectric thin film resonators, which is promising for the acoustic microsensors operating in liquid, were introduced. Theoretical predictions of electromechanical coupling and tilt of wave displacement as functions of c-axis tilt angle showed that pure shear mode excitation by using c-axis parallel oriented wurtzite piezoelectric films expected to achieve high-Q and high-coupling sensor. Fabrication of c-axis parallel oriented films by ion beam orientation control technique and characterization of the film by a conversion loss of the as-deposited resonator structure were discussed.

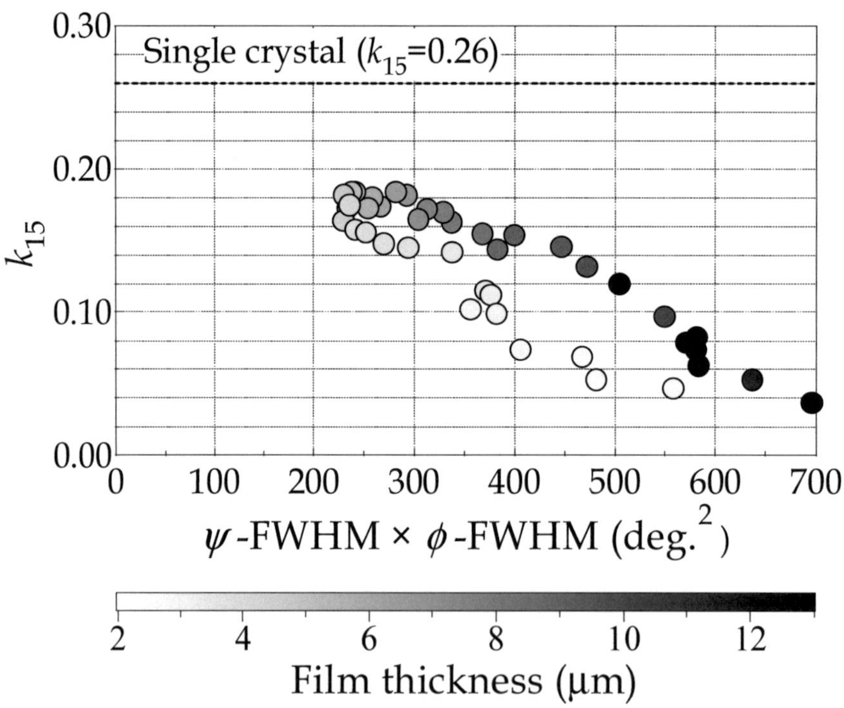

Fig. 10. k_{15} values of the ZnO piezoelectric layers as a function of multiplication of ψ-scan and ϕ-scan profile curve FWHM values extracted from XRD pole figure (indicating the degree of crystalline orientation in out-of-plane and in-plane) (Yanagitani et al., 2007b)

5. References

Auld, B. A. Acoustic Fields and Waves in Solid. (1973). Vol. 1, pp. 73-76, A Wiley-Interscience Publication.

Bond, W. (1943). The Mathmatics of the Physical Properties of Crystals. *Bell System Technical Journal*, Vol. 22, pp. 1–72.

Bradley R. M.; Harper J. M. E. & Smith, D. A. (1986). Theory of Thin-film Orientation by Ion bombardment during deposition. *J. Appl. Phys.*, Vol. 60, No. 12, pp. 4160–4164.

Corso, C. D.; Dickherber, A. & Hunt, W. D. (2007). Lateral Field Excitation of Thickness Shear Mode Waves in a Thin Film ZnO Solidly Mounted Resonator. *J. Appl. Phys.* Vol. 101, pp. 054514-1–054514-7.

Curie, P. (1885). *Bull. Soc. Franc. Miner. Crist.* Vol. 8 p. 145.

Dobrev, D. (1982). Ion-beam-induced Texture Formation in Vacuum-condensed Thin Metal Films. *Thin Solid Films*, Vol. 92, No. 1-2, pp. 41–53.

Dong, L. & Srolovitz, D. J. (1999). Mechanism of Texture Development in Ion-beam-assisted Deposition. *J. Appl. Phys. Lett.*, Vol. 75, No. 4, pp. 584–586.

Dong, L.; Srolovitz, D. J.; Was, G. S.; Zhao, Q. & Rollett, A. D. (2001). Combined Out-of-plane and In-plane Texture Control in Thin Films using Ion Beam Assisted Deposition. *J. Mater. Res.*, Vol. 16, No. 1, pp. 210–216.

Ensinger, W. (1995). On the Mechanism of Crystal Growth Orientation of Ion Beam Assisted Deposited Thin Films. *Nucl. Instrum. Meth. Phys. Res. B*, Vol. 106, pp. 142–146.

Foster, N. F.; Coquin, G. A.; Rozgonyi, G. A. & Vannatta, F. A. (1968). Cadmium Sulphide and Zinc Oxide Thin-Film Transducers. *IEEE Trans. Sonic. Ultrason.*, Vol. 15, pp. 28–41.

Fraser, D. B.; Krause, J. T. & Meitzler, A. H.; (1967). Physical Limitations on the Performance of Vitreous Silica in High-frequency Ultrasonic and Acousto-optical Devices. *Appl. Phys. Lett.* Vol. 11, No. 10, pp. 308–310.

Fujimura, N.; Nishihara, T.; Goto, S.; Xu, J. & Ito, T. (1993). Control of Preferred Orientation for ZnOx Films: Control of Self-texture. *J. Crystal Growth*, Vol. 130, pp. 269–279.

Harper, J. M. E.; Rodbell, K. P.; Colgan, E. G. & Hammond, R. H. (1997). Control of In-plane Texture of Body Centered Cubic Metal Thin Films. *J. Appl. Phys.* Vol. 82, pp. 4319–4326.

Hickernell, F. S. (1996). Measurement Techniques for Evaluating Piezoelectric Thin Films. *Proc. IEEE Ultrason. Symp.*, pp. 235–242.

Iijima, Y.; Tanabe, N.; Kohno, O. & Ikeno, Y. (1992). In-plane Aligned $YBa_2Cu_3O_{7-x}$ Thin Films Deposited on Polycrystalline Metallic Substrates. *Appl. Phys. Lett.*, Vol. 60, No. 6, pp. 769–771.

Kaufman, D. Y.; DeLuca, P. M.; Tsai, T. & Barnett, S. A. (1999). High-rate Deposition of Biaxially Textured Yttria-stabilized zirconia by Dual Magnetron Oblique Sputtering. *J. Vac. Sci. Technol. A*, Vol. 17, pp. 2826–2829.

Kawamoto, T.; Yanagitani T.; Matsukawa, M.; Watanabe, Y.; Mori1, Y.; Sasaki, S. & Oba M. (2010). Large-Area Growth of In-Plane Oriented (11-20) ZnO Films by Linear Cathode Magnetron Sputtering. *Jpn. J. Appl. Phys.* Vol. 49 pp. 07HD16-1–07HD16-4.

Kushibiki, J-I.; Akashi, N.; Sannomiya, T.; Chubachi, N.; & Dunn, F. (1995). VHF/UHF Range Bioultrasonic Spectroscopy System and Method. *IEEE Trans. Ultrason., Ferroelect., Freq. Contr.*, Vol. 42 No. 6, pp. 1028–1039

Link, M.; Weber, J.; Schreiter, M.; Wersing, W.; Elmazria, O. & Alonot, P. (2007). Sensing Characteristic of High-freqency Shear Mode Resonators in Glycerol Solutions. *Sens. Actuators B*, Vol. 121, No.2, pp. 372–378.

Matsumoto, Y.; Ujiie, T. & Kushibiki, J-I. (2000). Measurement of Shear Acoustic Properties of Water using the Ultrasonic Reflectance Method in Pulse-mode Operation. *Spring Meeting of Acoustical Society of Japan* (in Japanese)

Meeker, T. R. (1996). IEEE Stabdard on Piezoelectricity (ANSI/IEEE Std. 176-1987). *IEEE Trans. Ultrason., Ferroelect., Freq. Contr.*, Vol. 43, No. 5, pp. 719–772..

Milyutin, E. & Mural, P. (2010). Electro-Mechanical Coupling in Shear-Mode FBAR with Piezoelectric Modulated Thin Film. *IEEE Trans. Ultrason., Ferroelect., Freq. Contr.*, Vol. 58, No. 4, pp. 685–688.

Milyutin, E.; Gentil, S. & Mural, P. (2008). Shear Mode Bulk Acoustic Wave Resonator Based on c-axis Oriented AlN Thin Film. *J. Appl. Phys.* Vol. 104, pp. 084508-1-084508-6.

Mitsuyu, T.; Ono S. & Wasa, K. (1980). Structures and SAW Properties of Rf-sputtered Single-crystal Films of ZnO on Sapphire. *J. Appl. Phys.*, Vol. 51, No. 5, pp. 2464–2470.

Naik, B. S.; Lutsky, J. J.; Rief R.; & Sodini, C. D. (1998). Electromechanical Coupling Constant Extraction of Thin-film Piezoelectric Materials Using a Bulk Acoustic Wave Resonator. *IEEE Trans. Ultrason., Ferroelect., Freq. Contr.*, Vol.45, No.1, pp. 257–263.

Nakamura, K.; Shoji, T. & Kang, H. (2000). ZnO Film Growth on $(01\bar{1}2)$ $LiTaO_3$ by Electron Cyclotron Resonance-assisted Molecular Beam Epitaxy and Determination of Its Polarity. *Jpn. J. Appl. Phys.*, Vol. 39, No. 6A, pp. L534–L536.

Ogi, H.; Hirao, M.; Honda T. & Fukuoka, H. (1995). Ultrasonic Diffraction from a Transducer with Arbitrary Geometry and Strength Distribution. *J. Acoust. Soc. Am.* Vol. 98, No. 2, pp. 1191–1198.

Paco, P.; Menéndez, Ó. & Corrales E. (2008) Equivalent Circuit Modeling of Coupled Resonator Filters. *IEEE Trans. Ultrason., Ferroelect., Freq. Contr.*, Vol. 55, No. 9, pp. 2030–2037.

Park, S. J.; Norton, D. P. & Selvamanickam, V. (2005). Ion-beam Texturing of Uniaxially Textured Ni Films. *Appl. Phys. Lett.*, Vol. 87, pp. 031907–031909.

Ressler, K. G.; Sonnenberg, N. & Cima, M. J. (1997). Mechanism of Biaxial Alignment of Oxide Thin Films during Ion-Beam-Assisted Deposition. *J. Am. Ceram. Soc.*, Vol. 80, No. 10, pp. 2637–2648.

Rosenbaum, J. F. (1988). Bulk Acoustic Waves: Theory and Devices. Artech House Boston London.

Sauerbrey, G. (1959) Verwendung von Schwingquarzen zur Wägung Dünner Schichten und zur Mikrowägung. *Z. Physik*, Vol. 155, pp. 206–222.

Smith, R. T. & Stubblefield, V. E. (1969). Temperature Dependence of the Electroacoustical Constants of Li-doped ZnO Single Crystals. *J. Acoust. Soc. Am.*, Vol. 46, pp. 105.

Takayanagi, S.; Yanagitani, T.; Matsukawa, M. & Watanabe Y. (2010). A Simple Technique for Obtaining (11-20) or (10-10) Textured ZnO Films by RF Bias Sputtering," *Proc. 2010 IEEE Ultrason. Symp.* pp. 1060–1063.

Van Wyk, G. N. & Smith, H. J. (1980). Crystalline Reorientation Due to Ion Bombardment. *Nucl. Instrum. Meth.*, Vol. 170, pp. 433–439.

Weber, J.; Albers, W. M.; Tuppurainen, J.; Link, M.; Gabl, R.; Wersing, W. & Schreiter, M. (2006). Shear Mode FBARs as Highly Sensitive Liquid Biosensors. *Sens. Actuators A*, Vol. 128, No. 2, pp. 84–88.

Wingqvist G.; Anderson H.; Lennartsson C.; Weissbach T.; Yantchev V. & Lloyd Spetz A. (2009). On the Applicability of High Frequency Acoustic Shear Mode Biosensing in View of Thickness Limitations Set by The Film Resonance. *Biosens. Bioelectron.*, Vol. 24 No. 11, pp. 3387–3390.

Wingqvist, G. (2010). AlN-based Sputter-deposited Shear Mode Thin Film Bulk Acoustic Resonator (FBAR) for Biosensor Applications. *Surf. Coat. Tech.*, Vol. 205, No. 5, pp. 1279–1286

Wingqvist, G.; Bjurstrom, J.; Liljeholm, L.; Yantchev, V. & Katardjiev, I. (2007). Shear mode AlN Thin Film Electro-acoustic Resonant Sensor Operation in Viscous Media. *Sens. Actuators B*, Vol. 123, No. 1, pp. 466–473.

Wittstruck, R. H.; Tong, X.; Emanetoglu, N. W.; Wu, P.; Chen, Y.; Zhu, J.; Muthukumar, S.; Lu, Y.; & Ballato, A. (2003) Characteristic of $Mg_xZn_{1-x}O$ Thin Film Bulk Acoustic Wave Devices. *IEEE Trans. Ultrason., Ferroelect., Freq. Contr.*, Vol. 50, pp. 1272–1277.

Yanagitani, T. & Kiuchi, M. (2007c). Control of In-plane and Out-of-plane Texture in Shear Mode Piezoelectric ZnO Films by Ion-beam Irradiation. *J. Appl. Phys.*, Vol. 102, pp. 044115-1–044115-7.

Yanagitani, T. & Kiuchi, M. (2007e). Highly Oriented ZnO Thin Films Deposited by Grazing Ion-beam Sputtering: Application to Acoustic Shear Wave Excitation in the GHz Range. *Jpn. J. Appl. Phys.* Vol. 46, pp. L1167–L1169.

Yanagitani, T. & Kiuchi, M. (2011b). Texture Modification of Wurtzite Piezoelectric Films by Ion Beam Irradiation. *Surf. Coat. Technol.*, in press.

Yanagitani, T., Matsukawa, M., Watanabe, Y., Otani, T. (2005). Formation of Uniaxially (11-2 0) Textured ZnO Films on Glass Substrates. *J. Cryst. Growth*, Vol. 276, No. 3-4, pp. 424-430.

Yanagitani, T.; Arakawa, K.; Kano, K.; Teshigahara, A.; Akiyama, M. (2010). Giant Shear Mode Electromechanical Coupling Coefficient k_{15} in c axis Tilted ScAlN Films. *Proc. 2010 IEEE Ultrason. Symp.*, pp. 2095–2098.

Yanagitani, T.; Kiuchi, M.; Matsukawa, M. & Watanabe, Y. (2007b). Shear Mode Electromechanical Coupling Coefficient k_{15} and Crystallites Alignment of (11-20) Textured ZnO Films. *J. Appl. Phys.*, Vol. 102, pp. 024110-1–024110-7.

Yanagitani, T.; Kiuchi, M.; Matsukawa, M. & Watanabe, Y. (2007d). Characteristics of Pure-shear Mode BAW Resonators Consisting of (11-20) Textured ZnO Films. *IEEE Trans. Ultrason., Ferroelect., Freq. Contr.*, Vol. 54, No. 8, pp. 1680–1686.

Yanagitani, T.; Mishima, N.; Matsukawa, M. & Watanabe, Y. (2007a). Electromechanical Coupling Coefficient k_{15} of Polycrystalline ZnO Films with the c-axes Lie in the Substrate Plane. *IEEE Trans. Ultrason., Ferroelect., Freq. Contr.*, Vol. 54, No. 4, pp. 701–704.

Yanagitani, T.; Morisato, N.; Takayanagi, S.; Matsukawa, M. & Watanabe, Y. (2011a) c-axis Zig-Zag ZnO Film Ultrasonic Transducers for Designing Longitudinal and Shear

Wave Resonant Frequencies and Modes. *IEEE Trans. Ultrason., Ferroelect., Freq. Contr.*, Vol. 58, No. 5, pp. 1062–1068.

Yu, L. S.; Harper, J. M. E.; Cuomo, J. J. & Smith, D. A. (1985). Alignment of Thin Films by Glancing Angle Ion Bombardment During Deposition. *Appl. Phys. Lett.*, Vol. 47, No. 9, pp. 932–933.

Zhang, Y.; Wang Z. & Cheeke, J. D. N. (2003). Resonant Spectrum Method to Characterize Piezoelectric Films in Composite Resonators. *IEEE Trans. Ultrason., Ferroelect., Freq. Contr.*, Vol. 50, No. 3, pp. 321–333.

23

Polymer Coated Rayleigh SAW and STW Resonators for Gas Sensor Applications

Ivan D. Avramov
Georgi Nadjakov Institute of Solid State Physics, Sofia
Bulgaria

1. Introduction

Polymer coated gas-phase sensors using the classical Rayleigh-type surface acoustic wave (RSAW) mode have enjoyed considerable interest worldwide over the last two decades [1-3]. This interest is motivated by their orders of magnitude higher sensitivity and larger dynamic range compared to bulk acoustic wave (BAW) sensors, fast response times, excellent overall stability, coming close to that of quartz crystal sensors, and low phase noise of the sensor system making high-resolution measurements possible [4]. Because of these features that are difficult to achieve with other technologies, RSAW based gas sensors have found successful application in a variety of industrial implementations such as electronic noses, systems for analysis of chemical and biological gases, medical diagnostics, environmental monitoring and protection, etc. [5-11]. On the other hand, surface transverse wave (STW) based gas sensors, even though sharing the same operation principle, have not been studied so extensively yet. The purpose of this article is to present and discuss systematic experimental data with both acoustic wave modes which will prove that STW based gas-phase sensors not only successfully compete with their RSAW counterparts but also complement them in applications where RSAW gas sensors reach their limits. Successful corrosion proof RSAW sensors using gold metallization for operation in highly reactive chemical environments will also be presented.

2. Operation principle of RSAW/STW based resonant gas phase sensors

Both RSAW and STW based gas sensitive resonant sensors share the same operation principle illustrated in Fig. 1. The sensor device typically is a two-port RSAW or STW resonator on a temperature compensated rotated Y cut of quartz whose geometry has been optimized in such manner that the resonator retains a well behaved single-mode resonance and suffers minimum loss increase and Q-degradation after the gas sensitive layer, (typically a solid, semisolid or soft polymer film with good sorption properties), is deposited on its surface. On the other hand, the sensor has to have maximum active area in the centre of its geometry where the magnitude of the standing wave and deformation are maximized. Thus, strong interaction with the gas adsorbed in the polymer film occurs and maximum gas sensitivity is obtained. The sensor operation principle according to Fig. 1 is fairly simple. If a gas-phase analyte of a certain concentration is applied to its surface, gas molecules are absorbed by the sensing layer until thermodynamic equilibrium is achieved; i. e. the number of adsorbed molecules becomes equal to the number of desorbed ones. Due to adsorption, the layer becomes heavier and this increases the mass loading on the sensor surface. As a result of that, the acoustic wave propagation velocity v decreases and causes a concentration proportional frequency down shift Δf of the sensor's resonance, called sensor signal. The resonance frequency shift of RSAW gas sensors coated with a polyisobutilene (PIB) polymer film is shown in Fig. 2 a) and b) for two different concentrations of tetrachloroethilene vapors. If the vapor concentration is small (0,1% in Fig. 2 a)) then the resonance shifts down by 83 ppm without degradation in loss or Q. At large concentrations of the gas vapors (0,7% in Fig. 2 b)), the 550 ppm of observed frequency down shift is accompanied by a 2 dB loss increase due to the heavy mass loading. However, the sensor device retains a high loaded Q, (above 2000 in Fig. 2 a) versus >4000 in Fig. 2 b)) and a steep phase slope in a well behaved single-mode resonance without distortion or excitation of undesired longitudinal modes.

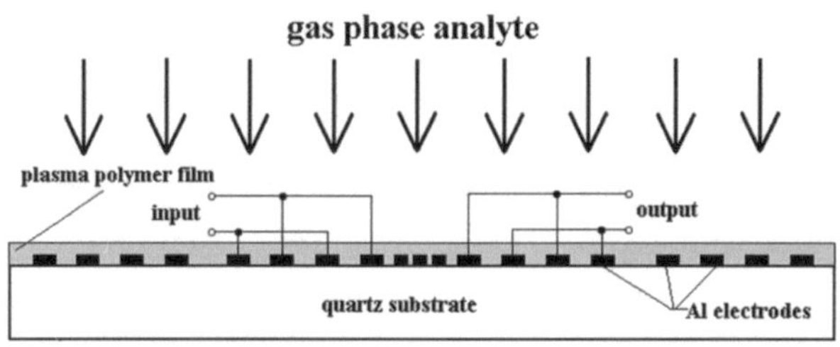

Fig. 1. Operation principle of RSAW/STW based resonant gas phase sensors

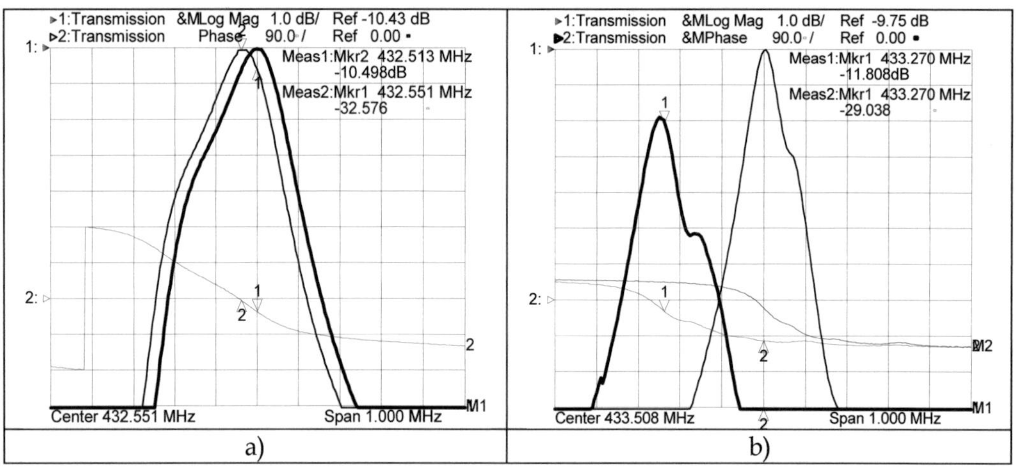

Fig. 2. Frequency (upper curves and phase (lower curves) responses of PIB coated RSAW sensors prior to (right) and after (left) tetrachloroethilene vapor probing at a) 0,1% and b) 0,7% concentration

3. Measurement resolution of RSAW/STW gas phase sensor systems

If a sensor device as the ones from Fig. 2 a) and b) is used as a frequency stabilizing element in the feedback loop of an oscillator circuit and its frequency f_0 is adjusted at the resonance frequency of the sensor (see marker positions in Fig. 2 a) and b)) then due to the high Q of the sensor device, low-noise oscillation with high short-term stability will be obtained. Any change in gas concentration will alter the resonance frequency and the output frequency f_0 of the sensor oscillator, accordingly. Thus Δf can be measured with a high precision using a high-resolution frequency counter, connected to the output of the sensor oscillator. At a given gas concentration C, measured in parts per million (ppm), the resolution R of the sensor system, also measured in ppm, will be limited only by the short-term stability of the sensor oscillator $\sigma_y(\tau)$, also called Allan's variation, for the measurement time τ. *The value of $\sigma_y(\tau)$ represents the flicker phase noise of the sensor oscillator in the time domain which is dominated by the actual flicker phase noise of the coated acoustic wave sensor.* The resolution R determines the minimum change in gas concentration that the system can detect and is, therefore, also called detection limit. It is calculated as follows:

$$R = [C\sigma_y(\tau) f_0 \tau] / \Delta f \tag{1}$$

To calculate R for a given gas concentration C, according to (1), it is sufficient to measure $\sigma_y(\tau)$ of the sensor oscillator for the time interval τ which is normally 1s for most frequency counters operating in the typical 0,3 to 1,0 GHz RSAW/STW sensor range with 1 Hz resolution. Then, according to [12], $\sigma_y(\tau)$ can be calculated from a finite number M of consecutive frequency measurements y_i of f_0 as:

$$\sigma_y(\tau) = \left[\frac{1}{2(M-1)} \sum_{i=1}^{M-1} (y_{i+1} - y_i)^2 \right]^{1/2}, \tag{2}$$

where i is an integer. In a well stabilized against thermal transients sensor oscillator typically 20 to 50 consecutive measurements of f_0 are enough to calculate $\sigma_y(\tau)$ with sufficient accuracy for practical sensor applications.

4. Chemosensitive layers for RSAW/STW based gas sensors

The correct choice of the sensing layer suitable for the chosen acoustic mode is the key to proper sensor operation and good sensitivity and dynamic range [13, 14]. A sensing layer is considered as "good" if it has an excellent adhesion to the surface of the acoustic device for proper interaction with the acoustic wave, can easily adsorb and restlessly desorb large amounts of probing gases without chemically reacting with them, has good temperature stability and low ageing and does not change its sensitivity and sorption characteristics over thousands of measurement cycles. It is also desirable that the layer provides some selectivity to a certain chemical compound, i. e. it absorbs larger amounts of that compound than other compounds. Finally, the layer should not significantly degrade the Q, loss and the shape of the resonance after deposition onto the acoustic device.

Because of their complicated net structure, many polymers feature excellent physical sorption, as required for reproducible sensor performance and this makes them appropriate for gas sensing applications [15-19]. If some of them have also appropriate viscoelastic properties for good interaction with the RSAW or STW mode, then they will provide the required performance of the acoustic wave sensor, accordingly. Layers with appropriate viscoelastic properties are those that follow the deformation of the surface as a result of the wave propagation without causing significant propagation loss and conversion of the acoustic energy into undesired modes that decay into the bulk of the substrate and may cause degradation of sensor performance.

An important parameter of the sensing film, except for its viscoelastic properties is its solidness. On one hand, the parameter "solidness" determines the sorption properties of the film and the amount of gas that the layer can accommodate before saturation is reached. On the other hand, it determines the way in which the polymer film interacts with the acoustic wave. Therefore, the film solidness will determine the sensitivity, dynamic range and detection limit of the sensor. Based on their solidness, there are three types of polymer films that are appropriate for RSAW/STW sensors:

a. *Solid polymer films.* In fact, these films are solid as glass. That is why, they are often called "glassy polymer films" and have a stiffness value close to that of the sensor's quartz substrate that they are deposited on. If used with the STW mode, due to the lower propagation velocity, these solid films trap the wave energy to the substrate surface and the acoustic wave propagates with low loss. That is why, solid films work much better with the STW mode than with the RSAW one. When their thickness becomes too high, a second slightly faster mode, called "Love mode" gets excited and multimoding occurs. Solid polymer films feature surface sorption and become easily saturated by the adsorbed gas but on the other hand, they feature very fast response times and are very sensitive if the sensor is operated far below saturation. That is why they are appropriate for high resolution measurements at low gas concentrations, (typically below 0,1%). A typical representative of the solid polymer family is the hexamethyldissiloxane (HMDSO), obtained in a glow-discharge plasma polymerization process [19].
b. *Soft polymer films.* These films are soft and elastic just like rubber. That is why they are referred to as "rubbery" or "jelly-like" films. Typically, they are deposited using spin coating or more advanced techniques such as airbrush or electro spray methods that provide good control over film thickness and uniformity. Since these soft polymers provide profound bulk sorption, they are capable of adsorbing large amounts of gas and are appropriate for measurements at high gas concentrations, (typically above 0,1%). They are well tolerated by the RSAW mode but do not work so well with STW. The reason is that they cause energy leakage of the STW into the bulk of the soft layer which results in increased loss and Q-degradation of the sensor resonator. Polymers like polyisobutilene (PIB), poly-(2-hydroxyethylmethacrylate) (PHEMA) and poly-(n-butylmethacrylate) (PBMA) are often used in RSAW based gas sensors.
c. *Semisolid polymer films.* These light and highly elastic films are also typically obtained in a plasma polymerization process [17, 18] for good reproducibility of the film parameters and have a structure very similar to polystyrene, the material used in plastic bags. They are highly resistant to almost all aggressive chemicals such as acids, bases and organic solvents and this makes them appropriate for environmental sensing applications. They are well tolerated by both, the RSAW and STW mode and often feature sensitivities comparable to those of the soft polymer films. The two semisolid films used in this study are styrene (ST) and allylalkohol (AA) synthesized in a plasma polymerization reactor.

5. Comparative characteristics of polymer coated RSAW and STW gas sensors operating at the same acoustic wave length

To identify the advantages and disadvantages of the STW mode versus its RSAW counterpart on quartz for gas sensor applications it is necessary to compare the sensor performance of both modes under identical real-life conditions. Such a performance comparison would be correct only if it is carried out with sensor devices of both modes operating on the same acoustic wave length for the following reason: If both types of devices are fabricated on the same piezoelectric material and cut orientation (AT-cut quartz in this case), use the same device geometry, are coated with the same sensing layer of the same thickness and are probed with identical gases and concentrations, then the only factors responsible for the differences in electrical and sensor performance would be the type of motion for each mode, (elliptical for the RSAW and shear horizontal for the STW) and the way the acoustic wave interacts with the sensing layer. The results presented in the next sections were performed with RSAW and STW sensors whose electrical characteristics in the uncoated state are summarized in Table 1.

5.1 Electrical performance of STW/RSAW sensor resonators coated with solid and semisolid sensing layers

The frequency and phase responses of the STW and RSAW sensor resonators from Table 1 prior to and after coating with the solid HMDSO are compared in Fig. 3. After film deposition, the frequency of the RSAW device shifts down by about 1,5 MHz (3500 ppm), its insertion loss increases by 5,7 dB and the loaded Q decreases from 6000 to about 2000.

Acoustic wave mode	STW	RSAW
Acoustic wave length	7,22 μm	7,22 μm
Sensor resonator frequency	433 MHz	701 MHz
Device insertion loss	5-7 dB	6-7 dB
Loaded Q-factor	3000-4000	5000-6000
Side lobe suppression	> 8 dB	>12 dB
Metallisation	Al	Al

Table 1. Electrical characteristics of the uncoated STW/RSAW sensor resonators used in the comparative studies

In addition, the RSAW device retains a well behaved single-mode resonance with excellent side lobe suppression as required for stable operation of the sensor oscillator. The STW device shows a different behavior. Its frequency shifts down by 4 MHz (6100 ppm) which accounts for about 2 times higher relative mass loading sensitivity than its RSAW counterpart. The insertion loss increases by just about 3 dB versus 5,7 dB for the RSAW mode which implies that the STW mode tolerates solid films better in terms of loss increase. On the other hand, excitation of a second higher-order Love wave mode [20] about 7 MHz higher than the main STW mode at 697 MHz is observed. Since a 180 deg. phase reversal at this Love mode occurs, (see the lower data plot in Fig. 3 b)), it is not very likely to degrade the performance of the sensor oscillator. A more serious problem, however, is the distortion at the main STW mode that indeed can cause the sensor oscillator to jump onto an adjacent peak during the measurement. That is why, coating STW sensor resonators with excessively thick solid films as the 190 nm HMDSO from Fig. 3 should be stopped before distortion and multiple peak behavior on the main STW mode occurs. As far as the higher-order Love mode at 704 MHz is concerned, we have noticed that its gas sensitivity is orders of magnitude lower than the STW mode on the right side. This lack of sensitivity is explained by the fact that the Love mode scatters its energy into the bulk of the sensing layer [20].

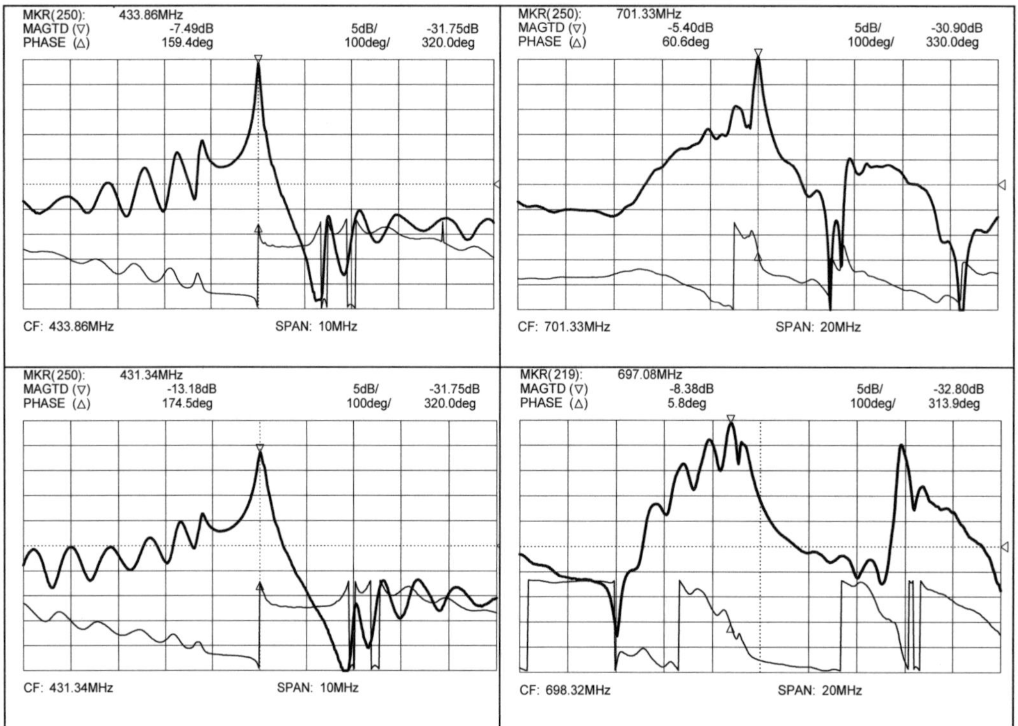

a) b)

Fig. 3. Frequency (upper curves) and phase responses (lower curves) of the a) RSAW and b) STW sensor resonators from Table 1 prior to (upper plots) and after (lower plots) 190 nm HMDSO solid film deposition

5.2 Electrical performance of STW/RSAW sensor resonators coated with soft polymer films

A similar comparison between both acoustic wave modes was performed by coating the devices from Table 1 with the soft polymer film PIB using the micro drop deposition method. The data obtained shows quite the opposite tendency compared to the solid film behavior from Section 5.1. The STW devices suffered a 5 dB increase in insertion loss and rather distorted frequency responses even at fairly thin soft layers. Only a moderate frequency down shift of 1330 ppm was obtain as a result of film coating. As evident from the frequency responses in Fig. 2 the RSAW devices were found to provide a much better performance at the same film thickness. They retain a high loaded Q and low insertion loss, as well as an undistorted single-mode resonance. These data imply that RSAW sensors will work better with soft polymer films while the STW mode will provide better performance with solid films as long as they are not excessively thick to cause distortion.

6. A practical method for film thickness optimization of RSAW/STW gas sensors coated with solid and semisolid sensing layers

The most important step in designing practical RSAW/STW resonant sensors is the selection of an optimum thickness of the sensing layer. It should be selected in such manner that maximum sensor sensitivity and dynamic range are obtained at minimum degradation of the electrical resonator performance (insertion loss, loaded Q, side lobe suppression and distortion) as required for stable low-noise operation of the sensor oscillator. A very efficient method for film thickness optimization using a controlled plasma deposition of the semisolid polymer Parylene C is described in [21]. This material has viscoelastic properties very similar to practical solid and semisolid layers but has the unique feature that it can polymerize directly on the surface of the acoustic devices at room temperature, thus avoiding undesired thermal frequency drifts. Also the actual deposition is performed in a chamber separate from the plasma reactor where the devices are protected from the high electric fields of the main plasma generator. This allows direct measurement of their frequency and phase responses with a network analyzer in the process of film deposition while the film thickness is measured with a quartz crystal microbalance (QCM). The results from a Parylene C deposition on a 433 MHz RSAW resonator and real time measurements of its electrical characteristics are shown in Fig. 4 for a polymer thickness ranging from 0 to 700 nm. From this measurement it is possible to identify the thickness range in which optimum sensor performance is expected and to extract information on the behavior of important sensor parameters in the process of polymer coating as follows:

- the down shift of the resonant frequency versus film thickness;
- the loss increase with film thickness;
- the loaded Q decrease with film thickness;
- the behavior of the adjacent longitudinal modes in the process of deposition;
- the thickness range over which the sensor demonstrates maximum mass sensitivity of frequency while retaining good electrical performance. In Fig. 4 this optimum thickness range is between 100 and 300 nm with an average of 200 nm.

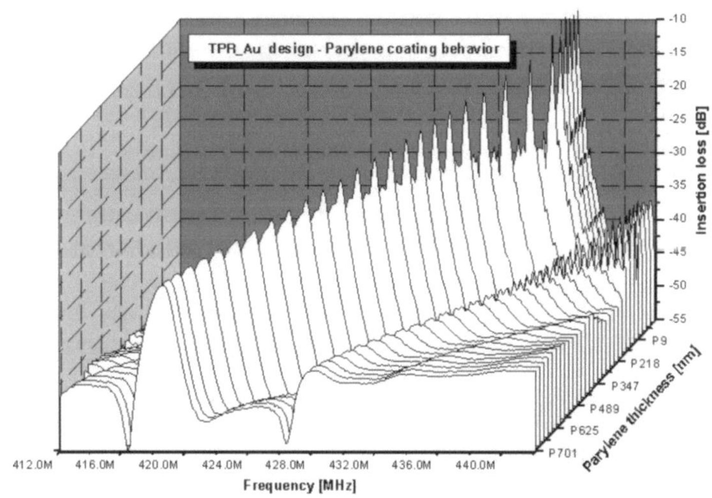

Fig. 4. Parylene C coating behavior of a 433 MHz RSAW sensor resonator in the 0 to 700 nm polymer thickness range

As evident from Fig. 4, in the optimum thickness range (100 to 300 nm) a maximum linear frequency shift, (maximum mass sensitivity), is accompanied by just about 7 dB loss increase. The Q remains high as shown by the sharp resonance while the first longitudinal mode on the left side of the resonance remains suppressed by at least 12 dB. If the sensor is intended for high-resolution measurements at low gas concentrations, then a thickness close to 100 nm should be chosen due to the highest Q and lowest loss. If measurements at higher gas concentrations are expected then a 300 nm thickness may be more appropriate since the thicker layer may adsorb larger amounts of gas without film saturation.

6.1 Critical thickness in RSAW/STW based sensor resonators coated with solid and semisolid sensing layers

As shown in the previous sections, the sensing layer does not only shift the resonant frequency down, increase the loss and decrease the loaded Q as a result of mass loading but it also influences the longitudinal modes supported by the resonator geometry that appear on the left side of the main resonance. In the uncoated resonator these modes are well enough suppressed (typically by 5 to 15 dB) and do not cause any problems when the resonator is operated in an oscillator circuit. As soon as a sensing layer is deposited on the surface, it will change the phase conditions along the device topology and this will cause the adjacent longitudinal modes to arise in magnitude at the expense of the main resonance. This situation gets worse at thick solid films for both, the STW and the RSAW mode. At a certain thickness which we call "critical thickness" the magnitude of the first adjacent low-frequency longitudinal mode on the left becomes equal to the magnitude of the higher-frequency main resonance. This creates a potential for instability in the sensor oscillator stabilized with this sensor since it can easily jump from the main resonance onto the left longitudinal mode during gas probing which will ruin the measurement. The critical thickness situation is illustrated in Fig. 5 a) and b) for a STW and a RSAW device from Table 1, accordingly, in the process of Parylene C deposition as described in Section 6. As evident from Fig. 5 a), at thickness values above 185 nm the first longitudinal mode on the left starts rapidly growing until its magnitude becomes equal to the main resonance. The critical thickness at which this happens is about 350 nm. At this thickness also a strong Love mode excitation on the right is observed. The RSAW device in Fig. 5 b) reaches its critical thickness at about 650 nm. From these data we can draw the conclusion that the devices from Table 1 can be usable as Parylene C coated sensors as long the film thickness is lower than 300 nm and 600 nm for the STW and RSAW devices, respectively. Comparing the coating behavior of both modes in Fig. 5 we see that the STW mode retains a much better behaved resonance than its RSAW counterpart until the critical thickness is reached. At that thickness the STW device has a loss of 14 dB (Fig. 5 a)), versus 35 dB for the RSAW device (Fig. 5 b)). Therefore, the STW mode tolerates solid and semisolid sensing films much better than the RSAW one and is more appropriate for operation with such films in practical gas sensors.

7. Gas sensing characteristics of RSAW/STW resonant sensors coated with solid and semisolid chemo sensitive films

In the following sections we present results from gas probing experiments on RSAW/STW sensors coated with solid HMDSO and semisolid ST and AA films. Four different chemical agents at different vapor concentrations are used for gas probing as follows:

- Dichloroethane 6500 ppm
- Ethylacetate 17600 ppm
- Tetrachloroethylene 2650 ppm
- Xylene 1400 ppm

The purpose of the gas probing tests is to identify which acoustic wave mode provides better performance in real-world gas sensing conditions.

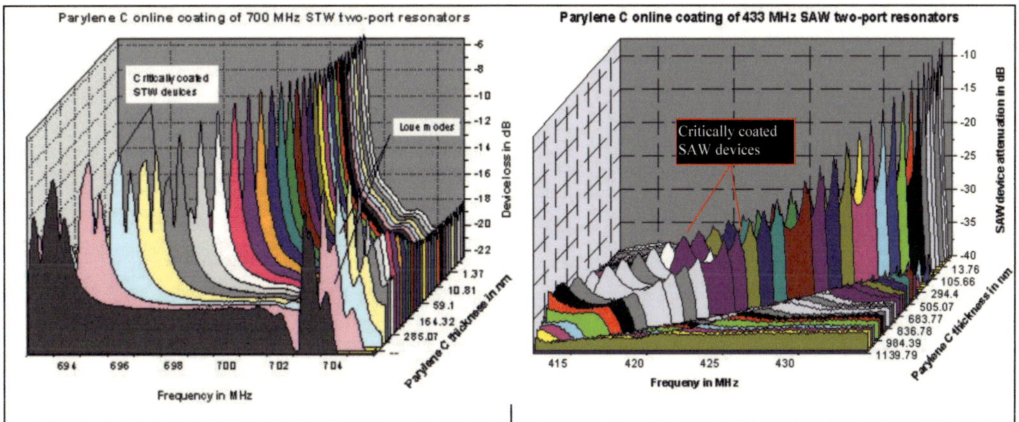

a) b)

Fig. 5. Critical thickness in a) STW and b) RSAW devices in the process of Parylene C coating

7.1 Computer controlled automatic system for gas probing measurements

The block diagram of the computer controlled system for measuring the gas sensing characteristics of the RSAW/STW polymer coated sensors is shown in Fig. 6. For correct comparison of the gas probing performance of both acoustic wave modes four pairs of devices (one RSAW and one STW sensor in each pair, coated with the same polymer to the same film thickness and in the same deposition process) are mounted in open TO 92 packages and placed in the sensor head which can accommodate a total of eight sensors. Each device is connected to one of the 8 sensor oscillator circuits in the head. During gas probing each of the 8 oscillators is turned on for a short period of time to take the measurement. The oscillators are operated one at a time and multiplexed consecutively to avoid possible injection locking. Their output frequency is down converted to an intermediate frequency in the 4-9 MHz range by means of a stable heterodyne reference oscillator to allow fast high-resolution measurements with a reciprocal frequency counter. The chemical compounds 1 through 4 used for gas probing are vapors from the 4 liquid-phase analytes in the 4 containers. A permeation cell is placed on top of each container to allow a defined vapor pressure which is controlled by the rotation speed of the pump. By a switch block of valves the vapors of each analyte are then consecutively fed to the sensor head where they interact with the sensors. After the measurements at each analyte are completed the sensors are flushed with dry air passing through a silica gel integrator which provides also a homogenous air and gas flow. The entire system is controlled by a computer which performs measurements in probe-flush cycles over time and provides real-time sensor data on the computer screen. The data in Fig. 7 is the gas probing performance of a 700 MHz STW styrene coated sensor probed with dichloroethane vapors at 6500 ppm concentration in 100 s probe-flush cycles. It should be noted that prior to this measurement the sensor was probed with a different compound (xylene) for 62 hours and 40 minutes. Note the excellent reproducibility of the noise free sensor signal with a magnitude

Δf=160KHz over time indicating that prolonged xylene treatment has not had any influence on the sensor performance. This is a clear indication that styrene has very good physical sorption properties to a variety of gas phase compounds.

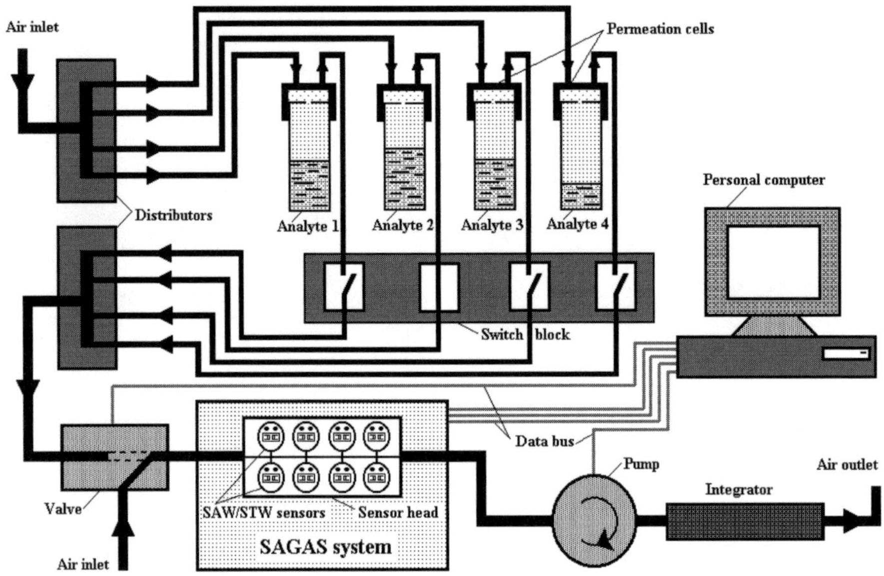

Fig. 6. Block diagram of the automated system for simultaneous gas sensitivity measurements on eight sensor devices

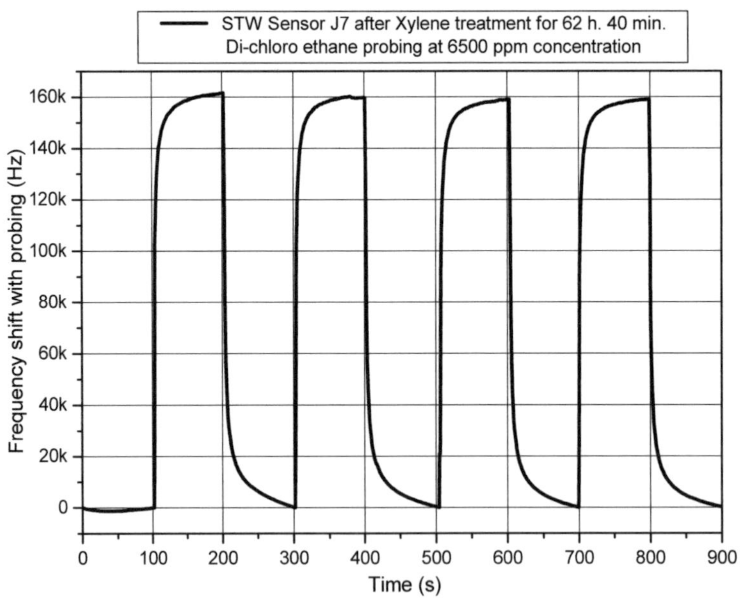

Fig. 7. Gas sensing performance of a 700 MHz styrene coated STW sensor probed with dichloroethane at 6500 ppm concentration in 100 s probe-flush cycles

7.2 Gas sensitivity comparison of RSAW/STW sensor resonators coated with solid HMDSO films

This study aims at finding out which of both acoustic wave modes provides better gas sensitivity when coated with solid HMDSO films and what is the optimum film thickness at which maximum sensitivity is achieved. For this purpose, 5 pairs of RSAW/STW devices according to Table 1 were coated at 5 different HMDSO thicknesses (50, 100, 190, 280 and 350 nm) each, in the same plasma deposition process for each pair. Figure 8 compares the gas sensing characteristics of both modes gas probed with tetrachloroethilene at 2650 ppm concentration. The results from all gas probing experiments on the 5 pairs of devices are summarized in Table 2. In this table the "sensitivity factor" is the ratio between the relative sensitivities (in ppm) for the two devices of each pair. It is given for each of the 5 film thicknesses and shows which mode is more sensitive and at which thickness. From the data in Fig. 8 and Table 2 the following important practical conclusions can be drawn:

1. *The HMDSO coated sensors have very short response times and reach adsorption-desorbtion equilibrium just a few seconds after the gas flow is applied. We attribute this behavior to the surface sorption of the HMDSO which is typical for solid sensing polymers.*

Sensor/Compound	Dichloroethane 6500 ppm	Ethylacetate 17600 ppm	Tetrachloroethylene 2650 ppm	Xylene 1400 ppm
700 MHz STW 50 nm HMDSO	2 KHz (2.9 ppm)	2.8 KHz (4 ppm)	3.5 KHz (5 ppm)	2.4 KHz (3.4 ppm)
433 MHz RSAW 50 nm HMDSO	1.5 KHz (3.5 ppm)	4 KHz (9.2 ppm)	3 KHz (6.9 ppm)	2.2 KHz (5 ppm)
Sensitivity factor (STW/ RSAW)	0.82	0.43	0.72	0.68
700 MHz STW 190 nm HMDSO	3 KHz (4.3 ppm)	4.8 KHz (6.9 ppm)	7.5 KHz (10.7 ppm)	3.7 KHz (5.3 ppm)
433 MHz SAW 190 nm HMDSO	1.8 KHz (4.2 ppm)	3.8 KHz (8.8 ppm)	3.5 KHz (8.1 ppm)	2.5 KHz (5.8 ppm)
Sensitivity factor (STW/RSAW)	**1.02**	0.78	**1.32**	0.91
700 MHz STW 280 nm HMDSO	8.3 KHz (11.9 ppm)	8.5 KHz (12.1 ppm)	7 KHz (10 ppm)	4 KHz (5.7 ppm)
433 MHz SAW 280 nm HMDSO	3.5 KHz (8 ppm)	6.5 KHz (15 ppm)	6 KHz (14 ppm)	4.3 KHz (10 ppm)
Sensitivity factor (HПAB/RSAW)	**1.49**	0.81	0.71	0.57
700 MHz STW 350 nm HMDSO	11 KHz (15.7 ppm)	14 KHz (20 ppm)	15 KHz (21.4 ppm)	9.5 KHz (13.6 ppm)
433 MHz SAW 350 nm HMDSO	1.8 KHz (4.2 ppm)	2.3 KHz (5.3 ppm)	5.1 KHz (11.8 ppm)	4.2 KHz (9.7 ppm)
Sensitivity factor (STW/RSAW)	**3.74**	**3.77**	**1.81**	**1.4**
700 MHz STW 100 nm HMDSO	11 KHz (16 ppm)	20 KHz (29 ppm)	**37 KHz** **(53 ppm)**	9 KHz (13 ppm)

Table 2. Gas sensitivity comparison of RSAW vs. STW devices coated with solid HMDSO film of 5 film thicknesses. Data on the 100 nm coated SAW device are not available

Compound Concentration	Dichloroethane 6500 ppm	Ethylacetate 17600 ppm	Tetrachloroethylene 2650 ppm	Xylene 1400 ppm
700 MHz STW, 100 nm HMDSO	11 KHz (16 ppm)	20 KHz (29 ppm)	37 KHz (53 ppm)	9 KHz (13 ppm)
433 MHz RSAW, 280 nm HMDSO	3.5 KHz (8 ppm)	6.5 KHz (15 ppm)	6 KHz (14 ppm)	4.3 KHz (10 ppm)
Sensitivity factor	*STW/RSAW 2.0*	*STW/RSAW 1.93*	*STW/RSAW 3.79*	*STW/RSAW 1.3*

Table 3. Gas sensitivity comparison of the RSAW and STW sensors coated at their optimum HMDSO thickness values

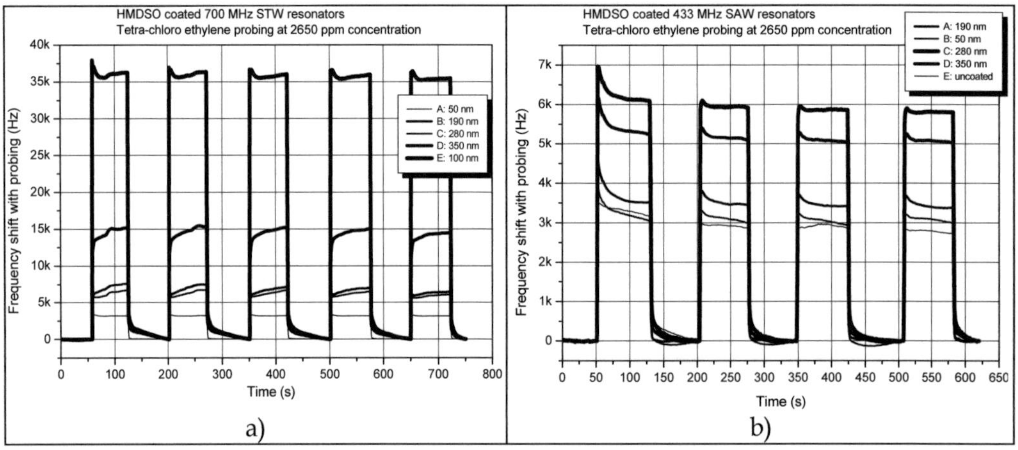

Fig. 8. Tetrachloroethilene probing data of HMDSO coated a) STW and b) RSAW sensors at 50, 100, 190, 280 and 350 nm film thicknesses

2. *Starting from very thin films (50 nm in this case) and increasing the thickness, the gas sensitivity increases in both the RSAW and STW devices, accordingly, until a thickness value is reached at which maximum sensitivity is achieved. This optimum thickness value is different for both modes (100 nm for the STW and 280 nm for the RSAW mode at the wavelength of 7,22 μm in this case). Further increase in film thickness beyond the optimum thickness value only reduces the relative frequency sensitivity and increases the loss of the gas sensor.*
3. *The optimum thickness values for both modes are far below critical thickness and are well within the thickness ranges in which the sensor devices demonstrate high mass sensitivity while retaining low insertion loss, high Q and a well behaved single-mode resonance in the Parylene C coating experiment from Fig. 5. Therefore, the practical film thickness optimization method described in Section 6 is well suited for identifying the optimum film thickness at which maximum gas sensitivity should be expected.*
4. *The relative sensitivities for both acoustic wave modes at their optimum thickness values, summarized in Table 3, demonstrate a 1,3 to 3,8 times higher sensitivity to all 4 gases of the STW mode versus its RSAW counterpart operating at the same acoustic wavelength. This suggests that the STW mode is much more appropriate for operation with solid chemo sensitive polymers.*

7.3 Gas sensitivity comparison of RSAW/STW sensor resonators coated with semisolid styrene (ST) and allylalcohol (AA) films

A similar comparative study was performed also when pairs of RSAW/STW devices from Table 1 were coated at three different thicknesses of the semisolid ST and AA polymer films in a glow discharge plasma reactor. Since no equipment was available to measure the thickness of semisolid layers directly we used the deposition time in seconds for each layer as a measure of the layer thickness. In this study the deposition times for all three thicknesses were 10, 15 and 20 s for the films with the lowest, medium and highest thicknesses, accordingly. The results from these gas probing tests are summarized in Tables 4 and 5 for the ST and AA coated sensors, accordingly. Note that in these experiments we had to reduce the concentrations of all four probing gases by a factor of 4 to avoid saturation of the sensing layers due to the much higher adsorption capacity of the ST and AA films compared to HDMSO. When comparing the data from Tables 4 and 5 an interesting behavior is observed. Styrene coated STW devices are up to 3 times more sensitive than their ST coated RSAW counterparts while with the AA coated sensors we see the opposite behavior - the AA coated RSAW devices are up to 3,6 times more sensitive than their AA coated STW counterparts. We attribute this behavior to the fact that AA is the softest of the three polymers we used in this work and this material behaves much more like a soft polymer than a semisolid one. This implies that the RSAW mode might be more suitable for operation with soft sensing films than the STW mode.

Compound/ Concentration	**Tetrachloroethylene (630 ppm)**		**Dichloroethane (1550 ppm)**		**Ethylacetate (4190 ppm)**		**Xylene (330 ppm)**	
Acoustic mode	STW	RSAW	STW	RSAW	STW	RSAW	STW	RSAW
Deposition time 10 s/styrene	119 ppm	74 ppm	100 ppm	92 ppm	94 ppm	76 ppm	71 ppm	44 ppm
Deposition time 15 s/styrene	254 ppm	85 ppm	206 ppm	171 ppm	190 ppm	95 ppm	140 ppm	65 ppm
Deposition time 20 s/styrene	239 ppm	108 ppm	219 ppm	201 ppm	206 ppm	122 ppm	111 ppm	76 ppm
Sensitivity factor 10s/15s/20s	*STW/RSAW 1.61/3.0/2.21*		*STW/RSAW 1.09/1.2/1.09*		*STW/RSAW 1.24/2.0/1.69*		*STW/RSAW 1.61/2.15/1.46*	

Table 4. Gas sensitivity comparison of semisolid ST coated RSAW and STW devices

Compound/ Concentration	**Tetrachloroethylene (630 ppm)**		**Dichloroethane (1550 ppm)**		**Ethylacetate (4190 ppm)**		**Xylene (330 ppm)**	
Acoustic mode	STW	RSAW	STW	RSAW	STW	RSAW	STW	RSAW
Deposition time 20 s/AA	12.3 ppm	20.8 ppm	22.8 ppm	66.4 ppm	32.6 ppm	69.9 ppm	7.1 ppm	25.2 ppm
Deposition time 25 s/AA	14.3 ppm	25.9 ppm	23.1 ppm	83.1 ppm	31.5 ppm	90.5 ppm	8.3 ppm	28.4 ppm
Sensitivity factor 20s/25s	*RSAW/STW 1.69/1.81*		*RSAW/STW 2.91/3.18*		*RSAW/STW 2.14/2.87*		*RSAW/STW 3.55/3.42*	

Table 5. Gas sensitivity comparison of semisolid AA coated RSAW and STW devices

For the semisolid film coated RSAW/STW sensors we can make the following conclusions:

1. *Semisolid sensing films improve gas sensitivity of the RSAW and STW modes dramatically compared to the solid films. ST coated devices demonstrate one to two orders of magnitude higher relative gas sensitivities compared to HMDSO coated ones.*
2. *As observed with the solid film, also semisolid layers seem to have an optimum film thickness at which maximum gas sensitivity is achieved. Here these optimum thicknesses are achieved at 15 s for ST and 20 s for AA with the STW mode, while the RSAW mode needs somewhat higher optimum thicknesses – 20 s for ST and 25 s for the AA film.*
3. *There is a significant film type dependent difference in gas sensitivities for both modes. ST provides much better gas sensitivity compared to AA and this is attributed to both – the different sorption properties and different viscoelastic properties of the films which determine how the wave interacts with the film and the gas sorbed in it.*
4. *The RSAW mode operates better with the relatively soft AA than with the semisolid ST.*

8. Noise and measurement resolution (detection limit) of RSAW/STW resonant sensors operating with gas sensing polymer layers

As shown in Section 3, if the sensor resonator is connected in the feedback loop of a sensor oscillator whose short-term stability over the time τ has been measured as $\sigma_y(\tau)$, then measuring the sensor signal Δf which is the response of the sensor oscillator to the gas with the concentration C, the measurement resolution R, also called detection limit of the sensor system, can readily be calculated with Equation (1). As an example, let us determine the measurement resolution of the sensor system with which the dichloroethane measurement at C=6500 ppm (6,5x10^{-3}) from Fig. 7 was performed. With $\sigma_y(\tau)$ measured as 1,17x10^{-9}/s at the oscillator frequency f_0=700 MHz (7x10^{8} Hz), for the sensor resolution over the measurement time τ=1 s we obtain R=33,3 ppb (parts per billion). This means that this sensor system can detect changes in the dichloroethane concentration as small as 33 ppb. Such high sensor resolutions are extremely difficult to achieve with other sensor technologies. They are attributed to the fact that RSAW/STW resonant sensors retain excellent resonance characteristics, low loss, high Q and low flicker phase noise when coated with solid and semisolid chemo sensitive polymer films at optimum thickness.

The data in Tables 6 and 7 represent the detection limits of the styrene coated RSAW and STW sensors, respectively, during the measurements on the 4 gas-phase analytes used in this work. The $\sigma_y(\tau)$ values for the sensor oscillators were measured as 5,6x10^{-9}/s and 3,04x10^{-9}/s, respectively. The "Resolution factor" in Table 7 indicates the resolution improvement of the STW mode versus its RSAW counterpart in these measurements. In

Compound/ Concentration	**Tetrachloroethylene (630 ppm)**	**Dichloroethane (1550 ppm)**	**Ethylacetate (4190 ppm)**	**Xylene (330 ppm)**
Acoustic mode/f_0	RSAW/433 MHz			
Dep. time/polymer	20 s/styrene			
Sensor signal Δf	47 KHz (108 ppm)	87 KHz (201 ppm)	53 KHz (122 ppm)	33 KHz (76 ppm)
$\sigma_y(1s)$	5,6x10^{-9}/s			
Resolution R	32,5 ppb	43,2 ppb	192 ppb	24 ppb

Table 6. Detection limits of the styrene coated RSAW sensors at the 4 gas-phase analytes

Compound/ Concentration	Tetra-chloro ethylene (630 ppm)	Di-chloro ethane (1550 ppm)	Ethyl acetate (4190 ppm)	Xylene (330 ppm)
Acoustic mode/*fo*	STW/700 MHz			
Dep. time/polymer	15 s/styrene			
Sensor signal Δf	178 KHz (254 ppm)	144 KHz (206 ppm)	133 KHz (190 ppm)	98 KHz (140 ppm)
$\sigma_y(1s)$	$3{,}04\times10^{-9}/s$			
Resolution *R*	7,5 ppb	22,9 ppb	67 ppb	7,2 ppb
Resolution factor $(STW/RSAW)^{-1}$	4,3	1,9	2,9	3,3

Table 7. Detection limits of the styrene coated STW sensors at the 4 gas-phase analytes

Table 7 this resolution improvement is by a factor of 1,9 to 4,3. We attribute this advantage of the STW mode to the following factors:

- *The styrene coated STW sensors feature lower flicker noise values than their RSAW counterparts. This is evident from the $\sigma_y(\tau)$ measurement ($5{,}6\times10^{-9}/s$ vs. $3{,}04\times10^{-9}/s$ for the RSAW and STW sensors, respectively). This suggests that the STW devices tolerate the styrene film better than the RSAW ones;*
- *The STW mode features better relative gas sensitivity than its RSAW counterpart at the same acoustic wave length and type of semisolid polymer film (styrene) as evident from Table 4.*

8.1 External factors that may degrade measurement resolution in practical RSAW/STW sensor systems

The data in Tables 6 and 7 represent the physical detection limits that can be achieved with practical RSAW/STW based sensor systems. If all other factors that may have a negative effect on the measurements are excluded then the only limiting quantity to the measurement resolution remains the electrical flicker phase noise of the sensor oscillator represented by its Allan's variation $\sigma_y(\tau)$. Unfortunately, in practical sensor systems there are several external factors that may seriously degrade sensor resolution and therefore care should be taken to eliminate them or to reduce their influence to acceptable values. The closer the system is brought to its $\sigma_y(\tau)$ limit, the better it has been designed.
The major factors that may degrade sensor resolution in practical sensor systems will be briefly discussed next.

a. *Gas flow homogeneity.*
This is one of the key disturbances that may seriously degrade sensor noise and should be eliminated first. If the gas flow applied to the sensor head from Fig. 6 is not homogeneous then the sensor devices will sense a variable gas concentration during the probe-flush cycles. Since the measurement cycle is much longer than the time over which inhomogeneities occur, amplitude fluctuation of the sensor signal at or close to equilibrium will occur. This situation is illustrated in Fig. 9 showing results from a tetrachloroethilene measurement with the setup from Fig. 6 when the integrators are removed. This causes a serious turbulence of the gas flow in the system which results in strong noise levels on top of the sensor signals. Even a periodicity on the noise signal is observed which is attributed to the pump rotation. Fortunately, gas flow

inhomogeneities are easily eliminated. After the integrators are placed back at the air inlet and outlet a noise free measurement similar to the one from Fig. 7 is obtained.

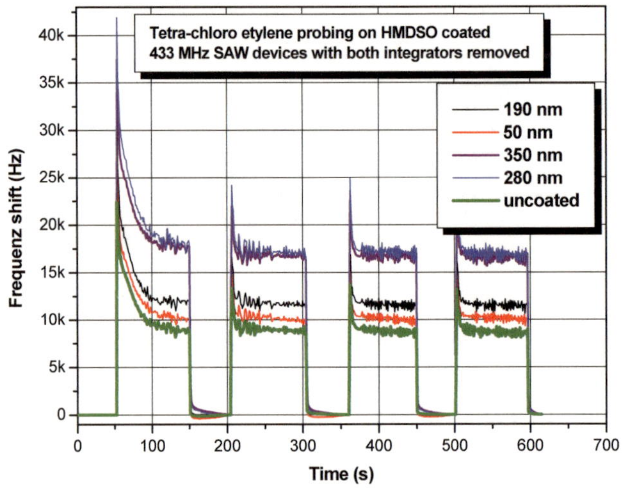

Fig. 9. Tetrachloroethilene probing with 5 HMDSO coated RSAW sensors with the integrators from the setup in Fig. 6 removed

b. *Gas saturation of the sensing films.*

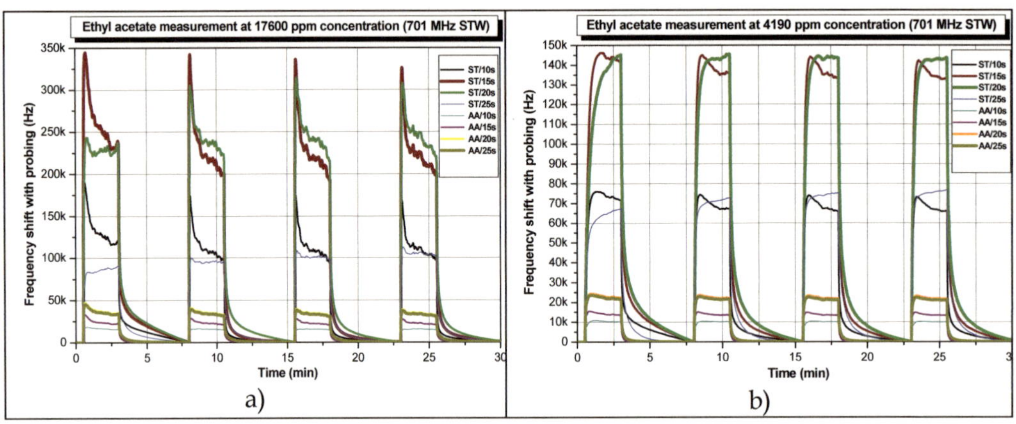

Fig. 10. Ethylacetate probing on ST and AA coated STW sensors at a) 17600 and b) 4190 ppm vapour concentration

Saturation of the sensing films occurs when gas concentrations become so high that sorption limit of the layer is reached. A situation like this during an ethylacetate measurement at 17600 ppm concentration is illustrated in Fig. 10 a). In this case strong peaks of overpressure in the analyte container as well as noise and distortion on the sensor signals are observed as a result of film saturation. When the gas concentration is reduced by a factor of 4 to 4190 ppm, the sensor signals become much more uniform and noise fluctuations are greatly reduced. The behaviour in Fig. 10 a) has the following explanation. When the films become saturated dynamic equilibrium is disturbed and a very turbulent adsorption-desorbtion process takes place, the films get lighter and heavier in a stochastic way and this generates noise on top of the sensor signals. Once the gas concentration is reduced, equilibrium occurs and the adsorption-desorbtion process returns back to normal (see Fig. 10 b)).

c. *Adsorption-desorbtion noise (ADN) in soft film coated RSAW sensors operated far below gas saturation.*

When RSAW sensors are coated with soft polymer films featuring profound bulk sorption these films can accommodate large amounts of gas without being driven into saturation. The larger the amount of adsorbed gas, the more turbulent the adsorption-desorbtion process at equilibrium becomes even though the film is operated far below its saturation limit. This results in ADN evident in Fig. 11 which presents results from tetrachloroethilene probing on RSAW/STW sensors coated with the soft PIB film. ADN is visible on top of all sensor signals regardless of how strong they are. It should be noted that ADN levels depend entirely on the sorption characteristics of the soft polymer films and this makes elimination of this type of noise very difficult. In practical sensor systems one should either cope with ADN or, in critical situations, a different type of polymer film with lower ADN should be used. For example, the magnitude of the sensor signals in Fig. 11 is comparable with those in Fig. 7 where ST was used and no measurable ADN levels were observed. Therefore, results as the ones from the PIB film coated sensors from Fig. 11 but free of ADN could readily be obtained with ST coated sensors.

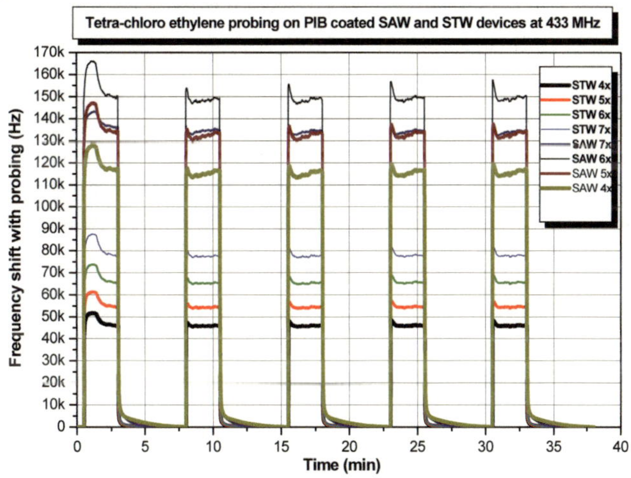

Fig. 11. Tetrachloroethilene probing on RSAW/STW sensors coated with the soft PIB polymer film and operated far below gas saturation

9. Corrosion proof RSAW resonant sensors using gold electrode structure

Typically, RSAW/STW sensors are fabricated with Al electrode structure using a well established photolithographic process. Al metallization is cheap and provides excellent electrical performance in almost all SAW devices fabricated to date. However, a major problem occurs if SAW devices with Al metallization are coated with sensing layers and used as gas sensors. Very often, the chemical gas-phase compounds to be detected form corrosive bases and alkalis with the ambient humidity and attack the Al electrode structure by entering into a chemical reaction with the Al film. The problem is further aggravated by the presence of the sensing layer which, by absorbing large amounts of gas, greatly increases the concentration of the aggressive analyte that comes in contact with the Al electrodes. As a result of corrosion, the sensors suffer irreversible performance degradation, provide inconsistent data and even dye within a limited number of measurement cycles. The solution to that problem is the implementation of SAW devices with corrosion proof gold (Au) metallization that can successfully stand severe corrosion attacks by chemically reactive substances. The design of RSAW sensor resonators with Au electrode structure is not so straight forward as with Al metallization. Due to the fact that Au has a 7 times higher density than Al and is much softer several side effects, such as excitation of strong SSBW modes and transverse waveguide modes occur that can cause serious loss and Q-degradation as well as distorted characteristics at the main resonance. However, by careful selection of the Au thickness and choosing proper device geometry, these side effects can be kept under control and very good resonance characteristics appropriate for gas sensor applications can be achieved [22], [23]. In the next sections we will discuss the performance of RSAW sensors with Au electrode structure intended for operation as gas sensors in highly reactive chemical environments. These sensors were designed to replace their predecessors using the problematic Al electrode structure in a practical sensor system operating at 433 MHz.

9.1 Electrical performance comparison of Au vs. Al RSAW sensor resonators

Generally two types of resonator devices are used in a practical resonator system – two-port resonators (TPR) featuring a single-mode resonance and coupled resonator filters (CRF) that have a two-pole resonance achieved with a coupling grating in the centre of the resonant cavity. As evident from Fig. 12 a) and b), the CRF devices have twice the phase slope of a TPR in their filter pass bands and generally provide better stabilization of the sensor oscillator than the TPR, especially in measurements at high gas concentrations. In a real-world sensor system, the sensor oscillator is designed to operate on the right CRF resonant mode since the left one vanishes when the device is coated with a polymer film [6]. Typically, the sensor oscillator provides stable oscillation on the right mode and never jumps onto the left one since a 180 deg. phase reversal makes oscillation impossible at that mode (see Fig. 12 b)). The frequency and group delay responses of the Al RSAW devices previously used in the sensor system are shown in Fig. 12 c) and d) while a) and b) represent the electrical performance of their Au substitutes designed in [23]. The insertion loss, Q and group delay data from these devices at resonance are compared in Table 8. Es evident from that data and also from Fig. 12 the Al and Au devices feature very similar frequency responses, insertion loss and loaded Q values and the replacement of the Al devices with their Au successors was made without any changes or adjustments of the sensor circuitry. The Au devices are slightly inferior to the Al ones in terms of loss and loaded Q. This is attributed to the loss of energy as a result of the heavy Au loading on the quartz surface.

Parameter / Device (433 MHz)	**Al-CRF**	**Au-CRF**	**Al-TPR**	**Au-TPR**
Unmatched insertion loss [dB]	6.5	10.5	6.5	7.5
Group delay (50Ω load) [μs]	4.01	3.44	3.94	2.83
Loaded Q	5450	4430	5350	3890
Unloaded Q	10400	6190	10160	6730

Table 8. Comparison of the insertion loss, Q and group delay data at resonance of the uncoated RSAW sensors with Al and Au metallization characterized in Fig. 12

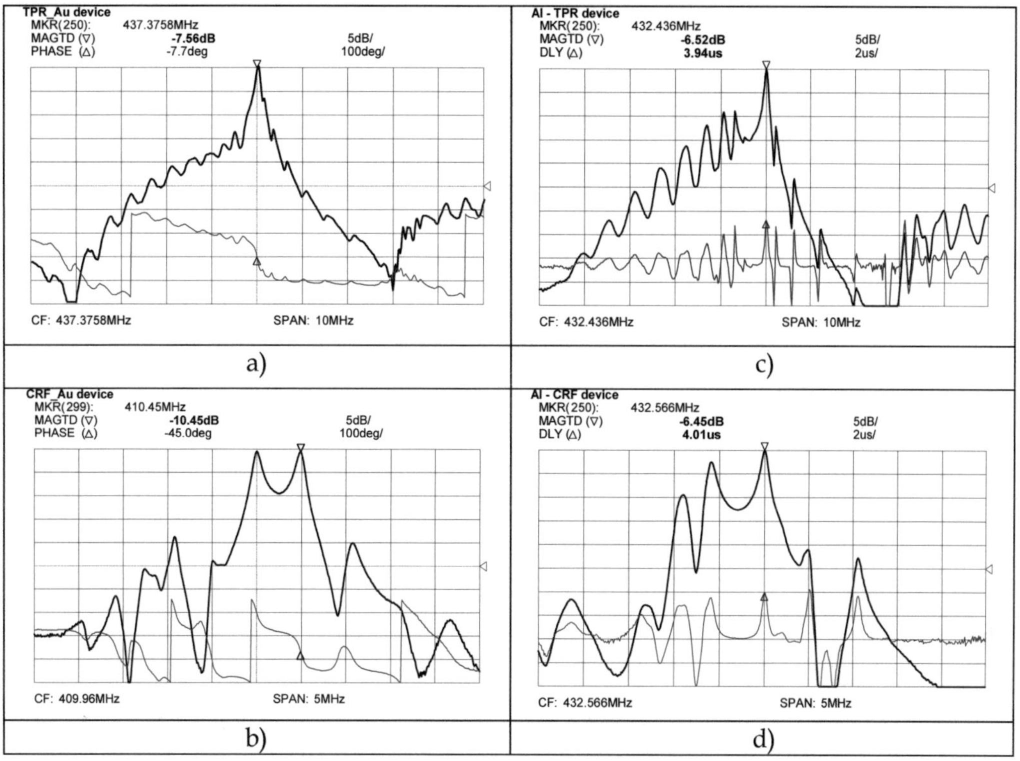

Fig. 12. Frequency (upper curves) and phase/group delay (lower curves) responses of RSAW single mode two-port resonators (upper row) and coupled resonator filter (lower row) using (a) and (b) Au and (c) and (d) Al electrode structure

9.2 Chemo sensitive polymer films and deposition methods used

This section compares the new Au vs. the old Al sensor devices in their sensor characteristics to find out if the replacement causes any performance degradation of the sensor system the Au devices were expected to operate in. To check the sensor performance we again used two types of polymer films: (A) solid Parylene C to simulate coating behaviour with solid and semisolid films as described in Section 6 and (B) a soft polymer called poly[chlorotrifluoroethylene-co-vinylidene fluoride] (PCFV) to test sensor performance at high gas concentrations. Since Parylene C coating was discussed in Section 6 already, here we will briefly discuss a relatively novel soft polymer deposition method, which is called electro spray method and is described in [24] in detail. We applied it successfully to all 4 devices from Fig. 12 to obtain very uniform high-quality soft PCFV films for reproducible sensor performance. According to this method, the holder with the SAW devices mounted on it, spins in a cloud of very small liquid-phase polymer droplets coming out from a narrow capillary tube and directed by an electrostatic field towards the sensor surface. The droplets settle onto the device surface, stick together and form a uniform film. Its thickness depends on the deposition time. Since the SAW device loss increases with film thickness, a certain insertion loss value, as necessary for optimum sensor performance, can be obtained simply by adjusting the deposition time. Except for excellent control over film thickness and uniformity [24], major advantage of this polymer coating method is that, even at 433 MHz, the droplets are much smaller than the acoustic wavelength of about 7 μm at this frequency. Because of the small droplet size, the electro spray films cause much less propagation loss for the SAW, compared to films of the same type and thickness, deposited in an older airbrush coating technique. In contrast to electro spray films, airbrush coatings have a rough textured structure and the droplet size typically exceeds the acoustic wavelength. The two optical microscope pictures in Fig. 13 a) and b) compare the film structures of PCFV deposited with the two methods on identical 433 MHz RSAW devices. Fig. 13 a) shows part of the electrode structure and bus bars at the centre of the electro spray coated SAW device. For better visibility of the textured film structure obtained in an airbrush technique, the picture in Fig. 13 b) has been taken with a slightly higher magnification and shows part of the reflector with some free substrate area on which the large drops are clearly visible.

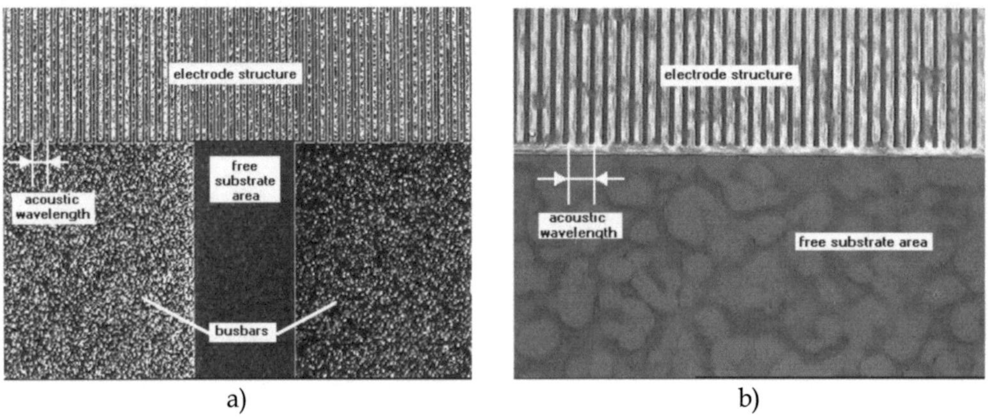

Fig. 13. Comparison of two identical RSAW devices PCFV coated with (a) the electrospray and (b) the airbrush method

9.3 Polymer coating behaviour of Au vs. Al sensor resonators

The three dimensional data plots in Fig. 14 compare the Parylene C coating behaviour of the Au and Al CRF devices from Fig. 12 b) and d) accordingly with a film thickness varying from 0 to 700 nm. In both cases the magnitude of the left longitudinal mode decreases with thickness until it disappears completely at about 300 nm for the Au and 450 nm for the Al device. Above these thickness ranges both devices demonstrate a smooth and well behaved single-mode resonance. In the 300 to 500 nm range the Au CRF reaches its highest mass sensitivity, accompanied with a gradual increase in insertion loss while the loss of the Al device decreases more rapidly. The critical for the system operation loss value of 20 dB is reached at 370 nm vs. 450 nm for the Au and Al devices, respectively.

An identical comparative Parylene C coating test, (not shown here), was performed also with the Au vs. Al TPR devices from Fig. 12 a) and c). In these tests again the Au devices were found to tolerate the solid Parylene C better than their Al counterparts. The increase in device insertion loss with Parylene C thickness for all four tested devices from Fig. 12 is shown in Fig. 15. At film thicknesses up to about 180 nm, all devices yield the same loss increase. Above 200 nm the loss behaviour starts to diverge. The two Al devices keep the same insertion loss up to about 550 nm thickness. Above 200 nm thickness, the loss of the Au devices increases at a much lower rate indicating that these devices can tolerate up to

40% thicker solid films than their Al counterparts for the same amount of loss increase. Finally, Fig. 16 compares the frequency sensitivities of the four tested devices with Parylene C film thickness which is also an indication of their gas probing sensitivity with solid films. Up to about 300 nm thickness, the sensitivity slope of the devices is nearly identical with a small advantage of the Al TPR device, followed by the Al CRF. Above 300 nm, the sensitivity slope of the Al TPR device increases but in view of its strong loss degradation, its sensitivity advantage gets lost. The other three devices keep a nearly constant sensitivity slope up to 500 nm thickness. Below the practical 370 nm thickness at which the critical for this particular system 20 dB of loss is reached for the Al devices, the sensitivity of all four devices differs by less than 20%. This difference is insignificant for practical sensor systems.

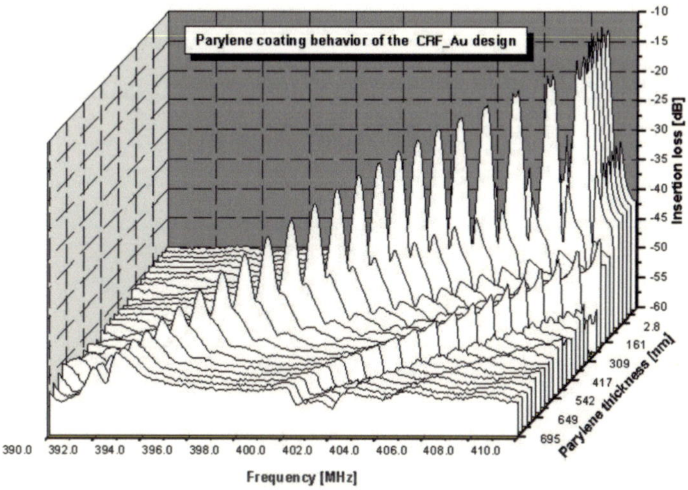

a)

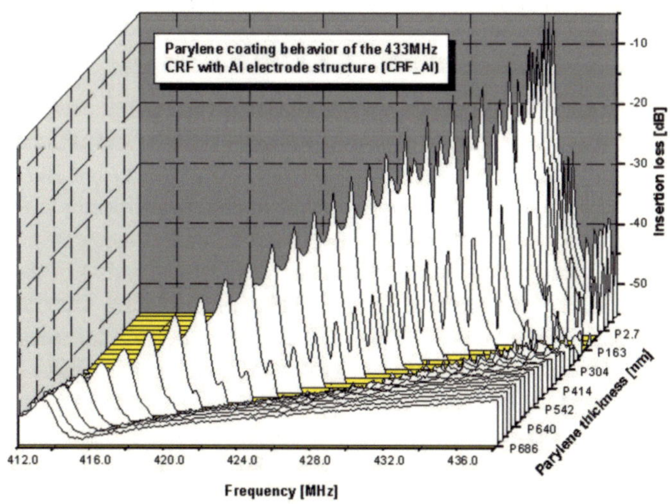

b)

Fig. 14. Parylene C coating behaviour of the RSAW CRF devices from Fig. 12 using (a) Au and (b) Al metallization

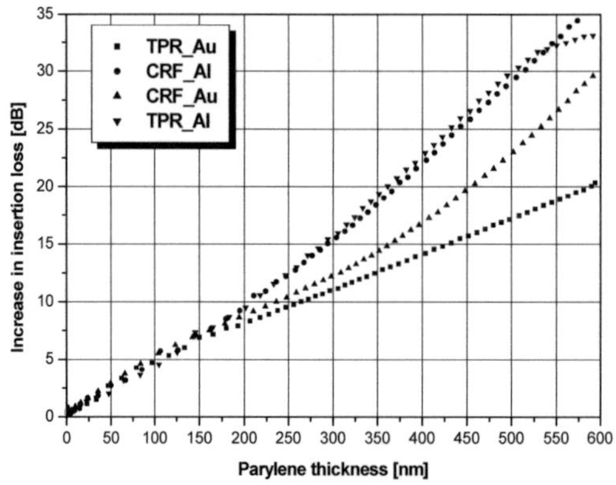

Fig. 15. Insertion loss behaviour of the Au and Al devices from Fig. 12 vs. Parylene C thickness

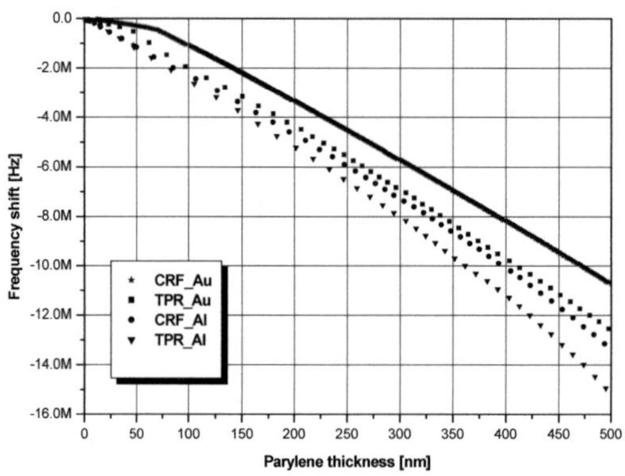

Fig. 16. Frequency (mass) sensitivity behaviour of the Au and Al devices from Fig. 12 vs. Parylene C thickness

The soft PCFV polymer coating experiments were performed on the two TPR devices from Fig. 12 a) and c) since they have about the same amount of loss prior to coating, as required by the sensor system. Since the devices are mounted on a spinning holder, monitoring of their electrical performance in the process of electro spray deposition is not possible. That is why we recorded the frequency and phase responses of each device prior to and after the deposition to obtain the frequency shift and loss increase as a function of the deposition time which we used as a measure for the thickness of the soft PCFV film. The data plots in Fig. 17 illustrate the PCFV coating behaviour of an Au TPR device in a 7,5 minutes deposition time. As a result of film loading the device loss increases by about 8 dB to 17,9 dB while its frequency shifts down by 3 MHz. The coated device on the left retains a well behaved single-mode resonance with a smooth phase response in the resonance region. The loss increase vs. thickness proportional deposition time for the Al and Au devices is shown in Fig. 18. For the Au device this dependence is linear while the Al device shows a small loss increase up to about 10 min. of deposition time and after that its loss degrades very rapidly. We attribute this behaviour again to the difference in Au vs. Al densities. Once the Au device has been optimized for operation under the heavy Au film load, it tolerates much better the polymer film which is much lighter than Au. Finally, Fig. 19 compares the frequency sensitivity of both devices to increased PCFV thickness. The curve for the Au device is steeper meaning that its mass sensitivity is higher than its Al counterpart.

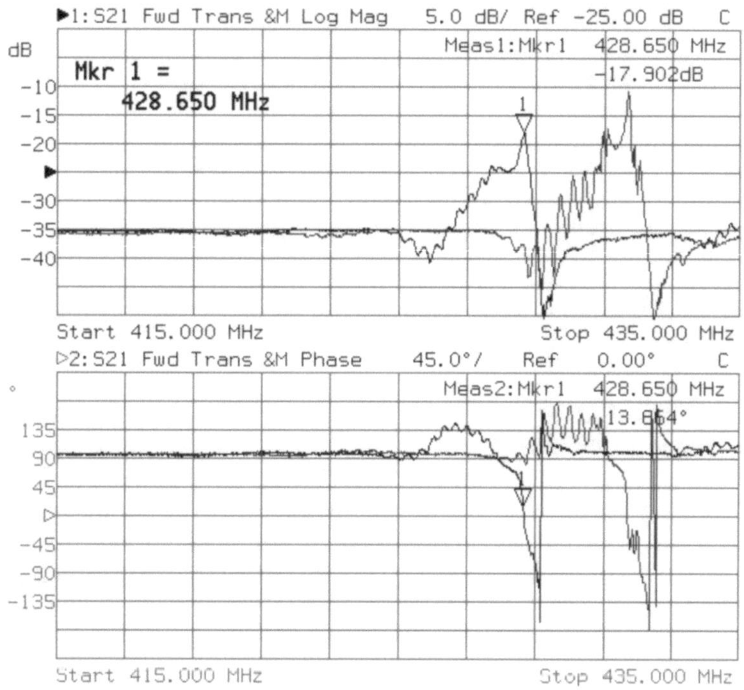

Fig. 17. Frequency responses (upper plots) and phase responses (lower plots) of an Au TPR device before (data on the right) and after (data on the left) 7.5 minutes of PCFV deposition using the electro spray method

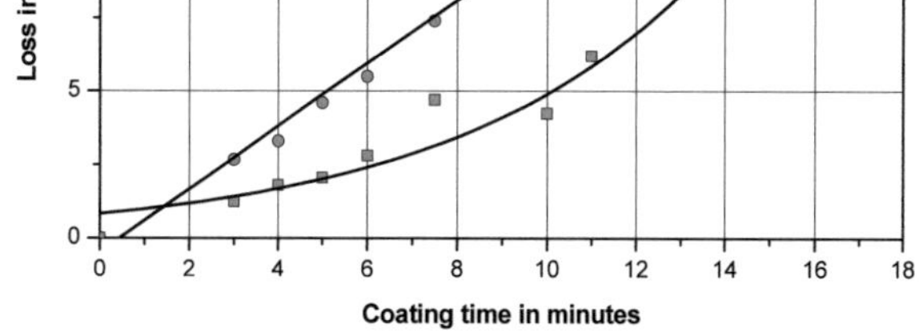

Fig. 18. Loss increase vs. deposition time for Au and Al devices electro spray coated with soft PCFV film

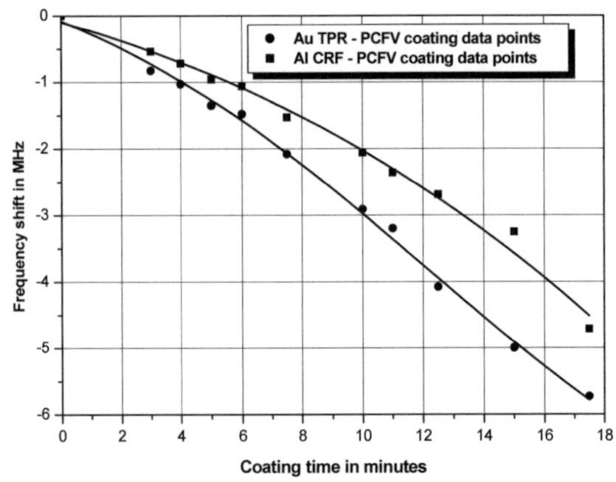

Fig. 19. Frequency downshift vs. deposition time for Au and Al devices electro spray coated with soft PCFV film

9.4 Gas probing behaviour of PCFV coated Au and Al TPR and CRF RSAW sensors

To compare the gas sensitivities of Au vs. Al sensors, pairs of devices of the same type according to Fig. 12 were PCFV coated in the same electro spray deposition method and probed with cooling agent, octane and tetrachloroethylene at different concentrations. Cooling agent and octane were among those gases that the sensors were intended to operate with in a specific application. After coating, the Au/Al pairs were selected to have nearly the same loss increase as a result of PCFV deposition to simulate identical mass loading. The gas probing results for 4 different loss increase values (thicknesses) are summarized in Table 9. As expected from the mass sensitivity data in Fig. 19, the Au devices demonstrate higher gas sensitivity than their Al counterparts. Another important conclusion evident from Table 9 is that soft polymer coating also requires an optimum film thickness for maximum gas sensitivity. In this case the optimum PCFV thickness is achieved when both the Au and Al devices are coated to about 6 dB loss increase, (numbers in bold in Table 9).

Probing gas → Loss increase values for each device pair ↓	**Cooling agent 40000 ppm concentration**		**Octane 1100 ppm concentration**		**Tetrachloroethylene 1000 ppm concentration**	
PCFV coated device→	Al CRF	Au TPR	Al CRF	Au TPR	Al CRF	Au TPR
Al_8/Au_8,7 dB	12,5 kHz	15 kHz	13 kHz	15 kHz	23 kHz	25 kHz
Al_6,2/Au_5,5 dB	**11 kHz**	**21 kHz**	**14 kHz**	**19 kHz**	**26 kHz**	**28 kHz**
Al_4,6/Au_4,6 dB	8 kHz	11 kHz	9 kHz	18 kHz	17 kHz	27 kHz
Al_3,5/Au_3,3 dB	6 kHz	8 kHz	11 kHz	13 kHz	17 kHz	17 kHz

Table 9. Summary of the gas probing performance of Au vs. Al device pairs PCFV coated to nearly identical loss increase values in the same electro spray deposition process

10. Summary and conclusions

This chapter has highlighted important practical aspects for the design and operation of chemical gas detection systems using STW and RSAW resonant devices. It has been shown that both acoustic wave modes provide excellent gas sensitivity and low detection limits, down to a few ppb when coated with solid, semisolid and soft polymer sensing layers. Furthermore, the RSAW and STW modes do not only compete but rather complement each other in different measurement tasks. The STW mode operates better with solid and semisolid sensing films featuring surface sorption and is better suited for high-resolution measurements at low gas concentrations (<1%) while the RSAW mode tolerates much better thick soft sensing layers with profound bulk sorption that operate better at high gas concentrations (>1%). Carefully designed RSAW sensors with Au metallization provide an excellent corrosion proof substitute of their Al counterparts when operated in highly reactive gas-phase environments, thereby greatly increasing system reliability and measurement reproducibility over time and a large number of measurement cycles. All gas sensors, regardless of acoustic wave mode, design, metallization and type of sensing polymer requires a careful thickness optimization to provide highest gas sensitivity, maximum dynamic range and lowest detection limit.

11. Acknowledgments

The author wishes to gratefully acknowledge Dr. E. Radeva from the Georgi Nadjakov Institute of Solid State Physics, Bulgarian Academy of Sciences in Sofia, Bulgaria for expert preparation of the HMDSO films used in this study as well Professor Shigeru Kurosawa and his research associates from the National Institute of Materials Chemistry in Tsukuba, Japan for the deposition of the semisolid ST and AA films. Special thanks are directed to Dr. Michael Rapp and his research team at the Research Centre Karlsruhe in Germany for the opportunity to perform a substantial part of this work at those laboratories.

12. References

[1] R. M. White, Acoustic sensors for physical, chemical and biochemical applications, *Proc. IEEE 1998 International Symposium on Frequency Control*, pp. 587-594.

[2] R. M. White, Surface acoustic wave sensors, *Proc. IEEE 1985 Ultrasonics Symposium*, pp. 490-494.

[3] H. Wohltjen, Mechanism of operation and design considerations for surface acoustic wave device vapour sensors, *Sens. Act.*, vol. 5, p. 307, 1984.

[4] R. Chung, R. A. McGill and P. Matthews, "Phase noise characterization of polymer coated SAW-gas sensors: Implications for the performance of an oscillator circuit", *Proc. 1997 IEEE Int. Freq. Control Symp.*, pp. 169-174.

[5] S. J. Martin, G. C. Frye, J. J. Spates, and M. A. Butler, Gas sensing with acoustic devices, *Proc. IEEE 1996 Ultrasonics Symposium*, pp. 423-434.

[6] M. Rapp, J. Reibel, S. Stier, A. Voigt, and J. Bahlo, SAGAS: Gas analyzing sensor systems based on surface acoustic wave devices—An issue of commercialization of SAW sensor technology, in *Proc. IEEE Int. Freq. Contr. Symp.*, 1997, pp. 129–132.

[7] E. J. Staples, Dioxin/furan detection and analysis using a SAW based electronic nose, *Proc. IEEE 1998 Ultrasonics Symposium*, pp. 521-524.

[8] E. J. Staples, T. Matsuda, and S Viswanathan, Real Time Environmental Screening of Air, Water and Soil Matrices Using a Novel Field Portable GC/SAW System, Environmental Strategies for for the 21st Century, *Asia Pacific Conference*, pp. 8-10 April 1998.

[9] United States Patent No. 5,289,715, Vapour Detection Apparatus and Method Using an Acoustic Interferometer.

[9] G. C. Frye, S. J. Martin, R. W. Cerenosek, K. B. Pfeifer, and J. S. Anderson, Portable acoustic wave sensor systems, in *Proc. IEEE Ultrason. Symp.*, 1991, pp. 311–316.

[10] G. C. Frye and S. J. Martin, Dual output acoustic wave sensor for molecular identification, in *Proc. IEEE Transducers*, 1991, pp. 566-569.

[11] T.Wessa, S. Kueppers, G. Mann, M. Rapp, and J. Reibel, "On-line monitoring of process HPLC by sensors," *Organ. Process Res. Develop.*, vol. 4, no. 2, pp. 102–106, 2000.

[12] D. A. Howe, D.W. Allan, and J. A. Barnes, Properties of Signal Sources and Measurement Methods, *U.S. Dept. Commerce, NIST Tech. Note 1337*, 1990.

[13] D. S. Ballantine, Jr. and H.Wohltjen, Elastic properties of thin polymer films investigated with surface acoustic wave devices, in *Chemical Sensors and Microinstrumentation*, R.W. Murry, R. E. Dessy,W. R. Heinemann, J. Janata, and W. R. Seitz, Eds. Washington, DC: Amer. Chem. Soc., 1989, pp. 222–236.

[14] J. W. Grate and E. T. Zellers, The fractional free volume of the sorbed vapor in modeling the viscoelastic contribution to polymer-coated surface acoustic wave vapor sensor responses, *Anal. Chem.*, vol. 72, no. 12, pp. 2861–2868, July 1, 2000.

[15] F. L. Dickert and A. Haunschild, Sensor materials for solvent vapor detection-donor-acceptor and host-guest interactions, *Adv. Mater.*, vol. 5, no. 12, pp. 887–895, Dec. 1993.

[16] N. Barie, M. Rapp, and H. J. Ache, UV crosslinked polysiloxanes as new coating materials for SAW devices with high long-term stability, *Sens. Actuators B, Chem.*, vol. B46, pp. 97–103, 1998.

[17] H. Yasuda, *Plasma Polymerization*. New York: Academic, 1985, p. 11 and 294.

[18] C. Hamann and G. Kampfrath, Glow discharge polymeric film: Preparation, structure, properties and applications, *Vacuum*, vol. 34, pp. 1053–1059, 1984.

[19] E. I. Radeva, Thin Plasma-Polymerized Layers of Hexamethyldisiloxane for Acoustoelectronic Humidity Sensors, *Sensors and Actuators*, B, vo1.44, pp.275-278, 1997.

[20] G. Kovach, G. W. Lubking, M. J. Vellekoop, and A. Venema, Love waves for (bio) chemical sensing in liquids, in *Proc. IEEE Ultrason. Symp.*, 1992, pp. 281–285.

[21] I. D. Avramov, M. Rapp, A. Voigt, U. Stahl and M. Dirschka, Comparative studies on polymer coated SAW and STW resonators for chemical gas sensing applications, *Proc. 2000 IEEE International Frequency Control Symposium*, pp. 58-65.

[22] I. D. Avramov, A. Voigt and M. Rapp, "Rayleigh SAW Resonators Using Gold Electrode Structure for Gas Sensor Applications in Chemically Reactive Environments", *Electronics Letters*, 31-st March 2005, Vol. 41, No. 7, pp. 450-452.

[23] I. D. Avramov, Design of Rayleigh SAW resonators for applications as gas sensors in highly reactive chemical environments, *Proc. 2006 IEEE International Frequency Control Symposium*, 5-7 June 2006, Miami, Florida, USA, pp. 381-388.

[24] F. Bender, L. Waechter, A. Voigt and M. Rapp, Deposition of High-Quality Coatings on SAW Sensors using Electrospray, *IEEE Sensors Conference*, 2003, Paper ID 1022 (on CD).

24

Ultrananocrystalline Diamond as Material for Surface Acoustic Wave Devices

Nicolas Woehrl and Volker Buck
University Duisburg-Essen and CeNIDE, Duisburg
Germany

1. Introduction

Diamond is one of the most promising materials for SAW applications due to the highest sound velocity and thermal conductivity. Unfortunately single crystals and epitaxial CVD-films are expensive and beyond that conventional CVD grown microcrystalline diamond films feature large facet structures with high roughness inapplicable for this application. Ultra-Nanocrystalline diamond (UNCD) films grown in a microwave plasma enhanced chemical vapor deposition (MPECVD) system on Si substrates possess a smooth surface making it an ideal material for SAW applications. Moreover, due to its nanocrystalline structure, the film properties of the UNCD material can be tailored in a wide range to adjust them to the specific needs of a SAW filter. For this task a profound understanding of the growth process of UNCD and the dependency of the film performance from the film properties is needed. In addition, a simple and quick method to characterize the properties of the UNCD films is necessary. Laser-induced SAW pulse method, which is fast and accurate, is demonstrated to measure the mechanical and structural properties of the UNCD films. AFM measurements were done to correlate the SAW pulse method results with the surface roughness of the deposited films.
Another advantage of the UNCD films is, that highly C-axis textured aluminum nitride (AlN) films can be grown directly on UNCD films by DC-sputtering. Using this technique a feasibility study for SAW devices has been successfully performed.

2. Saw filters

SAW devices are electromechanical products commonly used in high frequency applications such as filters, oscillators and transformers and are based on the transduction of acoustic waves. SAW filters are now widely used in mobile telephones applications for filtering and provide significant advantages in performance, cost, and size over other filter technologies such as quartz crystals (based on bulk waves), LC filters, and waveguide filters by offering a high degree of frequency selectivity with low insertion loss in compact size (Campbell, 1989). In SAW filters an Interdigital Transducer (IDT) that is attached to a piezoelectric material converts electrical signals to a mechanical wave. The piezoelectric effect and the electric field generated by the IDT are distorting the crystal close to its surface.

The oscillations of the crystal lattice can than add up by constructive wave interference and superimpose to a surface wave before being converted back to an electrical signal by further electrodes.
The passing frequency of a SAW filter can be calculated by

$$f_0 = \frac{v_0}{\lambda} \quad (1)$$

where λ represents the wavelength of the acoustic surface wave (corresponding to twice the distance of the fingers of the IDT) and v_0 is the crystal-dependent velocity oft the surface wave. The operation frequency of a SAW device is closely related to the spacing of the interdigital transducer (IDT) that is significantly limited by the photolithograph capability (Springer et al., 1999). Thus one way to achieve higher passing frequencies is to use crystals with a higher speed of sound, such as sapphire (Caliendo, 2003), SiC (Takagaki, 2002) or diamond (Yamanouchi et al., 1989)(Nakahata et al., 1992).

3. Diamond as SAW material

The material with the highest speed of sound is diamond with 18000 m/s. Besides the high speed of sound, diamond features other remarkable properties such as high thermal conductivity and high hardness to name only a few. Due to its extraordinary properties natural and HPHT diamond is used for a long time as a material for tools, especially for grinding or sawing of rocks. Since the 1980s the microcrystalline diamond deposited by thin film technology is increasingly used. One major problem with the microcrystalline diamond films deposited in CVD processes - especially for microelectronics and micromechanical applications with their decreasing structural sizes - is the high surface roughness (Malshe et. al., 1999). Moreover, high surface roughness results in large propagation loss, reducing the applicability of the material. Although Sumitomo Electrics developed SAW filters and resonators with various bandwidth in the 2-5 GHz range it turned out that the polishing of the rather rough CVD diamond surface was too expensive and time consuming due to the chemical inertness and highest hardness of diamond and the SAW filters were never produced in an industrial scale (Fujimori, 1996). Even if one solution to this problem was demonstrated by using the unpolished nucleation side of freestanding CVD diamond (Lamara et. al., 2004) this idea never went into production.
Another drawback of the microcrystalline diamond films is that the homogeneous deposition of such films on substrates with a high aspect ratio is difficult because the films consist of relatively large crystals.

4. Nanocrystalline diamond as SAW material

The growing interest in nanotechnology and nanostructured materials has encouraged the research of diamond films with reduced grain size. By reducing the grain size those films feature rather unique combinations of properties making them potential materials for emerging technological developments such as Nano/Micro- Electro-mechanical Systems (N/MEMS) (Auciello et. al., 2004) (Hernandez Guillen, 2004), optical coatings, bioelectronics (Yang et. al., 2002), tribological applications (Erdemir et. al., 1999) and also surface acoustic wave (SAW) filters (Bi et. al., 2002).

The nanostructured films differ from the microcrystalline films in grain size and in roughness of the surface as shown in Fig. 1.

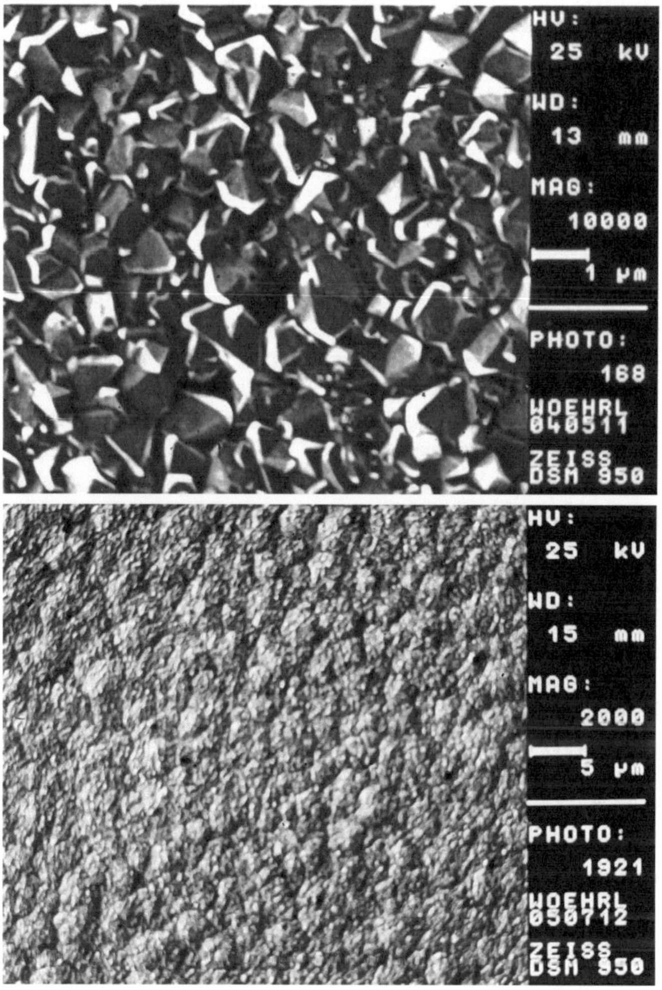

Fig. 1. Morphological comparison of microcrystalline diamond film (upper picture) and UNCD film (lower picture). The scale bar in the upper picture corresponds to 1 μm while the scale bar in the lower picture corresponds to 5 μm

The terms nanocrystalline (NCD) and ultrananocrystalline diamond (UNCD) were coined by the Argonne National Laboratory group that performed the pioneering work in this field. These terms were introduced to establish a differentiation to the microcrystalline diamond films that differ not only in film properties but also in the way they are deposited. The technology developed at Argonne National Laboratory started from deposition of hydrogen free plasmas using fullerenes in Ar (Ar/C_{60}) and was thereafter extended to hydrogen diluted plasmas using Ar/CH_4 and gas mixtures containing only about 1 % hydrogen (either added intentionally or through the thermal decomposition of CH_4) (Gruen, 1999). UNCD is grown from Argon-rich plasma giving it a very fine and uniform structure with grain sizes between 2 and 15 nm (Auciello et. al., 2004). The grain sizes are independent

from film thickness due to the high secondary nucleation of new growth sites during the whole deposition that is not taking place in the standard growth of diamond. This can be shown within the experimental errors when measuring the Young's modulus (GPa) as a function of the deposition time (Shen et. al., 2006).
UNCD consists of pure sp^3 crystalline grains that can be separated by atomically abrupt (0.5 nm) grain boundaries or embedded in an amorphous 3D matrix. By reducing the grain size of microcrystalline diamond films the amount of material between the grains is increased. This matrix in the films can contribute to a large fraction of the overall film, sometimes exceeding 10 % of the total volume, giving those films a great proportion of non-diamond or disordered carbon (Auciello et. al., 2004]. But also values down to 5 % sp^2-bonded carbon have been reported and determined by UV Raman spectroscopy and synchrotron based near-edge X-ray absorption fine structure measurements (NEXAFS) (Gruen, 1998).
In fact the overall volume and structure of the film matrix significantly determine the properties of nanocrystalline diamond films giving another degree of freedom for the material. The well-aimed use of an amorphous matrix for nanocrystalline diamond grains leads to an enormous field of new materials, because a whole class of carbon based materials (diamondlike carbon, DLC) can be used as matrix that may contain carbon solely (a-C) or carbon and hydrogen (a-C:H) as well as other components like metals (Me-C:H); additionally other dopants like silicon, oxygen, halogens or nitrogen may be added with considerable effect on the film properties. By combining soft matrix properties with the hard diamond crystals on the nanoscale it is possible to combine hard with elastic properties and get a material that is hard and tough at the same time. With tailoring the mechanical stress in the films or the coefficient of thermal expansion it was possible to tailor yet other very important mechanical properties for the application of UNCD films by adjusting the overall matrix fraction to the film volume (in the case of a 3D matrix surrounding the nanocrystals) (Woehrl & Buck, 2007) (Woehrl et. al., 2009).
Thus, when comprehensively characterizing UNCD films, one also has to analyze the matrix properties. Since the carbon atoms in the matrix have no crystalline configuration and are indeed amorphous, conventional techniques known from the analysis of amorphous carbon films can be used.

5. Deposition of UNCD films

It is well accepted that the initial nucleation is one decisive factor for the subsequent CVD diamond film growth. While a low nucleation density can lead to van-der-Drift growth – known as the "survival of the largest" – high initial nucleation leads to shorter coalescence time and lower surface roughness. Due to the fact that substrate pre-treatment can significantly increase initial nucleation, the pre-treatment is an important process step already predetermining the film properties (Liu & Dandy, 1995). Three effective seeding methods are known: Mechanical scratching of the substrate surface (see e.g. (Buck & Deuerler, 1998)), enhancing the nucleation by applying a bias voltage to the substrate in the early stages of deposition (Yugo, 1991), and ultrasonically activating the substrate in a suspension containing diamond powder (Lin et. al., 2006)(Sharma et. al., 2010). Nucleation densities of 10^{10} cm^{-2} or more were achieved with either of these methods. The latter method was used for the substrates in this work mainly because of the good reproducibility and uniformity even with larger substrates. Details on the pre-treatment and the deposition parameters used for the UNCD films deposited in this work are given below.

Ultra-Nanocrystalline diamond (UNCD) films were synthesized by microwave plasma enhanced chemical vapour deposition technique using a 2.45 GHz IPLAS CYRANNUS® I-6" plasma source. The nanocrystalline films were deposited from an $Ar/H_2/CH_4$ plasma.
As standard substrates in this work (100) oriented double side polished silicon wafers with a thickness of 425 µm were used. The substrates were usually cut from a wafer to a size of about 20 x 20 mm.
To enhance the nucleation of diamond the substrates were ultrasonically scratched for 30 min with a scratching solution consisting of diamond powder (~ 20 nm grain size), Ti powder (~ 5 nm particle size) and Ethanol in a weight percent ratio of 1:1:100 (wt%). Afterwards the substrates were ultrasonicated for 15 min in Acetone to clean the surface from any residues of the scratching solution (Lin et. al., 2006) (Buck, 2008). After the substrate pre-treatment they were immediately installed into the vacuum chamber placed on top of the molybdenum substrate holder and the recipient was pumped down to high-vacuum.
The plasma is ignited at ca. 1 mbar pressure with a process gas mixture of hydrogen (≈ 3 %) in argon (≈ 97 %) and a MW-power of 1 kW. After the ignition the pressure was slowly increased up to the deposition pressure (typically 200 mbar) during a 30 min period. This 30 min step is due to two reasons: Firstly the substrate surface is cleaned by the etching effect of the plasma. Secondly the temperature of the substrate is slowly increased in the process of the rising pressure. In doing so the substrate is already close to the targeted deposition temperature before switching to the deposition parameters and introducing the carbon carrier gas into the chamber. During the whole process of increasing the pressure the MW power coupling into the plasma is adjusted to the changing conditions. After reaching the desired deposition pressure the carbon carrier gas was introduced therewith starting the deposition process.
The nanocrystalline films shown here were deposited at a pressure of 200 mbar from an $Ar/H_2/CH_4$ plasma. To investigate the influence of the hydrogen admixture on the properties of the deposited films, the percentage of hydrogen in the process gas was varied between 2 % and 8 % as shown in Table 1.
The MW-power was kept constant at 1 kW and the films were deposited for 5 h.

Pressure	200 mbar
Gasflow	400 sccm
H_2 fraction	2 % - 8 %
CH_4 fraction	0,8 %
Ar fraction	91,2 % - 97,2 %
MW-power	1 kW
Deposition time	5 h
Substrate	Pretreated Silicon (100)

Table 1. CVD Deposition Parameters

6. Morphology of the films

The deposition parameters were systematically varied to investigate the influence on film structure and film properties with special attention to the speed of sound and the roughness of the films as most important properties for the application as SAW filters. Because of that the main focus was on deposition parameters that influence the diamond grain size and matrix. It is expected that both are directly influencing the elastic modulus of the films and thus the speed of sound. One important parameter that is influencing the crystal size is the admixture of hydrogen in the process gas. The higher the hydrogen fraction the bigger the crystals grow. (Woehrl & Buck, 2007)

In previous publications it was suggested that different species for the nucleation on the one hand and the growth of diamond grains on the other hand exist. The ratios of these species determine the macroscopic structure of the growing films by influencing the rate of secondary nucleation and therefore the matrix density and the grain size of the growing crystals. A higher amount of the nucleation species leads to smaller crystals and more material between the grains. A higher amount of growth species allows the grains to grow faster (thus a higher growth rate) suppressing the secondary nucleation. In the literature, C_2 was suggested to be the nucleation species (Gruen, 1999) as strong emission of the C_2 dimer could be found in the plasmas used for the deposition of fine-grained UNCD films. On the other hand the CH_3 radical is generally believed to be the growth species of diamond films (May & Mankelevich, 2008). Without taking part in the discussion concerning specific details of growth and nucleation species, previously published data can be interpreted in a way that these two competitive processes determine the structure of the deposited films.

Fig. 2. Morphology of UNCD films deposited with different hydrogen admixtures. The scale bars in all three pictures correspond to 2 µm

The atomic force microscope (AFM) is a scanning probe type microscope that offers a resolution of less than a nanometer that is by a factor of 1000 better than the optical diffraction limit. The AFM consists of a cantilever with a sharp tip with a radius of curvature in the order of nanometers at its end that is used to scan the sample surface. When the tip is brought close to the surface, atomic forces between the tip and the sample lead to a deflection of the cantilever. The deflection of the cantilever is then measured by a laser that is reflected from the cantilever onto an array of photodiodes. In comparison to the scanning electron microscope (SEM) that is measuring a two- dimensional image of a sample not necessarily corresponding to the morphological features, the AFM provides a true three-dimensional topographical image of the surface giving information about the roughness of the investigated surface. While specimens measured in SEM needs to be conducting and are therefore often coated with a thin metal film (e.g. gold) irreversibly alter the film properties, AFM measurements do not require such special treatments. While the SEM can easily measure an area in the order of square millimeters with a depth of field on the order of millimeters the AFM is usually restricted to a maximum scanning area of around 150 μm² with a depth of field in the order of micrometers. Another characteristic that has to be considered for high resolution AFM is the fact that the quality of an image is limited by the radius of curvature of the probe tip and can lead to image artifacts. (Sarid, 1991)

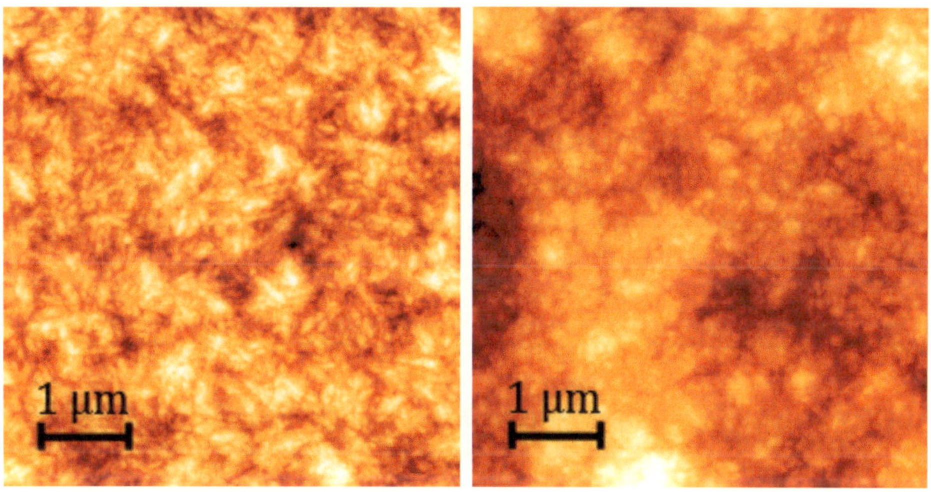

Fig. 3. AFM measurements of UNCD samples deposited with 2,5 % H_2 (left) and 6 % H_2 (right). Both images cover a 5 x 5 μm area

Fig. 2 shows UNCD films deposited with different admixtures of hydrogen to the process gas. It is clearly seen that the hydrogen is influencing the morphology of the deposited films. In fact the crystals are larger and the surface is rougher at hydrogen admixtures of 7% compared to the films deposited at lower admixtures.

Fig. 3 shows an AFM measurement of 5 μm thick UNCD films on a Si substrates. The measured area on the sample was 25 μm². The RMS-roughness (root-mean-squared roughness) of the surface is measured to be R_q= 21.1 nm for the sample deposited at 2.5 % H_2 (left picture) and R_q= 51.3 nm for the sample deposited at 6 % H_2 (right picture).

SEM as well as AFM measurements show that higher hydrogen admixture in the process gas lead to larger diamond crystals and rougher surfaces.

The RMS-roughness measurements as a function of the hydrogen admixture are shown in Fig. 4.

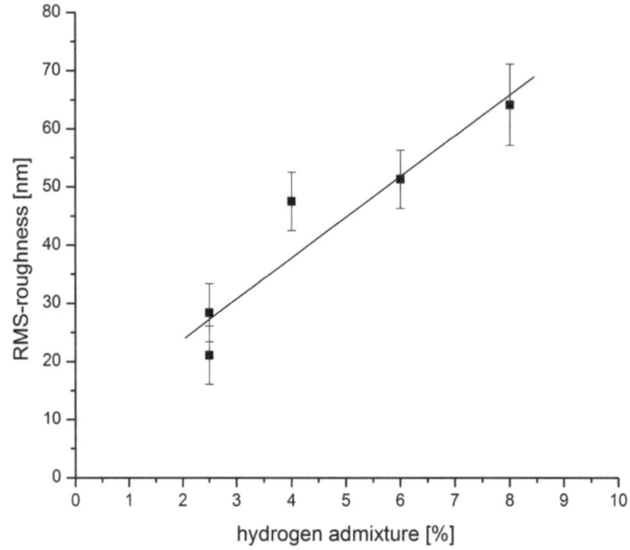

Fig. 4. RMS-roughness measurements as a function of hydrogen admixture in process gas

7. Influence of nitrogen admixture on morphology

An especially appealing field of application for UNCD is nitrogen doped semiconducting films. UNCD films are usually insulating, but n-doping is easily possible by admixture of nitrogen to the process gas. (Gruen, 2004)

To investigate the influence of the nitrogen admixture in the plasma on the film properties, more films were deposited at a pressure of 200 mbar with admixtures of nitrogen from 0 % to 7.5 %.

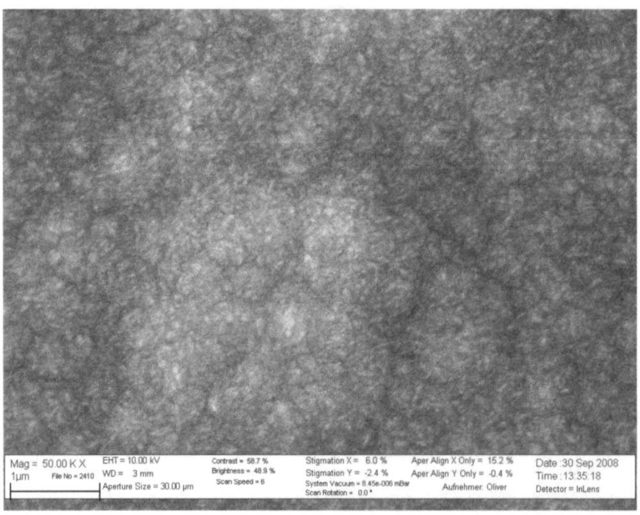

Fig. 5. High resolution SEM measurement of a UNCD film deposited with 2.5 % hydrogen and 2.5 % nitrogen admixture. The scale bar shown corresponds to 1 μm

High-resolution SEM pictures were taken to investigate the influence of hydrogen and nitrogen admixture on the morphology of the films. Fig. 5 shows a film deposited with 2.5 % hydrogen and 2.5 % nitrogen in the plasma. The diamond grains appear to be very fine. Increasing the nitrogen admixture to 7.5 % and keeping the hydrogen admixture at 2.5 % changes the shape of the diamond grains. They appear to be needle-shaped as shown in Fig. 6. These measurements show that the nitrogen admixture can influence the shape of the diamond grains.

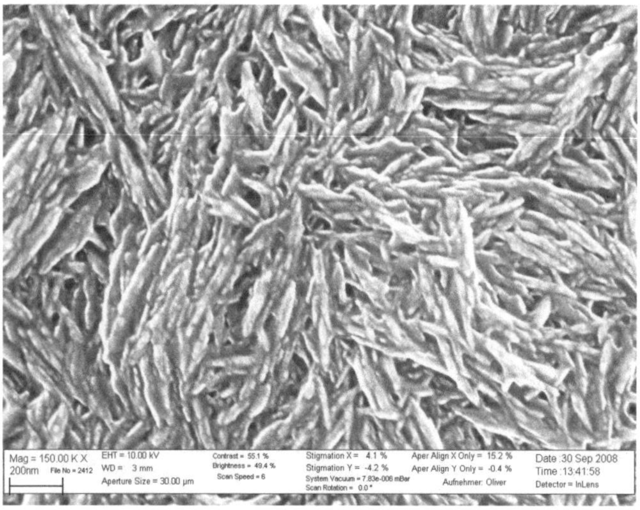

Fig. 6. High resolution SEM measurement of a UNCD film deposited with 2.5 % hydrogen and 7.5 % nitrogen admixture. The scale bar shown corresponds to 200 nm

It is expected that the change in the crystal shape will have a strong influence on the propagation speed of sound in the material giving yet another degree of freedom when designing the material for specific applications.

8. SAW pulse technique

The low surface roughness of UNCD films on the one hand and the high speed of sound in single crystalline diamond on the other hand are making UNCD a very promising material for SAW application. Yet the decisive question is whether the abundance of grain boundaries in the films or the amorphous matrix surrounding the grains will change this picture by e. g. damping the excellent propagation characteristics of surface acoustics waves. The laser-induced SAW pulse method is capable of measuring the SAW-related (i.e. mechanical and structural) properties of thin films (Weihnacht et. al, 1997) (Schenk et. al., 2001) and was used in this work. The applicability of this method for investigating the film properties of polycrystalline diamond films was demonstrated in previous publications (Lehmann et. al., 2001). This method allows measuring all necessary material constants and the wave excitation and propagation parameters decisive for the performance of the SAW material. The biggest advantage of this method is, that it is not necessary to prepare a piezoelectric layer or patterning an interdigital transducer (IDT) structure on the surface, and that rather thin films can also be measured without being disturbed by effects from the Si substrate. The method is a fast and accurate way to measure acoustic wave propagation

effects in thin film systems (Schneider et. al. 1997). Pioneering work on utilizing surface acoustic waves as a tool in material science has been done by P. Hess, a general overview can be found in (Hess, 2002).
A somewhat different setup has been used in this work and is schematically shown in Fig. 7. This setup is commercially available at Fraunhofer IWS Dresden[1].
A pulsed laser beam (N_2-laser at 337.1 nm, 0.5 ns pulse duration) is focused on the substrate by a cylindrical lens to excite a line-shaped broadband SAW pulse via a thermo-elastic mechanism. A piezoelectric PVDF polymer foil, pressed onto the sample surface by a sharp steel wedge (width around 5 µm), is used as a broadband sensor for detecting the SAW pulse propagated along the surface of the thin film system. SAW propagation measurements are performed for different propagation lengths between a few mm and some cm. The signal will then be amplified, digitized by an oscilloscope and converted to complex valued spectra (i.e. amplitude and phase spectra) by a fast Fourier transform algorithm. By doing so for different well-known propagation lengths on the one hand the SAW phase velocity dispersion can be determined accurately from the accompanying phase spectra. Knowledge of the velocity dispersion of a film system is decisive, because it gives the possibility to recover the materials parameters (e.g. elastic constants, mass density and film thickness). To derive the elastic properties, a theoretical approach, modeling the films as an isotropic layer but taking into account the anisotropy of the silicon substrate, was fitted to the measured dispersion data. The fact that we have a specimen that consists of a film on top of a substrate introduces a length scale, and thus generates the observed dispersion effect from that the elastic and mechanical properties can be derived.

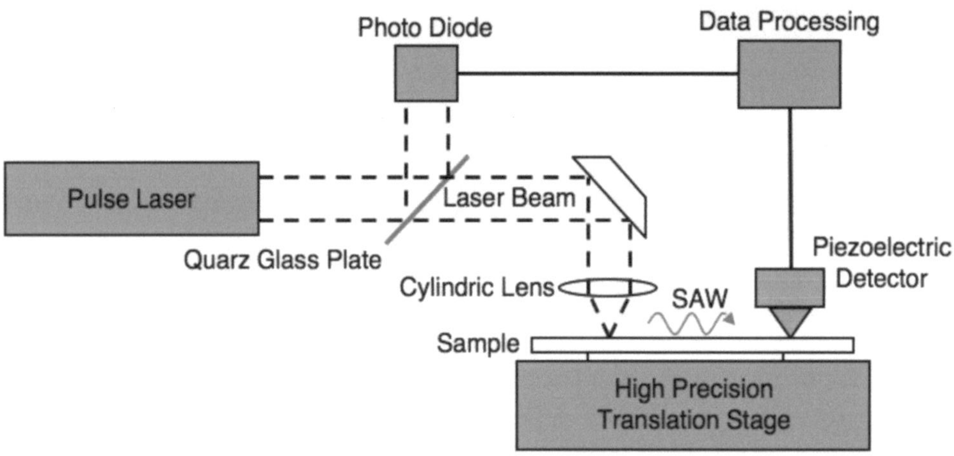

Fig. 7. Principle of SAW pulse technique

A measurement of the SAW phase velocity as a function of frequency as well as the fitted data is shown in Fig. 8. The phase velocity increases with frequency in the case of diamond on silicon substrate ('anomalous dispersion' or 'stiffening case'), because the smaller wavelengths, propagating predominantly in the film, have higher phase velocity.

[1] LAWave® (http://www.iws.fhg.de/projekte/062/e_pro062.html)

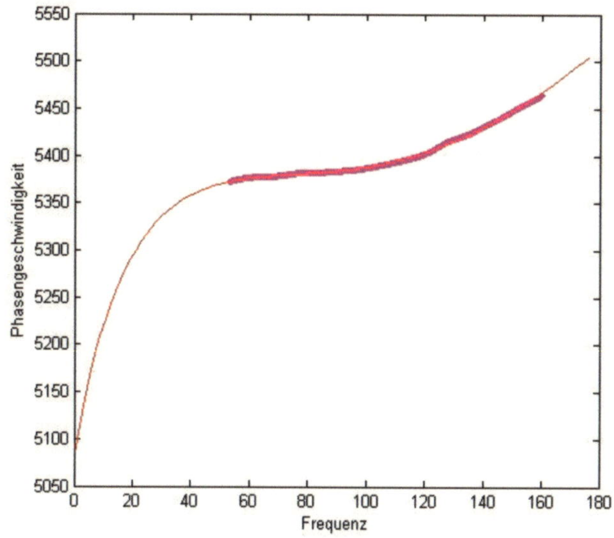

Fig. 8. Measured velocity dispersion and fitted data

Beyond that the damping of the amplitude spectra with increasing propagation length can deliver an estimation of SAW propagation losses due to scattering at thin film imperfections.

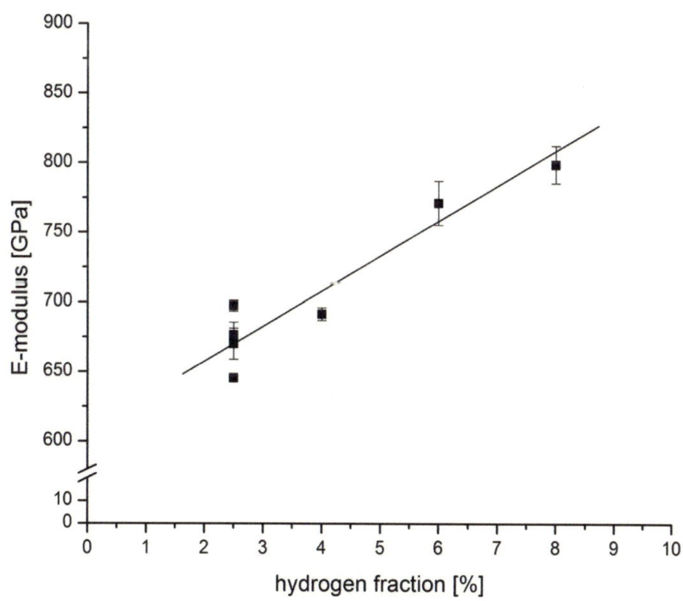

Fig. 9. E-modulus as a function of hydrogen admixture

As expected the elastic modulus is higher (the material is stiffer) for higher admixtures of hydrogen (Fig. 9). This can be explained by the larger diamond crystals and a smaller contribution of amorphous matrix and the fact that the elastic modulus of the amorphous matrix is significantly lower than the modulus of the diamond grains. While the elastic modulus for diamond is around 1220 GPa the elastic modulus of the deposited UNCD films can reach ca. 65 % of this value.

9. Influence of nitrogen and oxygen on mechanical properties

The influence of the nitrogen admixture on the elastic modulus of the deposited films was measured by nanoindentation.

The films that were deposited with additional nitrogen are less stiff compared to films where no additional nitrogen was used. The elastic modulus of the UNCD films deposited with 2.5 % nitrogen in the plasma was measured to be around 370 GPa and increasing the nitrogen admixture even higher to 7.5 % in the plasma resulted in UNCD films with values for the elastic modulus as low as 100 GPa. Thus it was shown that UNCD films deposited with additional nitrogen are unsuitable for the application as SAW device.

An opposite trend can be found when oxygen is used as admixture to the process gas. It was shown that the Young's modulus can be increased up to 950 GPa (ca. 75 % of single crystalline diamond). The reason can be found in the effective etching of sp^2-bonded carbon by the oxygen in the plasma and thus bigger diamond crystals (Shen et. al., 2006).

10. Feasibility study

As a feasibility study SAW resonators with sputtered AlN film as piezoelectric transducer have been produced. Fig. 10 shows the concept of the fabricated AlN-UNCD layered SAW resonator.

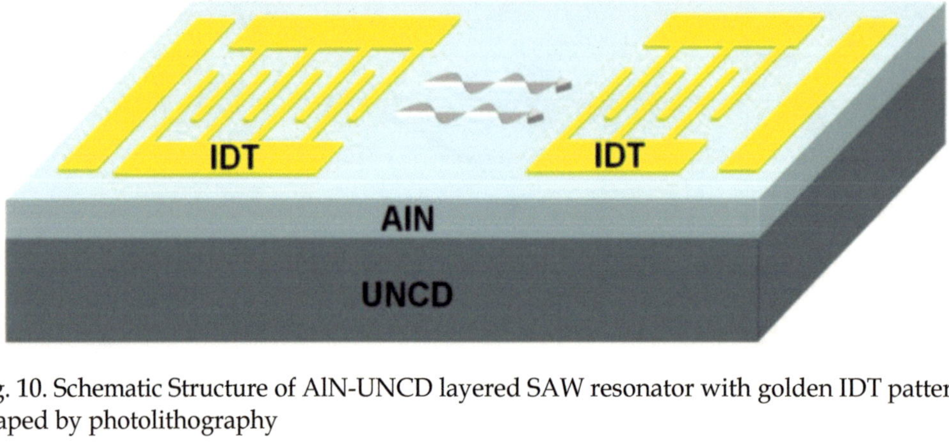

Fig. 10. Schematic Structure of AlN-UNCD layered SAW resonator with golden IDT patterns shaped by photolithography

In the previous chapters it was shown that UNCD films are very suitable as basic material for SAW applications. It was shown that the addition of hydrogen on the one hand improves the elastic constants (towards the value of diamond single crystals), and on the other hand increases the roughness (to values of microcrystalline diamond films), which leads to large propagation loss. Thus a compromise has to be made. The process parameters used for this feasibility study are given in table 2.

Pressure	240 mbar
Gasflow	400 sccm
H_2 fraction	2 %
CH_4 fraction	0,8 %
Ar fraction	97,2 %
MW-power	1 kW

Table 2. Deposition conditions

In order to induce a surface acoustic wave in the UNCD material, a piezoelectric layer is necessary. AlN was chosen for this feasibility study due to being the material with the highest phase velocity (6700 m/s) among piezoelectric materials (Ishihara et. al., 2002). The applicability of AlN thin films on various CVD diamond substrates was demonstrated before (Chalker et. al., 1999).

AlN is an intrinsic piezoelectric material; the wurtzite structure is thermodynamically stable. Several methods for deposition of AlN-films have been reported e.g. MOCVD (Tsubouchi & Mikoshiba, 1985), MBE (Weaver et. al., 1990) and reactive DC or RF sputtering (Akiyama et. al., 1998)(Karmann et. al., 1997). Reactive sputtering processes have the advantage of low substrate temperatures (Dubois & Muralt, 2001)(Naik et. al., 1999)(Tait & Mirfazli, 2001)(Assouar et. al., 2004). Here, magnetron sputtering processes was chosen, for being a common and reliable industrial process.

However, highly (002) oriented films with smooth surfaces are required. Thus deposition parameters (power, pressure, N_2 ratio and substrate temperature) have to be systematically optimized to reach this goal. The influence of oxygen on the film structure was demonstrated before (Vergara et. al., 2004) showing that a low residual gas pressure is crucial for the desired film properties. Therefore a vacuum chamber with turbo molecular pump and a load lock system was used in this work to assure clean conditions. By that, highly oriented AlN films with very smooth surface were deposited on UNCD films that turned out to possess good piezoelectric properties. (Lee et. al., 2007). DC power was 300 W at a pressure of 0.4 Pa and 50 sccm N_2 gas flow at 300°C. The film thickness of the AlN films was ca. 3.5 μm and structure, morphology and bonding structure were characterized by X-Ray diffractrometry (XRD), scanning electron microscopy (SEM), atomic force microscopy (AFM), Raman spectroscopy (Renishaw, RA100) and NEXAFS in synchrotron technique.

On top of the AlN film a gold film was deposited by sputtering which was shaped by conventional photolithography. The resonator consists of a central IDT with reflectors at each side (Fig. 11).

The produced SAW Resonators were analyzed due to their performance. Thickness of UNCD as well as AlN have been systematically varied (2 μm to 6.2 μm for UNCD, 1.4 μm to 3.5 μm for AlN). It was measured that with increasing thickness of AlN and UNCD films the resonance frequency increases as well and the resonance peak become clearer. The increase of resonance frequency and thus of SAW velocity is due to reduced influence of the low SAW velocity of the Si substrate. The clearer resonance peak means larger coupling coefficient, which is due to the relative thickness of AlN piezoelectric layer increasing.

Furthermore the influence of the IDT pair number on the SAW resonator performance was investigated (100 Pairs to 200 Pairs). It was measured that the resonance frequency and the resonance strength kept almost the same while doubling the IDT pair numbers.

This feasibility study indicates that the SAW velocity and coupling coefficient only depend on the relative thickness of ALN and UNCD films, but are not affected by IDT pattern.

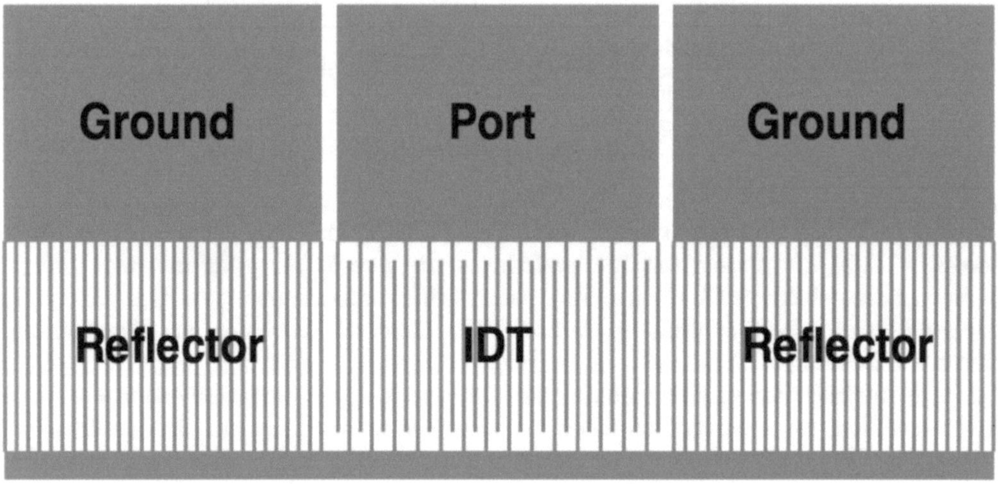

Fig. 11. Schematic Pattern design of SAW Resonator. The actual device consists of significant more lines

11. Acknowledgment

The authors like to thank Dr. Dieter Schneider at Fraunhofer IWS Dresden for the E-modulus measurements of the UNCD films and Hanna Bukowska, University Duisburg-Essen for the AFM measurements.

12. References

Akiyama M., Xu C.-N., Nonaka K., Shobu K., Watanabe T., *Thin Solid Films*, 315 (1998) 62

Assouar M. B., El Hakiki M., Elmazria O., Alnot P., Tiusan C., *Diamond Relat. Mater.* 13 (2004) 1111

Auciello O., Birrell J.,. Carlisle J. A, Gerbi J. E., Xiao X., Peng B., Espinosa H. D., *J. Phys.: Condens. Matter* 16 (2004) R539

Bi B., Huang W. -S., Asmussen J., Golding B., *Diamond Relat. Mater.* Vol. 11, Issues 3-6 (2002) 677-680

Buck V., Deuerler F., *Diamond Relat. Mater.* 7 (1998) 1544

Buck V., *J. Opto. Adv. Mater.* 10 (2008) 85

Caliendo C., *Appl. Phys. Lett.* 83 (2003) 4851

Campbell C., Academic Press (1989)

Chalker P. R., Joyce T. B., Johnston C., Crossley J. A. A., Huddlestone J., Whitfield M. D., Jackman R. B., *Diamond Relat. Mater.* 8 (1999) 309

Dubois M.-A., Muralt P., *J. Appl. Phys.*, 89 (2001) 6389

Erdemir A., Fenske G. R., Krauss A. R., Gruen D. M., McCauley T., Csencsits R. T., *Surf. Coat. Techn.*, 120-121 (1999) 565- 572

Fujimori N., Recent Progresses in Electrical Application of Diamond, *7th European Conference on Diamond, Diamond-like and Related Materials*, 8-13 September 1996

Gruen D. M., *MRS Bulletin*, Vol. 23, No. 9 (1998) 32

Gruen D. M., *Annu. Rev. Mater. Sci.* 29 (1999) 211

Gruen D. M., Krauss A. R., Auciello O. H., Carlisle J. A., (2004) US Patent No. 6793849

Hernandez Guillen F. J., Janischowsky K., Ebert W., Kohn E., *Phys. Stat. Sol.*, A, Appl. res., vol. 201, no11 (2004) 2553-2557

Hess P., *Physics Today*, March issue (2002) 42

Ishihara M., Nakamura T., Kokai F., Koga Y., *Diamond Relat. Mater.* 11 (2002) 408

Karmann S., Schenk H. P. D., Kaiser U., Fissel A., Richter W., *Mater. Sci. Eng.*, B50 (1997) 228

Lamara T., Belmahi M., Elmazria O., Le Brizoual L., Bougdira J., Rémy M., Alnot P., *Diamond Relat. Mater.* 13 (2004) 581

Lee Y. C., Lin S. J., Buck V., Kunze R., Schmidt H., Lin C. Y., Fang W. L., Lin I. N., *Diamond Relat. Mater.* 17 (2008) 446

Lehmann G., Schreck M., Hou L., Lambers J., Hess P., *Diamond Relat. Mater.* 10 (2001) 686

Lin I. N., Lee Y. C., Lin S. J., Lin C. Y., Yip M. C., Fang W., *Diamond Relat. Mater.*, Vol 15 No 11-12 (2006) 2046

Liu H., Dandy D. S., Diamond Chemical Vapor Deposition, Nucleation and early growth stages, Noyes Publications (1995)

Malshe A. P., Park B. S., Brown W. D., Naseem H. A., *Diamond Relat. Mater.* 8 (1999) 1198

May P. W., Mankelevich Yu. A., *J. Phys. Chem.* (2008)

Naik R. S., Reif R., Lutsky J. J., Sodini C.G., *J. Electrochem. Soc.*, 146 (1999) 691

Nakahata H., Hachigo A., Shikata S., Fujimori N., *IEEE Ultrason. Symp. Proc.* (1992) 377

Sarid D., Scanning Force Microscopy, Oxford Series in Optical and Imaging Sciences, Oxford University Press, New York (1991)

Schenk H. P. D., Feltin E., Vaille M., Gibart P., Kunze R., Schmidt H., M. Weihnacht, E. Dogheche, *Phys. Stat. Sol.*, A Appl. Res. 188 (2) (2001) 537

Schneider D., Schwarz T., Scheibe H.-J., Panzner M., *Thin Solid Films* 295 (1997) 107

Shen Z. H., Hess P., Huang J. P., Lin Y. C., Chen K. H., Chen L. C., Lin S. T., *J. Appl. Phys.*, 99 (2006) 124302

Sharma R., Woehrl N., Barhai P. K., Buck V., *J. Opt. Adv. Mater.*, Vol.12 Iss. 9 (2010) 1915

Springer A., Hollerweger F., Weigel R., Berek S., Thomas R., Ruile W., Ruppel C., Guglielmi M., *IEEE Trans. Microwave Theor. Tech.* 47 (1999) 2312

Tait R. N., Mirgazli A., J. *Vac. Sci. Technol.* A, 19 (2001) 1586

Takagaki Y., Santos P.V., Wiebicke E., Brandt O., Schoenherr H.P., Pliig K., *Appl. Phys. Lett.* 81 (2002) 2538

Tsubouchi K., Mikoshiba N., *IEEE Trans. Sonic Ultrason.*, SU-32 (1985) 634

Vergara L., Clement M., Iborra E., Sanz-Hervás A., García López J., Morilla Y., Sangrador J., Respaldiza M. A.

Weaver W., Timoshenko S. P.,. Young D. H, *Vibration Problems in Engineering*, John Wiley & Sons., (1990)

Weihnacht M., Franke K., Kaemmer K., Kunze R., Schmidt H., *IEEE Ultrason. Symp. Proc.* (1997) 217

Woehrl N., Buck V., *Diamond Relat. Mater.* 16 (4-7) (2007) 748

Woehrl N., Hirte T., Posth O., Buck V., *Diamond Relat. Mater.* 18 (2009) 224

Yamanouchi K., Sakurai N., Satoh T., *IEEE Ultrason. Symp. Proc* (1989) 351

Yang W. et al., *Nature Materials* 1, (2002) 253257
Yugo S., Kanai T., Kimura T., Muto T., *Appl. Phys. Lett.* 58, (1991) 1036

25

Aluminum Nitride (AlN) Film Based Acoustic Devices: Material Synthesis and Device Fabrication

Jyoti Prakash Kar[1] and Gouranga Bose[2]
[1]*Department of Electronics Engineering, University of Tor Vergata, Rome*
[2]*Department of Applied Electronics and Instrumentation Engineering, Institute of Technical Education and Research, Bhubaneswar, Orissa*
[1]*Italy*
[2]*India*

1. Introduction

Enormous growth has taken place in electronics, especially in the field of RF communications towards the beginning of 21st century and continuously striving for better communication performance. Presently, the key concerns of RF communications is bandwidth, in the range of low/medium GHz range, to avoid frequency crowding, especially for wireless communication mobile handsets and base stations (Kim et al., 2004). In addition, reduction in signal loss, low power consumption, scaling down device size, reduction in materials and fabrication costs, and packaging of the device are main issues today. Some of these issues can be resolved, if the new generation of electroacoustic devices can be monolithically integrated with integrated circuit (IC). Conventional electroacoustic devices, used in the communication e.g. Surface Acoustic Wave (SAW) and Bulk Acoustic Wave (BAW) based systems, are widely used for today's wireless communication. These devices are typically made on a single crystal piezoelectric substrate such as quartz, lithium niobate, and lithium tantalate (Assouar et al., 2004). Unfortunately, these substrates based electroacoustic devices are made separately and then it is wired with the signal processing chip, which has several limitations, in particular low acoustic wave velocity and high frequency device fabrication. To resolve these two core issues, thin film materials based electroacoustic devices are actively under consideration [Bender et al., 2003]. Where, a crystalline film is grown on a particular substrate, especially silicon wafer and electroacoustic device is made out of crystalline film. Thus, the electroacoustic device can be integrated with the signal processing circuit. Apart from the silicon wafer as a base material for crystalline film deposition, a variety of other substrates are also explored for academic and technology interests. Furthermore, to get electroacoustic devices of better quality in terms of high frequency and high quality factor (Q), the piezoelectric property of the film is also exploited with different type of device concept called "Micro-Electro-Mechanical Systems" (MEMS). Thin film bulk acoustic resonators (TFBAR) comes under this MEMS devices, where the crystalline film is made to resonate at RF frequency. These MEMS

devices have smaller size, lower insertion loss and higher-power handling capabilities than conventional SAW devices (Lee et al., 2004).
Generally, thin piezoelectric films, such as aluminum nitride (AlN), zinc oxide (ZnO) and lead zirconium titanate (PZT) are used for high frequency acoustic devices (Loebl et al., 2003; Yamada et al., 2004; Schreiter et al. 2004). AlN has higher SAW velocity, lower propagation loss, and higher thermal stability in comparison to ZnO; whereas, PZT thin films need selective substrates for deposition and thereafter, needs post-deposition poling to get specific cystal orientation. Thus, AlN seems to have edge over the ZnO and PZT films for electroacoustic devices. The critical factor of piezoelectric AlN thin film is its crystal orientation and morphology. Furthermore, to integrate with the signal processing chip, it is also essential that AlN film should be compatible to the complementary metal oxide semiconductor (CMOS) fabrication processes. In addition, AlN being a dielectric material, it can be used as an insulating material in integrated circuits as well as a piezoelectric material in electroacoustic device. Thus, it is imperative to study the presence of electrical charges and the nature of generation of defects in the AlN film along with its morphology. Usually, there are four types of electric charges present in the insulating film; namely, bulk charges (Q_{in}) and interface (D_{it}) charges, fixed charges (Q) and mobile charges (Q_m). In present IC processing, the presence of fixed charges (Q) and mobile charges (Q_m) are eliminated upto a large extent. Furthermore, the bulk charges (Q_{in}) and interface (D_{it}) charges are reduced further by the optimization of growth parameter and the post-deposition treatments. Reduction in the bulk charge (Q_{in}) and interface charge (D_{it}) density is most essential in cantilever beam based MEMS resonator, otherwise the electrostatic force produced by the these charges may stuck cantilever beam on the substrate (Luo et al., 2006). Most of the MEMS are made out of single crystal silicon substrate utilizing well-matured IC fabrication technology. This poses a challenge to be compatible with a new generation of functional materials. Apart from the electrical charges, the selective etching of piezoelectric materials and silicon for electroacoustic device fabrication is a key technology.

2. Properties of AlN film

AlN is a III-V family compound having hexagonal wurtzite crystal structure with lattice constants a = 3.112 Å and c = 4.982 Å (Yim et al., 1973). In this structure, each Al atom is surrounded by four N atoms, forming a distorted tetrahedron with three Al---$N_{(i)}$ (i = 1, 2,3) bonds named B_1 and one Al---N_0 bond in the direction of the c-axis, named B_2. The bond lengths of B_1 and B_2 are 1.885 Å and 1.917 Å, respectively. The bond angle N_0---Al---N_i is 107.7° and that for N_1---Al---N_2 is 110.5 ° (Xu et al., 2001).
AlN has gained ground in semiconductor industry because of its unique electrical, mechanical, piezoelectric and other properties (Table 1). Some of these noteworthy properties are wide bandgap, high thermal conductivity, high SAW velocity, moderately high electromechanical coupling coefficient, high temperature stability, chemical stability to atmospheric gases below 700 °C, high resistivity, low coefficient of thermal expansion (close to Si), high dielectric constant and mechanical hardness (Xu et al., 2001; Strite et al., 1992; Wang et al., 1994). Its high thermal conductivity (about 100 times that of SiO_2 and roughly equal to that of silicon) and electrical insulating property can prove to be a good dielectric layer for a new generation of integrated circuit devices, particularly in metal insulator semiconductor (MIS) devices. High heat dissipation of AlN can significantly enhance device lifetime and efficiency. AlN film with (002) preferred orientation (c-axis) has maximum

piezoelectricity among all other orientations of its crystal structure (Naik et al., 1999). Furthermore, its lattice matching is near to that of silicon and thus less stress is expected to be generated at the AlN/silicon interface. Owing to these properties, AlN films have received great interest as an electronic material for thermal dissipation, dielectric and passivation layers for ICs, acoustic devices, resonators and optoelectronic devices.

Bandgap	6.2 eV, direct
Thermal conductivity	2.85 $Wcm^{-1}K^{-1}$
Coefficient of thermal expansion	$4\text{-}5\times10^{-6}$ K^{-1}
Refractive index	1.8-2.2
Dielectric constant	8.5
Electrical resistivity	10^{11}-10^{13} Ω.cm
SAW velocity	6000 m/sec
Melting point	2490 °C
Hardness	9 Mhos

Table 1. Properties of AlN

3. Synthesis of AlN film

Depending on the intended application, various techniques have been implemented for synthesizing AlN films; namely, molecular beam epitaxy (MBE), reactive evaporation, pulsed laser deposition (PLD), chemical vapour deposition (CVD) and sputtering. Among these techniques, sputtering has the advantage of low-temperature deposition, ease of synthesis, less expensive, non-toxic, good quality films with a fairly smooth surface [Kar et al., 2006; Kar et al., 2007]. In addition, sputtering technique has also CMOS process compatibility. In sputtering technique, plasma is created between the two electrodes by applying high voltage in low pressure. The plasma region contains, positive ions, electrons and neutral sputtering gas, thus the plasma behaves like a conducting medium. Usually, argon gas is used as a sputtering gas. The material that is to be sputtered is called target and it is fixed to the negatively charged electrode. The other electrode is called anode, which is grounded so that the ratio of the target to anode area is significantly reduced. This electric configuration of the sputtering system makes high electric field at the target and that enhances the rate of sputtering. During sputtering process, the energetic ions strike the target and dislodge (sputter) the target atoms. These dislodged atoms travel through the plasma in a vapour state and stick to the surface of wafers, where they condense and form the film. AlN film can be deposited either by directly using the AlN target or by sputtering of aluminum metal in presence of argon and nitrogen gas. The sputtered aluminum atoms react with the nitrogen gas and form AlN film. This process of film deposition is called "reactive sputtering deposition". The sputtering parameters are required to be optimized for desired morphological and electrical properties. These deposition parameters are mainly sputtering pressure, wafer to target distance, sputtering power and wafer temperature. AlN film deposition by reactive sputter deposition technique requires nitrogen as a reactive gas, where it is introduced into the sputtering chamber along with inert argon gas. Argon ions produced in the plasma due to sputtering power and thereafter they strike to the aluminum target and sputter aluminum atoms. These aluminum atoms react with nitrogen and form AlN compound and that deposit on the wafer. Hence, the gas flow ratios need to be optimized. To increase sputtering rate, magnets are placed under the aluminum target, so that magnetic field and the electric field are perpendicular to each other. This configuration of sputtering system is called "magnetron sputtering technique". In the magnetron sputtering, electrons travel in spiral motion in the plasma region. This increases the collision of electrons to neutral argon atoms significantly and that increases argon ions in many folds, thus sputtering rate becomes high.

AlN film can be deposited by DC (direct current) and RF (radio frequency, 13.56 MHz) magnetron sputtering modes. In the DC mode of sputter deposition, the target material must be conductive, so that plasma can sustain. If trace of impurity is present in the system, the surface of the aluminum target becomes contaminated and target poisoning takes place. On the other hand, RF sputtering has the major advantages to produce good quality film, high deposition rate and less chance of target poisoning. For these reasons, RF sputtering technique is preferred than the DC sputtering technique. To obtain well oriented crystalline AlN films for SAW and MEMS structures, the RF sputtering parameters need to be optimized. The sputtering parameters are: RF power, substrate temperature, sputtering pressure, nitrogen concentration and target-substrate distances (D_{ts}). AlN films are deposited on CMOS IC compatibility silicon (100) wafer by the RF reactive magnetron sputtering. The change in morphological and electrical properties of the AlN films with the growth parameters are reported in following section.

3.1 RF power

Amorphous AlN film is found at lower RF sputtering power (100 W), but films became (002) oriented at a sputtering power of 200 W. Further increase of RF power to 400 W, a significant increase in (002) orientation has taken place. This is due to the increase of kinetic energy of atoms that leads to atomic movements on the substrate surface as a result of higher RF power. These newly arrived surface atoms are called "ad-atom". Higher sputtering power increases the AlN grain size that leads to increase in surface roughness as shown in scanning electron microscope (SEM) images (Fig. 1) (Kar et al., 2009).

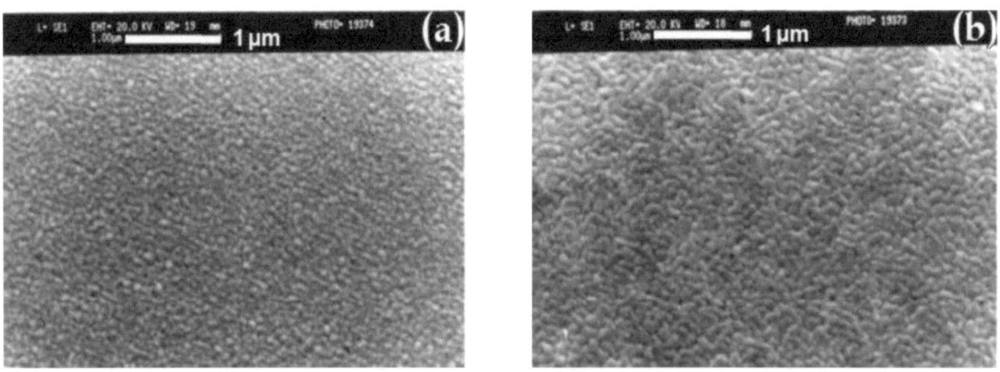

Fig. 1. SEM micrographs of AlN films deposited at (a) 200 W, and (b) 300 W

3.2 Substrate temperature

The structural and morphological properties of the deposited AlN films are strongly dependent on the kinetics of the sputtered atoms arrived at the substrate. The kinetics of sputtered atoms depends on the sputtering parameters. For instance, substrate temperature increases the ad-atom mobility and changes the film morphology significantly. One such illustrations of morphological change with temperature are seen from the X-Ray diffraction (XRD) studies. It is clearly seen from the XRD studies that the c-axis oriented AlN (002) peaks become prominent at moderate temperature range (200-300 °C), but degrades significantly at 400 °C (Kar et al., 2006). This can be attributed to the structural disorderness resulting from the incorporation of impurity atoms at higher temperature (Wang, 2000). The amount of contamination depends on the sputtering deposition system and process related factors, such as base pressure, temperature, gas purity and the partial pressure of moisture, etc. (Naik et al., 1999). Furthermore, smaller grain size with smoother surfaces is observed at lower deposition temperature, and that increases with temperature (Fig. 2). A possible reason may be that the smaller grains grow and merge to form bigger grain, due to the higher thermal energy.

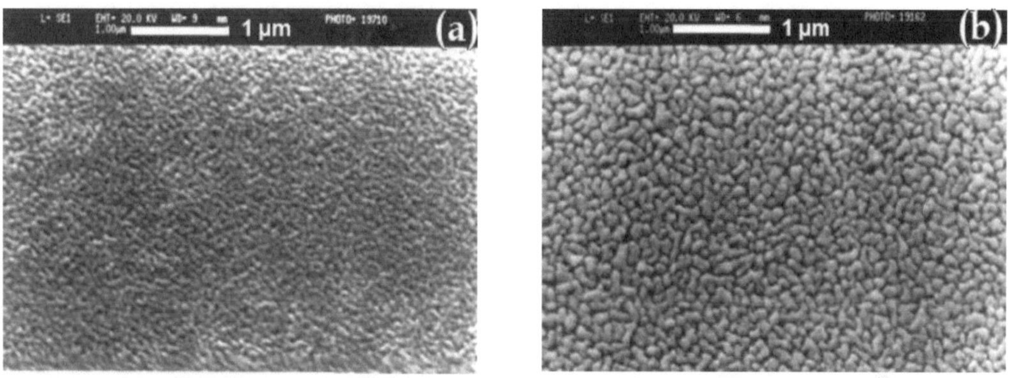

Fig. 2. SEM micrograph of AlN films deposited at (a) 100 °C, and (b) 400 °C

3.3 Sputtering pressure

The variation in crystal orientation with different sputtering pressure are observed from the XRD studies, where the intensity of (002) orientation increases with the deposition pressure and attained a maximum value at 6×10^{-3} mbar. On further increase to a deposition pressure of 8×10^{-3} mbar, the (002) crystal orientation of the AlN film is changed abruptly to the (100) orientation with lesser intensity. The deposited atoms may have altered their direction, energy, momentum and mobility due to the decrease in mean free path of the atoms with sputtering pressure. The hexagonal wurtzite structure of AlN has two kinds of Al-N bond named as B_1 and B_2. These bonds B_1 and B_2 together correspond to (110) and (002) planes, where B_1 corresponds to (100) plane. The formation energy of B_2 is relatively larger that of B_1 (Cheng et al., 2003). Hence, the energy required for sputtering species to orient along *c*-axis is larger than the other possible planes. At low pressure, sputtering species possess enough energy to form hexagonal wurtzite crystalline structure on substrate surface. It is also reported that the surface roughness of the film increases with the increase in deposition pressure. The grain size is increased till 6×10^{-3} mbar deposition pressure and then it reduced to 80 nm at 8×10^{-3} mbar (Kar et al., 2006). In addition, inhomogeneous patterns on the surface are also observed at this higher pressure (Fig. 3). It is also observed that the AlN film has changed its orientation with less Al-N bond density and reduction of grain size at 8×10^{-3} mbar sputtering pressure. Hence, it is inferred that the structural disorder and/or the change in the Al-N bond density/angles must have taken place at this particular sputtering pressure.

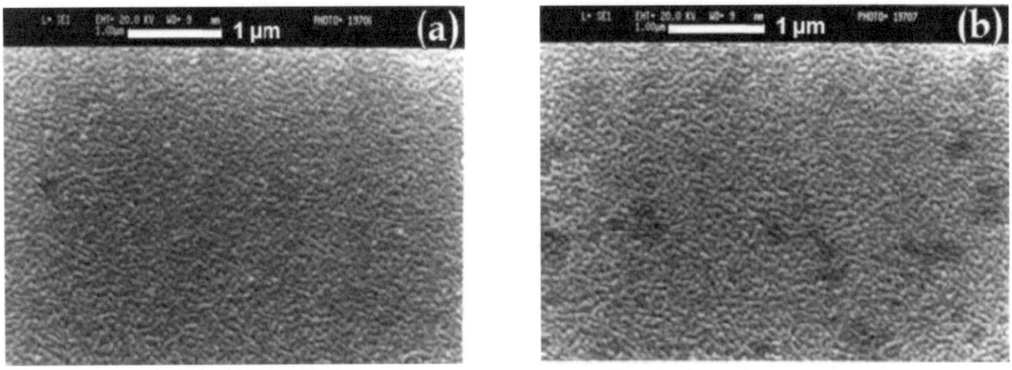

Fig. 3. SEM micrograph of the AlN films deposited at (a) 2×10^{-3} mbar, and (b) 8×10^{-3} mbar

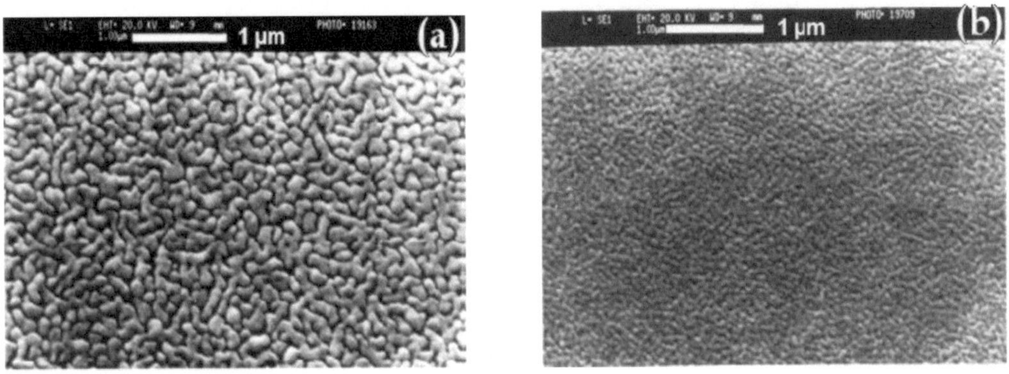

Fig. 4. SEM micrograph of AlN films deposited at (a) 20 % N_2, and (b) SEM image of AlN film for D_{ts} of 5 cm

3.4 Gas flow ratio

At lower nitrogen concentration, the intensity of (100) peak is relatively more prominent than (002), but the trend reverses with higher nitrogen concentration (Kar et al., 2006). At 80% N_2, a highly oriented (002) peak is observed without trace of (100) orientation. Lower argon and higher nitrogen gas concentration results slower aluminum sputtering rate. If the time interval for the arrival of Al species at the wafer surface is slower, the Al atom gets enough time to react with N_2. This increases the probability of Al-N bond formation and bonded Al-N molecules get more time to adjust themselves along (002) orientation on the substrate. On the other hand, at higher argon concentration, Al does not get enough time for complete nitridation due to higher sputtering rate. In addition, faster arrival of the Al at the substrate surface results not only in a poor AlN bond, but also provides less time for the newly formed AlN to arrange itself along c-axis. A surface texture of smaller grain size, smoother, homogeneous and dense granular microstructures has been observed at higher concentrations of nitrogen. This indicates a low surface mobility of the ad-atoms at high nitrogen concentration. In contrast, bigger grain size with increased roughness is observed at lower nitrogen concentration, where the newly formed smaller grain merges together with a previously formed grain and becomes bigger in size. The size and distribution of the micro-grains is quite uniform at 80% nitrogen concentration. At lower nitrogen concentrations, Ar^+ ions transfer more energy to the Al target during bombardment, generating more aluminum atoms that make clusters with incomplete nitridation of aluminum on the wafer surface. This leads to formation of fewer bonds, a poor c-axis orientated and a rough film (Fig 4 (a)).

3.5 Target-substrate distance

The kinetics of the sputtered species arriving at the substrate controls the ad-atom mobility and atomic rearrangement that governs the microstructure of the film. From the XRD studies, it is observed that the intensity of c-axis orientation of the film decreases with increase in target to substrate distance D_{ts} (Kar et al., 2008). At shorter D_{ts}, the Ar ions travel almost normal to the target due to the high electrical field and knock out Al atoms around perpendicular to the target. Because of short deposition path, the probability of collisions of the Al atom with gas atoms is low. Therefore, a good quality film is obtained at lower D_{ts} (5 cm). On the other hand, at larger D_{ts}, the chances of Al collision with gas molecules is increased. In this process Al atoms lose its kinetic energy significantly as well as alter deposition angles. These randomly arriving Al atoms, with lesser energy, cause self-shadowing effects and reduce atomic migration that leads to generation of voids in the film (Lee et al., 2003). The grain size of the AlN film increases with D_{ts}. For lower D_{ts}, smaller grain with minimum surface roughness is observed (Fig. 4(b)) and a coarser grain is found at the highest D_{ts} (8 cm). Surface roughness of the synthesized AlN films are also increases with D_{ts}. The kinetic energy of deposited species is considered to be a major factor for the grain size and the surface roughness of the film.

3.6 Variation of electrical properties with sputtering parameter

The AlN film can be used as a dielectric layer in IC; hence, the electric charges are essential to study with the sputter deposition parameters. Electric charges like Q_{in} and D_{it} are highly governed by the sputter deposition parameters. A decrease in the Q_{in} is observed with sputtering power, where as D_{it} is found to be minimum at moderate RF power. At higher temperature, better electrical properties in the bulk as well as the interface of sputtered AlN films are reported; this is mainly due to the formation of bigger grain size and its associated effects. It is reported that the defects produced by stress, voids and incorporation of gases are main responsible cause for the monotonic increase in Q_{in}. The D_{it} has a minimum value at 6×10^{-3} mbar sputtering pressure. The Q_{in} and D_{it} increases with nitrogen concentration. This will have a deleterious effect for silicon-based devices at higher nitrogen concentration. Rise in the Q_{in} and D_{it} with the increase in D_{ts} is also reported. It is seen that at larger D_{ts}, the morphological as well as the electrical properties of the AlN films deteriorates, whereas, at shorter D_{ts} the quality of the film comes out to be better (Kar et al., 2007). Apart from the electric charges, it is observed that better crystallinity posses AlN films of higher dielectric constant.

4. Post-deposition annealing effect

AlN film may see high temperature, if AlN film is monolithically integrated during IC fabrication. Post-deposition heat treatment significantly affects the morphology and electric charges of AlN film. The post-deposition thermal treatments (annealing) of AlN film are generally carried out by two distinguished modes; namely, Rapid Thermal Annealing (RTA) and conventional furnace annealing.

4.1 Rapid thermal annealing (RTA) process

The XRD studies show that the intensity of c-axis (002) orientation increases upto annealing temperature of 800 °C in nitrogen ambient and then it marginally decreases at 1000 °C (Kar et al., 2005). The shift in XRD diffraction peaks is reported at higher temperatures, which may be due to the generation of stress. Granular worm-like nanostructures are found in as-deposited AlN films (Fig. 5), whereas cracks are observed for annealing at 1000 °C. A short duration of heat pulse by RTA is barely sufficient to modulate the film surface, but not enough to activate the grains of the AlN films to merge themselves to form bigger grains. Appearances of cracks are due to the stress developed in the film. The thermal coefficient mismatch between the AlN and silicon substrate may be generated from the fast ramp up and ramp down annealing heat cycle during rapid thermal annealing. The position and density of cracks depend strongly on the defects, dislocations and the structural relaxation of grain boundaries. The surface roughness is considerably increased for the film annealed at higher temperatures due to the surface modulation.

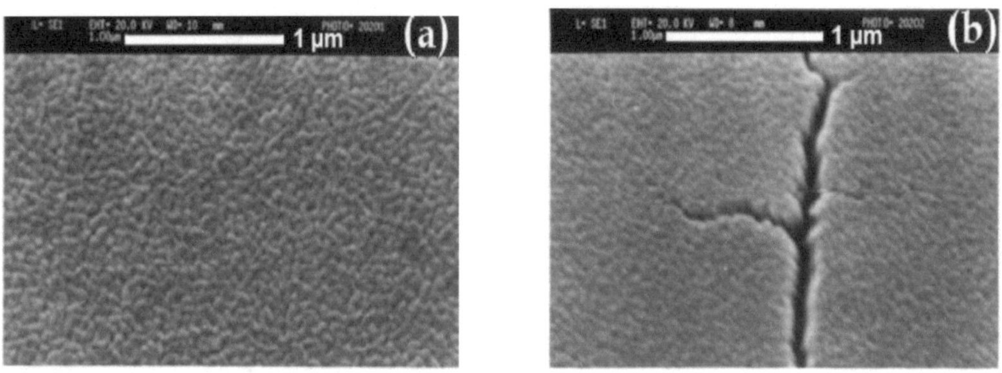

Fig. 5. SEM micrograph of AlN films RTA processed at (a) as-deposited, and (b) 1000 °C

4.2 Furnace annealing

The intensity of the (002) peak increases with furnace annealing temperature, where the atoms acquire adequate activation energy to become (002) oriented. Sometimes, many of the atoms may not be at the crystal lattice site in the as-deposited AlN film, which causes the lattice strain and the formation of microvoids. During conventional furnace annealing, atoms get enough time to acquire sufficient kinetic energy and occupy relative equilibrium positions that minimize the lattice strain and microvoids, which results in a better crystalline film. Furthermore, the furnace annealing process minimizes the dislocations and the other structural defects and forms a better stoichiometric material. From the SEM micrographs, it is observed that the granular worm-like textures grow bigger in size with increased surface roughness as a result of annealing (Fig. 6). The possible reasons for increase in the grain size and the surface roughness may be due to atomic migration in the film towards the lower surface energy with annealing temperature (Kar et al., 2009).

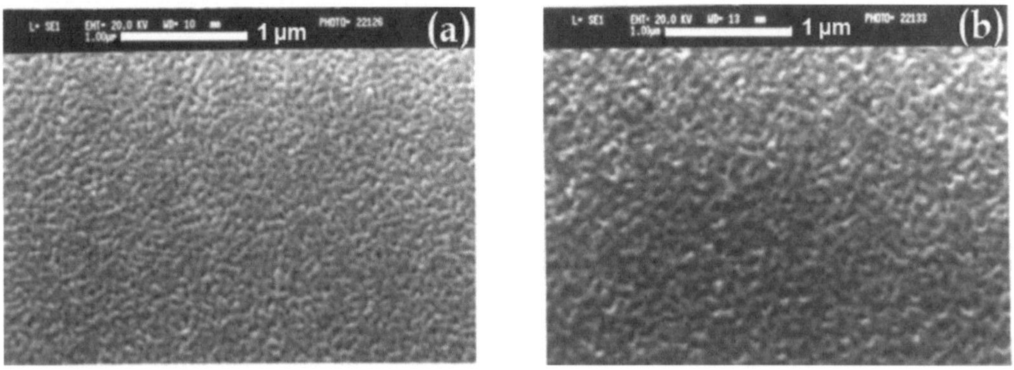

Fig. 6. SEM micrograph of AlN films annealed at (a) as-deposited, and (b) 800 °C

4.3 Effect of annealing on electrical properties

In both types of annealing, the Q_{in} is increased with temperature. The probable reasons for the rise in the Q_{in} may be due to the generation of trap centres with annealing in the nitrogen ambient. In RTA, the D_{it} is found to be strongly dependent on the annealing temperature and it significantly reduced at 600°C. With furnace annealing, the D_{it} decreases with increase in annealing temperature.

5. Growth of AlN films on different substrates

Many a times, AlN films are made on an insulator (SiO_2) for isolation or it is deposited over the metallic electrodes for thin film resonators (TFR). In future, AlN film on high speed semiconductor substrates such as GaAs, InP can be exploited for high speed signal processing and Micro-Opto-Electro-Mechanical Systems (MOEMS) applications. Hence, integration of AlN films on GaAs and InP substrates for a new generation of high-speed devices/subsystems, especially for telecommunications, and radar applications are required. Growth and surface morphology of a deposited film depends not only on the kinetics of the arriving species at the substrate, but also on the nature of the substrates chosen, even if they belong to the same family. In addition, substrate orientation, thermal conductivity and thermal expansion coefficients play vital roles in film growth and its morphology. C-axis oriented AlN films are deposited on Si and SiO_2/Si substrates by RF reactive magnetron sputtering, where the degree of orientation decreases with increase in oxide thickness. The surface roughness of the films deposited on SiO_2/Si is higher. AlN films are also deposited on GaAs and InP substrates by reactive magnetron sputtering technique under identical deposition conditions. c-axis (002) oriented films are observed on GaAs substrates; whereas, AlN (100), (002) and (102) oriented peaks are seen in case of InP substrates. Surface morphology of the films deposited on Si and InP substrates seems to be similar, but the films on InP are little rougher with the development of nano-pores. AlN films, grown on GaAs substrates, forms bump like structures (Kar et al., 2009), which may be due to thermal and/or lattice mismatch. It is important to note that the crystallinity and stochiometry of the initial layer of AlN film also plays a significant role in the creation of defects and mismatches (Ahmed et al., 1992). Crystal orientation of AlN films is also a strong function of the bottom metal electrodes. AlN films deposited on metals (Al, Cu, Cr, Au) are c-axis oriented, whereas the films deposited on Al and Cu are rough with larger grains.

6. AlN film based acoustic device

SAW and BAW are two import types of acoustic wave devices used in RF comunication. These devices can be realized either on a solid substrate/film or through micromachined suspended beam structures. In SAW devices, an elastic wave travels on the surface of a piezoelectric material and displaces the atoms about their equilibrium positions at the interface of piezoelectric film and solid substrate. The neighboring atoms at the interface then produce restoring forces to bring the displaced atoms back to their original positions. SAW can be generated by placing two inter digital transducers (IDT) either sides of the substrate. These IDTs have alternating periodic fingers (Fig. 7). RF signal is applied one of these alternating polarity fingers (IDT) that produces elastic mechanical wave in the substrate. This wave travels along the substrate and also collected by placing another IDT on the piezoelectric material, some distance away from the first IDT. The second IDT collects the RF signal, which can be retransformed into the electric signal. As the elastic-mechanical wave has the speed of acoustic wave, it introduces the delay of signal by 10^3 order. This is the prime use of SAW device. The periodicity p, (centre-to-centre spacing between neighbouring IDT fingers of same polarity) becomes the wavelength of the acoustic wave, and dictates its frequency $f = v/p$, where f and v represent acoustic wave frequency and acoustic propagation velocity, respectively.

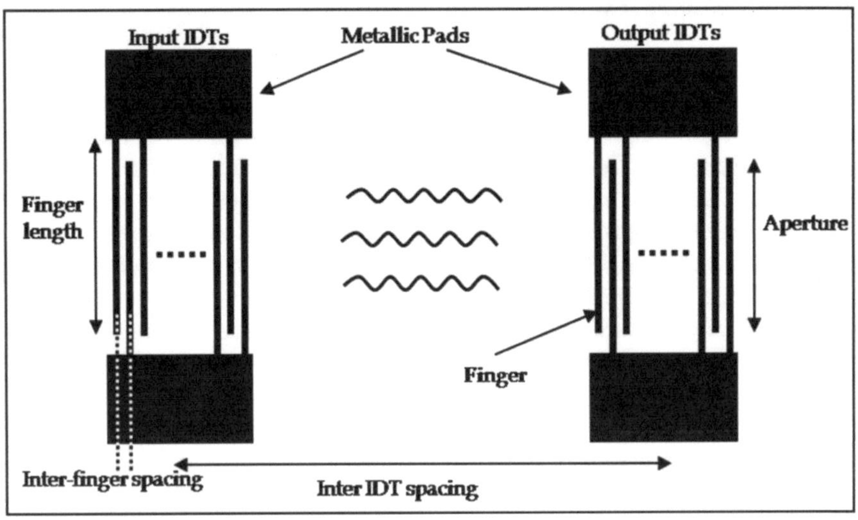

Fig. 7. Schematic diagram of SAW device

Thin Film Bulk Acoustic Resonator (FBAR) device consisting of a piezoelectric material sandwiched between two electrodes and is acoustically de-coupled from the surrounding medium. FBAR devices, using AlN piezoelectric with thickness ranging from tenth of micrometers to several micrometers, resonate in the cellular bands of cell phones and other wireless applications. On applying voltage across the electrodes, the piezoelectric thin film undergoes a shear deformation, and a BAW resonance occurs in the AlN film due to coherent reflection at the top and bottom boundaries of the metal film or plate electrodes. The frequency of resonance is dependent on the physical structures; hence, desired resonant frequency can be obtained by tailoring physical dimension of the structure. For RF frequency, physical dimension of resonators can be realized by using MEMS technology.

MEMS resonators are comprised of a microscale mechanical element, which converts mechanical to electrical signal and vice versa. One of the prominent resonator structures is MEMS cantilever, which is based on thin piezoelectric films. Film resonates when an ac voltage is applied across the film. Resonator can be made without piezoelectric material (electrostatic, capacitive resonator), but it suffers with large resistance, in the range of MΩ, and depends on driving voltage. On the other hand, piezoelectric resonators have smaller resistance of the order of KΩ and are more suitable for UHF device applications. (Lakin, 1999; Quandt et al., 2000; Humad et al., 2003). In addition, the output is easier to sense in a piezoelectric resonator. Furthermore, a piezoelectric resonator has certain advantages over the electrostatic resonator (capacitive resonators), such as low current consumption and lower actuation voltages (Olivares et al.; 2005). But the quality factor (Q) of piezoelectric resonator is smaller than that of a capacitive resonator. The quality factor of any resonator is proportional to the decay time, and is inversely proportional to the bandwidth around resonance. Higher Q represents higher frequency stability and accuracy capability of the resonator (De Los Santos, 1999).

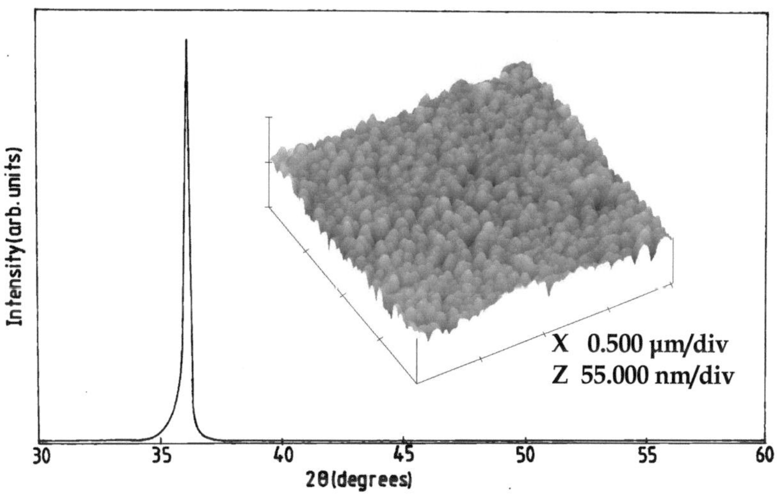

Fig. 8. XRD pattern, and AFM image (inset) of sputtered AlN film for SAW

6.1 Evaluation of AlN films through SAW devices

Higher RF power (400 W) and nitrogen concentration (80%), moderate substrate temperature (200 °C) and sputtering pressure (6×10^{-3} mbar), lower target-substrate distance (5 cm) is suitable for the growth of smooth, highly c-axis oriented AlN film with better electrical properties. A c-axis (002) oriented peak is recorded at 2θ value of 36.1° (Fig. 8). The atomic force micrograph of the film shows dense microstructure with continuous grain growth (inset of Fig. 8). This kind of film is suitable for SAW devices. In a typical case, each IDT consisted of 25 pairs of fingers/electrodes with 30 µm centre-to-centre spacing between the two neighbouring fingers comprising a pair (p/2). The width of each finger/electrode is designed to be 15 µm (p/4) with each of 6.0 mm length and 5.0 mm overlap, producing a SAW filter with an acoustic wavelength of 60 µm (Kar et al., 2009). The SAW device parameters are: AlN film thickness = 0.92 µm, acoustic wavelength = 60 µm, SAW velocity = 5058 m/sec, electromechanical coupling coefficient (K^2) = 0.34%.

Time response, due to acoustic frequency alone, is found after the gating out the response due to electromagnetic feed through. For the centre to centre distance between the two IDT's (d) = 7 mm, the main SAW signal centred at time (delay) = d/ V_{SAW} = 7/5058 = 1.384 μs is obtained. The central acoustic frequency (f_0) response after the gating out is observed at 84.304 MHz (Fig. 9).

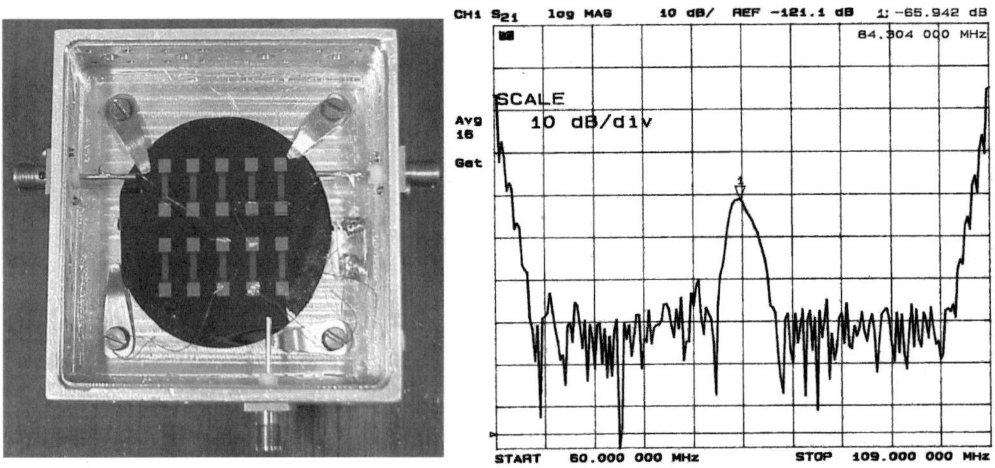

Fig. 9. (a) Optical image of AlN based SAW devices, and (b) Measured response of the SAW device

6.2 Fabrication of AlN film based MEMS

Anisotropic etching of silicon is a key technology for fabrication of various three-dimensional structures such as thin membranes and microbridges for MEMS. Generally, anisotropic silicon etching is done by potassium hydroxide (KOH) or ethylenediamine pyrochatechol (EDP) etchant (French, 2001; Ni et al., 2005). Another technique, which is more versatile, CMOS process compatible and nontoxic, provides better selective etching using doped tetramethylammonium hydroxide (TMAH) etchant (Biswas et al., 2006). The characteristics of these three etchants are listed in Table 2. Many AlN and Al based MEMS structures are also isolated from silicon by silicon dioxide. Hence, their selective etching is very important. Diluted tetramethylammonium hydroxide (TMAH, 5 wt %), doped with silicic acid (30.5 g/l) and ammonium persulphate (5.5 g/l), is suitable for CMOS silicon microprocessing. To protect Al and AlN, silicic acid has been chosen instead of pure silicon powder, because silicic acid dissolves quickly in TMAH solution. Ammonium persulphate (AP) is also added to the above-mentioned solution to reduce the surface roughness of etched silicon. The etch rate of silicon in doped TMAH is found to be 50 μm/hour. During silicon etching the Al, AlN and SiO_2 films are used as mask layers (Fig. 10). Low etch rates of Al and AlN (18-30 nm/hour) as well as SiO_2 (2.5 nm/hour) are found to be suitable for MEMS applications. Probable reason for low etch rate of Al may be the formation of a passivating layer during TMAH etching (Fujitsuka et al., 2004). Dilute TMAH is a well known etchant for AlN film (Kim et al., 2004) and Al as well. But doped TMAH shows significantly lower etch rates for AlN and Al, which are exploited for AlN based suspended microstructures. Fig. 10 (d) depicts suspended Cr/AlN/Cr/SiO_2 cantilevers fixed at one end, where in one of the microstructures is lifted up because of the stress (Kar et al., 2009).

Property	KOH	EDP	TMAH
Si etch rate (100), µm/h	150	30-35	40-60
Etch quality	high	high	medium
Selectivity (111)/(100)	1:30-100	1:20	1:10-50
Under-etch rate	0.5-1.5	1.4-1.5	0.2-1.7
CMOS compatible	no	yes	yes
Selectivity PECVD SiO_2/Si	1:100-300	1:10,000	1:100-1000
Selectivity PECVD SIN/Si	1:10,000	----------	1:150-200
Attack of aluminum	high	medium	low with Si
Etch stop	boron dope	boron dope	boron dope
Toxicity	low	high	low
Long-term stability	high	low	medium
Cost	low	high	medium

Table 2. Characteristics of important wet etchants used for silicon micromachining (French, 2001)

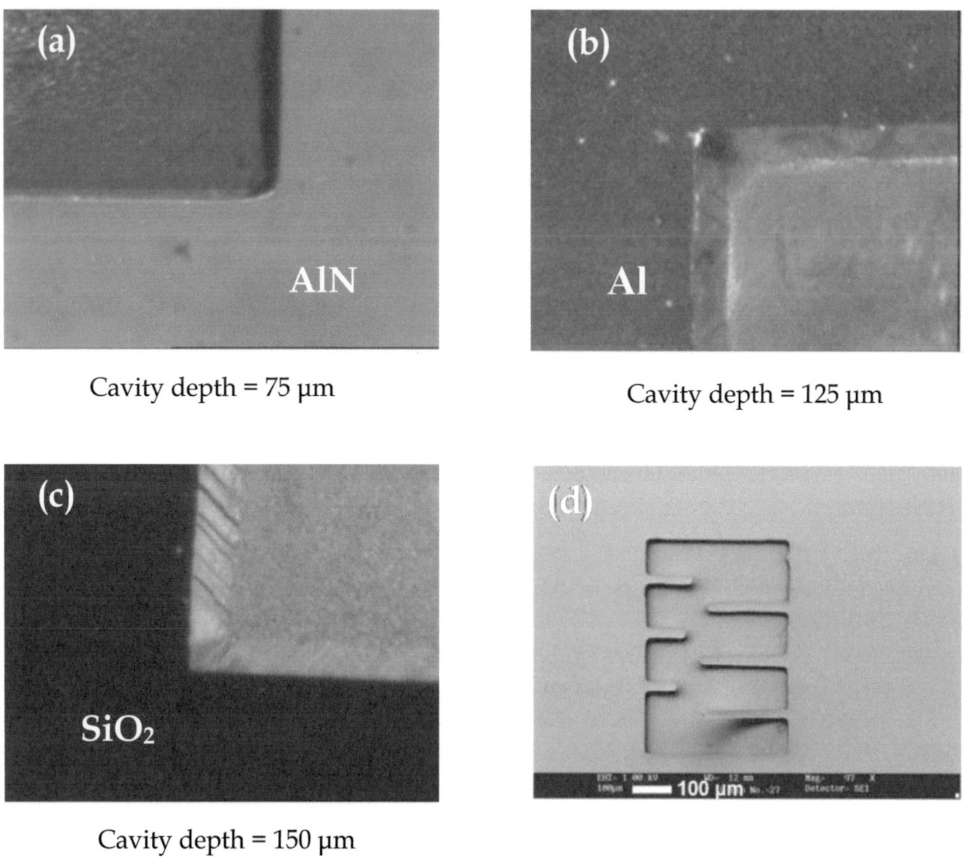

Fig. 10. Micrographs of etched silicon (a) AlN/Si, (b) Al/Si, (c) SiO_2/Si, (d) AlN based suspended microstructures

7. Conclusion

This chapter focuses on the study of RF sputtered AlN films in view of ICs, acoustic devices and MEMS applications. It is divided into two distinct parts; growth, characterization and optimization of device worthy AlN film (mostly on silicon), and demonstration of acoustic and MEMS device applications. The morphological and electrical properties of RF sputtered AlN films are studied with sputtering power, deposition temperature, sputtering pressure, gas flow ratio and target to substrate spacing (D_{ts}). Higher RF power (400 W) and nitrogen concentration (80%), moderate substrate temperature (200 °C) and sputtering pressure (6×10^{-3} mbar), lower target-substrate distance (5 cm) is suitable for the growth of smooth, highly c-axis oriented film with better electrical properties. Post-deposition (RTA and furnace) annealing has a significant impact on the morphology as well as the electrical properties. The RTA processed AlN films have relatively high c-axis (002) orientation films at 800 °C, where microcracks are appeared during RTA process at 1000 °C. Bulk charge density is increased with annealing temperature both types of annealing. A significant reduction in interface charge density is found at 600°C with RTA process, whereas it decreases with furnace annealing temperature. AlN films are deposited and characterized on different substrates by RF reactive magnetron sputtering. On SiO_2/Si substrates, (002) orientation is deteriorated and surface roughness of the films is increased with the increase in oxide thickness. c-axis (002) oriented films are observed on Si and GaAs substrates, whereas AlN (100), (002) and (102) oriented peaks are seen on InP substrates. AlN films, deposited on GaAs substrate, show bump like structures. c-axis oriented AlN films are also observed on metallic films. The AlN films, deposited on Al and Cu, are found to be rough with larger grains. Piezoelectric nature of RF deposited AlN films is ascertained from the performance of a SAW device. This device is centred around a frequency of 84.3 MHz and acoustic phase velocity is inferred to be 5058 m/ sec with K^2 of 0.34 %. The TMAH solution, doped with an appropriate ratio of silicic acid and ammonium persulphate, is developed for micromachining of AlN based structures. The etch rate of silicon is around 50 µm/hour. On the otherhand, the doped solution has negligible impact on Al, AlN and SiO_2 films. The growth of highly (002) oriented AlN films, post deposition process and micromachining method will provide an appropriate platform for the fabrication of futuristic electronic devices.

8. References

Ahmed, A.U., Rys, A., Singh, N., Edgar, J.H. & Yu, Z.J. (1992). The Electrical and Compositional Properties of AlN-Si Interfaces, *J. Electrochem Soc*, Vol. 139(4), pp. 1146-1151.

Assouar, M.B., Hakiki, M.E., Elmazria, O., Alnot, P. & Tiusan, C. (2004). Synthesis and microstructural characterisation of reactive RF magnetron sputtering AlN films for surface acoustic wave filters, *Diamond and Related Materials*, Vol. 13, pp. 1111-1115.

Belyanin, A.F., Boulov, L.L., Zhirnov, V.V., Kamenev, A.I., Kovalskij, K.A. & Spitsyn, B.V. (1999). Application of aluminum nitride films for electronic devices, *Diamond and Related Materials*, Vol. 8, pp. 369-372.

Bender, S., Dickert, F.L., Mokwa, W. & Pachatz, P. (2003). Investigations on temperature controlled monolithic integrated surface acoustic wave (SAW) gas sensors, *Sensors and Actuators B*, Vol. 93, pp. 164-168.

Biswas, K., Das, S., Maurya, D. K., Kal, S. & Lahiri S.K. (2006). Bulk micromachining of silicon in TMAH-based etchants for aluminum passivation and smooth surface, *Microelectronics Journal*, Vol. 37(4), pp. 321-327.

Cheng, H., Sun, Y. & Hing, P. (2003). The influence of deposition conditions on structure and morphology of aluminum nitride films deposited by radio frequency reactive sputtering, *Thin Solid Films*, Vol. 434, pp. 112-120.

De Los Santos, H.J. (1999). Introduction to Microelectromechanical (MEM) Microwave Systems, Artech House, pp. 83.

French, P.J. (2001). Integration of silicon MEMS devices: Materials and processing considerations, *Smart Materials Bulletin*, January, pp. 7-13.

Fujitsuka, N., Hamaguchi, K., Funabashi, H., Kawasaki, E. & Fukada, T. (2004). Silicon anisotropic etching without attacking aluminum with Si and oxidizing agent dissolved in TMAH solution, *Sensors and Actuators A*, Vol. 114, pp. 510-515.

Humad, S., Abdolvand, R., Ho, G.K. & Ayazi F. (2003). Micromechanical Piezo-on-Silicon Block Resonators, *in Proc. IEEE International Electron Devices Meeting (IEDM'03)*, Washington DC, Dec., pp. 957-960.

Kar, J.P., Mukherjee, S., Bose, G., Tuli, S. & Myoung, J.M. (2009), Impact of post-deposition annealing on the surface, bulk and interface properties of RF sputtered AlN films, *Materials Science and Technology*, Vol. 25, pp. 1023-1027.

Kar, J.P., Bose, G., Tuli, S., Myoung, J.M. & Mukherjee, S. (2009). Morphological investigation of AlN films on various substrates for MEMS applications, *Surface Engineering*, Vol. 25, pp. 526-530.

Kar, J.P., Bose, G., Tuli, S., Dangwal, A. & Mukherjee, S. (2009). Growth of AlN films and its process development for the fabrication of acoustic devices and micromachined structures, *Journal of Materials Engineering and Performance*, Vol. 18, pp. 1046-1051.

Kar, J.P., Mukherjee, S., Bose, G. & Tuli, S. (2008). Effect of inter-electrode spacing on structural and electrical properties of AlN films, *Journal of Materials Science: Materials in Electronics*, Vol. 19, pp. 261-265.

Kar, J.P. (2007). *Growth and characterization of aluminum nitride (AlN) films for electroacoustic and MEMS applications*, Ph. D. thesis, Indian Institute of Technology Delhi, India

Kar, J.P., Bose, G. & Tuli, S. (2006). A study on the interface and bulk charge density of AlN films with sputtering pressure, *Vacuum*, Vol. 81, pp. 494-498.

Kar, J.P., Bose, G. & Tuli, S. (2006). Correlation of electrical and morphological properties of sputtered aluminum nitride films with deposition temperature, *Current Applied Physics*, Vol. 6, pp. 873-876.

Kar, J.P., Bose, G. & Tuli, S. (2006). Influence of nitrogen concentration on grain growth, structural and electrical properties of sputtered aluminum nitride films, *Scripta Materialia*, Vol. 54, pp. 1755-1759.

Kar, J.P., Bose, G. & Tuli, S. (2005). Influence of rapid thermal annealing on morphological and electrical properties of RF sputtered AlN films, , *Materials Science in Semiconductor Processing*, Vol. 8, pp. 646-651.

Kim, H.H., Ju, B.K., Lee, Y.H., Lee S.H., Lee J.K. & Kim, S.W. (2004). Fabrication of suspended thin film resonator for application of RF bandpass filter, *Microelectronics Reliability*, Vol. 44, pp. 237-243.

Lakin, K. (1999). Thin film resonators and filters, *IEEE Ultrasonics Symposium*, pp. 895-906.

Lee, H.C., Park, J.Y., Lee, K.H. & Bu, J.U. (2004). Preparation of highly textured Mo and AlN films using a Ti seed layer for integrated high-*Q* film bulk acoustic resonators, *J. Vac. Sci. Technol. B*, Vol. 22(3), pp. 1127-1133.

Lee, S.H., Yoon, K.H., Cheong, D. S. & Lee, J.K. (2003). Relationship between residual stress and structural properties of AlN films deposited by r.f. reactive sputtering, *Thin Solid Films*, Vol. 435, pp. 193-198.

Loebl, H.P., Klee, M., Metzmacher, C., Brand, W., Milsom, R. & Lok, P. (2003). Piezoelectric thin AlN films for bulk acoustic wave (BAW) resonators, *Materials Chemistry and Physics*, Vol. 79, pp. 143-146.

Luo, J.K., Lin, M., Fu, Y.Q., Wang, L., Flewitt, A.J., Spearing, S.M., Fleck, N.A. & Milne, W.I. (2006).MEMS based digital variable capacitors with a high-*k* dielectric insulator, *Sensors and Actuators A*, Vol. 132 (1), pp. 139-146.

Naik, R.S., Reif, R., Lutsky J.J. & Sodini, C.G. (1999). Low-Temperature Deposition of Highly Textured Aluminum Nitride by Direct Current Magnetron Sputtering for Applications in Thin-Film Resonators, *J. the Electrochemical Society*, Vol. 146(2), pp. 691-696.

Ni, H., Lee H.J. & Ramirez, A.G. (2005). A robust two-step etching process for large-scale microfabricated SiO_2 and Si_3N_4 MEMS membranes, *Sensors and Actuators A*, Vol. 119, pp. 553-558.

Olivares, J., Iborra E., Clement M., Vergara, L., Sangrador, J. & Sanz-Herv´as, A. (2005). Piezoelectric actuation of microbridges using AlN, *Sensors and Actuators A*, Vol. 123–124, pp. 590-595.

Quandt, E. & Ludwig, A. (2000). Magnetostrictive actuation in microsystems, *Sensors and Actuators A*, Vol. 81, pp. 275-280.

Schreiter, M., Gabl, R., Pitzer, D., Primig, R. & Wersing, W. (2004). Electro-acoustic hysteresis behaviour of PZT thin film bulk acoustic resonators, *J. the European Ceramic Society*, Vol. 24, pp. 1589-1592.

Strite, S. & Morko, H. (1992). *GaN, AlN and InN: A review, J. Vac. Sci. Technol. B*, Vol. 10(4), pp. 1237-1266.

Wang, H.H. (2000). Properties and preparation of AlN thin films by reactive laser ablation with nitrogen discharge, *Modern Physics Letters B*, Vol. 14, 523-530.

Wang, X.D., Jiang, W., Norton, M.G. & Hipps, K.W. (1994). Morphology and orientation of nanocrystalline AlN thin films, *Thin Solid Films*, Vol. 251(2), pp. 121-126.

Xu, X.H., Wu, H.S., Zhang, C.J. & Jin Z.H. (2001). Morphological properties of AlN piezoelectric thin films deposited by DC reactive magnetron sputtering, *Thin Solid Films*, Vol. 388, pp. 62-67.

Yamada, H., Ushimi, Y., Takeuchi, M., Yoshino, Y., Makino, T. & Arai, S. (2004). Improvement of crystallinity of ZnO thin film and electrical characteristics of film bulk acoustic wave resonator by using Pt buffer layer, *Vacuum*, Vol. 74, pp. 689-692.

Yim, W.M., Stofko, E.J., Zanzucchi, P.J., Pankov, J.I., Ettenberg, M. & Gilbert, S.L. (1973). Epitaxially grown AlN and its optical band gap, *J. Appl. Phys.*, Vol. 44(1), pp. 292-296.

26

Surface Acoustic Wave Devices for Harsh Environment

Cinzia Caliendo
Istituto dei Sistemi Complessi, ISC-CNR, Area della Ricerca Roma 2, Rome
Italy

1. Introduction

There is an increasing demand of electronic components for aerospace, aircraft industries, sensors, automotive, chemical and material processing applications, to name just a few, able to operate reliably and for long time at high-temperature. Measurements reliability requires the electronic components to be placed directly inside the extreme environment, and to withstand temperatures of several centigrade degrees with lifetimes of several hours. The device mounting and packaging, but first of all the device materials must be stable with the working temperature, otherwise temperature-induced stress may result in device's failures. Electroacoustic devices based on surface and bulk acoustic wave (SAW and BAW) technology must satisfy the requirements of low cost, high frequency, high-Q, low loss, large piezoelectric coupling and zero temperature coefficient of delay (TCD) to be key devices in the communication and sensor fields. The temperature stability of the piezoelectric crystal is an essential characteristic because of its direct link with the temperature sensitivity of the electroacoustic device operation frequency. The high operation frequency is an essential characteristic for SAW and BAW devices to be used in mobile phones, cordless headphones, alarm and security systems, military equipment, sensors, etc. The temperature stability and the high operation frequency demands can be met through a proper choice of the piezoelectric substrate crystal cut, new piezoelectric materials and/or multilayer configurations. The use of temperature stable cuts of single crystal bulk piezoelectric materials or temperature compensated multilayers represents two possible solutions to the temperature stability requirement. The use of high-resolution lithography techniques and/or of high SAW velocity materials is required in order to extend the upper limit of the electroacoustic device frequency range. Submicron feature sized interdigital transducers (IDTs) are required to implement GHz range SAW devices on *slow* piezoelectric materials, while micron feature sized IDTs can still be used on *fast* materials, since the SAW device centre frequency, $f = v/\lambda$, depends on both the phase velocity of the propagating medium, v, and on the acoustic wavelength λ, being the IDT's period $p = \lambda/2$. Conventional piezoelectric substrates, such as quartz, lithium niobate ($LiNbO_3$), and lithium tantalate ($LiTaO_3$) crystals, cannot be used above 500°C. Quartz ST cut is a temperature stable material but it shows an alpha-beta transition at 573°C, which causes the loss of piezoelectricity, and results in a non-operable device. SAW devices implemented on $LiNbO_3$ have been studied for a temporary usage at 400°C [1]; however the $LiNbO_3$ acoustic wave properties are highly dependent on temperature since it is a pyroelectric material and has a TCD as high as ~75 ppm/°C. $LiTaO_3$ shows properties, such as a low Curie temperature (607°C), sensitivity to temperature variations (TCD = 22 ppm/°C for the X-112°Y cut) and a strong pyroelectricity, which limit its operation at elevated temperatures. Piezoelectric bulk single crystals such as $GaPO_4$, LGS ($La_5Ga_3SiO_{14}$) and its isomorphs (called LGX family group) substrates are widely investigated for the realization of SAW-based devices able to work at high temperature. LGS belongs to the trigonal class 32 group as quartz but it has no α-β transitions and can operate up to its melting temperature of 1470°. It shows zero or very low TCD cuts with zero power flow angle and higher electromechanical coupling coefficient than that of quartz [2]. Langasite based SAW devices are not suitable for operation in the GHz range as a consequence of their low phase velocity and high acoustic losses (from 1 to 0.01 dB/wavelength [3]). $GaPO_4$ has twice the sensitivity of quartz and many its physical constants are stable up to about 900°C, but the accessible frequencies are limited to values of 1 GHz as a consequence of quite high acoustic losses.

The technology of thin piezoelectric films (such as AlN) offers the opportunity of combining the properties of the substrate with those of the film: thus a composite arrangement of fast materials with opposite sign *TCD*s and a proper design of the electroacoustic configuration enable achieving a thermally stable SAW device operating in the GHz range.

Aluminium nitride (AlN) is a piezoelectric material that shows interesting properties, such as excellent thermal conductivity (180W/mK), low coefficient of thermal expansion (CTE, 4.1 x 10^{-6} °C^{-1}), and good resistance to thermal shock and caustic chemicals [4], that make it useful as protective coating and guarantee the stability of the AlN-based devices when they are in contact with extreme environments. It is currently being investigated due to its promising potentialities for high-temperature, high-power, and high-frequency electronics. It has demonstrated to be an ideal candidate for packaging SiC-devices for high-temperature applications [5] thanks to its CTE that closely matches those of Si (3.5 x 10^{-6} °C^{-1}) and SiC (3.7 x 10^{-6} °C^{-1}), high electrical resistivity, high mechanical strength, and chemical inertness. Reactively sputtered AlN films have been used as an effective encapsulant for GaN [6] at an annealing temperature of 1100°C substituting the standard dielectric encapsulants, such as SiO_2 and Si_3N_4, that are not viable at so high temperatures. AlN maintains its piezoelectricity up to 1200°C in vacuum and shows very high BAW and SAW velocities (~6000 m/s and 11300 m/s for transversal and longitudinal BAWs propagating along the z direction, 5607 m/s for SAWs propagating in the *z* plane) that make it the ideal candidate for microwave electroacoustic devices implementation. Furthermore, AlN can be grown in thin film form onto non piezoelectric substrates by techniques as simple as the rf reactive sputtering. Both the structural properties of the substrate and the experimental sputtering parameters (such as the reactive gas flow rate, the partial pressure of reactive and inert gasses, the substrate temperature, the rf power, and the substrate-target distance) affect the morphological and structural properties of the sputtered thin films. The requirements for a suitable substrate include also a thermal coefficient-of-expansion compatible with that of the film, high-temperature stability, machinability, good adherence of the AlN film: among the available substrates, silicon, platinum and sapphire satisfy these requirements. Al_2O_3, substrates have a wide range of industrial applications as structural ceramic and optical materials. Al_2O_3 is extensively used as a high temperature, corrosion resistant refractory material due to its hardness, chemical durability, abrasive resistance, mechanical strength, and good electrical insulation. Al_2O_3 shows good thermal conductivity (24 W/mK), high SAW velocity (in the range 5555 to 5706 m/s in the c-plane), positive TCD (~70 ppm/°C) and a CTE that closely matches that of AlN. The AlN/Al_2O_3 -based multilayers can withstand temperatures up to

900°C, thus allowing the realization of high frequency, temperature compensated dispersive electroacoustic devices for high temperature applications. Electroacoustic devices implemented on Si substrates offer the opportunity to integrate the device with the surrounding electronic circuitry on the same chip. Moreover, the opposite TCD of Si (~30 ppm/°C) and AlN (~ -30 ppm/°C) allows the realisation of zero-temperature-coefficient acoustic devices at the proper film thickness to be used as sensors and actuators where low loss, low thermal drift, high sensitivity and high signal-to-noise ratio are demanded [7, 8]. Pt is the material of choice for metallic components that have to withstand oxidation, thank to its high temperature coefficient of resistance: Pt can be grown in thin film form and both the IDTs and ground electrodes can be easily defined by lift off technique.
In the present chapter the sustainability of Pt and AlN films on sapphire and Si substrates for high temperature applications is assessed.

2. AlN-based SAW devices

Bulk piezoelectric single crystals, such as LGS and $GaPO_4$, can be used for the implementation of non dispersive SAW devices, such as delay lines, filters and resonators, and the SAW propagation characteristics, such as phase velocity, electroacoustic coupling efficiency K^2 and TCD, depend on the crystal cut and SAW propagation direction, as well as on the geometry of the IDTs . The SAW propagation is excited by IDTs located at the free surface of the piezoelectric substrate and directly exposed to the surrounding environment, as shown in figure 1.

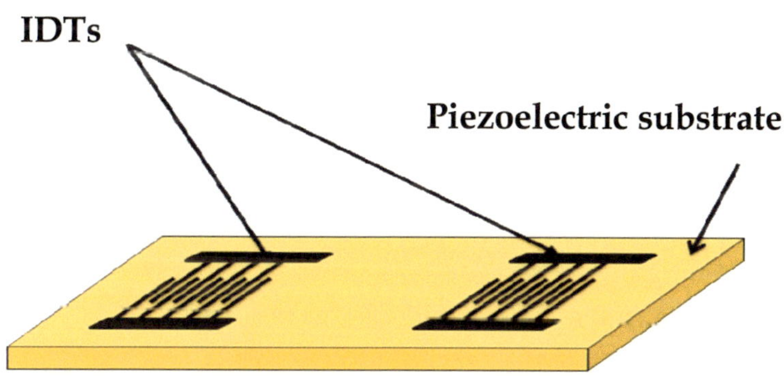

Fig. 1. SAW delay line on a piezoelectric substrate

AlN can be grown in thin film form onto non piezoelectric substrates, such as silicon or sapphire, thus allowing the realization of dispersive electroacoustic devices. Moreover, if the AlN film is sandwiched between the IDTs and the ground electrode, four piezoelectric coupling configurations can be obtained by placing the IDTs at the substrate/film interface or at the film surface, with and without the floating electrode opposite the IDTs. These four structures will be mentioned hereafter as substrate/film/IDT (SFT), substrate/IDT/film/metal (STFM), substrate/IDT/film (STF) and substrate/metal/film/IDT (SMFT), respectively. Figure 2a shows the top view of a dispersive SAW delay line, and figure 2b shows the cross sections of the four coupling configurations.
When the IDTs are located at the substrate/film interface, the piezoelectric film plays the role of both the acoustic wave transductor and protective layer of the underlying IDTs.

These four configurations show frequency dispersive SAW propagation characteristics, that are no longer solely determined by the geometry of the IDTs, the crystals cut, and the SAW propagation direction, but also by the film thickness and the electrical boundary conditions. For SAW propagating along layered structures, the achievable K^2 value is sometimes larger than that of the individual piezoelectric materials; it is frequency dispersive and depends on the type and orientation of the piezoelectric material, and it is drastically affected by the location of the IDTs and counter electrode with respect to the piezolectric layer. As an example, figures 3a and 3b show the K^2 vs film thickness to wavelength ratio, h/λ, for SAW propagation along *zx*- and *zy*-Al_2O_3/AlN for the four coupling configurations: the highest K^2 values obtainable are about 0.50 and 0.67% for STF x and y propagation (at $h/\lambda \sim 0.65$ and 0.60), 0.49 and 0.64 % for STFM x and y propagation (at $h/\lambda \sim 0.67$ and 0.62), being 0.3% the AlN K^2 value.

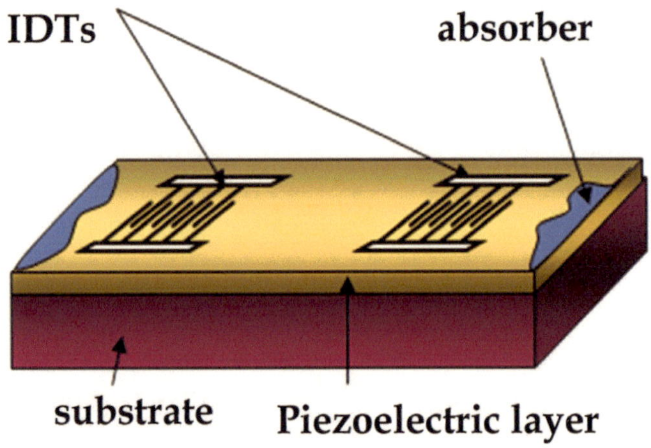

Fig. 2a. Dispersive SAW delay line

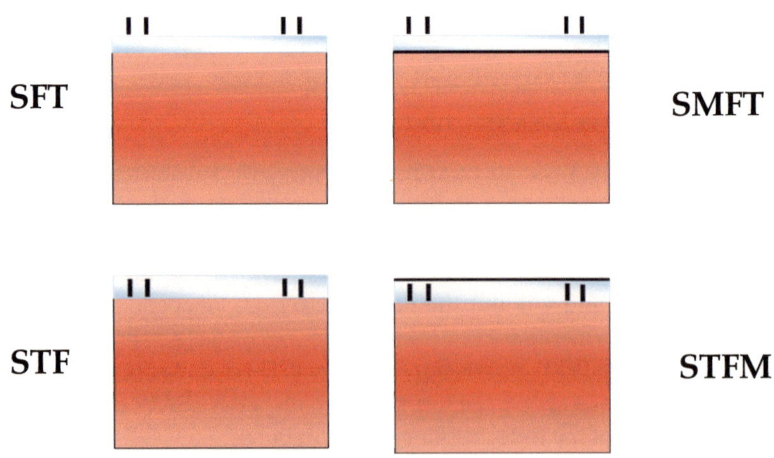

Fig. 2b. The four electroacoustic coupling configurations

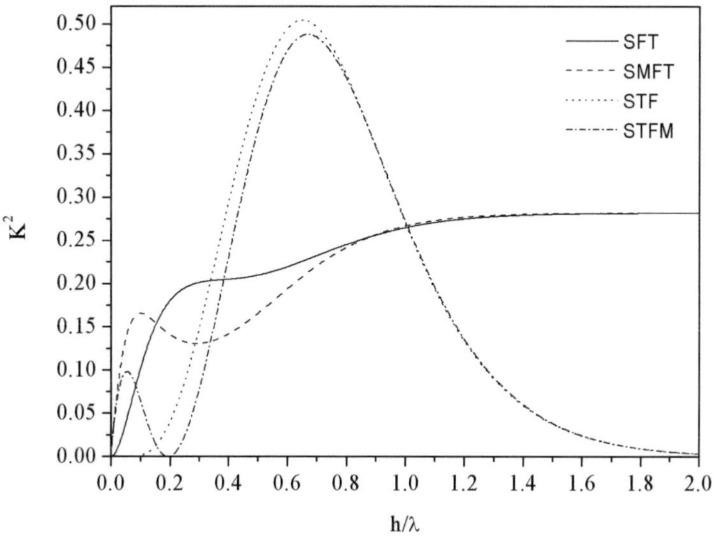

Fig. 3a. The K^2 vs h/λ for SAW propagating along zx-Al_2O_3/AlN for the four coupling configurations

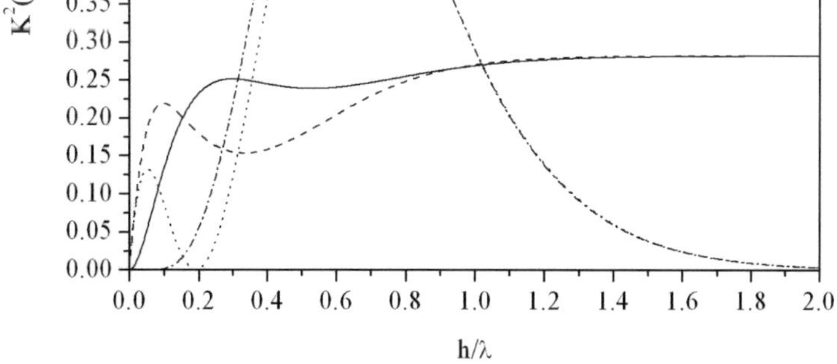

Fig. 3b. K^2 vs h/λ for SAW propagating along zy-Al_2O_3/AlN for the four coupling structures

Figures 4a and 4b show the K^2 vs h/λ for SAW propagation along Si/Pt/AlN and Si/AlN/Pt, being the Pt thickness the running parameter.

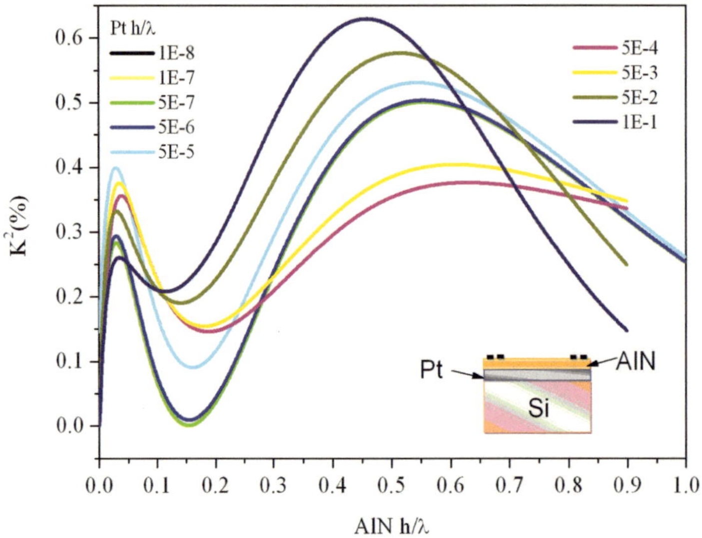

Fig. 4a. The K^2 vs AlN h/λ for SAW propagation along Si/Pt/AlN, for different Pt thickness values normalized to the acoustic wavelength

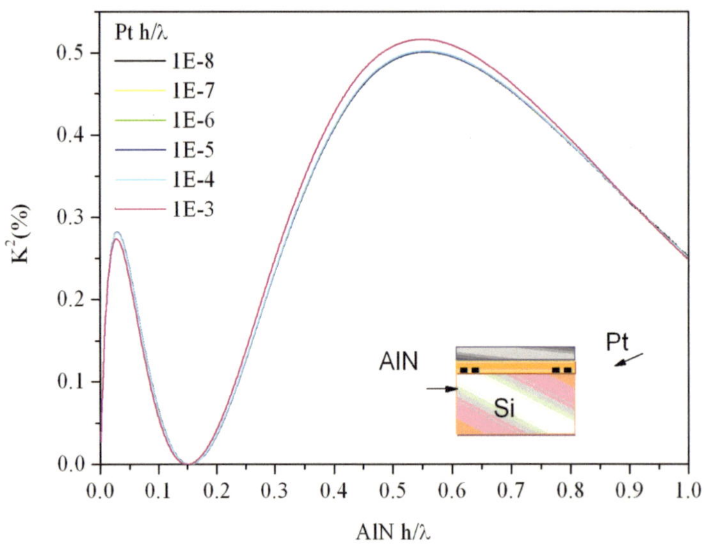

Fig. 4b. The K^2 vs AlN h/λ for SAW propagation along Si/AlN/Pt, for different Pt thickness values normalized to the acoustic wavelength

In figure 4a the highest K^2 values obtainable are in the range 0.5 to 0.63% at ~h/λ = 0.55 for Pt h/λ = 10^{-8} to 10^{-1}. In figure 4b the highest K^2 values obtainable are in the range 0.5 to

0.52% at ~h/λ = 0.555 for Pt h/λ = 10^{-8} to 10^{-3}. The theoretical data shown in figures 3 and 4 have been evaluated using the PC SAW software developed by Mc Gill University [9]. The theoretical K^2 has been approximated as $2 \cdot \frac{v_{ph}^f - v_{ph}^m}{v_{ph}^f}$ where v_{ph}^f and v_{ph}^m are the SAW phase velocities along the free and electrically short-circuited surfaces of the AlN film. The phase velocity v_{ph}^m is obtained by the insertion of a perfectly conductive and infinitesimally thin film at the interfaces where the IDTs and the ground plane are located in each of the four coupling structure. The physical data relative to the elastic, piezoelectric and dielectric constants of AlN film are extracted from [10] and [11] and refer to single crystal AlN thin films grown on the basal plane of Al_2O_3(0001) by metalorganic vapor deposition. The *TCD* of the bulk piezoelectric crystal depends only on the crystal cut and the SAW propagation direction, while that of a layered structures is frequency dispersive. Because the SAW penetration depth inside the propagating medium is about one wavelength, in a layered medium, for $h/\lambda << 1$, the most of the SAW energy is confined to the substrate, while, as h increases with respect to λ, more and more of the SAW energy is confined to the film. For small AlN film thickness ($h/\lambda < 1$) the *TCD* value of the multilayer corresponds approximately to that of the substrate. With increasing the AlN film thickness respect to the acoustic wavelength ($h/\lambda \geq 1$) the *TCD* reaches the AlN *TCD* value. If the film and the substrate show opposite sign TCD values, there will be a h/λ value at which $TCD = 0$ ppm/°C (the temperature compensated point, TCP). This h/λ value represents the film thickness for which the two opposite sign *TCD*s of the film and of the substrate equilibrate to form a thermally compensated structure. Thus high-frequency, enhanced coupling, and thermally compensated elctroacoustic devices can be designed at the proper AlN films thickness values [12].

3. Materials and methods

Highly c-axis oriented AlN films were grown at 180°C and at 200°C by rf reactive sputtering technique on the polished surface of (0001) oriented single crystal Al_2O_3 substrate, on SiO_2/Si(100) and Pt/SiO_2/Si(100) substrates. The AlN deposition process parameters were the following: gas atmosphere of 100% of N_2, high purity (99.999%) 4" diameter Al target disc, RF power 200 watt, background vacuum $5 \cdot 10^{-8}$ Torr and pressure during the deposition process 3×10^{-3} Torr. Before starting the sputtering process, a 30 minute pre-sputtering was performed. The substrate temperature was held at 180 °C during the deposition process. The optimized sputtering parameters ensure AlN films showing a high adhesion to the substrate, c-axis orientation, a columnar growth, smooth surface, and high piezoelectricity; the films are also uniform, stress-free and extremely adhesive to the substrates. The Pt sputtering process parameters were the following: substrate temperature 200°C, gas atmosphere of 100% of Ar, high purity (99.99%) 4" diameter Pt target disc, RF power 150 watt, background vacuum 10^{-7} Torr and pressure during the deposition process 5×10^{-3} Torr. The deposition process of both Pt and AlN films is performed subsequently without breaking the vacuum in order to avoid any oxidation effects of the layers. Then the obtained samples were heated at 900°C in air at ambient pressure by a quartz tube furnace, for different lengths of time. The *cold* (20°C) sample was abruptly put inside the furnace pre-heated at 900°C and the annealing time was measured from the set temperature was reached; then the sample was removed from the furnace and brought abruptly to room temperature. A temperature ramp of 1°C/s was measured by a thermocouple after the insertion of the *cold* sample inside the furnace. The furnace tube was not hermetically sealed, so ambient air was present during the loading and unloading of samples [14]. The structural properties of annealed Pt and AlN films were investigated by by X-ray diffraction measurements (XRD) a Seifert XRD 3003P performed on a a Seifert XRD 3003P diffractometer operating in the Bragg-Brentano geometry using Cu-Kα radiation (λ = 1.5418 Å) and the diffracted intensities were collected in θ-2θ scan mode in the range 20° <2θ < 80° with step size 0.04°. The reflection peaks of the diffractograms were compared with the standards of the JPCDS database. The crystallite size D was calculated from the Scherrer formula $D = \frac{0.9 \cdot \lambda}{B \cdot \cos\vartheta}$ where λ is the wavelength, B the θ-2θ full width at half maximum (FWHM) of the (0002) peak in rad and θ the Bragg angle. The *c* and *a* lattice parameters of the AlN and Pt films were calculated from the angular position of the AlN (002) and Pt (111) diffraction peaks of the θ-2θ scan and compared to the value from ref. 13 for Pt and powder AlN (*a* = 3.9231Å, *c* = 4.979Å). Since the electrical resistivity of the thin conducting films influences the device characteristics (such as insertion loss and Joule heating), the electrical resistance and the surface morphology of the outer Pt electrode were investigated at room temperature after each thermal annealing. The annealing effects on the piezoelectric constant d_{33} of the AlN films were also estimated [14].

3.1 Pt/AlN/Pt/SiO_2/Si

AlN films, 3.15 μm thick, were sputtered on bare and Pt (2200 Å thick) -covered SiO_2/Si(100) substrates, being ~2 μm the silicon oxide thickness; a Pt film (2200 Å thick) was sputtered on the AlN free surface and then the Pt/AlN/Pt/SiO_2/Si multilayers were heated at 900°C in air for lengths of time ranging from 1 to 32 hours. Figure 5 shows the XRD patterns of the as grown and annealed multilayers [14].

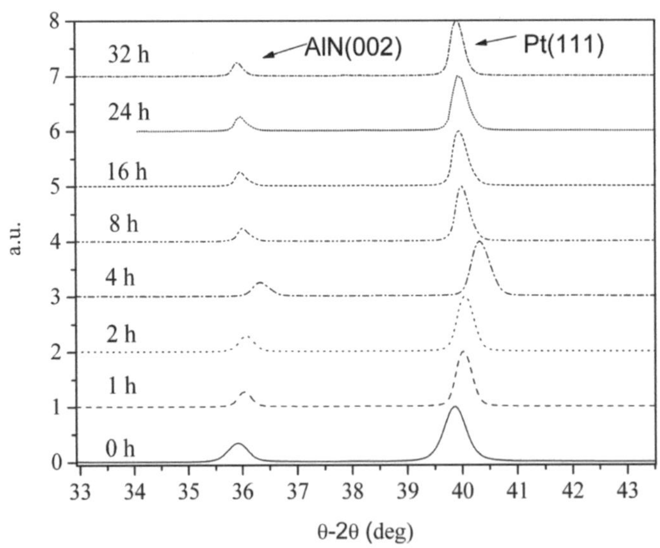

Fig. 5. XRD patterns of Pt/AlN/Pt/SiO_2/Si structures: the running parameter represents the annealing time

The piezoelectric AlN film is *c* axis oriented perpendicularly to the growth plane: in all the samples the AlN (002) and (004) peaks, at ~ 36° and at ~ 76°, are visible even after 32 hours annealing. The peak at 2ϑ ~40° corresponds to the Pt film strongly oriented along the (111) plane; with increasing the annealing time, a very small stress-induced shift in the Pt (111) peak position can be observed, while the same peak becomes narrower indicating a growing (111) fiber texture. As a consequence of the temperature-induced improvement of the Pt local epitaxy, the alignment precision of the AlN film crystalline planes also improves. The XRD data of the outer Pt film showed two peaks at 2θ~ 40° and 46.3° corresponding to the (111) and (200) orientations, and also a small platinum oxide (220) peak, at ~66°, clearly visible after the 1st annealing, that does not increase in percentage after the successive thermal cycles [14]. The AlN and Pt FWHM of the $\vartheta-2\vartheta$ scan before and after the annealing [14] resulted improved, starting from 0.391 and 0.415° of the as grown AlN and Pt films, up to 0.24 and 0.28° after 32 hours annealing. The *c* and *a* lattice parameter of AlN and Pt films of the as-grown samples, calculated from the angular position of the (002) and (111) diffraction peaks, are respectively larger (4.997 Å) and smaller (3.914Å) than the values reported for bulk single crystal AlN and Pt [13], as a consequence of the lattice mismatch between the two materials. Figures 6 and 7 show the FWHM of the AlN(002) and Pt(111) peaks, and the *c* and *a* lattice parameters of AlN and Pt films vs the annealing time.

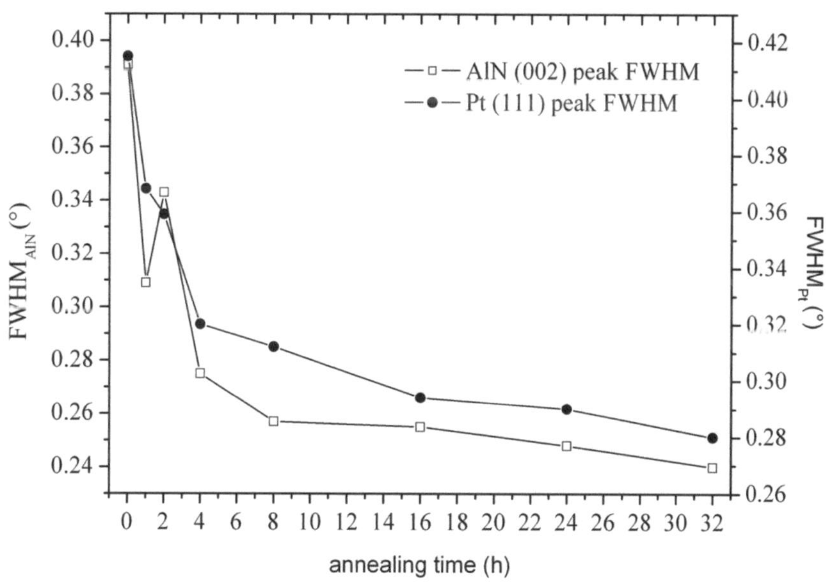

Fig. 6. FWHM of the AlN(002) and Pt(111) peaks vs annealing time

After the first annealings the AlN *c* parameter relaxes to the bulk while the Pt *a* parameter slightly decreases. Further annealings result in a permanent in-plane compressive stress for both the AlN and Pt films. The AlN(002) and Pt(111) peaks' FWHM decreases with increasing the annealing time indicating a decrease of inhomogeneous strain distribution.

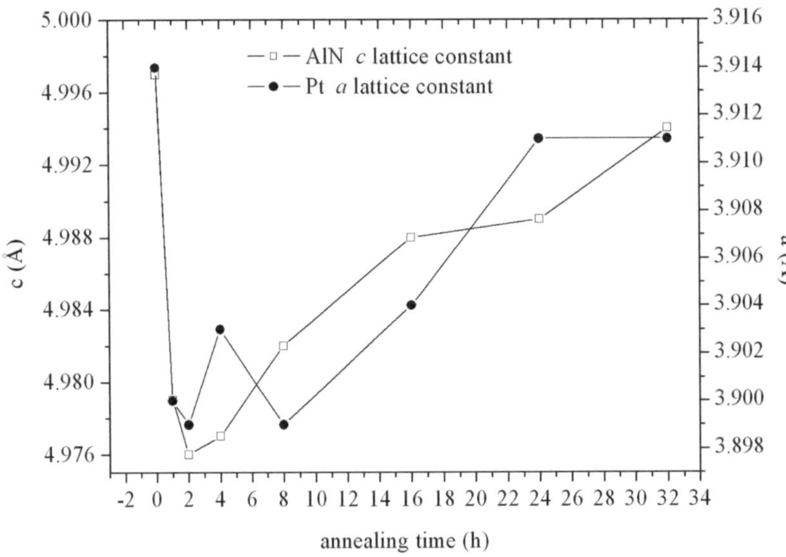

Fig. 7. *c* and *a* lattice parameters of AlN and Pt films vs the annealing time

The measurement of the longitudinal piezoelectric coefficient d_{33f} of the AlN film was done at ambient temperature on the same type multilayer without the outer Pt films, before and after the thermal annealing, with a method described in ref. 15 and based on the direct piezoelectric effect: a longitudinal acoustic wave perturbs the sample via a special probe and the electrical voltage induced in the piezoelectric film is measured. The probe consisted of a metal rod in contact with a $Pb(Zr,Ti)O_3$ (PZT)-based low frequency transducer that was connected to a pulse generator (pulse width 0.1-1.0 ns) to produce longitudinal bursts propagating along the metal rod. The contact between the rod and the piezoelectric film surface resulted in the application of a stress on the surface of AlN films. Stress-induced electrical charges were collected at the piezoelectric film surfaces by electrodes (the metal rod and the conducting substrate) and observed on a scope. The piezoelectric strain constant, d_{33f}, of the tested films was evaluated comparing the film response with the response of a thin single crystal reference sample, whose d_{33} was known. All the tested films showed to be piezoelectric with a difference in the d_{33f} obtained values not appreciable with this measurement technique because of an error of about 15-20 %. The estimated mean value is in the range from 6.2 and 7.4 pC/N, for both the as grown and all the annealed samples: these values well agree with the corresponding value reported in the available literature [16] that is about 6 pC/N.

The sheet resistivity of the Pt layer deposited on the $AlN/Pt/SiO_2/Si$ multilayer free surface was measured at ambient temperature by the four point method, before and after the thermal annealing. It was observed that the sheet resistance values decreases with increasing the annealing time, starting from ~0.6 Ω/sq, refereed to the unannealed samples, to ~0.5 Ω/sq, referred to the 32 hours annealed samples [14]. Scanning electron microscopy (SEM) investigations revealed an average grain size increased with increasing the annealing lasting. The as deposited Pt films have small grain size (about 400 nm), as shown in figure 8, and this high density of grain-boundary affects the high resistivity of the films.

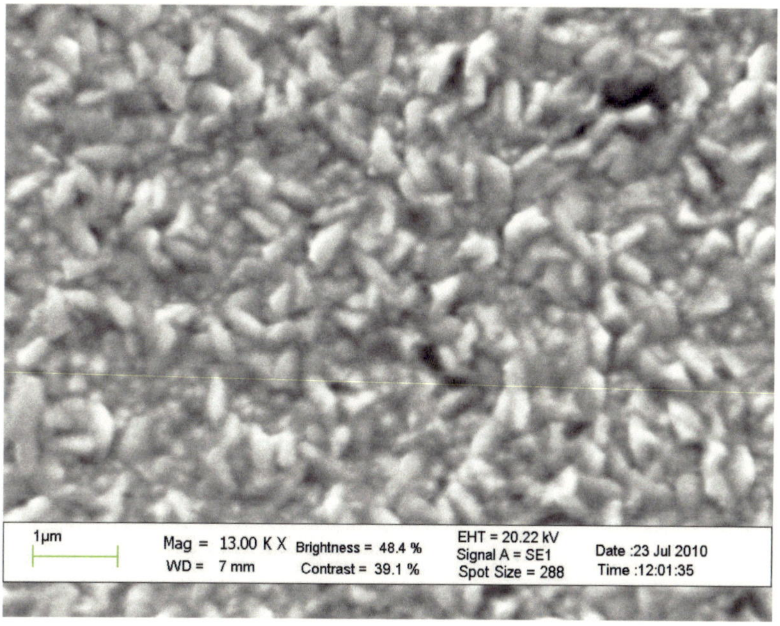

Fig. 8. SEM photo of the as deposited Pt film

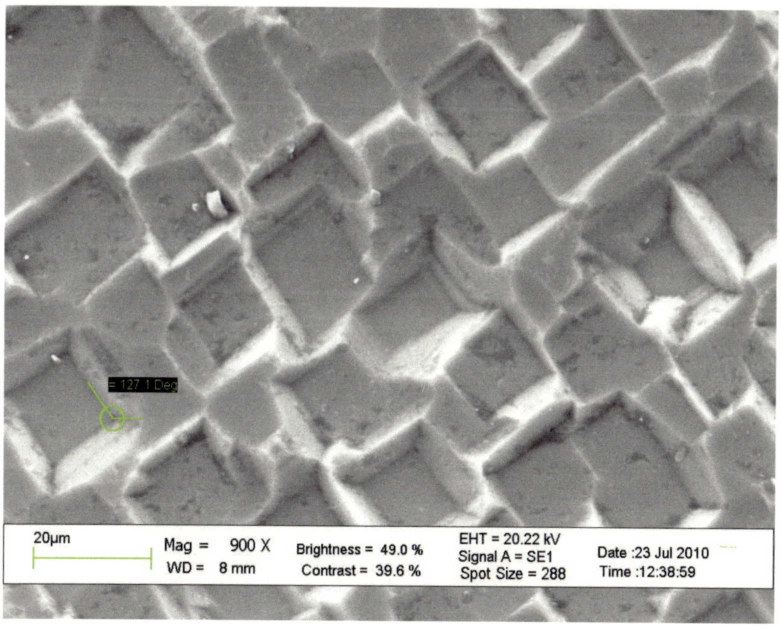

Fig. 9. SEM photo of the 32 hours annealed Pt film

Moreover, the as-grown films contain a number of structural defects that anneal out when heat treatment is carried out and, as a result, the film's resistivity decreases. The Pt annealing results in the relaxation of intrinsic stresses, as well as in the redistribution of structural imperfections: grains coalescence take place and the film sheet resistivity

decreases. SEM photo of the 32 hours annealed sample, shown in figure 9, demonstrates that recrystallization of Pt surface occurs [14].
The unannealed outer Pt film was silvery while the annealed films were matt: this color change can be explained with the increase of the Pt grain size. The adhesion strength of the Pt films was good enough to pass a rudimentary "tape test" with a transparent tape after each anneal as well.

3.2 c-AlN/(0001)-Al_2O_3

Highly c-axis oriented AlN films were grown by rf reactive sputtering technique on the polished surface of (0001) oriented single crystal Al_2O_3 substrate. Then the as grown films, 0.26 to 4.7 μm thick, were thermally annealed at 900°C in air for 1 to 18 hours and the AlN structural characteristics were evaluated after each thermal cycle. References are available in the literature concerning the high thermal annealing (HTA) of AlN on Si performed at 700 – 1200 °C for 2 to 12 hours in controlled atmosphere (in oxygen or nitrogen flux) or in high vacuum; AlN films on Al_2O_3 were heated at 950 °C in air for 30 minutes [20] to 1 hour [21], at 900 to 1200 °C for 10 s in flowing N_2[22], and at 800 °C for 90 minutes in air [23]; bulk AlN was annealed in oxygen at 900 – 1150 °C for 6 hours [24]. In the present work, the HTA of the AlN films on Si and on Al_2O_3 substrates have been performed up to 32 and 18 hours, respectively. No damage was observed on the surface of the AlN/Al_2O_3 film even after 18 hours annealing: the AlN films were still clear, uniform, and extremely adhesive to the substrate. The impact of the annealing on the films structural properties was investigated by XRD before and after undergoing the thermal annealing. The D and *c* parameter of the AlN films thermally annealed for different time periods, are listed in tab. 1.

AlN thickness (μm)	Time (hours)	$FWHM_{\theta-2\theta}$ (deg)	D (Å)	c (Å)
4.7	0	0.384	218	4.995
	4	0.281	297	4.977
	8	0.289	288	4.978
	12	0.290	290	4.978
4	0	0.270	540	4.976
	1	0.269	548	5.024
	3	0.264	555	5.026
	5	0.262	593	5.027
	9	0.261	611	5.037
	18	-	-	-
1.5	0	0.289	252	4.978
	4	0.332	288	4.984
	8	0.350	281	4.989
	10	0.410	303	5.001
0.26	0	0.500	220	4.986
	3	0.422	198	5.048
	4	-	-	-

Table 1. The AlN (002) FWHM, the D and *c* parameter of the AlN films thermally annealed for different time periods

Figures 10, 11 and 12 show the diffraction patterns of the AlN films, 4.0, 1.5 and 4.7 μm thick, annealed for different time.

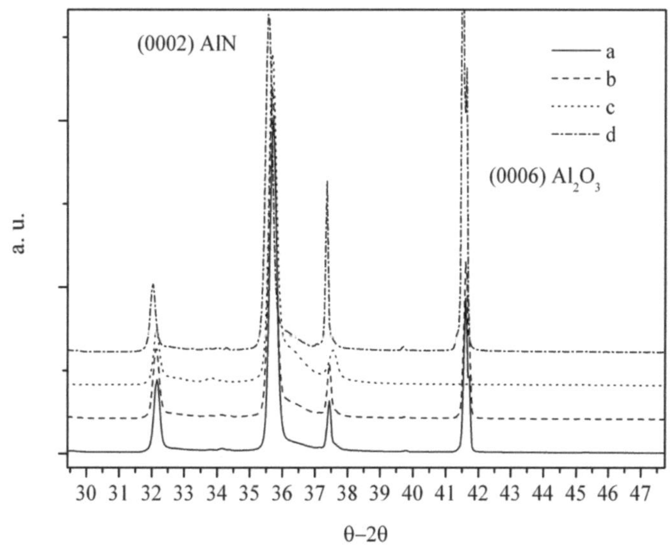

Fig. 10. XRD pattern of the AlN, 4 μm thick, annealed for 1 (a), 3 (b), 5 (c), and 9 (d) hours

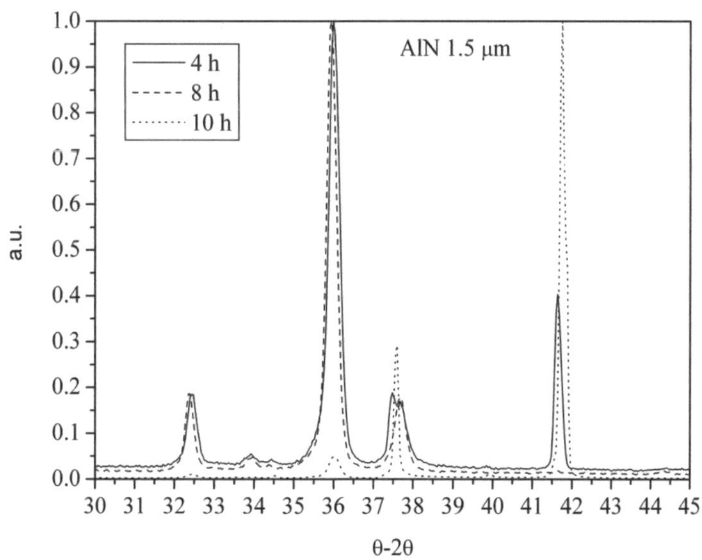

Fig. 11. XRD pattern of the AlN, 1.5 μm thick, annealed for 4, 8 and 10 hours

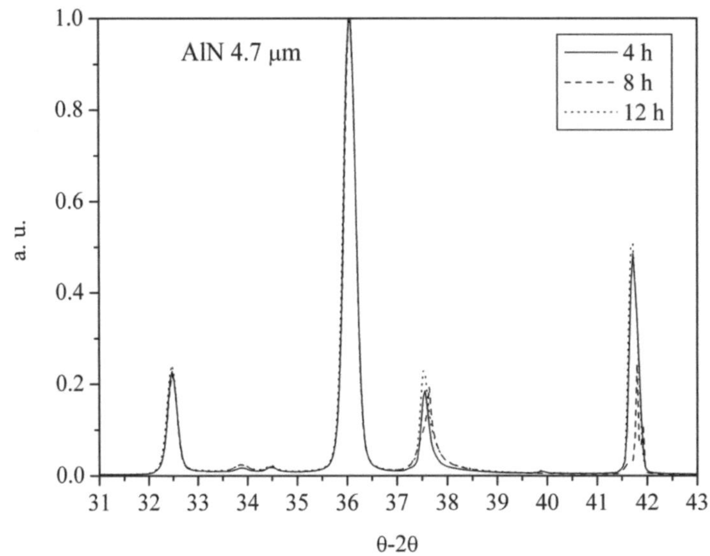

Fig. 12. XRD pattern of the AlN, 4.7 μm thick, annealed for 4, 8 and 12 hours

The AlN (0002) peak at approximately 2θ = 36° of the as deposited films, 0.26 to 4.7 μm thick, showed a FWHM (from θ-2θ scan) in the range 0.5° to 0.27°. Both the as grown and the annealed samples exhibit a strong peak at $2\theta \sim 42°$ due to the (0006) reflection of the Al_2O_3 substrate, and one small peak at 76°, corresponding to the (0004) reflections of the wurtzite AlN structure, not shown in figures 10 to 12. The 2nd order peaks of AlN(0002) and Al_2O_3 (00001) are clearly visible at ~32.4° and ~37.5°. After 4 hours at 900°C in air, the (0002) peak of the thin AlN film (0.26 μm thick) broadens completely as well as that of the 4 μm thick AlN film after 18 hours annealing. No other AlN phases are present, nor AlN-oxide traces are evident in the spectra after the annealing; only a decrease in the 2θ value of the (0002) peak is observed, resulting in an increase in the c lattice parameter whose values are larger than the bulk lattice parameter [13], indicating the presence of compressive stress in the surface plane. This lattice elongation, perpendicular to the growth plane, is usually associated with the intrinsic compressive stress caused by the lattice mismatch between the AlN and Al_2O_3, $(a_{sapphire}-a_{AlN})/a_{sapphire}$, that is approximately equal to 30%. Our previous results [25] have shown that the *c*-axis of the as grown AlN films on sapphire relaxes to the bulk value with increasing the film thickness: an interface layer, strained because of the large lattice mismatch, is formed on the sapphire surface and is followed by a columnar AlN layer which runs through the entire thickness of the film. The thinner the AlN film and more is strained and unable to survive to the HTA.

4. Conclusions

Highly c-axis oriented AlN films and thin film stacks of Pt/AlN/Pt were sputtered at 180°C on (0001)Al_2O_3 substrates and at 200°C on oxidized Si substrates. The multilayers were heated at 900 °C in air up to 32 hours to test their resistance to high temperature. The structural investigation of the annealed films showed that the thermal annealing improved the crystal quality of the AlN films sandwiched between the Pt films, as confirmed by the decreased FWHM of the rocking curves, and the film piezoelectric d_{33f} coefficient resulted unaffected by the temperature. The study of the electrical, morphological and structural characteristics of the Pt electrode revealed a dense surface with a hillock-free morphology, confirming that Pt is the material of choice when a high oxidation resistance is required for metallic components within devices operating at elevated temperatures. The AlN films on sapphire show a lattice elongation, perpendicular to the growth plane, that is usually associated with the compressive stress caused in the growth plane by the lattice mismatch between the film and the substrate. After the first few hours annealing, the FWHM of the (002) AlN peak decreases showing an improvement in the film texture; further annealing results in the FWHM broadening, whose magnitude depends on the film thickness. The structural investigations of thin (0.26 µm) and thick (from 2 to 4.7 µm) annealed AlN films revealed that the film behaviour in harsh environment is strongly affected by the film properties. Thick films, whose structure is more relaxed than the thin one, is able to survive to high temperature without suffering significant deterioration for longer annealing times than the thin one. The obtained results confirm that, during the HTA, the Pt film on the substrate surface is protected by the AlN film, while the Pt film directly exposed to the ambient conditions acts as a protective layer with respect to the AlN film; thus the AlN-based STFM coupling configuration to be implemented on sapphire or silicon substrates is an attractive alternative to langasite and $GaPO_4$ for the development of microwave electroacoustic devices able to work at high temperatures.

5. Acknowledgements

The author wishes to thank Mr. P.M. Latino for his technical support in the development of the technological processes and the HTA.

6. References

[1] Hauser, R. Reindl, L. Biniasch, J., ú *Ultrasonics Symposium*, Vol. 1, 192 (2003).
[2] M. Pereira da Cunha, M.P. Saulo de A. Fagundes, ú *Ultrasonics Symposium*, Vol. 1, 283 (1998).
[3] Maurício Pereira da Cunha, Eric L. Adler and Donald C. Malocha, ú *Ultrasonics Symposium*, Vol.1 169 (1999).
[4] Kar, J.P., Mukherjee, S., Bose, G., Tuli, S., Myoung, J.M., *Materials Science and Technology*, 25 1023 (2009).
[5] Zhigang Lin and Yoon, R.J., International Symposium on *Advanced Packaging Materials: Processes, Properties and Interfaces*, 156 (2005).
[6] C. Zolper, D. J. Rieger, and A. G. Baca S. J. Pearton and J. W. Lee R. A. Stall, *Appl. Phys. Lett.* 69, 538 (1996).
[7] S.M.Middelhoek, S.A.Audet, *Silicon sensors*, Academic Press, London 1989.
[8] M.J.Vellekoop, E.Nieuwkoop, J.C.Hsaartsen, A.Venema, ú *Ultrasonics Symposium*,Vol. 1, 375 (1981).
[9] E.L. Adler, G.W. Farnell, J. Slaboszewicz, C.K. Jen, ú *Ultrasonics Symposium*, Vol. 1, 103 (1982).
[10] K. Tsubouchi, K. Sugai, N. Mikoshiba, ú *Ultrasonics Symposium*, Vol. 1, 375 (1981).

[11] J.G. Gualtieri, J.A. Kosinski, A.Ballato, ú *Ultrasonics Symposium,* Vol. 1 403 (1992).
[12] C. Caliendo, *Appl. Phys. Letters* 92 103501 (2008).
[13] JPCDS cards No. 25-1133 Powder Diffraction File, Joint Committee on Powder Diffraction Standards, ASTM, Philadelphia, PA, 1967, Card 25-1133
[14] C. Caliendo, P. M. Latino, *Thin Solid Films* 519 (2011), pp. 6326-6329
[15] C.K.Xu, V.N.Umashev, I.B.Yakovkin, *Sov. Phys.* PTE6, 192 (1986).
[16] Landolt-Bornstein, *Numerical Data and Functional Relationships in Science and Technology, Group III: Crystal and Solid State Physics,* (Springer-Verlag Berlin, Heidelberg-New York, 1979), Vol. 11 (1979).
[17] C.-Y. Lin, F.-H. Lu, *J. Eur. Ceram. Soc.* 28 691 (2008).
[18] E. A. Chowdhury, J. Kolodzey, J. O. Olowolafe, G. Qiu, G. Katulka, D. Hits, M. Dashiell, D. van der Weide, C. P. Swann, K. M. Unruhm, *Appl. Phys. Lett.* 70 2732 (1997).
[19] F. Jose, R Ramaseshan, S. Dash, S. Bera, A. K. Tyagi, B. Raj, *J. Phys. D: Appl. Phys.* 43 075304 (2010).
[20] T. Aubert, O. Elmazria, B. Assouar, L. Bouvot, M. Oudich, *Appl. Phys. Lett.* 96 2035031 (2010).
[21] S.-K. Tien, C.-H. Lin, Y.-Z. Tsai and J.-G. Duh, *J. Alloys Compd.* 489 237 (2010).
[22] Z. Gu, J.H. Edgar, S.A. Spekman, D. Blom, J. Perrin, J. Chaudhuri, *J. Electron. Mater.* 34 1271 (2005).
[23] T. Aubert, M. B. Assouar, O. Legrani, O. Elmazria, C. Tiusan, S. Robert, *J. Vac. Sci. Technol. A* 29 021010 (2011).
[24] B. Liu, J. Gao, K.M. Wu, C. Liu, *Solid State Commun.* 149 715 (2009).
[25] C. Caliendo, P. Imperatori, *J. Appl. Phys.* 96, 2610 (2004).

27

Applications of In–Fiber Acousto–Optic Devices

C. Cuadrado-Laborde[1,2], A. Díez[1], M. V. Andrés[1],
J. L. Cruz[1], M. Bello-Jimenez[1], I. L. Villegas[1,3],
A. Martínez-Gámez[3] and Y. O. Barmenkov[1,3]

1. Introduction

Nowadays, in-fiber acousto-optic devices are increasingly used as frequency shifters, multiplexers, modulators, and tunable filters. They can be easily spliced into optical fiber systems, and the consequent low insertion loss, make them an attractive alternative to bulk optics devices. Our group, established at the Institute of Materials Science, Department of Applied Physics, of the Valencia University (ICMUV, Valencia, Spain), has been involved in this field for the last ten years, and our fabrication facilities allow us the development of new in-fiber acousto-optic devices for novel applications in different fields such as sensors, microwave photonics, lasers, and optical communications. Our aim here is to present the great potential shown by in-fiber acousto-optic devices for different photonic applications. Although this chapter is focused in our latest developments, the reader will also find discussed the work done by other research groups in the field. This chapter is divided in two main sections: Section 2 is focused on novel applications of acousto-optic fiber devices based on flexural acoustic waves, and Section 3 is focused on applications of the acousto-optic devices based on the interaction of longitudinal acoustic waves with fiber Bragg gratings (FBG). Finally, our conclusions are shown in Section 4.

2. Applications of acousto-optic devices based on flexural acoustic waves

In a standard single mode fiber, when a flexural acoustic wave propagates along an optical fiber a periodic perturbation is introduced in its refractive index, and it can induce coupling between the fundamental mode guided by the core and the modes supported by the cladding (Kim et al., 1997). This acousto-optic interaction can be seen as the dynamic counterpart of a long period grating (LPG). LPGs are usually fabricated by creating a periodic perturbation of the refractive index by UV radiation; this of course fixes its spectral characteristics. On the other hand, when the perturbation is introduced by an acoustic wave, its period and strength can be controlled through the frequency and amplitude of the acoustic wave, respectively. Thus, the spectral properties of the optical device can be controlled dynamically through the characteristics of the acoustic perturbation. The optical

[1]*Departamento de Física Aplicada y Electromagnetismo, ICMUV,*
Universidad de Valencia, Burjassot, Valencia, Spain
[2]*CONICET La Plata, Buenos Aires, Argentina,*
[3]*Centro de Investigaciones en Óptica, León, Guanajuato, Mexico*

coupling is resonant in wavelength; it takes place at the optical wavelength that verifies the phase-matching condition between the beat length of the two optical modes and the acoustic wavelength. At the output of the acousto-optic device, only the light that remains guided by the core mode is transmitted. Thus, the coupling of power from the fundamental mode to a cladding mode results in the appearance of an attenuation notch in the spectrum, whose amplitude is fixed when travelling acoustic waves are used. The characteristics of the fundamental flexural acoustic mode of an optical fiber fulfill the above requirements for the development of efficient in-fiber acousto-optic devices. In subsection 2.1 we will describe a *Q*-switched fiber laser where cavity loss modulation is reached by using this operation principle. As an interesting alternative, if standing –instead of travelling– acoustic waves are used, light transmitted at the resonance wavelength will experience an amplitude modulation at twice the frequency of the acoustic wave. In subsection 2.2 we describe a mode-locked fiber laser based on the use of this operation principle.

2.1 *Q*-switching by intermodal acousto-optic modulation in an optical fiber

Q-switched fiber lasers have attractive applications in different fields, such as in remote sensing and material processing. The mechanism of short pulse emission is based on the modulation of the *Q* factor of the cavity, which can be done either passively or actively (Siegman, 1986). In the former the setups are simpler, but the repetition rate only varies with the pump power of the medium gain. Further, they usually show long-term instability, and frequently the amplitude of the pulses is randomly modulated in time (Vicente et al., 2004). On the other hand, active *Q*-switching is independently and accurately controlled by an electrical signal, which triggers the modulator. The bulk approach, such as electro-optic (Kee et al, 1998) or acousto-optic modulators (Álvarez-Chávez et al., 2000), is not adapted to compact fiber laser systems required nowadays; further they have large optical coupling losses and stringent alignment requirements. For these reasons, the all-fiber approach is of permanent interest, being advantageous in terms of cost, loss, packaging, robustness, and simplicity. One solution widely investigated has been enabled by the use of FBGs as cavity mirrors. In this way, the tuning of the wavelength of one of the FBGs has been used to achieve active *Q*-switching (Andrés et al., 2008). Among the typical tuning methods, we could mention the stretching of fiber Bragg gratings by magnetostrictive materials (Pérez-Millán et al., 2005a; Andersen et al., 2006), by piezoelectric actuators (Imai et al., 1997; Russo et al., 2002), or by the interaction of longitudinal acoustic waves with the FBG (Delgado-Pinar et al., 2006; Cuadrado-Laborde et al., 2007). Other technique used for active *Q*-switching relied in the cavity loss modulation by the core-to-cladding mode-coupling in optical fibers induced by travelling flexural acoustic waves (Huang et al., 2000; Zalvidea et al., 2005). As opposed to the last two referred works –which were focused at the erbium lasing wavelength–, here we experimentally demonstrate the possibility of using this modulating technique at the technologically relevant ytterbium lasing wavelength (Tuchin, 1993). Very few actively *Q*-switched ytterbium-doped lasers with an all-fiber configuration have been reported (Andersen et al., 2006). The main difficulties arise from a relatively long time response and a limited modulation depth of all-fiber amplitude modulators. Here, we present the experimental results that we have obtained in a research work focused on the exploitation of in-fiber acousto-optics. In our proposed *Q*-switched fiber laser, when the acoustical signal is switched-off, the optical power losses within the cavity are reduced, and then a laser pulse is emitted (Villegas et al., 2011b).

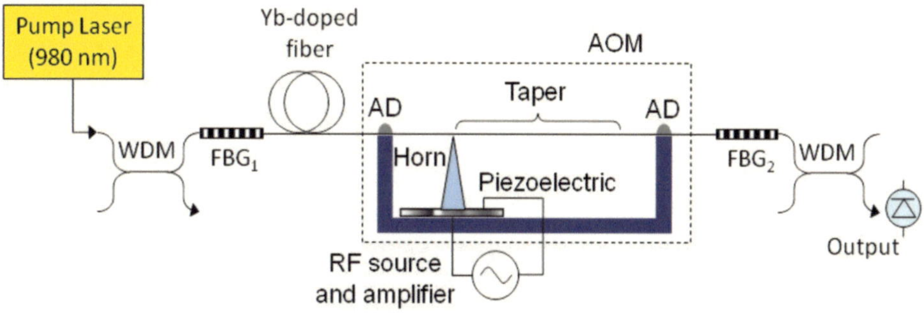

Fig. 1. *Q*-switched fiber-laser setup; the acousto-optic modulator is defined by the elements inside the dashed line; AD stands for acoustic dumper

The setup used for our *Q*-switched fiber laser is schematically illustrated in Fig. 1. The gain was provided by 0.65 m of a heavily-doped ytterbium-doped single-mode fiber –Nufern SM-YSF-HI, cut-off wavelength of 860 ± 70 nm, numerical aperture of 0.11, and fiber absorption of 250 dB/m at 975 nm–. The active fiber was pumped through a WDM coupler by a pigtailed laser diode emitting at 980 nm, providing a maximum pump power of 110 mW. The in-fiber acousto-optic modulator (AOM) was spliced between FBG_1 –Bragg wavelength at 1064 nm, FWHM of 0.23 nm, and 99.6 % of maximum reflectivity– and FBG_2 –Bragg wavelength at 1064 nm, FWHM of 70 pm, and 44 % of maximum reflectivity–, defining in this way a Fabry-Perot cavity. The AOM in turn is composed of an RF source, a transversal-mode piezoelectric disk, an aluminum horn, and a tapered single-mode optical fiber –Fibercore SM980 of low numerical aperture (0.13-0.15) –. The optical fiber was tapered down by the fusion and pulling technique using a travelling flame, which produces a taper waist with a uniform diameter of 76 μm and 0.1 m length, for this specific case. The tip of the aluminum cone –with the piezoelectric disc fixed to its base–, was glued to an uncoated section of fiber near the taper, see Fig. 1. Finally, the optical fiber in the AOM was acoustically dumped in both extremes; see Fig. 1, in order to prevent unwanted acoustical reflections.

When an RF signal is applied to the piezoelectric disc, a travelling flexural acoustic wave is launched through the taper. If the acoustic wavelength matches the beat-length between the fundamental mode guided by the core and one of the optical modes supported by the cladding, then light coupled to the later remains in the cladding downstream of the taper, being finally absorbed by the fiber coating (Birks et al., 1994). Thus, the coupling of power from the fundamental mode to a cladding mode results in the appearance of an attenuation notch in the spectrum. When the acoustic frequency is varied, the periodicity of the perturbation also does, and hence the phase-matching condition is shifted to a different optical wavelength. Figure 2(a) shows the tunability of the notches caused by the coupling between the fundamental core mode and the first three cladding modes LP_{1m}. The selected operating point was at an optical wavelength of 1064.1 nm for an RF signal applied to the piezoelectric of 825 kHz. These measurements were made by illuminating the taper with a broadband light source and detecting the light transmitted through the taper with an optical spectrum analyzer. As an example, Fig. 2(b) shows one of the measured transmission spectra for an applied voltage to the piezoelectric of 26 V and a frequency of 885 kHz. The transfer of optical power from the fundamental core mode to one of the cladding modes behaves periodically as a function of the acoustic power, which in turn is a function of the applied voltage to the piezoelectric. Figure 2(c) shows this effect for the selected operation point marked in Fig. 2(a). As it can be observed, the transmittance decays by a maximum of 12 dB for an applied voltage to the piezoelectric of 50 V (peak-to-peak measurement). Beyond this point, further increment in the applied voltage raises the transmittance again. Therefore, and bearing in mind its use as a *Q*-switching device, cavity loss modulation between 0 dB and 12 dB can be achieved by applying to the piezoelectric a sinusoidal signal at the frequency of 825 kHz fully-modulated by a rectangular signal, see the inset of Fig. 2(c). This modulation produces on–off periods of the acoustic wave travelling down the fiber, which results in a modulation of the cavity losses at the resonance wavelength. In passing, we should mention that several tapers were fabricated in order to better optimize this laser system, by using Fibercore SM980 of high numerical aperture (0.17-0.19), and Nufern SM-YSF-LO (a moderately ytterbium-doped fiber), and in turn with different taper waists. Despite this, the minimal decay in transmittance was always around the reported values, i.e. between 10 and 16 dB.

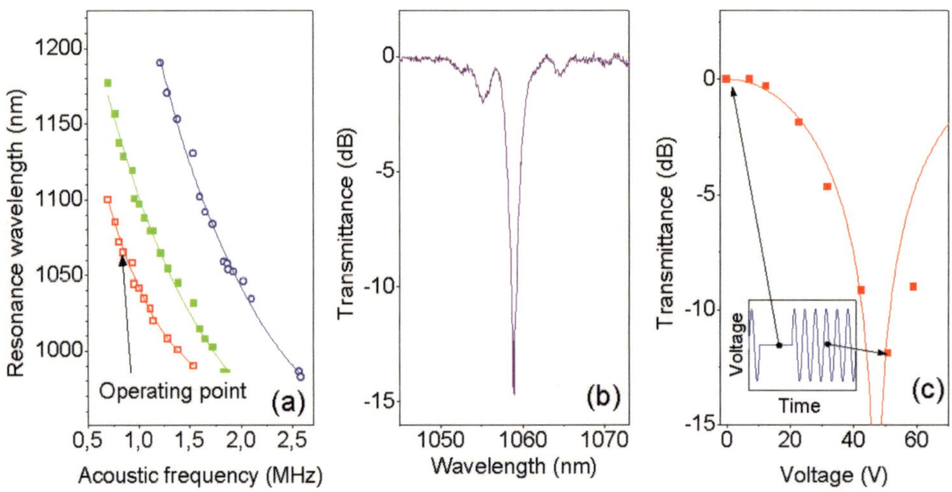

Fig. 2. (a) Resonant optical wavelengths as a function of the acoustic frequency for the first three mode-couplings LP_{01} -LP_{1m}. (b) Typical transmission notch caused by the first mode-coupling by applying to the piezoelectric a sinusoidal signal at 885 kHz and 26 V. (c) Transmittance at the selected operation point marked in (a) –i.e. 1064.1 nm and 825 kHz– as a function of the applied voltage to the piezoelectric (solid scatter points), the curve represents a theoretical fitting according to a $\sin^2$ function

The switching time is one of the key parameters for any modulator intended to be used as a *Q*-switching device in a fiber laser. Preferably, it should be as short as possible. For this reason we measured the temporal response of this device, by detecting the transmitted light, while simultaneously registering the modulating signal in an oscilloscope. It takes 25 μs to increase/decrease the transmitted optical power through the taper when the voltage applied to the piezoelectric is switched-off/on, respectively. This time corresponds reasonably well with the time it takes a flexural acoustical wave to travel down the taper, i.e. 0.1 m / 3764 m/s ≈ 26 μs. This value is short enough to allow the use of this modulator as a *Q*-switching device in a fiber laser, as it will be demonstrated in the following section. In principle, and for a given taper waist, shorter switching times could be achieved by decreasing the interaction length. Unfortunately, this simultaneously increases the acoustical power needed to reach the same optical coupling power, therefore there is a trade-off between both parameters.

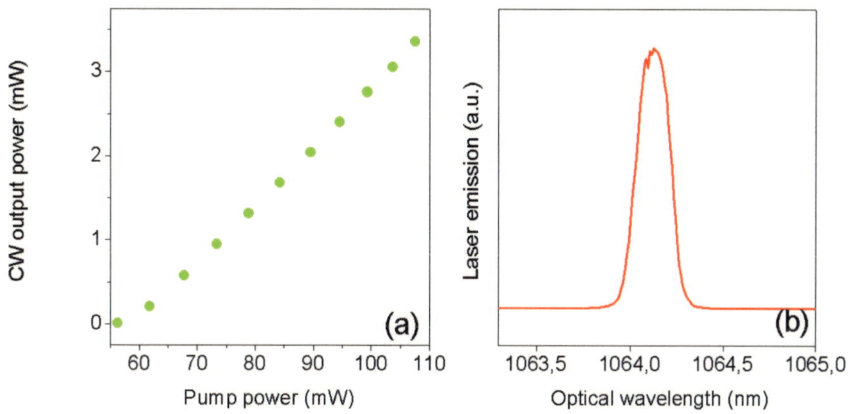

Fig. 3. (a) Output optical power in CW emission as a function of the pump power. (b) Emission spectrum

In this setup, both gratings are permanently tuned to the same wavelength by using two translational stages, one for each FBG; see the setup in Fig. 1. In this way, for zero voltage applied to the piezoelectric, this laser emits in CW, since then the cavity losses are minimal, see the transmittance in Fig. 2(c) for 0 V. Figure 3(a) shows the optical power in CW emission as a function of the pump power. There is a pump power threshold of 56 mW, about which the laser starts to emit, reaching a maximum output power of 3.4 mW, which in turn is determined by the maximum pump power available in our setup (110 mW). Figure 3(b) shows the spectrum of CW emission; its linewidth is of 0.21 nm at a center optical wavelength of 1064.1 nm.

Now, we discuss the *Q*-switched operation of this laser. To this end, we modulated the cavity losses by applying to the piezoelectric a fully-modulated sinusoidal signal at the frequency of 825 kHz and 50 V. A rectangular wave was used to modulate the RF voltage that generates the acoustic wave. At a given frequency of the modulating signal, we found always a maximal duty cycle –i.e. the fraction of time that the signal is in its high level– able to perform *Q*-switch correctly. If we decrease this duty cycle, then the cavity would stay in its high *Q* state longer, and more than one *Q*-switched pulse would be emitted in each time slot. The *Q*-switch repetition rate becomes determined by the frequency of the modulating signal; with this configuration we reached continuous tuning of the *Q*-switch repetition rate in the range 1-10 kHz. Figure 4(a) shows, as an example, a *Q*-switched optical pulses train at 1 kHz, together with the corresponding modulating signal, for a pump power of 59 mW. Figure 4(b) shows a detail of a single *Q*-switched optical pulse of the train shown in Fig. 4(a) with a time width (FWHM) of 3.72 μs. The pulses have a quasi-Gaussian profile; the fitting by this function is also shown in Fig. 4(b). The effect of pump power on the *Q*-switched pulses, for different repetition rates in the range 1-10 kHz, is shown in Fig. 5. For each *Q*-switching frequency there is a pump power threshold. Above threshold, the peak power increases with pump power, and there is a corresponding reduction of pulse width. For a given frequency there is also a pump power level beyond which extra pulses appear, the curves are truncated at that point. In order to overtake this limitation, a modulator with improved modulation depth is required. In addition, our laser produces easily multiple pulse emission, which is a well-known problem in actively *Q*-switched lasers with relatively large time responses. A large time response makes critical the adjustment of the duty cycle of the modulation voltage when the pump power is increased, particularly at low repetition rates. Consequently, a faster switching response of the acousto-optic modulator is required in order to improve the operation of the laser.

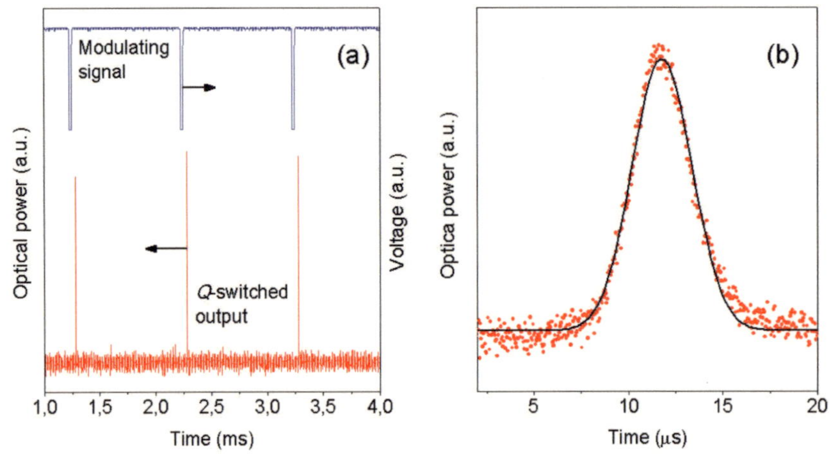

Fig. 4. (a) Modulating signal (above) at 1 kHz repetition rate together with the generated *Q*-switched laser output (below). (b) Detail of a single *Q*-switched pulse of (a) (scatter points) together with its corresponding fitting by a Gaussian function (solid curve)

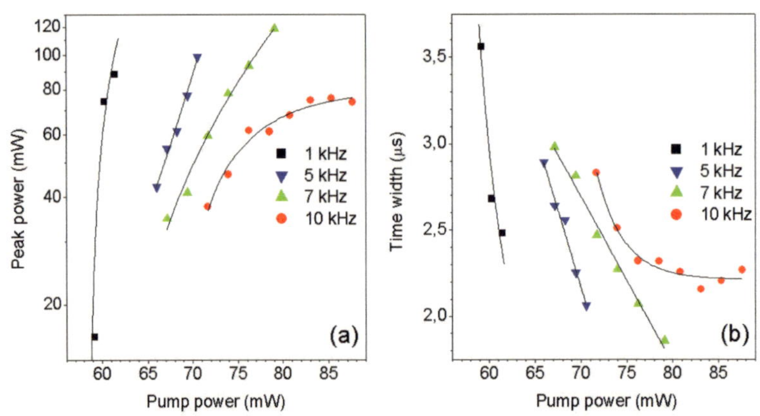

Fig. 5. Peak power and time width in *Q*-switched operation as a function of the pump power, (a) and (b), respectively, for several repetition rates

2.2 Mode-locking by intermodal acousto-optic modulation in an optical fiber

Mode-locking lasers have a vast number of applications, ranging from the telecommunications industry to medical surgery (Bonadeo et al., 2000; Haus, 2000; Yu et al., 2000; Schaffer et al., 2003). The shortest pulses are obtained by passive schemes. However, they also have some drawbacks, among them, they are generally unstable, they have higher timing-jitter between pulses and usually self-starting cannot be assured. An alternative to using passive schemes that reduces jitter and allows for synchronization is to use an active scheme. To this end, a modulator is driven at the cavity's fundamental repetition rate; therefore, timing-jitter is reduced since each pulse is triggered by the modulator, which in turn is accurately driven by an electrical signal. In this way, in active mode-locking one has the possibility to synchronize a laser to an external clock or electronic signal (French, 1995). Among all these lasers, the all-fiber solution has been the subject of intensive research due to their compact structure, free alignment problems, and low cost, between other merits. Earlier and subsequent active modelocked fiber lasers were not strictly all-fiber lasers, since they currently included a bulk modulator (Geister & Ulrich, 1988; Phillips et al., 1989a; Hudson et al., 2005). This not only induces extra losses into the laser cavity but of course also destroys the benefits of an all-fiber configuration. On the contrary a strictly all-fiber modulator can overcome these limitations and it can be used for actively mode-lock rare-earth-doped silica fiber lasers (Phillips et al., 1989b; Culverhouse et al., 1995; Jeon et al., 1998; Costantini et al., 2000; Myrén & Margulis, 2005; Cuadrado-Laborde et al., 2009a, 2010a, 2010b; Bello-Jiménez et al., 2010, 2011). Table 1 shows the major achievements towards the obtaining of all-fiber actively mode-locked lasers. As it can be observed, acousto-optic devices have been by far the preferred technique, with exception of the Myrén and Marguli`s work relying in a electro-optic effect. Our group was involved in works described in the last two rows of this table, which will be discussed in detail in this chapter.

In Jeon`s et al. work (1998), mode-locking is achieved by frequency shifting based on travelling flexural acoustic waves in a two-mode optical fiber; such a frequency shifter could be considered an improved version of the one reported in (Kim et al., 1986). Unlike Jeon (1998) and Kim (1986), here we propose using a standard single-mode optical fiber in which core-to-cladding mode-coupling is induced by a standing flexural acoustic wave. In this way, the device does not work as a frequency shifter, but as an amplitude modulator, which is used to actively mode-lock a fiber laser. In the following, we demonstrate that this simple acousto-optic modulator proves to be particularly well-suited for active mode-locking purposes (Bello-Jiménez et al., 2010, 2011).

The setup of the actively modelocked fiber ring laser is schematically illustrated in Fig. 6. The medium gain was provided by 4.36 m of erbium-doped fiber (EDF) containing 300 ppm Er^{3+}, with a cut-off wavelength of 939 nm, and a numerical aperture of 0.24. The active fiber was pumped through a WDM by a 980 nm pigtailed laser diode, providing a maximum pump power of 600 mW. Next –and following a clockwise direction– it was inserted the acousto-optic modulator between two polarization controllers (PCs), followed by a delay line fusion-spliced to port one of a polarization-independent three-port optical circulator (OC). The OC not only forces the unidirectional operation within the ring but simultaneously incorporates the FBG filtering in the cavity by the second port. Finally, the ring cavity is closed by connecting port three of the OC to the WDM. The output of this laser is obtained by transmission through the FBG –Bragg wavelength at 1549.5 nm, FWHM of 0.45 nm, and 50 % of maximum reflectivity–. In this way, the entire laser is spliced. The AOM in turn is composed of an RF source, a piezoelectric disk, an aluminum conical horn, and a 0.2 m long standard optical fiber stripped of its polymer coating; in order to prevent the attenuation of the acoustic wave. The horn is attached to the piezoelectric, and it focuses the vibrations into the fiber through its tip, which is glued to the uncoated fiber. In order to allow the generations of a standing flexural acoustic wave, the uncoated optical fiber of the AOM was firmly clamped at one extreme, whereas in the other it was damped, see Fig. 6.

Year	Group	Mode-locking mechanism	Major achievements
1989	Phillips et al.	Acousto-optically induced intra-fiber phase modulation	First all-fiber actively mode-locked laser, 200 ps time width pulses
1995	Culverhouse et al.	Acousto-optically induced frequency shifting in an all-fiber four-port null coupler	18 ps time width pulses (chirped)
1998	Jeon et al.	Acousto-optically induced mode-coupling in a two-mode optical fiber	Harmonic mode-locking, 3.8 ps time width pulses
2000	Costantini et al.	Acousto-optically induced intra-fiber phase modulation	Visible up-conversion laser, 550 ps time width pulses
2005	Myrén and Margulis	Electro-optic effect in an all-fiber Mach-Zehnder interferometer	1 ns time width pulses
2009	Cuadrado-Laborde et al.	Acousto-optic super-lattice effect in a fiber Bragg grating	630 ps un-chirped time width pulses. First to demonstrate double-active all-fiber *Q*-switching and mode-locking
2010	Bello-Jiménez et al.	Acousto-optically induced mode-coupling in a single-mode optical fiber	34 ps time width pulses

Table 1. Major achievements in the development of all-fiber actively mode-locked lasers

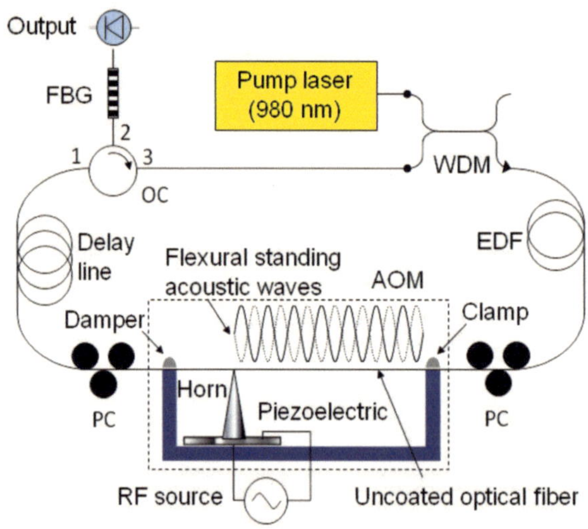

Fig. 6. Modelocked fiber-laser setup; where OC stands for optical circulator, EDF for erbium doped fiber, PC for polarization controller, FBG for fiber Bragg grating, and WDM for wavelength division multiplexer. The acousto-optic modulator is defined by the elements inside the dashed line

Figure 7(a) shows the modulation depth as a function of light wavelength for an acoustic frequency of 2.37315 MHz. The voltage applied to the piezoelectric was 18 V (whenever we refer to voltages, it is a peak-to-peak measurement). The measurement was performed by illuminating the AOM with a tunable laser and detecting the transmitted light at each wavelength. The modulation is maximal at the phase-matching wavelength of 1551 nm, and symmetrically decreases for longer and shorter wavelengths. It should be emphasized the high modulation depth achieved in this technique, reaching a maximum of 72 %, together with the relatively broad operative bandwidth of 1.5 nm (FWHM) –i.e. 187 GHz at the operation wavelength–. Moreover, the insertion loss is as low as 0.7 dB, corresponding to the maximum transmittance value attained, which is 86 %. When the acoustic frequency is varied, the periodicity of the perturbation also does, and hence the phase-matching condition is shifted to a different wavelength. Figure 7(b) illustrates this effect; the rate of change for the wavelength shift is of –0.169 nm/kHz; the maximum modulation reached at each specific operation point is also shown to the right. The differences between the modulation depths at different acoustic frequencies are mainly originated from the combination of the non-flat frequency response of the piezoelectric and the acoustic fiber resonator.

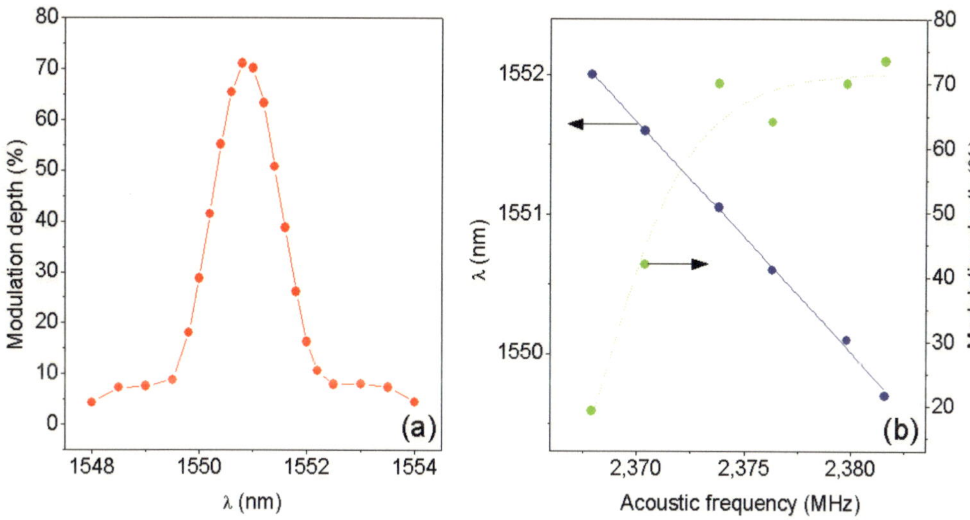

Fig. 7. (a) Modulation depth as a function of the optical wavelength for an RF signal applied to the piezoelectric of 2.37315 MHz and 18 V. (b) Resonant optical wavelengths as a function of the frequency of the RF signal, its corresponding modulation depth is also shown to the right ordinate

Now we discuss the operation of the modelocked laser shown in Fig. 6. The delay line length must be selected to match the round-trip time with the reciprocal of the optical modulation frequency, which in turn is twice the electrical frequency applied to the piezoelectric. The selected operation point for the piezoelectric is the same as in Fig. 7(a), i.e. 2.37315 MHz, and then the cavity length results in 43.7 m. However, fine tuning is always necessary to match the frequency of the modulation signal to the inverse of the round-trip time of the cavity. A translation stage was also used for matching the reflection band of the FBG to the operative wavelength (1551 nm). Regarding the dispersion of this laser; we measured the dispersion of each fiber and component within the cavity by the frequency-domain modulated-carrier method. The laser cavity is a mixed-dispersion fiber ring with an average cavity dispersion of –1 ps/nm/km (Bello-Jímenez et al., 2010). We measured also both the dispersion of the OC and WDM, both resulted normal also with a group delay per wavelength unit of –0.3 ps/nm and –0.04 ps/nm, respectively. In this way, we can assure that this laser works under the normal dispersion regime. Figure 8(a) shows the RF signal used to drive the piezoelectric together with the modelocked train of pulses generated (50 GHz bandwidth oscilloscope). As it can be observed, the frequency of the optical train (4.756 MHz) is twice the frequency of the signal used to drive the piezoelectric (2.37315 MHz). The optical pulses were best fitted by a sech^2 function; neither Gaussian nor Lorentzian functions improved the fitting. This is an expected profile for the optical pulses in this type of dispersion-mixed cavities, in which both kind of dispersion coexist, i.e. normal and anomalous (Bélanger, 2005). On the other hand, according to active mode-locking theory, the time width τ of the optical pulse can be expressed as (Kuizenga & Siegman, 1970):

$$\tau = \frac{\sqrt{\sqrt{2}\ln 2}}{\pi}\sqrt[4]{\frac{g_0}{\delta_m}}\frac{1}{\sqrt{f_m \Delta f_m}}, \tag{1}$$

where f_m is the modulation frequency, g_0 is the gain of the active medium, Δf_m is the modulator bandwidth, and δ_m is the modulation depth. As a rough estimation, let us compare these results with the results reported in Cuadrado-Laborde et al. (2009a), where optical pulses of 780 ps were obtained by using also AM mode-locking, but with a different modulator and setup. Thus, by comparing them, here there is a higher modulation depth (×7), modulation bandwidth (×240), and gain (×3), but a lower modulation frequency (×0.5) –where the additional assumption that gain scales with the EDF length was made–. Therefore, there is a narrowing factor for the time width of $(3/7)^{1/4}\times(0.5\times240)^{-1/2} = 7.4\times10^{-2}$. In this way, comparatively, it should be possible to reach optical pulses around $7.4\times10^{-2}\times780$ ps ≈ 58 ps in this laser, which could be taken roughly as a lower limit for this configuration. One key element, out of the preceding discussion, is the FBG bandwidth, which acts as a filter –see the setup in Fig. 6 –whose bandwidth must be selected narrower than the AOM bandwidth; otherwise modes not amplitude modulated would interfere in the mode-locking process. This evidently limits the effective bandwidth and raises the minimum pulse width able to be reached with this configuration, since modulator bandwidth and time width are inversely related. The optical pulse's peak power and temporal width as a function of the pump power are shown in Fig. 8(b). The time width as well as the peak power monotonically increases with the pump power; at the lowest pump powers, pulses as narrow as 95 ps were obtained with a pump power of 110 mW; see the inset of Fig. 8(b). Pump powers below 110 mW precludes mode-locking; whereas the upper limit is reached by the maximum pump power available in our setup.

Next, we analyzed the dependence of the pulse´s parameters with the applied voltage to the piezoelectric, see Fig. 9(a). Both the transmittance and modulation depth are functions of the applied voltage to the piezoelectric through a periodic relationship. However, we can consider that the relationship is linear within the range of voltages of Fig. 9(a). When the voltage increases, the modulation also does, and pulses become narrower, as expected from AM mode-locking theory, see Eq. (1). Finally, it is of practical interest to quantify the maximum allowable detuning ($\Delta\nu_{max}$), i.e. the maximum difference between the reciprocal of the cavity round trip´s time and modulation frequency that sustain mode-locking. Figure 9(b) shows the variation of the pulse´s parameters under frequency detuning, resulting in $\Delta\nu_{max}$ = ± 300 Hz at half the maximum peak power. According to theory of detuning in AM mode-locking (Li et al., 2001), the maximum allowable detuning can be expressed as:

$$\Delta \nu_{\max} = \frac{c\sqrt{m}}{4n\Delta f_m L}, \tag{2}$$

with c the speed of light in vacuum, L is the cavity length, and n the modal effective index. Thus, the minimum locking range is attained when the whole modulation bandwidth is available, by replacing we obtain $\Delta\nu_{max}$ = 42 Hz. This value represents a lower limit in this setup, since we are not using the whole modulator bandwidth –due to the reflection bandwidth of the FBG–. Thus, a higher measured locking range is an expected result.

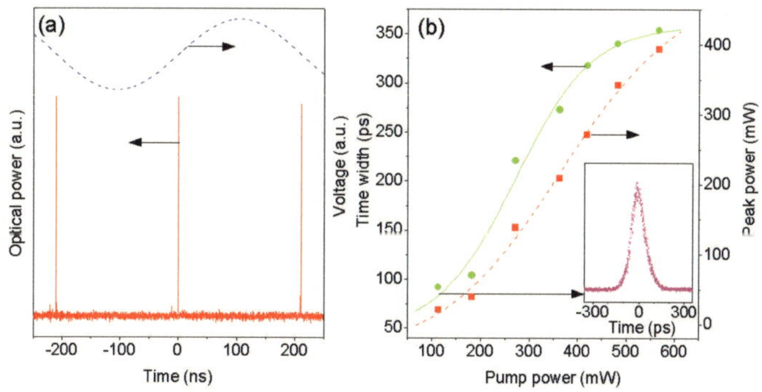

Fig. 8. (a) Voltage signal used to drive the piezoelectric and mode-locked train of pulses generated at 4.75 MHz repetition rate with 271 mW of pump power (dashed and solid curves, respectively). (b) Time width (FWHM) and peak power of the optical pulses as a function of the pump power (solid and open scatter points, respectively). The inset shows a single pulse of 95 ps time width

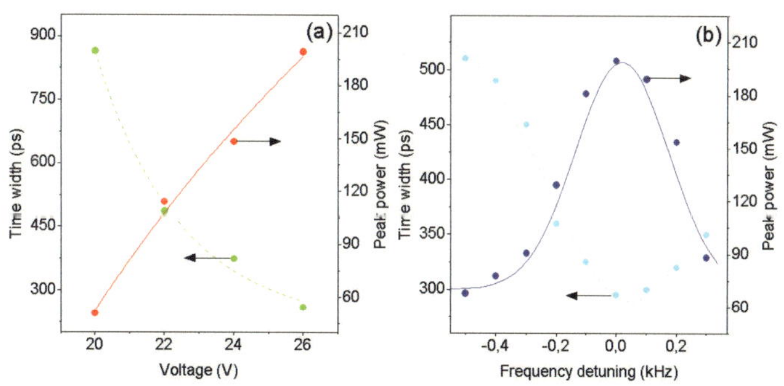

Fig. 9. (a) Time width (FWHM) and peak power as a function of the applied voltage; for a fixed frequency of 2.3731 MHz (open and solid scatter points, respectively). (b) Same as before, but as a function of the frequency detuning, for a fixed voltage of 26 V. In both cases, a pump power of 350 mW was used

Next, we carried out an experimental study of this actively mode-locked all-fiber ring laser, looking towards an improvement in its performance. To this end, we slightly modified the setup shown in Fig. 6, by replacing the FBG used before with a near 100 % reflective FBG, with a FWHM bandwidth of 0.3 nm, which should be compared with the preceding FBG with FWHM of 0.45 nm, and 50 % of maximum reflectivity. Together with this, the EDF length was shortened up to 2.85 m. As a result of these changes, output light pulses were obtained by one of the ports of a 3 dB coupler incorporated within the ring (Bello-Jímenez et al., 2011). With this new configuration, output light pulses were obtained with a maximum peak power of 380 mW and pulse width of 90 ps. By comparing the narrowest pulses obtained with this configuration with the narrowest pulses reported before (95 ps) where the same type of AOM was used, we find nearly no difference. However, from the laser operation point of view, there is an important improvement, since the present arrangement is rather more stable, and the laser can be adjusted more easily. We attribute this improvement to the FBG used, which has a flat spectral reflectivity, making less critical the spectral matching between the FBG reflectivity with the resonant dip of the AOM. On the contrary, the FBG used before, was of nearly the same FWHM but with a lower reflectivity (50 %). As a consequence, any spectral detuning resulted in a larger variation in reflectivity, which in turn induced a higher difficulty to stabilize the laser operation. Next, we analyzed the change in the pulse`s parameters as the EDF length was varied, see Fig. 10 (a). Changes in the EDF length induced a slight shortening in the output light pulses, which is in accordance with Eq. 1, if we assume that gain scales with the EDF length. By this trend, the shortest light pulses (65 ps) were obtained at 2.2 m of EDF length. Further narrowing by this trend was not possible, since no mode-locked output pulses were observed for EDF lengths shorter than 2.15 m.

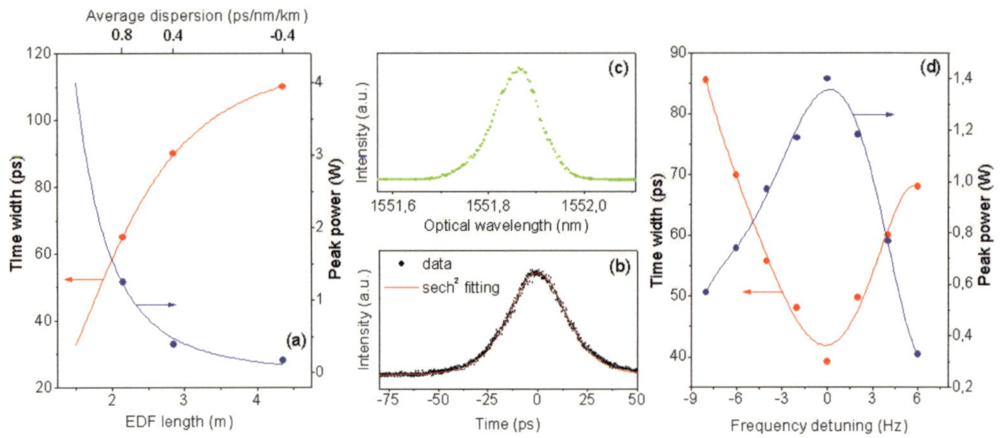

Fig. 10. (a) Time width (FWHM) and peak power as a function of the EDF length, using a FBG with 0.3 nm bandwidth. Using a 0.7 nm bandwidth: (b) single mode-locked pulse at 4.73 MHz repetition frequency, together with its corresponding fitting by a $sech^2$ function; (c) optical spectrum of the laser; (d) time width and peak power versus the frequency detuning

An important parameter of our laser arrangement is the FBG bandwidth, which acts as a spectral filter. Its bandwidth has to be selected narrower than the AOM bandwidth; in order to insure that the modes are properly amplitude modulated. For this reason we used first a 0.3 nm bandwidth FBG, which fulfills this requirement, since the AOM bandwidth is of 1.5 nm. However, it can be easily understood that a larger number of modes locked -i.e. amplitude modulated- produce the narrowest mode-locked train of pulses, which is in accordance with Eq. 1. In this respect, there is margin for improvement, since the bandwidth of the FBG used is far to be close to the AOM bandwidth. For this reason we replace the flat-top FBG with another FBG, which also has a flat-unit reflectivity, but a broader optical bandwidth of 0.7 nm. Fig. 10(b) shows the shortest pulse obtained with this configuration, by using an EDF length of 2.45 m at a pump power of 265 mW. The measured temporal width was 34 ps (FWHM), with a maximum peak power of 1.4 W. For this pulse the spectral linewidth was measured to be 110 pm (i.e. 13.7 Ghz at 1551.9 nm), using a 50 pm resolution optical spectrum analyzer, see Fig. 10(c). On the other hand, a Fourier-transform limited $sech^2$ pulse should have a spectral linewidth of 9.3 GHz. From the comparison between this last value and the linewidth measurement, we conclude that the optical pulses of our mode-locked laser could have some moderate degree of chirp. Finally, we present the variation of the pulse parameters as a function of the frequency detuning for this configuration, see Fig. 10(d); the measured allowable frequency detuning was of 8 Hz. According to theory of detuning in AM mode locking, the maximum allowable detuning is inversely proportional to the modulator optical bandwidth, in this case replaced by the FBG bandwidth, see Eq. 2. Since the later has been increased noticeably, i.e. from 0.3 nm to 0.7 nm, a reduction in the allowable detuning is an expected result.

3. Applications of acousto-optic devices based on the interaction of longitudinal acoustic waves with fiber Bragg gratings

When a travelling axially-propagating longitudinal acoustic wave is launched through a FBG, the periodic strain field of the acoustic wave perturbs the grating in two different ways. First, the average index changes in response to the stress-optical effect and second, the otherwise uniform Bragg grating pitch changes being modulated by the acoustical signal (Russell & Liu, 2000). As a consequence of both effects the reflectivity changes, the main features of these changes depend on the ratio between the acoustical wavelength and the grating length (Andrés et al., 2008). Thus, we can distinguish two well-different situations: the long-wavelength and the short-wavelength regimes. In the first, the Bragg grating pitch is homogeneously perturbed along its length. The successive cycles of compression and expansion generated by the longitudinal wave will shift periodically in time the spectral response of the grating as a whole to longer and shorter wavelengths (Cuadrado-Laborde et al., 2007; Andrés et al., 2008). On the contrary, in the short-wavelength regime the acoustic wave generates many compressed and expanded sections, which gives rise to a superstructure within the grating. In this case, the spectral response of the original FBG shows new and narrow reflection bands symmetrically at both sides of the original Bragg wavelength (Liu et al., 1997, 1998). The position and strength of these sidebands can be controlled by varying the frequency and voltage applied to the piezoelectric, respectively. Subsections 3.1 and 3.2 show the use of these effects to modulate the *Q*-factor of a fiber laser and as an active mode-locker; i.e. long and short-wavelength regimes, respectively. Subsection 3.3 shows the use of the acousto-optic interaction in a FBG in the short-wavelength regime as a tunable photonic true-time-delay line based on the group delay change of the light reflected from the grating sidebands. Finally, when the acoustic perturbation is not a harmonic wave but a single pulse, its passage through the grating creates a defect in the FBG, which can be used to actively *Q*-switch a distributed feedback fiber laser, which is discussed in subsection 3.4.

3.1 *Q*-switching of an all-fiber laser by acousto-optic interaction in a fiber Bragg grating in the long-wavelength regime

As an example of the use of the long-wavelength regime in FBGs, we present an all-fiber acousto-optic modulator suitable for *Q*-switching applications (Cuadrado-Laborde et al.,

2007). It consists of a short-length fiber Bragg grating modulated by a standing longitudinal acoustic wave, whose wavelength is much longer than the grating length. Periodic stretching and compression of the grating due to the elastic wave, causes that the Bragg wavelength is harmonically, continuously and repeatedly tuned in time over a given wavelength range. Figure 11 illustrates the operation principle. In the long-wavelength regime, the acoustic wavelength is much larger than the grating length.
However, the amplitude of the acoustic waves that can be achieved in a realistic arrangement is rather small to produce a significant perturbation of the FBG. We found that this limitation could be overcome by using an acoustic cavity. Thereby, if the grating is placed in the proper position within the acoustic cavity, a low-frequency standing longitudinal elastic wave will produce periodic stretching and compression of the grating. As a consequence, the reflection band will shift periodically in wavelength around the original Bragg wavelength, sweeping a given wavelength range in a harmonic way. The elastic wave may also introduce some chirp in the grating due to non-uniformity of the strain induced along the grating. To achieve efficient modulations this effect must be minimized. Thus, the grating length must be short compared with the acoustic wavelength, and the positioning of the grating within the cavity must match an anti-node of the standing acoustic wave.

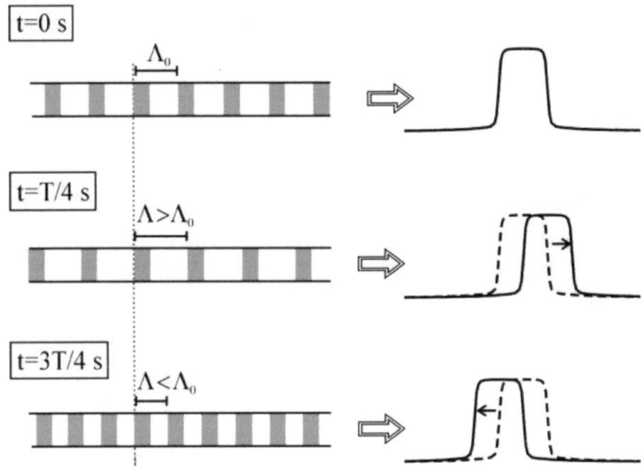

Fig. 11. Interaction of acoustic waves in the long-wavelength regime and FBGs; T is the period of the acoustic wave, Λ is the instantaneous pitch of the Bragg grating, and Λ_0 is the pitch of the Bragg grating in absence of acoustic wave. The effect on the reflection band is shown qualitatively. The standing acoustic wave stretches and compresses the grating periodically. During the time that the grating is stretched, the reflection band of the grating shifts towards longer wavelengths, whereas it shifts towards shorter wavelengths when it is compressed

Figure 12(a) shows a diagram of the laser setup, together with a detail of the acousto-optic modulator. The Fabry-Perot cavity becomes defined by FBG_1 (which also acts as modulator) and FBG_2. A piezoelectric disk (15 mm in diameter and 2 mm thickness) was used to generate the longitudinal acoustic waves, which were launched to FBG_1 through a silica horn. The base of the horn was glued to the piezoelectric and the tip reduced to ~125 μm and then fusion-spliced to FBG_1. At a distance of 210 mm from the piezoelectric disk, the fiber was clamped allowing standing acoustic waves to be created in the fiber section between the horn and the clamp. FBG_1 was written in the core of a photosensitive fiber by

UV irradiation using a phase-mask technique. The photosensitive fiber was previously tapered to enhance the acousto-optic interaction; the taper waist was 50 mm long, 90 µm in diameter, and the transition lengths were 13 mm. The grating was recorded in the taper waist (uniform diameter section); the FBG_1 length was 22 mm. The highest frequency used in our experiments was 61 kHz, and according to the velocity of sound in silica fibers for longitudinal elastic waves (5760 m/s), the wavelength for this frequency is 94 mm. Thus, we can verify that FBG_1 length is short enough to guaranty a standard long-wavelength regime. An apodization profile was introduced also during the grating inscription to reduce side-lobes in the spectrum. Figure 12(b) shows the transmission spectrum for both FBGs, the bandwidth of FBG_1 was 0.15 nm and the reflectivity was higher than 99.9%. On the short-wavelength side, the transmission drops sharply (~ 0.72 dB/pm). The side-lobes have been reduced entirely on this side, while a significant side-lobe still remains on the long-wavelength side of the transmission band. On the other hand, FBG_2 acted as the output coupler, its Bragg wavelength was 1543.27 nm, the reflectivity 50%, and the bandwidth was 40 pm.

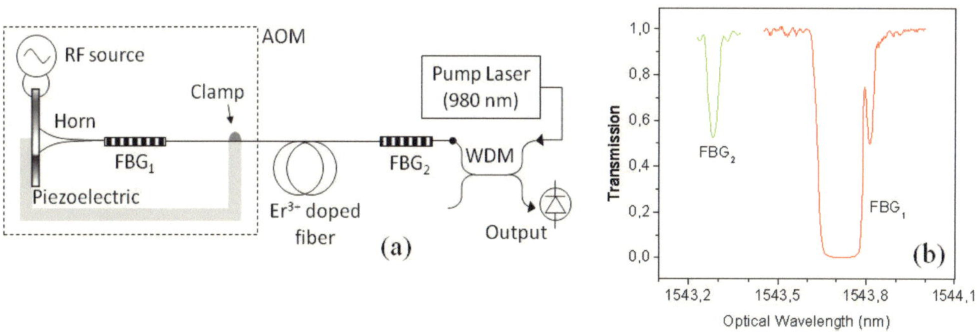

Fig. 12. (a) Laser setup, the dashed line defines the acousto-optic modulator. (b) Transmission spectra of the two fiber Bragg gratings

Next, we discuss the modulator´s performance. We detect the intensity of the light reflected by the mid-reflection point of the short-wavelength band-edge, when an AC electric signal of a given frequency was applied to the piezoelectric. To this end a tunable laser diode was used. The modulation of the grating caused by the elastic wave generates periodic modulation of the light intensity reflected by the grating. An example is shown in Fig. 13(a). At low voltage, a sinusoidal modulation was observed since the grating shifts little and the laser reflects at the band-edge at any time during the period of the electric signal. As the electric signal amplitude increases, the grating shifts further and then, the laser reflects also at the top and bottom of the band-edge, generating a square signal. We observed strong acoustic resonances at frequencies of 18 kHz, 37 kHz and 61 kHz. The peak-to-peak amplitude was measured for these frequencies as a function of the voltage applied to the piezoelectric, see Fig. 13(b). The most efficient frequency was 61 kHz, and the responses at 18 kHz and 37 kHz were quite similar. Taking into account the results shown in Fig. 13(b) and the width of the band-edge, 42 pm, the modulation efficiency (i.e. wavelength shift per unit volt) was estimated, giving 7 pm/V at 18 kHz and 37 kHz, and 16 pm/V at 61 kHz.
The switching speed is one of the key parameters of any modulator intended to be utilized as Q-switching element in a fiber laser, being desirable to be as high as possible. In this modulator, the switching time depends on the AC voltage amplitude, and it becomes shorter as the voltage is increased, as shown in Fig. 13(a). Switching times of the order of a

few microseconds were achievable. Finally, although the acoustic generator was designed for the excitation of longitudinal waves, we also observed flexural waves in the resonator when large voltages were applied to the piezoelectric. A large number of flexural standing waves, spaced fractions of kHz, were observed in the range from 0 to 100 kHz. These resonances gave rise to light modulated at double frequency of the electrical excitation and exhibited relatively low amplitude. The aforementioned modulator uses an acoustic resonator, which enhances the interaction between acoustic waves and the FBG. The fundamental mechanical resonance at 18 kHz had a Q factor of 1300. With respect to an unclamped configuration based on traveling acoustic waves, the use of an acoustic cavity increases the modulation efficiency, but it restricts the operation of the modulator at the resonant frequencies of the cavity. A modulator based on traveling acoustic waves would allow, in principle, tuning continuously the modulation frequency, but a large voltage would be required and important thermal effects would damage the acoustic transducer.

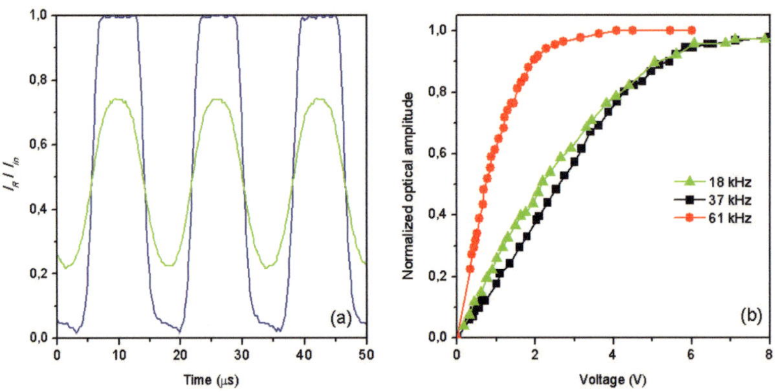

Fig. 13. (a) Light reflected from the band-edge of the grating when an AC electric signal of 61 kHz and peak-to peak amplitude of 4.6 V (green) and 14 V (blue) were applied to the piezoelectric. (b) Light modulation amplitude as a function of voltage for three different resonant frequencies

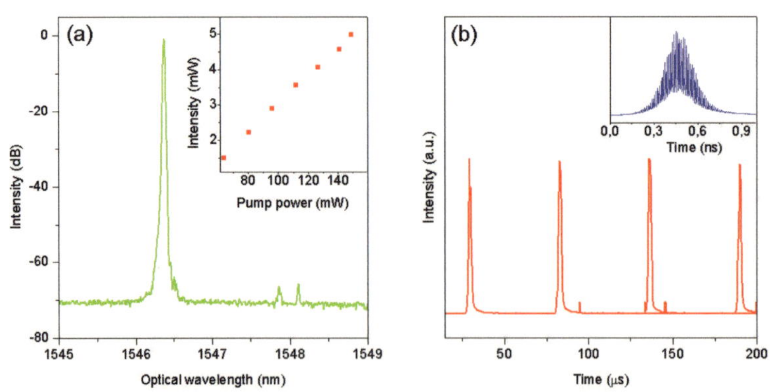

Fig. 14. (a) CW emission spectrum at 135 mW pump power; the inset shows the output CW power versus pump power. (b) Pulse train at 18 kHz repetition rate; the inset shows a single light pulse measured with a 1 GHz bandwidth photodetector, pump power of 80 mW

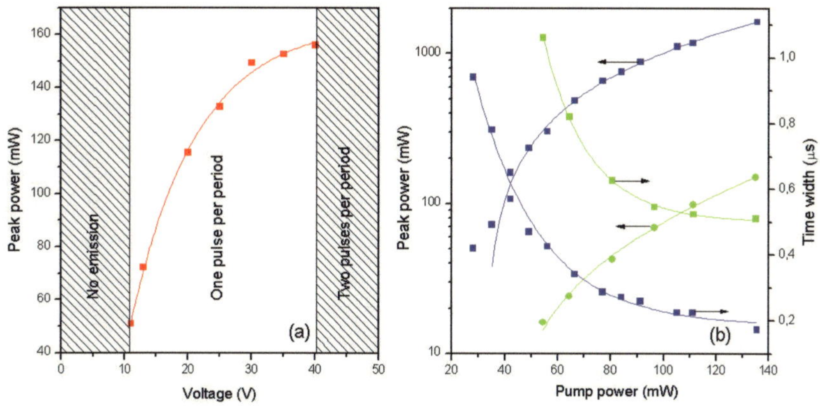

Fig. 15. (a) Pulse peak power as a function of the applied voltage to the piezoelectric (75 mW pump power, 18 kHz repetition rate). (b) Pulse peak power and pulse width as a function of pump power, for repetition rates of 18 kHz (blue) and 37 kHz (green). Voltage applied to the piezoelectric 24 V

The gain of the *Q*-switched all-fiber laser was provided by 0.3 m of an erbium-doped fiber containing 1000 p.p.m. Er^{3+}, with a cut-off wavelength of 965 nm, and a numerical aperture of 0.23. The active fiber was pumped through a WDM coupler by a pigtailed laser diode emitting at 979 nm, providing a maximum pump power of 140 mW. The Fabry–Perot cavity length was of 1.2 m. A translation stage was used for tuning the reflection band of FBG_2 closer to the short-wavelength band-edge of FBG_1, allowing full overlapping between both bands with moderate voltages being applied to the piezoelectric. In this arrangement, the longitudinal standing elastic wave causes periodic optimization of the *Q* factor of the laser cavity and, as a result, strong pulses are emitted at the repetition frequency of the acousto-optic modulation, when the fiber is pumped. The CW features of the fiber laser are shown in Fig. 14(a). The laser line was narrower than 20 pm (resolution limit of our measurement system), a signal-to-noise level as high as 70 dB was obtained. The efficiency and the threshold were 3 % and 25 mW, respectively. Figure 14(b) shows a pulse train emitted by the laser at a repetition rate of 18 kHz. At high pump powers each *Q*-switched pulse breaks into a train of narrower, 8 ns wide, sharp pulses (see inset). The frequency of this modulation, 83 MHz, coincides with the frequency mode-spacing of the laser cavity, which indicates beating between longitudinal cavity modes. We observed that the modulation amplitude of this high frequency component was affected by variations in pump level and repetition rate. This apparent passive mode-locking response has been previously reported repeatedly in both, passively and actively *Q*-switched fiber lasers (Philippov et al., 2004; Andersen et al., 2006). Figure 15(a) shows the peak power of the *Q*-switched pulse as a function of the applied voltage to the piezoelectric. Three regions can be clearly distinguished. For low voltages applied to the piezoelectric there is not emission, since the wavelength shift of FBG_1 is not enough to overlap the reflection band to FBG_2. As the applied voltage increases, there is a *Q*-switched pulse per period. For voltages close to the lower limit, the overlapping between both gratings is only partial and the laser is only a small amount above threshold, so weak pulses are emitted. As the voltage is increased, the gratings overlap further and the peak power increases. A steady state is reached when the voltage amplitude is high enough to obtain complete overlapping between the two reflection bands. If the applied voltage is increased further, the cavity stays in the high *Q* state longer and more than one Q-switched pulse is emitted per period. When changing the pump power and/or repetition rate, the upper and lower AC voltage limits vary, but the trend is preserved. Finally, the effect of pump power on the Q-switched pulses, for repetition rates of 18 and 37 kHz, is shown in Fig. 15(b). The pump power threshold at each frequency was 27 mW and 53 mW, respectively. Above threshold, the peak power increases with pump power, and there is a corresponding reduction of the pulse width. No evidences of peak power saturation were observed within the pump power range of our experiments. Pulses of 1.6 W peak power and 172 ns width were obtained at 18 kHz repetition rate and 135 mW pump power.

3.2 Mode-locking and *Q*-switching mode-locking by acousto-optic interaction in a fiber Bragg grating in the short-wavelength regime

In this subsection we discuss another mechanism developed by our group to mode-lock an all-fiber laser (Cuadrado-Laborde et al., 2009a, 2010a). This is based on the acousto-optic super-lattice modulation, and it is an example of the interaction of longitudinal acoustic waves and Bragg gratings in the short-wavelength regime (Liu et al., 1997, 1998). In the following, we start by reviewing the behavior of the acousto-optic super-lattice modulator, subsection 3.2.1. Then, we show the use of this device as mode-locker, by showing two different mode-locked lasers, either by driving the acousto-optic modulator by standing or travelling acoustic waves, subsections 3.2.2 and 3.2.3, respectively. The possibility to reach simultaneous Q-switching and mode-locking is also discussed in these subsections.

3.2.1 The fiber Bragg grating based acousto-optic modulator

The spectral response of the original FBG in this acoustical regime shows new –and narrow– reflection bands symmetrically at both sides of the original Bragg wavelength (Liu et al., 1997, 1998). The position of these sidebands can be controlled by varying the frequency at a slope of 0.15 nm /MHz and 0.30 nm /MHz for the first and second order sideband, respectively. The strength of the reflection bands, on the other hand, can be controlled independently by varying the voltage applied to the piezoelectric. Since these sidebands can be regarded as weak ghosts of the strong permanent Bragg grating, its FWHM bandwidth is that of a weak Bragg grating of the same length (Liu et al., 1997). Figure 16 shows the setup for a typical reflectivity measurement on an AOSLM. The AOSLM in turn is composed of an RF source, an electrical RF amplifier, a piezoelectric disk, a silica horn, and a FBG. The tip of the silica horn was reduced by chemical etching to the same diameter of FBG –125 µm– and subsequently fusion-spliced to the fiber. The uniform and non-apodized grating was written in photosensitive fiber using a doubled argon laser and a uniform period mask; the FBG was 120 mm long. The reflection properties of the AOSLM were investigated by illuminating the FBG through an optical circulator with a broadband light source, and detecting the reflected light with an optical spectrum analyzer (OSA). Figure 17(a) shows the reflectivity of the unperturbed FBG –i.e. without electrical signal applied to the piezoelectric– and with an electrical signal applied to the piezoelectric of 4.55 MHz and 16 V (whenever we refer to voltages throughout this chapter, it is a peak-to-peak measurement). The presence of the sidebands symmetrically positioned around the Bragg wavelength is clearly discernible. Further, these sidebands are produced either by standing or travelling acoustic waves. However, the light reflected by the sidebands in each case behaves differently. When travelling acoustic waves are used –by dumping the end of the FBG opposite to the silica horn for example with a drop of oil–, the light reflected on these sidebands is completely downshifted or upshifted by the frequency of the acoustical signal, depending if the reflection was in the long or short-wavelength sideband, respectively (Liu et al., 1997, 1998). Together with this and as a function of the instantaneous phase of the acoustic signal, there is also present an amplitude modulation at the frequency of the acoustical signal. Figure 17(b) shows the measurement performed by tuning a laser diode to the center of the short-wavelength sideband and measuring the reflected light. As expected, the amplitude modulation is at the same frequency of the electrical signal used to drive the piezoelectric. On the other hand, when standing acoustic waves are used –by clamping the end of the FBG opposite to the silica horn– the sidebands raise and fall at twice the frequency of the electrical signal (Cuadrado-Laborde et al., 2009a). Further, in principle, for a perfect acoustical reflection, the light reflected by the sidebands does not experiment any Doppler shift, as opposed to the previous case when travelling acoustic waves are used. Figure 17(c) shows the measurement performed by tuning a laser diode to the center of the short-wavelength sideband and measuring the reflected light. Now the optical modulation frequency is two times the acoustical signal frequency. To summarize, an AOSLM can be driven in two different regimes, either by using travelling or standing acoustic waves. In both cases we can use it as an amplitude modulator. However in the first case –travelling acoustic waves– the light reflected on the sidebands is modulated at the same frequency of the acoustical signal, whereas in the second case –standing acoustic waves– it is modulated at two times this frequency, respectively.

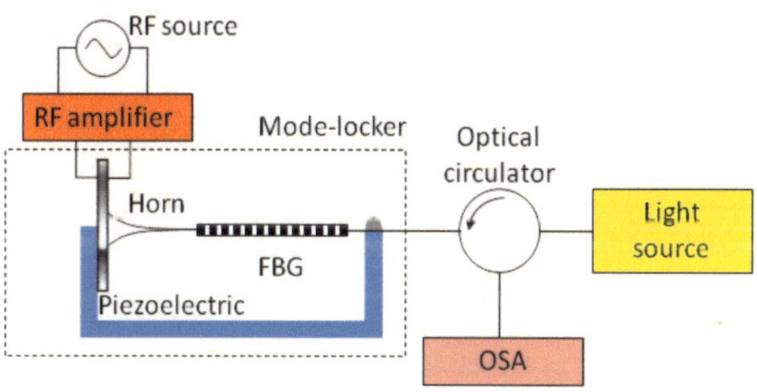

Fig. 16. Setup for the characterization of the acousto-optic super-lattice modulator; OSA stands for optical spectrum analyzer, FBG for fiber Bragg grating and RF for radio-frequency

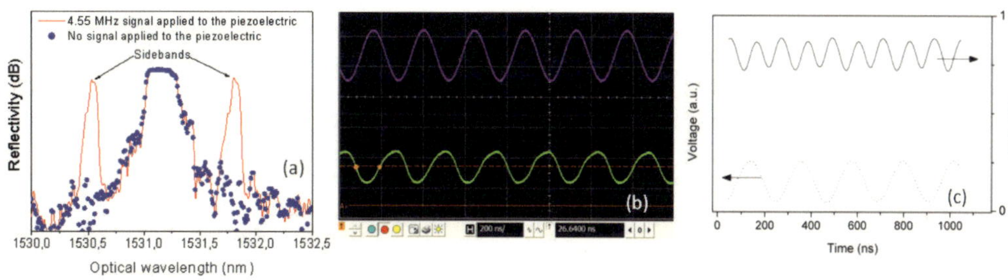

Fig. 17. (a) Reflectivity of the fiber Bragg grating with and without electrical signal applied to the piezoelectric (4.55 MHz and 16 V peak-to-peak). (b) Optical signal reflected by the short-wavelength sideband when travelling acoustic waves are used (lower trace) and RF voltage applied to the piezoelectric (4.11 MHz and 16 V, upper trace). (c) Same as in (b), but for standing acoustic waves (4.55 MHz and 16 V, dotted curve)

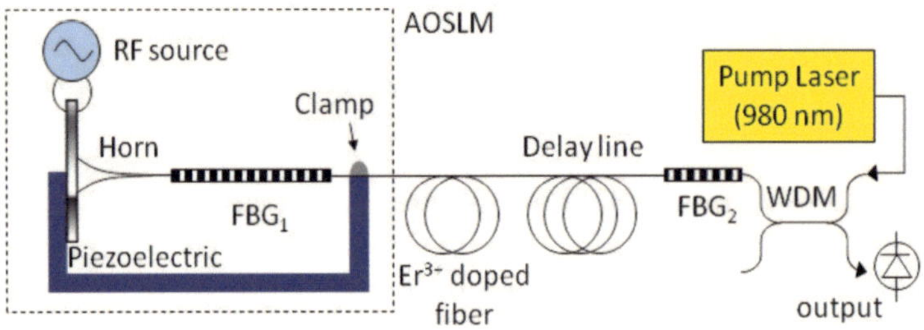

Fig. 18. Setup of the mode-locked all-fiber laser by using an acousto-optic supper-lattice modulator (AOSLM) as mode-locker

3.2.2 All-fiber actively mode-locked laser with a fiber Bragg grating based acousto-optic modulator driven by standing acoustic waves

Now, we show the use of the amplitude modulation induced by standing acoustic waves in the AOSLM to mode-lock an all-fiber laser (Cuadrado-Laborde et al., 2009a, 2010a). The setup proposed for the standard mode-locking laser is schematically illustrated in Fig. 18. The gain was provided by an erbium-doped fiber (EDF) containing 300 parts per million (ppm) of Er^{3+}, with a cut-off wavelength of 939 nm, and a numerical aperture of 0.24. The active fiber was pumped through a WDM coupler by a pigtailed laser diode emitting at 976 nm, providing a maximum pump power of 160 mW. The acousto-optic super-lattice modulator and a short delay line followed by a second fiber Bragg grating FBG_2 were fusion-spliced at each end of the active fiber. FBG_2 (10 mm long and with a Bragg wavelength of 1530.2 nm) was written with a uniform period in a photosensitive fiber using a doubled argon laser and a uniform period mask. The Fabry-Perot cavity was established once the reflection band of FBG_2 is made to match the short wavelength sideband of the FBG_1 of the AOSLM, i.e. 1530.5 nm, by straining it with a translational stage (see Fig. 18). The delay line length must be selected to match the round-trip time with the reciprocal of the optical modulation frequency, which in turn is twice the electrical frequency applied to the piezoelectric. Since the selected piezoelectric operation point for the piezoelectric was 4.55 MHz, this results in a cavity length of 11.4 m. However, fine tuning of the electrical frequency was required to achieve mode-locking.

Figure 19(a) exemplifies the laser behavior showing the train of optical pulses generated, at a frequency rate of 9 MHz (50 GHz bandwidth oscilloscope). Figure 19(b) shows a single optical pulse; the timing jitter was measured to be 40 ps (RMS). The pulses of this laser are best fitted by a $sech^2$ function, which is in accordance with earlier amplitude modulated mode-locking theory. Further, the dispersion measurements of the different fibers we used to form the laser cavity show that both kind of dispersion coexist, from the anomalous dispersion of the delay line to the normal dispersion of the EDF. For this type of dispersion-mixed cavities, the analytical steady-state solution is expected to be also of the $sech^2$ type (Belanger, 2005), which adds further evidence in the same direction. The emission linewidth was measured using a high resolution optical spectrum analyzer (BOSA-C, Aragón Photonics, resolution of 80 fm). Figure 19(c) shows the spectrum of the optical pulses shown in Fig. 19(b). The high resolution of the optical spectrum analyzer permits a direct observation of the individual cavity modes separated by 9 MHz, see the inset of Fig. 19(c). The measured 3-dB linewidth results in 2.8 pm, i.e. 360 MHz at 1530.5 nm. On the other side, the optical pulses have a temporal width of 780 ps, so according to the Fourier-transform-limited relation for a sech^2 pulse, time-bandwidth product of 0.315 (Wada et al., 2008), its bandwidth cannot be lower than 380 MHz. From the comparison between this last value and the linewidth measurement, we conclude that the optical pulses of our modelocked laser are transform-limited, i.e. un-chirped. A brief digression is in order here, the Fourier-transform-limited relation for Gaussian pulses with a time-bandwidth product of 0.44 (Kuizenga & Siegman, 1970; Wada et al., 2008), would result in a much higher linewidth (564 MHz), which is in contradiction with our spectral measurements. As a consequence, the fitting of the temporal pulses, see Fig. 19(b), the application of the Fourier-transform-limited relations, and dispersion measurements together with previous theoretical work, seems to corroborate a sech^2 nature for the emitted pulses.

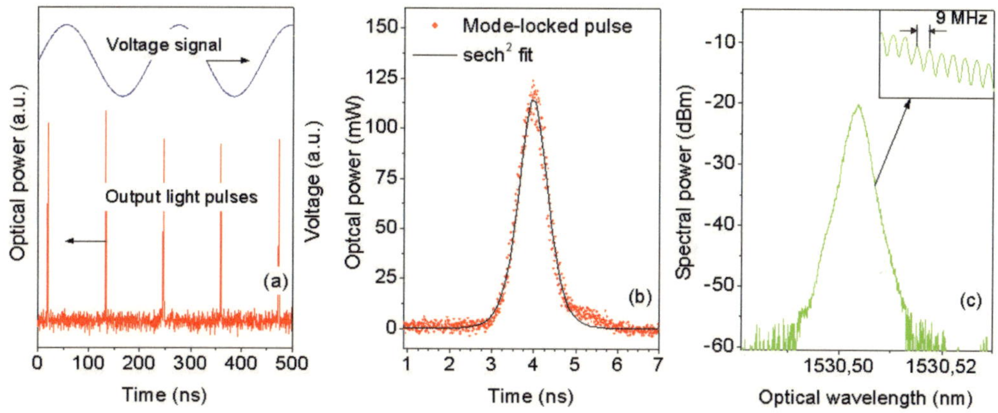

Fig. 19. (a) Voltage signal applied to the piezoelectric at 4.55 MHz and 16 V, and its corresponding modelocked train of pulses generated at twice this frequency with a pump power of 160 mW. (b) A single optical pulse and its corresponding fitting by a sech^2 function. (c) Optical spectrum of the output light pulses shown in (b); the inset shows the cavity modes distant by 9 MHz

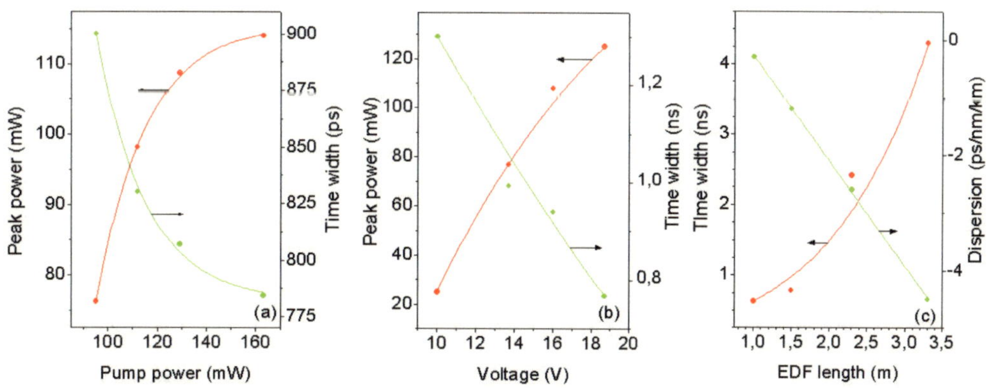

Fig. 20. (a) Peak power and pulse width of the optical pulses as a function of the pump power. (b) Peak power and pulse width of the output light pulses as a function of the voltage applied to the piezoelectric, for a pump power of 160 mW. (c) Pulse width of the output light pulses, as a function of the EDF length, for a fixed pump power of 160mW, additionally, the average dispersion of the cavity is also shown for each EDF length

Now we will show the variation in the pulse`s parameter as typical variables are changed. First we analyze the optical pulse's peak power and temporal width as a function of the pump power, which is shown in Fig. 20(a). A smooth variation of the pulse width with pump power can be observed, before reaching the gain saturation. A higher available pump power would allow additional axial modes to contribute, increasing the spectral bandwidth, and thereby decreasing the pulse´s width. Next, we analyzed the variation in the pulse`s parameters as a function of the modulation voltage. The intensity of the reflection sidebands of the AOSLM can be dynamically controlled by varying the voltage applied to the piezoelectric as we explained before in subsection 3.2.1. This voltage controls the amplitude of the propagating acoustic wave and the amplitude of the standing wave, as a function of the reflection at the clamp. Thus, as opposed to mode-locking by bulk acousto-optic or electro-optic modulators, here the reflectivity and the modulation depth are intrinsically linked, i.e. when the applied voltage increases, both the reflectivity and the modulation amplitude increase. Figure 20(b) shows the peak power and time width, respectively, as a function of the voltage applied to the piezoelectric. At lower reflectivities and modulation amplitudes, the peak power diminishes, whereas the time width increases. According to Eq. (1), the pulse width and the modulation depth are inverse parameters, and in turn the later rises with the applied voltage. However, the pulses shorten very slowly with the increased modulation strength; hence this shortening procedure is generally discouraged (Kuizenga & Siegman, 1970). One can conclude that further increase of the voltage might improve the pulse parameters, but it is not likely to lead to a great enhancement in terms either of peak power or temporal width. We analyzed also the influence that dispersion has on the pulse`s parameters, when the average dispersion of the cavity is changed. The dispersion for each type of optical fiber used in the setup was measured by the frequency-domain modulated-carrier method. The resulting average dispersion of this cavity is normal and its value is –1.2 ps/nm/km. Next, the optical fiber of the delay line –the Corning LEAF fiber– was replaced by a normal dispersion fiber –Fibercore SM980– leading to a lower (normal) overall dispersion of –5.8 ps/nm/km. With the new delay line, a higher pump reaches the EDF, since a better effective area compatibility is insured throughout the different fibers of the system –in fact, unlike LEAF fiber, the SM980 insures single-mode propagation of the pump power–. As a result, output pulses have higher peak power and shorter time width than in the previous configuration. In this way, optical pulses with temporal width and peak power of 640 ps and 160 mW, respectively, were obtained. These pulse parameters represent an improvement of 18% and 28% with respect to the previous configuration. Finally, in order to reverse the sign of the average dispersion of the cavity, we used an SMF28 optical fiber for the delay line, resulting in an average –anomalous– dispersion of 9.8 ps/nm/km. In this case, we did not observe mode-locking lasing. One reason for this could be motivated by the large average dispersion introduced within the cavity, since modelocked fiber lasers usually works with close-to-zero average dispersion cavities. Finally, the change on the pulse parameters was measured as a function of the EDF length –ranging from 0.5 to 3.3 m–, but keeping a constant cavity length of 11.4 m, by adjusting each time the length of LEAF fiber. The average dispersion remains normal regardless of the amount of EDF used, ranging from –0.3 ps/nm/km up to –4.5 ps/nm/km. The temporal pulse width as a function of the EDF length is shown in Fig. 20(c), for a fixed pump power of 160 mW. This figure includes the average dispersion of the cavity as a function of the EDF length. A direct relationship can be observed between the EDF length, the dispersion, and the time width of the pulses, with a minimum of 630 ps obtained for an EDF length of 1 m. This kind of interplay between gain and time width agrees with Eq. (1). Further narrowing of the optical pulses could not be reached by this trend, since no mode-locking lasing was observed with 0.5 m of EDF and the available pump power, as a result of insufficient cavity gain. Finally, we also developed the ytterbium-doped fiber version of this laser, with equally satisfactory results (Villegas et al., 2011).

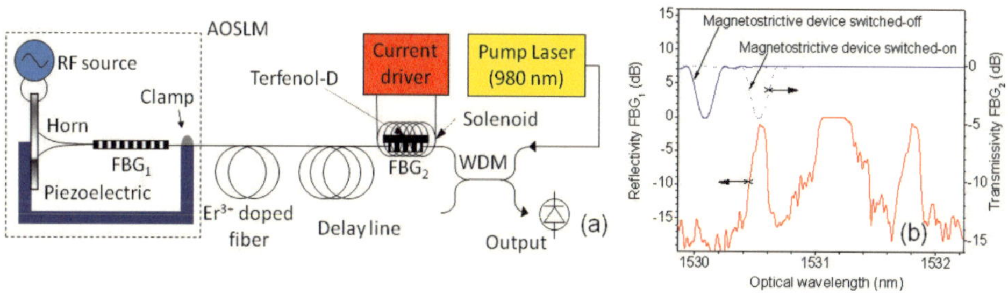

Fig. 21. (a) Q-switched mode-locked all-fiber laser setup when the AOSLM is driven by standing acoustic waves. (b) Reflection spectrum of the AOSLM when an RF signal of 4.55 MHz and 16 V is applied to the piezoelectric (left ordinate). Transmittance of FBG_2 (solid curve) when the magnetostrictive device is switched-off; the dotted curve shows the new spectral position of FBG_2 when the magnetostrictive device is switched-on (right ordinate)

Until now we have discussed the standard mode-locking operation. Now we demonstrate the possibility to simultaneously Q-switch and mode-lock, actively and independently. By combining both operations in a single laser, a superior performance is achieved with higher peak powers compared with ordinary mode-locked lasers, but almost the same pulse width as in a mode-locked laser is retained. Thus, the peak power of the central pulses of the mode-locked train, underneath the Q-switched envelope, can be greatly enhanced. The increased peak power provided by the Q-switching technique can be advantageous in applications such as wavelength conversion or super-continuum generation. It is worth to say, that any passive mechanism is used in this setup, neither for Q-switching nor for mode-locking. To this end, we design a solution for actively Q-switching, while simultaneously mode-locking, preserving the all-fiber configuration. This solution is based on fiber Bragg grating fast modulation using a magnetostrictive device. This feature provides a direct control of the repetition rate and a fully-modulated train of modelocked pulses. To the best of our knowledge, this was the first doubly active Q-switching mode-locking strictly all-fiber laser presented (Cuadrado-Laborde et al., 2009b). The setup of our doubly-active Q-switching mode-locking all-fiber laser is shown in Fig. 21(a). It is basically the same as it was described before for the standard mode-locking and shown in Fig. 18, with the same length of EDF and delay line, except by the magnetostrictive device controlling the FBG_2, see Fig. 21(a). It is composed of a 15 mm long (1 mm^2 cross-section) magnetostrictive rod of Terfenol-D bonded to the FBG_2. The rod and the fiber were placed inside a small coil driven by an electronic circuit designed to drive square current pulses with amplitudes up to 260 mA of any required duty cycle. Figure 21(b) shows the spectral positions of both FBGs –i.e. the FBG of the AOSLM and FBG_2, which helps understand the operation principle of this laser also. When the magnetic pulses generated in the solenoid stretch FBG_2, its central wavelength is brought to match the short-wavelength sideband of FBG_1 for a short period of time, which results in an increased Q-value. In this way by modulating the coil current with a given Q-switched frequency, the Q-factor is actively modulated at the same frequency. The magnetostriction has a low pass frequency response and, consequently, presents the advantage to permit a continuous tuning of both the Q-switched repetition rate and the duty cycle of the modulation pulses. The dependence of the wavelength shift with the coil current was measured by illuminating FBG_2 with a tunable laser source and detecting the reflected light. A quasi-linear behavior was observed, as a result of using moderate magnetic fields.

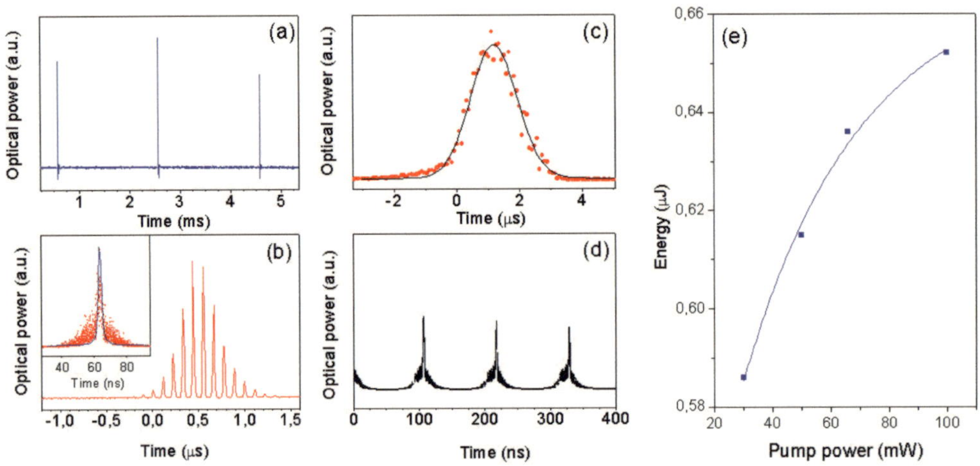

Fig. 22. (a) Q-switched mode-locked train of pulses at a Q-switching frequency of 500 Hz. (b) Single Q-switched pulse enveloping a train of 12–14 modelocked pulses; the inset shows a single modelocked pulse. (c) Single Q-switched pulse with no mode-locked. (d) Mode-locking operation of the laser without Q-switching. In all cases, the pump power was 70mW. (e) Energy of the Q-switched mode-locked pulses as a function of the pump power for a Q-switching repetition rate of 500 Hz

Now we discuss the Q-switching mode-locking laser operation. The laser emission wavelength was in 1530.55 nm, since the overlapping between the short-wavelength sideband of FBG_1 and the shifted position of FBG_2 takes place at this wavelength, see Fig. 21(b). Figure 22(a) exemplifies the Q-switched mode-locked laser behavior showing the train of optical pulses generated for a pump power of 70 mW. The optical Q-switched mode-locked pulses are 2 ms apart, since the repetition rate of the train of voltage pulses applied to the current driver was of 500 Hz. Figure 22(b) shows a single Q-switched envelope with a FWHM of 550 ns; it has between 12 and 14 fully-modulated modelocked pulses, with the expected temporal separation of 110 ns. However, the individual short pulses are broadened due to the limited bandwidth and sampling rate of the oscilloscope. In order to measure the FWHM of the individual short pulses, a 50 GHz sampling oscilloscope was used. In this case, the large difference between the Q-switching and mode-locking frequencies make difficult to trigger properly the oscilloscope. Even so, relatively good traces were recorded as the one depicted in the inset of Fig. 22(b). As we discussed before when we presented the standard mode-locking regime, the tuning sensitivity between the RF signal applied to the piezoelectric and the cavity round trip is not too critical in this setup. Only when the detuning is considerably –e.g. by a few kHz–, the pulse clearly deteriorates. Figure 22(c) shows the Q-switched pulse obtained by avoiding the mode-locking pulse formation, just by detuning the applied voltage to the piezoelectric by 10 kHz, in this way only the Q-switching operation is allowed within the cavity. As expected, the Q-switched pulse reproduces the waveform of the envelope of the *Q*-switched mode-locked pulses shown in Fig. 22(b), but with a much lower peak power (by a factor of ~4×10^{-3}) and an increased temporal width. The transition from the fully-modulated *Q*-switched mode-locked pulses to the pure *Q*-switching operation –i.e. from Fig. 22(b) to 22(c)– is progressive as we detuned the voltage applied to the piezoelectric. Finally, Fig. 22(d) shows the behavior of this laser when turning off the active *Q*-switching, i.e. standard mode-locking operation of the laser. To this end, a DC current was driven to the coil of the magnetostrictive device, in order to achieve a stationary overlap with the short wavelength sideband of the AOSLM. The trace of Fig. 22(d) was recorded with the 1 GHz oscilloscope. In this case, the thermal effects produced by the DC current made difficult an optimum and stable adjustment of FBG_2. The comparison of Figs. 22(b) and 22(d), illustrates the improved performance of the laser in terms of peak power when both *Q*-switching and mode-locking are operating. The energy of a *Q*-switched train of pulses, as that reported in Fig. 22(b), as a function of the pump power is shown in Fig. 22(e). This energy was measured directly with a pyroelectric detector. At high pump powers, the energy of the *Q*-switched mode-locked train of pulses reach gain saturation. Thus, a peak power higher than 250 W can be calculated for the central pulses of a train with energy of 0.65 μJ, assuming that the mode-locked pulses are 1 ns width. This result demonstrate a dramatic enhancement in comparison with the peak power achieved when the laser was operated in standard mode-locking regime (Cuadrado-Laborde et al., 2009a), since the ratio is higher than 2×10^3.

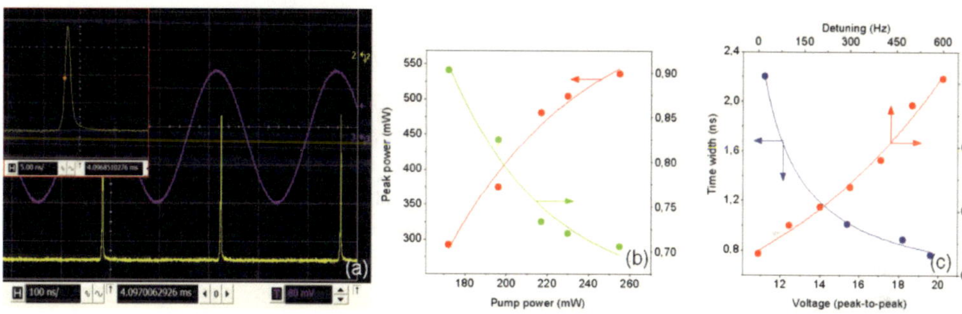

Fig. 23. (a) Oscilloscope traces of the sinusoidal electrical signal applied to the piezoelectric at 4.1 MHz repetition rate and 16 V (upper trace), and the train of optical pulses generated (lower trace); a pump power of 230 mW was used. The inset shows the detail of a single pulse of the optical train, with a FWHM of 720 ps and a peak power of 500 mW. (b) Peak power and pulse width (FWHM) of the optical pulses as a function of the pump power (solid and dotted curves, respectively). (c) Pulse width of the optical pulses when the applied voltage to the piezoelectric is varied, and when the frequency of the electrical signal is varied

3.2.3 All-fiber actively mode-locked laser with a fiber Bragg grating based acousto-optic modulator driven by travelling acoustic waves

The setup of the mode-locking laser driven by travelling acoustic waves is essentially the same described in Fig. 18, but with two important differences. First, the pigtailed laser diode emitting at 976 nm provides now a –higher– maximum pump power of 410 mW. Second, the modulation frequency is now at the same frequency of the acoustical signal (Cuadrado-Laborde et al., 2010b), since now we use travelling acoustic waves, see Subsection 3.2.1. Then, the delay line length is now much larger, because we must match the round-trip time with this frequency, which is lower. Since the selected piezoelectric operation point was of 4.1 MHz; then it results in a cavity length of 25 m –compare this with the 11.4 m of cavity length used before. Apart from the 1.4 m of EDF, the cavity was constructed entirely with Fibercore SM980 fiber, this result in an average dispersion of –4.9 ps/nm/km (i.e. normal). Figure 23(a) exemplifies the laser behavior showing the sinusoidal electrical signal applied to the piezoelectric at a frequency rate of 4.1 MHz and the train of optical pulses generated at the same frequency. The inset shows a detail of a single optical pulse with a temporal width (FWHM) of 710 ps. Once mode-locking was reached, the polarization controllers were adjusted in order to obtain the minimum pulse width.

The emission linewidth was measured using a classical heterodyne configuration (Galtarossa et al., 1993). To this end, a tunable laser was used as local oscillator, with a 100 kHz linewidth. The output of this laser was superimposed to the optical pulses of the modelocked laser, through a 1550 nm 50/50 coupler. The beat signal at the coupler output was detected with a 45 GHz bandwidth photodetector, and analyzed with a 2.5 GHz oscilloscope. This results in a linewidth for the mode-locked laser of 560 MHz for a 900 ps (FWHM) pulse, measured directly in the spectrum at –3 dB. Again, we found that the output pulses of this laser are best fitted by a sech^2 function rather than with a Gaussian function. From the Fourier-transform-limited relation for a sech^2 pulse, its time-bandwidth product cannot be lower than 0.315 (FWHM) (Wada et al., 2008). Since we obtain 560 MHz × 900 ps = 0.504, we conclude that the optical pulses of our modelocked laser could have some moderate degree of chirp.

According to Eq. (1), the pulse duration is limited by a variety of factors in an active modelocked laser. From these parameters, we believe the narrow spectral bandwidth of the AOSLM $\Delta\lambda$ plays a key role in this setup. As we explained before, these sidebands can be regarded as weak ghosts of the strong permanent Bragg grating, its FWHM bandwidth is that of a weak Bragg grating of the same length (Liu et al., 1997), namely $\Delta\lambda = 1.39\lambda^2/(\pi L n_0)$, where L is the fiber length and n_0 is the modal effective index. For the grating used in this chapter, with L = 120 mm, at λ = 1530 nm, this translates into $\Delta\lambda$ = 6 pm, which is equivalent to 770 MHz at the operation wavelength of this laser (1530.5 nm). As a rough estimation, since the cavity modes are distant 4.1 MHz, it is easily seen that this AOSLM only is able to lock a few percent of the axial modes available by the medium gain, namely 770 MHz /4.1 MHz = 192 modes (FWHM). If we assume for the output pulses of this laser a sech^2 envelope, then this parameter alone determines a lower limit for the pulse width around 0.315 /770 MHz = 410 ps. Therefore, we believe that narrower pulses can be reached by broadening the sidebands of the AOSLM in order to lock additional axial modes. The optical pulse's peak power and temporal width as a function of the pump power are shown in Fig. 23(b). A smooth variation of the pulse width with pump power can be observed, before reaching the gain saturation. Fig 23(c) shows the time width as a function of the frequency detuning. The behavior was asymmetric, i.e. it depends if the detuning was positive or negative. Once mode-locking is reached and the PC adjusted to get the minimum pulse width, a small detuning in either direction does not modify the pulse width –e.g. up to a few tens of Hertz, as expected for amplitude modulation mode-locking (Kuizenga & Siegman, 1970). However, when the detuning is considerably higher, a positive detuning continuously broadens the time width, whereas for a negative detuning, the mode-locking is rapidly lost, and the pulses drop out. We believe this asymmetry could be caused by the non-flat frequency response of the piezoelectric. We conclude this subsection by analyzing the behavior of this laser when the voltage applied to the piezoelectric changes. Figure 23(c) shows the influence on the time width of the optical pulses with the voltage applied to the piezoelectric. For higher voltages, the time width decreases; further narrowing by this trend appears to finish when the reflectivity of the sideband approaches the maximum (in addition, 20 V was the maximum voltage provided by the electrical amplifier used to drive the piezoelectric). Within the range of voltage available in our experiments, both the reflectivity and the modulation amplitude increase continuously. This is consistent with Eq. (1), by which the time width is inversely related with both, the modulation depth and the reflectivity.

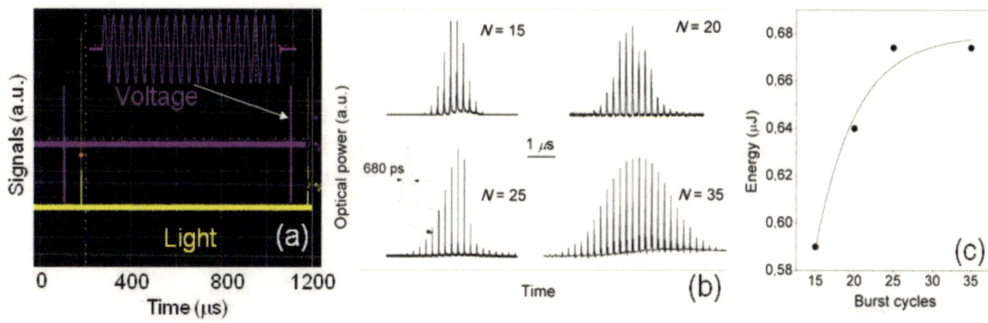

Fig. 24. (a) Voltage signal applied to the piezoelectric and *Q*-switched mode-locked train of pulses generated for a 1 kHz *Q*-switching repetition rate; the inset shows a detail of the burst signal. (b) Single *Q*-switched mode-locked pulses enveloping a train of modelocked pulses for different burst cycles; the inset shows a single-shot capture of a single modelocked pulse with a FWHM of 680 ps. (c) Energy of the *Q*-switched mode-locked pulses as a function of the cycles contained in the burst for a fixed *Q*-switching frequency of 500 Hz and a pump power of 270 mW

The most significant advantage of using travelling instead of standing acoustic waves is determined by the possibility to actively *Q*-switch this modelocked laser simply by launching to the grating a burst acoustical wave (Cuadrado-Laborde et al. 2010b). Therefore, the setup is exactly the same as the reported above for the standard mode-locking regime driven by travelling acoustic waves. The burst signal consists of an integer number of sinusoidal acoustical cycles *N*. Figure 24(a) shows the burst voltage signal applied to the piezoelectric together with the *Q*-switched mode-locked train of pulses generated, at a frequency rate of 1 kHz. Figure 24(b) shows several *Q*-switched mode-locked pulses as a function of *N*. When *N* is low, *Q*-switched mode-locked is not reached. For higher *N*, *Q*-switched mode-locked is reached and the emission is allowed. For example, the left bottom corner of Fig. 24(b) shows a single *Q*-switching envelope with a FWHM of 1.4 μs for *N* = 25; it has between 16 and 18 fully-modulated modelocked pulses, with the expected temporal separation given by 1/4.1 MHz = 244 ns. The inset shows a single-shot capture of a modelocked pulse within this train with a temporal width of 680 ps. As observed, the *Q*-switched mode-locked pulses of this laser have excellent inter-pulse characteristics. The energy of a *Q*-switched mode-locked train of pulses, as a function of *N* is shown in Fig. 24(c). This energy was measured directly with a pyroelectric detector. As expected, for a longer burst signal, the energy of the *Q*-switched mode-locked train of pulses increases, until it reaches saturation. Thus, a peak power higher than 200 W can be calculated for the central pulses of a train with energy of 0.68 μJ, assuming that the modelocked pulses are 680 ps width. This result clearly demonstrates the degree of enhancement in comparison with the peak power achieved when the laser was operated in the standard mode-locking regime, since the ratio is higher than 4×10^2.

3.3 Photonic true-time delay-line by acousto-optic interaction in a fiber Bragg grating in the short-wavelength regime

A photonic true-time-delay (PTTD) line is an optical device that permits to vary the group delay of an optical signal. The ideal PTTD line should exhibit fast tuning within a broad range of time delay values. PTTD lines attract much attention due to an increasing number of applications such as the control of phased array antennas (Liu et al., 2002; Perez-Millán et al., 2004; Italia et al., 2005;), tunable microwave band-pass filters (Capmany et al., 2005), buffering and packet synchronization (Li et al., 2007; Caucheteur et al., 2010). The advantages of PTTD lines are their broad bandwidth, high frequency operation and immunity to electromagnetic interference. In addition, fiber-based PTTD lines are readily compatible with fiber systems, robust, compact and lightweight, and they present low insertion loss.

Recently several types of fiber-based PTTD lines were proposed. For example, an optically controlled PTTD line based on stimulated Brillouin scattering demonstrates a fast time response within a relatively small range of the time delay (230 ps were reported using a fiber of 3.5 km length) (Zadok et al., 2007). Another example is the tunable PTTD line based on a uniform or chirped FBGs perturbed by heating or mechanical stress (Ortega et al., 2000; Liu et al., 2002; Perez-Millán et al., 2004), which permits simple experimental realization and easy tuning. However, such system has a very slow response (below 1 ms). In a number of PTTD lines tunable light sources are required to adjust the time delay (Liu et al., 2003; Blais et al., 2009), but this approach is not easily scalable when a large number of independent delays need to be adjusted.

Here we discuss a tunable PTTD line based on longitudinal ultra-sound modulation of a FBG written in a standard fiber similar to SMF-28. This device operates with a fixed optical carrier, demonstrates a wide time delay range (400 ps) and a fast response time [μs range, see Delgado-Pinar et al. (2006)].

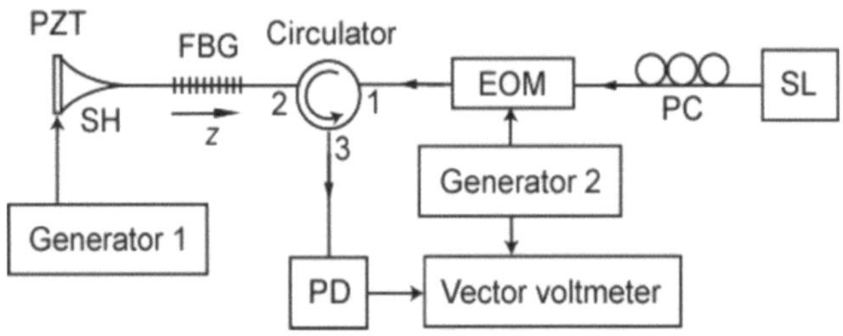

Fig. 25. Experimental setup. SL: semiconductor laser; PC: polarization controller; EOM: electro-optical amplitude modulator; FBG: fiber Bragg grating; PZT: piezoelectric transducer; SH: fused silica horn; PD: photodetector

The setup of the PTTD line is shown in Fig. 25. Light from a single-frequency semiconductor laser (SL) modulated by an electro-optical amplitude modulator (EOM) with a frequency of f_m = 400 MHz is sent through the optical circulator to a long uniform FBG written in a standard photosensitive fiber; the RF modulation depth is chosen to be comparatively low (~ 10%), which maintains a practically linear regime of light modulation. The FBG is 12 cm long, with a

Bragg wavelength of 1530.8 nm and reflectivity of 99.99% (40 dB transmission attenuation measured at the Bragg wavelength) that corresponds to a coupling coefficient $\kappa = 0.442$ cm^{-1}.
In the experimental arrangement, a longitudinal acoustic wave is generated by a piezoelectic (PZT) disk attached to a heat sink. This wave is launched along the single-mode optical fiber containing the uniform FBG using a fused silica horn (SH). The PZT is driven by Generator 1 at the PZT resonance frequency (2.08 MHz in our case). The light reflected by the FBG is registered by a photodetector connected to a vector voltmeter. AC voltage from Generator 2 drives the EOM; it is also used as a reference signal for the RF phase measurements.
The operation principle of such a PTTD line is based on the reflection of narrow-line light by one of the FBG side bands generated by the longitudinal acoustical wave. This wave modulates both the period and the refractive index of the grating producing thereby a sinusoidal chirp that moves along the fiber synchronously with the acoustic wave, and will be described by an effective phase modulation of the grating with amplitude $\Delta\Phi$. Under these conditions, the refractive index for the light mode propagating through the single-mode fiber core (so called LP_{01} mode) is written as follows:

$$\begin{aligned} n(z,t) &= n_0 + \Delta n \cos\left[Kz - \Delta\Phi \cos\left(k_s z + \Omega t\right)\right] = n_0 + \Delta n J_0(\Delta\Phi)\cos(Kz) \\ &+ 2\Delta n \cos(Kz)\sum_{m=1}^{\infty}(-1)^m J_{2m}(\Delta\Phi)\cos\left[2m\left(k_s z + \Omega t\right)\right] \\ &+ 2\Delta n \sin(Kz)\sum_{m=1}^{\infty}(-1)^{m-1} J_{2m-1}(\Delta\Phi)\cos\left[(2m-1)\left(k_s z + \Omega t\right)\right], \end{aligned} \tag{3}$$

where n_0 is the effective refractive index for LP_{01} mode, Δn is the grating amplitude, $K = 2\pi/\Lambda$ is the unperturbed grating wavenumber (Λ is the grating period), z is the distance along the fiber, $\Delta\Phi$ is the phase modulation amplitude, k_s is the acoustic wavenumber, Ω is the angular acoustic frequency, m is a natural number corresponding to the order of the FBG side band, and J_m is the Bessel function of the m-th order. From Eq. 3 one can see that the longitudinal acoustic wave produces a series of the traveling FBGs with amplitudes depending on $\Delta\Phi$.
Using the coupled-wave theory (Erdogan, 1997), one can obtain the following set of the coupled-wave equations for two contra-propagating optical waves (one probe wave, $A_{\pm l}$, and one reflected wave, $B_{\pm l}$) falling into the $\pm l$ sideband induced by the acoustic wave and interacting with the perturbed grating described by Eq. 3:

$$\frac{dA_{\pm l}}{dz} = -i\Delta\beta_{\pm l}A_{\pm l} + i^{(l-1)}\kappa_l B_{\pm l}, \tag{4}$$

$$\frac{dB_{\pm l}}{dz} = i\Delta\beta_{\pm l}B_{\pm l} + (-i)^{(l-1)}\kappa_l A_{\pm l}, \tag{5}$$

where $\pm l$ is the sideband number, $\Delta\beta_{\pm l} = \beta - (\beta_0 \mp lk_s/2) = 2\pi n_0(\lambda^{-1} - \lambda_0^{-1}) \pm l\pi/\lambda_s$ is the detuning from the $\pm$ l-sideband Bragg wavelength, λ_0 is the Bragg wavelength of the unperturbed FBG, λ is the probe light wavelength, β and β_0 are the light wavenumbers that correspond to λ and λ_0, and $\kappa_l = \kappa_0 J_l(\Delta\Phi)$ is the coupling coefficient for the l-th sideband (κ_0 is the coupling coefficient of the unperturbed FBG).
The analysis based on the theory developed by Barmenkov et al. (2006, 2010) permits one to write the equations for the sideband reflectivity, R_l, and the sideband effective length, $L_{\pm l}^{eff}$, in the following form:

$$R_{\pm l} = \left(\tanh\left(\kappa_{\pm l} L\right)\right)^2, \tag{6}$$

$$L_{\pm l}^{eff} = \frac{1}{2\kappa_{\pm l}} \tanh\left(\kappa_{\pm l} L\right) = \frac{L\sqrt{R_{\pm l}}}{2\operatorname{atanh}\left(\sqrt{R_{\pm l}}\right)}, \tag{7}$$

where L is the FBG length. The formula for the group delay is found as

$$\tau_{\pm l} = 2n_0 L_{\pm l}^{eff} / c. \tag{8}$$

From the last three equations one can conclude that the sideband diffraction efficiency, the effective length and the group delay may be controlled by the amplitude of the phase modulation of the grating induced by the acoustic wave, which, in turn, depends on the acoustic wave magnitude. The sideband Bragg wavelength can be tuned by adjusting the acoustical frequency within the PZT resonance in the case of a slight tuning, or by replacing the PZT with another one having the necessary resonance frequency.

The diffraction efficiency of the FBG sidebands can be measured using a wide-spectrum LED connected directly to the circulator port 1, and an optical spectrum analyzer (OSA) connected, in turn, to the port 3 instead of a photodetector (see Fig. 25); The Generator 2 is switched off. Fig. 26(a) shows the FBG spectra at different AC voltages applied to the PZT. At low voltages, the ±1 sidebands have amplitudes much higher than that of the high-order sidebands. Reflectivity of the ±1 sidebands did not reach 100% because the OSA resolution was not enough to resolve comparatively narrow reflection peaks (3 dB spectrum width is about 40 pm). Note that a proper FBG design based on the grating apodization and chirping should permit increasing the sideband width.

As it is seen from Fig. 26 (b), the dependence of the +1 sideband efficiency on the voltage applied to the PZT is in good agreement with Eq. (6), assuming that the amplitude of the phase modulation of the grating is proportional to the PZT voltage; a slight difference observed at high voltage amplitudes (> 4V) could be explained by a small heating of the PZT, which slightly changes the PZT electromechanical coupling factor. Note that the ±1 sideband allows operation of AOSLM at low PZT voltages in comparison with high-order sidebands, the –1 sideband has the same properties as the +1 sideband.

The RF modulation envelope phase, Ψ_{+1}, is related directly to the +1 sideband group delay τ_{+1} and the effective length L_{+1}^{eff}:

$$\Psi_{+1} = 2\pi\tau_{+1} f_m = 4\pi n_0 L_{+1}^{eff} f_m / c \tag{9}$$

Thus, the dependence of Ψ_{+1} on +1 sideband efficiency is the basic feature that permits to implement a dynamic control of the group delay and the effective length of the grating by means of the AC voltage. The same is for the –1 sideband.

Fig. 27(a) shows the experimental relationship between Ψ_{+1} and the amplitude of the AC voltage applied to the PZT. One can see that the phase decreases as voltage amplitude increases, which is explained by decreasing of the group delay and the effective length. Figure 27(b) plots these two parameters versus the +1 sideband diffraction efficiency. The experimental data show a good agreement with Eqs. (7) and (8). Since the optical path between the EOM and the photodetector was relatively long (the fiber length was ~ 2 m), producing an additional phase shift into the measured Ψ_{+1} values, the experimental data were corrected by a phase offset (the right scale in Fig. 27(a) that permits to compensate the fiber length excess. The correcting parameter was chosen so that the +1 sideband effective length is equal to a half of FBG physical length at low sideband efficiency (low PZT voltage), as the Eq. (7) predicts.

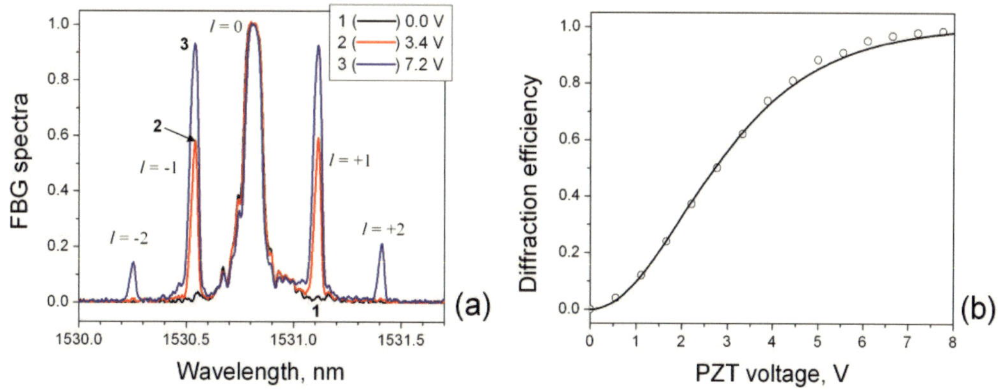

Fig. 26. (a) Spectra of FBG perturbed by ultra-sound wave at different PZT voltages (the voltage values are shown in the upper right corner); *l* indicates the sideband number. The OSA resolution is 20 pm. (b) Dependence of the +1 sideband efficiency on PZT voltage. Circles: experimental data; solid line: fitting

From Fig. 27(b) one can conclude that the group delay and the effective length for the +1 sideband depend on its efficiency that is controlled electrically by the AC voltage applied to the PZT. In the experiments, the group delay was adjustable from 150 ps to 550 ps, i.e., covering a range of 400 ps. Within the 3 dB range of the sideband efficiency, the group delay could be continuously adjusted from 150 to 450 ps (a range of 300 ps). The implementation of an automatic SL power control would permit to adjust the group delay at constant optical power, compensating for the reflectivity changes of the sideband. The response time of this PTTD line is approximately 20 μs and is determined by the acoustic wave speed and the grating length.

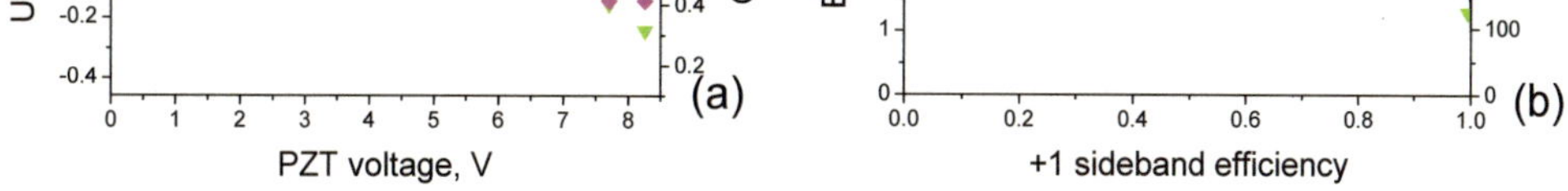

Fig. 27. (a) Dependence of the RF modulation envelope phase of light reflected from the FBG +1 sideband on amplitude of voltage applied to PZT. The left scale: uncorrected values, the right scale: corrected values. (b) FBG +1 sideband effective length and group delay versus the diffraction efficiency. Symbols: experimental data; solid line: curve calculated by Eq. (7). In both figures different symbols correspond to different experimental series

In summary, the experimental and theoretical study of the phase and group delay response of a FBG modulated by a longitudinal acoustic wave permits the implementation of electrically-tuned PTTD line controlled by the AC voltage applied to the piezo-electric transducer that generates the acoustic wave. The proposed PTTD line permits to vary the group delay in the range of 400 ps.

3.4 *Q*-switching in a distributed feedback fiber laser

When the acoustic perturbation is not a harmonic wave but a single pulse, its passage through the FBG creates a defect which can be used to control the *Q*-factor in a DFB all-fiber laser. Single-mode narrow-linewidth lasers are of great interest as they have a wide applicability in fields such as high-resolution interferometry, distributed Brillouin sensing, optical coherent communications, etc. DFB lasers based on semiconductor technology have linewidths typically in the MHz domain; due to the short cavity length and low thermal dissipation its output spectrum is typically broad. Fiber lasers usually operate in a multi-mode regime and have a broad spectrum; to make efficient single-mode narrow-linewidth fiber lasers different approaches have been implemented. Fiber rings lasers fulfill the requirement and produce narrow-linewidth outputs; most importantly, they can be made widely tunable. However, they are more complex and –because of their long cavities– susceptible to mode hopping. Distributed Bragg reflector (DBR) fiber lasers is another option, however, temperature stabilization is still required to prevent mode hopping and, in the short laser cavity, the pump wave absorption is low, as a consequence the DBR fiber laser is not efficient and external optical amplifiers are generally needed (Babin et al., 2007). Distributed feedback fiber lasers can overcome part of these problems. They have the simplest, robust and compact design providing operation without mode hopping. Its fabrication is relatively simple, and involves the writing of a grating structure –i.e. a fiber Bragg grating– with ultraviolet light into an active fiber. Single-mode pump leads to an alignment free resonator with optimum overlap of pump and signal light. For these FBG-based DFB lasers, the distributed reflection occurs in the grating when a phase shift has been generated within it. A number of techniques have been proposed for this; however, statics phase shifts only allow CW operation, which translates into low power emission. Recently, some approaches have been reported to obtain single-frequency pulsed all-fiber lasers, based on active *Q*-switching of DFB fiber cavities. The pulsed operation of a DFB laser is interesting because for certain applications (e.g. Brillouin sensing), it is not only required a narrow linewidth but also a peak power about a minimum threshold, which otherwise could not be reached by a CW operation (Cuadrado-Laborde et al., 2008).

Figure 28(a) shows the scheme of the proposed DFB all-fiber laser. The FBG was 100 mm long, and was written in a 1500 ppm erbium hydrogen-loaded fiber (codoped with germanium and aluminum) of the same length using a doubled argon laser and a uniform period mask. The FBG shows more than 30 dB attenuation at the Bragg wavelength of 1532.45 nm and a 3 dB bandwidth of 88 pm. The FBG was pumped through a 980/1550 nm WDM with a 980 nm semiconductor laser, providing a maximum pump power of 130 mW. A square shape rod of a magnetostrictive material (Terfenol-D, 15 mm long and 1 mm^2 section) was bonded outside the FBG to a free section of fiber at 88 mm from the center of the grating, and placed inside a small coil, see Fig. 28(a).

The acoustic pulse generated by using a magnetostrictive device has a superior performance as compared with piezoelectric devices which have a frequency-dependent response characterized by strong mechanical resonances. On the contrary, a magnetostrictive device has a lower frequency range but its frequency response is basically flat. The magnetostrictive rod can be bonded directly over the FBG, but this has some detrimental consequences. First, any external actuator bonded directly on the fiber is likely to exhibit long-term instabilities (Andrés et al., 2008). Second, once the magnetostrictive rod is fixed to the fiber, any temperature change generates a differential expansion between the fiber and the magnetostrictive rod, producing a local static perturbation of the FBG that may cause CW emission. Because of this, we avoided these drawbacks by attaching he magnetostrictive rod outside the FBG and generating a dynamic defect in the FBG through an acoustic axially-propagating pulse. Thereby, when a pulse of electric current is applied to the coil, the small rod lengthens and stretches the section of fiber attached to it, generating in this way a longitudinal acoustic pulse that propagates toward the FBG. The pulse propagating along the grating generates a phase shift opening a transmission peak within the reflection band of the grating; as a consequence a high Q resonance is produced and a laser pulse is emitted (Pérez-Millán et al., 2005b; Delgado-Pinar et al., 2007). Otherwise, if no perturbation is present within the FBG, there is no efficient feedback for the optical signal, and the laser emission is not allowed. Since the defect is not fixed, but induced by a travelling acoustic pulse with a time-varying position along the FGB, one could think that this might have important consequences on the spectral position of the transmission peak. However, this is not the case, as it was recently demonstrated (Andrés et al., 2008). The spectral position of the resonance is constant, no matter the actual spatial location of the acoustic pulse, although some short transients are produced when the pulse overlaps the extremes of the FBG. This property insures that the laser will emit with a narrow linewidth, preserving one of the most attractive properties of DFB fiber lasers. The transmission properties of the passive FBG –i.e. without pumping– interacting with the acoustic pulses were investigated by illuminating the FBG with a tunable laser at the Bragg wavelength and detecting the reflected signal. When the coil current is zero, the reflectance is maximal; but when a current pulse of 220 mA and 5 µs temporal width is applied to the coil, a transmission peak opens the reflection band, being as narrow as 2 µs (FWHM), see Fig. 28(b). Figures 29(a) to 29(c) shows the voltage pulses applied to the current driver, the backward output train of the DFB laser, a detail of the optical pulse, and a single current pulse applied to the coil, respectively. The observed delay time between the voltage signal and the optical pulses is mainly due to the distance that pulses have to travel from the magnetostrictive device to the FBG, see Fig. 28(a).

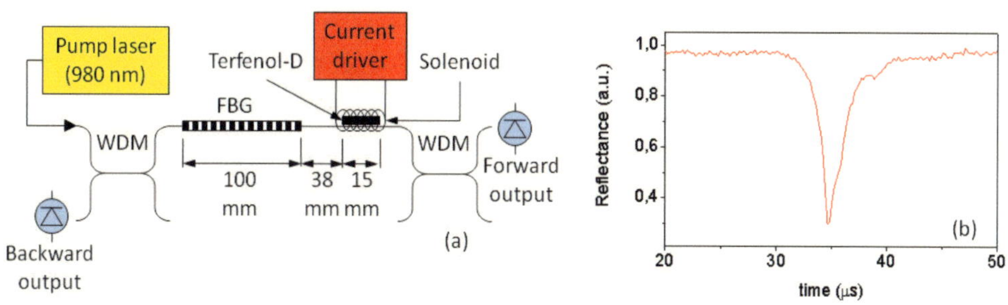

Fig. 28. (a) Q-switched distributed feedback-fiber laser setup. (b) Reflection at the Bragg wavelength when an acoustic pulse travels along the FBG

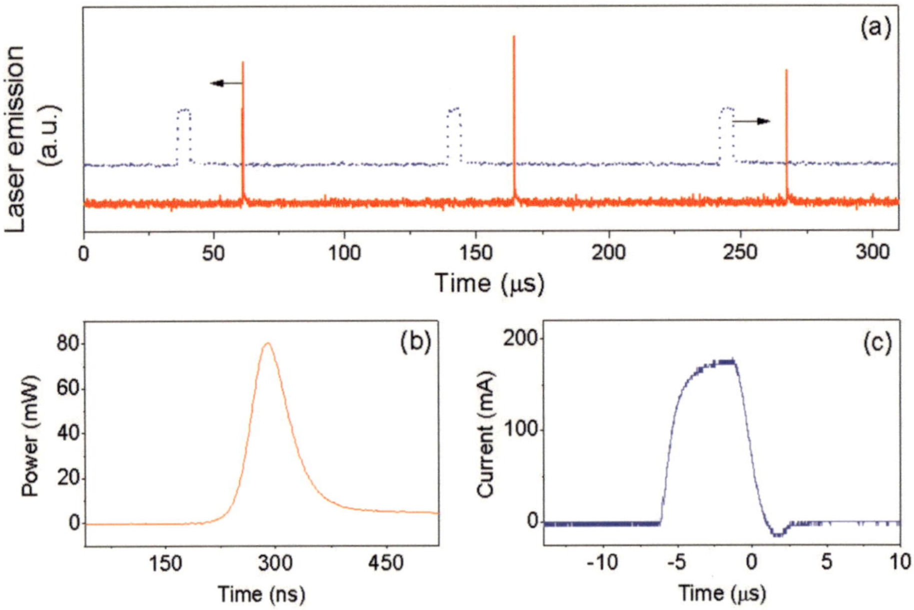

Fig. 29. Q-switched DFB behavior at 10 kHz repetition rate and 80 mW of pump power. (a) Emitted optical train pulses and voltage signal applied to the current driver, (b) detail of a single optical pulse of the train, and (c) detail of a single current pulse applied to the coil

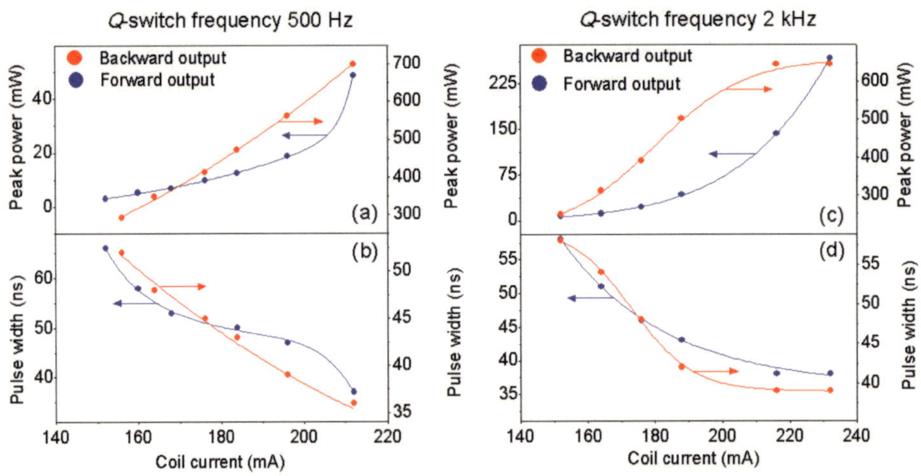

Fig. 30. Peak power and pulse width for backward and forward outputs as a function of the coil current for two different Q-switched repetition rates: 500 Hz (a, b) and 2 kHz (c, d), respectively, and 55mW of pump power

One key characteristic in this setup is its versatility, which is given by the possibility of selecting a variety of peak powers and time widths just by varying the coil current. Figure 30 shows the peak power and pulse width for backward and forward outputs as a function of the coil current for two different repetition rates: 500 Hz and 2 kHz, Figs. 30(a-b) and 30(cd), respectively. There is a low threshold value for the coil current (I), bellow this value; there is not laser emission, since the Q value is not high enough (i.e. I < 150 mA). Above this value, the laser emission is allowed and one optical pulse per cycle is emitted; at higher electric currents (i.e., I > 230 mA) there are two optical pulses per current pulse. At different repetition rates the peak power and time width values change but the general trend is preserved. The effect of pump power on the optical pulses, for repetition rates ranging from 200 Hz to 20 kHz, is shown in Figs. 31(a,b) and 31(c,d), for the backward and forward outputs, respectively. A fixed coil current of 200 mA amplitude was used in all cases. At each different frequency, there is a corresponding pump power threshold. Above threshold, the peak power increases with pump power and there is a corresponding reduction of the pulse width. Pulses of 800 mW peak power and 32 ns time width were obtained at 500 Hz repetition rate and 46 mW pump power for the backward output. If pump power is high enough, the laser emits more than one pulse per cycle, defining in this way an upper limit also. It can be observed the differences in peak powers between backward and forward outputs, being higher in the first case. Temporal widths and peak power jitters were measured to be below 5 %.

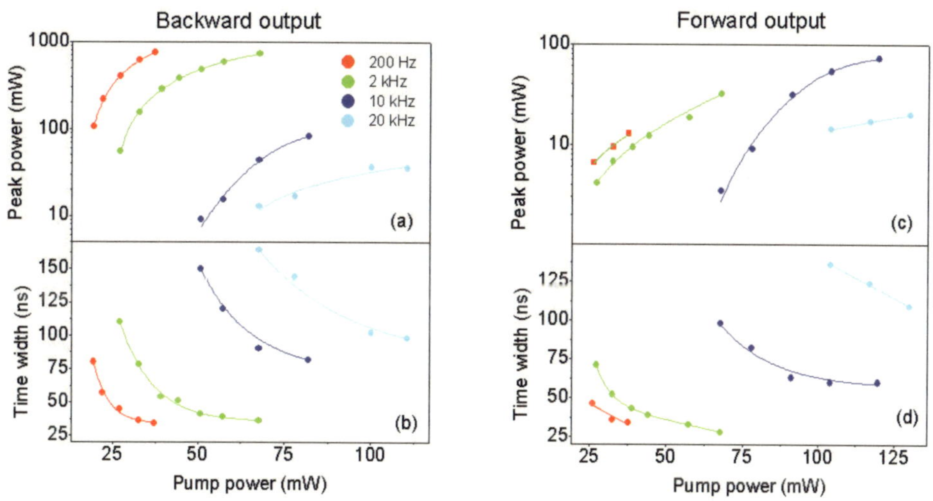

Fig. 31. Peak power and pulse width for the backward (a, b) and forward (c, d) laser outputs, as a function of the pump power, for several repetition rates

The emission linewidth was measured, using a classical heterodyne configuration (Galtarossa et al., 1993). A tunable laser was used as local oscillator with spectral linewidth of 100 kHz. The output of this laser was superimposed to the optical pulses of the DFB laser (backward output), through a 1550 nm 20 dB coupler. The beat signal at the coupler output was detected with a 1 GHz bandwidth photodetector, and analyzed with a 500 MHz bandwidth oscilloscope. From the spectrum of the beating signals, resulted a DFB laser linewidth of 6 MHz. The optical pulses from the DFB all-fiber laser had a temporal width of 80 ns, so according to the time-frequency uncertainty principle (Agrawal, 2001), its bandwidth cannot be lower than 5.5 MHz. From the comparison between this last value and the linewidth measurement, we conclude that the optical pulses of our DFB fiber laser are transform-limited.
Finally, as an example of application of this compact laser source, we demonstrate the possibility to generate Brillouin scattering for sensing purposes. Brillouin scattering

essentially refers to the scattering of a light wave by an acoustic wave (Agrawal, 2001). When a coherent pulse of light propagates through a medium, part of its energy is backscattered due to a non elastic interaction with the acoustic phonons. This back-scattered light is composed of a frequency down-shifted Stokes light and an up-shifted anti-Stokes light, whose spectral positions are dependent on temperature and strain of the fiber, in this way allowing its use as a sensing mechanism (Culverhouse et al., 1989a, 1989b; Bao et al., 1995; Parker et al., 1998; Y. Li et al., 2003). Figure 32(b) shows the backscattered light spectrum after illuminating a 10.5 km length optical fiber spool (Corning SMF-28) with the backward output of our DFB all-fiber laser (4 kHz repetition rate and 74 mW pump power). The Brillouin spectrum was registered with an optical spectral analyzer (resolution of 20 pm). The extreme of the fiber optic spool was terminated with a matching refractive index liquid (n = 1.46). The central (highest) peak corresponds to the proper laser beam reflections after successive connections and splices together with Rayleigh scattering. Peaks symmetrically positioned at both sides correspond to the Brillouin backscattering by Stokes and anti-Stokes processes (Parker et al., 1998). The measured Brillouin shift results in 88 pm (i.e. 11.24 GHz at 1532.4 nm), which correspond to the expected value in this fiber (Agrawal, 2001). Fiber optics distributed temperature, and/or strain, sensors have becoming very attractive for applications requiring sensing lengths of many kilometers, principally due to its inexpensiveness and availability. Optical fiber based distributed sensor systems normally make use of the principle of optical time domain reflectometry (Parker et al., 1998). Therefore, an optical pulse is launched into one end of the fiber system and the variation of the scattered light is detected as a function of time, giving in this way information of temperature or strain as a function of distance. A key requirement in this measurement system is a stable light source with a narrow enough spectral linewidth. In addition, for time domain reflectometry applications, the sensor spatial resolution proportionally depends on the optical pulse width, so it must be considered also. In order to fulfill all these requirements, solid state lasers with external cavities –plus amplifiers and amplitude modulators– are currently used in these systems (Parker et al., 1998). Here we propose using this compact all-fiber pulsed light source as a relatively simpler alternative.

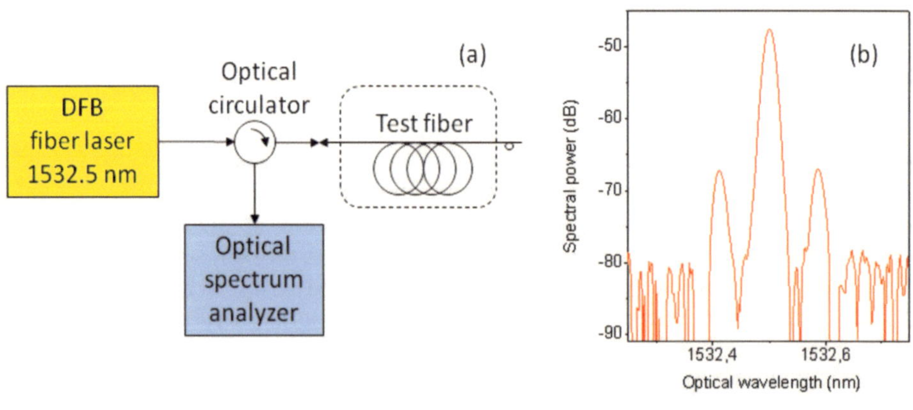

Fig. 32. Schematic diagram of the setup used for the Brillouin backscattering measurements. Brillouin spectra at room temperature for a 10.5-km length Corning SMF-28 optical fiber spool, for a backward output at 4 kHz repetition rate and 74 mW pump power

4. Conclusions

In this chapter we have discussed the use of acoustic waves to control all-fiber devices for different applications. In Section 2 we focused on novel applications of acousto-optic fiber devices based on flexural acoustic waves. As an example of application, we described in subsection 2.1 an actively *Q*-switched ytterbium-doped strictly all-fiber laser. *Q*-switching modulation was achieved by intermodal modulation induced by flexural acoustic waves travelling in a tapered optical fiber. *Q*-switched light pulses at 1064.1 nm were successfully obtained at repetition rates in the range 1-10 kHz, with pump powers between 59 mW and 88 mW. Best results were for laser pulses of 118 mW peak power, 1.8 μs of time width, with a pump power of 79 mW, at 7 kHz repetition rate.
Next, in subsection 2.2 we described an actively mode-locked fiber ring laser. As mode-locker was proposed an acousto-optic modulator driven by standing flexural acoustic waves, which couples core-to-cladding modes in a standard single-mode optical fiber. Among the remarkable features of the modulator, we mention its high modulation depth (72%), broad bandwidth (187 GHz), easy tunability in the optical wavelength, and low insertion losses (0.7 dB). The narrowest optical pulses obtained were of 95 ps time width, 21 mW peak power, repetition rate of 4.758 MHz, and 110 mW of pump power. We also characterized this laser as a function of the RF voltage that controls the modulator, the length of the active fiber, and the optical bandwidth of an intracavity filter implemented with a fiber Bragg grating. Best results were for output light pulses of 34 ps time width, and 1.4 W peak power.
In Section 3 we focused on novel applications of acousto-optic devices based on the interaction of longitudinal acoustic waves with fiber Bragg gratings. From the comparison of the acoustic wavelength with the periodicity of the FBG, two well-different situations could be distinguished: the long-wavelength and the short-wavelength regimes. As an example of application of the former, we described in subsection 3.1 an in-fiber resonant acousto-optic modulator suitable for *Q*-switching applications. The modulator consists of a short-length FBG modulated by a long-wavelength standing longitudinal acoustic wave. The acoustic wave shifts periodically in time the reflection band of the grating along a given wavelength range; by using this modulator, we demonstrated an actively *Q*-switched all-fiber laser. Output light pulses of 1.6 W peak power and 172 ns time width were obtained at 18 kHz repetition rate.
Next, in section 3.2 and 3.3 we focused on applications of acousto-optic devices based on the interaction of longitudinal acoustic waves with fiber Bragg gratings in the short-wavelength regime. Thus, in subsection 3.2 we proposed the use of an acousto-optic supper-lattice modulator as mode-locker. Since an acousto-optic supper-lattice modulator can be driven in two different ways either by standing or traveling acoustic waves; we discussed first the construction of a mode-locked laser driven by standing acoustic waves. In this configuration we obtained transform-limited optical light pulses of up to 120 mW peak power and 780 ps pulse width generated at a fixed repetition rate of 9 MHz, with an emission linewidth of 2.8 pm at 1530.5 nm. We also study the influence of different parameters on the mode-locking process such as: frequency detuning, EDF length, amplitude modulation, and dispersion. In this case, narrower pulses were obtained at higher modulation depths, normal dispersion, and shorter lengths of active fiber. Best results were reached when the laser was optimized according to these variables (160mW peak power and 630 ps pulse width). On the other hand, we showed that by slightly modifying the cavity, it is possible to operate the laser in a double-active *Q*-switching mode-locking regime. It is worth to mention that this approach was unique, being this double-active all-fiber laser the first of its kind. To this end, we attached a magnetostrictive rod to the output FBG to modulate the *Q* factor of the Fabry–

Perot cavity. Fully modulated Q-switched mode-locked trains of optical pulses were obtained for a wide range of pump powers and repetition rates. For a Q-switched repetition rate of 500 Hz and a pump power of 100 mW, the laser generates trains of 12–14 modelocked pulses of about 1 ns each, within an envelope of 550 ns, an overall energy of 0.65 µJ, and a peak power higher than 250 W for the central pulses of the train. Then, we discussed the construction of a mode-locked laser when the mode-locker is driven by travelling acoustic waves. In this case, the modulation frequency is half the frequency obtained when standing acoustic waves are used. Optical pulses were obtained of 530 mW peak power, 700 ps pulse width, at a repetition rate of 4.1 MHz. The variation of the pulses parameters under frequency detuning and applied voltage was also studied. Finally, we demonstrated that it is not necessary to modify the setup in order to reach double active Q-switching and mode-locking, when travelling acoustic waves were used to drive the mode-locker. In this case the commutation between mode-locking and Q-switching mode-locking is remarkably simple; it just needs use of a different electrical signal to drive the piezoelectric of the mode-locker, i.e. from a sinusoidal to a burst-sinusoidal electrical signal. In this case, fully modulated 10–25 modelocked pulses around 700 ps each within a Q-switching envelope around 1 µs and a maximum overall energy of 0.68 µJ were obtained.

In subsection 3.3 we shown another example of application of acousto-optic devices based on the interaction of longitudinal acoustic waves with fiber Bragg gratings in the short-wavelength regime. Thus, we carried out an experimental and theoretical study of the phase and group delay response of the acousto-optic supper-lattice modulator. The phase properties of the first sidebands permit the implementation of electrically-tuned photonic true-time delay line controlled by the AC voltage applied to the piezoelectric transducer that generates the acoustic wave. The proposed photonic true-time delay line permits to vary the group delay up to 400 ps.

Finally, when the acoustic perturbation is not a harmonic wave but a single pulse, its passage through the FBG creates a defect which can be used to control the Q-factor in a DFB all-fiber laser, subsection 3.4. Thus, we shown a single-mode, transform-limited, actively Q-switched distributed-feedback fiber laser. Optical pulses of 800 mW peak power, 32 ns temporal width, and up to 20 kHz repetition rates were obtained. The measured linewidth demonstrates that these pulses are transform limited: 6 MHz for a train of pulses of 10 kHz repetition rate, 80 ns temporal width, and 60 mW peak power. Efficient excitation of spontaneous Brillouin scattering was demonstrated.

In summary, photonic devices can benefit highly of a strictly all-fiber configuration which provides them a series of attractive advantages. Among all the proposed in-fiber solutions, devices controlled by acoustic waves have been by far the most employed, especially in mode-locked lasers, providing a broad range of alternatives. The recent advances in acoustically controlled photonic systems positioned them as a promising candidate for commercially available systems in the near future.

5. Acknowledgments

This work has been financially supported by the *Ministerio de Ciencia e Innovación* and the *Generalitat Valenciana* of Spain (projects TEC2008-05490 and PROMETEO/2009/077, respectively). C. Cuadrado-Laborde acknowledges the *Secretaría de Estado de Universidades e Investigación del Ministerio de Ciencia e Innovación* (Spain) and ANPCyT (project PICT 2008-1506, Argentina).

6. References

Agrawal, G. P. (2001). *Nonlinear Fiber Optics*, Academic Press, New York, 2001.

Álvarez-Chávez, J. A.; Offerhaus, H. L.; Nilsson, J.; Turner, P. W.; Clarkson, W. A.; Richardson, D. J. (2000). High-energy, high-power ytterbium-doped Q-switched fiber laser. *Optics Letters*, 25 (1): 37-39 Jan 1 2000.

Andersen, T. V.; Pérez-Millán, P.; Keiding, S. R.; Agger, S.; Duchowicz, R.; Andrés, M. V. (2006). All-fiber actively Q-switched Yb-doped laser. *Optics Communications*, 260 (1): 251-256 Apr 1 2006.

Andres, M. V.; Cruz, J. L.; Díez, A.; Pérez-Millán, P.; Delgado-Pinar, M. (2008). Actively Q-switched all-fiber lasers. *Laser Physics Letters*, 5 (2): 93-99 Feb 2008.

Babin, S. A.; Churkin, D. V.; Ismagulov, A. E.; Kablukov, S. I.; Nikulin, M. A. (2007). Single frequency single polarization DFB fiber laser. *Laser Physics Letters*, 4 (6): 428-432 Jun 2007.

Bao, X.; Dhliwayo, J.; Heron, N.; Webb, D. J.; Jackson, D. A. (1995). Experimental and theoretical-studies on a distributed temperature sensor-based on Brillouin-scattering. *Journal of Lightwave Technology*, 13 (7): 1340-1348 Jul 1995.

Barmenkov, Y. O.; Zalvidea, D.; Torres-Peiró, S.; Cruz, J. L.; Andrés, M. V. (2006). Effective length of short Fabry-Perot cavity formed by uniform fiber Bragg gratings. *Optics Express*, 14 (14): 6394-6399 Jul 10 2006.

Barmenkov, Y. O.; Cruz, J. L.; Díez, A.; Andrés, M. V. (2010). Electrically tunable photonic true-time-delay line. *Optics Express*, 18 (17): 17859-17864 Aug 16 2010.

Belanger, P. A. (2005). On the profile of pulses generated by fiber lasers. *Optics Express*, 13 (20): 8089-8096 Oct 3 2005.

Bello-Jiménez, M.; Cuadrado-Laborde, C.; Sáez-Rodriguez, D.; Díez, A.; Cruz, J. L.; Andrés, M. V. (2010). Actively mode-locked fiber ring laser by intermodal acousto-optic modulation. *Optics Letters*, 35 (22): 3781-3783 Nov 15 2010.

Bello-Jiménez, M.; Cuadrado-Laborde, C.; Díez, A.; Cruz, J. L.; Andrés, M. V. (2011). Experimental study of an actively mode-locked fiber ring laser based on in-fiber amplitude modulation. *Applied Physics B*, in press (2011).

Birks, T. A.; Russell, P. S. J.; Pannell, C. N. (1994). Low-power acoustooptic device based on a tapered single-mode fiber. *IEEE Photonics Technology Letters*, 6 (6): 725-727 Jun 1994.

Blais, S.; Yao, J. P. (2009). Photonic true-time delay beamforming based on superstructured fiber Bragg gratings with linearly increasing equivalent chirps. *Journal of Lightwave Technology*, 27 (9): 1147-1154 May 1 2009.

Bonadeo, N. H.; Knox, W. H.; Roth, J. M.; Bergman, K. (2000). Passive harmonic mode-locked soliton fiber laser stabilized by an optically pumped saturable Bragg reflector. *Optics Letters*, 25 (19): 1421-1423 Oct 1 2000.

Capmany, J.; Ortega, B.; Pastor, D.; Sales, S. (2005). Discrete-time optical processing of microwave signals. *Journal of Lightwave Technology*, 23 (2): 702-723 Feb 2005.

Caucheteur, C.; Mussot, A.; Bette, S.; Kudlinski, A.; Douay, M.; Louvergneaux, E.; Megret, P.; Taki, M.; Gonzalez-Herraez, M. (2010). All-fiber tunable optical delay line. *Optics Express*, 18 (3): 3093-3100 Feb 1 2010.

Costantini, D. M.; Limberger, H. G.; Lasser, T.; Muller, C. A. P.; Zellmer, H.; Riedel, P.; Tunnermann, A. (2000). Actively mode-locked visible upconversion fiber laser. *Optics Letters*, 25 (19): 1445-1447 Oct 1 2000.

Cuadrado-Laborde, C.; Delgado-Pinar, M.; Torres-Peiró, S.; Díez, A.; Andrés, M. V. (2007). Q-switched all-fibre laser using a fibre-optic resonant acousto-optic modulator. *Optics Communications*, 274 (2): 407-411 Jun 15 2007.

Cuadrado-Laborde, C.; Perez-Millán, P.; Andres, M. V.; Díez, A.; Cruz, J. L.; Barmenkov, Y. O. (2008). Transform-limited pulses generated by an actively Q-switched distributed fiber laser. *Optics Letters,* 33 (22): 2590-2592 Nov 15 2008.

Cuadrado-Laborde, C.; Díez, A.; Delgado-Pinar, M.; Cruz, J. L.; Andres, M. V. (2009a). Mode locking of an all-fiber laser by acousto-optic superlattice modulation. *Optics Letters,* 34 (7): 1111-1113 Apr 1 2009.

Cuadrado-Laborde, C.; Diez, A.; Cruz, J. L.; Andres, M. V. (2009b). Doubly active Q switching and mode locking of an all-fiber laser. *Optics Letters,* 34 (18): 2709-2711 Sep 15 2009.

Cuadrado-Laborde, C.; Diez, A.; Cruz, J. L.; Andres, M. V. (2010a). Experimental study of an all-fiber laser actively mode-locked by standing-wave acousto-optic modulation. *Applied Physics B-Lasers and Optics,* 99 (1-2): 95-99 Apr 2010.

Cuadrado-Laborde, C.; Diez, A.; Cruz, J. L.; Andres, M. V. (2010b). Actively Q-switched and modelocked all-fiber lasers. *Laser Physics Letters,* 7 (12): 870-875 Dec 2010.

Culverhouse, D. O.; Farahi, F.; Pannell, C. N.; Jackson, D. A. (1989a). Potential of stimulated Brillouin-scattering as sensing mechanism for distributed temperature sensors. *Electronics Letters,* 25 (14): 913-915 Jul 6 1989.

Culverhouse, D. O.; Farahi, F.; Pannell, C. N.; Jackson, D. A. (1989b). Stimulated Brillouin-scattering - a means to realize tunable microwave generator or distributed temperature sensor. *Electronics Letters,* 25 (14): 915-916 Jul 6 1989.

Culverhouse, D. O.; Richardson, D. J.; Birks, T. A.; Russell, P. S. J. (1995). All-fiber sliding-frequency Er^{3+}/Yb^{3+} soliton laser. *Optics Letters,* 20 (23): 2381-2383 Dec 1 1995.

Delgado-Pinar, M.; Zalvidea, D.; Diez, A.; Perez-Millan, P.; Andres, M. V. (2006). *Q*-switching of an all-fiber laser by acousto-optic modulation of a fiber Bragg grating. *Optics Express,* 14 (3): 1106-1112 Feb 6 2006.

Delgado-Pinar, M.; Diez, A.; Cruz, J. L.; Andres, M. V. (2007). Single-frequency active Q-switched distributed fiber laser using acoustic waves. *Applied Physics Letters,* 90 (17): Art. No. 171110 Apr 23 2007.

Erdogan, T. (1997). Fiber crating spectra. *Journal of Lightwave Technology,* 15 (8): 1277-1294 Aug 1997.

French, P. M. W. (1995). The generation of ultrashort laser pulses. *Reports on Progress in Physics* 58 (2): 169-267 Feb 1995.

Galtarossa, A.; Nava, E.; Valentini, G. (1993). *Single-Mode Optical Fiber Measurement: Characterization and Sensing,* Ed. G. Cancellieri, Artech Pub. 1993.

Geister, G.; Ulrich, R. (1988). Neodymium-fiber laser with integrated-optic mode locker. *Optics Communications,* 68 (3): 187-189 Oct 1 1988.

Haus, H. A. (2000). Mode-locking of lasers. *IEEE Journal of Selected Topics in Quantum Electronics,* 6 (6): 1173-1185 Nov-Dec 2000.

Huang, D. W.; Liu, W. F.; Yang, C. C. (2000). *Q*-switched all-fiber laser with an acoustically modulated fiber attenuator. *IEEE Photonics Technology Letters,* 12 (9): 1153-1155 Sep 2000.

Hudson, D. D.; Holman, K. W.; Jones, R. J.; Cundiff, S. T.; Ye, J.; Jones, D. J. (2005). Mode-locked fiber laser frequency-controlled with an intracavity electro-optic modulator. *Optics Letters,* 30 (21): 2948-2950 Nov 1 2005.

Imai, T.; Komukai, T.; Yamamoto, T.; Nakazawa, M. (1997). A wavelength tunable Q-switched fiber laser using fiber Bragg gratings. *Electronics and Communications in Japan Part II-Electronics,* 80 (11): 12-21 Nov 1997.

Italia, V.; Pisco, M.; Campopiano, S.; Cusano, A.; Cutolo, A. (2005). Chirped fiber Bragg gratings for electrically tunable time delay lines. *IEEE Journal of Selected Topics in Quantum Electronics,* 11 (2): 408-416 Mar-Apr 2005.

Jeon, M. Y.; Lee, H. K.; Kim, K. H.; Lee, E. H.; Oh, W. Y.; Kim, B. Y.; Lee, H. W.; Koh, Y. W. (1998). Harmonically mode-locked fiber laser with an acousto-optic modulator in a Sagnac loop and Faraday rotating mirror cavity. *Optics Communications,* 149 (4-6): 312-316 Apr 15 1998.

Kee, H. H.; Lees, G. P.; Newson, T. P. (1998). Narrow linewidth CW and Q-switched erbium-doped fibre loop laser. *Electronics Letters,* 34 (13): 1318-1319 Jun 25 1998.

Kim, B. Y.; Blake, J. N.; Engan, H. E.; Shaw, H. J. (1986). All-fiber acoustooptic frequency shifter. *Optics Letters,* 11 (6): 389-391 Jun 1986.

Kim, H. S.; Yun, S. H.; Kwang, I. K.; Kim, B. Y. (1997). All-fiber acousto-optic tunable notch filter with electronically controllable spectral profile. *Optics Letters,* 22 (19): 1476-1478 Oct 1 1997.

Kuizenga, D. J.; Siegman, A. E. (1970). FM and AM mode locking of homogeneous laser .1. Theory. *IEEE Journal of Quantum Electronics,* QE 6 (11): 694-& 1970.

Li, Y. H.; Lou, C. Y.; Han, M.; Gao, Y. Z. (2001). Detuning characteristics of the AM mode-locked fiber laser. *Optical and Quantum Electronics,* 33 (6): 589-597 Jun 2001.

Li, Y. Q.; Zhang, F. C.; Yoshino, T. (2003). Wide-range temperature dependence of Brillouin shift in a dispersion-shifted fiber and its annealing effect. *Journal of Lightwave Technology,* 21 (7): 1663-1667 Jul 2003.

Li, X. W.; Peng, L. M.; Wang, S. B.; Kim, Y. C.; Chen, J. P. (2007). A novel kind of programmable 3(n) feed-forward optical fiber true delay line based on SOA. *Optics Express,* 15 (25): 16760-16766 Dec 10 2007.

Liu, W. F.; Russell, P. S. J.; Dong, L. (1997). Acousto-optic superlattice modulator using a fiber Bragg grating. *Optics Letters,* 22 (19): 1515-1517 Oct 1 1997.

Liu, W. F.; Russell, P. S. J.; Dong, L. (1998). 100% efficient narrow-band acoustooptic tunable reflector using fiber Bragg grating. *Journal of Lightwave Technology,* 16 (11): 2006-2009 Nov 1998.

Liu, Y. Q.; Yang, J. L.; Yao, J. P. (2002). Continuous true-time-delay beamforming for phased array antenna using a tunable chirped fiber grating delay line. IEEE *Photonics Technology Letters,* 14 (8): 1172-1174 Aug 2002.

Liu, Y. Q.; Yao, J. P.; Yang, J. L. (2003). Wideband true-time-delay beam former that employs a tunable chirped fiber grating prism. *Applied Optics,* 42 (13): 2273-2277 May 1 2003.

Myren, N.; Margulis, W. (2005). All-fiber electrooptical mode-locking and tuning. *IEEE Photonics Technology Letters,* 17 (10): 2047-2049 Oct 2005.

Ortega, B.; Cruz, J. L.; Capmany, J.; Andres, M. V.; Pastor, D. (2000). Analysis of a microwave time delay line based on a perturbed uniform fiber Bragg grating operating at constant wavelength. *Journal of Lightwave Technology,* 18 (3): 430-436 Mar 2000.

Parker, T. R.; Farhadiroushan, M.; Feced, R.; Handerek, V. A.; Rogers, A. J. (1998). Simultaneous distributed measurement of strain and temperature from noise-initiated Brillouin scattering in optical fibers. *IEEE Journal of Quantum Electronics,* 34 (4): 645-659 Apr 1998.

Pérez-Millán, P.; Torres-Peiró, S.; Mora, J.; Díez, A.; Cruz, J. L.; Andres, M. V. (2004). Electronic tuning of delay lines based on chirped fiber gratings for phased arrays powered by a single optical carrier. *Optics Communications,* 238 (4-6): 277-280 Aug 15 2004.

Pérez-Millán, P.; Díez, A.; Andres, M. V.; Zalvidea, D.; Duchowicz, R. (2005a). Q-switched all-fiber laser based on magnetostriction modulation of a Bragg grating. *Optics Express*, 13 (13): 5046-5051 Jun 27 2005.

Pérez-Millán, P.; Cruz, J. L.; Andres, M. V. (2005b). Active Q-switched distributed feedback erbium-doped fiber lasers. *Applied Physics Letters*, 87 (1): Art. No. 011104 Jul 4 2005.

Philippov, V. N.; Kiryanov, A. V.; Unger, S. (2004). Advanced configuration of erbium fiber passively Q-switched laser with Co^{2+}: ZnSe saturable absorber. *Ieee Photonics Technology Letters*, 16 (1): 57-59 Jan 2004.

Phillips, M. W.; Ferguson, A. I.; Hanna, D. C. (1989a). Frequency-modulation mode-locking of a Nd^{3+}-doped fiber laser. *Optics Letters*, 14 (4): 219-221 Feb 15 1989.

Phillips, M. W.; Ferguson, A. I.; Kino, G. S.; Patterson, D. B. (1989b). Mode-locked fiber laser with a fiber phase modulator. *Optics Letters*, 14 (13): 680-682 Jul 1 1989.

Russell, P. S. J.; Liu, W. F. (2000). Acousto-optic superlattice modulation in fiber Bragg gratings. *Journal of the Optical Society of America A-Optics Image Science and Vision*, 17 (8): 1421-1429 Aug 2000.

Russo, N. A.; Duchowicz, R.; Mora, J.; Cruz, J. L.; Andres, M. V. (2002). High-efficiency Q-switched erbium fiber laser using a Bragg grating-based modulator. *Optics Communications*, 210 (3-6): 361-366 Sep 15 2002.

Schaffer, C. B.; Garcia, J. F.; Mazur, E. (2003). Bulk heating of transparent materials using a high-repetition-rate femtosecond laser. *Applied Physics A-Materials Science & Processing*, 76 (3): 351-354 Mar 2003.

Siegman, E. (1986). *Lasers*, University Science, Mill Valley, CA, 1986.

Tuchin, V. V. (1993). Lasers and fiber optics in biomedicine. *Laser Physics*, 3 (4): 767-820 1993.

Vicente, S. G. C.; Gámez, M. A. M.; Kiryanov, A. V.; Barmenkov, Y. O.; Andrés, M. V. (2004). Diode-pumped self-Q-switched erbium-doped all-fibre laser. *Quantum Electronics*, 34 (4): 310-314 Apr 2004.

Villegas, I. L.; Cuadrado-Laborde, C.; Abreu-Afonso, J.; Díez, A.; Cruz, J. L.; Martinez-Gámez, M. A.; Andrés, M. V. (2011a). Mode-locked Yb-doped all-fiber laser based on in-fiber acoustooptic modulation. *Laser Physics Letters*, 8 (3): 227-231 Mar 2011.

Villegas, I. L.; Cuadrado-Laborde, C.; Díez, A.; Cruz, J. L.; Martinez-Gámez, M. A.; Andrés, M. V. (2011b). Yb-doped strictly all-fiber laser actively Q-switched by intermodal acousto-optic modulation. *Laser Physics, in press* 2011.

Wada, K.; Fujita, J.; Yamada, J.; Matsuyama, T.; Horinaka, H. (2008). Simple method for estimating shape functions of optical spectra. *Optics Communications*, 281 (3): 368-373 Feb 1 2008.

Yu, C. X.; Haus, H. A.; Ippen, E. P.; Wong, W. S.; Sysoliatin, A. (2000). Gigahertz-repetition-rate mode-locked fiber laser for continuum generation. *Optics Letters*, 25 (19): 1418-1420 Oct 1 2000.

Zadok, A.; Raz, O.; Eyal, A.; Tur, M. (2007). Optically controlled low-distortion delay of GHz-wide radio-frequency signals using slow light in fibers. *IEEE Photonics Technology Letters*, 19 (5-8): 462-464 Mar-Apr 2007.

Zalvidea, D.; Russo, N.A.; Duchowicz, R.; Delgado-Pinar, M.; Díez, A.; Cruz, J. L.; Andrés, M. V. (2005). High-repetition rate acoustic-induced Q-switched all-fiber laser. *Optics Communications*, 244 (1-6): 315-319 Jan 3 2005.

28

Surface Acoustic Waves and Nano–Electromechanical Systems

Dustin J. Kreft and Robert H. Blick
University of Wisconsin – Madison
Department of Electrical and Computer Engineering
U.S.A.

1. Introduction

Surface acoustic waves (SAW) follow the industrial trend of reducing the size, enhancing the speed, while enhancing the efficiency of energy coupling. Integrating this with micro-electromechanical systems (MEMS) and nano-electromechanical systems (NEMS) offers a wide variety of applications such as touch screens, gas and biological sensors, and embedded RFID devices. With modern lithographic techniques, allowing the fabrication of smaller SAW devices, we now use SAWs to probe the mechanical interactions of nano structures. In particular, SAWs can be used to actuate NEMS which gives rise to many interesting phenomena including anomalous acoustoelectric currents, shock waves in suspended devices, and few electron transport, to only name a few, (Beil et al., 2008; Talyanskii et al., 1997). Today, SAWs are also used to generate a quantized current for use as a current standard. In practice two counter-propagating SAWs are used to observe a quantized acoustoelectric current. This leads to population and depopulation of discrete states (Kataoka et al., 2007). In the following we want to give an overview of the state of the art of applying SAWs to nanomechanical devices. We will also give a brief introduction to recent nanoelectromechanical systems with integrated low-dimensional electron gases, which have the potential to reveal insights into quantized acoustoelectric states.

2. Fabrication

The focus for generating SAWs in this chapter will involve the fabrication of interdigitated transducers (IDT). An IDT is simple in concept but can be very involved when fine tuning a structure for engineering applications. Such topics as electronic impedance matching to RF lines, effects of bulk waves in contrast to SAWs, and increasing bandwidth will not be covered; though, this is simply a shortened list of things to consider when designing a proper IDT for engineering applications, they do fall outside the scope of this chapter. Nevertheless, another fabrication step we will consider is the use of acoustic waveguides for acoustic impedance matching of IDTs to nanomechanical devices.

2.1 Interdigitated Transducer Design

The main equation to consider when designing an IDT is:

$$v = \lambda f \quad (1)$$

Where v is the velocity of sound in the material, λ is the SAW wavelength or pitch of the IDT, and f is the frequency of the propagating SAW. The pitch of the IDT fingers is the same as the SAW wavelength, λ, which will propagate across the sample. Fig. 1 shows a simple IDT schematic along with a scanning electron micrograph (SEM) image and optical image.

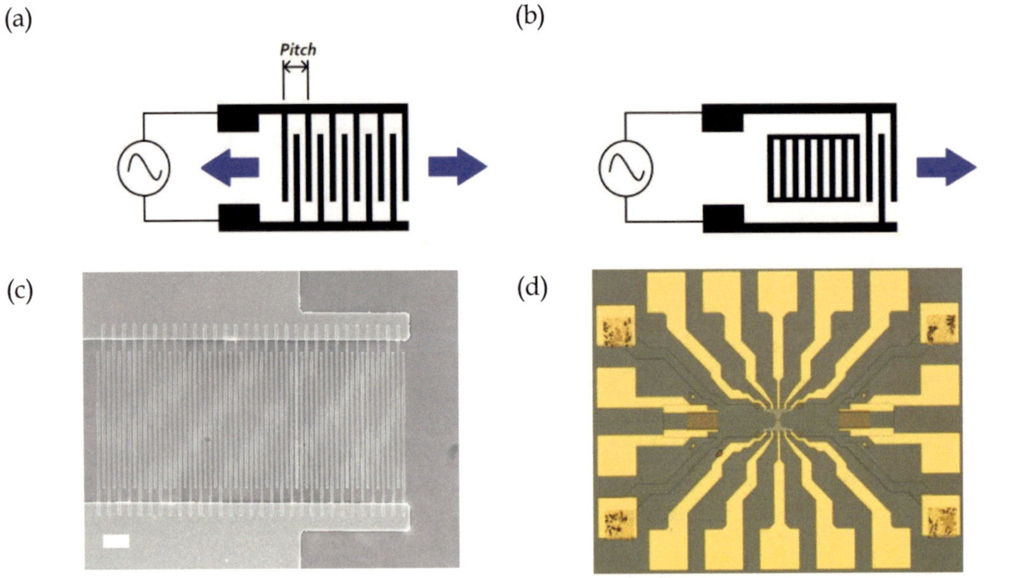

Fig. 1. Schematics and scanning electron microscope images of IDTs. (a) Schematic of bidirectional IDT, (b) schematic of unidirectional IDT through the use of a reflection grating, (c) SEM image of a bidirectional IDT on GaAs with a pitch of 4 μm and a center frequency of ~715 MHz. The scale bar is 10 μm. (d) Optical image of device with an IDT on both ends and a nanomechanical device placed in the center

In Fig. 1 the arrows indicate the SAW propagation directions. It can be easily seen that the SAW will propagate in both directions away from the IDT. A singly propagating SAW direction can be achieved by placing a reflection grating on one end of the SAW. The reflection grating will have the same geometry as the IDT; that is, the same finger spacing and width. The grating distance from the IDT is ~ $\lambda/2$. The IDT fingers are typically chosen with evenly spaced fingers, where the spacing between the fingers is equal to the finger width. The finger width and finger spacing is $\lambda/4$ in this scenario and gives a metallization ratio of $\eta = 0.5$, which will generate only odd harmonics with no response of the third harmonic (Campbell, 1998). The bandwidth of an IDT is defined as $BW\% = f/N_{Pairs}$. Where f, again, is the center frequency and N_{Pairs} is the number of finger pairs of the IDT. The IDT in Fig. 1a has five pairs of fingers and is symmetric about a center point along the axis of SAW propagation. It is typical to fabricate two IDTs, one on each end of the device, so that their propagating waves can interfere either constructively or destructively across the center region of the device; this is typically the region where a Quantum Point Contact (QPC) or other structures resides. This can be seen in Fig. 1d where in the center a double quantum dot (DQD) is placed and the IDTs are placed to the left and right.

2.2 Acoustic Waveguide Design

Another detail of fabrication is the use of waveguides for the SAW. A waveguide for SAWs is a pattern that is fabricated onto the device so the SAWs can be guided into a certain region allowing a stronger SAW signal or amplitude; it is analogous to a coplanar waveguide for RF signals. When fabricating IDTs, and a device as a whole, using waveguides does not have much of a use in the areas of RFID, cellular delay lines, sensors, and other non-region specific devices. However, this chapter will focus on NEMS and the use of SAWs to interact with these devices. Since the NEMS device is orders of magnitude smaller than the IDT aperture very little SAW power can interact with the NEMS structure; it can be beneficial to include a waveguide to focus the SAW onto the nanostructure.

Waveguides can be modeled using standard acoustic, or horn, waveguides with some minor tuning. There are several shapes which can be used and each shape offers its own benefits regarding a particular application. We will start by first mentioning the base equation for hyperbolic horns used in speaker systems, these horns can also be referred to as hyperbolic-exponential horns or hypex horns (Salmon, 1946).

$$A = A_t[\mathrm{Cosh}(kx) + T \cdot \mathrm{Sinh}(kx)]^2 \quad (2)$$

$$z = \frac{\rho_0{}^{v}}{A_t}\left(\sqrt{1-\left(\frac{f_c}{f}\right)^2} + i\frac{f_c T}{f}\right)\left(1-\frac{f_c{}^2(1-T^2)}{f^2}\right)^{-1} \quad (3)$$

Eq. 2 gives the wave front area expansion of the wave as propagating through the waveguide or horn. A_t is the area of the throat, or base, of the waveguide, T is a factor describing the shape of waveguide; for T = 1 the waveguide becomes exponential in shape, and becomes canonical as $T \rightarrow \infty$, k is given as $(2\pi f_c)/v$; where v is the velocity of sound, and f_c is the cutoff frequency of the waveguide. For hyperbolic and exponential waveguides there is no transmission below the cutoff frequency, f_c, since at this point the impedance is purely reactive. Eq. 3 gives the impedance of the waveguide; all variables are defined previously with the addition of ρ_0 which is the density of the material.

Fig. 2 contains three SEM images of an exponential waveguide with a tube waveguide in the center. A tube waveguide is simply a small or narrow SAW delay line. The material is a GaAs/AlGaAs heterostructure containing a two-dimensional electron gas (2DEG) and a sacrificial layer. We use a 2DEG because the high electron mobility makes this an ideal candidate for nanostructures, such as quantum dot (QD) and DQD systems. The waveguide was defined using electron-beam lithography to open up an etch region through the PMMA. The material was then wet etched with H_3PO_4:H_2O_2:H_20 into the sacrificial layer. Then diluted HCl was used to remove the sacrificial layer which resulted in a suspended region in the center and around the waveguide. This suspended region is visible as a darkened shadow indicating a height difference due to strain relaxation of the material. The suspended region forms the actual nanomechanical device. Since this device contains a DQD in the center region of the structure it is beneficial to use a waveguide to ensure a higher SAW power density coupled into that region. Since the IDT aperture of this device is 50 µm and the DQD region is roughly 600 nm at the widest QPCs.

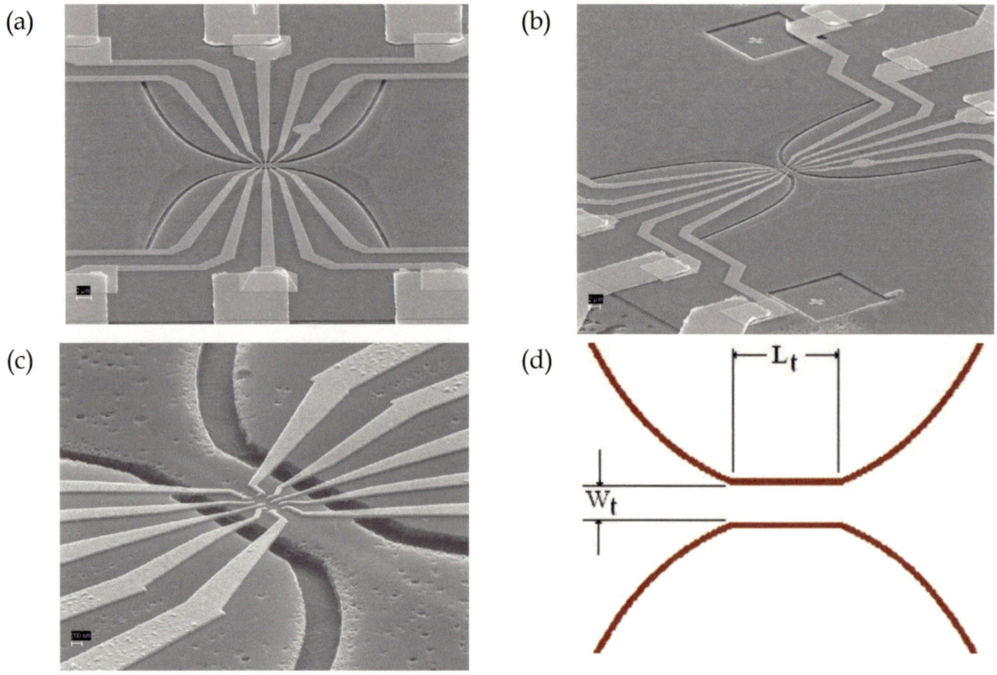

Fig. 2. SEM images of an exponential waveguide with a cutoff frequency of 75 MHz and throat width of ~1.5 μm. The darker region around the waveguide shows that the area is suspended, which was achieved by an HCl etch to remove the sacrificial layer below the 2DEG, (a), (b) and (c). (d) Schematic of waveguide, center region is the throat (pipe waveguide) with width W_t and length L_t. The outer curves are the exponential portion of the waveguide

3. Surface Acoustic Waves in Two Dimensional Electron Gases

3.1 Surface Acoustic Wave Basics

When applying an RF signal to the IDTs an alternating electric potential is seen by the electrodes, this will create an electric field distribution within the piezoelectric substrate, see Fig. 3a. This field distribution, in turn, will cause a mechanical deformation of the material through the inverse piezoelectric effect. As stated before, this mechanical deformation will propagate away from the IDTs and continue to move along the surface of the device, see Fig. 3d. Here we neglect other forms of waves being produced such as bulk waves.

Once the RF signal of the IDT is coupled into the piezoelectric substrate a Rayleigh-wave is produced. The elliptical wave propagating (Fig. 3b, d) along the surface can be described quite accurately by a traveling wave:

$$U = |U|\, e^{i(\omega t - kz)} e^{-k|y|} \tag{4}$$

The electric potential being created by the propagating wave can also be described in a like manner:

$$\Phi = |\Phi|\, e^{i(\omega t - kz)} e^{-k|y|} \tag{5}$$

In both Eqs. 4 and 5, ω is the angular frequency and k is the phase constant. The SAW penetrates into the depth of the material by about one wavelength, ~λ. This value is different in the suspended region. Hence, the actuation of nanomechanical resonators is considerably enhanced. Another property of the SAW is that the electric field created from the induced electric potential does not terminate at the surface of the material but can extend beyond by λ.

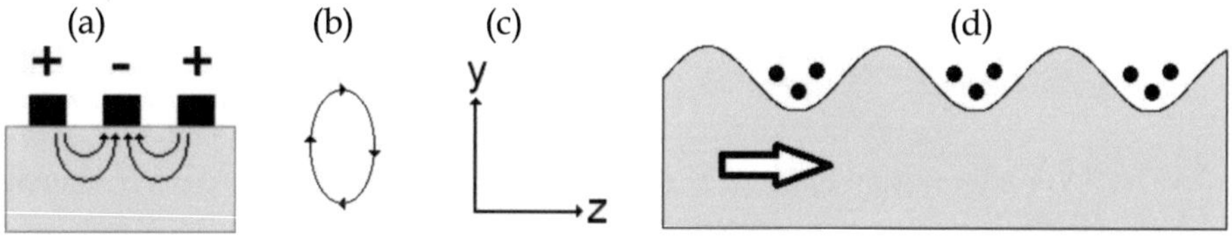

Fig. 3. (a) Side view of IDT with an applied voltage, the electric field couples into the piezoelectric substrate causing a deformation. (b) The elliptical motion of the Rayleigh-wave. (c) Axis used for reference in equations as it applies to the orientation of the piezoelectric substrate. (d) Side view of piezoelectric material as it is deformed causing a SAW to propagate

The set of base equations used for describing SAW phenomena are listed.

$$E_i = -\frac{\partial \Phi}{\partial x_i} \tag{6}$$

$$T_{ij} = c_{ijkl} S_{kl} + e_{nij} E_n \tag{7}$$

$$S_{kl} = \frac{1}{2}\frac{\partial u_k}{\partial x_l} + \frac{\partial u_l}{\partial x_k} \tag{8}$$

Eq. 6 is the electric field intensity that is produced from the deformed piezoelectric material from the SAW. Eq. 7 is the piezoelectric mechanical stress and Eq. 8 is the linear strain displacement.

3.1.1 Attenuation

SAW attenuation can be described by the following equations (Wixworth et al., 1989):

$$\Gamma = k\frac{K^2{}_{eff}}{2}\frac{\sigma_s / \sigma_M}{1 + (\sigma_s / \sigma_M)^2} \tag{9}$$

$$\frac{\Delta v}{v} = k\frac{K^2{}_{eff}}{2}\frac{1}{1 + (\sigma_s / \sigma_M)^2} \tag{10}$$

Here the attenuation occurs because part of the longitudinal electric field of the propagating wave couples into the electrons of the 2DEG. This not only causes a current to flow but pulls power from the SAW due to ohmic losses. This attenuation is described by Eq. 9. A SAW velocity shift is also observed due to the piezoelectric stiffening of the substrate, see Eq. 10 (Wixworth et al., 1989). Below are the recreated graphs from (Wixworth et al., 1989) to show the relationship of the attenuation and sound velocity shift due to a change in conductivity.

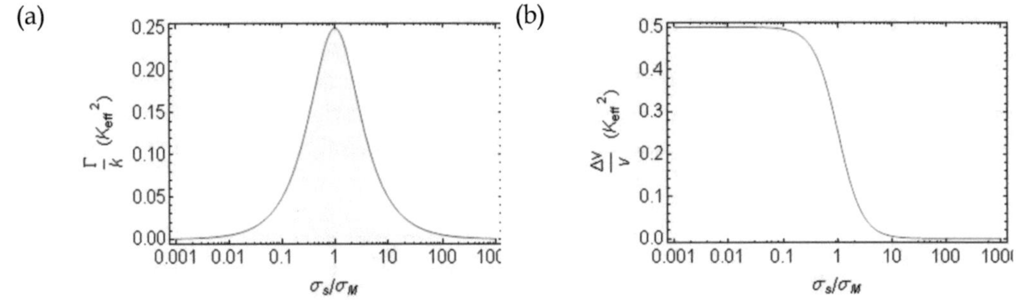

Fig. 4. (a) The SAW attenuation in units of K_{eff}^2, see Eq. 9. (b) The change in SAW velocity in units of K_{eff}^2, see Eq. 10

Where k is the SAW wave vector, K^2_{eff} is the effective piezoelectric coupling coefficient, σ_s is the 2DEG sheet conductivity, and $\sigma_M = v_0(\varepsilon_1 + \varepsilon_2)$. Again, where v_0 is the sound velocity and ε_1 and ε_2 are the dielectric constants of the piezoelectric substrate and half space above it.

3.2 Acoustoelectric Current

As the SAW propagates across the material it creates two types of currents, one is the normal acoustoelectric current and the other being the anomalous acoustoelectric current. The normal acoustoelectric current is created by electrons being "dragged" across the material and can be described by Eq. 11 where I is the current, n is the number of electrons, e is the charge of an electron, and f is the frequency of the SAW or RF signal to the IDT. This current always flows in the direction of the SAW and is produced as a DC current despite an oscillating RF signal being applied to the IDTs. Eq. 11 shows that at higher frequencies the normal acoustoelectric current becomes quantized.

$$I = nef \tag{11}$$

$$j(z) = \frac{a\omega}{2\pi}\int_0^{2\pi/\omega} \sigma_{zz}(z,t)E_z(z,t)dt \tag{12}$$

Once this deformation occurs the energy bands in the material bend as well causing the electrons to fall into the created quantum wells and are dragged along with the SAW, see Fig. 3d. As the frequency increases the wavelength, and pitch of the IDT, decreases causing fewer energy states to be available within the wells. This idea is also implemented to generate QDs within the SAW, (Barnes et al., 2000).

The anomalous acoustoelectric current is produced from the deformation of the material and flows as a DC current. The difference is that the anomalous current always flows in one direction regardless of which IDT, left or right, produced the SAW. The current is smaller than the normal acoustoelectric current and is detected by different methods. The anomalous acoustoelectric current can be obtained by sweeping the RF signal of the IDTs and typically appears at an off-center frequency. Since the normal acoustoelectric current is much smaller, the anomalous current can be more easily detected, as shown in Fig. 5b. One such method for direct detection of the anomalous acoustoelectric current is to apply an RF signal to both the left and right IDTs, while phase locking the two signals (Beil et al., 2008). By phase locking the IDT signals a standing SAW can be created. Thus, a surface deformation is produced without a net direction of propagation. As seen in Fig. 5a the phase is shifted between the two IDTs as the current is measured. At a phase difference of 180° the current tends towards zero.

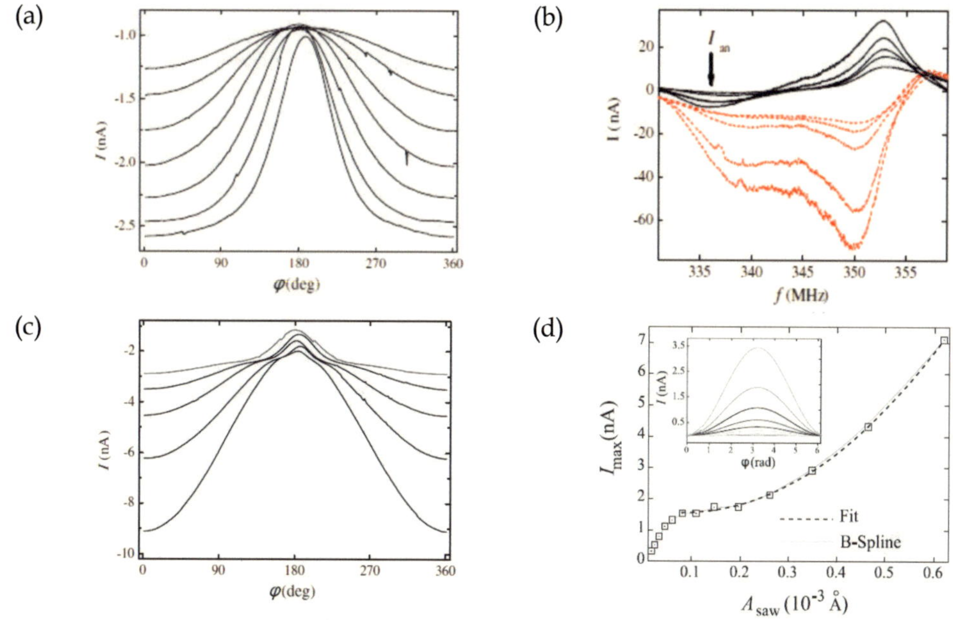

Fig. 5. Results for a suspended beam that is 1.2 μm long and 300 nm wide. (a) Anomalous acoustoelectric current as a result of phase adjusting two counter propagating SAWs. (b) Acoustoelectric current vs. applied frequency of the IDTs. There is always a negative current value to the left and is indicative of the anomalous current. (c) Formation of a shock wave seen by its Sinc(x) shape. The device's response jumps to a higher order mode and returns to Sin(x). (d) Derived anomalous acoustoelectric current amplitude as a result of the calculated SAW amplitude. All images are taken from Beil et al., 2008, Copyright (2008) by The American Physical Society

This zero current occurs because the two SAWs interfere destructively and create a nearly smooth surface with no deformation. At 0°, and also 360°, the current is at a maximum since the two SAWs interfere constructively allowing a maximum in the surface deformation.
The anomalous acoustoelectric current can be used to probe the SAW amplitude of a suspended beam or nanostructure. The accompanying graphs show the acoustoelectric current from the device. It has been shown that the anomalous acoustoelectric current through the device in relation to the SAW amplitude can be described by Eqs. 12-14; where Eq. 12 is the general equation for an anomalous acoustoelectric current, 13 is the derived equation, and 14 is the strain equation of the beam.

$$j_{an}(L/2) = -\frac{\sigma_0 e_{z4}}{2\varepsilon_{GaAs}}(\Pi_{zzzy} S_{zy}(L/2)^2) \quad (13)$$

$$S_{zy}(z,t) = 2A_t\left(\left(\frac{6z^2}{L^3} - \frac{6z}{L^2}\right)\right)(\mathrm{Cos}(\varphi/2) - \mathrm{Cos}(k_{SAW}L + \varphi/2)) \quad (14)$$

Where σ_0 is the unperturbed conductivity, ε_{GaAs} is the permittivity of GaAs, Π is a tensor which describes the effect of strain S, and A_t is the transverse component of the SAW.
An interesting phenomenon that occurs is the formation of shock waves in the suspended structure. From Figs. 5a and 5c it is seen that the device exhibits a Sin(x) type behavior in the anomalous acoustoelectric current at low RF powers. Once the RF power starts to increase it can be seen that the shape of the current starts to exhibit more of a Sinc(x) shape. This transition from a linear to a non-linear response indicates the formation of a shock wave in the suspended beam. As the power increases, the beam jumps into the next higher order mode and the current trace returns back to a linear response again. The shock wave formation is an indication that the beam will transition from one mode to another.

3.3 Surface Acoustic Waves in Magnetic Fields

We can use SAWs to probe the characteristics of a 2DEG under the presence of a magnetic field. Since a high mobility 2DEG is subject to Shubnikov-de Haas (SdH) oscillations, which creates changes in the conductivity, and SAWs are more sensitive to a conducting plane or surface, and the conductance of that plane, makes this combination an ideal candidate to investigate quantum effects of the 2DEG. The real interesting features are seen when integral Landau level filling factors are observed which causes a drop in conductivity of the 2DEG. At these quantized values the SAW responds strongly to the conductivity, σ.
In Fig. 6 it can be seen that the SdH oscillations and the acoustoelectric current oscillations are nearly identical. The SdH measurement was measured using a standard four-point lock-in technique, where as the acoustoelectric current was created by a SAW and was taken from two ohmic contacts on opposite sides of the sample. The peaks and valleys of the two measurements line up, for the most part, but there is a small offset. This offset is due to the fact that the SAW attenuation is not a linear with respect to the 2DEG conductivity, or magneto conductivity in this case, see Eq. 9.

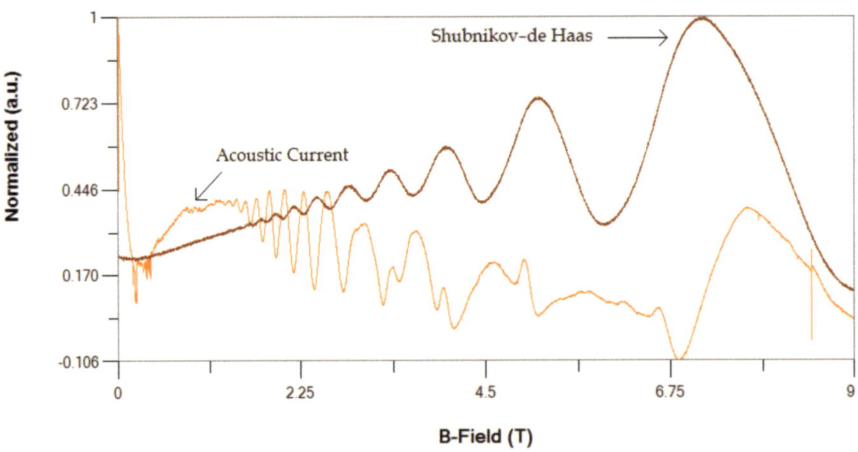

Fig. 6. A magnetic field, of maximum value 9 T, was applied perpendicular to the surface of the device. The device is shown in figures 1d and 2a-2d and contains a 2DEG 40 nm below the surface. Upper trace is the Shubnikov-de Haas oscillations measured using standard lock-in techniques and normalized to 713.4 Ω. The lower trace is the acoustoelectric current generated from a SAW with -12 dBm applied to the right IDT at 1.488 GHz and is normalized to 10.47 nA

Special attention has to be given to the acoustic current trace of Fig. 6 around 3 - 4 Tesla. There is a splitting of the peaks that occurs. This information does not appear in the SdH oscillation and is only isolated to a SAW effect. During this splitting the conductivity of the 2DEG is very low and σ_s becomes much smaller than σ_M, see Eq. 9. When this happens a maximum in SAW attenuation will occur which results in a reduced SAW amplitude. The center of a SAW current split is a minimum in the 2DEG conductivity; further details are discussed in Wixworth et al., 1989.

From Fig. 6 there are two observations that need to be explained. One is the spike-like feature or discontinuity around the 8.5 T mark of the acoustic trace. This is caused from SAW reflections on the sample. This spike-like feature can be seen on other data plots as well with SAW currents. The final feature is the negative acoustic current that was measured. Since at this point the SAW was highly attenuated, causing little or no current to flow, caused the sample itself to heat. The negative current may be a combination of a thermal current and a small offset in the measuring equipment.

4. Surface Acoustic Waves in Quantum Electronics

SAWs are used to produce a current in low dimensional electronic systems and NEMS. This can be used in various applications and for numerous different designs. The information presented in this section will deal with piezoelectric materials with an embedded two dimensional electron gas (2DEG); mainly GaAs/AlGaAs heterostructures unless otherwise stated. These measurements are carried out at liquid helium temperatures or lower, ≤ 4.2 K.

4.1 Quantum Point Contacts and Low Dimensional Channels

SAWs can be used to create a quantized current based upon Eq. 7. The advantage of the SAW inducing a quantized current is that this process can be used to populate and depopulate QDs and DQDs at higher frequencies then what can be used by applying an oscillating signal to the source-drain of the 2DEG. When using a SAW as the current source the acoustoelectric current can be pinched off in a Coulomb blockade just like a standard source-drain current can be. This pinch off is done via a quantum point contact (QPC) or a set of QPCs, which can be used to form QDs, see Fig. 7.

4.1.1 Quantum Point Contact Fabrication

Fabricating QPCs is done with the same methods as fabricating IDTs, see Sec. 2. Since QPCs are have small dimensions it is most common to use Electron Beam Lithography, or e-beam lithography, to create the structures. The exact dimensions of the QPC pair depends on what works best for the user and there is no set rules for design like there is for IDTs. When viewing Fig. 7 it can be seen that there are five sets of QPCs. A single QPC is seen as an electrode with another electrode opposite its position. All of the QPCs shown have a few common features; the tip of the QPC is small when compared to the rest of the electrode and the gap between the electrode tips is small as well. The tip is small so the electric field being emitted from the QPC is very localized, and the majority of the electrode is made wider so that it covers a wider portion of the 2DEG so the electrons are repelled. The gap between electrodes is small so pinch-off can be achieved with small voltages, more on this in Sec. 4.1.2.

4.1.2 Quantum Point Contact Operation

QPCs work by applying a negative voltage to them, this negative voltage produces an electric field that penetrates into the 2DEG causing electrons to be repelled at or around the QPC region; this can be seen if Fig 7(c). Here the black region represents the metallic electrode while the blue region represents the electric field and the green region represents the 2DEG or substrate. It can be taken as a good estimate that the electric field penetrates into the 2DEG from the QPC electrode at a 45° angle. As a larger negative voltage is applied to the QPC the effective screening electric field becomes larger causing the electron path within the 2DEG to become more constricted until single electron transport is achieved; this leads to a Coulomb Blockade, see Sec. 4.2. Then finally, the current is completely pinched off, see Figs 9 and 10.

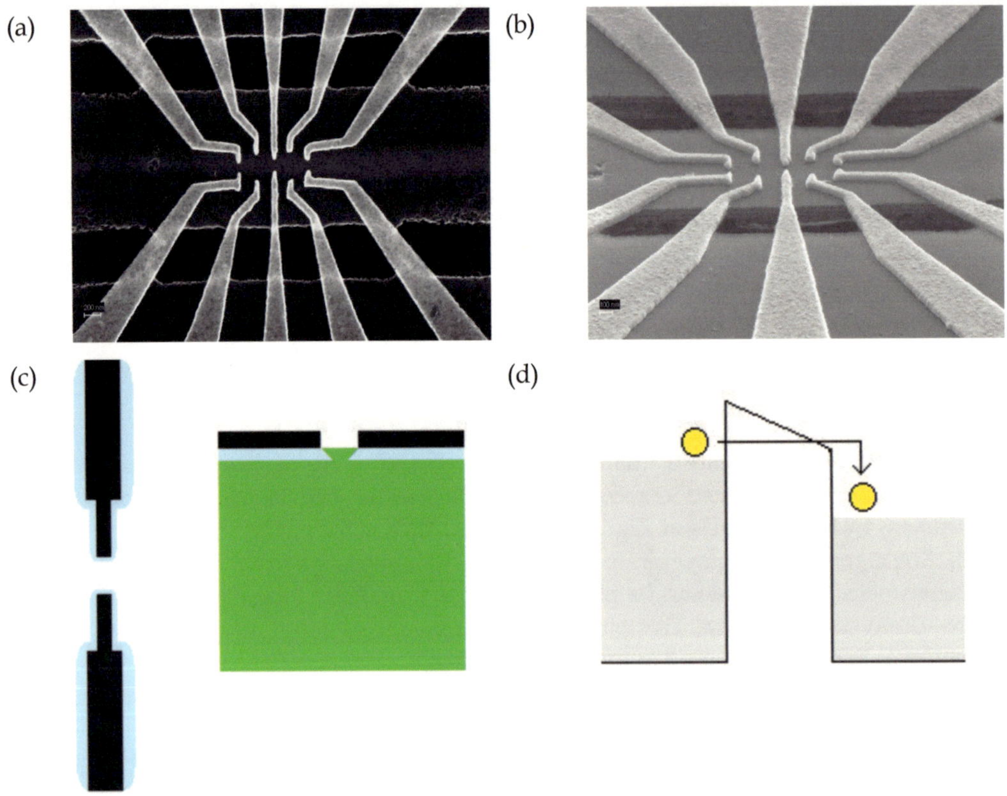

Fig. 7. Two SEM images of the same DQD structure on top of a suspended tube section of an acoustic waveguide. The DQD is made up of five pairs of QPCs. The material is a GaAs/AlGaAs heterostructure with an embedded 2DEG about 40 nm below the surface. (a) Top view with a scale bar of 200 nm, (b) angled view with a scale bar of 100 nm. (c) schematic view of QPC with applied voltage, the blue area represents the electric field surrounding the QPC and penetrating into the 2DEG (green). (d) Tunneling of electrons across potential barrier created by applied voltage from the QPC

The QPC creates a tunneling barrier which separates the source and drain regions of the sample. As the voltage magnitude becomes larger the strength, or height, or the tunneling barrier is increased and the width of the barrier becomes larger. When plotted a step like feature can be seen which corresponds to a single conductance step which has a value of $G = 2e^2/h$, where h is Plank's constant and e is the charge of an electron. Now the temperature must be low so the thermal energy, $E = k_bT$, of the background is smaller than the tunneling energy needed for the electrons to "jump" across the barrier, where k_b is the Boltzmann constant. As seen in Figs 9 and 10 this step like feature can be seen by doing an I-V measurement. By changing the temperature of the system the phonon energy is increased and causes scattering events to increase, or increase the electron-phonon interaction, and the steepening or smoothing of the step like feature is a direct measure of this.

4.1.3 Usage of Quantum Point Contacts and Surface Acoustic Waves

The use of QPCs offers a benefit of determining which SAW mode(s) are propagating in the sample. Different SAW modes, such as bulk, longitudinal, and transverse with propagate at different frequencies due to the fact that they have different sound velocities, see Eq. 1. Another factor, which affects the sound velocity, is the propagation direction of the SAW with respect to the crystal orientation of the material. In Fig. 8 a QPC had an applied voltage of -0.8 V, which puts the QPC into pinch-off mode. Since it is in pinch-off higher RF power is required to create a sufficiently strong SAW that will overcome the potential barrier. As the power is increased from -18 dBm to -10 dBm, three peaks emerge as transferring current through the tunnel barrier. From Eq. 7 we can calculate the electron count to be 6, 3, and 2 for RF powers of -10 dBm, -12 dBm, and -14 dBm, respectively (some rounding is taken into account, due to thermal errors in measurement).

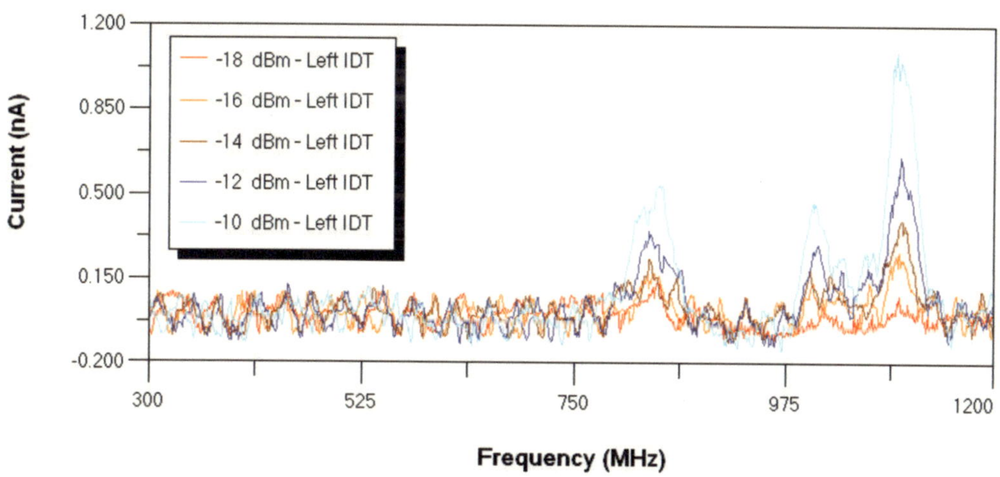

Fig. 8. A frequency sweep of varying RF powers while the center QPC of the sample in Fig. 4 is held at -0.8 V. The first peak is at 840 MHz with a current of 540 pA and a velocity of 3,368 m/s, the second peak at 1.005 GHz with a current of 472 pA and a velocity of 4,020 m/s, the third peak is at 1.095 GHz with a current of 1.098 nA and a velocity of 4,380 m/s

Now the three peaks represent different SAW modes. The highest frequency peak of Fig. 8 of 1.095 GHz represents a longitudinal wave with an acoustic velocity of 4,380 m/s and an angle of about 10° off from the (110) direction (Kuok et al., 2001). This small angle variation is due to a small misalignment during the lithography process. When viewing the lower peak of 840 MHz at a velocity of 3,368 m/s, this coincides with a fast transverse wave with, again, a 10° difference between the SAW direction and the (110) GaAs crystal orientation (Kuok et al., 2001).

4.2 Coulomb Blockade of Acoustoelectric Current

When looking over Fig. 9, we see that single steps are observed just before the total acoustoelectric current becomes zero. Focusing on Fig. 9a, we see that the last step is at about 0.5 nA. When solving with Eq. 11 the first step yields an answer of n = 1, i.e. a single electron is being transferred. We can look at all of the remaining steps and solve for them as well which will reveal integer numbers for n and will increase by 1 for each step, as is expected. Fig. 9a also shows that an applied voltage across the source and drain contacts, or ohmics, has very little effect on the final quantized acoustoelectric current. We do see, however, that a larger, or smaller, gate voltage is required for final pinch-off in the system but this is due to the small offset in the Fermi energy because of the applied bias. The only real difference is the shift from negative current to a positive current which is easy explained by the fact that the source-drain bias is producing a current that is opposite in direction, when $V_{ds} < 0$ mV, to that of the acoustoelectric current. Another aspect is the RF power dependence on the quantized current. When viewing Fig. 9b there is a similar trend to that of Fig. 9a. The small change in RF power has only a small effect on the acoustoelectric current. This is because the RF power is directly proportional to the SAW amplitude and we do not really identify any significant difference in the number of electrons being transferred, or Coulomb steps, until a drop of about 2 dBm.

So far an acoustoelectric current behaves in the same way as a standard source-drain bias current when a QPC is near pinch off. There is, however, another effect which can arise. As seen in Fig 9c a negative current arises and still exhibits the step like behavior. This negative current is the negative anomalous acoustoelectric current. This is said to be produced as an effect of SAW reflections, this is mostly seen in a two IDT system. The second IDT acts as a reflector much as the same way a reflection gradient is used for a unidirectional SAW, see Fig 1b. This can cause a standing wave to occur in the sample in such a way that the reflections effectively reduce the potential of the SAW and cause fewer electrons to be transported.

With slight phase shifts added the standing wave with respect to the QPC a net negative potential in the energy landscape can exist on one side of the QPC which will cause the current to change direction. These steps are best observed for QPCs that are long when compared to the SAW wavelength. This makes it possible to observe the acoustoelectric step current without applying a magnetic field (Shilton et al, 1996). As mentioned earlier, the high mobility of the 2DEG will screen the acoustoelectric current but this effect is not as prevalent when inside of a long QPC channel. The current screening is reduced around the QPC region and thus electrons can be transported through the channel. A long channel can allow ballistic transport of the electron causing it to shuttle from one side of the QPC to another. Since the screening is minimal, the electron will be "dragged" through the channel and the current steps of Fig. 9 can be observed.

Another phenomenon is oscillations in the acoustoelectric current once approaching the QPC pinch-off limit. As described in Shilton et al, 1996, and Maaø & Galperin, 1997, the acoustoelectric current oscillations occur at the same positions as the Coulomb steps. This oscillation is described as interference effects near the QPC at high frequencies which may be attributed to impurities in the 2DEG channel. The theory presented suggests that these oscillations are due to state transitions, both propagating and reflecting (Maaø & Galperin, 1997). The SAW may reflect from the QPC and create interference patterns; these patterns will affect the electron transmission probability through the QPC simply by changing the potential landscape. This mechanism is also sensitive to scattering events near the QPC, where energy quanta are emitted and absorbed between two waves. Others have contributed to the theory and more is being added.

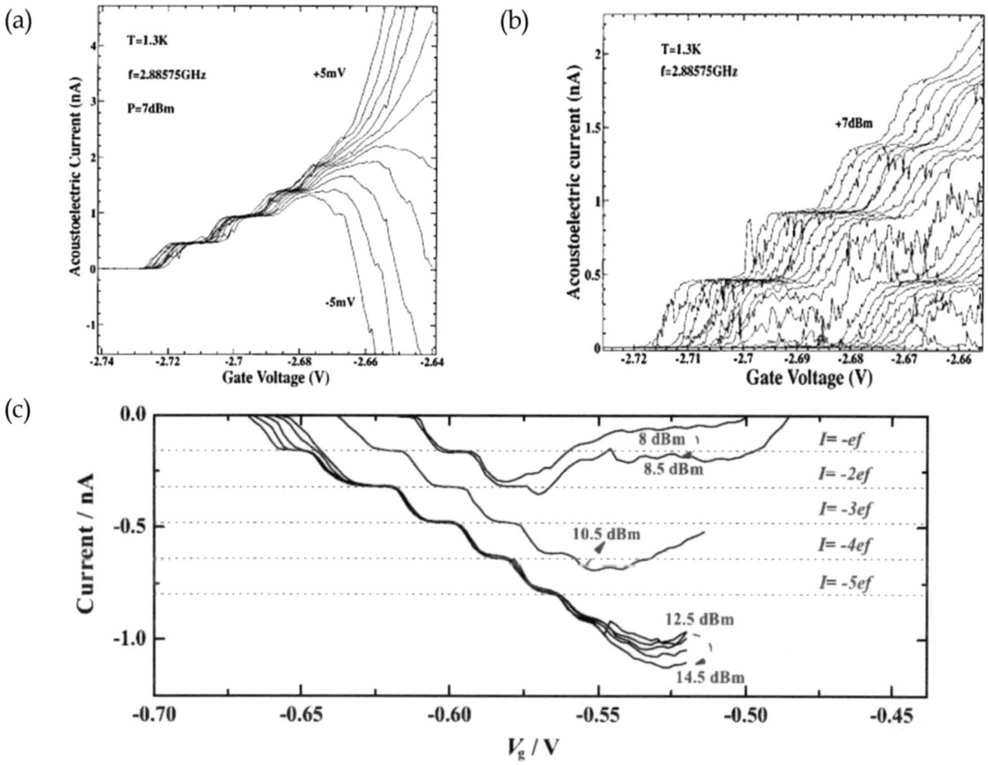

Fig. 9. QPC pinch-off of acoustoelectric current: (a) pinch-off with varying source-drain voltages applied at the SAW, changes are in 1 mV increments (Talyanskii et al., 1997) Copyright (1997) by The American Physical Society. (b) Pinch-off with different SAW power levels, the left most trace is 7 dBm and decrements by 0.2 dBm (Talyanskii et al., 1997) Copyright (1997) by The American Physical Society. (c) A negative anomalous acoustoelectric current for different gate voltages and SAW powers; f = 1007.426 MHz, V_{ds} = 0 V, T = 1.7 K, (Song et al., 2010)

Fig. 10 shows the oscillations of the acoustoelectric current as it is being driven through a QPC (orange trace) as it approaches pinch-off. It can be seen that the step like features observed in Fig. 9 are have been replaced by oscillations. These oscillations are negative, or on the lower cycle, when the current from the source-drain bias is flat or non-changing. We then see that as the current starts to decrease the acoustoelectric current has a positive value, or is on the upper cycle of the oscillation. The inset graph of Fig. 10 shows the entire sweep of the QPC gate voltage. As one conduction channel pinches off at the V_g = -0.65 V we see that there is a large, nearly single, oscillation in the acoustoelectric current. Since SAWs are very sensitive to 2DEGs we get what seems to be an amplified signal when compared to a source-drain bias. This can be used to identify information that may otherwise have been too weak or nearly washed out from thermal effects.

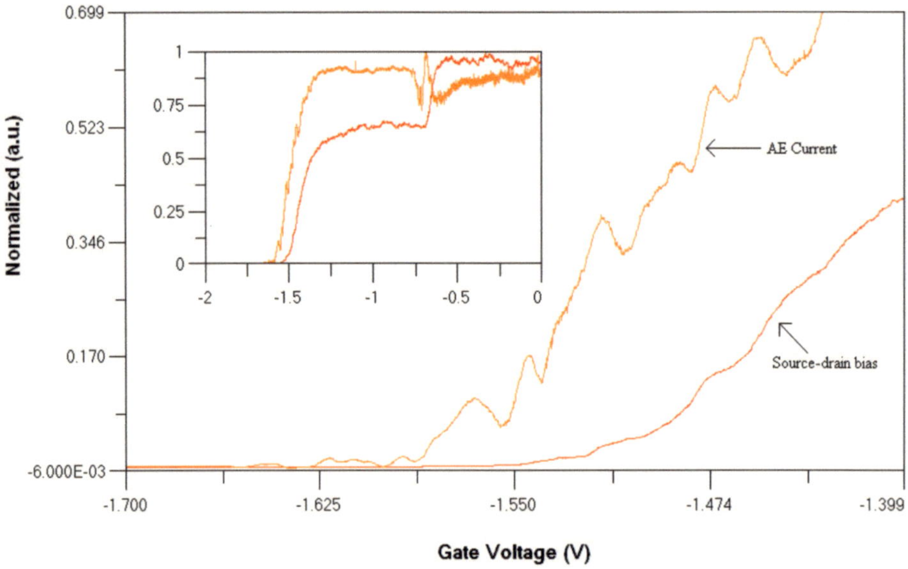

Fig. 10. Acoustoelectric current oscillations as the QPC nears pinch-off. The data plot has been normalized to show the relationship between the acoustoelectric current oscillations (orange trace) and the pinch-off when using a voltage bias across the source-drain electrodes (red trace). The acoustoelectric current trace has been normalized to -8.832 nA and the source-drain bias trace has been normalized to 4.692 nA. Measurements were taken at 4.2 K with the sample shown in Fig. 7. Inset plot shows entire gate voltage sweep range

4.3 Population and Depopulation of Quantum Dots

SAWs have been used widely as a mechanism to control the population and depopulation of QDs and DQDs. The high speed, or frequency, of operation combined with quantized current production make SAWs suitable for use with quantum systems. In a traditional QD system the gate and source-drain voltages are changed to allow single electron population and depopulation. This has several drawbacks; one being it is hard to decouple the dot from the surrounding environment due to the precise control of the voltages. There is always some noise either in the system or from the electronics. Another problem is if one gate is changed then the others must be changed to ensure a constant electron count in each dot, if applicable.

A SAW can be used as an alternative and has gotten more attention lately for the use in a QD or DQD system. The biggest reason for the attention is that SAWs can be used to interact with QD systems at much higher frequencies then what has been previously achieved by pulsing the gate electrodes. Since the acoustoelectric current can be well defined a QD system can use very high gate voltages to ensure no leakage current. We will take a look at the work done in Kataoka et al., 2007. This work describes very well the use of SAWs as a way to populate and depopulate a QD.

When the QD needs to be populated, the center plunger gate voltage value is lowered; causing the Fermi energy to lower below the Fermi energy of the surrounding 2DEG. The gate electrodes are still held at a high voltage value so the electrons cannot tunnel in or out of the QD, see Fig. 11e. The IDT is pulsed and as the SAW enters the QD region it changes the tunneling barrier created by the electrodes. The barrier is changed just enough that a single electron can tunnel into the dot, this is population; see Figs. 11a-b. The IDT is pulsed many times since the tunneling barrier is so high that the additional pulses are needed to increase the tunneling probability of the electron.

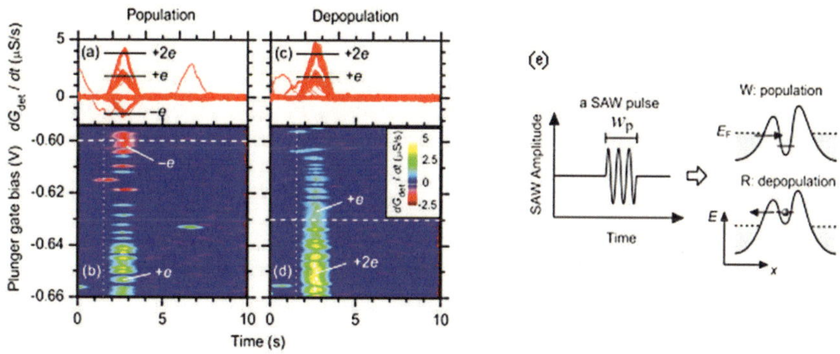

Fig. 11. Results from the paper Kataoka et al., 2007, Copyright (2007) by The American Physical Society. (a) and (b) The population of the QD, the –e line is an electron entering the dot. (c) and (d) Depopulation of the QD. (e) Schematic of SAW for population and depopulation of the QD

Likewise, as shown in Figs. 11c-d, the QD is depopulated by a similar mechanism. Here the plunger gate voltage is raised, made to be more negative, which causes the Fermi energy of the QD to be raised in comparison to the 2DEG Fermi energy outside of the dot, Fig. 11e. At this time the IDT is pulsed and the SAW enters the QD region. The SAW alters the tunneling barrier created by the gate electrodes when the potential of the SAW is superimposed onto the barrier, see Eq. 5. When the barrier is raised nothing happens since this will simply ensure the electron stays in the system, but when the negative cycle of the SAW superimposes with the barrier it decreases it which causes the electron to tunnel out.

This method of operation has proven, quite nicely, that SAWs and high potential QPCs offer a more robust method of single electron population and depopulation. As shown in Figs. 11a-d the QD can maintain the electron confinement for a long period of time which can be difficult in a traditional setup due to noise reduction of the measurement electronics. The IDTs can be accessed from an outside system such as an antenna, for example, and interact with the QD allowing the system another degree of freedom from the traditional closed electronics. However, this may come at the cost of additional noise reduction equipment and filtering.

5. Conclusion

This chapter has shown all of the basic properties and uses of SAWs in nano-structures and nano-systems. The reader is shown the parameters required for fabrication, theory of operation, real results, and application. The use of SAWs is quite limitless in the area of

NEMS. As people continue to refine the research through experiment and by continuing to add to the theory more applications will emerge.

6. Acknowledgment

We like to thank the National Science Foundation for support under a NIRT grant (No. ECCS-0708759) and the Air Force Office for Scientific Research for support under a MURI grant (No. FA9550-08-1-0337).

7. References

Barnes, C. H. W., Shilton, J. M., & Robinson, A. M. (2000). Quantum computing using electrons trapped by surface acoustic waves. *Physical Review B*, Vol. 62, No. 12, (September 2000), pp. 8410-8419, ISSN 0163-1829

Beil, F. W., Wixforth, A., Wegscheider, W., Schuh, D., Bichler, M., & Blick, R. H. (2008). Shock Waves in Nanomechanical Resonators. *Physical Review Letters*, Vol. 100, No. 2, (January 2008), pp. 026801 1-4, ISSN 0031-9007

Campbell, C. (1998). *Surface Acoustic Wave Devices for Mobile and Wireless Communications*, Academic Press, Inc, ISBN 0-12-157340-0, San Diego, CA, USA

Kataoka, M., Schneble, R. J., Thorn, A. L., Barnes, C. H. W., Ford, C. J. B., Anderson, D., Jones, G. A. C., Farrer, I., Ritchie, D. A., & Pepper, M. (2007). Single-Electron Population and Depopulation of an Isolated Quantum Dot Using a Surface-Acoustic-Wave Pulse. *Physical Review Letters*, Vol. 98, No. 4, (January 2007), pp. 46801-1 - 46801-4, ISSN 0031-9007

Kuok, M., Ng, S. C., & Zhang, V. L. (2001). Angular dispersion of surface acoustic waves on (001), (110), and (111) GaAs. *Journal of Applied Physics*, Vol. 89, No. 12, (June 2001), pp. 7899-7902, ISSN 0021-8979

Maaø, F. A. & Galperin, Y. (1997). Acoustoelectric effects in quantum constrictions. *Physical Review B*, Vol. 56, No. 7, (August 1997), pp. 4028-4036, ISSN 0163-1829

Salmon, V. (1946). A New Family of Horns. *Jounal of the Acoustical Society of America*, Vol. 17, No. 3, (January 1946), pp. 212-218, ISSN 0001-4966

Shilton, J. M., Mace, D. R., Talyanskii, V. I., Galperin, Y., Simmons, M. Y., Pepper, M., & Ritchie, D. A. (1996). On the acoustoelectric current in a one-dimensional channel. *Journal of Condensed Matter Physics*, Vol. 8, (March 1996), pp. L337-L343, ISSN 0953-8984

Song, L., Chen, S. W., He, J. H., Zhang, C. Y., Lu, C., & Gao, J. (2010). The anomalous negative acoustoelectric current in single-electron transport. *Solid State Communications*, Vol. 150, (2010), pp. 292-296, ISSN 0038-1098

Talyanskii, V. I., Shilton, J. M., Pepper, M., Smith, C. G., Ford, C. J. B., Linfield, E. H., Ritchie, D. A., & Jones, G. A. C. (1997). Single-electron transport in a one-dimensional channel by high-frequency surface acoustic waves. *Physical Review B*, Vol. 56, No. 23, (December 1997), pp. 15180-15184, ISSN 0163-1829

Wixforth, A., Scriba, J., Wassermeier, M., Kothaus, J. P., Weimann, G., & Schlapp, W. (1989). Surface acoustic waves on GaAs/$Al_xGa_{1-x}As$ heterostructures. *Physical Review B*, Vol. 40, No. 11, (October 1989), pp. 7874-7887